Applied Probability and Statistics (*Continued*)

CHAKRAVARTI, LAHA and ROY · [illegible] Methods of Applied Statistics, Vol. I
CHAKRAVARTI, LAHA and ROY · Handbook of Methods of Applied Statistics, Vol. II
CHERNOFF and MOSES · Elementary Decision Theory
CHIANG · Introduction to Stochastic Processes in Biostatistics
CLELLAND, deCANI, BROWN, BURSK, and MURRAY · Basic Statistics with Business Applications
COCHRAN · Sampling Techniques, *Second Edition*
COCHRAN and COX · Experimental Designs, *Second Edition*
COX · Planning of Experiments
COX and MILLER · The Theory of Stochastic Processes
DAVID · Order Statistics
DEMING · Sample Design in Business Research
DODGE and ROMIG · Sampling Inspection Tables, *Second Edition*
DRAPER and SMITH · Applied Regression Analysis
GOLDBERGER · Econometric Theory
GUTTMAN and WILKS · Introductory Engineering Statistics
HALD · Statistical Tables and Formulas
HALD · Statistical Theory with Engineering Applications
HOEL · Elementary Statistics, *Second Edition*
HUANG · Regression and Econometric Methods
JOHNSON and LEONE · Statistics and Experimental Design: In Engineering and the Physical Sciences, Volumes I and II
LANCASTER · The Chi Squared Distribution
MILTON · Rank Order Probabilities: Two-Sample Normal Shift Alternatives
PRABHU · Queues and Inventories: A Study of Their Basic Stochastic Processes
SARHAN and GREENBERG · Contributions to Order Statistics
SEAL · Stochastic Theory of a Risk Business
WILLIAMS · Regression Analysis
WOLD and JURÉEN · Demand Analysis
WONNACOTT and WONNACOTT · Econometrics
YOUDEN · Statistical Methods for Chemists

Tracts on Probability and Statistics

BILLINGSLEY · Ergodic Theory and Information
BILLINGSLEY · Convergence of Probability Measures
CRAMÉR and LEADBETTER · Stationary and Related Stochastic Processes
RIORDAN · Combinatorial Identities
TAKÁCS · Combinatorial Methods in the Theory of Stochastic Processes

An Introduction to Probability Theory and Its Applications

An Introduction to Probability Theory and Its Applications

WILLIAM FELLER (1906-1970)

Eugene Higgins Professor of Mathematics

Princeton University

VOLUME II

SECOND EDITION

John Wiley & Sons, Inc.

New York · London · Sydney · Toronto

Library of Congress Catalogue Card Number: 57-10805

ISBN 0-471-257095

Printed in the United States of America

10 9 8 7 6 5 4 3 2 1

To

O. E. Neugebauer

o et praesidium et dulce decus meum

Preface to the First Edition

AT THE TIME THE FIRST VOLUME OF THIS BOOK WAS WRITTEN (BETWEEN 1941 and 1948) the interest in probability was not yet widespread. Teaching was on a very limited scale and topics such as Markov chains, which are now extensively used in several disciplines, were highly specialized chapters of pure mathematics. The first volume may therefore be likened to an all-purpose travel guide to a strange country. To describe the nature of probability it had to stress the mathematical content of the theory as well as the surprising variety of potential applications. It was predicted that the ensuing fluctuations in the level of difficulty would limit the usefulness of the book. In reality it is widely used even today, when its novelty has worn off and its attitude and material are available in newer books written for special purposes. The book seems even to acquire new friends. The fact that laymen are not deterred by passages which proved difficult to students of mathematics shows that the level of difficulty cannot be measured objectively; it depends on the type of information one seeks and the details one is prepared to skip. The traveler often has the choice between climbing a peak or using a cable car.

In view of this success the second volume is written in the same style. It involves harder mathematics, but most of the text can be read on different levels. The handling of measure theory may illustrate this point. Chapter IV contains an informal introduction to the basic ideas of measure theory and the conceptual foundations of probability. The same chapter lists the few facts of measure theory used in the subsequent chapters to formulate analytical theorems in their simplest form and to avoid futile discussions of regularity conditions. The main function of measure theory in this connection is to justify formal operations and passages to the limit that would never be questioned by a non-mathematician. Readers interested primarily in practical results will therefore not feel any need for measure theory.

To facilitate access to the individual topics the chapters are rendered as self-contained as possible, and sometimes special cases are treated separately ahead of the general theory. Various topics (such as stable distributions and renewal theory) are discussed at several places from different angles. To avoid repetitions, the definitions and illustrative examples are collected in chapter VI, which may be described as a collection of introductions to the subsequent chapters. The skeleton of the book consists of chapters V, VIII, and XV. The reader will decide for himself how much of the preparatory chapters to read and which excursions to take.

Experts will find new results and proofs, but more important is the attempt to consolidate and unify the general methodology. Indeed, certain parts of probability suffer from a lack of coherence because the usual grouping and treatment of problems depend largely on accidents of the historical development. In the resulting confusion closely related problems are not recognized as such and simple things are obscured by complicated methods. Considerable simplifications were obtained by a systematic exploitation and development of the best available techniques. This is true in particular for the proverbially messy field of limit theorems (chapters XVI–XVII). At other places simplifications were achieved by treating problems in their natural context. For example, an elementary consideration of a particular random walk led to a generalization of an asymptotic estimate which had been derived by hard and laborious methods in risk theory (and under more restrictive conditions independently in queuing).

I have tried to achieve mathematical rigor without pedantry in style. For example, the statement that $1/(1 + \xi^2)$ is the characteristic function of $\frac{1}{2}e^{-|x|}$ seems to me a desirable and legitimate abbreviation for the logically correct version that the function which at the point ξ assumes the value $1/(1 + \xi^2)$ is the characteristic function of the function which at the point x assumes the value $\frac{1}{2}e^{-|x|}$.

I fear that the brief historical remarks and citations do not render justice to the many authors who contributed to probability, but I have tried to give credit wherever possible. The original work is now in many cases superseded by newer research, and as a rule full references are given only to papers to which the reader may want to turn for additional information. For example, no reference is given to my own work on limit theorems, whereas a paper describing observations or theories underlying an example is cited even if it contains no mathematics.[1] Under these circumstances the index of authors gives no indication of their importance for probability theory. Another difficulty is to do justice to the pioneer work to which we owe new directions of research, new approaches, and new methods. Some theorems which were considered strikingly original and deep now appear with simple proofs among more refined results. It is difficult to view such a theorem in its historical perspective and to realize that here as elsewhere it is the first step that counts.

[1] This system was used also in the first volume but was misunderstood by some subsequent writers; they now attribute the methods used in the book to earlier scientists who could not have known them.

ACKNOWLEDGMENTS

Thanks to the support by the U.S. Army Research Office of work in probability at Princeton University I enjoyed the help of J. Goldman, L. Pitt, M. Silverstein, and, in particular, of M. M. Rao. They eliminated many inaccuracies and obscurities. All chapters were rewritten many times and preliminary versions of the early chapters were circulated among friends. In this way I benefited from comments by J. Elliott, R. S. Pinkham, and L. J. Savage. My special thanks are due to J. L. Doob and J. Wolfowitz for advice and criticism. The graph of the Cauchy random walk was supplied by H. Trotter. The printing was supervised by Mrs. H. McDougal, and the appearance of the book owes much to her.

WILLIAM FELLER

October 1965

THE MANUSCRIPT HAD BEEN FINISHED AT THE TIME OF THE AUTHOR'S DEATH but no proofs had been received. I am grateful to the publisher for providing a proofreader to compare the print against the manuscript and for compiling the index. J. Goldman, A. Grunbaum, H. McKean, L. Pitt, and A. Pittenger divided the book *among* themselves to check on the mathematics. Every mathematician knows what an incredible amount of work that entails. *I express my deep gratitude to these men and extend my heartfelt thanks for their labor of love.*

May 1970 CLARA N. FELLER

Introduction

THE CHARACTER AND ORGANIZATION OF THE BOOK REMAIN UNCHANGED, BUT the entire text has undergone a thorough revision. Many parts (Chapter XVII, in particular) have been completely rewritten and a few new sections have been added. At a number of places the exposition was simplified by streamlined (and sometimes new) arguments. Some new material has been incorporated into the text.

While writing the first edition I was haunted by the fear of an excessively long volume. Unfortunately, this led me to spend futile months in shortening the original text and economizing on displays. This damage has now been repaired, and a great effort has been spent to make the reading easier. Occasional repetitions will also facilitate a direct access to the individual chapters and make it possible to read certain parts of this book in conjunction with Volume 1.

Concerning the organization of the material, see the introduction to the first edition (repeated here), starting with the second paragraph.

I am grateful to many readers for pointing out errors or omissions. I especially thank D. A. Hejhal, of Chicago, for an exhaustive and penetrating list of errata and for suggestions covering the entire book.

January 1970
Princeton, N.J.

WILLIAM FELLER

Abbreviations and Conventions

Iff is an abbreviation for *if and only if.*

Epoch. This term is used for points on the time axis, while time is reserved for intervals and durations. (In discussions of stochastic processes the word "times" carries too heavy a burden. The systematic use of "epoch," introduced by J. Riordan, seems preferable to varying substitutes such as moment, instant, or point.)

Intervals are denoted by bars: $\overline{a, b}$ is an open, $\overset{|\!\!-\!\!-\!\!|}{a, b}$ a closed interval; half-open intervals are denoted by $\overset{-\!\!-\!\!|}{a, b}$ and $\overset{|\!\!-\!\!-}{a, b}$. *This notation is used also in higher dimensions.* The pertinent conventions for vector notations and order relations are found in V,1 (and also in IV,2). The symbol (a, b) is reserved for pairs and for points.

$\mathfrak{R}^1, \mathfrak{R}^2, \mathfrak{R}^r$ stand for the line, the plane, and the r-dimensional Cartesian space.

1 refers to volume one, Roman numerals to chapters. Thus **1**; XI,(3.6) refers to section 3 of chapter XI of volume **1**.

▶ indicates the end of a proof or of a collection of examples.

$\mathfrak{n}$ *and* $\mathfrak{N}$ denote, respectively, the normal density and distribution function with zero expectation and unit variance.

O, o, *and* $\sim$. Let u and v depend on a parameter x which tends, say, to a. Assuming that v is positive we write

$$\left.\begin{array}{l} u = O(v) \\ u = o(v) \\ u \sim v \end{array}\right\} \qquad \text{if} \quad \frac{u}{v} \quad \left\{\begin{array}{l} \text{remains bounded} \\ \to 0 \\ \to 1. \end{array}\right.$$

$f(x)\, U\{dx\}$. For this abbreviation see V,3.

Regarding Borel sets and Baire functions, see the introduction to chapter V.

Contents

* Starred sections are not required for the understanding of the sequel and should be omitted at first reading.

An Introduction to Probability Theory and Its Applications

CHAPTER I

The Exponential and the Uniform Densities

1. INTRODUCTION

In the course of volume **1** we had repeatedly to deal with probabilities defined by sums of many small terms, and we used approximations of the form

$$\mathbf{P}\{a < \mathbf{X} < b\} \approx \int_a^b f(x)\, dx. \tag{1.1}$$

The prime example is the normal approximation to the binomial distribution.[1] An approximation of this kind is usually formulated in the form of a limit theorem involving a succession of more and more refined discrete probability models. In many cases this passage to the limit leads conceptually to a new sample space, and the latter may be intuitively simpler than the original discrete model.

Examples. (*a*) *Exponential waiting times.* To describe waiting times by a discrete model we had to quantize the time and pretend that changes can occur only at epochs[2] $\delta, 2\delta, \ldots.$ The simplest waiting time $\mathbf{T}$ is the waiting time for the first success in a sequence of Bernoulli trials with probability p_δ for success. Then $\mathbf{P}\{\mathbf{T} > n\delta\} = (1-p_\delta)^n$ and the expected waiting time is $\mathbf{E}(\mathbf{T}) = \delta/p_\delta$. Refinements of this model are obtained by letting δ grow smaller in such a way that the expectation $\delta/p_\delta = \alpha$ remains

[1] *Further examples from volume* **1**: The arc sine distribution, chapter III, section 4; the distributions for the number of returns to the origin and first passage times in III,7; the limit theorems for random walks in XIV; the uniform distribution in problem 20 of XI,7.

[2] Concerning the use of the term *epoch*, see the list of abbreviations at the front of the book.

fixed. To a time interval of duration t there correspond $n \approx t/\delta$ trials, and hence for small δ

$$\mathbf{P}\{\mathbf{T} > t\} \approx (1 - \delta/\alpha)^{t/\delta} \approx e^{-t/\alpha} \tag{1.2}$$

approximately, as can be seen by taking logarithms. This model considers the waiting time as a geometrically distributed discrete random variable, and (1.2) states that "in the limit" one gets an exponential distribution. From the point of view of intuition it would seem more natural to start from the sample space whose points are real numbers and to introduce the exponential distribution directly.

(*b*) *Random choices.* To "choose a point at random" in the interval[3] $\overline{0, 1}$ is a conceptual experiment with an obvious intuitive meaning. It can be described by discrete approximations, but it is easier to use the whole interval as sample space and to assign to each interval its length as probability. The conceptual experiment of making two independent random choices of points in $\overline{0, 1}$ results in a pair of real numbers, and so the natural sample space is a unit square. In this sample space one equates, almost instinctively, "probability" with "area." This is quite satisfactory for some elementary purposes, but sooner or later the question arises as to what the word "area" really means. ▶

As these examples show, a continuous sample space may be conceptually simpler than a discrete model, but the definition of probabilities in it depends on tools such as integration and measure theory. In denumerable sample spaces it was possible to assign probabilities to *all* imaginable events, whereas in general spaces this naïve procedure leads to logical contradictions, and our intuition has to adjust itself to the exigencies of formal logic. We shall soon see that the naïve approach can lead to trouble even in relatively simple problems, but it is only fair to say that many probabilistically significant problems do not require a clean definition of probabilities. Sometimes they are of an analytic character and the probabilistic background serves primarily as a support for our intuition. More to the point is the fact that complex stochastic processes with intricate sample spaces may lead to significant and comprehensible problems which do not depend on the delicate tools used in the analysis of the whole process. A typical reasoning may run as follows: if the process can be described at all, the random variable $\mathbf{Z}$ must have such and such properties, and its distribution must therefore satisfy such and such an integral equation. Although probabilistic arguments can greatly influence the analytical treatment of the equation in question, the latter is in principle independent of the axioms of probability.

[3] Intervals are denoted by bars to preserve the symbol (a, b) for the coordinate notation of points in the plane. See the list of abbreviations at the front of the book.

Specialists in various fields are sometimes so familiar with problems of this type that they deny the need for measure theory because they are unacquainted with problems of other types and with situations where vague reasoning did lead to wrong results.[4]

This situation will become clearer in the course of this chapter, which serves as an informal introduction to the whole theory. It describes some analytic properties of two important distributions which will be used throughout this book. Special topics are covered partly because of significant applications, partly to illustrate the new problems confronting us and the need for appropriate tools. It is not necessary to study them systematically or in the order in which they appear.

Throughout this chapter probabilities are *defined* by elementary integrals, and the limitations of this definition are accepted. The use of a probabilistic jargon, and of terms such as random variable or expectation, may be justified in two ways. They may be interpreted as technical aids to intuition based on the formal analogy with similar situations in volume **1**. Alternatively, everything in this chapter may be interpreted in a logically impeccable manner by a passage to the limit from the discrete model described in example 2(*a*). Although neither necessary nor desirable in principle, the latter procedure has the merit of a good exercise for beginners.

2. DENSITIES. CONVOLUTIONS

A *probability density on the line* (or $\mathcal{R}^1$) is a function f such that

$$f(x) \geq 0, \qquad \int_{-\infty}^{+\infty} f(x)\,dx = 1. \tag{2.1}$$

For the present we consider only piecewise continuous densities (see V,3 for the general notion). To each density f we let correspond its *distribution function*[5] F defined by

$$F(x) = \int_{-\infty}^{x} f(y)\,dy. \tag{2.2}$$

---

[4] The roles of rigor and intuition are subject to misconceptions. As was pointed out in volume **1**, natural intuition and natural thinking are a poor affair, but they gain strength with the development of mathematical theory. Today's intuition and applications depend on the most sophisticated theories of yesterday. Furthermore, strict theory represents economy of thought rather than luxury. Indeed, experience shows that in applications most people rely on lengthy calculations rather than simple arguments because these appear risky. [The nearest illustration is in example 5(*a*).]

[5] We recall that by "distribution function" is meant a right continuous non-decreasing function with limits 0 and 1 at $\pm\infty$. Volume **1** was concerned mainly with distributions whose growth is due entirely to jumps. Now we focus our attention on distribution functions defined as integrals. General distribution functions will be studied in chapter V.

It is a monotone continuous function increasing from 0 to 1. We say that f and F are *concentrated on the interval* $a \leq x \leq b$ if f vanishes outside this interval. The density f will be considered as an assignment of probabilities to the intervals of the line, the interval $\overline{a, b} = \{a < x < b\}$ having probability

$$F(b) - F(a) = \int_a^b f(x)\, dx. \tag{2.3}$$

Sometimes this probability will be denoted by $\mathbf{P}\{\overline{a, b}\}$. Under this assignment an individual point carries probability zero, and the closed interval $a \leq \times \leq b$ has the same probability as $\overline{a, b}$.

In the simplest situation the real line serves as "*sample space*," that is, the outcome of a conceptual experiment is represented by a number. (Just as in volume **1**, this is only the first step in the construction of sample spaces representing sequences of experiments.) *Random variables* are functions defined on the sample space. For simplicity we shall for the time being accept as random variable only a function $\mathbf{U}$ such that for each t the event $\{\mathbf{U} \leq t\}$ consists of finitely many intervals. Then

$$G(t) = \mathbf{P}\{\mathbf{U} \leq t\} \tag{2.4}$$

is well defined as the integral of f over these intervals. The function G defined by (2.4) is called the *distribution function* of $\mathbf{U}$. If G is the integral of a function g, then g is called the *density* of the distribution G or (interchangeably) the density of the variable $\mathbf{U}$.

The basic random variable is, of course, the coordinate variable[6] $\mathbf{X}$ as such, and all other random variables are functions of $\mathbf{X}$. The distribution function of $\mathbf{X}$ is identical with the distribution F by which probabilities are defined. Needless to say, any random variable $\mathbf{Y} = g(\mathbf{X})$ can be taken as coordinate variable on a new line.

As stated above, these terms may be justified by mere analogy with the situation in volume **1**, but the following example shows that our model may be obtained by a passage to the limit from discrete models.

Examples. (*a*) *Grouping of data.* Let F be a given distribution function. Choose a fixed $\delta > 0$ and consider the discrete random variable $\mathbf{X}_\delta$ which for $(n-1)\delta < x \leq n\delta$ assumes the constant value $n\delta$. Here $n = 0, \pm 1, \pm 2, \ldots$. In volume **1** we would have used the multiples of δ as sample

[6] As far as possible we shall denote random variables (that is, functions on the sample space) by capital boldface letters, reserving small letters for numbers or location parameters. This holds in particular for the coordinate variable $\mathbf{X}$, namely the function defined by $\mathbf{X}(x) = x$.

space, and described the probability distribution of $\mathbf{X}_\delta$ by saying that

$$\mathbf{P}\{\mathbf{X}_\delta = n\delta\} = F(n\delta) - F((n-1)\delta). \tag{2.5}$$

Now $\mathbf{X}_\delta$ becomes a random variable in an enlarged sample space, and its distribution function is the function that for $n\delta \leq x < (n+1)\delta$ equals $F(n\delta)$. In the continuous model, $\mathbf{X}_\delta$ serves as an approximation to $\mathbf{X}$ obtained by identifying our intervals with their right-hand endpoints (a procedure known to statisticians as grouping of data). In the spirit of volume **1** we should treat $\mathbf{X}_\delta$ as the basic random variable and δ as a free parameter. Letting $\delta \to 0$ we would obtain limit theorems stating, for example, that F is the limit distribution of $\mathbf{X}_\delta$.

(*b*) For $x > 0$, the event $\{\mathbf{X}^2 \leq x\}$ is the same as $\{-\sqrt{x} \leq \mathbf{X} \leq \sqrt{x}\}$; the random variable $\mathbf{X}^2$ has a distribution concentrated on $\overline{0, \infty}$ and given there by $F(\sqrt{x}) - F(-\sqrt{x})$. By differentiation it is seen that *the density* g of $\mathbf{X}^2$ is given by

$$g(x) = \tfrac{1}{2}[f(\sqrt{x}) + f(-\sqrt{x})]/\sqrt{x} \quad \text{for} \quad x > 0 \qquad g(x) = 0 \quad \text{for} \quad x < 0.$$

The distribution function of $\mathbf{X}^3$ is given for all x by $F(\sqrt[3]{x})$ and has density $\frac{1}{3}f(\sqrt[3]{x})/\sqrt[3]{x^2}$.

The *expectation of* $\mathbf{X}$ is defined by

$$\mathbf{E}(\mathbf{X}) = \int_{-\infty}^{+\infty} xf(x)\,dx, \tag{2.6}$$

provided the integral converges absolutely. The expectations of the approximating discrete variables $\mathbf{X}_\delta$ of example (*a*) coincide with Riemann sums for this integral, and so $\mathbf{E}(\mathbf{X}_\delta) \to \mathbf{E}(\mathbf{X})$. If u is a bounded continuous function the same argument applies to the random variable $u(\mathbf{X})$, and the relation $\mathbf{E}(u(\mathbf{X}_\delta)) \to \mathbf{E}(u(\mathbf{X}))$ implies

$$\mathbf{E}(u(\mathbf{X})) = \int_{-\infty}^{+\infty} u(x)f(x)\,dx; \tag{2.7}$$

the point here is that this formula makes no explicit use of the distribution of $u(\mathbf{X})$. Thus the knowledge of the distribution of a random variable $\mathbf{X}$ suffices to calculate the expectation of functions of it.

The *second moment* of $\mathbf{X}$ is defined by

$$\mathbf{E}(\mathbf{X}^2) = \int_{-\infty}^{+\infty} x^2 f(x)\,dx, \tag{2.8}$$

provided the integral converges. Putting $\mu = \mathbf{E}(\mathbf{X})$, *the variance of* $\mathbf{X}$ is again defined by

$$\operatorname{Var}(\mathbf{X}) = \mathbf{E}((\mathbf{X}-\mu)^2) = \mathbf{E}(\mathbf{X}^2) - \mu^2. \tag{2.9}$$

Note. If the variable $\mathbf{X}$ is *positive* (that is, if the density f is concentrated on $\overline{0, \infty})$ and if the integral in (2.6) diverges, it is harmless and convenient to say that $\mathbf{X}$ *has an infinite expectation* and write $\mathbf{E}(\mathbf{X}) = \infty$. By the same token one says that $\mathbf{X}$ has an infinite variance when the integral in (2.8) diverges. For variables assuming positive and negative values the expectation remains undefined when the integral (2.6) diverges. A typical example is provided by the density $\pi^{-1}(1+x^2)^{-1}$. ▶

The notion of density carries over to higher dimensions, but the general discussion is postponed to chapter III. Until then we shall consider only the analogue to the product probabilities introduced in definition 2 of **1**; V,4 to describe combinations of independent experiments. In other words, in this chapter we shall be concerned only with product densities of the form $f(x)\,g(y)$, $f(x)\,g(y)\,h(z)$, etc., where $f, g, \ldots$ are densities on the line. Giving a density of the form $f(x)\,g(y)$ in the plane $\mathfrak{R}^2$ means identifying "probabilities" with integrals:

$$\mathbf{P}\{A\} = \iint_A f(x)\,g(y)\,dx\,dy. \tag{2.10}$$

Speaking of "*two independent random variables* $\mathbf{X}$ *and* $\mathbf{Y}$ *with densities* f *and* g" is an abbreviation for saying that probabilities in the $(\mathbf{X}, \mathbf{Y})$-plane are assigned in accordance with (2.10). This implies the multiplication rule for intervals, for example $\mathbf{P}\{\mathbf{X} > a, \mathbf{Y} > b\} = \mathbf{P}\{\mathbf{X} > a\}\mathbf{P}\{\mathbf{Y} > b\}$. The analogy with the discrete case is so obvious that no further explanations are required.

Many new random variables may be defined as functions of $\mathbf{X}$ and $\mathbf{Y}$, but the most important role is played by the sum $\mathbf{S} = \mathbf{X} + \mathbf{Y}$. The event $A = \{\mathbf{S} \leq s\}$ is represented by the half-plane of points (x, y) such that $x + y \leq s$. Denote the distribution function of $\mathbf{Y}$ by G so that one has $g(y) = G'(y)$. To obtain the *distribution function of* $\mathbf{X} + \mathbf{Y}$ we integrate in (2.10) over $y \leq s - x$ with the result

$$\mathbf{P}\{\mathbf{X}+\mathbf{Y} \leq s\} = \int_{-\infty}^{+\infty} G(s-x) f(x)\,dx. \tag{2.11}$$

For reasons of symmetry the roles of F and G can be interchanged without affecting the result. By differentiation it is then seen that *the density of* $\mathbf{X} + \mathbf{Y}$ *is given by either of the two integrals*

$$\int_{-\infty}^{+\infty} f(s-y)\,g(y)\,dy = \int_{-\infty}^{+\infty} f(y)\,g(s-y)\,dy. \tag{2.12}$$

The operation defined in (2.12) is a special case of the convolutions to be introduced in V,4. For the time being we use the term convolution only for densities: *The convolution of two densities* f *and* g *is the function defined by* (2.12). *It will be denoted by* $f * g$.

Throughout volume **1** we dealt with convolutions of discrete distributions, and the rules are the same. According to (2.12) we have $f * g = g * f$. Given a third density h we can form $(f * g) * h$ and this is the density of a sum $\mathbf{X} + \mathbf{Y} + \mathbf{Z}$ of three independent variables with densities f, g, h. The fact that summation of random variables is commutative and associative implies the same properties for convolutions, and so $f * g * h$ is independent of the order of the operations.

Positive random variables play an important role, and it is therefore useful to note that *if* f *and* g *are concentrated on* $\overline{0, \infty}$ *the convolution* $f * g$ *of* (2.12) *reduces to*

$$f * g(s) = \int_0^s f(s-y)\, g(y)\, dy = \int_0^s f(x)\, g(s-x)\, dx. \tag{2.13}$$

Example. (*c*) Let f and g be concentrated on $\overline{0, \infty}$ and defined there by $f(x) = \alpha e^{-\alpha x}$ and $g(x) = \beta e^{-\beta x}$. Then

$$f * g(x) = \alpha\beta\, \frac{e^{-\alpha x} - e^{-\beta x}}{\beta - \alpha} \qquad x > 0. \tag{2.14}$$

(Continued in problem 12.) ▶

Note *on the notion of random variable.* The use of the line or the Cartesian spaces $\mathcal{R}^n$ as sample spaces sometimes blurs the distinction between random variables and "ordinary" functions of one or more variables. In volume **1** a random variable $\mathbf{X}$ could assume only denumerably many values and it was then obvious whether we were talking about a function (such as the square or the exponential) defined on the line, or the random variable $\mathbf{X}^2$ or $e^{\mathbf{X}}$ defined in the sample space. Even the outer appearance of these functions was entirely different inasmuch as the "ordinary" exponential assumes all positive values whereas $e^{\mathbf{X}}$ had a denumerable range. To see the change in this situation, consider now "two independent random variables $\mathbf{X}$ and $\mathbf{Y}$ with a common density f." In other words, the plane $\mathcal{R}^2$ serves as sample space, and probabilities are defined as integrals of $f(x)f(y)$. Now every function of two variables *can* be defined *in* the sample space, and then it becomes a random variable, but it must be borne in mind that a function of two variables can be defined also without reference to our sample space. For example, certain statistical problems compel one to introduce the random variable $f(\mathbf{X})f(\mathbf{Y})$ [see example VI,12(*d*)]. On the other hand, in introducing our sample space $\mathcal{R}^2$ we have evidently referred to the "ordinary" function f defined independently of the sample space. This "ordinary" function induces many random variables, namely $f(\mathbf{X})$, $f(\mathbf{Y})$, $f(\mathbf{X} \pm \mathbf{Y})$, etc. Thus the same f may serve either as a random variable or as an ordinary function.

As a rule (and in each individual case) it will be clear whether or not we are concerned with a random variable. Nevertheless, in the general theory there arise situations in which functions (such as conditional probabilities and expectations) can be considered either as free functions or as random variables, and this is somewhat confusing if the freedom of choice is not properly understood.

Note on *terminology and notations.* To avoid overburdening of sentences it is customary to call $\mathbf{E}(\mathbf{X})$, interchangeably, expectation of the variable $\mathbf{X}$, or of the density f, or of the distribution F. Similar liberties will be taken for other terms. For example, convolution really signifies an operation, but the term is applied also to the result of the operation and the function $f \star g$ is referred to as "the convolution."

In the older literature the terms distribution and frequency function were applied to what we call densities; our distribution functions were described as "cumulative," and the abbreviation c.d.f. is still in use.

3. THE EXPONENTIAL DENSITY

For arbitrary but fixed $\alpha > 0$ put

$$(3.1) \qquad f(x) = \alpha e^{-\alpha x}, \qquad F(x) = 1 - e^{-\alpha x}, \quad \text{for} \quad x \geq 0$$

and $F(x) = f(x) = 0$ for $x < 0$. Then f is an exponential density, F its distribution function. A trite calculation shows that *the expectation equals* α^{-1}, *the variance* α^{-2}.

In example 1(*a*) the exponential distribution was derived as the limit of geometric distributions, and the method of example 2(*a*) leads to the same result. We recall that in stochastic processes the geometric distribution frequently governs waiting times or lifetimes, and that this is due to its "lack of memory," described in **1**; XIII,9: *whatever the present age, the residual lifetime is unaffected by the past and has the same distribution as the lifetime itself.* It will now be shown that this property carries over to the exponential limit and to no other distribution.

Let $\mathbf{T}$ be an arbitrary positive variable to be interpreted as life- or waiting time. It is convenient to replace the distribution function of $\mathbf{T}$ by its *tail*

$$(3.2) \qquad U(t) = \mathbf{P}\{\mathbf{T} > t\}.$$

Intuitively, $U(t)$ is the "probability at birth of a lifetime exceeding t." Given an age s, the event that the residual lifetime exceeds t is the same as $\{\mathbf{T} > s+t\}$ and the conditional probability of this event (given age s) equals the ratio $U(s+t)/U(s)$. This is the residual lifetime distribution, and it coincides with the total lifetime distribution iff

$$(3.3) \qquad U(s+t) = U(s)\,U(t), \qquad s, t > 0.$$

It was shown in **1**; XVII,6 that a positive solution of this equation is necessarily of the form $U(t) = e^{-\alpha t}$, and hence *the lack of aging described above in italics holds true if the lifetime distribution is exponential.*

We shall refer to this lack of memory as the *Markov property* of the exponential distribution. Analytically it reduces to the statement that only for the exponential distribution F do the tails $U = 1-F$ satisfy (3.3), but this explains the constant occurrence of the exponential distribution in Markov processes. (A stronger version of the Markov property will be described in section 6.) Our description referred to temporal processes, but the argument is general and the Markov property remains meaningful when time is replaced by some other parameter.

Examples. (*a*) *Tensile strength.* To obtain a continuous analogue to the proverbial finite chain whose strength is that of its weakest link denote by $U(t)$ the probability that a thread of *length* t (of a given material) can sustain a certain fixed load. A thread of length $s+t$ does not snap iff the two segments individually sustain the given load. Assuming that there is no interaction, the two events must be considered independent and U must satisfy (3.3). Here the length of the thread takes over the role of the time parameter, and the length at which the thread will break is an exponentially distributed random variable.

(*b*) *Random ensembles of points in space* play a role in many connections so that it is important to have an appropriate definition for this concept. Speaking intuitively, the first property that perfect randomness should have is a lack of interaction between different regions: the observed configuration within region A_1 should not permit conclusions concerning the ensemble in a non-overlapping region A_2. Specifically, the probability p that both A_1 and A_2 are empty should equal the product of the probabilities p_1 and p_2 that A_1 and A_2 be empty. It is plausible that this product rule cannot hold for *all* partitions unless the probability p depends only on the volume of the region A but not on its shape. Assuming this to be so, we denote by $U(t)$ the probability that a region of volume t be empty. These probabilities then satisfy (3.3) and hence $U(t) = e^{-\alpha t}$; the constant α depends on the density of the ensemble or, what amounts to the same, on the unit of length. It will be shown in the next section that the knowledge of $U(t)$ permits us to calculate the probabilities $p_n(t)$ that a region of volume t contains exactly n points of the ensemble; they are given by the Poisson distribution $p_n(t) = e^{-\alpha t}(\alpha t)^n/n!$. We speak accordingly of *Poisson ensembles* of points, this term being less ambiguous than the term random ensemble which may have other connotations.

(*c*) *Ensembles of circles and spheres.* Random ensembles of particles present a more intricate problem. For simplicity we assume that the particles are of a spherical or circular shape, the radius ρ being fixed. The configuration is then completely determined by the centers and it is tempting to assume that these centers form a Poisson ensemble. This, however, is impossible in the strict sense since the mutual distances of centers necessarily exceed 2ρ. One feels nevertheless that for small radii ρ the effect of the finite size should be negligible in practice and hence the model of a Poisson ensemble of centers should be usable as an approximation.

For a mathematical model we postulate accordingly that the centers form a Poisson ensemble and accept the implied possibility that the circles or spheres intersect. This idealization will have no practical consequences if the radii ρ are small, because then the theoretical frequency of intersections will be negligible. Thus astronomers treat the stellar system as a Poisson ensemble and the approximation to reality seems excellent. The next two examples show how the model works in practice.

(*d*) *Nearest neighbors.* We consider a Poisson ensemble of spheres (stars) with density α. The probability that a domain of volume t contains no center equals $e^{-\alpha t}$. Saying that the nearest neighbor to the origin has a distance $>r$ amounts to saying that a sphere of radius r contains no star center in its interior. The volume of such a ball equals $\frac{4}{3}\pi r^3$, and hence in a Poisson ensemble of stars the probability that the nearest neighbor has a distance $>r$ is given by $e^{-\frac{4}{3}\pi\alpha r^3}$. The fact that this expression is independent of the radius ρ of the stars shows the approximative character of the model and its limitations.

In the plane, spheres are replaced by circles and the distribution function for the distance of nearest neighbors is given by $1 - e^{-\alpha\pi r^2}$.

(*e*) *Continuation: free paths.* For ease of description we begin with the two-dimensional model. The random ensemble of circular disks may be interpreted as the cross section of a thin forest. I stand at the origin, which is not contained in any disk, and look in the direction of the positive x-axis. The longest interval $\overline{0, t}$ not intersecting any disk represents the *visibility* or free path in the x-direction. It is a random variable and we denote it by $\mathbf{L}$.

Denote by A the region formed by the points at a distance $\leq\rho$ from a point of the interval $\overline{0, t}$ on the x-axis. The boundary of A consists of the segments $0 \leq x \leq t$ on the lines $y = \pm\rho$ and two semicircles of radii ρ about the origin and the point $(t, 0)$ on the x-axis. Thus the area of A equals $2\rho t + \pi\rho^2$. The event $\{\mathbf{L} > t\}$ occurs iff no disk center is contained within A, but it is known in advance that the circle of radius ρ about the origin is empty. The remaining domain has area $2\rho t$ and we conclude that the distribution of the visibility $\mathbf{L}$ is exponential:

$$\mathbf{P}\{\mathbf{L} > t\} = e^{-2\alpha\rho t}.$$

In space the same argument applies and the relevant region is formed by rotating our A about the x-axis. The rectangle $0 < x < t$, $|y| < \rho$ is replaced by a cylinder of volume $\pi\rho^2 t$. We conclude that in a Poisson ensemble of spherical stars the *free path* $\mathbf{L}$ *in any direction has an exponential distribution:* $\mathbf{P}\{\mathbf{L} > t\} = e^{-\pi\alpha\rho^2 t}$. The mean free path is given by $\mathbf{E}(\mathbf{L}) = 1/(\pi\alpha\rho^2)$. ▶

The next theorem will be used repeatedly.

Theorem. *If* $\mathbf{X}_1, \ldots, \mathbf{X}_n$ *are mutually independent random variables with the exponential distribution* (3.1), *then the sum* $\mathbf{X}_1 + \cdots + \mathbf{X}_n$ *has a density* g_n *and distribution function* G_n *given by*

$$g_n(x) = \alpha \frac{(\alpha x)^{n-1}}{(n-1)!} e^{-\alpha x} \qquad x > 0 \tag{3.4}$$

$$G_n(x) = 1 - e^{-\alpha x}\left(1 + \frac{\alpha x}{1!} + \cdots + \frac{(\alpha x)^{n-1}}{(n-1)!}\right) \qquad x > 0. \tag{3.5}$$

Proof. For $n = 1$ the assertion reduces to the definition (3.1). The density g_{n+1} is defined by the convolution

$$g_{n+1}(t) = \int_0^t g_n(t-x)\, g_1(x)\, dx, \tag{3.6}$$

and assuming the validity of (3.4) this reduces to

$$g_{n+1}(t) = \frac{\alpha^{n+1}}{(n-1)!} e^{-\alpha t} \int_0^t x^{n-1}\, dx = \alpha \frac{(\alpha t)^n}{n!} e^{-\alpha t}. \tag{3.7}$$

Thus (3.4) holds by induction for all n. The validity of (3.5) is seen by differentiation. ▶

The densities g_n are among the *gamma densities* to be introduced in II,2. They represent the continuous analogue of the negative binomial distribution found in **1**; VI,8 for the sum of n variables with a common geometric distribution. (See problem 6.)

4. WAITING TIME PARADOXES. THE POISSON PROCESS

Denote by $\mathbf{X}_1, \mathbf{X}_2, \ldots$ mutually independent random variables with the common exponential distribution (3.1), and put $\mathbf{S}_0 = 0$,

$$\mathbf{S}_n = \mathbf{X}_1 + \cdots + \mathbf{X}_n, \qquad n = 1, 2, \ldots. \tag{4.1}$$

We introduce a family of new random variables $\mathbf{N}(t)$ as follows: $\mathbf{N}(t)$ *is the number of indices* $k \geq 1$ *such that* $\mathbf{S}_k \leq t$. The event $\{\mathbf{N}(t) = n\}$

occurs iff $\mathbf{S}_n \leq t$ but $\mathbf{S}_{n+1} > t$. As $\mathbf{S}_n$ has the distribution G_n the probability of this event equals $G_n(t) - G_{n+1}(t)$ or

$$\mathbf{P}\{\mathbf{N}(t) = n\} = e^{-\alpha t}\frac{(\alpha t)^n}{n!}. \tag{4.2}$$

In words, *the random variable* $\mathbf{N}(t)$ *has a Poisson distribution with expectation* αt.

This argument looks like a new derivation of the Poisson distribution but in reality it merely rephrases the original derivation of **1**; VI,6 in terms of random variables. For an intuitive description consider chance occurrences (such as cosmic ray bursts or telephone calls), which we call "arrivals." Suppose that there is no aftereffect in the sense that the past history permits no conclusions as to the future. As we have seen, this condition requires that the waiting time $\mathbf{X}_1$ to the first arrival be exponentially distributed. But at each arrival the process starts from scratch as a probabilistic replica of the whole process: the successive waiting times $\mathbf{X}_k$ between arrivals must be independent and must have the same distribution. The sum $\mathbf{S}_n$ represents the epoch of the nth arrival and $\mathbf{N}(t)$ *the number of arrivals* within the interval $\overline{0, t}$. In this form the argument differs from the original derivation of the Poisson distribution only by the use of better technical terms.

(In the terminology of stochastic processes the sequence $\{\mathbf{S}_n\}$ constitutes a *renewal process* with exponential *interarrival times* $\mathbf{X}_k$; for the general notion see VI,6.)

Even this simple situation leads to apparent contradictions which illustrate the need for a sophisticated approach. We begin by a naïve formulation.

Example. *Waiting time paradox.* Buses arrive in accordance with a Poisson process, the expected time between consecutive buses being α^{-1}. I arrive at an epoch t. What is the expectation $\mathbf{E}(\mathbf{W}_t)$ of my waiting time $\mathbf{W}_t$ for the next bus? (It is understood that the epoch t of my arrival is independent of the buses, say noontime sharp.) Two contradictory answers stand to reason:

(*a*) The lack of memory of the Poisson process implies that the distribution of my waiting time should not depend on the epoch of my arrival. In this case $\mathbf{E}(\mathbf{W}_t) = \mathbf{E}(\mathbf{W}_0) = \alpha^{-1}$.

(*b*) The epoch of my arrival is "chosen at random" in the interval. between two consecutive buses, and for reasons of symmetry my expected waiting time should be half the expected time between two consecutive buses, that is $\mathbf{E}(\mathbf{W}_t) = \frac{1}{2}\alpha^{-1}$.

Both arguments appear reasonable and both have been used in practice. What to do about the contradiction? The easiest way out is that of the formalist, who refuses to see a problem if it is not formulated in an impeccable manner. But problems are not solved by ignoring them.

We now show that *both* arguments are substantially, if not formally, *correct*. The fallacy lies at an unexpected place and we now proceed to explain it.[7] ▶

We are dealing with interarrival times $\mathbf{X}_1 = \mathbf{S}_1$, $\mathbf{X}_2 = \mathbf{S}_2 - \mathbf{S}_1, \ldots$. By assumption the $\mathbf{X}_k$ have a common exponential distribution with expectation α^{-1}. Picking out "any" particular $\mathbf{X}_k$ yields a random variable, and one has the intuitive feeling that its expectation should be α^{-1} provided the choice is done without knowledge of the sample sequence $\mathbf{X}_1, \mathbf{X}_2, \ldots$. But this is not true. In the example we chose that element $\mathbf{X}_k$ for which

$$\mathbf{S}_{k-1} < t \leq \mathbf{S}_k,$$

where t is fixed. This choice is made without regard to the actual process, but it turns out that the $\mathbf{X}_k$ so chosen has the *double* expectation $2\alpha^{-1}$. Given this fact, the argument (*b*) of the example postulates an expected waiting time α^{-1} and the contradiction disappears.

This solution of the paradox came as a shock to experienced workers, but it becomes intuitively clear once our mode of thinking is properly adjusted. Roughly speaking, a long interval has a better chance to cover the point t than a short one. This vague feeling is supported by the following

Proposition. *Let* $\mathbf{X}_1, \mathbf{X}_2, \ldots$ *be mutually independent with a common exponential distribution with expectation* α^{-1}. *Let* $t > 0$ *be fixed, but arbitrary. The element* $\mathbf{X}_k$ *satisfying the condition* $\mathbf{S}_{k-1} < t \leq \mathbf{S}_k$ *has the density*

$$v_t(x) = \begin{cases} \alpha^2 x e^{-\alpha x} & \text{for } 0 < x \leq t \\ \alpha(1+\alpha t)e^{-\alpha x} & \text{for } x > t. \end{cases} \tag{4.3}$$

The point is that the density (4.3) *is not the common density of the* $\mathbf{X}_k$. Its explicit form is of minor interest. [The analogue for arbitrary waiting time distributions is contained in XI,(4.16).]

Proof. Let k be the (chance-dependent) index such that $\mathbf{S}_{k-1} < t \leq \mathbf{S}_k$ and put $\mathbf{L}_t$ equal to $\mathbf{S}_k - \mathbf{S}_{k-1}$. We have to prove that $\mathbf{L}_t$ has density (4.3). Suppose first $x < t$. The event $\{\mathbf{L}_t \leq x\}$ occurs iff $\mathbf{S}_n = y$ and $t-y < \mathbf{X}_{n+1} \leq x$ for some combination n, y. This necessitates

$$t-x \leq y \leq t.$$

Summing over all possible n and y we obtain

$$\mathbf{P}\{\mathbf{L}_t \leq x\} = \sum_{n=1}^{\infty} \int_{t-x}^{t} g_n(y) \cdot [e^{-\alpha(t-y)} - e^{-\alpha x}]\, dy. \tag{4.4}$$

[7] For a variant of the paradox see example VI,7(*a*). The paradox occurs also in general renewal theory where it caused serious trouble and contradictions before it was properly understood. For the underlying theory see XI,4.

But $g_1(y) + g_2(y) + \cdots = \alpha$ identically, and so

$$\mathbf{P}\{\mathbf{L}_t \le x\} = 1 - e^{-\alpha x} - \alpha x e^{-\alpha x}. \tag{4.5}$$

By differentiation we get (4.3) for $x < t$. For $x > t$ a similar argument applies except that y ranges from 0 to t and we must add to the right side in (4.4) the probability $e^{-\alpha t} - e^{-\alpha x}$ that $0 < t < \mathbf{S}_1 < x$. This completes the proof. ►

The break in the formula (4.3) at $x = t$ is due to the special role of the origin as the starting epoch of the process. Obviously

$$\lim_{t\to\infty} v_t(x) = \alpha^2 x e^{-\alpha x}, \tag{4.6}$$

which shows that the special role of the origin wears out, and for an "old" process the distribution of $\mathbf{L}_t$ is nearly independent of t. One expresses this conveniently by saying that the "*steady state*" *density* of $\mathbf{L}_t$ is given by the right side in (4.6).

With the notations of the proof, the waiting time $\mathbf{W}_t$ considered in the example is the random variable $\mathbf{W}_t = \mathbf{S}_k - t$. The argument of the proof shows also that

$$\begin{aligned}\mathbf{P}\{\mathbf{W}_t \le x\} &= e^{-\alpha t} - e^{-\alpha(x+t)} + \sum_{n=1}^{\infty} \int_0^t g_n(y)[e^{-\alpha(t-y)} - e^{-\alpha(x+t-y)}]\, dy \\ &= 1 - e^{-\alpha x}\end{aligned} \tag{4.7}$$

Thus $\mathbf{W}_t$ *has the same exponential distribution as the* $\mathbf{X}_k$ in accordance with the reasoning (*a*). (See problem 7.)

Finally, a word about the *Poisson process.* The Poisson variables $\mathbf{N}(t)$ were introduced as functions on the sample space of the infinite sequence of random variables $\mathbf{X}_1, \mathbf{X}_2, \ldots$. This procedure is satisfactory for many purposes, but a different sample space is more natural. The conceptual experiment "observing the number of incoming calls up to epoch t" yields for each positive t an integer, and the result is therefore a step function with unit jumps. The appropriate sample space has these step functions as sample points; the sample space is a function space—the space of all conceivable "paths." In this space $\mathbf{N}(t)$ is defined as the value of the ordinate at epoch t and $\mathbf{S}_n$ as the coordinate of the nth jump, etc. Events can now be considered that are not easily expressible in terms of the original variables $\mathbf{X}_n$. A typical example of practical interest (see the ruin problem in VI,5) is the event that $\mathbf{N}(t) > a + bt$ for some t. The individual path (just as the individual infinite sequence of ± 1 in binomial trials) represents the natural and unavoidable object of probabilistic inquiry. Once one gets used to the new phraseology, the space of paths becomes most intuitive.

Unfortunately the introduction of probabilities in spaces of sample paths is far from simple. By comparison, the step from discrete sample spaces to the line, plane, etc., and even to infinite sequences of random variables, is neither conceptually nor technically difficult. Problems of a new type arise in connection with function spaces, and the reader is warned that we shall not deal with them in this volume. We shall be satisfied with an honest treatment of sample spaces of sequences (denumerably many coordinate variables). Reference to stochastic processes in general, and to the Poisson process in particular, will be made freely, but only to provide an intuitive background or to enhance interest in our problems.

Poisson Ensembles of Points

As shown in **1**; VI,6, the Poisson law governs not only "points distributed randomly along the time axis," but also ensembles of points (such as flaws in materials or raisins in a cake) distributed randomly in plane or space, provided t is interpreted as area or volume. The basic assumption was that the probability of finding k points in a specified domain depends only on the area or volume of the domain, but not on its shape, and that occurrences in non-overlapping domains are independent. In example 3(*b*) we used the same assumption to show that the probability that a domain of volume t be empty is given by $e^{-\alpha t}$. This corresponds to the exponential distribution for the waiting time for the first event, and we see now that the Poisson distribution for the number of events is a simple consequence of it. The same argument applies to random ensembles of points in space, and we have thus a new proof for the fact that the number of points of the ensemble contained in a given domain is a Poisson variable. Easy formal calculations may lead to interesting results concerning such random ensembles of points, but the remarks about the Poisson process apply equally to Poisson ensembles; a complete probabilistic description is complex and beyond the scope of the present volume.

5. THE PERSISTENCE OF BAD LUCK

As everyone knows, he who joins a waiting line is sure to wait for an abnormally long time, and similar bad luck follows us on all occasions. How much can probability theory contribute towards an explanation? For a partial answer we consider three examples typical of a variety of situations. They illustrate unexpected general features of chance fluctuations.

Examples. (*a*) *Record values.* Denote by $\mathbf{X}_0$ my waiting time (or financial loss) at some chance event. Suppose that friends of mine expose themselves to the same type of experience, and denote the results by $\mathbf{X}_1, \mathbf{X}_2, \ldots$. To exclude bias we assume that $\mathbf{X}_0, \mathbf{X}_1, \ldots$ are mutually independent random variables with a common distribution. The nature of the latter really does not matter but, since the exponential distribution serves as a model for randomness, we assume the $\mathbf{X}_j$ exponentially distributed in accordance with (3.1). For simplicity of description we treat the sequence $\{\mathbf{X}_j\}$ as infinite.

To find a measure for my ill luck I ask how long it will take before a friend experiences worse luck (we neglect the event of probability zero that $\mathbf{X}_k = \mathbf{X}_0$). More formally, we introduce the waiting time $\mathbf{N}$ *as the value of the first subscript* n *such that* $\mathbf{X}_n > \mathbf{X}_0$. The event $\{\mathbf{N} > n-1\}$ occurs iff the maximal term of the n-tuple $\mathbf{X}_0, \mathbf{X}_1, \ldots, \mathbf{X}_{n-1}$ appears at the initial place; for reasons of symmetry the probability of this event is n^{-1}. The event $\{\mathbf{N} = n\}$ is the same as $\{\mathbf{N} > n-1\} - \{\mathbf{N} > n\}$, and hence for $n = 1, 2, \ldots,$

$$\mathbf{P}\{\mathbf{N} = n\} = \frac{1}{n} - \frac{1}{n+1} = \frac{1}{n(n+1)}. \tag{5.1}$$

This result fully confirms that I have indeed very bad luck: *The random variable* $\mathbf{N}$ *has infinite expectation*! It would be bad enough if it took on the average 1000 trials to beat the record of my ill luck, but the actual waiting time has infinite expectation.

It will be noted that the argument does not depend on the condition that the $\mathbf{X}_k$ are exponentially distributed. It follows that whenever the variables $\mathbf{X}_j$ are independent and have a common continuous distribution function F the first record value has the distribution (5.1). The fact that this distribution is independent of F is used by statisticians for tests of independence. (See also problems 8–11.)

The striking and general nature of the result (5.1) combined with the simplicity of the proof are apt to arouse suspicion. The argument is really impeccable (except for the informal presentation), but those who prefer to rely on brute calculation can easily verify the truth of (5.1) from the direct definition of the probability in question as the $(n+1)$-tuple integral of $\alpha^{n+1}e^{-\alpha(x_0+\cdots+x_n)}$ over the region defined by the inequalities $0 < x_0 < x_n$ and $0 < x_j < x_0$ for $j = 1, \ldots, n-1$.

An *alternative derivation* of (5.1) is an instructive exercise in conditional probabilities; it is less simple, but leads to additional results (problem 8). Given that $\mathbf{X}_0 = x$, the probability of a greater value at later trials is $p = e^{-\alpha x}$, and we are concerned with the waiting time for the first "success" in Bernoulli trials with probability p. The conditional probability that $\mathbf{N} = n$ given $\mathbf{X}_0 = x$ is therefore $p(1-p)^{n-1}$. To obtain $\mathbf{P}\{\mathbf{N} = n\}$ we have to multiply by the density $\alpha e^{-\alpha x}$ of the hypothesis $\mathbf{X}_0 = x$ and integrate with respect to x. The substitution $1 - e^{-\alpha x} = t$ reduces the integrand to $t^{n-1}(1-t)$, the integral of which equals $n^{-1} - (n+1)^{-1}$ in agreement with (5.1).

(*b*) *Ratios.* If $\mathbf{X}$ and $\mathbf{Y}$ are two independent variables with a common exponential distribution, the ratio $\mathbf{Y}/\mathbf{X}$ is a new random variable. Its distribution function is obtained by integrating $\alpha^2 e^{-\alpha(x+y)}$ over $0 < y < tx$, $0 < x < \infty$. Integration with respect to y leads to

$$\mathbf{P}\left\{\frac{\mathbf{Y}}{\mathbf{X}} \le t\right\} = \int_0^\infty \alpha e^{-\alpha x}(1-e^{-\alpha t x})\,dx = \frac{t}{1+t}. \tag{5.2}$$

The corresponding *density* is given by $(1+t)^{-2}$. It is noteworthy that *the variable* $\mathbf{Y}/\mathbf{X}$ *has infinite expectation.*

We find here a new confirmation for the persistence of bad luck. Assuredly Peter has reason for complaint if he has to wait three times as long as Paul, but the distribution (5.2) attributes to this event probability $\frac{1}{4}$. It follows that, on the average, in one out of two cases either Paul or Peter has reason for complaint. The observed frequency increases in practice because very short waiting times naturally pass unnoticed.

(*c*) *Parallel waiting lines.* I arrive in my car at the car inspection station (or at a tunnel entrance, car ferry, etc.). There are two waiting lines to choose from, but once I have joined a line I have to stay in it. Mr. Smith, who drove behind me, occupies the place that I might have chosen and I keep watching whether he is ahead of or behind me. Most of the time we stand still, but occasionally one line or the other moves one car-length forward. To maximize the influence of pure chance we assume the two lines stochastically independent; also, the time intervals between successive moves are independent variables with a common exponential distribution. Under these circumstances the successive moves constitute Bernoulli trials in which "success" means that I move ahead, "failure" that Mr. Smith moves. The probability of success being $\frac{1}{2}$, we are, in substance, dealing with a symmetric random walk, and the curious properties of fluctuations in random walks find a striking interpretation. (For simplicity of description we disregard the fact that only finitely many cars are present.) Am I ever going to be ahead of Mr. Smith? In the random walk interpretation the question is whether a first passage through $+1$ will ever take place. As we know, this event has probability one, but the expected waiting time for it is infinite. Such waiting gives ample apportunity to bemoan my bad luck, and this only grows more irritating by the fact that Mr. Smith argues in the same way. ►

6. WAITING TIMES AND ORDER STATISTICS

An ordered n-tuple $(x_1, \ldots, x_n)$ of real numbers, may be reordered in increasing order of magnitude to obtain the new n-tuple

$$(x_{(1)}, x_{(2)}, \ldots, x_{(n)}) \quad \text{where} \quad x_{(1)} \le x_{(2)} \le \cdots \le x_{(n)}.$$

This operation applied to all points of the space $\mathcal{R}^n$ induces n well-defined functions, which will be denoted by $\mathbf{X}_{(1)}, \ldots, \mathbf{X}_{(n)}$. If probabilities are defined in $\mathcal{R}^n$ these functions become random variables. We say that $(\mathbf{X}_{(1)}, \ldots, \mathbf{X}_{(n)})$ is obtained by reordering $(\mathbf{X}_1, \ldots, \mathbf{X}_n)$ according to increasing magnitude. The variable $\mathbf{X}_{(k)}$ is called kth-*order statistic*[8] of the given sample $\mathbf{X}_1, \ldots, \mathbf{X}_n$. In particular, $\mathbf{X}_{(1)}$ and $\mathbf{X}_{(n)}$ are the *sample extremes*; when $n = 2\nu + 1$ is odd, $\mathbf{X}_{(\nu+1)}$ is the *sample median.*

We apply this notion to the particular case of independent random variables $\mathbf{X}_1, \ldots, \mathbf{X}_n$ with the common exponential density $\alpha e^{-\alpha x}$.

Examples. (*a*) *Parallel waiting lines.* Interpret $\mathbf{X}_1, \ldots, \mathbf{X}_n$ as the lengths of n service times commencing at epoch 0 at a post office with n counters. The order statistics represent the successive epochs of terminations or, as one might say, the *epochs of the successive discharges* (the "output process"). In particular, $\mathbf{X}_{(1)}$ is the waiting time for the first discharge. Now if the assumed lack of aftereffect is meaningful, the waiting time $\mathbf{X}_{(1)}$ must have the Markov property, that is, $\mathbf{X}_{(1)}$ must be exponentially distributed. As a matter of fact, the event $\{\mathbf{X}_{(1)} > t\}$ is the simultaneous realization of the n events $\{\mathbf{X}_k > t\}$, each of which has probability $e^{-\alpha t}$; because of the assumed independence the probabilities multiply and we have indeed

$$\mathbf{P}\{\mathbf{X}_{(1)} > t\} = e^{-n\alpha t}. \tag{6.1}$$

We can now proceed a step further and consider the situation at epoch $\mathbf{X}_{(1)}$. The assumed lack of memory seems to imply that the original situation is restored except that now only $n - 1$ counters are in operation; the continuation of the process should be *independent of* $\mathbf{X}_{(1)}$ and a replica of the whole process. In particular, the waiting time for the next discharge, namely $\mathbf{X}_{(2)} - \mathbf{X}_{(1)}$, should have the distribution

$$\mathbf{P}\{\mathbf{X}_{(2)} - \mathbf{X}_{(1)} > t\} = e^{-(n-1)\alpha t} \tag{6.2}$$

analogous to (6.1). This reasoning leads to the following general proposition concerning the order statistics for independent variables with a common exponential distribution.

[8] Strictly speaking the term "sample statistic" is synonymous with "function of the sample variables," that is, with random variable. It is used to emphasize linguistically the different role played *in a given context* by the primary variable (the sample) and some derived variables. For example, the "sample mean" $(\mathbf{X}_1 + \cdots + \mathbf{X}_n)/n$ is called a statistic. Order statistics occur frequently in the statistical literature. We conform to the standard terminology except that the extremes are usually called extreme "*values*."

Proposition[9] *The* n *variables* $\mathbf{X}_{(1)}, \mathbf{X}_{(2)} - \mathbf{X}_{(1)}, \ldots, \mathbf{X}_{(n)} - \mathbf{X}_{(n-1)}$ *are independent and the density of* $\mathbf{X}_{(k+1)} - \mathbf{X}_{(k)}$ *is given by* $(n-k)\alpha e^{-(n-k)\alpha t}$.

Before verifying this proposition formally let us consider its implications. When $n = 2$ the difference $\mathbf{X}_{(2)} - \mathbf{X}_{(1)}$ is the *residual waiting time* after the expiration of the shorter of two waiting times. The proposition asserts that this residual waiting time has the same exponential distribution as the original waiting time and is independent of $\mathbf{X}_{(1)}$. This is an extension of the Markov property enunciated for *fixed epochs* t to the chance-dependent stopping time $\mathbf{X}_{(1)}$. It is called the *strong Markov property*. (As we are dealing with only finitely many variables we are in a position to *derive* the strong Markov property from the weak one, but in more complicated stochastic processes the distinction is essential.)

The *proof of the proposition* serves as an example of formal manipulations with integrals. For typographical simplicity we let $n = 3$. As in many similar situations we use a symmetry argument. With probability one, no two among the variables $\mathbf{X}_j$ are equal. Neglecting an event of probability zero the six possible orderings of $\mathbf{X}_1, \mathbf{X}_2, \mathbf{X}_3$ according to magnitude therefore represent six mutually exclusive events of equal probability. To calculate the distribution of the order statistics it suffices therefore to consider the contingency $\mathbf{X}_1 < \mathbf{X}_2 < \mathbf{X}_3$. Thus

$$\begin{aligned} \mathbf{P}\{\mathbf{X}_{(1)} > t_1, \mathbf{X}_{(2)} - \mathbf{X}_{(1)} > t_2, \mathbf{X}_{(3)} - \mathbf{X}_{(2)} > t_3\} &= \\ = 6\mathbf{P}\{\mathbf{X}_1 > t_1, \mathbf{X}_2 - \mathbf{X}_1 > t_2, \mathbf{X}_3 - \mathbf{X}_2 > t_3\}&. \end{aligned} \tag{6.3}$$

(Purely analytically, the space $\mathfrak{R}^3$ is partitioned into six parts congruent to the region defined by $x_1 < x_2 < x_3$, each contributing the same amount to the integral. The boundaries where two or more coordinates are equal have probability zero and play no role.) To evaluate the right side in (6.3) we have to integrate $\alpha^3 e^{-\alpha(x_1+x_2+x_3)}$ over the region defined by the inequalities

$$x_1 > t_1, \qquad x_2 - x_1 > t_2, \qquad x_3 - x_2 > t_3.$$

A simple integration with respect to x_3 leads to

$$\begin{aligned} 6e^{-\alpha t_3} \int_{t_1}^{\infty} \alpha e^{-\alpha x_1}\, dx_1 \int_{x_1+t_2}^{\infty} \alpha e^{-2\alpha x_2}\, dx_2 &= \\ = 3e^{-\alpha t_3 - 2\alpha t_2} \int_{t_1}^{\infty} \alpha e^{-3\alpha x_1}\, dx_1 = e^{-\alpha t_3 - 2\alpha t_2 - 3\alpha t_1}&. \end{aligned} \tag{6.4}$$

[9] This proposition has been discovered repeatedly for purposes of statistical estimation but the usual proofs are computational instead of appealing to the Markov property. See also problem 13.

Thus the joint distribution of the three variables $\mathbf{X}_{(1)}$, $\mathbf{X}_{(2)}-\mathbf{X}_{(1)}$, $\mathbf{X}_{(3)}-\mathbf{X}_{(2)}$ is a product of three exponential distributions, and this proves the proposition.

It follows in particular that $\mathbf{E}(\mathbf{X}_{(k+1)}-\mathbf{X}_{(k)}) = 1/(n-k)\alpha$. Summing over $k = 0, 1, \ldots, \nu-1$ we obtain

$$\mathbf{E}(\mathbf{X}_{(\nu)}) = \frac{1}{\alpha}\left(\frac{1}{n} + \frac{1}{n-1} + \cdots + \frac{1}{n-\nu+1}\right). \tag{6.5}$$

Note that this expectation was calculated without knowledge of the distribution of $\mathbf{X}_{(\nu)}$ and we have here another example of the advantage to be derived from the representation of a random variable as a sum of other variables. (See **1**; IX,3.)

(*b*) *Use of the strong Markov property.* For picturesque language suppose that at epoch 0 three persons A, B, and C arrive at a post office and find two counters free. The three service times are independent random variables $\mathbf{X}, \mathbf{Y}, \mathbf{Z}$ with the same exponential distribution. The service times of A and B commence immediately, but that of C starts at the epoch $\mathbf{X}_{(1)}$ when either A or B is discharged. We show that the Markov property leads to simple answers to various questions.

(i) What is the probability that C will not be the last to leave the post office? The answer is $\frac{1}{2}$, because epoch $\mathbf{X}_{(1)}$ of the first departure establishes symmetry between C and the other person being served.

(ii) What is the distribution of the time $\mathbf{T}$ spent by C at the post office? Clearly $\mathbf{T} = \mathbf{X}_{(1)} + \mathbf{Z}$ is the sum of two independent variables whose distributions are exponential with parameters 2α and α. The convolution of two exponential distributions is given by (2.14), and it is seen that $\mathbf{T}$ has density $u(t) = 2\alpha(e^{-\alpha t} - e^{-2\alpha t})$ and $\mathbf{E}(\mathbf{T}) = 3/(2\alpha)$.

(iii) What is the distribution of the epoch of the *last* departure? Denote the epochs of the successive departures by $\mathbf{X}_{(1)}$, $\mathbf{X}_{(2)}$, $\mathbf{X}_{(3)}$. The difference $\mathbf{X}_{(3)} - \mathbf{X}_{(1)}$ is the sum of the two variables $\mathbf{X}_{(3)} - \mathbf{X}_{(2)}$ and $\mathbf{X}_{(2)} - \mathbf{X}_{(1)}$. We saw in the preceding example that these variables are independent and have exponential distributions with parameters 2α and α. It follows that $\mathbf{X}_{(3)} - \mathbf{X}_{(1)}$ has the same density u as the variable $\mathbf{T}$. Now $\mathbf{X}_{(1)}$ is independent of $\mathbf{X}_{(3)} - \mathbf{X}_{(1)}$ and has density $2\alpha e^{-2\alpha t}$. The convolution formula used in (ii) shows therefore that $\mathbf{X}_{(3)}$ has density

$$4\alpha[e^{-\alpha t} - e^{-2\alpha t} - \alpha t e^{-2\alpha t}]$$

and $\mathbf{E}(\mathbf{X}_{(3)}) = 2/\alpha$.

The advantage of this method becomes clear on comparison with direct calculations, but the latter apply to arbitrary service time distributions (problem 19).

(*c*) *Distribution of order statistics.* As a final exercise we derive the distribution of $\mathbf{X}_{(k)}$. The event $\{\mathbf{X}_{(k)} \leq t\}$ signifies that at least k among the n variables $\mathbf{X}_j$ are $\leq t$. This represents at least k "successes" in n independent trials, and hence

$$(6.6) \qquad \mathbf{P}\{\mathbf{X}_{(k)} \leq t\} = \sum_{j=k}^{n} \binom{n}{j} (1-e^{-\alpha t})^j e^{-(n-j)\alpha t}.$$

By differentiation it is seen that the *density of* $\mathbf{X}_{(k)}$ is given by

$$(6.7) \qquad n\binom{n-1}{k-1}(1-e^{-\alpha t})^{k-1} e^{-(n-k)\alpha t} \cdot \alpha e^{-\alpha t}.$$

This result may be obtained directly by the following loose argument. We require (up to terms negligible in the limit as $h \to 0$) the probability of the joint event that one among the variables $\mathbf{X}_j$ lies between t and $t + h$ and that $k - 1$ among the remaining $n - 1$ variables are $\leq t$, while the other $n - k$ variables are $>t + h$. Multiplying the number of choices and the corresponding probabilities leads to (6.7). Beginners are advised to formalize this argument, and also to derive (6.7) from the discrete model. (Continued in problems 13, 17.) ▶

7. THE UNIFORM DISTRIBUTION

The random variable $\mathbf{X}$ *is distributed uniformly in the interval* $\overline{a, b}$ if its density is constant $= (b-a)^{-1}$ for $a < x < b$ and vanishes outside this interval. In this case the variable $(\mathbf{X}-a)(b-a)^{-1}$ is distributed uniformly in $\overline{0, 1}$, and we shall usually use this interval as standard. Because of the appearance of their graphs the densities of the uniform distribution function are called "*rectangular*."

With the uniform distribution the interval $\overline{0, 1}$ becomes a sample space in which probabilities of intervals are identical with their lengths. The sample space corresponding to two independent variables $\mathbf{X}$ and $\mathbf{Y}$ that are uniformly distributed over $\overline{0, 1}$ is the unit square in $\mathcal{R}^2$, and probabilities in it are defined by their area. The same idea applies to triples and n-tuples.

A uniformly distributed random variable is often called a "*point* $\mathbf{X}$ *chosen at random.*" The result of the conceptual experiment "n independent random choices of a point in $\overline{0, 1}$" requires an n-dimensional hypercube for its probabilistic description, but the experiment as such yields n points $\mathbf{X}_1, \ldots, \mathbf{X}_n$ *in the same interval.* With unit probability no two of them are equal, and hence they partition $\overline{0, 1}$ into $n + 1$ subintervals. Reordering the n points $\mathbf{X}_1, \ldots, \mathbf{X}_n$ in their natural order from left to right we get n new random variables which will be denoted by $\mathbf{X}_{(1)}, \ldots, \mathbf{X}_{(n)}$. These are the *order statistics* defined in the last section. The subintervals of the partition are now $\overline{0, \mathbf{X}_{(1)}}$, then $\overline{\mathbf{X}_{(1)}, \mathbf{X}_{(2)}}$, etc.

The notion of a *point chosen at random on a circle* is self-explanatory. To visualize the result of n independent choices on the circle we imagine the circle *oriented* anticlockwise, so that intervals have left and right endpoints and may be represented in the form $\overline{a, b}$. Two points $\mathbf{X}_1$ and $\mathbf{X}_2$ chosen independently and at random divide the circle into the two intervals $\overline{\mathbf{X}_1, \mathbf{X}_2}$ and $\overline{\mathbf{X}_2, \mathbf{X}_1}$. (We disregard again the zero-probability event that $\mathbf{X}_1 = \mathbf{X}_2$.)

Examples. (*a*) *Empirical interpretations.* The *roulette wheel* is generally thought of as a means to effect a "random choice" on the circle. In numerical calculations to six decimals the *rounding error* is usually treated as a random variable distributed uniformly over an interval of length 10^{-6}. (For the error committed by dropping the last two decimals, the discrete model with 100 possible values is more appropriate, though less convenient in practice.) The waiting time of a passenger arriving at the *bus station* without regard to the schedule may be regarded as uniformly distributed over the interval between successive departures. Of wider theoretical interest are the applications to *random splittings* discussed in section 8. In many problems of mathematical statistics (such as non-parametric tests) the uniform distribution enters in an indirect way: given an arbitrary random variable $\mathbf{X}$ with a continuous distribution F the random variable $F(\mathbf{X})$ *is distributed uniformly over* $\overline{0, 1}$. (See section 12.)

(*b*) *The induced partition.* We prove the following proposition: *n independently and randomly chosen points $\mathbf{X}_1, \ldots, \mathbf{X}_n$ partition $\overline{0, 1}$ into $n + 1$ intervals whose lengths have the common distribution given by*

$$\mathbf{P}\{\mathbf{L} > t\} = (1-t)^n, \qquad 0 < t < 1. \tag{7.1}$$

This result is surprising, because intuitively one might expect that at least the two end intervals should have different distributions. That all $n + 1$ intervals have the same distribution becomes clear on considering the equivalent situation on the (oriented) circle of unit length.[10] Here $n + 1$ points $\mathbf{X}_1, \ldots, \mathbf{X}_{n+1}$ chosen independently and at random partition the circle into $n + 1$ intervals, and for reasons of symmetry these intervals must have the same distribution. Imagine now the circle cut at the point $\mathbf{X}_{n+1}$ to obtain an interval in which $\mathbf{X}_1, \ldots, \mathbf{X}_n$ are chosen independently

[10] For a computational verification note that the probability of the event

$$\{\mathbf{X}_{(k+1)} - \mathbf{X}_{(k)} > t\}$$

equals the integral of the constant function 1 over the union of the $n!$ congruent regions defined either by the string of inequalities $x_1 < \cdots < x_k < x_k + t < x_{k+1} < \cdots < x_n$ or by similar strings obtained by permuting the subscripts. A more streamlined calculation leading to a stronger result is contained in example III, 3(*c*).

and at random. The lengths of the $n + 1$ intervals of the induced partition are the same, and they have a common distribution. That this distribution is given by (7.1) may be seen by considering the leftmost interval $\overline{0, \mathbf{X}_{(1)}}$. Its length exceeds t iff all n points $\mathbf{X}_1, \ldots, \mathbf{X}_n$ are in $\overline{t, 1}$, and the probability of this event is $(1-t)^n$.

It is a good exercise to verify the proposition in the special case $n = 2$ by inspection of the three events in the unit square representing the sample space. (Continued in problems 22–26.)

(*c*) *A paradox* (related to the waiting time paradox of section 4). Let two points $\mathbf{X}_1$ and $\mathbf{X}_2$ be chosen independently and at random on the circle of unit length. Then *the lengths of the two intervals* $\overline{\mathbf{X}_1, \mathbf{X}_2}$ *and* $\overline{\mathbf{X}_2, \mathbf{X}_1}$ *are uniformly distributed, but the length* $\boldsymbol{\lambda}$ *of the one containing the arbitrary point* P *has a different distribution* (*with density* $2x$).

In particular, each of the two intervals has expected length $\frac{1}{2}$, but the one containing P has expected length $\frac{2}{3}$. The point P being fixed, but arbitrary, one has the feeling that the interval covering P is chosen "without advance knowledge of its properties" (to borrow a phrase from the philosophers of probability). Certainly naïve intuition is not prepared for the great difference between covering or not covering an arbitrary point, but after due reflection this difference becomes "intuitively obvious." In fact, however, rather experienced writers have fallen into the trap.

For a proof imagine the circle cut at P leaving us with two points chosen independently and at random in $\overline{0, 1}$. Using the same notation as before the event $\{\boldsymbol{\lambda} < t\}$ occurs iff $\mathbf{X}_{(2)} - \mathbf{X}_{(1)} > 1 - t$ and by (7.1) the probability for this equals t^2. The variable $\boldsymbol{\lambda}$ has therefore density $2t$, as asserted. (Beginners are advised to try a *direct* computational verification.)

(*d*) *Distribution of order statistics.* If $\mathbf{X}_1, \ldots, \mathbf{X}_n$ are independent and distributed uniformly in $\overline{0, 1}$, the number of variables satisfying the inequality $0 < \mathbf{X}_j \leq t < 1$ has a binomial distribution with probability of "success" equal to t. Now the event $\{\mathbf{X}_{(k)} \leq t\}$ occurs iff at least k among the variables are $\leq t$ and hence

$$\mathbf{P}\{\mathbf{X}_{(k)} \leq t\} = \sum_{j=k}^{n} \binom{n}{j} t^j (1-t)^{n-j}. \tag{7.2}$$

This gives us the distribution function of the kth-order statistics. By differentiation it is found that *the density of* $\mathbf{X}_{(k)}$ *is given by*

$$n\binom{n-1}{k-1} t^{k-1}(1-t)^{n-k}. \tag{7.3}$$

This may be seen directly as follows: The probability that one among the $\mathbf{X}_j$ lies between t and $t + h$, and that $k - 1$ among the remaining ones are less than t while $n - k$ are greater than $t + h$, equals

$$n\binom{n-1}{k-1}t^{k-1}(1-t-h)^{n-k}h.$$

Divide by h and let $h \to 0$ to obtain (7.3).

(*e*) *Limit theorems.* To see the nature of the distribution of $\mathbf{X}_{(1)}$ when n is large it is best to introduce $\mathbf{E}(\mathbf{X}_{(1)}) = (n+1)^{-1}$ as a new unit of measurement. As $n \to \infty$ we get then for the tail of the distribution function

$$\mathbf{P}\{n\mathbf{X}_{(1)} > t\} = \left(1 - \frac{t}{n}\right)^n \to e^{-t}. \tag{7.4}$$

It is customary to describe this relation by saying that *in the limit* $\mathbf{X}_{(1)}$ *is exponentially distributed with expectation* n^{-1}. Similarly

$$\mathbf{P}\{n\mathbf{X}_{(2)} > t\} = \left(1 - \frac{t}{n}\right)^n + \binom{n}{1}\frac{t}{n}\left(1 - \frac{t}{n}\right)^{n-1} \to e^{-t} + te^{-t}, \tag{7.5}$$

and on the right one recognizes the tail of the gamma distribution G_2 of (3.5). In like manner it is easily verified that for every fixed k as $n \to \infty$ *the distribution of* $n\mathbf{X}_{(k)}$ *tends to the gamma distribution* G_k (see problem 33).

Now G_k is the distribution of the sum of k independent exponentially distributed variables while $\mathbf{X}_{(k)}$ is the sum of the first k intervals considered in example (*b*). We can therefore say that the lengths of the successive intervals of our partition behave in the limit as if they were mutually independent exponentially distributed variables.

[In view of the obvious relation of (7.2) with the binomial distribution the central limit theorem may be used to obtain approximations to the distribution of $\mathbf{X}_{(k)}$ when both n and k are large. See problem 34.]

(*f*) *Ratios.* Let $\mathbf{X}$ be chosen at random in $\overline{0, 1}$ and denote by $\mathbf{U}$ the length of the *shorter* of the intervals $\overline{0, \mathbf{X}}$ and $\overline{\mathbf{X}, 1}$ and by $\mathbf{V} = 1 - \mathbf{U}$ the length of the longer. The random variable $\mathbf{U}$ is uniformly distributed between 0 and $\frac{1}{2}$ because the event $\{\mathbf{U} < t < \frac{1}{2}\}$ occurs iff either $\mathbf{X} < t$ or $1 - \mathbf{X} < t$ and therefore has probability $2t$. For reasons of symmetry $\mathbf{V}$ is uniformly distributed between $\frac{1}{2}$ and 1, and so $\mathbf{E}(\mathbf{U}) = \frac{1}{4}$, $\mathbf{E}(\mathbf{V}) = \frac{3}{4}$. What can we say about the ratio $\mathbf{V}/\mathbf{U}$? It necessarily exceeds 1 and it lies between 1 and $t > 1$ iff either

$$\frac{1}{1+t} \leq \mathbf{X} \leq \tfrac{1}{2} \qquad \text{or} \qquad \tfrac{1}{2} \leq \mathbf{X} \leq \frac{t}{1+t}.$$

For $t > 1$ it follows that

$$\mathbf{P}\left\{\frac{\mathbf{V}}{\mathbf{U}} \leq t\right\} = \frac{t-1}{t+1}, \tag{7.6}$$

and the density of this distribution is given by $2(t+1)^{-2}$. It is seen that $\mathbf{V}/\mathbf{U}$ *has infinite expectation.* This example shows how little information is contained in the observation that $\mathbf{E}(\mathbf{V})/\mathbf{E}(\mathbf{U}) = 3$. ▶

8. RANDOM SPLITTINGS

The problem of this section concludes the preceding parade of examples and is separated from them partly because of its importance in physics, and partly because it will serve as a prototype for general Markov chains.

Formally we are concerned with products of the form $\mathbf{Z}_n = \mathbf{X}_1\mathbf{X}_2 \cdots \mathbf{X}_n$ where $\mathbf{X}_1, \ldots, \mathbf{X}_n$ are mutually independent variables distributed uniformly in $\overline{0, 1}$.

Examples *for applications.* In certain collision processes a physical *particle* is split into two and its mass m divided between them. Different laws of partition may fit different processes, but it is frequently assumed that the fraction of parental mass received by each descendant particle is distributed uniformly in $\overline{0, 1}$. If one of the two particles is chosen at random and subject to a new collision then (assuming that there is no interaction so that the collisions are independent) the masses of the two second-generation particles are given by products $m\mathbf{X}_1\mathbf{X}_2$, and so on. (See problem 21.) With trite verbal changes this model applies also to splittings of mineral grains or pebbles, etc. Instead of masses one considers also *energy losses* under collisions, and the description simplifies somewhat if one is concerned with changes of energy of the *same* particle in successive collisions. As a last example consider the changes in the *intensity of light* when passing through matter. Example 10(a) shows that when a light ray passes through a sphere of radius R "in a random direction" the distance traveled through the sphere is distributed uniformly between 0 and $2R$. In the presence of uniform absorption such a passage would reduce the intensity of the incident ray by a factor that is uniformly distributed in an interval $\overline{0, a}$ (where $a < 1$ depends on the strength of absorption). The scale factor does not seriously affect our model and it is seen that n independent passages would reduce the intensity of the light by a factor of the form $\mathbf{Z}_n$. ▶

To find the distribution of $\mathbf{Z}_n$ we can proceed in two ways.

(i) *Reduction to exponential distributions.* Since sums are generally preferable to products we pass to logarithms putting $\mathbf{Y}_k = -\log \mathbf{X}_k$. The $\mathbf{Y}_k$ are mutually independent, and for $t > 0$

$$\text{(8.1)} \qquad \mathbf{P}\{\mathbf{Y}_k \geq t\} = \mathbf{P}\{\mathbf{X}_k \leq e^{-t}\} = e^{-t}.$$

Now the distribution function G_n of the sum $\mathbf{S}_n = \mathbf{Y}_1 + \cdots + \mathbf{Y}_n$ of n independent exponentially distributed variables was calculated in (3.5), and the distribution function of $\mathbf{Z}_n = e^{-\mathbf{S}_n}$ is given by $1 - G_n(\log t^{-1})$ where $0 < t < 1$. The density of this distribution function is $t^{-1}g_n(\log t^{-1})$ or

$$f_n(t) = \frac{1}{n-1}\left(\log\frac{1}{t}\right)^{n-1}, \qquad 0 < t < 1. \tag{8.2}$$

Our problem is solved explicitly. This method reveals the advantages to be derived from an appropriate transformation, but the success depends on the accidental equivalence of our problem with one previously solved.

(ii) *A recursive procedure* has the advantage that it lends itself also to related problems and generalizations. Let $F_n(t) = \mathbf{P}\{\mathbf{Z}_n \leq t\}$ and $0 < t < 1$. By definition $F_1(t) = t$. Suppose F_{n-1} known and note that $\mathbf{Z}_n = \mathbf{Z}_{n-1}\mathbf{X}_n$ is the product of two independent variables. Given $\mathbf{X}_n = x$ the event $\{\mathbf{Z}_n \leq t\}$ occurs iff $\mathbf{Z}_{n-1} \leq t/x$ and has probability $F_{n-1}(t/x)$. Summing over all possible x we obtain for $0 \leq t \leq 1$

$$F_n(t) = \int_0^1 F_{n-1}(t/x)\,dx = \int_t^1 F_{n-1}(t/x)\,dx + t. \tag{8.3}$$

This formula permits us in principle to calculate successively $F_2, F_3, \ldots$. In practice it is preferable to operate with the corresponding densities f_n. By assumption f_1 exists. Assume by induction the existence of f_{n-1}. Recalling that $f_{n-1}(s) = 0$ for $s > 1$ we get by differentiation from (8.3)

$$f_n(t) = \int_t^1 f_{n-1}\left(\frac{t}{x}\right)\frac{dx}{x}, \qquad 0 < t < 1, \tag{8.4}$$

and trite calculations show that f_n is indeed given by (8.2).

9. CONVOLUTIONS AND COVERING THEOREMS

The results of this section have a mild amusement value in themselves and some obvious applications. Furthermore, they turn up rather unexpectedly in connection with seemingly unrelated topics, such as significance tests in harmonic analysis [example III,3(*f*)], Poisson processes [XIV,2(*a*)], and random flights [example 10(*e*)]. It is therefore not surprising that all formulas, as well as variants of them, have been derived repeatedly by different methods. The method used in the sequel is distinguished by its simplicity and applicability to related problems.

Let $a > 0$ be fixed, and denote by $\mathbf{X}_1, \mathbf{X}_2, \ldots$ mutually independent random variables distributed uniformly over $\overline{0, a}$. Let $\mathbf{S}_n = \mathbf{X}_1 + \cdots + \mathbf{X}_n$. Our first problem consists in finding the distribution U_n of $\mathbf{S}_n$ and its density $u_n = U'_n$.

By definition $u_1(x) = a^{-1}$ for $0 < x < a$ and $u_1(x) = 0$ elsewhere (rectangular density). The higher u_n are defined by the convolution formula (2.13) which in the present situation reads

$$u_{n+1}(x) = \frac{1}{a}\int_0^a u_n(x-y)\,dy = \frac{1}{a}\,[U_n(x) - U_n(x-a)]. \tag{9.1}$$

It is easily seen that

$$u_2(x) = \begin{cases} xa^{-2} & 0 \leq x \leq a \\ (2a-x)a^{-2} & a \leq x \leq 2a, \end{cases} \tag{9.2}$$

and, of course, $u_2(x) = 0$ for all other x. The graph of u_2 appears as an isosceles triangle with basis $\overline{0, 2a}$, and hence u_2 is called *triangular density.* Similarly u_3 is concentrated on $\overline{0, 3a}$ and is defined by three different quadratic polynomials in the three thirds of this interval. For a general formula we introduce the following

Notation. *We write*

$$x_+ = \frac{x + |x|}{2} \tag{9.3}$$

for the positive part of the real number x. In the following the ambiguous symbol x_+^n stands for $(x_+)^n$, namely the function that vanishes for $x \leq 0$ and equals x^n when $x \geq 0$. Note that $(x-a)_+$ is zero for $x < a$ and a linear function when $x > a$. With this notation the uniform distribution may be written in the form

$$U_1(x) = (x_+ - (x-a)_+)a^{-1}. \tag{9.4}$$

Theorem 1. *Let* $\mathbf{S}_n$ *be the sum of* n *independent variables distributed uniformly over* $\overline{0, a}$. *Let* $U_n(x) = \mathbf{P}\{\mathbf{S}_n \leq x\}$ *and denote by* $u_n = U'_n$ *the density of this distribution. Then for* $n = 1, 2, \ldots$ *and* $x \geq 0$

$$U_n(x) = \frac{1}{a^n n!}\sum_{\nu=0}^{n}(-1)^\nu\binom{n}{\nu}(x-\nu a)_+^n; \tag{9.5}$$

$$u_{n+1}(x) = \frac{1}{a^{n+1} n!}\sum_{\nu=0}^{n+1}(-1)^\nu\binom{n+1}{\nu}(x-\nu a)_+^n. \tag{9.6}$$

(These formulas remain true also for $x < 0$ and for $n = 0$ provided x_+^0 is defined to equal 0 on the negative half-axis, and 1 on the positive.)

Note that for a point x between $(k-1)a$ and ka only k terms of the sum are different from zero. In practical calculations it is convenient to disregard the limits of summation and to pretend that ν varies from $-\infty$

to ∞. This is possible, because with the standard convention the binomial coefficients in (9.5) vanish for $\nu < 0$ and $\nu > n$ (see **1**; II,8).

Proof. For $n = 1$ the assertion (9.5) reduces to (9.4) and is obviously true. We now prove the two assertions simultaneously by induction. Assume (9.5) to be true for some $n \geq 1$. Substituting into (9.1) we get u_{n+1} as the difference of two sums. Changing the summation index ν in the second sum to $\nu - 1$ we get

$$u_{n+1}(x) = \frac{1}{a^{n+1}n} \sum (-1)^\nu \left[\binom{n}{\nu} + \binom{n}{\nu-1} \right] (x-\nu a)_+^n$$

which is identical with (9.6). Integrating this relation leads to (9.5) with n replaced by $n + 1$, and this completes the proof. ▶

(An alternative proof using a passage to the limit from the discrete model is contained in problem 20 of **1**; XI,7.)

Let $a = 2b$. The variables $\mathbf{X}_k - b$ are then distributed uniformly over the symmetric interval $\overline{-b, b}$, and hence the sum of n such variables has the same distribution as $\mathbf{S}_n - nb$. It is given by $U_n(x+nb)$. Our theorem may therefore be reformulated in the following equivalent form.

Theorem 1a. *The density of the sum of* n *independent variables distributed uniformly over* $\overline{-b, b}$ *is given by*

$$(9.7) \qquad u_n(x+nb) = \frac{1}{(2b)^n(n-1)!} \sum_{\nu=0}^{n} (-1)^\nu \binom{n}{\nu} (x + (n-2\nu)b)_+^{n-1}.$$

We turn to a theorem which admits of two equivalent formulations both of which are useful in many special problems arising in applications. By unexpected good luck the required probability can be expressed simply in terms of the density u_n. We prove this analytically by a method of wide applicability. For a proof based on geometric arguments see problem 23.

Theorem 2. *On a circle of length* t *there are given* $n \geq 2$ *arcs of length* a *whose centers are chosen independently and at random. The probability* $\varphi_n(t)$ *that these* n *arcs cover the whole circle is*

$$(9.8) \qquad \varphi_n(t) = a^n(n-1)!u_n(t)\frac{1}{t^{n-1}},$$

which is the same as

$$(9.9) \qquad \varphi_n(t) = \sum_{\nu=0}^{n} (-1)^\nu \binom{n}{\nu} \left(1 - \nu\frac{a}{t}\right)_+^{n-1}$$

Before proving it, we reformulate the theorem in a form to be used later. Choose one of the n centers as origin and open the circle into an interval of length t. The remaining $n-1$ centers are randomly distributed in $\overline{0, t}$ and theorem 2 obviously expresses the same thing as

Theorem 3. *Let the interval* $\overline{0, t}$ *be partitioned into* n *subintervals by choosing independently at random* $n-1$ *points* $\mathbf{X}_1, \ldots, \mathbf{X}_{n-1}$ *of division. The probability* $\varphi_n(t)$ *that none of these subintervals is of length exceeding* a *equals* (9.9).

Note that $\varphi_n(t)$, considered for fixed t as a function of a, represents *the distribution function of the maximal length among the* n *intervals* into which $\overline{0, t}$ is partitioned. For related questions see problems 22–27.

Proof. It suffices to prove theorem 3. We prove the recursion formula

$$\varphi_n(t) = (n-1)\int_0^a \varphi_{n-1}(t-x)\left(\frac{t-x}{t}\right)^{n-2}\frac{dx}{t}. \tag{9.10}$$

Its truth follows directly from the definition of φ_n as an $(n-1)$-tuple integral, but it is preferable to read (9.10) probabilistically as follows. The smallest among $\mathbf{X}_1, \ldots, \mathbf{X}_{n-1}$ must be less than a, and there are $n-1$ choices for it. Given that $\mathbf{X}_j = x$, the probability that $\mathbf{X}_j$ is leftmost equals $[(t-x)/t]^{n-2}$. The remaining variables are distributed uniformly over $\overline{x, t}$ and the conditional probability that they satisfy the conditions of the theorem is $\varphi_{n-1}(t-x)$. Summing over all possibilities we get (9.10).[11]

Let us for the moment *define* u_n by (9.8). Then (9.10) reduces to

$$u_n(t) = a^{-1}\int_0^a u_{n-1}(t-x)\,dx \tag{9.11}$$

which is exactly the recursion formula (9.1) which served to define u_n. It suffices therefore to prove the theorem for $n = 2$. But it is obvious that $\varphi_2(t) = 1$ for $0 < t < a$ and $\varphi_2(t) = (2a-t)/t$ for $a < t < 2a$, in agreement with (9.8). ▶

10. RANDOM DIRECTIONS

Choosing a random direction in the plane $\mathfrak{R}^2$ is the same as choosing at random a point on the circle. If one wishes to specify the direction by its angle with the right x-axis, the circle should be referred to its arc length θ with $0 \leq \theta < 2\pi$. For random directions in the space $\mathfrak{R}^3$ the unit sphere serves as sample space; each domain has a probability equal to its area divided by 4π. Choosing a random direction in $\mathfrak{R}^3$ is equivalent to

---

[11] Readers who feel uneasy about the use of conditional probabilities in connection with densities should replace the hypothesis $\mathbf{X}_j = x$ by the hypothesis $x - h < \mathbf{X}_j < x$, which has positive probability, and pass to the limit as $h \to 0$.

choosing at random a point on this unit sphere. As this involves a pair of random variables (the longitude and latitude), consistency would require postponing the discussion to chapter III, but it appears more naturally in the present context.

Propositions. (i) *Denote by* $\mathbf{L}$ *the length of the projection of a unit vector with random direction in* $\mathfrak{R}^3$ *on a fixed line, say the* x*-axis. Then* $\mathbf{L}$ *is uniformly distributed over* $\overline{0, 1}$, *and* $\mathbf{E}(\mathbf{L}) = \frac{1}{2}$.

(ii) *Let* $\mathbf{U}$ *be the length of the projection of the same vector on a fixed plane, say the* x,y*-plane. Then* $\mathbf{U}$ *has density* $t/\sqrt{1-t^2}$ *for* $0 < t < 1$, *and* $\mathbf{E}(\mathbf{U}) = \frac{1}{4}\pi$.

The important point is that the two projections have different distributions. That the first is uniform is not an attribute of randomness, but depends on the number of dimensions. The counterpart to (i) in $\mathfrak{R}^2$ is contained in

Proposition. (iii) *Let* $\mathbf{L}$ *be the length of the projection of a random unit vector in* $\mathfrak{R}^2$ *on the* x*-axis. Then* $\mathbf{L}$ *has density* $2/(\pi\sqrt{1-x^2})$, *and* $\mathbf{E}(\mathbf{L}) = 2/\pi$.

Proofs. (iii) If θ is the angle between our random direction and the y-axis, then $\mathbf{L} = |\sin\theta|$ and hence for $0 < x < 1$ we get by symmetry

$$\mathbf{P}\{\mathbf{L} \le x\} = \mathbf{P}\{0 < \theta < \arcsin x\} = \frac{2}{\pi}\arcsin x. \tag{10.1}$$

The assertion now follows by differentiation.

(i), (ii). Recall the elementary theorem that the area of a spherical zone between two parallel planes is proportional to the height of the zone. For $0 < t < 1$ the event $\{\mathbf{L} \le t\}$ is represented by the zone $|x_1| \le t$ of height $2t$, whereas $\{\mathbf{U} \le t\}$ corresponds to the zones $|x_3| \ge \sqrt{1-t^2}$ of total height $2 - 2\sqrt{1-t^2}$. This determines the two distribution functions up to numerical factors, and these follow easily from the condition that both distributions equal 1 at $t = 1$ ►

Examples. (*a*) *Passage through spheres.* Let Σ be a sphere of radius r and N a point on it. A line drawn through N in a random direction intersects Σ in P. Then: *The length of the segment* NP *is a random variable distributed uniformly between* 0 *and* $2r$.

To see this consider the axis NS of the sphere and the triangle NPS which has a right angle at P and an angle Θ at N. The length of NP is then $2r\cos\Theta$. But $\cos\Theta$ is also the projection of a unit vector in the line NP into the diameter NS, and therefore $\cos \Theta$ is uniformly distributed in $\overline{0, 1}$.

In physics this model is used to describe the passage of light through "randomly distributed spheres." The resulting *absorption of light* was used as one example for the random-splitting process in the last section. (See problem 28.)

(*b*) *Circular objects under the microscope.* Through a microscope one observes the projection of a cell on the x_1, x_2-plane rather than its actual shape. In certain biological experiments the cells are lens-shaped and may be treated as circular disks. Only the horizontal diameter of the disk projects in its natural length, and the whole disk projects into an ellipse whose minor axis is the projection of the steepest radius. Now it is generally assumed that the orientation of the disk is random, meaning that the direction of its normal is chosen at random. In this case the projection of the unit normal on the x_3-axis is distributed uniformly in $\overline{0, 1}$. But the angle between this normal and the x_3-axis equals the angle between the steepest radius and the x_1, x_2-plane and hence the ratio of the minor to the major axis is distributed uniformly in $\overline{0, 1}$. Occasionally the evaluation of experiments was based on the erroneous belief that the angle between the steepest radius and the x_1, x_2-plane should be distributed uniformly.

(*c*) *Why are two violins twice as loud as one?* (The question is serious because the loudness is proportional to the *square* of the amplitude of the vibration.) The incoming waves may be represented by random *unit* vectors, and the superposition effect of two violins corresponds to the addition of two independent random vectors. By the law of the cosines the square of the length of the resulting vector is $2 + 2 \cos \Theta$. Here Θ is the angle between the two random vectors, and hence $\cos \Theta$ is uniformly distributed in $\overline{-1, 1}$ and has zero expectation. The expectation of the square of the resultant length is therefore indeed 2.

In the plane $\cos \theta$ is not uniformly distributed, but for reasons of symmetry its expectation is still zero. Our result therefore holds in any number of dimensions. See also example V,4(*e*). ▶

By a *random vector in* $\mathfrak{R}^3$ is meant a vector drawn in a random direction with a length $\mathbf{L}$ which is a random variable independent of its direction. The probabilistic properties of a random vector are completely determined by those of its projection on the x-axis, and using the latter it is frequently possible to avoid analysis in three dimensions. For this purpose it is important to know the relationship between the distribution function V of the true length $\mathbf{L}$ and the distribution F of the length $\mathbf{L}_x$ of the projection on the x-axis. Now $\mathbf{L}_x = \mathbf{XL}$, where $\mathbf{X}$ is the length of the projection of a *unit* vector in the given direction. Accordingly, $\mathbf{X}$ is distributed uniformly over $\overline{0,1}$ and is independent of $\mathbf{L}$. Given $\mathbf{X} = x$, the event $\{\mathbf{L}_x \leq t\}$ occurs iff $\mathbf{L} \leq t/x$, and so[12]

$$F(t) = \int_0^1 V(t/x)\,dx \qquad t > 0. \tag{10.2}$$

For the corresponding *densities* we get by differentiation

$$f(t) = \int_0^1 v\left(\frac{t}{x}\right)\frac{dx}{x} = \int_t^\infty v(y)\,\frac{dy}{y}, \tag{10.3}$$

and differentiation leads to

$$v(t) = -tf'(t), \qquad t > 0. \tag{10.4}$$

We have thus found *the analytic relationship between the density* v *of the length of a random vector in* $\mathfrak{R}^3$ *and the density* f *of the length of its projection on a fixed direction.* The relation (10.3) is used to find f when v is known, and (10.4) in the opposite direction. (The asymmetry between the two formulas is due to the fact that the direction is *not* independent of the length of the projection.)

Examples. (*d*) *Maxwell distribution for velocities.* Consider random vectors in space whose projections on the x-axis have the normal density with zero expectation and unit variance. Since length is taken positive we have

$$f(t) = 2\mathfrak{n}(t) = \sqrt{2/\pi}\, e^{-\frac{1}{2}t^2}, \qquad t > 0. \tag{10.5}$$

From (10.4) then

$$v(t) = \sqrt{2/\pi}\, t^2 e^{-\frac{1}{2}t^2}, \qquad t > 0. \tag{10.6}$$

This is the Maxwell density for velocities in statistical mechanics. The usual derivation combines the preceding argument with a proof that f must be of the form (10.5). (For an alternative derivation see III,4.)

(*e*) *Lord Rayleigh's random flights in* $\mathfrak{R}^3$. Consider n *unit* vectors whose directions are chosen independently and at random. We seek the distribution of *the length* $\mathbf{L}_n$ of their resultant (or vector sum). Instead of studying this resultant directly we consider its projection on the x-axis. This projection is obviously the sum of n independent random variables distributed uniformly over $\overline{-1,1}$. The density of this sum is given by (9.7) with $b = 1$. Substituting into (10.4) one sees that *the density of the length* $\mathbf{L}_n$ *is given by*[13]

$$v_n(x) = \frac{-x}{2^{n-1}(n-2)!}\sum_{\nu=0}^{n}(-1)^\nu\binom{n}{\nu}(x+n-2\nu)_+^{n-2}, \qquad x > 0. \tag{10.7}$$

[12] This argument repeats the proof of (8.3).

[13] The standard reference is to a paper by S. Chandrasekhar [reprinted in Wax (1954)] who calculated v_3, v_4, v_6 and the Fourier transform of v_n. Because he used polar coordinates, his $W_n(x)$ must be multiplied by $4\pi x^2$ to obtain our v_n.

This problem occurs in physics and chemistry (the vectors representing, for example, plane waves or molecular links). The reduction to one dimension seems to render this famous problem trivial.

The same method applies to random vectors with arbitrary length and thus (10.4) enables us to *reduce random-walk problems in* $\mathfrak{R}^3$ *to simpler problems in* $\mathfrak{R}^1$. Even when explicit solutions are hard to get, the central limit theorem provides valuable information [see example VIII,4(*b*)]. ▶

Random vectors in $\mathfrak{R}^2$ are defined in like manner. The distribution V of the true length and the distribution F of the projection are related by the obvious analogue to (10.2), namely

$$F(x) = \frac{2}{\pi}\int_0^{\pi/2} V\left(\frac{x}{\sin\theta}\right) d\theta. \tag{10.8}$$

However, the inversion formula (10.4) has no simple analogue, and to express V in terms of F we must depend on the relatively deep theory of Abel's integral equation. [14] We state without proof that if F has a continuous density f, then

$$1 - V(x) = x\int_0^{\pi/2} f\left(\frac{x}{\sin\theta}\right)\frac{d\theta}{\sin^2\theta}. \tag{10.9}$$

(See problems 29–30.)

Example. (*f*) *Binary orbits.* In observing a spectroscopic binary orbit astronomers can measure only the projections of vectors onto a plane perpendicular to the line of sight. An ellipse in space projects into an ellipse in this plane. The major axis of the true ellipse lies in the plane determined by the line of sight and its projection, and it is therefore reasonable to assume that the angle between the major axis and its projection is uniformly distributed. Measurements determine (in principle) the distribution of the projection. The distribution of the true major axis is then given by the solution (10.9) of Abel's integral equation. ▶

11. THE USE OF LEBESGUE MEASURE

If a set A in $\overline{0,1}$ is the union of finitely many non-overlapping intervals $I_1, I_2, \ldots$ of lengths $\lambda_1, \lambda_2, \ldots,$ the uniform distribution attributes to it probability

$$\mathbf{P}\{A\} = \lambda_1 + \lambda_2 + \cdots. \tag{11.1}$$

The following examples will show that some simple, but significant, problems

---

[14] The transformation to Abel's integral equation is by means of the change of variables

$$F_1(x) = F\left(\frac{1}{\sqrt{x}}\right), \qquad V_1(x) = V\left(\frac{1}{\sqrt{x}}\right), \quad \text{and} \quad x\sin^2\theta = y.$$

Then (10.8) takes on the form

$$F_1(t) = \int_0^t \frac{V_1(y)}{\sqrt{y(t-y)}}\, dy.$$

lead to unions of infinitely many non-overlapping intervals. The definition (11.1) is still applicable and identifies $\mathbf{P}\{A\}$ with the Lebesgue measure of A. It is consistent with our program to identify probabilities with the integral of the density $f(x) = 1$, except that we use the Lebesgue integral rather than the Riemann integral (which need not exist). Of the Lebesgue theory we require only the fact that if A is the union of possibly overlapping intervals $I_1, I_2, \ldots$ the measure $\mathbf{P}\{A\}$ exists and does not exceed the sum $\lambda_1 + \lambda_2 + \cdots$ of the lengths. For non-overlapping intervals the equality (11.1) holds. The use of Lebesgue measure conforms to uninhibited intuition and simplifies matters inasmuch as many formal passages to the limit are justified. A set N is called a *null set* if it is contained in sets of arbitrarily small measure, that is, to each ϵ there exists a set $A \supset N$ such that $\mathbf{P}\{A\} < \epsilon$. In this case $\mathbf{P}\{N\} = 0$.

In the following $\mathbf{X}$ stands for a random variable distributed uniformly in $\overline{0, 1}$.

Examples. (*a*) *What is the probability of* $\mathbf{X}$ *being rational*? The sequence $\frac{1}{2}, \frac{1}{3}, \frac{2}{3}, \frac{1}{4}, \frac{3}{4}, \frac{1}{5}, \ldots$ contains *all* the rationals in $\overline{0, 1}$ (ordered according to increasing denominators). Choose $\epsilon < \frac{1}{2}$ and denote by J_k an interval of length ϵ^{k+1} centered at the kth point of the sequence. The sum of the lengths of the J_k is $\epsilon^2 + \epsilon^3 + \cdots < \epsilon$, and their union covers the rationals. Therefore by our definition *the set of all rationals has probability zero*, and so $\mathbf{X}$ is irrational with probability one.

It is pertinent to ask why such sets should be considered in probability theory. One answer is that nothing can be gained by excluding them and that the use of Lebesgue theory actually simplifies matters without requiring new techniques. A second answer may be more convincing to beginners and non-mathematicians; the following variants lead to problems of undoubted probabilistic nature.

(*b*) *With what probability does the digit* 7 *occur in the decimal expansion of* $\mathbf{X}$? In the decimal expansion of each x in the open interval between 0.7 and 0.8 the digit 7 appears at the first place. For each n there are 9^{n-1} intervals of length 10^{-n} containing only numbers such that the digit 7 appears at the nth place but not before. (For $n = 2$ their endpoints are 0.07 and 0.08, next 0.17 and 0.18, etc.) These intervals are non-overlapping, and their total length is $\frac{1}{10}(1 + \frac{9}{10} + (\frac{9}{10})^2 + \cdots) = 1$. Thus our event has *probability* 1.

Notice that certain numbers have two expansions, for example $0.7 = 0.6999\ldots$. To make our question unequivocal we should therefore specify whether the digit 7 must or may occur in the expansion, but our argument is independent of the difference. The reason is that only rationals can have two expansions, and the set of all rationals has probability zero.

(*c*) *Coin tossing and random choice.* Let us now see how a "random choice of a point $\mathbf{X}$ between 0 and 1" can be described in terms of discrete random variables. Denote by $\mathbf{X}_k(x)$ the kth decimal of x. (To avoid ambiguities let us use terminating expansions when possible.) The random variable $\mathbf{X}_k$ assumes the values $0, 1, \ldots, 9$, each with probability $\frac{1}{10}$, and the $\mathbf{X}_k$ are mutually independent. By the definition of a decimal expansion, we have the identity

$$\mathbf{X} = \sum_{k=1}^{\infty} 10^{-k}\mathbf{X}_k. \tag{11.2}$$

This formula reduces the random choice of a point $\mathbf{X}$ to successive choices of its decimals.

For further discussion we switch from decimal to *dyadic* expansions, that is, we replace the basis 10 by 2. Instead of (11.2) we have now

$$\mathbf{X} = \sum_{k=1}^{\infty} 2^{-k}\mathbf{X}_k \tag{11.3}$$

where the $\mathbf{X}_k$ are mutually independent random variables assuming the values 0 and 1 with probability $\frac{1}{2}$. These variables are defined on the interval $\overline{0, 1}$ on which probability is equated with Lebesgue measure (length). This formulation brings to mind the coin-tossing game of volume **1**, in which the sample space consists of infinite sequences of heads and tails, or zeros and ones. A new interpretation of (11.3) is now possible in this sample space. In it, the $\mathbf{X}_k$ are coordinate variables, and $\mathbf{X}$ is a random variable defined by them; its distribution function is, of course, uniform. Note that the second formulation contains two distinct sample points 0111111 and 1000000 even though the corresponding dyadic expansions represent the same point $\frac{1}{2}$. Nevertheless, the notion of zero probability enables us to *identify the two sample spaces.* Stated in more intuitive terms, neglecting an event of probability zero *the random choice of a point* $\mathbf{X}$ *between* 0 *and* 1 *can be effected by a sequence of coin tossings*; conversely, the result of an infinite coin-tossing game may be represented by a point x of $\overline{0, 1}$. Every random variable of the coin-tossing game may be represented by a function on $\overline{0, 1}$, etc. This convenient and intuitive device has been used since the beginning of probability theory, but it depends on neglecting events of zero probability.

(*d*) *Cantor-type distributions.* A distribution with unexpected properties is found by considering in (11.3) the contribution of the even-numbered terms or, what amounts to the same, by considering the random variable

$$\mathbf{Y} = 3\sum_{\nu=1}^{\infty} 4^{-\nu}\mathbf{X}_\nu. \tag{11.4}$$

(The factor 3 is introduced to simplify the discussion. The contribution of the odd-numbered terms has the same distribution as $\frac{2}{3}\mathbf{Y}$.) The distribution function $F(x) = \mathbf{P}\{\mathbf{Y} \leq x\}$ will serve as example for so-called singular distributions.

In the calculation we refer to $\mathbf{Y}$ as the gain of a gambler who receives the amount $3 \cdot 4^{-k}$ if the kth toss of a fair coin results in tails. This gain lies between 0 and $3(4^{-1}+4^{-2}+\cdots) = 1$. If the first trial results in 1 the gain is $\geq\frac{3}{4}$, while in the contrary case $\mathbf{Y} \leq 3(4^{-2}+4^{-3}+\cdots) = 4^{-1}$. Thus the inequality $\frac{1}{4} < \mathbf{Y} < \frac{3}{4}$ cannot be realized under any circumstances, and so $F(x) = \frac{1}{2}$ in this interval of length $\frac{1}{2}$. It follows that F can have no jump exceeding $\frac{1}{2}$.

Next notice that up to a factor $\frac{1}{4}$ the contribution of the trials number 2, 3, ... constitute a replica of the whole sequence, and so the graph of F in the interval $\overline{0, \frac{1}{4}}$ differs from the whole graph only by a similarity transformation

$$F(x) = \tfrac{1}{2}F(4x), \qquad 0 < x < \tfrac{1}{4}. \tag{11.5}$$

It follows that $F(x) = \frac{1}{4}$ throughout an interval of length $\frac{1}{8}$ centered at $x = \frac{1}{8}$. For reasons of symmetry, $F(x) = \frac{3}{4}$ throughout an interval of length $\frac{1}{8}$ centered at $x = \frac{7}{8}$. We have now found three intervals of total length $\frac{1}{2} + \frac{2}{8} = \frac{3}{4}$ in each of which F assumes a constant value, namely $\frac{1}{4}, \frac{1}{2}$, or $\frac{3}{4}$. Consequently, F can have no jump exceeding $\frac{1}{4}$. There remain four intervals of length $\frac{1}{16}$ each, and in each of them the graph of F differs from the whole graph only by a similarity transformation. Each of the four intervals therefore contains a subinterval of half its length in which F assumes a constant value (namely $\frac{1}{8}, \frac{3}{8}, \frac{5}{8}, \frac{7}{8}$, respectively). Continuing in like manner we find in n steps $1 + 2 + 2^2 + \cdots + 2^{n-1}$ intervals of total length $2^{-1} + 2^{-2} + 2^{-3} + \cdots + 2^{-n} = 1 - 2^{-n}$ in each of which F assumes a constant value.

Thus F *is a continuous function increasing from* $F(0) = 0$ *to* $F(1) = 1$ *in such a way that the intervals of constancy add up to length* 1. Roughly speaking, the whole increase of F takes place on a set of measure 0. We have here a continuous distribution function F without density f. ▶

12. EMPIRICAL DISTRIBUTIONS

The "*empirical distribution function*" F_n of n points $a_1, \ldots, a_n$ on the line is the step function with jumps $1/n$ at $a_1, \ldots, a_n$. In other words, $n\,F_n(x)$ equals the number of points a_k in $\overline{-\infty, x}|$, and F_n is a distribution function. Given n random variables $\mathbf{X}_1, \ldots, \mathbf{X}_n$, their values at a particular point of the sample space form an n-tuple of numbers and its empirical distribution function is called the empirical sample distribution. For each

x, the value $\mathbf{F}_n(x)$ of the empirical sample distribution defines a new random variable, and *the empirical distribution of* $(\mathbf{X}_1, \ldots, \mathbf{X}_n)$ represents a whole family of random variables depending on the parameter x. (In technical language we are concerned with a stochastic process with x as time parameter.) No attempt will be made here to develop the theory of empirical distributions, but the notion may be used to illustrate the occurrence of complicated random variables in simple applications. Furthermore, the uniform distribution will appear in a new light.

Let $\mathbf{X}_1, \ldots, \mathbf{X}_n$ stand for mutually independent random variables with a common continuous distribution F. The probability that any two variables assume the same value is zero, and we can therefore restrict our attention to samples of n distinct values. For fixed x the number of variables $\mathbf{X}_k$ such that $\mathbf{X}_k \leq x$ has a binomial distribution with probability of "success" $p = F(x)$, and so *the random variable* $\mathbf{F}_n(x)$ *has a binomial distribution* with possible values $0, 1/n, \ldots, 1$. For large n and x fixed, $\mathbf{F}_n(x)$ is therefore likely to be close to $F(x)$ and the central limit theorem tells us more about the probable deviations. More interesting is the (chance-dependent) graph of $\mathbf{F}_n$ as a whole and how close it is to F. A measure for this closeness is the *maximum discrepancy*, that is,

$$\mathbf{D}_n = \sup_{-\infty < x < \infty} |\mathbf{F}_n(x) - F(x)|. \tag{12.1}$$

This is a new random variable of great interest to statisticians because of the following property. The *probability distribution of the random variable* $\mathbf{D}_n$ *is independent of* F (provided, of course, that F is continuous).

For the proof it suffices to verify that the distribution of $\mathbf{D}_n$ remains unchanged when F is replaced by a uniform distribution. We begin by showing that *the variables* $\mathbf{Y}_k = F(\mathbf{X}_k)$ *are distributed uniformly in* $\overline{0, 1}$. For that purpose we restrict t to the interval $\overline{0, 1}$, and in this interval we define v as the inverse function of F. The event $\{F(\mathbf{X}_k) \leq t\}$ is then identical with the event $\{\mathbf{X}_k \leq v(t)\}$ which has probability $F(v(t)) = t$. Thus $\mathbf{P}\{\mathbf{Y}_k \leq t\} = t$ as asserted.

The variables $\mathbf{Y}_1, \ldots, \mathbf{Y}_n$ are mutually independent, and we denote their empirical distribution by $\mathbf{G}_n$. The argument just used shows also that for fixed t the random variable $\mathbf{G}_n(t)$ is identical with $\mathbf{F}_n(v(t))$. Since $t = F(v(t))$ this implies that at every point of the sample space $\mathfrak{R}^n$

$$\sup |\mathbf{G}_n(t) - t| = \sup |\mathbf{F}_n(v(t)) - F(v(t))| = \mathbf{D}_n.$$

This proves the proposition.

The fact that the distribution of $\mathbf{D}_n$ is independent of the underlying distribution F enables statisticians to devise tests and estimation procedures applicable in situations when the underlying distribution is unknown. In this connection other variables related to $\mathbf{D}_n$ are of even greater practical use.

Let $\mathbf{X}_1, \ldots, \mathbf{X}_n, \mathbf{X}_1^\#, \ldots, \mathbf{X}_n^\#$ be $2n$ mutually independent random variables with the common continuous distribution F, and denote the empirical distributions of $(\mathbf{X}_1, \ldots, \mathbf{X}_n)$ and $(\mathbf{X}_1^\#, \ldots, \mathbf{X}_n^\#)$ by $\mathbf{F}_n$ and $\mathbf{F}_n^\#$, respectively. Put

$$\mathbf{D}_{n,n} = \sup_x |\mathbf{F}_n(x) - \mathbf{F}_n^\#(x)|. \tag{12.2}$$

This is the *maximum discrepancy between the two empirical distributions.* It shares with $\mathbf{D}_n$ the property that it does not depend on the distribution F. For this reason it serves in statistical tests of "the hypothesis that $(\mathbf{X}_1, \ldots, \mathbf{X}_n)$ and $(\mathbf{X}_1^\#, \ldots, \mathbf{X}_n^\#)$ are random samples from the same population."

The distribution of $\mathbf{D}_{n,n}$ was the object of cumbersome calculations and investigations but in 1951 B. V. Gnedenko and V. S. Koroljuk showed that the whole question reduces to a random-walk problem with a well-known solution. Their argument is pleasing by its elegance and we use it as illustration of the power of simple combinatorial methods.

Theorem. $\mathbf{P}\{\mathbf{D}_{n,n} < r/n\}$ *equals the probability in a symmetric random walk that a path of length* $2n$ *starting and terminating at the origin does not reach the points* $\pm r$.

Proof. It suffices to consider integral r. Order the $2n$ variables $\mathbf{X}_1, \ldots, \mathbf{X}_n^\#$ in order of increasing magnitude and put $\epsilon_k = 1$ or $\epsilon_k = -1$, according to whether the kth place is occupied by an $\mathbf{X}_j$ or an $\mathbf{X}_j^\#$. The resulting arrangement contains n plus ones and n minus ones, and all $\binom{2n}{n}$ orderings are equally likely. The resulting $2n$-tuples $(\epsilon_1, \ldots, \epsilon_{2n})$ are therefore in a one-to-one correspondence with the paths of length $2n$ starting and terminating at the origin. Now if $\epsilon_1 + \cdots + \epsilon_j = k$ the first j places contain $(j+k)/2$ unsuperscripted and $(j-k)/2$ superscripted variables, and so there exists a point x such that $\mathbf{F}_n(x) = (j+k)/2n$ and $\mathbf{F}_n^\#(x) = (j-k)/2n$. But then $|\mathbf{F}_n(x) - \mathbf{F}_n^\#(x)| = |k|/n$ and hence $\mathbf{D}_{n,n} \geq |k|/n$. The same argument in reverse completes the proof. ▶

An *explicit expression* for the probability in question is contained in **1**, XIV,(9.1). In fact

$$\mathbf{P}\{\mathbf{D}_{n,n} < r/n\} = w_{r,n}$$

is the probability that a particle starting at the origin returns at epoch $2n$ to the origin without touching $\pm r$. The last condition can be realized by putting absorbing barriers at $\pm r$, and so $w_{r,n}$ is the probability of a return to the origin at epoch $2n$ when $\pm r$ are absorbing barriers. [In **1**; XIV,(9.1) the interval is $\overline{0, a}$ rather than $\overline{-r, r}$. Our $w_{r,n}$ is identical with $u_{r,2n}(r)$.]

It was shown in **1**; XIV that a limiting procedure leads from random walks to diffusion processes, and in this way it is not difficult to see that the distribution of $\sqrt{n}\mathbf{D}_{n,n}$ tends to a limit. Actually this limit was discovered by N. V. Smirnov as early as 1939 and the similar limit for $\sqrt{n}\mathbf{D}_n$ by A. Kolmogorov in 1933. Their calculations are very intricate and do not explain the connection with diffusion processes, which is inherent in the Gnedenko-Koroljuk approach. On the other hand, they have given impetus to fruitful work on the convergence of stochastic processes (P. Billingsley, M. F. Donsker, Yu. V. Prohorov, A. V. Skorohod, and others).

It may be mentioned that the Smirnov theorems apply equally to discrepancies $\mathbf{D}_{m,n}$ of the empirical distributions of samples of different sizes m and n. The random-walk approach carries over, but loses much of its elegance and simplicity (B. V. Gnedenko, E. L. Rvateva). A great many variants of $\mathbf{D}_{m,n}$ have been investigated by statisticians. (See problem 36.)

13. PROBLEMS FOR SOLUTION

In all problems it is understood that the given variables are *mutually independent.*

1. Let $\mathbf{X}$ and $\mathbf{Y}$ have densities $\alpha e^{-\alpha x}$ concentrated on $\overline{0, \infty}$. Find the densities of

(i) $\mathbf{X}^3$ (ii) $3 + 2\mathbf{X}$
(iii) $\mathbf{X} - \mathbf{Y}$ (iv) $|\mathbf{X} - \mathbf{Y}|$
(v) The smaller of $\mathbf{X}$ and $\mathbf{Y}^3$ (vi) The larger of $\mathbf{X}$ and $\mathbf{Y}^3$.

2. Do the same problem if the densities of $\mathbf{X}$ and $\mathbf{Y}$ equal $\frac{1}{2}$ in $\overline{-1, 1}$ and 0 elsewhere.

3. Find the densities for $\mathbf{X} + \mathbf{Y}$ and $\mathbf{X} - \mathbf{Y}$ if $\mathbf{X}$ has density $\alpha e^{-\alpha x}(x > 0)$ and the density of $\mathbf{Y}$ equals h^{-1} for $0 < x < h$.

4. Find the probability that $\lambda^2 - 2a\lambda + b$ has complex roots if the coefficients a and b are random variables whose common density is

(i) uniform, that is, h^{-1} for $0 < x < h$
(ii) exponential, that is, $\alpha e^{-\alpha x}$ for $x > 0$.

5. Find the distribution functions of $\mathbf{X}+\mathbf{Y}/\mathbf{X}$ and $\mathbf{X}+\mathbf{Y}/\mathbf{Z}$ if the variables $\mathbf{X}$, $\mathbf{Y}$, and $\mathbf{Z}$ have a common exponential distribution.

6. Derive the convolution formula (3.6) for the exponential distribution by a direct passage to the limit from the convolution formula for the "negative binomial" distribution of **1**; VI,(8.1).

7. In the Poisson process of section 4, denote by $\mathbf{Z}$ the time between epoch t and the last preceding arrival or 0 (the "age" of the current interarrival time). Find the distribution of $\mathbf{Z}$ and show that it tends to the exponential distribution as $t \to \infty$.

8. In example 5(*a*) show that the probability of the first record value occurring at the nth place and being $\leq x$ equals

$$\frac{1}{n(n+1)}(1 - e^{-\alpha x})^{n+1}.$$

Conclude that the probability distribution of the first record value is $1 - (1 + \alpha x)e^{-\alpha x}$.

[More generally, if the $\mathbf{X}_j$ are positive and subject to an arbitrary continuous distribution F, the first probability equals $[n(n+1)]^{-1}F^{n+1}(x)$ and the distribution of the first record value is $F - (1-F)\log(1-F)^{-1}$.]

9. *Downward runs.* The random variable $\mathbf{N}$ is defined as the unique index such that $\mathbf{X}_1 \geq \mathbf{X}_2 \geq \cdots \geq \mathbf{X}_{\mathbf{N}-1} < \mathbf{X}_{\mathbf{N}}$. If the $\mathbf{X}_j$ have a common continuous distribution F prove that $\mathbf{P}\{\mathbf{N} = n\} = (n-1)/n!$ and $\mathbf{E}(\mathbf{N}) = e$.

Hint: Use the method of example 5(*a*) concerning record values.

10. *Platoon formation in traffic.*[15] Cars start successively at the origin and travel at different but constant speeds along an infinite road on which no passing is possible. When a car reaches a slower car it is compelled to trail it at the same speed. In this way platoons will be formed whose ultimate size depends on the speeds of the cars but not on the times between successive departures.

Consider the speeds of the cars as independent random variables with a common continuous distribution. Choose a car at random, say the next car to depart. Using the combinatorial method of example 5(*a*) show that:

(*a*) The probability that the given car does not trail any other car tends to $\frac{1}{2}$.

(*b*) The probability that it leads a platoon of total size n (with exactly $n - 1$ cars trailing it) tends to $1/(n+1)(n+2)$.

(*c*) The probability that the given car is the last in a platoon of size n tends to the same limit.

11. *Generalization*[16] *of the record value example* 5(*a*). Instead of taking the single preliminary observation $\mathbf{X}_0$ we start from a sample $(\mathbf{X}_1, \ldots, \mathbf{X}_m)$ with order statistics $(\mathbf{X}_{(1)}, \ldots, \mathbf{X}_{(m)})$. (The common distribution F plays no role as long as it is continuous.)

(*a*) If $\mathbf{N}$ is the first index n such that $\mathbf{X}_{m+n} \geq \mathbf{X}_{(m)}$ show that $\mathbf{P}\{\mathbf{N} > n\} = m/(m+n)$. [In example 5(*a*) we had $m = 1$.]

(*b*) If $\mathbf{N}$ is the first index n such that $\mathbf{X}_{m+n} \geq \mathbf{X}_{(m-r+1)}$ show that

$$\mathbf{P}\{\mathbf{N} > n\} = \binom{m}{r} \Big/ \binom{m+n}{r}.$$

For $r \geq 2$ we have $\mathbf{E}(\mathbf{N}) < \infty$ and

$$\mathbf{P}\{\mathbf{N} \leq mx\} \to 1 - \frac{1}{(1+x)^r}, \qquad m \to \infty.$$

(*c*) If $\mathbf{N}$ is the first index such that $\mathbf{X}_{m+n}$ falls outside the interval between $\mathbf{X}_{(1)}$ and $\mathbf{X}_{(m)}$ then

$$\mathbf{P}\{\mathbf{N} > n\} = \frac{m(m-1)}{(m+n)(m+n-1)}, \quad \text{and} \quad \mathbf{E}(\mathbf{N}) < \infty.$$

12. (*Convolutions of exponential distributions*). For $j = 0, \ldots, n$ let $\mathbf{X}_j$ have density $\lambda_j e^{-\lambda_j x}$ for $x > 0$ where $\lambda_j \neq \lambda_k$ unless $j = k$. Put

$$\psi_{k,n} = [(\lambda_0 - \lambda_k) \cdots (\lambda_{k-1} - \lambda_k)(\lambda_{k+1} - \lambda_k) \cdots (\lambda_n - \lambda_k)]^{-1}.$$

Show that $\mathbf{X}_0 + \cdots + \mathbf{X}_n$ has a density given by

$$(*) \qquad P_n(t) = \lambda_0 \cdots \lambda_{n-1}[\psi_{0,n} e^{-\lambda_0 t} + \cdots + \psi_{n,n} e^{-\lambda_n t}].$$

Hint: Use induction, a symmetry argument, and (2.14). No calculations are necessary.

---

[15] G. F. Newell, Operations Research, vol. 7 (1959), pp. 589–598.

[16] S. S. Wilks, J. Australian Math. Soc., vol. 1 (1959) pp. 106–112.

13. (*Continuation*). If $\mathbf{Y}_j$ has the density je^{-jx}, the density of the sum $\mathbf{Y}_1 + \cdots + \mathbf{Y}_n$ is given by

$$f_n(x) = n \sum_{k=1}^{n} (-1)^{k-1} \binom{n-1}{k-1} e^{-kx}, \qquad x > 0.$$

Using the proposition of example 6(*b*) conclude that f_{n-1} is the density of the spread $\mathbf{X}_{(n)} - \mathbf{X}_{(1)}$ of a sample $\mathbf{X}_1, \ldots, \mathbf{X}_n$ if the $\mathbf{X}_j$ have the common density e^{-x}.

14. *Pure birth processes*. In the pure birth process of **1**; XVII,3 the system passes through a sequence of states $E_0 \to E_1 \to \cdots$, staying at E_k for a sojourn time $\mathbf{X}_k$ with density $\lambda_k e^{-\lambda_k x}$. Thus $\mathbf{S}_n = \mathbf{X}_0 + \cdots + \mathbf{X}_n$ is the epoch of the transition $E_n \to E_{n+1}$. Denote by $P_n(t)$ the probability of E_n at epoch t. Show that $\mathbf{P}_n(t) = \mathbf{P}\{\mathbf{S}_n > t\} - \mathbf{P}\{\mathbf{S}_{n-1} > t\}$ and hence that $\mathbf{P}_n$ is given by formula (*) of problem 12. The differential equations of the process, namely

$$P_0'(t) = -\lambda_0 P_0(t), \qquad P_n'(t) = -\lambda_n P_n(t) + \lambda_{n-1} P_{n-1}(t), \qquad n \geq 1,$$

should be derived (*a*) from (1), and (*b*) from the properties of the sums $\mathbf{S}_n$.

Hint: Using inductively a symmetry argument it suffices to consider the factor of $e^{-\lambda_0 t}$.

15. In example 6(*a*) for parallel waiting lines we say that the system is in state k if k counters are free. Show that the birth process model of the last example applies with $\lambda_k = (n-k)\alpha$. Conclude that

$$P_k(t) = \binom{n}{k} (1 - e^{-\alpha t})^k e^{-(n-k)\alpha t}.$$

From this derive the distribution of $\mathbf{X}_{(k)}$.

16. Consider two independent queues of m and $n > m$ persons respectively, assuming the same exponential distribution for the service times. Show that the probability of the longer queue finishing first equals the probability of obtaining n heads before m tails in a fair coin-tossing game. Find the same probability also by considering the ratio $\mathbf{X}/\mathbf{Y}$ of two variables with gamma distributions G_m and G_n given in (3.5).

17. *Example of statistical estimation*. It is assumed that the lifetimes of electric bulbs have an exponential distribution with an unknown expectation α^{-1}. To estimate α a sample of n bulbs is taken and one observes the lifetimes

$$\mathbf{X}_{(1)} < \mathbf{X}_{(2)} < \cdots < \mathbf{X}_{(r)}$$

of the first r bulbs to fail. The "best unbiased estimator" of α^{-1} is a linear combination $\mathbf{U} = \lambda_1 \mathbf{X}_{(1)} + \cdots + \lambda_r \mathbf{X}_{(r)}$ such that $E(\mathbf{U}) = \alpha^{-1}$ and $\operatorname{Var}(\mathbf{U})$ is the smallest possible. Show that

$$\mathbf{U} = (\mathbf{X}_{(1)} + \cdots + \mathbf{X}_{(r)}) \frac{1}{r} + \mathbf{X}_{(r)}(n-r)\frac{1}{r}, \quad \text{and then} \quad \operatorname{Var}(\mathbf{U}) = \frac{1}{r}\alpha^{-2}.$$

Hint: Do the calculations in terms of the *independent* variables $\mathbf{X}_{(k)} - \mathbf{X}_{(k-1)}$ (see example 6(*b*)).

18. If the variables $\mathbf{X}_1, \ldots, \mathbf{X}_n$ are distributed uniformly in $\overline{0,1}$ show that the spread $\mathbf{X}_{(n)} - \mathbf{X}_{(1)}$ has density $n(n-1)x^{n-2}(1-x)$ and expectation $(n-1)/(n+1)$. What is the probability that all n points lie within an interval of length t?

19. Answer the questions of example 6(*b*) when the three service times are distributed uniformly in $\overline{0, 1}$. (*Note:* The problem involves tedious calculations, but may provide a useful exercise in technical manipulations.)

20. Four points are chosen independently and at random on a circle. Find the probability that the chords $\mathbf{X}_1\mathbf{X}_2$ and $\mathbf{X}_3\mathbf{X}_4$ intersect: (*a*) without calculation using a symmetry argument; (*b*) from the definition by an integral.

21. In the random-splitting process of section 8 denote by $\mathbf{X}_{11}, \mathbf{X}_{12}, \mathbf{X}_{21}, \mathbf{X}_{22}$ the masses of the four fragments of the second generation, the subscript 1 referring to the smaller and 2 to the larger part. Find the densities and expectations of these variables.

Note. The next few problems contain new theorems concerning *random partitions of an interval* [see example 7(*b*)]. The variables $\mathbf{X}_1, \ldots, \mathbf{X}_n$ are supposed independent and uniformly distributed over $\overline{0, t}$. They induce a partition of this interval into $n + 1$ subintervals whose lengths, taken in proper order, will be denoted by $\mathbf{L}_1, \ldots, \mathbf{L}_{n+1}$. [In the notation of order statistics we have

$$\mathbf{L}_1 = \mathbf{X}_{(1)}, \qquad \mathbf{L}_2 = \mathbf{X}_{(2)} - \mathbf{X}_{(1)}, \ldots, \qquad \mathbf{L}_{n+1} = t - \mathbf{X}_{(n)}.]$$

22. Denote by $p_n(t)$ the probability that all $n + 1$ intervals are longer than h. [In other words, $p_n(t) = \mathbf{P}\{\min \mathbf{L}_k > h\}$, which is the tail of the distribution function of the shortest among the intervals.] Prove the recurrence relation

$$(*) \qquad p_n(t) = \frac{n}{t^n} \int_0^{t-h} x^{n-1} p_{n-1}(x)\, dx.$$

Conclude that $p_n(t) = t^{-n}(t - (n+1)h)_+^n$.

23. From a recurrence relation analogous to (*) prove without calculations that *for arbitrary* $x_1 \geq 0, \ldots, x_{n+1} \geq 0$

$$(**) \qquad \mathbf{P}\{\mathbf{L}_1 > x_1, \ldots, \mathbf{L}_{n+1} > x_{n+1}\} = t^{-n}(t - x_1 - \cdots - x_{n+1})_+^n.$$

[This elegant result was derived by B. de Finetti[17] from geometrical considerations. It contains many interesting special cases. When $x_j = h$ for all j we get the preceding problem. Example 7(*b*) corresponds to the special case where exactly one among the x_j is different from zero. The covering theorem 3 of section 9 follows from (**) and the formula **1**; IV,(1.5) for the realization of at least one among $n + 1$ events.]

24. Denote by $q_n(t)$ the probability that all mutual distances of the $\mathbf{X}_k$ exceed h. (This differs from problem 22 in that no restrictions are imposed on the end intervals $\mathbf{L}_1$ and $\mathbf{L}_{n+1}$.) Find a relation analogous to (*) and hence derive $q_n(t)$.

25. *Continuation.* Without using the solution of the preceding problems show a priori that $p_n(t) = (t-2h)^n t^{-n} q_n(t-2h)$.

26. Formulate the analogue to problem 24 for a circle and show that problem 23 furnishes its solution.

---

27. An isosceles triangle is formed by a unit vector in the x-direction and another in a random direction. Find the distribution of the length of the third side (i) in $\mathfrak{R}^2$ and (ii) in $\mathfrak{R}^3$.

[17] Giornale Istituto Italiano degli Attuari, vol. 27 (1964) pp. 151–173, in Italian.

28. A unit circle (sphere) about 0 has the north pole on the positive x-axis. A ray enters at the north pole and its angle with the x-axis is distributed uniformly over $\overline{-\frac{1}{2}\pi, \frac{1}{2}\pi}$. Find the distribution of the length of the chord within the circle (sphere).

Note. In $\mathfrak{R}^2$ the ray has a random direction and we are concerned with the analogue to example 10(a). In $\mathfrak{R}^3$ the problem is new.

29. The ratio of the *expected lengths* of a random vector and of its projection on the x-axis equals 2 in $\mathfrak{R}^3$ and $\pi/2$ in $\mathfrak{R}^2$. *Hint:* Use (10.2) and (10.8).

30. The length of a random vector is distributed uniformly over $\overline{0, 1}$. Find the density of the length of its projection on the x-axis (a) in $\mathfrak{R}^3$, and (b) in $\mathfrak{R}^2$. *Hint:* Use (10.4) and (10.9).

31. Find the distribution function of the projection on the x-axis of a randomly chosen direction in $\mathfrak{R}^4$.

32. Find the analogue in $\mathfrak{R}^4$ to the relation (10.2) between the distributions of the lengths of a random vector and that of its projection on the x-axis. Specialize to a unit vector to verify the result of problem 31.

33. *A limit theorem for order statistics.* (a) Let $\mathbf{X}_1, \ldots, \mathbf{X}_n$ be distributed uniformly in $\overline{0, 1}$. Prove that for k fixed and $n \to \infty$

$$\mathbf{P}\left\{\mathbf{X}_{(k)} \leq \frac{x}{n}\right\} \to G_{k-1}(x), \qquad x > 0,$$

where G_k is the gamma distribution (3.5) [see example 7(e)].

(b) If the $\mathbf{X}_k$ have an arbitrary continuous distribution function F, the same limit exists for $\mathbf{P}\{\mathbf{X}_{(k)} \leq \Phi(x/n)\}$ where Φ is the inverse function of F. (Smirnov.)

34. *A limit theorem for the sample median.* The nth-order statistic $\mathbf{X}_{(n)}$ of $(\mathbf{X}_1, \ldots, \mathbf{X}_{2n-1})$ is called the sample median. If the $\mathbf{X}_j$ are independent and uniformly distributed over $\overline{0, 1}$ show that

$$\mathbf{P}\{\mathbf{X}_{(n)} - \tfrac{1}{2} < t/\sqrt{8n}\} \to \mathfrak{N}(t)$$

where $\mathfrak{N}$ stands for the standard normal distribution.

35. *Continuation.* Let the $\mathbf{X}_j$ have a common distribution F with a continuous density f. Let m be the theoretical median, that is, let $F(m) = \frac{1}{2}$. Show that

$$\mathbf{P}\{\mathbf{X}_{(n)} < x\} = (2n-1)\binom{2n-2}{n-1}\int_{-\infty}^{x} F^{n-1}(y)[1-F(y)]^{n-1} f(y)\, dy$$

whence, using the preceding problem,

$$\mathbf{P}\left\{\mathbf{X}_{(n)} - m < \frac{t}{f(m)\sqrt{8n}}\right\} \to \mathfrak{N}(t).$$

36. Prove the following variant of the Gnedenko-Koroljuk theorem in section 12:

$$\mathbf{P}\left\{\sup_x\,[\mathbf{F}_n(x) - \mathbf{F}_n^{\#}(x)] \geq \frac{r}{n}\right\} = \binom{2n}{n-r}\bigg/\binom{2n}{n},$$

where $r = 1, 2, \ldots, n$. (In contrast to the original formulation the absolute values on the left are omitted and so only one absorbing barrier at r occurs in the associated random walk.)

37. *Generation of exponentially distributed variables from uniform ones.*[18] Let $\mathbf{X}_1, \mathbf{X}_2, \ldots$ be independent and uniformly distributed in $\overline{0, 1}$. Define the random variable $\mathbf{N}$ as the index such that $\mathbf{X}_1 \geq \mathbf{X}_2 \geq \cdots \geq \mathbf{X}_{\mathbf{N}-1} < \mathbf{X}_\mathbf{N}$ (see problem 9). Prove that

$$\mathbf{P}\{\mathbf{X}_1 \leq x, \mathbf{N} = n\} = \frac{x^{n-1}}{(n-1)!} - \frac{x^n}{n!},$$

whence $\mathbf{P}\{\mathbf{X}_1 \leq x, \mathbf{N} \text{ even}\} = 1 - e^{-x}$.

Define $\mathbf{Y}$ as follows: A "trial" is a sequence $\mathbf{X}_1, \ldots, \mathbf{X}_\mathbf{N}$; it is a "failure" if $\mathbf{N}$ is odd. We repeat independent trials as long as necessary to produce a "success." Let $\mathbf{Y}$ equal the number of failures plus the first variable in the successful trial. Prove that $\mathbf{P}\{\mathbf{Y} < x\} = 1 - e^{-x}$.

---

[18] J. von Neumann, National Bureau of Standards, Appl. Math. Series, No. 12 (1951) pp. 36–38.

CHAPTER II

Special Densities. Randomization

The main purpose of this chapter is to list for reference the densities that will occur most frequently in the following chapters. The randomization procedure described in the second part is of general use. Its scope is illustrated by deriving certain distributions connected with Bessel functions which occur in various applications. It turns out that this simple probabilistic approach replaces involved calculations and hard analysis.

1. NOTATIONS AND CONVENTIONS

We say that a density f and its distribution F are *concentrated*[1] *on an interval* $I = \overline{a, b}$ if $f(x) = 0$ for all x outside I. Then $F(x) = 0$ for $x < a$ and $F(x) = 1$ for $x > b$. Two distributions F and G, and also their densities f and g, are said to be of the *same type* if they stand in the relationship

$$G(x) = F(ax+b), \qquad g(x) = af(ax+b), \tag{1.1}$$

where $a > 0$. We shall frequently refer to b as a *centering* parameter, to a as a *scale* parameter. These terms are readily understood from the fact that when F serves as distribution function of a random variable $\mathbf{X}$ then G is the distribution function of

$$\mathbf{Y} = \frac{\mathbf{X} - b}{a}. \tag{1.2}$$

In many contexts only the type of a distribution really matters.

---

[1] According to common usage the closed interval I should be called the *support* of f. A new term is introduced because it will be used in the more general sense that a distribution may be concentrated on the set of integers or rationals.

The *expectation* m and *variance* σ^2 of f (or of F) are defined by

(1.3)
$$m = \int_{-\infty}^{+\infty} xf(x)\,dx, \qquad \sigma^2 = \int_{-\infty}^{+\infty} (x-m)^2 f(x)\,dx = \int_{-\infty}^{+\infty} x^2 f(x)\,dx - m^2,$$

provided the integrals converge absolutely. It is clear from (1.2) that in this case g has expectation $(m-b)/a$ and variance σ^2/a^2. It follows that for each type there exists at most one density with zero expectation and unit variance.

We recall from I,(2.12) that the *convolution* $f = f_1 * f_2$ of two densities f_1 and f_2 is the probability density defined by

$$(1.4) \qquad f(x) = \int_{-\infty}^{+\infty} f_1(x-y) f_2(y)\,dy.$$

When f_1 and f_2 are concentrated on $\overline{0, \infty}$ this formula reduces to

$$(1.5) \qquad f(x) = \int_0^x f_1(x-y) f_2(y)\,dy, \qquad x > 0.$$

The former represents the density of the sum of two independent random variables with densities f_1 and f_2. Note that for $g_i(x) = f_i(x+b_i)$ the convolution $g = g_1 * g_2$ is given by $g(x) = f(x+b_1+b_2)$ as is obvious from (1.2).

Finally we recall the standard *normal distribution* function and its density defined by

$$(1.6) \qquad \mathfrak{n}(x) = \frac{1}{\sqrt{2\pi}}\, e^{-\frac{1}{2}x^2}, \qquad \mathfrak{N}(x) = \frac{1}{\sqrt{2\pi}} \int_{-\infty}^{x} e^{-\frac{1}{2}y^2}\,dy.$$

Our old acquaintance, the *normal density with expectation* m *and variance* σ^2, is given by

$$\frac{1}{\sigma}\,\mathfrak{n}\left(\frac{x-m}{\sigma}\right) \qquad \sigma > 0.$$

Implicit in the central limit theorem is the basic fact that the *family of normal densities is closed under convolutions;* in other words, the convolution of two normal densities with expectations m_1, m_2 and variances σ_1^2, σ_2^2 is the normal density with expectation $m_1 + m_2$ and variance $\sigma^2 = \sigma_1^2 + \sigma_2^2$. In view of what has been said it suffices to prove it for $m_1 = m_2 = 0$. It is asserted that

$$(1.7) \quad \frac{1}{\sqrt{2\pi}\,\sigma} \exp\left[-\frac{x^2}{2\sigma^2}\right] = \frac{1}{2\pi\sigma_1\sigma_2} \int_{-\infty}^{+\infty} \exp\left[-\frac{(x-y)^2}{2\sigma_1^2} - \frac{y^2}{2\sigma_2^2}\right] dy$$

and the truth of this assertion becomes obvious by the change of variables $z = y(\sigma/\sigma_1\sigma_2) - x(\sigma_2/\sigma\sigma_1)$ where x is fixed. (See problem 1.)

2. GAMMA DISTRIBUTIONS

The *gamma function* Γ is defined by

$$\Gamma(t) = \int_0^\infty x^{t-1}e^{-x}\,dx, \qquad t > 0. \tag{2.1}$$

[See **1**; II,(12.22).] It interpolates the factorials in the sense that

$$\Gamma(n+1) = n! \quad \text{for} \quad n = 0, 1, \ldots .$$

Integration by parts shows that $\Gamma(t) = (t-1)\,\Gamma(t-1)$ for all $t > 0$. (Problem 2.)

The *gamma densities* concentrated on $\overline{0, \infty}$ are defined by

$$f_{\alpha,\nu}(x) = \frac{1}{\Gamma(\nu)}\,\alpha^\nu x^{\nu-1}e^{-\alpha x}, \qquad \nu > 0, \quad x > 0. \tag{2.2}$$

Here $\alpha > 0$ *is* the trivial scale parameter, but $\nu > 0$ is essential. The special case $f_{\alpha,1}$ represents the *exponential* density, and the densities g_n of I,(3,4) coincide with $f_{\alpha,n}$ $(n = 1, 2, \ldots)$. A trite calculation shows that *the expectation* of $f_{\alpha,\nu}$ *equals* ν/α, *the variance* ν/α^2.

The family of gamma densities is closed under convolutions:

$$f_{\alpha,\mu} * f_{\alpha,\nu} = f_{\alpha,\mu+\nu} \qquad \mu > 0, \quad \nu > 0. \tag{2.3}$$

This important property generalizes the theorem of I,3 and will be in constant use; the *proof* is exceedingly simple. By (1.5) the left side equals

$$\frac{\alpha^{\mu+\nu}}{\Gamma(\mu)\,\Gamma(\nu)}\,e^{-\alpha x}\int_0^x (x-y)^{\mu-1}y^{\nu-1}\,dy. \tag{2.4}$$

After the substitution $y = xt$ this expression differs from $f_{\alpha,\mu+\nu}(x)$ by a numerical factor only, and this equals unity since both $f_{\alpha,\mu+\nu}$ and (2.4) are probability densities.

The value of the last integral for $x = 1$ is the so-called *beta integral* $B(\mu, \nu)$, and as a by-product of the proof we have found that

$$B(\mu, \nu) = \int_0^1 (1-y)^{\mu-1}y^{\nu-1}\,dy = \frac{\Gamma(\mu)\,\Gamma(\nu)}{\Gamma(\mu+\nu)} \tag{2.5}$$

for all $\mu > 0$, $\nu > 0$. [For integral μ and ν this formula is used in **1**; VI,(10.8) and (10.9). See also problem 3 of the present chapter.]

As to the *graph of* $f_{1,\nu}$, it is clearly monotone if $\nu \le 1$, and unbounded near the origin when $\nu < 1$. For $\nu > 1$ the graph of $f_{1,\nu}$ is bell-shaped, attaining at $x = \nu - 1$ its maximum $(\nu-1)^{\nu-1}e^{-(\nu-1)}/\Gamma(\nu)$ which is close to $[2\pi(\nu-1)]^{-\frac{1}{2}}$ (Stirling's formula, problem 12 of **1**; II, 12). It follows from the central limit theorem that

$$\frac{\sqrt{\nu}}{\alpha} f_{\alpha,\nu}\left(\frac{\nu + \sqrt{\nu}\, x}{\alpha}\right) \to \mathfrak{n}(x), \qquad \nu \to \infty. \tag{2.6}$$

*3. RELATED DISTRIBUTIONS OF STATISTICS

The gamma densities play a crucial, though sometimes disguised, role in mathematical statistics. To begin with, in the classical (now somewhat outdated) system of densities introduced by K. Pearson (1894) the gamma densities appear as "type III." A more frequent appearance is due to the fact that for a random variable $\mathbf{X}$ with normal density $\mathfrak{n}$ the square $\mathbf{X}^2$ has density $x^{-\frac{1}{2}}\mathfrak{n}(x^{\frac{1}{2}}) = f_{\frac{1}{2},\frac{1}{2}}(x)$. In view of the convolution property (2.3) it follows that:

If $\mathbf{X}_1, \ldots, \mathbf{X}_n$ *are mutually independent normal variables with expectation* 0 *and variance* σ^2, *then* $\mathbf{X}_1^2 + \cdots + \mathbf{X}_n^2$ *has density* $f_{1/2\sigma^2, n/2}$.

To statisticians $\chi^2 = \mathbf{X}_1^2 + \cdots + \mathbf{X}_n^2$ is the "sample variance from a normal population" and its distribution is in constant use. For reasons of tradition (going back to K. Pearson) in this connection $f_{\frac{1}{2},\frac{1}{2}n}$ is called *chi-square density with* n *degrees of freedom.*

In statistical mechanics $\mathbf{X}_1^2 + \mathbf{X}_2^2 + \mathbf{X}_3^2$ appears as the square of the speed of particles. Hence $v(x) = 2xf_{\frac{1}{2},\frac{1}{2}3}(x^2)$ represents the density of the speed itself. This is the *Maxwell density* found by other methods in I,(10.6). (See also the example in III,4.)

In queuing theory the gamma distribution is sometimes called *Erlangian.*

Several random variables (or "statistics") of importance to statisticians are of the form $\mathbf{T} = \mathbf{X}/\mathbf{Y}$, where $\mathbf{X}$ and $\mathbf{Y}$ are independent random variables, $\mathbf{Y} > 0$. Denote their distributions by F and G, respectively, and their densities by f and g. As $\mathbf{Y}$ is supposed positive, g is concentrated on $\overline{0, \infty}$ and so

$$\mathbf{P}\{\mathbf{T} \leq t\} = \mathbf{P}\{\mathbf{X} \leq t\mathbf{Y}\} = \int_0^\infty F(ty)\, g(y)\, dy. \tag{3.1}$$

By differentiation it is found that *the ratio* $\mathbf{T} = \mathbf{X}/\mathbf{Y}$ *has density*

$$w(t) = \int_0^\infty f(ty) y\, g(y)\, dy. \tag{3.2}$$

Examples. (*a*) *If* $\mathbf{X}$ *and* $\mathbf{Y}$ *have densities* $f_{\frac{1}{2},\frac{1}{2}m}$ *and* $f_{\frac{1}{2},\frac{1}{2}n}$, *then* $\mathbf{X}/\mathbf{Y}$ *has density*

$$w(t) = \frac{\Gamma(\frac{1}{2}(m+n))}{\Gamma(\frac{1}{2}m)\,\Gamma(\frac{1}{2}n)} \frac{t^{\frac{1}{2}m-1}}{(1+t)^{\frac{1}{2}(m+n)}}, \qquad t > 0. \tag{3.3}$$

* This section treats special topics and is not used in the sequel.

In fact, the integral in (3.2) equals

$$\frac{t^{\frac{1}{2}m-1}}{2^{\frac{1}{2}(m+n)}\,\Gamma(\frac{1}{2}m)\,\Gamma(\frac{1}{2}n)}\int_0^\infty y^{\frac{1}{2}(m+n)-1}e^{-\frac{1}{2}(1+t)y}\,dy \tag{3.4}$$

and the substitution $\frac{1}{2}(1+t)y = s$ reduces it to (3.3).

In the analysis of variance one considers the special case

$$\mathbf{X} = \mathbf{X}_1^2 + \cdots + \mathbf{X}_m^2 \quad \text{and} \quad \mathbf{Y} = \mathbf{Y}_1^2 + \cdots + \mathbf{Y}_n^2$$

where $\mathbf{X}_1, \ldots, \mathbf{X}_m$, $\mathbf{Y}_1, \ldots, \mathbf{Y}_n$ are mutually independent variables with the common normal density $\mathfrak{n}$. The random variable $\mathbf{F} = (n\mathbf{X}/m\mathbf{Y}/)$ is called Snedecor's statistic and its density $(m/n)\, w((m/n)\,x)$ is *Snedecor's density*, or the *F*-density. The variable $\mathbf{Z} = \log \frac{1}{2}\mathbf{F}$ is Fisher's **Z**-statistic, and its density *Fisher's* **Z**-*density*. The two statistics are, of course, merely notational variants of each other.

(*b*) *Student's* **T**-*density*. Let $\mathbf{X}, \mathbf{Y}_1, \ldots, \mathbf{Y}_n$ be independent with the common normal density $\mathfrak{n}$. The variable

$$\mathbf{T} = \frac{\mathbf{X}\sqrt{n}}{\sqrt{\mathbf{Y}_1^2 + \cdots + \mathbf{Y}_n^2}} \tag{3.5}$$

is known to statisticians as Student's **T**-statistic. We show that its density is given by

$$w(t) = \frac{C_n}{(1+t^2/n)^{\frac{1}{2}(n+1)}}, \quad \text{where} \quad C_n = \frac{1}{\sqrt{\pi n}}\,\frac{\Gamma(\frac{1}{2}(n+1))}{\Gamma(\frac{1}{2}n)}. \tag{3.6}$$

In fact, the numerator in (3.5) has a normal density with zero expectation and variance n, while the density of the denominator is given by $2xf_{\frac{1}{2},\frac{1}{2}n}(x^2)$. Thus (3.2) takes on the form

$$\frac{1}{\sqrt{\pi n}\; 2^{(n-1)/2}\,\Gamma(n/2)}\int_0^\infty e^{-\frac{1}{2}(1+t^2/n)y^2}y^n\,dy. \tag{3.7}$$

The substitution $s = \frac{1}{2}(1+t^2/n)y^2$ reduces the integral to a gamma integral and yields (3.6). ▶

4. SOME COMMON DENSITIES

In the following it is understood that all densities vanish identically outside the indicated interval.

(*a*) *The bilateral exponential* is defined by $\frac{1}{2}\alpha e^{-\alpha|x|}$ where α is a scale parameter. It has zero expectation and variance $2\alpha^{-2}$. This density is *the convolution of the exponential density* $\alpha e^{-\alpha x}$ $(x > 0)$ *with the mirrored density* $\alpha e^{\alpha x}$ $(x < 0)$. In other words, the bilateral exponential is the density of $\mathbf{X}_1 - \mathbf{X}_2$ when $\mathbf{X}_1$ and $\mathbf{X}_2$ are independent and have the common exponential density $\alpha e^{-\alpha x}$ $(x > 0)$. In the French literature it is usually referred to as the "second law of Laplace," the first being the normal distribution.

(*b*) The *uniform* (or *rectangular*) density ρ_a and the *triangular* density τ_a concentrated on $\overline{-a, a}$ are defined by

$$(4.1) \qquad \rho_a(x) = \frac{1}{2a}, \qquad \tau_a(x) = \frac{1}{a}\left(1 - \frac{|x|}{a}\right), \qquad |x| < a.$$

It is easily seen that $\rho_a \star \rho_a = \tau_{2a}$. In words: the sum of two uniformly distributed variables in $\overline{-a, a}$ has a triangular density in $\overline{-2a, 2a}$. [The repeated convolutions $\rho_a \star \cdots \star \rho_a$ are described in I,(9.7).]

(*c*) *Beta densities* in $\overline{0, 1}$ are defined by

$$(4.2) \qquad \beta_{\mu,\nu}(x) = \frac{\Gamma(\mu+\nu)}{\Gamma(\mu)\,\Gamma(\nu)}(1-x)^{\mu-1}\,x^{\nu-1}, \qquad 0 < x < 1,$$

where $\mu > 0$ and $\nu > 0$ are free parameters. That (4.2) indeed defines a probability density follows from (2.5). By the same formula it is seen that $\beta_{\mu,\nu}$ has expectation $\nu/(\mu+\nu)$, and variance $\mu\nu/[(\mu+\nu)^2(\mu+\nu+1)]$. If $\mu < 1$, $\nu < 1$, the graph of $\beta_{\mu,\nu}$ is U-shaped, tending to ∞ at the limits. For $\mu > 1$, $\nu > 1$ the graph is bell-shaped. For $\mu = \nu = 1$ we get the *uniform* density as a special case.

A simple variant of the beta density is defined by

$$(4.3) \qquad \frac{1}{(1+t)^2}\,\beta_{\mu,\nu}\left(\frac{1}{1+t}\right) = \frac{\Gamma(\mu+\nu)}{\Gamma(\mu)\,\Gamma(\nu)} \cdot \frac{t^{\mu-1}}{(1+t)^{\mu+\nu}}, \qquad 0 < t < \infty.$$

If the variable $\mathbf{X}$ has density (4.2) then $\mathbf{Y} = \mathbf{X}^{-1} - 1$ has density (4.3).

In the Pearson system the densities (4.2) and (4.3) appear as types I and VI. The Snedecor density (3.3) is a special case of (4.3). The densities (4.3) are sometimes called after the economist *Pareto*. It was thought (rather naïvely from a modern statistical standpoint) that income distributions should have a tail with a density $\sim Ax^{-\alpha}$ as $x \to \infty$, and (4.3) fulfills this requirement.

(*d*) *The so-called arc sine density*

$$(4.4) \qquad \frac{1}{\pi\sqrt{x(1-x)}}, \qquad 0 < x < 1$$

is actually the same as the beta density $\beta_{\frac{1}{2},\frac{1}{2}}$, but deserves special mention because of its repeated occurrence in fluctuation theory. (It was introduced in **1**; III,4 in connection with the unexpected behavior of sojourn times.) The misleading name is unfortunately in general use; actually the distribution function is given by $2\pi^{-1}$ arc sin $\sqrt{x}$. (The beta densities with $\mu + \nu = 1$ are sometimes referred to as "generalized arc sine densities.")

(*e*) *The Cauchy density* centered at the origin is defined by

$$(4.5)\qquad \gamma_t(x) = \frac{1}{\pi}\cdot\frac{t}{t^2+x^2}, \qquad -\infty < x < \infty,$$

where $t > 0$ is a scale parameter. The corresponding distribution function is $\frac{1}{2} + \pi^{-1}\arctan(x/t)$. The graph of γ_t resembles that of the normal density but approaches the axis so slowly that an *expectation does not exist*.

The importance of the Cauchy densities is due to the convolution formula

$$(4.6)\qquad \gamma_s \star \gamma_t = \gamma_{s+t}.$$

It states that *the family of Cauchy densities* (4.5) *is closed under convolutions*. Formula (4.6) can be proved in an elementary (but tedious) fashion by a routine decomposition of the integrand into partial fractions. A simpler proof depends on Fourier analysis.

The convolution formula (4.6) has the amazing consequence that for independent variables $\mathbf{X}_1, \ldots, \mathbf{X}_n$ with the common density (4.5) *the average* $(\mathbf{X}_1 + \cdots + \mathbf{X}_n)/n$ *has the same density as the* $\mathbf{X}_j$.

Example. Consider a laboratory experiment in which a vertical mirror projects a horizontal light ray on a wall. The mirror is free to rotate about a vertical axis through A. We assume that the direction of the reflected ray is chosen "at random," that is, the angle $\boldsymbol{\varphi}$ between it and the perpendicular AO to the wall is distributed uniformly between $-\frac{1}{2}\pi$ and $\frac{1}{2}\pi$. The light ray intersects the wall at a point at a distance

$$\mathbf{X} = t \cdot \tan \boldsymbol{\varphi}$$

from O (where t is the distance AO of the center A from the wall). It is now obvious that the random variable $\mathbf{X}$ has density (4.5).[2] If the experiment is repeated n times the average $(\mathbf{X}_1+\cdots+\mathbf{X}_n)/n$ has the same density and so the averages do not cluster around 0 as one should expect by analogy with the law of large numbers. ▶

The Cauchy density has the curious property that if $\mathbf{X}$ has density γ_t then $2\mathbf{X}$ has density $\gamma_{2t} = \gamma_t \star \gamma_t$. Thus $2\mathbf{X} = \mathbf{X} + \mathbf{X}$ *is the sum of two dependent variables, but its density is given by the convolution formula*. More generally, if $\mathbf{U}$ and $\mathbf{V}$ are two independent variables with common density γ_t and $\mathbf{X} = a\mathbf{U} + b\mathbf{V}$, $\mathbf{Y} = c\mathbf{U} + d\mathbf{V}$, then $\mathbf{X} + \mathbf{Y}$ has density $\gamma_{(a+b+c+d)t}$ which is the convolution of the densities $\gamma_{(a+b)t}$ of $\mathbf{X}$ and $\gamma_{(c+d)t}$

---

[2] A simple reformulation of this experiment leads to physical interpretation of the convolution formula (4.6). Our argument shows that if a unit light source is situated at the origin then γ_t represents the distribution of the intensity of light along the line $y = t$ of the x,y-plane. Then (4.6) expresses *Huygens' principle*, according to which the intensity of light along $y = s + t$ is the same as if the source were distributed along the line $y = t$ following the density γ_t. (I owe this remark to J. W. Walsh.)

of **Y**; nevertheless, **X** and **Y** are not independent. (For a related example see problem 1 in III,9.)

[The Cauchy density corresponds to the special case $n = 1$ of the family (3.5) of Student's **T**- densities. In other words, if **X** and **Y** are independent random variables with the normal density $\mathfrak{n}$, then $\mathbf{X}/|\mathbf{Y}|$ has the Cauchy density (4.5) with $t = 1$. For some related densities see problems 5–6.]

The convolution property (2.3) of the gamma densities looks exactly like (4.6) but there is an important difference in that the parameter ν of the gamma densities is essential whereas (4.6) contains only a scale parameter. With the Cauchy density the *type* is stable. This stability under convolutions is shared by the normal and the Cauchy densities; the difference is that the scale parameters compose according to the rules $\sigma^2 = \sigma_1^2 + \sigma_2^2$ and $t = t_1 + t_2$ respectively. There exist other *stable densities* with similar properties, and with a systematic terminology we should call the normal and Cauchy densities "symmetric, stable of exponent 2 and 1." (See VI,1.)

(*f*) *One-sided stable distribution of index* $\frac{1}{2}$. If $\mathfrak{N}$ is the normal distribution of (1.6), then

$$F_\alpha(x) = 2[1 - \mathfrak{N}(\alpha/\sqrt{x})], \qquad x > 0, \tag{4.7}$$

defines a distribution function with density

$$f_\alpha(x) = \frac{\alpha}{\sqrt{2\pi}} \cdot \frac{1}{\sqrt{x^3}} e^{-\frac{1}{2}\alpha^2/x}, \qquad x > 0. \tag{4.8}$$

Obviously no expectation exists. This distribution was found in **1**; III,(7.7) and again in **1**; X,1 as limit of the distribution of recurrence times, and this derivation implies the *composition rule*

$$f_\alpha \star f_\beta = f_\gamma \quad \textit{where} \quad \gamma = \alpha + \beta. \tag{4.9}$$

(A verification by elementary, but rather cumbersome, integrations is possible. The Fourier analytic proof is simpler.) If $\mathbf{X}_1, \ldots, \mathbf{X}_n$ are independent random variables with the distribution (4.7), then (4.9) implies that $(\mathbf{X}_1 + \cdots + \mathbf{X}_n)n^{-2}$ has the same distribution, and so the averages $(\mathbf{X}_1 + \cdots + \mathbf{X}_n)n^{-1}$ are likely to be of the order of magnitude of n; instead of converging they increase over all bounds. (See problems 7 and 8.)

(*g*) Distributions of the form $e^{-x^{-\alpha}}$ $(x > 0, \alpha > 0)$ appear in connection with order statistics (see problem 8). Together with the variant $1 - e^{-x^\alpha}$ they appear (rather mysteriously) under the name of *Weibull* distributions in statistical reliability theory.

(*h*) The *logistic distribution* function

$$F(t) = \frac{1}{1 + e^{-\alpha t - \beta}}, \qquad \alpha > 0 \tag{4.10}$$

may serve as a warning. An unbelievably huge literature tried to establish a transcendental "law of logistic growth"; measured in appropriate units, practically all growth processes were supposed to be represented by a function of the form (4.10) with t representing time. Lengthy tables, complete with chi-square tests, supported this thesis for human populations, for bacterial colonies, development of railroads, etc. Both height *and* weight of plants and animals were found to follow the logistic law even though it is theoretically clear that these two variables cannot be subject to the same distribution. Laboratory experiments on bacteria showed that not even systematic disturbances can produce other results. Population theory relied on logistic extrapolations (even though they were demonstrably unreliable). The only trouble with the theory is that not only the logistic distribution but also the normal, the Cauchy, and other distributions can be fitted to the *same material with the same or better goodness of fit.*[3] In this competition the logistic distribution plays no distinguished role whatever; most contradictory theoretical models can be supported by the same observational material.

Theories of this nature are short-lived because they open no new ways, and new confirmations of the same old thing soon grow boring. But the naïve reasoning as such has not been superseded by common sense, and so it may be useful to have an explicit demonstration of how misleading a mere goodness of fit can be.

5. RANDOMIZATION AND MIXTURES

Let F be a distribution function depending on a parameter θ, and u a probability density. Then

$$W(x) = \int_{-\infty}^{+\infty} F(x, \theta)\, u(\theta)\, d\theta \tag{5.1}$$

is a monotone function of x increasing from 0 to 1 and hence a distribution function. If F has a continuous density f, then W has a density w given by

$$w(x) = \int_{-\infty}^{+\infty} f(x, \theta)\, u(\theta)\, d\theta. \tag{5.2}$$

Instead of integrating with respect to a density u we can sum with respect to a discrete probability distribution: if $\theta_1, \theta_2, \ldots$ are chosen arbitrarily and if $p_k \geq 0$, $\Sigma p_k = 1$, then

$$w(x) = \sum_k f(x, \theta_k)\, p_k \tag{5.3}$$

defines a new probability density. The process may be described probabilistically as *randomization;* the parameter θ is treated as random variable and a new probability distribution is defined in the x, θ-plane, which serves as sample space. Densities of the form (5.3) are called *mixtures*, and the term is now used generally for distributions and densities of the form (5.1) and (5.2).

We do not propose at this juncture to develop a general theory. Our aim is rather to illustrate by a few examples the scope of the method and its

[3] W. Feller, *On the logistic law of growth and its empirical verifications in biology*, Acta Biotheoretica, vol. 5 (1940) pp. 51–66.

probabilistic content. The examples serve also as preparation for the notion of conditional probabilities. The next section is devoted to examples of discrete distributions obtained by randomization of a continuous parameter. Finally, section 7 illustrates the construction of continuous processes out of random walks; as a by-product we shall obtain distributions occurring in many applications and otherwise requiring hard calculations.

Examples. (*a*) *Ratios.* If $\mathbf{X}$ is a random variable with density f, then for fixed $y > 0$ the variable $\mathbf{X}/y$ has density $f(xy)y$. Treating the parameter y as random variable with density g we get the new density

$$w(x) = \int_{-\infty}^{+\infty} f(xy)y\, g(y)\, dy. \tag{5.4}$$

This is the same as formula (3.2) on which the discussion in section 3 was based.

In probabilistic language randomizing the denominator y in $\mathbf{X}/y$ means considering the random variable $\mathbf{X}/\mathbf{Y}$, and we have merely rephrased the derivation of the density (3.2) of $\mathbf{X}/\mathbf{Y}$. In this particular case the terminology is a matter of taste.

(*b*) *Random sums.* Let $\mathbf{X}_1, \mathbf{X}_2, \ldots$ be mutually independent random variables with a common density f. The sum $\mathbf{S}_n = \mathbf{X}_1 + \cdots + \mathbf{X}_n$ has the density f^{n*}, namely the n-fold convolution of f with itself. [See I,2.] The number n of terms is a parameter which we now randomize by a probability distribution $\mathbf{P}\{\mathbf{N} = n\} = p_n$. The density of the resulting sum $\mathbf{S}_\mathbf{N}$ with the random number $\mathbf{N}$ of terms is

$$w = \sum_1^\infty p_n f^{n*}. \tag{5.5}$$

As an example take for $\{p_n\}$ the geometric distribution $p_n = qp^{n-1}$, and for f an exponential density. Then $f^{n*} = g_n$ is given by (2.2) and

$$w(x) = q\alpha e^{-\alpha x} \sum_{n=1}^\infty p^{n-1} \frac{(\alpha x)^{n-1}}{(n-1)!} = q\alpha e^{-\alpha q x}. \tag{5.6}$$

(*c*) *Application to queuing.* Consider a single server with exponential servicing time distribution (density $f(t) = \mu e^{-\mu t}$) and assume the incoming traffic to be Poisson, that is, the inter-arrival times are independent with density $\lambda e^{-\lambda t}$, $\lambda < \mu$. The model is described in **1**; XVII,7(*b*). Arriving customers join a (possibly empty) "waiting line" and are served in order of arrival without interruption.

Consider a customer who on his arrival finds $n \geq 0$ other customers in the line. The total time that he spends at the server is the sum of the service times of these n customers plus his own service time. This is a random variable with density $f^{(n+1)*}$. We saw in **1**; XVII,(7.10) that in the steady state the probability of finding exactly n customers in the waiting line equals qp^n with $p = \lambda/\mu$. Assuming this steady state we see that *the total time* **T** *spent by a customer at the server is a random variable with density*

$$\sum_{n=0}^{\infty} qp^n f^{(n+1)*}(t) = q\mu\, e^{-\mu t} \sum_{n=0}^{\infty} (p\mu t)^n/n! = (\mu-\lambda)\, e^{-(\mu-\lambda)t}.$$

Thus $\mathbf{E}(\mathbf{T}) = 1/(\mu-\lambda)$. (See also problem 10.)

(*d*) *Waiting lines for buses*. A bus is supposed to appear every hour on the hour, but is subject to delays. We treat the successive delays $\mathbf{X}_k$ as independent random variables with a common distribution F and density f. For simplicity we assume $0 \leq \mathbf{X}_k \leq 1$. Denote by $\mathbf{T}_x$ the waiting time of a person arriving at epoch $x < 1$ after noon. The probability that the bus scheduled for noon has already departed is $F(x)$, and it is easily seen that

$$(5.7)\qquad \mathbf{P}\{\mathbf{T}_x \leq t\} = \begin{cases} F(t+x) - F(x) & \textit{for} \quad 0 < t < 1-x \\ 1 - F(x) + F(x)\, F(t+x-1) & \textit{for} \quad 1-x < t < 2-x \end{cases}$$

and, of course, $\mathbf{P}\{\mathbf{T}_x \leq t\} = 1$ for all greater t. The corresponding density is given by

$$(5.8)\qquad \begin{cases} f(t+x) & \textit{for} \quad 0 < t < 1-x \\ F(x) f(t+x-1) & \textit{for} \quad 1-x < t < 2-x. \end{cases}$$

Here the epoch x of arrival is a free parameter and it is natural to randomize it. For example, for a person arriving "at random" the epoch of arrival is a random variable distributed uniformly in $\overline{0, 1}$. The expected waiting time in this case equals $\frac{1}{2} + \sigma^2$ where σ^2 is the variance of the delay. In other words, *the expected waiting time is smallest if the buses are punctual* and increases with the variance of the delay. (See problems 11–12.) ▶

6. DISCRETE DISTRIBUTIONS

This section is devoted to a quick glance at some results of randomizing binomial and Poisson distributions.

The number $\mathbf{S}_n$ of successes in Bernoulli trials has a distribution depending on the probability p of success. Treating p as a random variable with density u leads to the new distribution

$$(6.1)\qquad \mathbf{P}\{\mathbf{S}_n = k\} = \binom{n}{k} \int_0^1 p^k (1-p)^{n-k}\, u(p)\, dp \qquad k = 0, \ldots, n.$$

Example. (*a*) When $u(p) = 1$ an integration by parts shows (6.1) to be independent of k, and (6.1) reduces to the discrete *uniform distribution*

$\mathbf{P}\{\mathbf{S}_n = k\} = (n+1)^{-1}$. More illuminating is an argument due to Bayes. Consider $n + 1$ independent variables $\mathbf{X}_0, \ldots, \mathbf{X}_n$ distributed uniformly between 0 and 1. The integral in (6.1) (with $u = 1$) equals the probability that exactly k among the variables $\mathbf{X}_1, \ldots, \mathbf{X}_n$ will be $<\mathbf{X}_0$ or, in other words, that in an enumeration of the points $\mathbf{X}_0, \ldots, \mathbf{X}_n$ in order of magnitude $\mathbf{X}_0$ appears at the $(k+1)$st place. But for reasons of symmetry all positions are equally likely, and so the integral equals $(n+1)^{-1}$. ►

In gambling language (6.1) corresponds to the situation when a skew coin is picked by a chance mechanism and then trials are performed with this coin of unknown structure. To a gambler the trials do not look independent; indeed, if a long sequence of heads is observed it becomes likely that for our coin p is close to 1 and so it is safe to bet on further occurrences of heads. Two formal examples may illustrate estimation and prediction problems of this type.

Examples. (*b*) Given that n trials resulted in k successes (= hypothesis H), what is the probability of the event that $p < \alpha$? By the definition of conditional probabilities

$$\mathbf{P}\{A \mid H\} = \frac{\mathbf{P}\{AH\}}{\mathbf{P}\{H\}} = \frac{\int_0^\alpha p^k(1-p)^{n-k}\, u(p)\, dp}{\int_0^1 p^k(1-p)^{n-k}\, u(p)\, dp}. \tag{6.2}$$

This type of estimation with $u(p) = 1$ was used by Bayes. Within the framework of our model (that is, if we are really concerned with a mixed population of coins *with known* density u) there can be no objection to the procedure. The trouble is that it used to be applied indiscriminately to judge "probabilities of causes" when there was no randomization in sight; this point was fully discussed in example 2(*e*) of **1**; V in connection with a so-called probability that the sun will rise tomorrow.

(*c*) A variant may be formulated as follows. Given that n trials resulted in k successes, what is the probability that the next m trials will result in j successes? The preceding argument leads to the answer

$$\frac{\binom{m}{j}\int_0^1 p^{j+k}(1-p)^{m+n-j-k}\, u(p)\, dp}{\int_0^1 p^k(1-p)^{n-k}\, u(p)\, dp}. \tag{6.3}$$

(See problem 13.) ►

Turning to the *Poisson* distribution let us interpret it as regulating the number of "arrivals" during a time interval of duration t. The expected number of arrivals is αt. We illustrate two conceptually different randomization procedures.

Examples. (*d*) *Randomized time.* If the duration of the time interval is a random variable with density u, the probability p_k of exactly k arrivals becomes

$$p_k = \int_0^\infty e^{-\alpha t} \frac{(\alpha t)^k}{k!} u(t)\, dt. \tag{6.4}$$

For example, if the time interval is exponentially distributed, the probability of $k = 0, 1, \ldots$ new arrivals equals

$$p_k = \int_0^\infty e^{-(\alpha+\beta)t} \frac{(\alpha t)^k}{k!} \beta\, dt = \frac{\beta}{\alpha + \beta} \cdot \left(\frac{\alpha}{\alpha + \beta}\right)^k \tag{6.5}$$

which is a geometric distribution.

(*e*) *Stratification.* Suppose there are several independent sources for random arrivals, each source having a Poisson output, but with different parameters. For example, *accidents* in a plant during a fixed exposure time t may be assumed to represent Poisson variables, but the parameter will vary from plant to plant. Similarly, *telephone calls* originating at an individual unit may be Poissonian with the expected number of calls varying from unit to unit. In such processes the parameter α appears as random variable with a density u, and the probability of exactly n arrivals during time t is given by

$$P_n(t) = \int_0^\infty e^{-\alpha t} \frac{(\alpha t)^n}{n!} u(\alpha)\, d\alpha. \tag{6.6}$$

For the special case of a gamma density $u = f_{\beta,\nu+1}$ we get

$$P_n(t) = \binom{n+\nu}{n} \left(\frac{\beta}{\beta + t}\right)^{\nu+1} \left(\frac{t}{\beta + t}\right)^n, \tag{6.7}$$

which is the limiting form of the *Polya distribution* as given in problem 24 of **1**; V,8 and **1**; XVII,(10.2) (setting $\beta = a^{-1}$, $\nu = a^{-1} - 1$). ►

Note on spurious contagion. A curious and instructive history attaches to the distribution (6.7) and its dual nature.

The Polya urn model and the Polya process which lead to (6.7) are models for true contagion where every accident effectively increases the probability of future accidents. This model enjoyed great popularity, and (6.7) was fitted empirically to a variety of phenomena, a good fit being taken as an *indication of true contagion.*

By coincidence, the same distribution (6.7) has been derived previously (in 1920) by M. Greenwood and G. U. Yule with the intent that a good fit should *disprove presence of contagion.* Their derivation is roughly equivalent to our stratification model, which starts from the assumption underlying the Poisson process, namely, that there is no aftereffect whatever. We have thus the curious fact that a good fit of the same distribution may be interpreted in two ways diametrically opposite in their nature as well as in their practical implications. This should serve as a warning against too hasty interpretations of statistical data.

The explanation lies in the phenomenon of *spurious* contagion, described in **1**; V,2(*d*) and above in connection with (6.1). In the present situation, having observed m accidents during a time interval of length s one may estimate the probability of n accidents during a future exposure of duration t by a formula analogous to (6.3). The result will depend on m, but this dependence is due to the method of sampling rather than to nature itself; the information concerning the past enables us to make better predictions concerning the future behavior of our sample, and this should not be confused with the future of the whole population.

7. BESSEL FUNCTIONS AND RANDOM WALKS

Surprisingly many explicit solutions in diffusion theory, queuing theory, and other applications involve Bessel functions. It is usually far from obvious that the solutions represent probability distributions, and the analytic theory required to derive their Laplace transforms and other relations is rather complex. Fortunately, the distributions in question (and many more) may be obtained by simple randomization procedures. In this way many relations lose their accidental character, and much hard analysis can be avoided.

By the Bessel function of order $\rho > -1$ we shall understand the function I_ρ defined for all real x by[4]

$$I_\rho(x) = \sum_{k=0}^{\infty} \frac{1}{k!\Gamma(k+\rho+1)} \left(\frac{x}{2}\right)^{2k+\rho}. \tag{7.1}$$

We proceed to describe three procedures leading to three different types of distributions involving Bessel functions.

(a) Randomized Gamma Densities

For fixed $\rho > -1$ consider the gamma density $f_{1,\rho+k+1}$ of (2.2). Taking the parameter k as an integral-valued random variable subject to a Poisson distribution we get in accordance with (5.3) the new density

$$w_\rho(x) = e^{-t}\sum_{k=0}^{\infty} \frac{t^k}{k!} f_{1,\rho+k+1}(x) = e^{-t-x}\sum_{k=0}^{\infty} \frac{t^k x^{\rho+k}}{k!\Gamma(\rho+k+1)}. \tag{7.2}$$

Comparing terms in (7.1) and (7.2) one sees that

$$w_\rho(x) = e^{-t-x}\sqrt{(x/t)^\rho}\, I_\rho(2\sqrt{tx}), \qquad x > 0. \tag{7.3}$$

---

[4] According to standard usage I_ρ is the "modified" Bessel function or Bessel function "with imaginary argument." The "ordinary" Bessel function, always denoted by J_ρ, is defined by inserting $(-1)^k$ on the right in (7.1). Our use of the term Bessel function should be understood as abbreviation rather than innovation.

If $\rho > -1$ then w_ρ *is a probability density* concentrated on $\overline{0, \infty}$. (For $\rho = -1$ the right side is not integrable with respect to x.) Note that t is not a scale parameter, so that these densities are of different types.

Incidentally, from this construction and the convolution formula (2.3) for the gamma densities it is clear that

$$w_\rho \star f_{1,\nu} = w_{\rho+\nu}. \tag{7.4}$$

(b) Randomized Random Walks

In discussing random walks one pretends usually that the successive jumps occur at epochs $1, 2, \ldots$. It should be clear, however, that this convention merely lends color to the description and that the model is entirely independent of time. An honest continuous-time stochastic process is obtained from the ordinary random walk by postulating that the *time intervals between successive jumps correspond to independent random variables with the common density* e^{-t}. In other words, the epochs of the jumps are regulated by a Poisson process, but the jumps themselves are random variables assuming the values $+1$ and -1 with probabilities p and q independent of each other and of the Poisson process.

To each distribution connected with the random walk there corresponds a distribution for the continuous-time process, which is obtained formally by randomization of the number of jumps. To see the procedure in detail consider the position at a given epoch t. In the basic random walk the nth step leads to the position $r \geq 0$ iff among the first n jumps $\frac{1}{2}(n+r)$ are positive and $\frac{1}{2}(n-r)$ negative. This is impossible unless $n - r = 2\nu$ is even. In this case the probability of the position r just after the nth jump is

$$\binom{n}{\frac{1}{2}(n+r)} p^{\frac{1}{2}(n+r)} q^{\frac{1}{2}(n-r)} = \binom{r+2\nu}{r+\nu} p^{r+\nu} q^{\nu}. \tag{7.5}$$

In our Poisson process the probability that up to epoch t exactly $n = 2\nu + \gamma$ jumps occur is $e^{-t}t^n/n!$ and so in our time-dependent process the probability of the position $r \geq 0$ at epoch t equals

$$e^{-t} \sum_{\nu=0}^{\infty} \frac{t^{r+2\nu}}{(r+2\nu)!} \binom{r+2\nu}{r+\nu} p^{r+\nu} q^{\nu} = \sqrt{(p/q)^r}\, e^{-t} I_r(2\sqrt{pq}\, t) \tag{7.6}$$

and we reach two conclusions.

(i) If we define $I_{-r} = I_r$ for $r = 1, 2, 3, \ldots$ then *for fixed* $t > 0, p, q$,

$$a_r(t) = \sqrt{(p/q)^r}\, e^{-t} I_r(2\sqrt{pq}\, t), \quad r = 0, \pm 1, \pm 2, \ldots, \tag{7.7}$$

represents a probability distribution (that is, $a_r \geq 0$, $\sum a_r = 1$).

(ii) In our time-dependent random walk $a_r(t)$ *equals the probability of the position* r *at epoch* t.

Two famous formulas for Bessel functions are immediate corollaries of this result. First, with the change of notations $2\sqrt{pq}\,t = x$ and $p/q = u^2$, the identity $\sum a_r(t) = 1$ becomes

$$e^{\frac{1}{2}x(u+u^{-1})} = \sum_{-\infty}^{+\infty} u^r I_r(x). \tag{7.8}$$

This is the so-called generating function for Bessel functions or *Schlömilch's formula* (which sometimes serves as definition for I_r).

Second, it is clear from the nature of our process that the probabilities $a_r(t)$ must satisfy the Chapman-Kolmogorov equation

$$a_r(t+\tau) = \sum_{k=-\infty}^{\infty} a_k(t)a_{r-k}(\tau), \tag{7.9}$$

which expresses the fact that at epoch t the particle must be at some position k and that a transition from k to r is equivalent to a transition from 0 to $r-k$. We shall return to this relation in XVII,3. [It is easily verified directly from the representation (7.6) and the analogous formula for the probabilities in the random walk.] The Chapman-Kolmogorov relation (7.9) is equivalent to

$$I_r(t+\tau) = \sum_{k=-\infty}^{\infty} I_k(t)I_{r-k}(\tau) \tag{7.10}$$

which is known as *K. Neumann's identity*.

(c) First Passages

For simplicity let us restrict our attention to symmetric random walks, $p = q = \frac{1}{2}$. According to **1**; III,(7.5), the probability that the *first passage* through the point $r > 0$ occurs at the jump number $2n - r$ is

$$\frac{r}{2n-r}\binom{2n-r}{n}2^{-2n+r} \qquad n \geq r. \tag{7.11}$$

The random walk being recurrent, such a first passage occurs with probability one, that is, for fixed r the quantities (7.11) add up to unity. In our time-dependent process the epoch of the kth jump has the gamma density $f_{1,k}$ of (2.2). It follows that the epoch of the first passage through $r > 0$ has density

$$\begin{aligned}\sum_n \frac{r}{2n-r}\binom{2n-r}{n}2^{-2n+r} f_{1,2n-r}(t) &= \\ = e^{-t}\sum \frac{t^{2n-r-1}}{(2n-r-1)!}\left(\frac{r}{2n-r}\right)\cdot\frac{(2n-r)!}{n!(n-r)!}\,2^{-2n+r} &= e^{-t}\frac{r}{t}I_r(t).\end{aligned} \tag{7.12}$$

Thus: (i) *for fixed* $r = 1, 2, \ldots$

$$v_r(t) = e^{-t}\frac{r}{t}I_r(t) \tag{7.13}$$

defines a probability density concentrated on $\overline{0, \infty}$.

(ii) *The epoch of the first passage through* $r > 0$ *has density* v_r. (See problem 15.)

This derivation permits another interesting conclusion. A first passage through $r + \rho$ at epoch t presupposes a previous first passage through r at some epoch $s < t$. Because of the independence of the jumps in the time intervals $\overline{0, s}$ and $\overline{s, t}$ and the lack of memory of the exponential waiting times we must have

$$v_r * v_\rho = v_{r+\rho}. \tag{7.14}$$

[A computational verification of this relation from (7.12) is easy if one uses the corresponding convolution property for the probabilities (7.11).]

Actually the proposition (i) and the relation (7.14) are true for all positive values of the parameters r and ρ.[5]

8. DISTRIBUTIONS ON A CIRCLE

The half-open interval $\overline{0, 1}$ may be taken as representing the points of a circle of unit length, but it is preferable to wrap the whole line around the circle. The circle then receives an orientation, and the arc length runs from $-\infty$ to ∞ but x, $x \pm 1$, $x \pm 2, \ldots$ are interpreted as the same point. Addition is modulo 1 just as addition of angles is modulo 2π. *A probability density on the circle is a periodic function* $\varphi \geq 0$ *such that*

$$\int_0^1 \varphi(x)\, dx = 1. \tag{8.1}$$

Examples. (*a*) *Buffon's needle problem* (1777). The traditional formulation is as follows. A plane is partitioned into strips of unit width parallel to the y-axis. A needle of unit length is thrown at random. What is the probability that it lies athwart two strips? To state the problem formally consider first the *center* of the needle. Its position is determined by two coordinates, but y is disregarded and x is reduced modulo 1. In this way "the center of the needle" becomes a random variable $\mathbf{X}$ on the circle with a uniform distribution. The *direction* of the needle may be described by the angle (measured clockwise) between the needle and the y-axis. A turn through π restores the position of the needle and hence the angle is determined only up to a multiple of π. We denote it by $\mathbf{Z}\pi$. In Buffon's needle problem it is implied that $\mathbf{X}$ and $\mathbf{Z}$ are independent and uniformly distributed variables[6] on the circle with unit length.

---

[5] W. Feller, *Infinitely divisible distributions and Bessel functions associated with random walks*, J. Soc. Indust. Appl. Math., vol. 14 (1966), pp. 864–875.

[6] The sample space of the pair $(\mathbf{X}, \mathbf{Z})$ is a torus.

If we choose to represent $\mathbf{X}$ by values between 0 and 1 and $\mathbf{Z}$ by values between $-\frac{1}{2}$ and $\frac{1}{2}$ the needle crosses a boundary iff $\frac{1}{2}\cos \mathbf{Z}\pi > \mathbf{X}$ or $\frac{1}{2}\cos \mathbf{Z}\pi > 1 - \mathbf{X}$. For a given value z between $-\frac{1}{2}$ and $\frac{1}{2}$ the probability that $\mathbf{X} < \frac{1}{2}\cos z\pi$ is the same as the probability that $1 - \mathbf{X} < \frac{1}{2}\cos z\pi$, namely $\frac{1}{2}\cos z\pi$. Thus the required probability is

$$\int_{-\frac{1}{2}}^{\frac{1}{2}} \cos z\pi \cdot dz = \frac{2}{\pi}. \tag{8.2}$$ ►

A random variable $\mathbf{X}$ on the line may be reduced modulo 1 to obtain a variable ${}^0\mathbf{X}$ on the circle. *Rounding errors* in numerical calculations are random variables of this kind. If $\mathbf{X}$ has density f the density of ${}^0\mathbf{X}$ is given by[7]

$$\varphi(x) = \sum_{-\infty}^{+\infty} f(x+n). \tag{8.3}$$

Every density on the line thus induces a density on the circle. [It will be seen in XIX,5 that the same φ admits of an entirely different representation in terms of Fourier series. For the special case of normal densities see example XIX,5(*e*).]

Examples. (*b*) *Poincaré's roulette problem.* Consider the number of rotations of a roulette wheel as a random variable $\mathbf{X}$ with a density f concentrated on the positive half-axis. The observed net result, namely the point ${}^0\mathbf{X}$ at which the wheel comes to rest, is the variable $\mathbf{X}$ reduced modulo 1. Its density is given by (8.3).

One feels instinctively that "under ordinary circumstances" the density of ${}^0\mathbf{X}$ should be nearly uniform. In 1912 H. Poincaré put this vague feeling on the solid basis of a limit theorem. We shall not repeat this analysis because a similar result follows easily from (8.3). The tacit assumption is, of course, that the given density f is spread out effectively over a long interval so that its *maximum* m *is small.* Assume for simplicity that f increases up to a point a where it assumes its maximum $m = f(a)$, and that f decreases for $x > a$. For the density φ of the reduced variable ${}^0\mathbf{X}$ we have then

$$\varphi(x) - 1 = \sum_n f(x+n) - \int_{-\infty}^{+\infty} f(s)\, ds. \tag{8.4}$$

For fixed x denote by x_k the unique point of the form $x+n$ such that

[7] Readers worried about convergence should consider only densities f concentrated on a finite interval. The uniform convergence is obvious if f is monotone for x and $-x$ sufficiently large. Without any conditions on f the series may diverge at some points, but φ always represents a density because the partial sums in (8.2) represent a *monotone* sequence of functions whose integrals tend to 1. (See IV,2.)

$a+k \leq x_k < a+k+1$. Then (8.4) may be rewritten in the form

$$\varphi(x) - 1 = \sum_{k=-\infty}^{+\infty} \int_{a+k}^{a+k+1} [f(x_k) - f(s)]\, ds. \tag{8.5}$$

For $k < 0$ the integrand is ≤ 0, and so

$$\varphi(x) - 1 \leq \sum_{k=0}^{\infty} [f(a+k) - f(a+k+1)] = f(a) = m.$$

A similar argument shows that $\varphi(x) - 1 \geq -m$. Thus $|\varphi(x) - 1| < m$ and so φ is indeed nearly constant.

The monotonicity conditions were imposed only for the sake of exposition and can be weakened in many ways. [Neat sufficient conditions can be obtained using Poisson's summation formula, XIX,5(2).]

(*c*) *Distribution of first significant digits.* A distinguished applied mathematician was extremely successful in bets that a number chosen at random in the *Farmer's Almanac*, or the *Census Report* or a similar compendium, would have the first significant digit less than 5. One expects naïvely that all 9 digits are equally likely, in which case the probability of a digit ≤ 4 would be $\frac{4}{9}$. In practice[8] it is close to 0.7.

Consider the discrete probability distribution attributing to the digit k probability $p_k = \text{Log}\,(k+1) - \text{Log}\,k$ (where Log denotes the logarithm to the basis 10 and $k = 1, \ldots, 9$). These probabilities are approximately

$$p_1 = 0.3010 \qquad p_2 = 0.1761 \qquad p_3 = 0.1249 \qquad p_4 = 0.0969$$
$$p_5 = 0.0792 \qquad P_6 = 0.0669 \qquad p_7 = 0.0580 \qquad p_8 = 0.0512 \qquad p_9 = 0.0458,$$

and it is seen that the distribution $\{p_k\}$ differs markedly from the uniform distribution with weights $\frac{1}{9} = 0.111 \cdots$.

We now show (following R. S. Pinkham) that $\{p_k\}$ is plausible for the empirical distribution of the first significant digit for numbers taken at random from a large body of physical or observational data. Indeed, such a number may be considered as a random variable $\mathbf{Y} > 0$ with some unknown distribution. The first significant digit of $\mathbf{Y}$ equals k iff $10^n k \leq \mathbf{Y} < 10^n(k+1)$ for some n. For the variable $\mathbf{X} = \text{Log}\,\mathbf{Y}$ this means

$$n + \text{Log}\,k \leq \mathbf{X} < n + \text{Log}\,(k+1). \tag{8.6}$$

If the spread of $\mathbf{Y}$ is very large the reduced variable ${}^0\mathbf{X}$ will be approximately uniformly distributed, and the probability of (8.6) is then close to $\text{Log}\,(k+1) - \text{Log}\,k = p_k$.

---

[8] For empirical material see F. Benford, *The law of anomalous numbers*, Proc. Amer. Philos. Soc., vol. 78 (1938) pp. 551–572.

The convolution formula (1.5) and the argument leading to it remain valid when addition is taken modulo 1. Accordingly, *the convolution of two densities on the circle of length* 1 *is the density defined by*

$$w(x) = \int_0^1 f_1(x-y)f_2(y)\,dy. \tag{8.7}$$

If $\mathbf{X}_1$ and $\mathbf{X}_2$ are independent variables with densities f_1 and f_2 then $\mathbf{X}_1 + \mathbf{X}_2$ has the density w. Since these densities are periodic, *the convolution of the uniform density with any other density is uniform.* (See problem 16.)

9. PROBLEMS FOR SOLUTION

1. Show that the normal approximation to the binomial distribution established in **1**; VII implies the convolution formula (1.7) for the normal densities.

2. Using the substitution $x = \frac{1}{2}y^2$ prove that $\Gamma(\frac{1}{2}) = \sqrt{\pi}$.

3. *Legendre's duplication formula.* From (2.5) for $\mu = \nu$ conclude that

$$\Gamma(2\nu) = \frac{1}{\sqrt{\pi}}\, 2^{2\nu-1}\Gamma(\nu)\Gamma(\nu+\tfrac{1}{2}).$$

Hint: Use the substitution $4(y-y^2) = s$ in $0 < y < \frac{1}{2}$.

4. If $g(x) = \frac{1}{2}e^{-|x|}$ find the convolutions $g * g$ and $g * g * g$ as well as g^{4*}.

5. Let $\mathbf{X}$ and $\mathbf{Y}$ be independent with the common Cauchy density $\gamma_1(x)$ of (4.5). Prove that the product $\mathbf{XY}$ has density $2\pi^{-2}\quad (x-1)^{-1}\,g|x|$
Hint: No calculations are required beyond the observation that

$$\frac{a-1}{(1+s)(a+s)} = \frac{1}{1+s} - \frac{1}{a+s}.$$

6. Prove that if

$$f(x) = \frac{2}{\pi}\frac{1}{e^x + e^{-x}} \quad \text{then} \quad f * f(x) = \frac{4}{\pi^2}\frac{x}{e^x - e^{-x}}$$

(*a*) by considering the variables $\log|\mathbf{X}|$ and $\log|\mathbf{Y}|$ of the preceding problem; (*b*) directly by the substitution $e^{2y} = t$ and a partial fraction decomposition. (See problem 8 of XV,9.)

7. If $\mathbf{X}$ has the normal density $\mathfrak{n}$ then obviously $\mathbf{X}^{-2}$ has the stable density (4.8). From this conclude that if $\mathbf{X}$ and $\mathbf{Y}$ are independent and normal with zero expectations and variances σ_1^2 and σ_2^2, then $\mathbf{Z} = \mathbf{XY}/\sqrt{\mathbf{X}^2+\mathbf{Y}^2}$ *is normal with variance* σ_3^2 such that $1/\sigma_3 = 1/\sigma_1 + 1/\sigma_2$ (L. Shepp).

8. Let $\mathbf{X}_1, \ldots, \mathbf{X}_n$ be independent and $\mathbf{X}_{(n)}$ the largest among them. Show that if the $\mathbf{X}_j$ have:

(*a*) the Cauchy density (4.5), then

$$\mathbf{P}\{n^{-1}\mathbf{X}_{(n)} \le x\} \to e^{-t/(\pi x)}, \qquad x > 0$$

(*b*) the stable density (4.8), then

$$\mathbf{P}\{n^{-2}\mathbf{X}_{(n)} \le x\} \to e^{-\alpha\sqrt{2/(\pi x)}}, \qquad x > 0.$$

9. Let **X** and **Y** be independent with densities f and g concentrated on $\overline{0, \infty}$. If $\mathbf{E}(\mathbf{X}) < \infty$ the ratio $\mathbf{X}/\mathbf{Y}$ has a finite expectation iff

$$\int_0^1 y^{-1}g(y)\,dy < \infty.$$

10. In example 5(*c*) find the density of the waiting time to the *next discharge* (*a*) if at epoch 0 the server is empty, (*b*) under steady-state conditions.

11. In example 5(*d*) show that

$$\mathbf{E}(\mathbf{T}_x) = F(x)(\mu+1-x) + \int_0^{1-x} tf(t+x)\,dt,$$

where μ is the expectation of F. From this verify the assertion concerning $\mathbf{E}(\mathbf{T})$ when x is uniformly distributed.

12. In example 5(*d*) find the waiting time distribution when $f(t) = 1$ for $0 < t < 1$.

13. In example 6(*c*) assume that u is the beta density given by (4.2). Evaluate the conditional probability (6.3) in terms of binomial coefficients.

14. Let **X** and **Y** be independent with the common Poisson distribution $\mathbf{P}\{\mathbf{X} = n\} = e^{-t}t^n/n!$ Show that

$$\mathbf{P}\{\mathbf{X}-\mathbf{Y} = r\} = e^{-2t}I_{|r|}(2t), \qquad r = 0, \pm 1, \pm 2, \ldots.$$

[See problem 9 of V,11.]

15. The results of section 7.*c* remain valid for unsymmetric random walks provided the probability of a first passage through $r > 0$ equals one, that is, provided $p \geq q$. Show that the only change in (7.11) is that 2^{-2n+r} is replaced by $p^n q^{n-r}$, and the conclusion is that for $p \geq q$ and $r = 1, 2, \ldots,$

$$\sqrt{(p/q)^r}e^{-t}\frac{r}{t}I_r(2\sqrt{pq}\,t)$$

defines a probability density concentrated on $t > 0$.

16. Let **X** and **Y** be independent variables and ${}^0\mathbf{X}$ and ${}^0\mathbf{Y}$ be the same variables reduced modulo 1. Show that ${}^0\mathbf{X}+{}^0\mathbf{Y}$ is obtained by reducing $\mathbf{X}+\mathbf{Y}$ modulo 1. Verify the corresponding formula for convolutions by direct calculation.

CHAPTER III

Densities in Higher Dimensions. Normal Densities and Processes

For obvious reasons multivariate distributions occur less frequently than one-dimensional distributions, and the material of this chapter will play almost no role in the following chapters. On the other hand, it covers important material, for example, a famous characterization of the normal distribution and tools used in the theory of stochastic processes. Their true nature is best understood when divorced from the sophisticated problems with which they are sometimes connected.

1. DENSITIES

For typographical convenience we refer explicitly to the Cartesian plane $\mathfrak{R}^2$, but it will be evident that the number of dimensions is immaterial. We refer the plane to a fixed coordinate system with coordinate variables $\mathbf{X}_1, \mathbf{X}_2$. (A more convenient single-letter notation will be introduced in section 5.)

A non-negative integrable function f defined in $\mathfrak{R}^2$ and such that its integral equals one is called a *probability density*, or density for short. (All the densities occurring in the chapter are piecewise continuous, and so the concept of integration requires no comment.) The density f attributes to the region Ω the probability

$$\mathbf{P}\{\Omega\} = \iint_{\Omega} f(x_1, x_2)\, dx_1\, dx_2 \tag{1.1}$$

provided, of course, that Ω is sufficiently regular for the integral to exist. All such probabilities are uniquely determined by the probabilities of rectangles parallel to the axes, that is, by the knowledge of

$$\mathbf{P}\{a_1 < \mathbf{X}_1 \le b_1, a_2 < \mathbf{X}_2 \le b_2\} = \int_{a_1}^{b_1}\int_{a_2}^{b_2} f(x_1, x_2)\, dx_1\, dx_2 \tag{1.2}$$

for all combinations $a_i < b_i$. Letting $a_1 = a_2 = -\infty$ we get the *distribution function* F of f, namely

$$F(x_1, x_2) = \mathbf{P}\{\mathbf{X}_1 \leq x_1, \mathbf{X}_2 \leq x_2\}. \tag{1.3}$$

Obviously $F(b_1, x_2) - F(a_1, x_2)$ is the probability of a semi-finite strip of width $b_1 - a_1$ and, the rectangle appearing in (1.2) being the difference of two such strips, the probability (1.2) equals the so-called mixed difference

$$F(b_1, b_2) - F(a_1, b_2) - F(b_1, a_2) + F(a_1, a_2).$$

It follows that the knowledge of the distribution function F uniquely determines all probabilities (1.1). Despite the formal analogy with the situation on the line, the concept of distribution function F is much less useful in the plane and it is best to concentrate on the assignment of probabilities (1.1) in terms of the density itself. This assignment differs from the joint probability distribution of two discrete random variables (**1**; IX,1) in two respects. First, integration replaces summation and, second, probabilities are now assigned only to "sufficiently regular regions" whereas in discrete sample spaces all sets had probabilities. As the present chapter treats only simple examples in which the difference is hardly noticeable, the notions and terms of the discrete theory carry over in a self-explanatory manner. Just as in the preceding chapters we employ therefore a probabilistic language without any attempt at a general theory (which will be supplied in chapter V).

It is apparent from (1.3) that[1]

$$\mathbf{P}\{\mathbf{X}_1 \leq x_1\} = F(x_1, \infty). \tag{1.4}$$

Thus $F_1(x) = F(x, \infty)$ defines the distribution function of $\mathbf{X}_1$, and its *density* f_1 is given by

$$f_1(x) = \int_{-\infty}^{+\infty} f(x, y)\, dy. \tag{1.5}$$

When it is desirable to emphasize the connection between $\mathbf{X}_1$ and the pair $(\mathbf{X}_1, \mathbf{X}_2)$ we again speak of F_1 as *marginal* distribution[2] and of f_1 as marginal density.

The *expectation* μ_1 and *variance* σ_1^2 of $\mathbf{X}_1$—if they exist—are given by

$$\mu_1 = \mathbf{E}(\mathbf{X}_1) = \int_{-\infty}^{+\infty}\int_{-\infty}^{+\infty} x_1 f(x_1, x_2)\, dx_1\, dx_2 \tag{1.6}$$

and

$$\sigma_1^2 = \operatorname{Var}(\mathbf{X}_1) = \int_{-\infty}^{+\infty}\int_{-\infty}^{+\infty} (x_1 - \mu_1)^2 f(x_1, x_2)\, dx_1\, dx_2. \tag{1.7}$$

---

[1] Here and in the following $U(\infty) = \lim U(x)$ as $x \to \infty$ and the use of the symbol $U(\infty)$ implies the existence of the limit.

[2] *Projection* on the axes is another accepted term.

By symmetry these definitions apply also to $\mathbf{X}_2$. Finally, the *covariance* of $\mathbf{X}_1$ and $\mathbf{X}_2$ is

$$\text{(1.8)} \qquad \operatorname{Cov}(\mathbf{X}_1, \mathbf{X}_2) = \int_{-\infty}^{+\infty}\int_{-\infty}^{+\infty} (x_1-\mu_1)(x_2-\mu_2)f(x_1, x_2)\, dx_1\, dx_2.$$

The normalized variables $\mathbf{X}_i\sigma_i^{-1}$ are dimensionless and their covariance, namely $\rho = \operatorname{Cov}(\mathbf{X}_1, \mathbf{X}_2)\sigma_1^{-1}\sigma_2^{-1}$, is the *correlation coefficient* of $\mathbf{X}_1$ and $\mathbf{X}_2$ (see **1**; IX,8).

A *random variable* $\mathbf{U}$ is a function of the coordinate variables $\mathbf{X}_1$ and $\mathbf{X}_2$; again we consider for the present only functions such that the probabilities $\mathbf{P}\{\mathbf{U} \leq t\}$ can be evaluated by integrals of the form (1.1). Thus each random variable will have a unique distribution function, each pair will have a joint distribution, etc.

In many situations it is expedient to *change the coordinate variables*, that is, to let two variables $\mathbf{Y}_1, \mathbf{Y}_2$ play the role previously assigned to $\mathbf{X}_1, \mathbf{X}_2$. In the simplest case the $\mathbf{Y}_j$ are defined by a linear transformation

$$\text{(1.9)} \qquad \mathbf{X}_1 = a_{11}\mathbf{Y}_1 + a_{12}\mathbf{Y}_2, \qquad \mathbf{X}_2 = a_{21}\mathbf{Y}_1 + a_{22}\mathbf{Y}_2,$$

with determinant $\Delta = a_{11}a_{22} - a_{12}a_{21} > 0$. Generally a transformation of the form (1.9) may be described either as a mapping from one plane to another or as a change of coordinates in the same plane. Introducing the change of variables (1.9) into the integral (1.1) we get

$$\text{(1.10)} \qquad \mathbf{P}\{\Omega\} = \iint_{\Omega_*} f(a_{11}y_1+a_{12}y_2,\, a_{21}y_1+a_{22}y_2)\cdot \Delta\, dy_1\, dy_2$$

the region Ω_* containing all points (y_1, y_2) whose image (x_1, x_2) is in Ω. Since the events $(\mathbf{X}_1, \mathbf{X}_2) \in \Omega$ and $(\mathbf{Y}_1, \mathbf{Y}_2) \in \Omega_*$ are identical it is seen that *the joint density of* $(\mathbf{Y}_1, \mathbf{Y}_2)$ *is given by*

$$\text{(1.11)} \qquad g(y_1, y_2) = f(a_{11}y_1+a_{12}y_2,\, a_{21}y_1+a_{22}y_2)\cdot \Delta.$$

All this applies equally to higher dimensions.

A similar argument applies to more general transformations, except that the determinant Δ is replaced by the Jacobian. We shall use explicitly only the *change to polar coordinates*

$$\text{(1.12)} \qquad \mathbf{X}_1 = \mathbf{R}\cos\mathbf{\Theta}, \qquad \mathbf{X}_2 = \mathbf{R}\sin\mathbf{\Theta}$$

with $(\mathbf{R}, \mathbf{\Theta})$ restricted to $\mathbf{R} \geq 0$, $-\pi < \mathbf{\Theta} \leq \pi$. Here the density of $(\mathbf{R}, \mathbf{\Theta})$ is given by

$$\text{(1.13)} \qquad g(r, \theta) = f(r\cos\theta, r\sin\theta)r.$$

In three dimensions one uses the geographic longitude φ and latitude θ (with $-\pi < \varphi \leq \pi$ and $-\frac{1}{2}\pi \leq \theta \leq \frac{1}{2}\pi$). The coordinate variables in the polar system are then defined by

(1.14) $\mathbf{X}_1 = \mathbf{R}\cos\boldsymbol{\Phi}\cos\boldsymbol{\Theta}, \qquad \mathbf{X}_2 = \mathbf{R}\sin\boldsymbol{\Phi}\cos\boldsymbol{\Theta}, \qquad \mathbf{X}_3 = \mathbf{R}\sin\boldsymbol{\Theta}.$

For their joint density one gets

(1.15) $g(r, \varphi, \theta) = f(r\cos\varphi\cos\theta,\ r\sin\varphi\cos\theta,\ r\sin\theta)r^2\cos\theta.$

In the transformation (1.14) the "planes" $\boldsymbol{\Theta} = -\frac{1}{2}\pi$ and $\boldsymbol{\Theta} = \frac{1}{2}\pi$ correspond to the half axes in the x_3-direction, but this singularity plays no role since these half axes have zero probability. A similar remark applies to the origin for polar coordinates in the plane.

Examples. (*a*) *Independent variables.* In the last chapters we considered *independent* variables $\mathbf{X}_1$ and $\mathbf{X}_2$ with densities f_1 and f_2. This amounts to defining a bivariate density by $f(x_1, x_2) = f_1(x_1)f_2(x_2)$, and the f_i represent the marginal densities.

(*b*) "*Random choice.*" Let Γ be a bounded region; for simplicity we assume Γ convex. Denote the area of Γ by γ and put f equal to γ^{-1} within Γ and equal to 0 outside Γ. Then f is a density, and the probability of any region $\Omega \subset \Gamma$ equals the ratio of the areas of Ω and Γ. By obvious analogy with the one-dimensional situation we say that the pair $(\mathbf{X}_1, \mathbf{X}_2)$ is *distributed uniformly over* Γ. The marginal density of $\mathbf{X}_1$ at the abscissa x_1 equals the width of Γ at x_1 in the obvious sense of the word. (See problem 1.)

(*c*) *Uniform distribution on a sphere.* The unit sphere Σ in three dimensions may be represented in terms of the geographic longitude φ and latitude θ by the equations

(1.16) $x_1 = \cos\varphi\cos\theta, \qquad x_2 = \sin\varphi\cos\theta, \qquad x_3 = \sin\theta.$

To each pair (φ, θ) such that $-\pi < \varphi \leq \pi$, $-\frac{1}{2}\pi < \theta < \frac{1}{2}\pi$ there corresponds exactly one point on the sphere and, except for the two poles, each point of Σ is obtained in this way. The exceptional role of the poles need not concern us since they will have probability 0. A region Ω on the sphere is defined by its image in the φ, θ-plane, and the area of Ω equals the integral of $\cos\theta\, d\varphi\, d\theta$ over this image [see (1.15)]. For the conceptual experiment "random choice of a point on Σ" we should put $4\pi\mathbf{P}\{\Omega\} =$ area of Ω. This is equivalent to defining in the φ, θ-plane a density

(1.17) $$g(\varphi, \theta) = \begin{matrix} (4\pi)^{-1}\cos\theta & \textit{for} \quad -\pi < \varphi \leq \pi, \quad |\theta| < \frac{1}{2}\pi \\ 0 & \textit{elsewhere.} \end{matrix}$$

With this definition the coordinate variables are independent, the longitude being distributed uniformly over $\overline{-\pi, \pi}|$.

The device of referring the sphere Σ to the φ, θ-plane is familiar from geographic maps and useful for probability theory. Note, however, that the coordinate variables are largely arbitrary and their expectations and variances meaningless for the original conceptual experiment.

(*d*) *The bivariate normal density*. Normal densities in higher dimensions will be introduced systematically in section 6. The excuse for anticipating the bivariate case is to provide an easy access to it. An obvious analogue to the normal density $\mathfrak{n}$ of II,(2.1) is provided by densities of the form $c \cdot e^{-q(x_1, x_2)}$ where $q(x_1, x_2) = a_1 x_1^2 + 2bx_1 x_2 + a_2 x_2^2$. It is easily seen that e^{-q} will be integrable iff the a_j are positive and $a_1 a_2 - b^2 > 0$. For purposes of probability theory it is preferable to express the coefficients a_i and b in terms of these variances and to *define the bivariate normal density centered at the origin by*

(1.18)
$$\varphi(x_1, x_2) = \frac{1}{2\pi\sigma_1\sigma_2\sqrt{1-\rho^2}} \exp\left[-\frac{1}{2(1-\rho^2)}\left(\frac{x_1^2}{\sigma_1^2} - 2\rho\frac{x_1 x_2}{\sigma_1\sigma_2} + \frac{x_2^2}{\sigma_2^2}\right)\right]$$

where $\sigma_1 > 0$, $\sigma_2 > 0$, and $-1 < \rho < 1$. The integration with respect to x_2 is easily performed by the substitution $t = x_2/\sigma_2 - \rho\, x_1/\sigma_1$ (completing squares), and it is seen that φ indeed represents a density in $\mathcal{R}^2$. Furthermore, it becomes obvious that the *marginal* distributions for $\mathbf{X}_1$ and $\mathbf{X}_2$ are again normal[3] and that $\mathbf{E}(\mathbf{X}_i) = 0$, $\text{Var}(\mathbf{X}_i) = \sigma_i^2$, $\text{Cov}(\mathbf{X}_1, \mathbf{X}_2) = \rho\sigma_1\sigma_2$. In other words, ρ is the *correlation* coefficient of $\mathbf{X}_1$ and $\mathbf{X}_2$. Replacing x_i by $x_i - c_i$ in (1.18) leads to a normal density centered at the point (c_1, c_2).

It is important that *linear transformations* (1.9) *change a normal distribution into another normal distribution*. This is obvious from the definition and (1.11). [Continued in example 2(*a*).]

(*e*) *The symmetric Cauchy distribution in* $\mathcal{R}^2$. Put

$$u(x_1, x_2) = \frac{1}{2\pi} \cdot \frac{1}{\sqrt{(1+x_1^2+x_2^2)^3}}. \tag{1.19}$$

To see that this is a density note[4] that

$$\int_{-\infty}^{+\infty} u(x_1, y)\, dy = \frac{1}{2\pi} \cdot \frac{1}{1+x_1^2} \cdot \frac{y}{\sqrt{1+x_1^2+y^2}}\Bigg|_{-\infty}^{+\infty} = \frac{1}{\pi} \cdot \frac{1}{1+x_1^2}. \tag{1.20}$$

[3] Contrary to a widespread belief *there exist non-normal bivariate* densities with normal marginal densities (two types are described in problems 2, 3; two more in problems 5 and 7 of V,12). In the desire to deal with normal densities, statisticians sometimes introduce a pair of new coordinate variables $\mathbf{Y}_1 = g_1(\mathbf{X}_1)$, $\mathbf{Y}_2 = g_2(\mathbf{X}_2)$ which are normally distributed. Alas, this does *not* make the joint distribution of $(\mathbf{Y}_1, \mathbf{Y}_2)$ normal.

[4] The substitution $y = \sqrt{1 + x_1^2}\tan t$ makes the calculation easy.

It follows that u is a density and that the marginal density of $\mathbf{X}_1$ is the *Cauchy density* γ_1 of II,(4.5). Obviously $\mathbf{X}_1$ has no expectation.

Switching to polar coordinates [as in (1.12)] $\mathbf{R}$ gets a density independent of θ and so the variables $\mathbf{R}$ and $\mathbf{\Theta}$ are stochastically independent. In the terminology of I,10 we can therefore say that with the symmetric Cauchy distribution $(\mathbf{X}_1, \mathbf{X}_2)$ *represents a vector in a randomly chosen direction with a length* $\mathbf{R}$ *whose density is given by* $r\sqrt{(1+r^2)^{-3}}$, whence $\mathbf{P}\{\mathbf{R} \leq r\} = 1 - \sqrt{(1+r^2)^{-1}}$. [Continued in example 2(*c*).]

(*f*) *The symmetric Cauchy distribution in* $\mathcal{R}^3$. Put

$$v(x_1, x_2, x_3) = \frac{1}{\pi^2} \cdot \frac{1}{(1+x_1^2+x_2^2+x_3^2)^2}. \tag{1.21}$$

It is easily seen[5] that the marginal density of $(\mathbf{X}_1, \mathbf{X}_2)$ is the symmetric Cauchy density u of (1.19). The marginal density of $\mathbf{X}_1$ is therefore the Cauchy density γ_1. (Continued in problem 5.) ▶

Although it will not play an explicit role in the sequel it should be mentioned that we can define *convolutions* just as in one dimension. Consider two pairs $(\mathbf{X}_1, \mathbf{X}_2)$ and $(\mathbf{Y}_1, \mathbf{Y}_2)$ with joint densities f and g, respectively. Saying that *the two pairs are independent* means that we take the four-dimensional space with coordinate variables $\mathbf{X}_1, \mathbf{X}_2, \mathbf{Y}_1, \mathbf{Y}_2$ as sample space and define in it a density given by the product $f(x_1, x_2)\, g(y_1, y_2)$. Just as in $\mathcal{R}^1$ it is then easily seen that the joint density v of the sum $(\mathbf{X}_1+\mathbf{Y}_1, \mathbf{X}_2+\mathbf{Y}_2)$ is given by the convolution formula

$$v(z_1, z_2) = \int_{-\infty}^{+\infty}\int_{-\infty}^{+\infty} f(z_1-x_1, z_2-x_2)\, g(x_1, x_2)\, dx_1\, dx_2 \tag{1.22}$$

which is the obvious analogue to I,(2.12). (See problems 15–17.)

2. CONDITIONAL DISTRIBUTIONS

Suppose that the pair $(\mathbf{X}_1, \mathbf{X}_2)$ has a continuous density f and that the marginal density f_1 of $\mathbf{X}_1$ is strictly positive. Consider the conditional probability of the event $\mathbf{X}_2 \leq \eta$ given that $\xi < \mathbf{X}_1 \leq \xi + h$, namely

$$\mathbf{P}\{\mathbf{X}_2 \leq \eta \mid \xi < \mathbf{X}_1 \leq \xi + h\} = \frac{\displaystyle\int_\xi^{\xi+h} dx \int_{-\infty}^{\eta} f(x, y)\, dy}{\displaystyle\int_\xi^{\xi+h} f_1(x)\, dx}. \tag{2.1}$$

Dividing numerator and denominator by h, one sees that as $h \to 0$ the

[5] Use the substitution $z = \sqrt{1 + x_1^2 + x_2^2} \tan t$.

right side tends to

$$U_\xi(\eta) = \frac{1}{f_1(\xi)} \int_{-\infty}^{\eta} f(\xi, y)\, dy. \tag{2.2}$$

For fixed ξ this is a distribution function in η with density

$$u_\xi(\eta) = \frac{1}{f_1(\xi)} f(\xi, \eta). \tag{2.3}$$

We call u_ξ the *conditional density of* $\mathbf{X}_2$ *given that* $\mathbf{X}_1 = \xi$. *The conditional expectation of* $\mathbf{X}_2$ *given that* $\mathbf{X}_1 = \xi$ *is defined by*

$$\mathbf{E}(\mathbf{X}_2 \mid \mathbf{X}_1 = \xi) = \frac{1}{f_1(\xi)} \int_{-\infty}^{+\infty} y\, f(\xi, y)\, dy \tag{2.4}$$

provided that the integral converges absolutely. With ξ considered as a variable the right side becomes a function of it. In particular, we may identify ξ with the coordinate variable $\mathbf{X}_1$ to obtain a random variable called *the regression of* $\mathbf{X}_2$ *on* $\mathbf{X}_1$ and denoted by $\mathbf{E}(\mathbf{X}_2 \mid \mathbf{X}_1)$. The appearance of $\mathbf{X}_2$ should not obscure the fact that this random variable is a function of the single variable $\mathbf{X}_1$ [its values being given by (2.4)].

So far we have assumed that $f_1(\xi) > 0$ for all ξ. The expression (2.4) is meaningless at any place where $f_1(\xi) = 0$, but the set of such points has probability zero and we agree to interpret (2.4) as zero at all points where f_1 vanishes. Then $\mathbf{E}(\mathbf{X}_2 \mid \mathbf{X}_1)$ is defined whenever the density is continuous. (In V,9–11 conditional probabilities will be introduced for arbitrary distributions.)

Needless to say, the regression $\mathbf{E}(\mathbf{X}_1 \mid \mathbf{X}_2)$ of $\mathbf{X}_1$ on $\mathbf{X}_2$ is defined in like manner. Furthermore, a *conditional variance* $\mathrm{Var}\,(\mathbf{X}_2 \mid \mathbf{X}_1)$ is defined by obvious analogy with (2.4).

These definitions carry over to higher dimensions, except that a density in $\mathcal{R}^3$ gives rise to three bivariate and three univariate conditional densities (See problem 6.)

Examples. (*a*) *The normal density.* For the density (1.18) obviously

$$u_\xi(y) = \frac{1}{\sqrt{2\pi(1-\rho^2)\sigma_2^2}} \exp\left[-\frac{(y-\rho(\sigma_2/\sigma_1)\xi)^2}{2(1-\rho^2)\sigma_2^2}\right] \tag{2.5}$$

which is a normal density with expectation $\rho(\sigma_2/\sigma_1)\xi$ and variance $(1-\rho^2)\sigma_2^2$. Thus

$$\mathbf{E}(\mathbf{X}_2 \mid \mathbf{X}_1) = \rho(\sigma_2/\sigma_1)\mathbf{X}_1, \qquad \mathrm{Var}\,(\mathbf{X}_2 \mid \mathbf{X}_1) = (1-\rho^2)\sigma_2^2. \tag{2.6}$$

It is one of the pleasing properties of the normal distribution that the regressions are *linear* functions.

Perhaps the earliest application of these relations is due to Galton, and one of his examples may illustrate their *empirical meaning*. Imagine that $\mathbf{X}_1$ and $\mathbf{X}_2$ represent the heights (measured in inches from their respective expectations) of fathers and sons in a human population. The height of a randomly chosen son is then a normal variable with expectation 0 and variance σ_2^2. However, in the subpopulation of sons whose fathers have a fixed height ξ, the height of the sons is a normal variable with expectation $\rho(\sigma_2/\sigma_1)\xi$ and variance $\sigma_2^2(1-\rho^2) < \sigma_2^2$. Thus the regression of $\mathbf{X}_2$ on $\mathbf{X}_1$ indicates how much statistical information about $\mathbf{X}_2$ is contained in observation of $\mathbf{X}_1$.

(*b*) Let $\mathbf{X}_1$ and $\mathbf{X}_2$ be independent and uniformly distributed in $\overline{0, 1}$. Denote by $\mathbf{X}_{(1)}$ the smaller and by $\mathbf{X}_{(2)}$ the larger among these variables. The pair $(\mathbf{X}_{(1)}, \mathbf{X}_{(2)})$ has a density equal to the constant 2 within the triangle $0 \leq x_1 \leq x_2 \leq 1$, and vanishing elsewhere. Integration over x_2 shows that the marginal density of $\mathbf{X}_{(1)}$ is given by $2(1-x_1)$. The conditional density of $\mathbf{X}_{(2)}$ for given $\mathbf{X}_{(1)} = x_1$ therefore equals the constant $1/1-x_1$ within the interval $\overline{x_1, 1}$ and zero elsewhere. In other words, given the value x_1 of $\mathbf{X}_{(1)}$ the variable $\mathbf{X}_{(2)}$ is uniformly distributed over $\overline{x_1, 1}$.

(*c*) *Cauchy distribution in* $\mathcal{R}^2$. For the bivariate density (1.19) the marginal density for $\mathbf{X}_1$ is given in (1.20), and so the conditional density of $\mathbf{X}_2$ for given $\mathbf{X}_1$ is

$$u_\xi(y) = \frac{1}{2} \cdot \frac{1 + \xi^2}{\sqrt{(1+\xi^2+y^2)^3}}. \tag{2.7}$$

Note that u_ξ differs only by the scale factor $\sqrt{1 + \xi^2}$ from the density $u_0(y)$ and so all the densities u_ξ are of the same type. Conditional expectations do not exist in this example. (See problem 6.) ►

In terms of the conditional densities (2.3) the distribution function of $\mathbf{X}_2$ takes on the form

$$\mathbf{P}\{\mathbf{X}_2 < y\} = \int_{-\infty}^{y} \int_{-\infty}^{+\infty} u_\xi(\eta) \cdot f_1(\xi) \, d\xi \, d\eta. \tag{2.8}$$

In other words, the distribution of $\mathbf{X}_2$ is obtained by *randomization* of the parameter ξ in the conditional densities u_ξ, and so *every*[6] *distribution may be represented as mixture*. Despite this theoretical universality there is a great difference in emphasis. In some situations [such as example (*a*)] one *starts* from a bivariate distribution for $(\mathbf{X}_1, \mathbf{X}_2)$ and derives conditional distributions, whereas in true randomization the conditional probabilities

[6] We have so far considered only continuous densities, but the general case will be covered in V,9. The notion of randomization was discussed in II,5.

u_x are the primary notion and the density $f(x, y)$ is actually *defined* by $u_x(y)f_1(x)$. (This procedure of defining probabilities in terms of conditional probabilities was explained in an elementary way in **1**; V,2.)

3. RETURN TO THE EXPONENTIAL AND THE UNIFORM DISTRIBUTIONS

The object of this section is to provide illustrative examples to the preceding sections and at the same time to supplement the theory of the first chapter.

Examples. (*a*) *A characteristic property of the exponential distribution.* Let $\mathbf{X}_1$ and $\mathbf{X}_2$ be two *independent* random variables with densities f_1 and f_2, and denote the density of their sum $\mathbf{S} = \mathbf{X}_1+\mathbf{X}_2$ by g. The pairs $(\mathbf{X}_1, \mathbf{S})$ and $(\mathbf{X}_1, \mathbf{X}_2)$ are related by the linear transformation $\mathbf{X}_1 = \mathbf{X}_1$, $\mathbf{X}_2 = \mathbf{S} - \mathbf{X}_1$ with determinant 1 and by (1.11) the *joint density* of the pair $(\mathbf{X}_1, \mathbf{S})$ is given by $f_1(x)f_2(s-x)$. Integrating over all x we obtain the marginal density g of $\mathbf{S}$. *The conditional density* u_s *of* $\mathbf{X}_1$ *given that* $\mathbf{S} = s$ satisfies

$$u_s(x) = \frac{f_1(x)f_2(s-x)}{g(s)}. \tag{3.1}$$

In the special case of exponential densities $f_1(x) = f_2(x) = \alpha e^{-\alpha x}$ (where $x > 0$) we get $u_s(x) = s^{-1}$ for $0 < x < s$. In other words, *given that* $\mathbf{X}_1 + \mathbf{X}_2 = s$, *the variable* $\mathbf{X}_1$ *is uniformly distributed* over the interval $\overline{0, s}$. Intuitively speaking, the knowledge that $\mathbf{S} = s$ gives us no clue as to the possible position of the random point $\mathbf{X}_1$ within the interval $\overline{0, s}$. This result conforms with the notion of complete randomness inherent in the exponential distribution. (A stronger version is contained in example (*d*). See also problem 12.)

(*b*) *Random partitions of an interval.* Let $\mathbf{X}_1, \ldots, \mathbf{X}_n$ be n points chosen independently and at random in the (one-dimensional) interval $\overline{0, 1}$. As before we denote by $\mathbf{X}_{(1)}, \mathbf{X}_{(2)}, \ldots, \mathbf{X}_{(n)}$ the random points $\mathbf{X}_1, \ldots, \mathbf{X}_n$ rearranged in increasing order. These points divide the interval $\overline{0, 1}$ into $n + 1$ subintervals which we denote by $I_1, I_2, \ldots, I_{n+1}$ numbering them from left to right so that $\mathbf{X}_{(j)}$ is the right endpoint of I_j. Our first aim is to calculate the joint density of $(\mathbf{X}_{(1)}, \ldots, \mathbf{X}_{(n)})$.

The sample space corresponding to $(\mathbf{X}_1, \ldots, \mathbf{X}_n)$ is the n-dimensional hypercube Γ defined by $0 < x_k < 1$, and probabilities equal the n-dimensional volume. The natural sample space with the $\mathbf{X}_{(k)}$ as coordinate variables is the subset Ω of Γ containing all points such that

$$0 < x_1 \leq \cdots \leq x_n < 1.$$

The volume of Ω is $1/n!$ Evidently the hypercube Γ contains $n!$ congruent replicas of the set Ω and in each the ordered n-tuple $(\mathbf{X}_{(1)}, \ldots, \mathbf{X}_{(n)})$ coincides with a fixed permutation of $\mathbf{X}_1, \ldots, \mathbf{X}_n$. (Within Γ, in particular, $\mathbf{X}_{(k)} = \mathbf{X}_k$.) The probability that $\mathbf{X}_j = \mathbf{X}_k$ for some pair $j \neq k$ equals zero, and only this event causes overlaps among the various replicas. It follows that for any subset $A \subset \Omega$ the probability that $(\mathbf{X}_{(1)}, \ldots, \mathbf{X}_{(n)})$ lies in A equals the probability that $(\mathbf{X}_1, \ldots, \mathbf{X}_n)$ lies in one of the $n!$ replicas of A, and this probability in turn equals $n!$ times the volume of A. Thus $\mathbf{P}\{(\mathbf{X}_{(1)}, \ldots, \mathbf{X}_{(n)}) \in A\}$ equals the ratio of the volumes of A and of Ω, which means that *the n-tuple* $(\mathbf{X}_{(1)}, \ldots, \mathbf{X}_{(n)})$ *is distributed uniformly over the set* Ω. The joint density of our n-tuple equals $n!$ within Ω and 0 outside.

From the joint density of $(\mathbf{X}_{(1)}, \ldots, \mathbf{X}_{(n)})$ the density of $\mathbf{X}_{(k)}$ may be calculated by keeping x_k fixed and integrating over the remaining variables. The result is easily seen to agree with the density calculated by other methods in I,(7.2).

This example was treated in detail as an exercise in handling and computing multivariate densities.

(*c*) *The distribution of the lengths.* In the random partition of the preceding example denote the length of the kth interval I_k by $\mathbf{U}_k$. Then

$$(3.2) \qquad \mathbf{U}_1 = \mathbf{X}_{(1)}, \qquad \mathbf{U}_k = \mathbf{X}_{(k)} - \mathbf{X}_{(k-1)} \qquad \textit{for} \quad k = 2, 3, \ldots, n$$

This is a linear transformation of the form (1.9) with determinant 1. The set Ω of points $0 < x_1 \leq \cdots \leq x_n < 1$ is mapped into the set Ω^* of points such that $u_j \geq 0$, $u_1 + \cdots + u_n < 1$, and hence $(\mathbf{U}_1, \ldots, \mathbf{U}_n)$ *is distributed uniformly over this region.* This result is stronger than the previously established fact that the $\mathbf{U}_k$ have a common distribution function [example I,7(*b*) and problem in I,13.]

(*d*) *Once more the randomness of the exponential distribution.* Let $\mathbf{X}_1, \ldots, \mathbf{X}_{n+1}$ be independent with the common density $\alpha e^{-\alpha x}$ for $x > 0$. Put $\mathbf{S}_j = \mathbf{X}_1 + \cdots + \mathbf{X}_j$. Then $(\mathbf{S}_1, \mathbf{S}_2, \ldots, \mathbf{S}_{n+1})$ is obtained from $(\mathbf{X}_1, \ldots, \mathbf{X}_{n+1})$ by a linear transformation of the form (1.9) with determinant 1. Denote by Ω the "octant" of points $x_j > 0$. The density of $(\mathbf{X}_1, \ldots, \mathbf{X}_{n+1})$ is concentrated on Ω and is given by

$$\alpha^{n+1} e^{-\alpha(x_1 + \cdots + x_{n+1})}$$

if $x_j > 0$. The variables $\mathbf{S}_1, \ldots, \mathbf{S}_{n+1}$ map Ω onto the region Ω^* defined by $0 < s_1 \leq s_2 \leq \cdots \leq s_{n+1} < \infty$, and [see (1.11)] within Ω^* the density of $(\mathbf{S}_1, \ldots, \mathbf{S}_{n+1})$ is given by $\alpha^{n+1} e^{-\alpha s_{n+1}}$. The marginal density of $\mathbf{S}_{n+1}$ is known to be the gamma density $\alpha^{n+1} s^n e^{-\alpha s}/n!$ and hence the *conditional density of the n-tuple* $(\mathbf{S}_1, \ldots, \mathbf{S}_n)$ *given that* $\mathbf{S}_{n+1} = s$ *equals* $n! s^{-n}$ *for* $0 < s_1 < \cdots < s_n < s$ (and zero elsewhere). In other words, given that

$\mathbf{S}_{n+1} = s$ the variables $(\mathbf{S}_1, \ldots, \mathbf{S}_n)$ are uniformly distributed over their possible range. Comparing this with example (*b*) we may say that *given* $\mathbf{S}_{n+1} = s$, *the variables* $(\mathbf{S}_1, \ldots, \mathbf{S}_n)$ *represent* n *points chosen independently and at random in the interval* $\overline{0, s}$ numbered in their natural order from left to right.

(*e*) *Another distribution connected with the exponential.* With a view to a surprising application we give a further example of a transformation. Let again $\mathbf{X}_1, \ldots, \mathbf{X}_n$ be independent variables with a common exponential distribution and $\mathbf{S}_n = \mathbf{X}_1 + \cdots + \mathbf{X}_n$. Consider the variables $\mathbf{U}_1, \ldots, \mathbf{U}_n$ defined by

$$\mathbf{U}_k = \mathbf{X}_k/\mathbf{S}_n \quad for \quad k = 1, \ldots, n-1, \qquad \mathbf{U}_n = \mathbf{S}_n, \tag{3.3}$$

or, what amounts to the same,

$$\mathbf{X}_k = \mathbf{U}_k\mathbf{U}_n \quad for \quad k < n, \qquad \mathbf{X}_n = \mathbf{U}_n(1-\mathbf{U}_1-\cdots-\mathbf{U}_{n-1}). \tag{3.4}$$

The Jacobian of (3.4) equals $\mathbf{U}_n^{n-1}$. The joint density of $(\mathbf{X}_1, \ldots, \mathbf{X}_n)$ is concentrated on the region Ω defined by $x_k > 0$, and in it this density is given by $\alpha^n e^{-\alpha(x_1+\cdots+x_n)}$. It follows that the joint density of $(\mathbf{U}_1, \ldots, \mathbf{U}_n)$ is given by $\alpha^n u_n^{n-1} e^{-\alpha u_n}$ in the region Ω^* defined by

$$u_1 + \cdots + u_{n-1} < 1, \qquad u_k > 0 \qquad k = 1, \ldots, n$$

and that it vanishes outside Ω^*. An integration with respect to u_n shows that the joint density for $(\mathbf{U}_1, \ldots, \mathbf{U}_{n-1})$ equals $(n-1)!$ in Ω^* and 0 elsewhere. Comparing with example (*c*) we see that $(\mathbf{U}_1, \ldots, \mathbf{U}_{n-1})$ *has the same distribution as if* $\mathbf{U}_k$ *were the length of the* kth *interval in a random partition of* $\overline{0, 1}$ *by* $n - 1$ *points.*

(*f*) *A significance test in periodogram analysis and the covering theorem.* In practice, any continuous function of time t can be approximated by a trigonometric polynomial. If the function is a sample function of a stochastic process the coefficients become random variables, and the approximating polynomial may be written in the form

$$\sum_{\nu=1}^{n}(\mathbf{X}_\nu \cos \omega_\nu t + \mathbf{Y}_\nu \sin \omega_\nu t) \equiv \sum_{\nu=1}^{n} \mathbf{R}_\nu \cos (\omega_\nu t - \mathbf{\Phi}_\nu) \tag{3.5}$$

where $\mathbf{R}_\nu^2 = \mathbf{X}_\nu^2 + \mathbf{Y}_\nu^2$ and $\tan \mathbf{\Phi}_\nu = \mathbf{Y}_\nu/\mathbf{X}_\nu$. Conversely, reasonable assumptions on the random variables $\mathbf{X}_\nu$, $\mathbf{Y}_\nu$ lead to a stochastic process with sample functions given by (3.5). For a time it was fashionable to introduce models of this form and to detect "hidden periodicities" for sunspots, wheat prices, poetic creativity, etc. Such hidden periodicities used to be discovered as easily as witches in medieval times, but even strong faith must be fortified by a statistical test. The method is roughly as follows. A trigonometric polynomial of the form (3.5) with well-chosen frequencies $\omega_1, \ldots, \omega_n$ is fitted to some observational data, and a particularly large amplitude $\mathbf{R}_\nu$ is observed. One wishes to prove that this cannot be due to chance and hence that ω_ν is a true period. To test this conjecture one asks whether the large observed value of $\mathbf{R}_\nu$ is plausibly compatible with the hypothesis that all n components play the same role. For a test one assumes, accordingly, that the coefficients $\mathbf{X}_1, \ldots, \mathbf{Y}_n$ are mutually independent with a common normal distribution with zero expectation and variance σ^2. In this case (see II,3) the $\mathbf{R}_\nu^2$ are mutually independent and have a common exponential distribution with expectation $2\sigma^2$. If an observed value $\mathbf{R}_\nu^2$ deviated "significantly" from this predicted expectation it was customary to jump to the conclusion that the hypothesis of equal weights was untenable, and $\mathbf{R}_\nu$ represented a "hidden periodicity."

The fallacy of this reasoning was exposed by R. A. Fisher (1929) who pointed out that the maximum among n independent observations does not obey the same probability distribution as each variable taken separately. The error of treating the worst case statistically as if it had been chosen at random is still common in medical statistics, but the reason for discussing the matter here is the surprising and amusing connection of Fisher's test of significance with covering theorems.

As only the ratios of the several components are significant we normalize the coefficients by letting

$$\mathbf{V}_j = \frac{\mathbf{R}_j^2}{\mathbf{R}_1^2 + \cdots + \mathbf{R}_n^2} \qquad j = 1, \ldots, n. \tag{3.6}$$

Since the $\mathbf{R}_j^2$ have a common exponential distribution we can use the preceding example with $\mathbf{X}_j = \mathbf{R}_j^2$. Then $\mathbf{V}_1 = \mathbf{U}_1, \ldots, \mathbf{V}_{n-1} = \mathbf{U}_{n-1}$, but $\mathbf{V}_n = 1 - \mathbf{U}_1 - \cdots - \mathbf{U}_{n-1}$. Accordingly, the n-tuple $(\mathbf{V}_1, \ldots, \mathbf{V}_n)$ *is distributed as the length of the* n *intervals into which* $\overline{0, 1}$ *is partitioned by a random distribution of* $n-1$ *points. The probability that all* $\mathbf{V}_j$ *be less than* a *is therefore given by formula* I,(9.9) *of the covering theorem.* This result illustrates the occurrence of unexpected relations between apparently unconnected problems.[7] ▶

*4. A CHARACTERIZATION OF THE NORMAL DISTRIBUTION

Consider a non-degenerate linear transformation of coordinate variables

$$\mathbf{Y}_1 = a_{11}\mathbf{X}_1 + a_{12}\mathbf{X}_2, \qquad \mathbf{Y}_2 = a_{21}\mathbf{X}_1 + a_{22}\mathbf{X}_2, \tag{4.1}$$

[7] Fisher derived the distribution of the maximal term among the $\mathbf{V}_j$ in 1929 without knowledge of the covering theorem, and explained in 1940 the equivalence with the covering theorem after W. L. Stevens had proved the latter. [See papers No. 16 and 37 in Fisher's *Contributions to Mathematical Statistics*, John Wiley, New York (1950).] For an alternative derivation using Fourier analysis see U. Grenander and M. Rosenblatt (1957).

* This section treats a special topic and is not used in the sequel.

and suppose (without loss of generality) that the determinant $\Delta = 1$. If $\mathbf{X}_1$ and $\mathbf{X}_2$ are independent normal variables with variances σ_1^2 and σ_2^2 the distribution of the pair $(\mathbf{Y}_1, \mathbf{Y}_2)$ is normal with covariance

$$a_{11}a_{21}\sigma_1^2 + a_{12}a_{22}\sigma_2^2$$

[see example 1(*d*)]. In this case there exist non-trivial choices of the coefficients a_{jk} such that $\mathbf{Y}_1$ and $\mathbf{Y}_2$ are independent. The following theorem shows that this property of the univariate normal distribution *is not shared by any other distribution*. We shall here prove it only for distributions with continuous densities, in which case it reduces to a lemma concerning the functional equation (4.3). By the use of characteristic functions the most general case is reduced to the *same* equation, and so our proof will really yield the theorem in its greatest generality (see XV,8). The elementary treatment of densities reveals better the basis of the theorem.

The transformation (4.1) is meaningful only if no coefficient a_{jk} vanishes. Indeed, suppose for example that $a_{11} = 0$. Without loss of generality we may choose the scale parameters so that $a_{12} = 1$. Then $\mathbf{Y}_1 = \mathbf{X}_2$, and a glance at (4.4) shows that in this case $\mathbf{Y}_2$ must have the same density as $\mathbf{X}_1$. In other words, such a transformation amounts to a mere renaming of the variables, and need not be considered.

Theorem. *Suppose that* $\mathbf{X}_1$ *and* $\mathbf{X}_2$ *are independent of each other, and that the same is true of the pair* $\mathbf{Y}_1$, $\mathbf{Y}_2$. *If no coefficient* a_{jk} *vanishes then all four variables are normal.*

The most interesting special case of (4.1) is presented by *rotations*, namely transformations of the form

$$\text{(4.2)} \qquad \mathbf{Y}_1 = \mathbf{X}_1 \cos \omega + \mathbf{X}_2 \sin \omega, \qquad \mathbf{Y}_2 = -\mathbf{X}_1 \sin \omega + \mathbf{X}_2 \cos \omega$$

where ω is not a multiple of $\frac{1}{2}\pi$. Applying the theorem to them we get

Corollary. *If* $\mathbf{X}_1$ *and* $\mathbf{X}_2$ *are independent and there exists one rotation* (4.2) *such that* $\mathbf{Y}_1$ *and* $\mathbf{Y}_2$ *are also independent, then* $\mathbf{X}_1$ *and* $\mathbf{X}_2$ *have normal distributions with the same variance. In this case* $\mathbf{Y}_1$ *and* $\mathbf{Y}_2$ *are independent for every* ω.

Example. *Maxwell distribution of velocities.* In his study of the velocity distributions of molecules in $\mathcal{R}^3$ Maxwell assumed that in *every* Cartesian coordinate system the three components of the velocity are mutually independent random variables with zero expectation. Applied to rotations leaving one axis fixed our corollary shows immediately that the three components are normally distributed with the same variance. As we saw in II,3 this implies the Maxwell distribution for velocities. ►

The theorem has a long history going back to Maxwell's investigations. Purely probabilistic studies were initiated by M. Kac (1940) and S. Bernstein (1941), who proved our corollary assuming finite variances. An impressive number of authors contributed improvements and variants, sometimes by rather deep methods. The development culminates in a result proved by V. P. Skitovič.[8]

Now to the proof in the case of continuous densities. We denote the densities of $\mathbf{X}_j$ and $\mathbf{Y}_j$ respectively by u_j and f_j. For abbreviation we put

$$(4.3)\qquad y_1 = a_{11}x_1 + a_{12}x_2, \qquad y_2 = a_{21}x_1 + a_{22}x_2.$$

Under the conditions of the theorem we must have

$$(4.4)\qquad f_1(y_1)\,f_2(y_2) = u_1(x_1)\,u_2(x_2).$$

We shall show that this relation implies that

$$(4.5)\qquad f_j(y) = \pm e^{\varphi_j(y)}, \qquad u_j(x) = \pm e^{\omega_j(x)}$$

where the exponents are polynomials of degree 2 or lower. The only probability densities of this form are the normal densities. For distributions with continuous densities the theorem is therefore contained in the following

Lemma. *Suppose that four continuous functions f_j and u_j are connected by the functional equation* (4.4), *and that no coefficient a_{jk} vanishes. The functions are then of the form* (4.5) *where the exponents are polynomials of degree ≤ 2.*

(It is, of course, assumed that none of the functions vanishes identically.)

Proof. We note first that none of our functions can have a zero. Indeed, otherwise there would exist a domain Ω in the x_1, x_2-plane in which the two members of (4.4) have no zeros and on whose boundary they vanish. But the two sides require on the one hand that the boundary consists of segments parallel to the axes, on the other hand of segments parallel to the lines $y_j = \text{const}$. This contradiction shows that no such boundary exists.

We may therefore assume our functions to be strictly positive. Passing to logarithms we can rewrite (4.4) in the form

$$(4.6)\qquad \varphi_1(y_1) + \varphi_2(y_2) = \omega_1(x_1) + \omega_2(x_2).$$

For fixed h_1 and h_2 define the mixed difference operator Δ by

$$(4.7)\qquad \Delta v(x_1, x_2) = v(x_1+h_1, x_2+h_2) - v(x_1+h_1, x_2-h_2) - \\ - v(x_1-h_1, x_2+h_2) + v(x_1-h_1, x_2-h_2).$$

[8] Izvestia Acad. Nauk SSSR, vol. 18 (1954) pp. 185–200. The theorem: Let $\mathbf{X}_1, \ldots, \mathbf{X}_n$ be mutually independent, $\mathbf{Y}_1 = \Sigma a_i\mathbf{X}_i$, and $\mathbf{Y}_2 = \Sigma b_i\mathbf{X}_i$ where no coefficient is 0. If $\mathbf{Y}_1$ and $\mathbf{Y}_2$ are independent the $\mathbf{X}_i$ are normally distributed.

Because each ω_j depends on the single variable x_j it follows that $\Delta\omega_j = 0$. Also

$$(4.8)\quad \Delta\varphi_1(y_1) = \varphi_1(y_1+t_1) - \varphi_1(y_1+t_2) - \varphi_1(y_1-t_2) + \varphi_1(y_1-t_1)$$

where we put for abbreviation

$$(4.9)\qquad t_1 = a_{11}h_1 + a_{12}h_2, \qquad t_2 = a_{11}h_1 - a_{12}h_2.$$

We have thus $\Delta\varphi_1 + \Delta\varphi_2 = 0$ with φ_j depending on the single variable y_j. Keeping y_2 fixed one sees that $\Delta\varphi_1(y_1)$ is a constant depending only on h_1 and h_2. We now choose h_1 and h_2 so that $t_1 = t$ and $t_2 = 0$, where t is arbitrary, but fixed. The relation $\Delta\varphi_1 = \text{const.}$ then takes on the form

$$(4.10)\qquad \varphi_1(y_1+t) + \varphi_1(y_1-t) - 2\varphi_1(y_1) = \lambda(t).$$

Near a point y_1 at which φ_1 assumes a minimum the left side is ≥ 0, and hence such a point can exist only if $\lambda(t) \geq 0$ for all t in some neighborhood of the origin. But in this case φ_1 cannot assume a maximum. Now a continuous function vanishing at three points assumes both a maximum and a minimum. We conclude that if a continuous solution of (4.10) vanishes at three distinct points, then it is identically zero.

Every quadratic polynomial $q(y_1) = \alpha y_1^2 + \beta y_1 + \gamma$ satisfies an equation of the form (4.10) (with a different right side), and hence the same is true of the difference $\varphi_1(y_1) - q(y_1)$. But q can be chosen such that this difference vanishes at three prescribed points, and then $\varphi_1(y_1)$ is identical with q. The same argument applies to φ_2, and this proves the assertion concerning f_1 and f_2. Since the variables $\mathbf{X}_j$ and $\mathbf{Y}_j$ play the same role, the same argument applies to the densities u_j. ▶

5. MATRIX NOTATION. THE COVARIANCE MATRIX

The notation employed in section 1 is messy and becomes more so in higher dimensions. Elegance and economy of thought may be achieved by the use of matrix notation.

For ease of reference we summarize the few facts of matrix theory and the notations used in the sequel. The basic rule is: first rows, then columns. Thus an α by β matrix A has α rows and β columns; its elements are denoted by a_{jk}, the first index indicating the row. If B is a β by γ matrix with elements b_{jk} the product AB is the α by γ matrix with elements $a_{j1}b_{1k} + a_{j2}b_{2k} + \cdots + a_{j\beta}b_{\beta k}$. No product is defined if the number of columns of A does not agree with the number of rows of B. The associative law $(AB)C = A(BC)$ holds, whereas in general $AB \neq BA$. The *transpose* A^T is the β by α matrix with elements $a_{jk}^T = a_{kj}$. Obviously $(AB^T) = B^TA^T$.

A one by α matrix with a single row is called a *row vector*; a matrix with a single column, a *column vector*.[9] A row vector $r = (r_1, \ldots, r_\alpha)$ is easily printed, but a column

[9] This is really an abuse of language. In a concrete case x_1 may represent pounds and x_2 cows; then (x_1, x_2) is no "vector" in the strict sense.

vector is better indicated by its transpose $c^T = (c_1, \ldots, c_\alpha)$. Note that *cr is an* α *by* α *matrix* (of the "multiplication table" type) *whereas rc is a one by one matrix*, or scalar. In the case $\alpha = 2$

$$cr = \begin{pmatrix} c_1r_1 & c_1r_2 \\ c_2r_1 & c_2r_2 \end{pmatrix}, \qquad rc = (r_1c_1 + r_2c_2).$$

The *zero vector* has all components equal to 0.

Matrices with the same number of rows and columns are called *square* matrices. With a square matrix A there is associated its *determinant*, a number which will be denoted by $|A|$. For our purposes it suffices to know that the determinants are multiplicative: if A and B are square matrices and $C = AB$, then $|C| = |A| \cdot |B|$. The transpose A^T has the same determinant as A.

By *identity matrix* is meant a square matrix with ones in the main diagonal and zeros at all other places. If I is the identity matrix with r rows and columns and A an r by r matrix, obviously $IA = AI = A$. By *inverse* of A is meant a matrix A^{-1} such that $AA^{-1} = A^{-1}A = I$. [Only square matrices can have inverses. The inverse is unique, for if B is any inverse of A we have $AB = I$ and by the associative law $A^{-1} = (A^{-1}A)B = B$.] A square matrix without inverse is called *singular*. The multiplicative property of determinants implies that a matrix with zero determinant is singular. The converse is also true if $|A| \neq 0$ then A is non-singular. In other words, a matrix A is singular iff there exists a non-zero vector x such that $xA = 0$.

A square matrix A is *symmetric* if $a_{jk} = a_{kj}$, that is, if $A^T = A$. The *quadratic form associated with a symmetric r by r matrix A* is defined by

$$xAx^T = \sum_{j,k=1}^{r} a_{jk}x_jx_k$$

where $x_1, \ldots, x_r$ are indeterminates. The matrix is *positive definite* if $xAx^T > 0$ for all non-zero vectors x. It follows from the last criterion that a positive definite matrix is non-singular.

Rotations in $\mathfrak{R}^\alpha$. For completeness we mention briefly a geometric application of matrix calculus although it will not be used in the sequel.

The *inner product* of two row vectors $x = (x_1, \ldots, x_\alpha)$ and $y = (y_1, \ldots, y_\alpha)$ is defined by

$$xy^T = yx^T = \sum_{j=1}^{\alpha} x_jy_j.$$

The *length* L of x is given by $L^2 = xx^T$. If x and y are vectors of unit length the *angle* δ between them is given by $\cos\delta = xy^T$.

An α by α matrix A induces a transformation mapping x into $\xi = xA$; for the transpose one has $\xi^T = A^Tx^T$. The matrix A is *orthogonal* if the induced transformation preserves lengths and angles, that is to say, if any two row vectors have the same inner product as their images: Thus A is orthogonal iff for any pair of row vectors x, y

$$xAA^Ty^T = xy^T.$$

This implies that AA^T is the identity matrix I as can be seen by choosing for x and y vectors with $\alpha - 1$ vanishing components. We have thus found that A *is orthogonal iff* $AA^T = I$. Since A and A^T have the same determinant it follows that it equals $+1$ or -1. An orthogonal matrix with determinant 1 is called a *rotation matrix* and the induced transformation is a rotation.

From now on we denote a point of the r-dimensional space $\mathcal{R}^r$ by a single letter to be interpreted as a *row vector*. Thus $x = (x_1, \ldots, x_r)$ and $f(x) = f(x_1, \ldots, x_r)$, etc. Inequalities are to be interpreted coordinatewise: $x < y$ iff $x_k < y_k$ for $k = 1, \ldots, r$ and similarly for other inequalities. In the plane $\mathcal{R}^2$ the relation $x < y$ may be read as "x lies southwest of y." A novel feature of this notation is that two points need not stand in either of the relations $x \leq y$ or $y < x$, that is, in higher dimensions the inequality $<$ introduces only a partial ordering.

We write $\mathbf{X} = (\mathbf{X}_1, \ldots, \mathbf{X}_r)$ for the row vector of the coordinate variables and use this notation for random variables in general (mainly for normally distributed variables).

If the variables $\mathbf{X}_1, \ldots, \mathbf{X}_r$ have expectations $\mathbf{E}(\mathbf{X}_j)$ we write $\mathbf{E}(\mathbf{X})$ *for the row vector with components* $\mathbf{E}(\mathbf{X}_j)$. The vector $\mathbf{X} - \mathbf{E}(\mathbf{X})$ has zero expectation. More generally, if $\mathbf{M}$ is a matrix whose elements $\mathbf{M}_{jk}$ are random variables we write $\mathbf{E}(\mathbf{M})$ for the matrix of elements $\mathbf{E}(\mathbf{M}_{jk})$ assuming that it exists.

Definition. *If* $E(\mathbf{X}) = 0$ *the covariance matrix* $\operatorname{Var}(\mathbf{X})$ *of* $\mathbf{X}$ *is the symmetric* r *by* r *matrix with elements* $\mathbf{E}(\mathbf{X}_j\mathbf{X}_k)$ (*provided they all exist*). *In other words*

$$\operatorname{Var}(\mathbf{X}) = \mathbf{E}(\mathbf{X}^T\mathbf{X}). \tag{5.1}$$

For arbitrary $\mathbf{X}$ *we define* $\operatorname{Var}(\mathbf{X})$ *to be the same as* $\operatorname{Var}(\mathbf{X}-\mathbf{E}(\mathbf{X}))$.

The use of row vectors necessitates writing a *linear transformation from* $\mathcal{R}^r$ *to* $\mathcal{R}^m$ in the form

$$\mathbf{Y} = \mathbf{X}A, \tag{5.2}$$

that is,

$$y_k = \sum_{j=1}^{r} a_{jk}x_j \qquad k = 1, \ldots, m \tag{5.3}$$

where A is an r by m matrix. Obviously $\mathbf{E}(\mathbf{Y}) = \mathbf{E}(\mathbf{X})A$ whenever $\mathbf{E}(\mathbf{X})$ exists. To find the variances we assume without loss of generality $\mathbf{E}(\mathbf{X}) = 0$. Then $\mathbf{E}(\mathbf{Y}) = 0$ and

$$\mathbf{E}(\mathbf{Y}^T\mathbf{Y}) = \mathbf{E}(A^T\mathbf{X}^T\mathbf{X}A) = A^T\mathbf{E}(\mathbf{X}^T\mathbf{X})A. \tag{5.4}$$

We thus have the important result that

$$\operatorname{Var}(\mathbf{Y}) = A^T \operatorname{Var}(\mathbf{Y})A. \tag{5.5}$$

Of particular interest is the special case $m = 1$ when

$$\mathbf{Y} = a_1\mathbf{X}_1 + \cdots + a_r\mathbf{X}_r \tag{5.6}$$

is an ordinary random variable. Here $\text{Var}(\mathbf{Y})$ is the (scalar) quadratic form

(5.7) $$\text{Var}(\mathbf{Y}) = \sum_{j,k=1}^{r} \mathbf{E}(\mathbf{X}_j\mathbf{X}_k)a_j a_k.$$

The linear form (5.6) vanishes with probability one if $\text{Var}(\mathbf{Y}) = 0$ and in this case every region outside the hyperplane $\sum a_k x_k = 0$ carries zero probability. The probability distribution is then concentrated on an $(r-1)$-dimensional manifold and is *degenerate* when considered in r dimensions. We have now proved that *the covariance matrix of any non-degenerate probability distribution is positive definite.* Conversely, every such matrix may serve as covariance matrix of a normal density (see theorem 4 of the next section).

6. NORMAL DENSITIES AND DISTRIBUTIONS

Throughout this section Q stands for a symmetric r by r matrix, and $q(x)$ for the associated quadratic form

(6.1) $$q(x) = \sum_{j,k=1}^{r} q_{jk}x_j x_k = xQx^T$$

where $x = (x_1, \ldots, x_r)$ is a *row* vector. Densities in $\mathcal{R}^r$ defined by an exponential with a quadratic form in the exponent are a natural counterpart of the normal density on the line, and we start therefore from the following

Definition. *A density φ in r dimensions is called normal*[10] *and centered at the origin of it is of the form*

(6.2) $$\varphi(x) = \gamma^{-1} \cdot e^{-\frac{1}{2}q(x)}$$

where γ is a constant. A normal density centered at $a = (a_1, a_2, \ldots, a_r)$ is given by $\varphi(x-a)$.

The special case of two dimensions was discussed in examples $1(d)$ and $2(a)$.

We take $\mathcal{R}^r$ with the probability distribution of (6.2) as sample space and denote by $\mathbf{X} = (\mathbf{X}_1, \ldots, \mathbf{X}_r)$ the row vector formed by the coordinate variables. Its covariance matrix will be denoted by M:

(6.3) $$M = \text{Var}(\mathbf{X}) = \mathbf{E}(\mathbf{X}^T\mathbf{X}).$$

Our problem consists in investigating the nature of the matrices Q and M, and the relationship between them.

First we observe that no diagonal element of Q can vanish. Indeed, if we had $q_{rr} = 0$, then for fixed values of $x_1, \ldots, x_{r-1}$ the density (6.2) would

[10] "*Degenerate*" normal distributions will be introduced at the end of this section.

take on the form $\gamma^{-1}e^{-ax_r+b}$ and the integral with respect to x_r would diverge. We now introduce the substitution $y = xA$ defined by

$$y_1 = x_1, \ldots, y_{r-1} = x_{r-1}, \qquad y_r = q_{1r}x_1 + \cdots + q_{rr}x_r. \tag{6.4}$$

It is seen by inspection that $q(x) - y_r^2/q_{rr}$ is a quadratic form in $x_1, \ldots, x_{r-1}$ not involving x_r. Thus

$$q(x) = \frac{1}{q_{rr}} y_r^2 + \bar{q}(y). \tag{6.5}$$

where $\bar{q}(y)$ is a quadratic form in $y_1, \ldots, y_{r-1}$. This shows that the vector $\mathbf{Y} = \mathbf{X}A$ has a normal density that factors into two normal densities for $\mathbf{Y}_r$ and $(\mathbf{Y}_1, \ldots, \mathbf{Y}_{r-1})$, respectively. The first conclusion to be drawn is the simple but important

Theorem 1. *All marginal densities of a normal density are again normal.*

Less expected is

Theorem 2. *There exists a matrix C with positive determinant such that $\mathbf{Z} = \mathbf{X}C$ is a row vector whose components $\mathbf{Z}_j$ are mutually independent normal variables.*

The matrix C is not unique; in fact, the theorem can be strengthened to the effect that C can be chosen as a *rotation matrix* (see problem 19).

Proof. We proceed by induction. When $r = 2$ the assertion is contained in the factorization (6.5). If the theorem is true in $r - 1$ dimensions, the variables $\mathbf{Y}_1, \ldots, \mathbf{Y}_{r-1}$ are linear combinations of independent normal variables $\mathbf{Z}_1, \ldots, \mathbf{Z}_{r-1}$ while $\mathbf{Y}_r$ itself is normal and independent of the remaining variables. Since $\mathbf{X} = \mathbf{Y}A^{-1}$ it follows also that the $\mathbf{X}_j$ are linear combinations of $\mathbf{Z}_1, \ldots, \mathbf{Z}_{r-1}$ and $\mathbf{Y}_r$. The determinant of A equals q_{rr}, and (6.5) implies that it is positive. The determinant of the transformation $\mathbf{X} \to \mathbf{Z}$ is the product of the determinants of A and the transformation $\mathbf{Y} \to \mathbf{Z}$ and hence it is positive. ▶

Theorem 3. *The matrices Q and M are inverses of each other and*

$$\gamma^2 = (2\pi)^r \cdot |M| \tag{6.6}$$

where $|M| = |Q|^{-1}$ is the determinant of M.

Proof. With the notations of the preceding theorem put

$$D = \mathbf{E}(\mathbf{Z}^T\mathbf{Z}) = C^TMC. \tag{6.7}$$

This is a matrix with diagonal elements $\mathbf{E}(\mathbf{Z}_j^2) = \sigma_j^2$ and zero elements outside the diagonal. The density of $\mathbf{Z}$ is the product of normal densities $\mathfrak{n}(x\sigma_j^{-1})\sigma_j^{-1}$ and hence induced by the matrix D^{-1} with diagonal elements

σ_j^{-2}. Now the density of $\mathbf{Z}$ is obtained from the density (6.2) of $\mathbf{X}$ by the substitution $x = zC^{-1}$ and multiplication by the determinant $|C^{-1}|$. Accordingly

$$zD^{-1}z^T = xQx^T \tag{6.8}$$

and

$$(2\pi)^r\,|D| = \gamma^2 \cdot |C|^2. \tag{6.9}$$

From (6.8) it is seen that

$$Q = CD^{-1}C^T, \tag{6.10}$$

and in view of (6.7) this implies $Q = M^{-1}$. From (6.7) it follows also that $|D| = |M| \cdot |C|^2$, and hence (6.9) is equivalent to (6.6). ▶

The theorem implies in particular that a factorization of M corresponds to an analogous factorization of Q and hence we have the

Corollary. *If* $(\mathbf{X}_1, \mathbf{X}_2)$ *is normally distributed then* $\mathbf{X}_1$ *and* $\mathbf{X}_2$ *are independent iff* $\text{Cov}\,(\mathbf{X}_1, \mathbf{X}_2) = 0$, *that is, iff* $\mathbf{X}_1$ *and* $\mathbf{X}_2$ *are uncorrelated.*

More generally, if $(\mathbf{X}_1, \ldots, \mathbf{X}_r)$ has a normal density then $(\mathbf{X}_1, \ldots, \mathbf{X}_n)$ and $(\mathbf{X}_{n+1}, \ldots, \mathbf{X}_r)$ are independent iff $\text{Cov}\,(\mathbf{X}_j, \mathbf{X}_k) = 0$ for $j \leq n, k > n$.

Warning. The corollary depends on the joint density of $(\mathbf{X}_1, \mathbf{X}_2)$ being normal and *does not apply if it is only known that the marginal densities of* $\mathbf{X}_1$ *and* $\mathbf{X}_2$ *are normal.* In the latter case the density of $(\mathbf{X}_1, \mathbf{X}_2)$ need not be normal and, in fact, need not exist. This fact is frequently misunderstood (see problems 2–3).

Theorem 4. *A matrix* M *is the covariance matrix of a normal density iff it is positive definite.*

Since the density is induced by the matrix $Q = M^{-1}$ an equivalent formulation is: *A matrix* Q *induces a normal density* (6.2) *iff it is positive definite.*

Proof. We saw at the end of section 5 that every covariance matrix of a density is positive definite. The converse is trivial when $r = 1$ and we proceed by induction. Assume Q positive definite. For $x_1 = \cdots = x_{r-1} = 0$ we get $q(x) = q_{rr}x_r^2$ and hence $q_{rr} > 0$. Under this hypothesis we saw that q may be reduced to the form (6.5). Choosing x_r such that $y_r = 0$ we see that the positive definiteness of Q implies $\bar{q}(x) > 0$ for all choices of $x_1, \ldots, x_{r-1}$. By the induction hypothesis therefore $\bar{q}$ corresponds to a normal density in $r - 1$ dimensions. From (6.5) it is now obvious that q corresponds to a normal density in r dimensions, and this completes the proof. ▶

We conclude this general theory by an interpretation of (6.5) in terms of *conditional densities* which leads to a general formulation of the regression theory explained for the two-dimensional case in example 2(*a*).

Put for abbreviation $a_k = -q_{kr}/q_{rr}$, so that

$$y_r = q_{rr}(x_r - a_1x_1 - \cdots - a_{r-1}x_{r-1}). \tag{6.11}$$

For a probabilistic interpretation of the coefficients a_k we recall that $\mathbf{Y}_r$ was found to be independent of $\mathbf{X}_1, \ldots, \mathbf{X}_{r-1}$. In other words, the a_k are numbers such that

$$\mathbf{T} = \mathbf{X}_r - a_1\mathbf{X}_1 - \cdots - a_{r-1}\mathbf{X}_{r-1} \tag{6.12}$$

is independent of $(\mathbf{X}_1, \ldots, \mathbf{X}_{r-1})$, and this property uniquely characterizes the coefficients a_k.

To obtain the conditional density of $\mathbf{X}_r$ for given $\mathbf{X}_1 = x_1, \ldots, \mathbf{X}_{r-1} = x_{r-1}$ we must divide the density of $(\mathbf{X}_1, \ldots, \mathbf{X}_r)$ by the marginal density for $(\mathbf{X}_1, \ldots, \mathbf{X}_{r-1})$. In view of (6.5) we get an exponential with exponent $-\frac{1}{2}y_r^2/q_{rr}$. It follows that the conditional density of $\mathbf{X}_r$ for given $\mathbf{X}_1 = x_1, \ldots, \mathbf{X}_{r-1} = x_{r-1}$ is normal with expectation $a_1x_1 + \cdots + a_{r-1}x_{r-1}$ and variance $1/q_{rr}$. Accordingly

$$\mathbf{E}(\mathbf{X}_r \mid \mathbf{X}_1, \ldots, \mathbf{X}_{r-1}) = a_1\mathbf{X}_1 + \cdots + a_{r-1}\mathbf{X}_{r-1}. \tag{6.13}$$

We have thus proved the following generalization of the two-dimensional regression theory embodied in (2.6).

Theorem 5. *If* $(\mathbf{X}_1, \ldots, \mathbf{X}_r)$ *has a normal density, the conditional density of* $\mathbf{X}_r$ *for given* $\mathbf{X}_1, \ldots, \mathbf{X}_{r-1}$ *is again normal. Furthermore, the conditional expectation* (6.13) *is the unique linear function of* $\mathbf{X}_1, \ldots, \mathbf{X}_{r-1}$ *making* $\mathbf{T}$ *independent of* $(\mathbf{X}_1, \ldots, \mathbf{X}_{r-1})$. *The conditional variance equals* $\mathrm{Var}(\mathbf{T}) = q_{rr}^{-1}$.

Example. *Sample mean and variance.* In statistics the random variables

$$\hat{\mathbf{x}} = \frac{1}{r}(\mathbf{X}_1 + \cdots + \mathbf{X}_r), \qquad \hat{\boldsymbol{\sigma}}^2 = \frac{1}{r}\sum_{k=1}^{r}(\mathbf{X}_k - \hat{\mathbf{x}})^2 \tag{6.14}$$

are called the sample mean and sample variance of $\mathbf{X} = (\mathbf{X}_1, \ldots, \mathbf{X}_r)$. It is a curious fact that *if* $\mathbf{X}_1, \ldots, \mathbf{X}_r$ *are independent normal variables with* $\mathbf{E}(\mathbf{X}_k) = 0$, $\mathbf{E}(\mathbf{X}_k^2) = \sigma^2$, *the random variables* $\hat{\mathbf{X}}$ *and* $\hat{\boldsymbol{\sigma}}^2$ *are independent.*[11]

The proof illustrates the applicability of the preceding results. We put $\mathbf{Y}_k = \mathbf{X}_k - \hat{\mathbf{X}}$ for $k \leq r-1$ but $\mathbf{Y}_r = \hat{\mathbf{X}}$. The transformation from $\mathbf{X}$ to $\mathbf{Y} = (\mathbf{Y}_1, \ldots, \mathbf{Y}_r)$ being linear and non-singular, $\mathbf{Y}$ has a normal density. Now $\mathbf{E}(\mathbf{Y}_k\mathbf{Y}_r) = 0$ for $k \leq r-1$ and so $\mathbf{Y}_r$ is independent of

[11] That this fact characterizes the normal distribution in $\mathcal{R}^1$ was shown by R. C. Geary and by E. Lukacs.

$(\mathbf{Y}_1, \ldots, \mathbf{Y}_{r-1})$. But

(6.15) $$r\hat{\mathbf{\sigma}}^2 = \mathbf{Y}_1^2 + \cdots + \mathbf{Y}_{r-1}^2 + (\mathbf{Y}_1 + \cdots + \mathbf{Y}_{r-1})^2$$

depends only on $\mathbf{Y}_1, \ldots, \mathbf{Y}_{r-1}$, and thus $\hat{\mathbf{\sigma}}^2$ is indeed independent of $\mathbf{Y}_r = \hat{\mathbf{X}}$. ►

General Normal Distributions

It follows from the lemma that if $\mathbf{X} = (\mathbf{X}_1, \ldots, \mathbf{X}_r)$ has a normal density, every non-zero linear combination $\mathbf{Y}_1 = a_1\mathbf{X}_1 + \cdots + a_r\mathbf{X}_r$ also has a normal density. The same is true of every pair $(\mathbf{Y}_1, \mathbf{Y}_2)$ provided that no linear relationship $c_1\mathbf{Y}_1 + c_2\mathbf{Y}_2 = 0$ holds. In this exceptional case the probability distribution of $(\mathbf{Y}_1, \mathbf{Y}_2)$ is concentrated on the line with the equation $c_1y_1 + c_2y_2 = 0$ and hence it is singular *if viewed as a two-dimensional distribution*. For many purposes it is desirable to preserve the term normal distribution also for degenerate distributions concentrated on a lower-dimensional manifold, say on a particular axis. The simplest general definition is as follows: *The distribution of* $\mathbf{Y} = (\mathbf{Y}_1, \ldots, \mathbf{Y}_\rho)$ *is normal if there exists a vector* $\mathbf{X} = (\mathbf{X}_1, \ldots, \mathbf{X}_r)$ *with normal r-dimensional density such that* $\mathbf{Y} = a + \mathbf{X}A$ where A is a (constant) r by ρ matrix and $a = (a_1, \ldots, a_\rho)$. If $\rho > r$ the distribution of $\mathbf{Y}$ is *degenerate in* ρ *dimensions*. For $\rho \leq r$ it is non-degenerate iff the ρ forms defining $\mathbf{Y}_k$ are linearly independent.

*7. STATIONARY NORMAL PROCESSES

The purpose of this section is partly to supply examples of normal distributions, partly to derive some relations of considerable use in the theory of discrete stochastic processes and time series. They are of an analytic character and easily separated from the deeper stochastic analysis. In fact, we shall be concerned only with finite-dimensional normal densities or, what amounts to the same, their covariance matrices. The reference to random variables is essential for probabilistic intuition and as a preparation for applications, but at the present stage we are concerned only with their joint distributions; the random variables themselves are used merely as a convenient way of describing all marginal densities by indicating the corresponding collections $(\mathbf{X}_{\alpha_1}, \ldots, \mathbf{X}_{\alpha_k})$. By the same token a reference to an infinite sequence $\{\mathbf{X}_k\}$ implies merely that the number of terms in $(\mathbf{X}_1, \ldots, \mathbf{X}_n)$ may be taken arbitrarily large.

We shall, in fact, consider a doubly infinite sequence $\{\ldots, \mathbf{X}_{-2}, \mathbf{X}_{-1}, \ldots\}$. By this we mean simply that corresponding to each finite collection

* Not used in the sequel. In particular, section 8 can be read independently. (See also XIX,8.)

$(\mathbf{X}_{n_1}, \ldots, \mathbf{X}_{n_r})$ we are given a *normal* density with the obvious consistency rules. The sequence is *stationary* if these distributions are invariant under time shifts, that is, if all r-tuples of the form $(\mathbf{X}_{n_1+\nu}, \ldots, \mathbf{X}_{n_r+\nu})$ with fixed $n_1, \ldots, n_r$ have a common distribution independent of ν. For $r = 1$ this implies that the expectations and variances are constant, and hence there is no loss in generality in assuming that $\mathbf{E}(\mathbf{X}_n) = 0$. The joint distributions are completely determined by the covariances $\rho_{jk} = \mathbf{E}(\mathbf{X}_j\mathbf{X}_k)$, and the stationarity requires that ρ_{jk} depends only on the difference $|k - j|$. Accordingly we put $\rho_{j,j+n} = r_n$. Thus

$$r_n = \mathbf{E}(\mathbf{X}_k\mathbf{X}_{k+n}) = \mathbf{E}(\mathbf{X}_{k-n}\mathbf{X}_k), \tag{7.1}$$

whence $r_n = r_{-n}$. In effect we are dealing only with sequences of numbers r_n that can serve as covariances for a stationary process.

Throughout this section $\{\mathbf{Z}_n\}$ stands for a doubly infinite sequence of *mutually independent normal variables* normed by

$$\mathbf{E}(\mathbf{Z}_n) = 0, \qquad \mathbf{E}(\mathbf{Z}_n^2) = 1. \tag{7.2}$$

Three methods of constructing stationary sequences in terms of a given sequence $\{\mathbf{Z}_n\}$ will be described. They are in constant use in time series analysis and may serve as an exercise in routine manipulations.

Examples. (*a*) *Generalized moving average processes.* With arbitrary constants $b_0, b_1, \ldots, b_N$ put

$$\mathbf{X}_n = b_0\mathbf{Z}_n + b_1\mathbf{Z}_{n-1} + \cdots + b_N\mathbf{Z}_{n-N}. \tag{7.3}$$

In the special case of equal coefficients $b_k = 1/(N+1)$ the variable $\mathbf{X}_n$ is an arithmetic average of the type used in time series analysis to "smooth data" (that is, to eliminate local irregularities). In the general case (7.3) represents a linear operator taking the stationary sequence $\{\mathbf{Z}_n\}$ into a new stationary sequence $\{\mathbf{X}_n\}$. The fashionable term for such operations is "filters." The sequence $\{\mathbf{X}_n\}$ has covariances

$$r_k = r_{-k} = \mathbf{E}(\mathbf{X}_n\mathbf{X}_{n+k}) = \sum_\nu b_\nu b_{\nu+k} \qquad (k \geq 0) \tag{7.4}$$

the series having finitely many terms only.

Since $2\,|b_\nu b_{\nu+k}| \leq b_\nu^2 + b_{\nu+k}^2$ the expression (7.4) makes sense also for infinite sequences such that $\sum b_\nu^2 < \infty$. It is easily seen that the limit of a sequence of covariance matrices is again a covariance matrix and, letting $N \to \infty$, we conclude that *for any sequence* $b_0, b_1, b_2, \ldots$ *such that* $\sum b_n^2 < \infty$ *the numbers* r_k *of* (7.4) *may serve as covariances of a stationary process* $\{\mathbf{X}_n\}$. Formally we get for the new process

$$\mathbf{X}_n = \sum_{k=0}^{\infty} b_k\mathbf{Z}_{n-k}. \tag{7.5}$$

It can be shown without difficulty that every stationary process with covariances (7.4) is of this form, but the relation (7.5) involves infinitely many coordinates and we cannot justify it at present. (See XIX,8.)

(*b*) *The auto-regression process.* Since the inception of time series analysis various theoretical models have been proposed to explain empirical phenomena such as economic time series, sunspots, and observed (or imagined) periodicities. The most popular model assumes that the variables $\mathbf{X}_n$ of the process are related to our sequence $\mathbf{Z}_n$ of independent normal variables of (7.2) by an *auto-regression equation* of the form

$$a_0\mathbf{X}_n + a_1\mathbf{X}_{n-1} + \cdots + a_N\mathbf{X}_{n-N} = \mathbf{Z}_n. \tag{7.6}$$

This model is based on the empirical assumption that the value of the variable $\mathbf{X}_n$ at epoch n (price, supply, or intensity) depends on its past development superimposed on a "random disturbance" $\mathbf{Z}_n$ which is not related to the past. As is frequently the case, the assumption of *linear* dependence serves to simplify (or make possible) a theoretical analysis. More general models may be obtained by letting $N \to \infty$ or by letting the $\mathbf{Z}_n$ be the variables of another stationary process.

If $a_0 \neq 0$ one may chosse $(\mathbf{X}_0, \ldots, \mathbf{X}_{N-1})$ in an arbitrary way and then calculate $\mathbf{X}_N, \mathbf{X}_{N+1}, \ldots$ and $\mathbf{X}_{-1}, \mathbf{X}_{-2}, \ldots$ recursively. In this sense (7.6) determines a process, but we ask whether there exists a *stationary* solution.

To answer this question we rewrite (7.6) in a form not involving the immediate predecessors of $\mathbf{X}_n$. Consider (7.6) with n replaced successively by $n-1$, $n-2, \ldots, n-\nu$. Multiply these equations by $b_1, b_2, \ldots, b_\nu$, respectively, and add to (7.6). The variables $\mathbf{X}_{n-1}, \ldots, \mathbf{X}_{n-\nu}$ will not appear in the new equation iff the b_j are such that

$$a_0b_1 + a_1b_0 = 0, \ldots, \qquad a_0b_\nu + a_1b_{\nu-1} + \cdots + a_\nu b_0 = 0 \tag{7.7}$$

with $b_0 = 1$. The resulting identity is then of the form

$$a_0\mathbf{X}_n = b_0\mathbf{Z}_n + b_1\mathbf{Z}_{n-1} + \cdots + b_\nu\mathbf{Z}_{n-\nu} + \mathbf{Y}_{n,\nu} \tag{7.8}$$

where $\mathbf{Y}_{n,\nu}$ is a linear combination of $\mathbf{X}_{n-\nu-1}, \ldots, \mathbf{X}_{n-N-\nu}$ (with coefficients that are of no interest). In (7.8) we have expressed the variable $\mathbf{X}_n$ as a resultant of the chance contributions at epochs $n, n-1, \ldots, n-\nu$ and a variable $\mathbf{Y}_{n,\nu}$ representing the influence of the time before epoch $n - \nu$. As $\nu \to \infty$ this time becomes the "infinitely remote past" and in most situations it will have no influence. In passing to the limit we shall (at least temporarily) assume this to be the case, that is, we are looking for a process satisfying a limiting relation of the form

$$a_0\mathbf{X}_n = \sum_{k=0}^{\infty} b_k\mathbf{Z}_{n-k}. \tag{7.9}$$

[Roughly speaking we assume that the residual variables $\mathbf{Y}_{n,\nu}$ tend to zero. Other possible limits will be indicated in example (*d*).]

Processes of the form (7.9) are the object of example (*a*) and we saw that a stationary solution exists whenever $\sum b_k^2 < \infty$. (If the series diverges, not even the expressions for the covariances make sense.) To solve the equations (7.7) for b_k we use the formal generating functions

$$A(s) = \sum a_k s^k, \qquad B(s) = \sum b_k s^k. \tag{7.10}$$

The equations (7.7) hold iff $A(s)\,B(s) = a_0 b_0$ and, A being a polynomial, B is rational. We can therefore use the theory of partial fractions developed in **1**; XI,4. If the polynomial $A(s)$ has distinct roots $s_1, \ldots, s_N$ we get

$$B(s) = \frac{A_1}{s_1 - s} + \cdots + \frac{A_N}{s_N - s} \tag{7.11}$$

and hence

$$b_n = A_1 s_1^{-n-1} + \cdots + A_N s_N^{-n-1}. \tag{7.12}$$

Obviously $\sum b_n^2 < \infty$ *iff all roots satisfy* $|s_j| > 1$, and it is easily verified that this remains true also in the presence of multiple roots. We have thus shown that a *stationary solution of the auto regression model* (7.6) *exists whenever all roots of the polynomial* $A(s)$ *lie outside the unit disk.* The covariances of our process are given by (7.4) and in the process the "infinitely remote past" plays no role.

Our solution $\{\mathbf{X}_n\}$ of the auto-regression equation (7.6) is *unique*. Indeed, the difference of two solutions would satisfy the homogeneous equation (7.13) and we shall now show that the condition $|s_j| > 1$ precludes the existence of a probabilistically meaningful solution of this equation.

(*c*) *Degenerate processes.* We turn to stationary sequences $\{\mathbf{Y}_n\}$ satisfying the *stochastic difference equation*

$$a_0\mathbf{Y}_n + a_1\mathbf{Y}_{n-1} + \cdots + a_N\mathbf{Y}_{n-N} = 0. \tag{7.13}$$

They represent an interesting counterpart to the auto-regression processes governed by (7.6). Typical examples are

$$\mathbf{Y}_n = \lambda(\mathbf{Z}_1 \cos n\omega + \mathbf{Z}_{-1} \sin n\omega) \tag{7.14}$$

and

$$\mathbf{Y}_n = \alpha_1\mathbf{Z}_1 + (-1)^n\alpha_2\mathbf{Z}_{-1} \tag{7.15}$$

where the coefficients and ω are constants, and $\mathbf{Z}_1$ and $\mathbf{Z}_{-1}$ independent normal variables normed by (7.2). These processes satisfy (7.13), the first with $a_0 = a_2 = 1$ and $a_1 = -2\cos\omega$, the second with $a_0 = -a_2 = 1$ and $a_1 = 0$. They are degenerate in the sense that the whole process is completely determined by two observations, say $\mathbf{Y}_{k-1}$ and $\mathbf{Y}_k$. These two observations can be taken as far back in the past as we please, and in this sense the process is completely determined by its "infinitely remote past." The same remark applies to any process satisfying a difference equation of the form (7.13), and hence these processes form the counterpart to example (*b*) where the infinitely remote past had no influence at all. ▶

These examples explain the general interest attaching to the stochastic difference equation (7.13). Before passing to its theory we observe that any process $\{\mathbf{Y}_n\}$ satisfying (7.13) satisfies also various difference equations of higher order, for example

$$a_0\mathbf{Y}_n + (a_1-a_0)\mathbf{Y}_{n-1} + \cdots + (a_N-a_{N-1})\mathbf{Y}_{n-N} - a_N\mathbf{Y}_{n-N-1}.$$

To render the problem meaningful we must suppose that (7.13) represents the difference equation of *lowest order* satisfied by $\{\mathbf{Y}_n\}$. This amounts to saying that the N-tuple $(\mathbf{Y}_1, \ldots, \mathbf{Y}_N)$ is non-degenerate with a normal density in N dimensions. It implies that $a_0 \neq 0$ and $a_N \neq 0$.

It will now be shown that the theory of stationary solutions of the difference equation (7.13) is intimately related to the "characteristic equation"

$$a_0\xi^N + a_1\xi^{N-1} + \cdots + a_N = 0. \tag{7.16}$$

To each quadratic factor of the polynomial on the left there corresponds a second-order stochastic difference equation, and through it a process of the form (7.14) or (7.15). Corresponding to the factorization of the characteristic polynomial we shall thus represent the general solution of (7.13) as a sum of components of the form (7.14) and (7.15).

As before we assume the centering $\mathbf{E}(\mathbf{Y}_n) = 0$. The whole theory depends on the following

Lemma 1. *A stationary sequence with* $\mathbf{E}(\mathbf{Y}_n\mathbf{Y}_{n+k}) = r_k$ *satisfies the stochastic difference equation* (7.13) *iff*

$$a_0r_n + a_1r_{n-1} + \cdots + a_Nr_{n-N} = 0. \tag{7.17}$$

Proof. Multiplying (7.13) by $\mathbf{Y}_0$ and taking expectations leads to (7.17). Squaring the left side in (7.13) and taking expectations yields $\sum a_j(\sum a_kr_{k-j})$, and so (7.17) implies that the left side in (7.13) has zero variance. This proves the lemma. ▶

We proceed to derive a canonical form for r_n. It is, of course, real, but it involves the roots of the characteristic equation (7.16), and we must therefore resort to a temporary use of complex numbers.

Lemma 2. *If* $\{\mathbf{Y}_n\}$ *satisfies* (7.13), *but no difference equation of lower order, then the characteristic equation* (7.16) *possesses* N *distinct roots*

$\xi_1, \ldots, \xi_N$ *of unit modulus. In this case*

$$r_n = c_1\xi_1^n + \cdots + c_N\xi_N^n \tag{7.18}$$

with $c_j > 0$ *for* $j = 1, \ldots, N$.

Proof. Suppose first that the characteristic equation (7.16) has N *distinct* roots $\xi_1, \ldots, \xi_N$. We solve (7.17) by the method of particular solutions which was used for similar purposes in volume 1. Inspection shows that (7.18) represents a formal solution depending on N free parameters $c_1, \ldots, c_N$. Now the r_n are completely determined by the N values $r_1, \ldots, r_N$, and to show that *every* solution of (7.17) is of the form (7.18) it suffices therefore to show that the c_j can be chosen so that the relations (7.18) yield prescribed values for $r_1, \ldots, r_N$. This means that the c_j must satisfy N linear equations whose matrix A has elements $a_{jk} = \xi_k^j$ ($j, k = 1, \ldots, N$). The determinant of A does not vanish,[12] and hence the desired solution exists.

We have thus established that (in the case of distinct roots) r_n is indeed of the form (7.18). Next we show that only roots of unit modulus can effectively appear in it. We know that $a_N \neq 0$, and hence 0 is not a root of the characteristic equation. Next we note that the covariances r_n are bounded by the common variance r_0 of the $\mathbf{Y}_n$. But if ξ_j is not of unit modulus then $|\xi_j|^n \to \infty$ either as $n \to \infty$ or as $n \to -\infty$. It follows that for each j either $|\xi_j| = 1$ or else $c_j = 0$.

Suppose now that ξ_1 and ξ_2 are a pair of conjugate roots and $c_1 \neq 0$. Then ξ_1 is of unit modulus and hence $\xi_2 = \xi_1^{-1}$. The symmetry relation $r_n = r_{-n}$ therefore requires that $c_2 = c_1$. Again, $\xi_1^n + \xi_2^n$ is real, and therefore c_1 must be real. Thus the complex roots appear in (7.18) in conjugate pairs with real coefficients, and if some coefficient c_j vanished, r_n would satisfy a difference equation of order less than N. Accordingly all roots are of unit modulus, all c_j are real and $c_j \neq 0$.

To show that the c_j are actually positive we introduce the covariance matrix R of $(\mathbf{Y}_1, \ldots, \mathbf{Y}_N)$. Its elements are given by r_{j-k}, and it is easily verified from (7.18) that

$$R = AC\bar{A}^T \tag{7.19}$$

where C is the diagonal matrix with elements c_j, A is the matrix introduced

[12] The determinant is usually called after Vandermonde. To show that it does not vanish replace ξ_1 by a free variable x. Inspection then shows that the determinant is of the form $xP(x)$ where P is a polynomial of degree $N-1$. Now $P(x) = 0$ for $x = \xi_2, \ldots, \xi_n$ because for these values of x two columns of the determinant become identical. The determinant can therefore not vanish for any other value of x, and in particular not for $x = \xi_1$.

above, and $\bar{A}$ is its conjugate (that is, it is obtained from A by replacing ξ_j by ξ_j^{-1}). Now R is real and positive definite, and therefore for any complex N-dimensional non-zero row vector $x = u + iv$

$$(7.20) \qquad x R \bar{x}^T = vRu^T + vRv^T > 0.$$

Letting $y = xA$ this reduces to

$$(7.21) \qquad y C \bar{y}^T = \sum_{j=1}^{N} c_j |y_j|^2 > 0.$$

Since the determinant of A does not vanish this inequality holds for arbitrary y, and thus $c_j > 0$ as asserted.

To complete the proof we have to show that the characteristic equation can not have multiple roots. Assume that $\xi_1 = \xi_2$ but the other roots are distinct. We get again a representation of the form (7.18) except that the term $c_1\xi_1^n$, is replaced by $c_1 n \xi_1^n$. The boundedness of r_n again necessitates that $c_1 = 0$. In the case of one double root we would therefore get a representation of the form (7.18) with fewer than N non-zero terms, and we have seen that this is impossible. The same argument shows more generally that no multiple roots are possible. ▶

We now state the final result for the case that N is an odd integer. The modifications required for even N should be obvious.

Theorem. *Suppose that the stationary sequence* $\{\mathbf{Y}_n\}$ *satisfies the difference equation* (7.13) *with* $N = 2\nu + 1$, *but no difference equation of lower order. The characteristic equation* (7.16) *possesses* ν *pairs of complex roots* $\xi_j = \cos\omega_j \pm i \sin\omega_j$ (*with* ω_j *real*), *and one real root* $\omega_0 = \pm 1$. *The sequence* $\{\mathbf{Y}_n\}$ *is of the form*

$$(7.22) \qquad \mathbf{Y}_n = \lambda_0 \mathbf{Z}_0 \cdot \omega_0^n + \sum_{j=1}^{\nu} \lambda_j [\mathbf{Z}_j \cos n\omega_j + \mathbf{Z}_{-j} \sin n\omega_j],$$

where the $\mathbf{Z}_j$ *are mutually independent normal variables with zero expectations and unit variances, and the* λ_j *are constants. For this sequence*

$$(7.23) \qquad r_n = \lambda_0^2 \omega_0^n + \sum_{j=1}^{\nu} \lambda_j^2 \cos n\omega_j.$$

Conversely, choose real $\lambda_\nu \neq 0$ *arbitrary and* $\omega_0 = \pm 1$, *and let* $\omega_1, \ldots, \omega_j$ *be distinct real numbers with* $0 < \omega_j < \pi$. *Then* (7.22) *defines a stationary process with covariances* (7.23) *and satisfying a difference equation of order* $2\nu + 1$ *but no difference equation of a lower order.*

Proof. Let the λ_j and ω_j be numbers, and the Z_j normal variables satisfying the conditions of the theorem. Define the variables $\mathbf{Y}_n$ by (7.22). A trite calculation shows that the covariances r_n of $\{\mathbf{Y}_n\}$ are given by

(7.23). There exists a real algebraic equation of the form (7.16) with the roots ξ_j described in the theorem. The r_j then satisfy the difference equation (7.17), and by lemma 1 this implies that the $\mathbf{Y}_n$ satisfy the stochastic difference equation (7.13). By construction this is the equation of lowest degree satisfied by the $\mathbf{Y}_n$.

Conversely, let $\{\mathbf{Y}_n\}$ stand for the solution of a given difference equation (7.13). The covariances r_n of $\{\mathbf{Y}_n\}$ determine the numbers λ_j and ω_j appearing in (7.22). Consider these equations for $n = 0, 1, \ldots, 2\nu$ as a linear transformation of an arbitrary N-tuple of normal variables $(\mathbf{Z}_{-\nu}, \ldots, \mathbf{Z}_\nu)$ into $(\mathbf{Y}_0, \ldots, \mathbf{Y}_N)$. This transformation is non-singular, and hence the covariance matrices of the two N-tuples determine each other uniquely. We have just shown that if the covariance matrix of the $\mathbf{Z}_j$ reduces to the identity matrix the $\mathbf{Y}_k$ will have the prescribed covariances r_n. The converse is therefore also true, and so there exist normal variables $\mathbf{Z}_j$ satisfying the conditions of the theorem and such that (7.22) holds for $n = 0, \ldots, N$. But both sides of these equations represent solutions of the stochastic difference equation (7.13), and since they agree for $0 \leq n \leq N$ they are necessarily identical. ►

8. MARKOVIAN NORMAL DENSITIES

We turn to a discussion of the particular class of normal densities occurring in Markov processes. Without loss of generality *we consider only densities centered at the origin.* Then $\mathbf{E}(\mathbf{X}_k) = 0$ and we use the usual abbreviations

$$\mathbf{E}(\mathbf{X}_k^2) = \sigma_k^2, \qquad \mathbf{E}(\mathbf{X}_j\mathbf{X}_k) = \sigma_j\sigma_k\rho_{jk}. \tag{8.1}$$

The ρ_{jk} are the *correlation coefficients* and $\rho_{kk} = 1$.

Definition. *The r-dimensional normal density of* $(\mathbf{X}_1, \ldots, \mathbf{X}_r)$ *is Markovian if for* $k \leq r$ *the conditional density of* $\mathbf{X}_k$ *for given* $\mathbf{X}_1, \ldots, \mathbf{X}_{k-1}$ *is identical with the conditional density of* $\mathbf{X}_k$ *for given* $\mathbf{X}_{k-1}$.

Roughly speaking, if we know $\mathbf{X}_{k-1}$ (the "present") then the additional knowledge of the "past" $\mathbf{X}_1, \ldots, \mathbf{X}_{k-2}$ does not contribute any relevant information about the "future," that is, about any $\mathbf{X}_j$ with $j \geq k$.

As usual in similar situations, we apply the term Markovian interchangeably to $(\mathbf{X}_1, \ldots, \mathbf{X}_r)$ and its density.

Theorem 1. *For* $(\mathbf{X}_1, \ldots, \mathbf{X}_r)$ *to be Markovian each of the following two conditions is necessary and sufficient:*

(i) *For* $k \leq r$

$$\mathbf{E}(\mathbf{X}_k \mid \mathbf{X}_1, \ldots, \mathbf{X}_{k-1}) = \mathbf{E}(\mathbf{X}_k \mid \mathbf{X}_{k-1}). \tag{8.2}$$

(ii) *For* $j \leq \nu \leq k \leq r$

$$\rho_{jk} = \rho_{j\nu}\rho_{\nu k}. \tag{8.3}$$

For (8.3) *to hold it suffices that*

$$\rho_{jk} = \rho_{j,k-1}\rho_{k-1,k}, \qquad j \leq k. \tag{8.4}$$

Proof. Identity of densities implies equality of expectations and so (8.2) is trivially necessary. On the other hand, if (8.2) is true, theorem 5 of section 6 shows that the conditional density of $\mathbf{X}_k$ for given $\mathbf{X}_1, \ldots, \mathbf{X}_{k-1}$ depends only on $\mathbf{X}_{k-1}$, but not on the preceding variables. Now the conditional density of $\mathbf{X}_k$ for given $\mathbf{X}_{k-1}$ is obtained by integrating out the variables $\mathbf{X}_1, \ldots, \mathbf{X}_{k-2}$, and hence the two conditional densities are identical. Thus (8.2) is necessary and sufficient.

Referring again to theorem 5 of section 6 it is clear that the variable

$$\mathbf{T} = \mathbf{X}_k - \mathbf{E}(\mathbf{X}_k \mid \mathbf{X}_{k-1}) \tag{8.5}$$

is identical with

$$\mathbf{T} = \mathbf{X}_k - \frac{\sigma_k}{\sigma_{k-1}}\,\rho_{k-1,k}\mathbf{X}_{k-1}. \tag{8.6}$$

because this is the only variable of the form $\mathbf{X}_k - c\mathbf{X}_{k-1}$ uncorrelated to $\mathbf{X}_{k-1}$. By the same theorem therefore (8.2) holds iff $\mathbf{T}$ is uncorrelated also to $\mathbf{X}_1, \ldots, \mathbf{X}_{k-2}$, that is, iff (8.4) holds. Thus (8.4) is necessary and sufficient. As it is a special case of (8.3) the latter condition is sufficient. It is also necessary, for repeated application of (8.4) shows that for $j < \nu < k \leq r$

$$\frac{\rho_{jk}}{\rho_{\nu k}} = \frac{\rho_{j,k-1}}{\rho_{\nu,k-1}} = \frac{\rho_{j,k-2}}{\rho_{\nu,k-2}} = \frac{\rho_{j\nu}}{\rho_{\nu\nu}} = \rho_{j\nu} \tag{8.7}$$

and so (8.4) implies (8.3). ►

Corollary. *If* $(\mathbf{X}_1, \ldots, \mathbf{X}_r)$ *is Markovian, so is every subset* $(\mathbf{X}_{\alpha_1}, \ldots, \mathbf{X}_{\alpha_\nu})$ *with* $\alpha_1 < \alpha_2 < \cdots < \alpha_\nu \leq r$.

This is obvious since (8.3) automatically extends to all subsets. ►

Examples. (*a*) *Independent increments.* A (finite or infinite) sequence $\{\mathbf{X}_k\}$ of normal random variables with $\mathbf{E}(\mathbf{X}_k) = 0$ is said to be a process with independent increments if for $j < k$ the increment $\mathbf{X}_k - \mathbf{X}_j$ is independent of $(\mathbf{X}_1, \ldots, \mathbf{X}_j)$. This implies, in particular, $\mathbf{E}(\mathbf{X}_j(\mathbf{X}_k - \mathbf{X}_j)) = 0$ or

$$\rho_{jk} = \frac{\sigma_j}{\sigma_k} \qquad j < k. \tag{8.8}$$

Comparing this with (8.3) one sees that a normal process with independent increments is automatically Markovian. Its structure is rather trite: $\mathbf{X}_k$ is the sum of the k mutually independent normal variables

$$\mathbf{X}_1, \mathbf{X}_2-\mathbf{X}_1, \ldots, \mathbf{X}_k-\mathbf{X}_{k-1}.$$

(*b*) *Autoregressive models.* Consider a normal Markovian sequence $\mathbf{X}_1, \mathbf{X}_2, \ldots$ with $\mathbf{E}(\mathbf{X}_k) = 0$. There exists a unique constant a_k making $\mathbf{X}_k - a_k\mathbf{X}_{k-1}$ independent of $\mathbf{X}_{k-1}$, and hence of $\mathbf{X}_1, \ldots, \mathbf{X}_{k-1}$. Put

$$\lambda_k^2 = \operatorname{Var}(\mathbf{X}_k - a_k\mathbf{X}_{k-1})$$

and, recursively,

$$\mathbf{X}_1 = \lambda_1\mathbf{Z}_1 \tag{8.9}$$

$$\mathbf{X}_k = a_k\mathbf{X}_{k-1} + \lambda_k\mathbf{Z}_k \qquad k = 2, 3, \ldots$$

The variables $\mathbf{Z}_k$ thus defined are easily seen to be independent and

$$\mathbf{E}(\mathbf{Z}_k) = 0, \qquad \mathbf{E}(\mathbf{Z}_k^2) = 1. \tag{8.10}$$

Now the converse is also true. If the $\mathbf{Z}_k$ are normal and satisfy (8.10), then (8.9) defines a sequence $\{\mathbf{X}_n\}$ and the very structure of (8.9) shows that $\{\mathbf{X}_n\}$ is Markovian. As an exercise we verify it computationally. Multiply (8.9) by $\mathbf{X}_j$ and take expectations. As $\mathbf{Z}_k$ is independent of $\mathbf{X}_1, \ldots, \mathbf{X}_{k-1}$ we get for $j < k$

$$a_k = \frac{\sigma_k}{\sigma_{k-1}} \frac{\rho_{jk}}{\rho_{j,k-1}}. \tag{8.11}$$

Now (8.4) is a simple consequence of this, and we know that it implies the Markovian character of the $\mathbf{X}_k$. Thus $(\mathbf{X}_1, \ldots, \mathbf{X}_r)$ *is Markovian iff relations of the form* (8.9) *hold* with normal variables $\mathbf{Z}_j$ satisfying (8.10). [This is a special case of example 7(*b*).] ▶

So far we have considered only finite sequences $(\mathbf{X}_1, \ldots, \mathbf{X}_r)$, but the number r plays no role and we may as well speak of infinite sequences $\{\mathbf{X}_n\}$. This does *not* involve infinite sequence spaces or any new theory, but is merely an indication that a distribution for $(\mathbf{X}_1, \ldots, \mathbf{X}_r)$ is defined for all r. Similarly, we speak of a *Markovian family* $\{\mathbf{X}(t)\}$ when any *finite* collection $\mathbf{X}_1 = \mathbf{X}(t_1), \ldots, \mathbf{X}_r = \mathbf{X}(t_r)$ is Markovian. The description depends on the functions

$$\mathbf{E}(\mathbf{X}^2(t)) = \sigma^2(t), \qquad \mathbf{E}(\mathbf{X}(s)\,\mathbf{X}(t)) = \sigma(s)\,\sigma(t)\,\rho(s, t). \tag{8.12}$$

In view of the criterion (8.3) it is obvious that *the family is Markovian iff for* $s < t < \tau$

$$\rho(s, t)\,\rho(t, \tau) = \rho(s, \tau). \tag{8.13}$$

Despite the fancy language we are really dealing only with families of finite-dimensional normal distributions with covariances satisfying (8.13).

As explained in greater detail at the beginning of section 7, the sequence $\{\mathbf{X}_n\}$ *is stationary* if for each fixed n-tuple $(\alpha_1, \ldots, \alpha_n)$ the distribution of $(\mathbf{X}_{\alpha_1+\nu}, \ldots, \mathbf{X}_{\alpha_n+\nu})$ is independent of ν. A finite section of such a sequence may be extended to both sides, and hence it is natural to consider only doubly infinite sequences $\{\ldots, \mathbf{X}_{-2}, \mathbf{X}_{-1}, \mathbf{X}_0, \mathbf{X}_1, \ldots\}$. These notions carry over trivially to families $\{\mathbf{X}(t)\}$.

For a stationary sequence $\{\mathbf{X}_n\}$ the variance σ_n^2 is independent of n and in the Markovian case (8.3) implies that $\rho_{jk} = \rho_{12}^{|k-j|}$. Thus *for a stationary Markovian sequence*

$$\mathbf{E}(\mathbf{X}_j\mathbf{X}_k) = \sigma^2\rho^{|k-j|} \tag{8.14}$$

where σ^2 and ρ are constants, $|\rho| \leq 1$. Conversely, a sequence with normal distributions satisfying (8.14) is Markovian and stationary.

In the case of a stationary family $\{\mathbf{X}(t)\}$ the correlation $\rho(s, t)$ depends only on the difference $|t-s|$ and (8.13) takes on the form

$$\rho(t)\,\rho(\tau) = \rho(t+\tau) \qquad \text{for} \quad t, \tau > 0.$$

Obviously $\rho(\tau) = 0$ would imply $\rho(t) = 0$ for all $t > \tau$ and also $\rho(\frac{1}{2}\tau) = 0$, and so ρ can have no zeros except if $\rho(t) = 0$ for all $t > 0$. Hence $\rho(t) = e^{-\lambda t}$ by the repeatedly used result of **1**; XVII, 6. Accordingly, *for a stationary Markovian family*

$$\mathbf{E}(\mathbf{X}(s)\,\mathbf{X}(s+t)) = \sigma^2 e^{-\lambda t}, \qquad t > 0 \tag{8.15}$$

except if $\mathbf{X}(s)$ and $\mathbf{X}(t)$ are uncorrelated for all $s \neq t$.

Example. (*c*) *Stationary sequences* may be constructed by the scheme of the last example. Because of (8.11) we must have

$$\mathbf{X}_k = \rho\mathbf{X}_{k-1} + \sigma\sqrt{1-\rho^2}\,\mathbf{Z}_k. \tag{8.16}$$

For each k it is possible to express $\mathbf{X}_k$ as a linear combination of $\mathbf{Z}_k$, $\mathbf{Z}_{k-1}, \ldots, \mathbf{Z}_{k-\nu}$, and $\mathbf{X}_{k-\nu}$. A formal passage to the limit would lead to the representation

$$\mathbf{X}_k = \sigma\sqrt{1-\rho^2}\sum_{j=0}^{\infty}\rho^j\mathbf{Z}_{k-j} \tag{8.17}$$

of $\{\mathbf{X}_k\}$ in terms of a doubly infinite sequence of independent normal variables $\mathbf{Z}_j$ normed by (8.10). Since $|\rho| < 1$ the convergence of the series is plausible, but the formula as such involves an infinite sequence space. [See the remarks concerning (7.5) of which (8.17) is a special case.]

It may be useful to discuss the relation of theorem 1 to the direct description of Markovian sequences in terms of densities. Denote by g_i the density of $\mathbf{X}_i$ and by $g_{ik}(x, y)$ the value at y of the conditional density of $\mathbf{X}_k$ given that $\mathbf{X}_i = x$. (In stochastic processes g_{ik} is called a transition density from $\mathbf{X}_i$ to $\mathbf{X}_k$.) For normal Markovian sequences g_i is the normal density with zero expectation and variance σ_i^2. As for the transition probabilities, it was shown in example 2(*a*) that

$$g_{ik}(x, y) = \frac{1}{\sigma_k\sqrt{1-\rho_{ik}^2}}\,\mathfrak{n}\left(\frac{y - \sigma_i^{-1}\rho_{ik}\sigma_k x}{\sigma_k\sqrt{1-\rho_{ik}^2}}\right) \tag{8.18}$$

where $\mathfrak{n}$ stands for the standard normal density. However, we shall not use this result and proceed to analyze the properties of g_{ik} by an independent method. As usual we interpret the subscripts as time parameters.

The joint density of $(\mathbf{X}_i, \mathbf{X}_j)$ is given by $g_i(x)\, g_{ij}(x, y)$. The joint density for $(\mathbf{X}_i, \mathbf{X}_j, \mathbf{X}_k)$ is the product of this with the conditional density for $\mathbf{X}_k$ for given $\mathbf{X}_j$ *and* $\mathbf{X}_i$, but in view of the Markovian character the index i drops out if $i < j < k$ and the density of $(\mathbf{X}_i, \mathbf{X}_j, \mathbf{X}_k)$ is given by

$$g_i(x)\, g_{ij}(x, y)\, g_{jk}(y, z). \tag{8.19}$$

In the Markovian case the density of every n-tuple $(\mathbf{X}_{\alpha_1}, \ldots, \mathbf{X}_{\alpha_n})$ is given by a product of the form (8.19), but the densities g_{jk} cannot be chosen arbitrarily. Indeed, integration of (8.19) with respect to y yields the marginal density for $(\mathbf{X}_i, \mathbf{X}_k)$ and so we have the consistency condition

$$g_{ik}(x, z) = \int_{-\infty}^{+\infty} g_{ij}(x, y)\, g_{jk}(y, z)\, dy \tag{8.20}$$

for all $i < j < k$. This is a special case of the *Chapman-Kolmogorov identity* for Markov processes.[13] Very roughly, it expresses that a transition from x at epoch i to z at epoch k takes place via an arbitrary intermediate position y, the transition from y to z being independent of the past. It is obvious that with any system of transition probabilities g_{ik} satisfying the Chapman-Kolmogorov identity the multiplication scheme (8.19) leads to a consistent system of densities for $(\mathbf{X}_1, \mathbf{X}_2, \ldots, \mathbf{X}_r)$ and the sequence is Markovian. We have thus the following analytic counterpart to theorem 1.

Theorem 2. *A family* $\{g_{ik}\}$ *can serve for transition densities in a normal Markovian process iff it satisfies the Chapman-Kolmogorov identity and* $g_{ik}(x, y)$ *represents for each fixed* x *a normal density in* y.

[13] Other special cases were encountered in **1**; XV,(13.3) and XVII,(9.1). Note that the system (8.19) is the analogue to the definition **1**; XV,(1.1) of probabilities for Markov chains, except that there summation replaces the integration and that only stationary transition probabilities were considered.

Both theorems contain necessary and sufficient conditions and they are therefore, in a sense, equivalent. They are, nevertheless, of different natures. The second is really not restricted to normal processes; applied to families $\{\mathbf{X}(t)\}$ it leads to differential and integral equations for the transition probabilities and in this way it serves to introduce new classes of Markovian processes. On the other hand, from theorem 2 one would not guess that the g_{ik} are necessarily of the form (8.18), a result implicit in the more special theorem 1.

For reference and later comparisons we list here the two most important Markovian families $\{\mathbf{X}(t)\}$.

Example. (*d*) *Brownian motion* or *Wiener-Bachelier process.* It is defined by the condition that $\mathbf{X}(0) = 0$, and that for $t > s$ the variable $\mathbf{X}(t) - \mathbf{X}(s)$ be independent of $\mathbf{X}(s)$ with a variance depending only on $t - s$. In other words, the process has independent increments [example (*a*)] and stationary transition probabilities [but it is not stationary since $\mathbf{X}(0) = 0$]. Obviously $\mathbf{E}(\mathbf{X}^2(t)) = \sigma^2 t$ and $\mathbf{E}(\mathbf{X}(s)\mathbf{X}(t)) = \sigma^2 s$ for $s < t$. For $\tau > t$ the transition densities from (t, x) to (τ, y) are normal with expectation x and variance $\sigma^2(\tau - t)$. They depend only on $(y-x)/(\tau-t)$, and the Chapman-Kolmogorov identity reduces to a convolution.

(*e*) *Ornstein-Uhlenbeck process.* By this is meant the most general normal *stationary Markovian process* with zero expectations. Its covariances are given by (8.15). In other words, for $\tau > t$ the transition density from (t, x) to (τ, y) is normal with expectation $e^{-\lambda(\tau-t)}x$ and variance $\sigma^2(1-e^{-2\lambda(\tau-t)})$. As $\tau \to \infty$ the expectation tends to 0 and the variance to σ^2. This process was considered by Ornstein and Uhlenbeck from an entirely different point of view. Its connection with diffusion will be discussed in X,4. ►

9. PROBLEMS FOR SOLUTION

1. Let Ω be the region of the plane (of area $\frac{1}{2}$) bounded by the quadrilateral with vertices $(0, 0)$, $(1, 1)$, $(0, \frac{1}{2})$, $(\frac{1}{2}, 1)$ and the triangle with vertices $(\frac{1}{2}, 0)$, $(1, 0)$, $(1, \frac{1}{2})$. (The unit square is the union of Ω and the region symmetric to Ω with respect to the bisector.) Let $(\mathbf{X}, \mathbf{Y})$ be distributed uniformly in Ω. Prove that the marginal distributions are uniform and the $\mathbf{X} + \mathbf{Y}$ has the same density as if $\mathbf{X}$ and $\mathbf{Y}$ were independent.[14]

Hint: A diagram renders calculations unnecessary.

2. *Densities with normal marginal densities.* Let u be an *odd* continuous function on the line, vanishing outside $\overline{-1, 1}$. If $|u| < (2\pi e)^{-\frac{1}{2}}$ then

$$\mathfrak{n}(x)\mathfrak{n}(y) + u(x)u(y)$$

[14] In other words, the distribution of a sum may be given by the convolution even if the variables are dependent. This intuitive example is due to H. E. Robbins. For another freak of the same type see II,4(*c*).

represents a bivariate density which *is not normal, but whose marginal densities are both normal.* (E. Nelson.)

3. *A second example.* Let φ_1 and φ_2 be two bivariate normal densities with unit variances but different correlation coefficients. The mixture $\frac{1}{2}(\varphi_1+\varphi_2)$ is *not normal, but its two marginal densities coincide with* $\mathfrak{n}$.

Note. In the sequel all random variables are in $\mathcal{R}^1$. *Vector variables are indicated by pairs* $(\mathbf{X}_1, \mathbf{X}_2)$, *etc.*

4. Let $\mathbf{X}_1, \ldots, \mathbf{X}_n$ be independent random variables with the common density f and distribution function F. If $\mathbf{X}$ is the smallest and $\mathbf{Y}$ the largest among them, the joint density of the pair $(\mathbf{X}, \mathbf{Y})$ is given by

$$n(n-1)f(x)f(y)[F(y)-F(x)]^{n-2}, \qquad y > x.$$

5. Show that the symmetric Cauchy distribution in $\mathcal{R}^3$ [defined in (1.21)] corresponds to a random vector whose length has the density $v(r) = 4\pi^{-1}r^2(1+r^2)^{-2}$ for $r > 0$. [*Hint:* Use polar coordinates and either (1.15) or else the general relation I,(10.4) for projections.]

6. For the Cauchy distribution (1.21) the conditional density of $\mathbf{X}_3$ for given $\mathbf{X}_1, \mathbf{X}_2$ is

$$v_{\xi_1,\xi_2}(z) = \frac{2}{\pi} \cdot \frac{\sqrt{(1+\xi_1^2+\xi_2^2)^3}}{(1+\xi_1^2+\xi_2^2+z^2)^2},$$

and the bivariate conditional density of $\mathbf{X}_2, \mathbf{X}_3$ for given $\mathbf{X}_1 = \xi$

$$v_\xi(y, z) = \frac{1}{\pi} \cdot \frac{1+\xi^2}{(1+\xi^2+y^2+z^2)^2}.$$

7. Let $0 < a < 1$ and $f(x, y) = [(1+ax)(1+ay) - a]e^{-x-y-axy}$ for $x > 0$, $y > 0$ and $f(x, y) = 0$ elsewhere.

(*a*) Prove that f is a density of a pair $(\mathbf{X}, \mathbf{Y})$. Find the marginal densities and the distribution function.

(*b*) Find the conditional density $u_x(y)$ and $\mathbf{E}(\mathbf{Y} \mid \mathbf{X})$, $\text{Var}(\mathbf{Y} \mid \mathbf{X})$.

8. Let f be a density concentrated on $\overline{0, \infty}$. Put $u(x, y) = f(x+y)/(x+y)$ for $x > 0, y > 0$ and $u(x, y) = 0$ otherwise. Prove that u is a density in $\mathcal{R}^2$ and find its covariance matrix.

9. Let $\mathbf{X}_1, \mathbf{X}_2, \mathbf{X}_3$ be mutually independent and distributed uniformly over $\overline{0, 1}$. Let $\mathbf{X}_{(1)}, \mathbf{X}_{(2)}, \mathbf{X}_{(3)}$ be the corresponding order statistics. Find the density of the pair

$$\left(\frac{\mathbf{X}_{(1)}}{\mathbf{X}_{(2)}}, \frac{\mathbf{X}_{(2)}}{\mathbf{X}_{(3)}}\right)$$

and show that the two ratios are independent. Generalize to n dimensions.

10. Let $\mathbf{X}_1, \mathbf{X}_2, \mathbf{X}_3$ be independent with a common exponential distribution. Find the density of $(\mathbf{X}_2-\mathbf{X}_1, \mathbf{X}_3-\mathbf{X}_1)$.

11. A particle of unit mass is split into two fragments with masses $\mathbf{X}$ and $1 - \mathbf{X}$. The density f of $\mathbf{X}$ is concentrated on $\overline{0, 1}$ and for reasons of symmetry $f(x) = f(1-x)$. Denote the smaller fragment by $\mathbf{X}_1$ the larger by $\mathbf{X}_2$. The two fragments are split independently in like manner resulting in four fragments with masses $\mathbf{X}_{11}, \mathbf{X}_{12}, \mathbf{X}_{21}, \mathbf{X}_{22}$. Find (*a*) the density of $\mathbf{X}_{11}$. (*b*) The joint density of $\mathbf{X}_{11}$ and $\mathbf{X}_{22}$. Use (*b*) to verify the result in (*a*).

12. Let $\mathbf{X}_1, \mathbf{X}_2, \ldots$ be independent with the common normal density $\mathfrak{n}$, and $\mathbf{S}_k = \mathbf{X}_1 + \cdots + \mathbf{X}_k$. If $m < n$ find the joint density of $(\mathbf{S}_m, \mathbf{S}_n)$ and the conditional density for $\mathbf{S}_m$ given that $\mathbf{S}_n = t$.

13. In the preceding problem find the conditional density of $\mathbf{X}_1^2 + \cdots + \mathbf{X}_m^2$ given $\mathbf{X}_1^2 + \cdots + \mathbf{X}_n^2$.

14. Let $(\mathbf{X}, \mathbf{Y})$ have a bivariate normal density centered at the origin with $\mathbf{E}(\mathbf{X}^2) = \mathbf{E}(\mathbf{Y}^2) = 1$, and $\mathbf{E}(\mathbf{XY}) = \rho$. In polar coordinates $(\mathbf{X}, \mathbf{Y})$ becomes $(\mathbf{R}, \boldsymbol{\Phi})$ where $\mathbf{R}^2 = \mathbf{X}^2 + \mathbf{Y}^2$. Prove that $\boldsymbol{\Phi}$ has a density given by

$$\frac{\sqrt{1-\rho^2}}{2\pi(1-2\rho \sin \varphi \cos \varphi)} \qquad 0 < \varphi < 2\pi$$

and is uniformly distributed iff $\rho = 0$. Conclude

$$\mathbf{P}\{\mathbf{XY} > 0\} = \tfrac{1}{2} + \pi^{-1} \arcsin \rho \quad \text{and} \quad \mathbf{P}\{\mathbf{XY} < 0\} = \pi^{-1} \arccos \rho.$$

15. Let f be the uniform density for the triangle with vertices $(0, 0)$, $(0, 1)$, $(1, 0)$ and g the uniform density for the symmetric triangle in the third quadrant. Find $f \star f$, and $f \star g$.

Warning. A tedious separate consideration of individual intervals is required.

16. Let f be the uniform density in the unit disk. Find $f \star f$ in polar coordinates.

17. Let u and v be densities in $\mathcal{R}^2$ of the form

$$u(x, y) = f(\sqrt{x^2 + y^2}), \qquad v(x, y) = g(\sqrt{x^2 + y^2}).$$

Find $u \star v$ in polar coordinates.

18. Let $\mathbf{X} = (\mathbf{X}_1, \ldots, \mathbf{X}_r)$ have a normal density in r dimensions. There exists a unit vector $a = (a_1, \ldots, a_r)$ such that

$$\operatorname{Var}(a_1\mathbf{X}_1 + \cdots + a_r\mathbf{X}_r) \geq \operatorname{Var}(c_1\mathbf{X}_1 + \cdots + c_r\mathbf{X}_r)$$

for all unit vectors $c = (c_1, \ldots, c_r)$. If $a = (1, 0, \ldots, 0)$ is such a vector then $\mathbf{X}_1$ is independent of the remaining $\mathbf{X}_j$.

19. Prove the

Theorem. *Given a normal density in* $\mathcal{R}^r$ *the coordinate axes can be rotated in such a way that the new coordinate variables are mutually independent normal variables.*

In other words: in theorem 2 of section 6 the matrix C may be taken as a rotation matrix.

Hint: Let $\mathbf{Y} = \mathbf{X}C$ and choose a rotation matrix C such that

$$\mathbf{Y}_r = a_1\mathbf{X}_1 + \cdots + a_r\mathbf{X}_r$$

where $a = (a_1, \ldots, a_r)$ is the maximizing vector of the preceding example. The rest is easy.

20. Find the general normal stationary process satisfying

(*a*) $\mathbf{X}_{n+2} + \mathbf{X}_n = 0$

(*b*) $\mathbf{X}_{n+2} - \mathbf{X}_n = 0$

(*c*) $\mathbf{X}_{n+3} - \mathbf{X}_{n+2} + \mathbf{X}_{n+1} - \mathbf{X}_n = 0.$

21. *A servo-stochastic process.* (H. D. Mills.) A servomechanism is exposed to random shocks, but corrections may be introduced at any time. Thus the

error $\mathbf{Y}_n$ at time n is (in proper units) of the form $\mathbf{Y}_{n+1} = \mathbf{Y}_n - \mathbf{C}_n + \mathbf{X}_{n+1}$, where $\mathbf{C}_n$ is the correction and the $\mathbf{X}_n$ are independent normal variables, $\mathbf{E}(\mathbf{X}_n) = 0$, $\mathbf{E}(\mathbf{X}_n^2) = 1$. The $\mathbf{C}_n$ are, in principle, arbitrary functions of the past observations, that is, of $\mathbf{Y}_k$ and $\mathbf{X}_k$ for $k \leq n$. One wishes to choose them so as to minimize $\text{Var}(\mathbf{Y}_n)$ (which is a measure of how *well* the mechanism works), and $\text{Var}(\mathbf{C}_n)$ (which is a measure of how *hard* it works).

(*a*) Discuss the covariance function of $\{\mathbf{Y}_n\}$ and show that $\text{Var}(\mathbf{Y}_n) \geq 1$.

(*b*) Assuming that $\text{Var}(\mathbf{C}_n) \to \alpha^2$, $\text{Var}(\mathbf{Y}_n) \to \sigma^2$ (tendency to stationarity) show that $\sigma \geq \frac{1}{2}(\alpha + \alpha^{-1})$.

(*c*) Consider, in particular, the *linear device* $\mathbf{C}_n = a + p(\mathbf{Y}_n - b)$, $0 < p \leq 1$. Find the covariance function and a representation of the form (7.8) for $\mathbf{Y}_n$.

22. *Continuation.* If there is a *time lag* in information or adjustment the model is essentially the same except that $\mathbf{C}_n$ is to be replaced by $\mathbf{C}_{n+N}$. Discuss this situation.

CHAPTER IV

Probability Measures and Spaces

As stated in the introduction, very little of the technical apparatus of measure theory is required in this volume, and most of the book should be readable without the present chapter.[1] It is nevertheless desirable to give a brief account of the basic concepts which form the theoretical background for this book and, for reference, to record the main theorems. The underlying ideas and facts are not difficult, but proofs in measure theory depend on messy technical details. For the beginner and outsider access is made difficult also by the many facets and uses of measure theory; excellent introductions exist, but of necessity they dwell on great generality and on aspects which are not important in the present context. The following survey concentrates on the needs of this volume and omits many proofs and technical details.[2] (It is fair to say that the simplicity of the theory is deceptive in that much more difficult measure theoretic problems arise in connection with stochastic processes depending on a continuous time parameter. The treatment of conditional expectations is deferred to V, 10–11; that of the Radon-Nikodym theorem to V,3.)

Formulas relating to Cartesian (or Euclidean) spaces $\mathfrak{R}^r$ are independent of the number of dimensions provided x is read as abbreviation for $(x_1, \ldots, x_r)$.

---

[1] This applies to readers acquainted with the rudiments of measure theory as well as to readers interested primarily in results and facts. For the benefit of the latter the definition of integrals is repeated in V,1. Beyond this they may rely on their intuition, because in effect measure theory justifies simple formal manipulations.

[2] An excellent source for Baire functions and Lebesgue-Stieltjes integration is found in E. J. McShane and T. A. Botts, *Real analysis*, D. Van Nostrand, Princeton, 1959. Widely used are presentations of general measure theory in P. R. Halmos, *Measure theory*, D. Van Nostrand, Princeton, 1950 and in N. Bourbaki, *Eléments de mathématiques* [livre VI, chapters 3–5] Hermann, Paris, 1952 and 1956. For presentations for the specific purposes of probability see the books of Doob, Krickeberg, Loève, Neveu, and Hennequin-Tortrat.

1. BAIRE FUNCTIONS

We shall have to decide on a class of sets for which probabilities are defined and on a class of functions acceptable as random variables. The two problems are not only related but their treatment is unified by a streamlined modern notation. We begin by introducing it and by recalling the definition of convergence in terms of monotone limits.

The *indicator*[3] *of a set* A is the function which assumes the value 1 at all points of A and the value 0 at all points of the complement A'. It will be denoted by $\mathbf{1}_A$: thus $\mathbf{1}_A(x) = 1$ if $x \in A$ and $\mathbf{1}_A(x) = 0$ otherwise. Every set has an indicator, and every function assuming only the values 1 and 0 is the indicator of some set. If f is an arbitrary function, the product $\mathbf{1}_A f$ is the function that equals f on A and vanishes elsewhere.

Consider now the intersection $C = A \cap B$ of two sets. Its indicator $\mathbf{1}_C$ equals 0 wherever either $\mathbf{1}_A$ or $\mathbf{1}_B$ vanishes, that is, $\mathbf{1}_C = \inf(\mathbf{1}_A, \mathbf{1}_B)$ equals the smaller of the two functions. To exploit this parallelism one writes $f \cap g$ instead of $\inf(f, g)$ for the function which at each point x equals the smaller of the values of $f(x)$ and $g(x)$. Similarly $f \cup g = \sup(f, g)$ denotes the larger of the two values.[4] The operators $\cap$ and $\cup$ are called *cap* and *cup* respectively. They apply to arbitrary numbers of functions, and one writes

$$f_1 \cap \cdots \cap f_n = \bigcap_{k=1}^{n} f_k, \qquad f_1 \cup \cdots \cup f_n = \bigcup_{k=1}^{n} f_k. \tag{1.1}$$

To repeat, at each point x these functions equal, respectively, the minimum and the maximum among the n values $f_1(x), \ldots, f_n(x)$. If f_k is the indicator of a set A_k then (1.1) exhibits the indicators of the intersection $A_1 \cap \cdots \cap A_n$ and of the union $A_1 \cup \cdots \cup A_n$.

Consider now an *infinite* sequence $\{f_n\}$. The functions defined in (1.1) depend monotonically on n, and hence the limits $\bigcup_{k=1}^{\infty} f_k$ and $\bigcup_{k=1}^{\infty} f_k$ are well defined though possibly infinite. For fixed j

$$w_j = \bigcap_{k=j}^{\infty} f_k \tag{1.2}$$

is the limit of the monotone sequence of functions $f_j \cap \cdots \cap f_{j+n}$, and the sequence $\{w_j\}$ itself is again monotone, that is, $w_n = w_1 \cup \cdots \cup w_n$. With our notations $w_n \to \bigcup_{k=1}^{\infty} w_k$. By definition $w_n(x)$ is the greatest lower bound (the infimum) of the numerical sequence $f_n(x), f_{n+1}(x), \ldots$.

[3] This term was introduced by Loève. The older term "characteristic function" is confusing in probability theory.

[4] Many writers prefer the symbols $\vee$ and $\wedge$ for functions and reserve $\cap$ and $\cup$ for sets. Within our context there is no advantage in the dual notation.

Hence the limit of w_n is the same as $\liminf f_n$ and thus

$$\liminf f_n = \bigcup_{j=1}^{\infty} \bigcap_{k=j}^{\infty} f_k. \tag{1.3}$$

In this way the lim inf is obtained by a succession of two passages to the limit in *monotone* sequences. For $\limsup f_n$ one gets (1.3) with $\cap$ and $\cup$ interchanged.

All these considerations carry over to sets. In particular, we write $A = \lim A_n$ iff $\mathbf{1}_A = \lim \mathbf{1}_{A_n}$. In words, the sequence $\{A_n\}$ of sets converges to the set A iff each point of A belongs to *all* A_n with finitely many exceptions, and each point of the complement A' belongs at most to finitely many A_n.

Example. (*a*) *The set* $\{A_n \text{ i. o.}\}$. As a probabilistically significant example of limiting operations among sets consider the event A defined as "the realization of infinitely many among a given sequence of events $A_1, A_2, \ldots$." [Special cases were considered in **1**; VIII,3 (Borel-Cantelli lemmas) and in **1**; XIII (recurrent events).] More formally, given a sequence $\{A_n\}$ of sets, a point x belongs to A iff it belongs to infinitely many A_k. Since 0 and 1 are the only possible values of indicators this definition is equivalent to saying that $\mathbf{1}_A = \limsup \mathbf{1}_{A_n}$. In standard notation therefore $A = \limsup A_n$, but the notation $\{A_n \text{ i. o.}\}$ (read "A_n infinitely often") is more suggestive. It is due to K. L. Chung.

Our next problem is to delimit the class of functions[5] in $\mathcal{R}^r$ with which we propose to deal. The notion of an arbitrary function is far too broad to be useful for our purposes, and a modernized version of Euler's notion of a function is more appropriate. Taking continuous functions as given, the only effective way of constructing new functions depends on taking limits. As it turns out, all our needs will be satisfied if we know how to deal with functions that are limits of sequences $\{f_n\}$ of continuous functions, or limits of sequences where each f_n is such a limit, and so on. In other words, we are interested in a class $\mathfrak{B}$ of functions with the following properties: (1) every continuous function belongs to $\mathfrak{B}$, and (2) if $f_1, f_2, \ldots$ belong to $\mathfrak{B}$ and a limit $f(x) = \lim f_n(x)$ exists for all x, then f belongs to $\mathfrak{B}$.

[5] We are, in principle, interested only in finite-valued functions, but it is sometimes convenient to permit $\pm\infty$ as values. For example, the simple theorem that every monotone sequence has a limit is false for finite-valued functions and without it many formulations become clumsy. For this reason we adhere to the usual convention that all functions are to the extended real line, that is, their values are numbers of $\pm\infty$. In practice the values $\pm\infty$ will play no role. To make sure that the sum and product of two functions are again functions one introduces for their values the conventions $\infty + \infty = \infty$, $\infty - \infty = 0$, $\infty \cdot \infty = \infty$, $0 \cdot \infty = 0$, etc.

Such a class is said to be *closed* under pointwise limits. There is no doubt that such classes exist, the class of *all* functions being one. The intersection of all such classes is itself a closed family, and obviously is the smallest such class. Prudence requires us to limit our considerations to this smallest class.

The smallest closed class of functions containing all continuous functions is called the Baire class and will be denoted by $\mathfrak{B}$. *The functions in* $\mathfrak{B}$ *are called Baire functions.*[6]

We shall use this notion not only for functions defined in the whole space but also for functions defined only on a subset (for example, $\sqrt{x}$ or $\log x$ in $\mathfrak{R}^1$).

It is obvious from the definition that the sum and the product of two Baire functions are again Baire functions, but much more is true. If w is a continuous function in r variables and $f_1, \ldots, f_r$ are Baire functions, then $w(f_1, \ldots, f_r)$ is again a Baire function. Replacing w by w_n and passing to a limit it can be shown that more generally *every Baire function of Baire functions is again a Baire function.* Fixing the value of one or more variables leads again to a Baire function, and so on. In short, none of the usual operations on Baire functions will lead outside the class, and therefore the class $\mathfrak{B}$ is a natural object for our analysis. It will turn out that no simplifications are possible by considering smaller classes.

2. INTERVAL FUNCTIONS AND INTEGRALS IN $\mathfrak{R}^r$

We shall use the word *interval*, and the indicated notation, for sets of points satisfying a double inequality of one of the following four types:

$$\overline{a, b}: \quad a < x < b \qquad\qquad \overline{a, b|}: \quad a < x \leq b$$

$$\overline{|a, b|}: \quad a \leq x \leq b \qquad\qquad \overline{|a, b}: \quad a \leq x < b.$$

In one dimension this covers all possible intervals, including the degenerate interval of length zero. In two dimensions the inequalities are interpreted coordinate-wise, and intervals are (possibly degenerate) rectangles parallel to the axes. Other types of partial closure are possible but are herewith *excluded.* The limiting case where one or more coordinates of either a or b are replaced by $\pm\infty$ is admitted; in particular, the whole space is interval the $\overline{-\infty, \infty}$.

A point function f assigns a value $f(x)$ to individual points. A *set function* F assigns values to sets or regions of the space. The volume in

[6] This definition depends on the notion of continuity but not on other properties of Cartesian spaces. It is therefore applicable to arbitrary topological spaces.

$\mathfrak{R}^3$, area in $\mathfrak{R}^2$, or length in $\mathfrak{R}^1$ are typical examples but there are many more, probabilities representing a special case of primary concern to us. We shall be interested only in set functions with the property that if a set A is partitioned into two sets A_1 and A_2, then $F\{A\} = F\{A_1\} + F\{A_2\}$ Such functions are called additive.[7]

As we have seen, it occurs frequently that probabilities $F\{I\}$ are assigned to all intervals of the r-dimensional space $\mathfrak{R}^r$ and it is desired to extend this assignment to more general sets. The same problem occurs in elementary calculus, where the area (content) is originally defined only for rectangles and it is desired to define the area of a more general domain A. The simplest procedure is first to define integrals for functions of two variables and then to equate "the area of A" with the integral of the indicator $\mathbf{1}_A$ (that is the function that equals 1 in A and vanishes outside A). In like manner we shall define the integral

$$\mathbf{E}(u) = \int_{\mathfrak{R}^r} u(x)\, F\{dx\} \tag{2.1}$$

of a point function u with respect to the interval function F. The probability of A will then be defined by $\mathbf{E}(\mathbf{1}_A)$. In the construction of the integral (2.1) the interpretation of F plays no role, and we shall actually describe the general notion of a Lebesgue-Stieltjes integral. With this program in mind we now start anew.

Let F be a function assigning to each interval I a finite value $F\{I\}$. Such a function is called (finitely) *additive* if for every partition of an interval I into finitely many non-overlapping intervals $I_1, \ldots, I_n$.

$$F\{I\} = F\{I_1\} + \cdots + F\{I_n\}. \tag{2.2}$$

Examples. (*a*) *Distributions in* $\mathfrak{R}^1$. In volume **1** we considered discrete probability distributions attributing probabilities $p_1, p_2, \ldots$ to the points $a_1, a_2, \ldots$. Here $F\{I\}$ is the sum of the weights p_n of all points a_n contained in I, and $\mathbf{E}(u) = \sum u(a_n)p_n$.

(*b*) If G is any continuous monotone function increasing from 0 at $-\infty$ to 1 at ∞ one may define $F\{\overline{a, b}\} = G(b) - G(a)$.

(*c*) *Random vectors in* $\mathfrak{R}^2$. A vector of unit length issues from the origin in a random direction. The probability that its endpoint lies in a two-dimensional interval I is proportional to the length of the intersection of I with the unit circle. This defines a continuous probability distribution without density. The distribution is *singular* in the sense that the whole

---

[7] Empirical examples for additive functions are the mass and amount of heat in a region, the land value, the wheat acreage and the number of inhabitants of a geographical region, the yearly coal production, the passenger miles flown or the kilowatt hours consumed during a period, the number of telephone calls, etc.

probability is carried by a circle. One may think that such distributions are artificial and that the circle rather than the plane should serve as natural sample space. The objection is untenable because the sum of two independent random vectors is capable of all lengths between 0 and 2 and has a positive density within the disk of radius 2 [see example V,4(*e*)]. For some problems involving random unit vectors the plane is therefore the natural sample space. Anyhow, the intention was only to show by a simple example what happens in more complicated situations.

(*d*) We conclude with an example illustrating the contingency that will be *excluded* in the sequel. In $\mathcal{R}^1$ put $F\{I\} = 0$ for any interval $I = \overline{a, b}$ with $b < \infty$ and $F\{I\} = 1$ when $I = \overline{a, \infty}$. This interval function is additive but weird because it violates the natural continuity requirement that $F\{\overline{a, b}\}$ should tend to $F\{\overline{a, \infty}\}$ as $b \to \infty$. ►

The last example shows the desirability of strengthening the requirement (2.2) of finite additivity. We shall say that *an interval function* F *is countably additive, or* σ*-additive, if for every partitioning of an interval* I *into countably many intervals* $I_1, I_2, \ldots,$

$$F\{I\} = \sum F\{I_k\}. \tag{2.3}$$

["Countably many" means finitely or denumerably many. The term completely additive is synonymous with countably additive. The condition (2.3) is manifestly violated in the last example.]

We shall restrict our attention entirely to countably additive set functions. This is justified by the success of the theory, but the restriction can be defended a priori on heuristic or pragmatic grounds. In fact, if $A_n = I_1 \cup \cdots \cup I_n$ is the union of the first n intervals, then $A_n \to I$. One could argue that "for n sufficiently large A_n is practically indistinguishable from I." If $F\{I\}$ can be found by experiments, $F\{A_n\}$ must be "practically indistinguishable" from $F\{I\}$, that is, $F\{A_n\}$ must tend to $F\{I\}$. The countable additivity (2.3) expresses precisely this requirement.

Being interested principally in probabilities we shall consider only non-negative interval functions F normed by the condition that $F\{\overline{-\infty, \infty}\} = 1$. This norming imposes no serious restriction when $F\{\overline{-\infty, \infty}\} < \infty$, but it excludes interval functions such as length in $\mathcal{R}^1$ or area in $\mathcal{R}^2$. To make use of the following theory in such cases it suffices to partition the line or the plane into unit intervals and treat them separately. This procedure is so obvious and so well known that it requires no further explanation.

A function on $\mathcal{R}^r$ is called a *step function* if it assumes only finitely many values, each on an interval. For a step function u assuming the values $a_1, \ldots, a_n$ on intervals $I_1, \ldots, I_n$ (that is, with probabilities

$F\{I_1\}, \ldots, F\{I_n\})$, respectively we put

$$\mathbf{E}(u) = a_1\, F\{I_1\} + \cdots + a_n\, F\{I_n\} \tag{2.4}$$

in analogy with the definition of expectation of discrete random variables. [It is true that the partioning of the space into intervals on which u is constant is not unique, but just as in the discrete case the definition (2.4) is easily seen to be independent of the partition.] This expectation $\mathbf{E}(u)$ satisfies the following conditions:

(*a*) *Additivity* for linear combinations:

$$\mathbf{E}(\alpha_1 u_1 + \alpha_2 u_2) = \alpha_1 \mathbf{E}(u_1) + \alpha_2 \mathbf{E}(u_2). \tag{2.5}$$

(*b*) *Positivity:*

$$u \geq 0 \quad \textit{implies} \quad \mathbf{E}(u) \geq 0. \tag{2.6}$$

(*c*) *Norming:* For the constant function

$$\mathbf{E}(1) = 1. \tag{2.7}$$

The last two conditions are equivalent to the *mean value thoerem:* $\alpha \leq u \leq \beta$ implies $\alpha \leq \mathbf{E}(u) \leq \beta$ and so the function $\mathbf{E}(u)$ represents a sort of *average*.[8]

The problem is to extend the definition of $\mathbf{E}(u)$ to larger classes of functions preserving the properties (*a*)–(*c*). The classical Riemann integration utilizes the fact that to each continuous function u on $\overline{0, 1}$ there exists a sequence of step functions u_n such that $u_n \to u$ uniformly on $\overline{0, 1}$. By definition then $\mathbf{E}(u) = \lim \mathbf{E}(u_n)$. It turns out that the uniformity of the convergence is unnecessary and the same definition for $\mathbf{E}(u)$ can be used whenever $u_n \to u$ pointwise. In this way it is possible to extend $\mathbf{E}(u)$ to *all bounded Baire functions*, and the extension is *unique*. When it comes to unbounded functions divergent integrals are unavoidable, but at least for *positive* Baire functions it is possible to define $\mathbf{E}(u)$ either as a number or as

[8] When F represents probabilities $\mathbf{E}(u)$ may be interpreted as the expected gain of a gambler who can gain the amounts $a_1, a_2, \ldots$. To grasp the intuitive meaning in other situations consider three examples in which $u(x)$ represents, respectively, the temperature at time x, the number of telephone conversations at time x, the distance of a mass point from the origin, while F represents, respectively, the duration of a time interval, the value (cost of conversation) of a time interval, and mechanical mass. In each case integration will be extended over a finite interval only, and $\mathbf{E}(u)$ will represent the accumulated "temperature hours," the accumulated gain, and a static moment. These examples will show our integration with respect to arbitrary set functions to be simpler and more intuitive than Riemann integration where the independent variable plays more than one role and the "area under the curve" is of no help to the beginner. One should beware of the idea that the concept of expectation occurs only in probability theory.

the symbol ∞ (indicating divergence). No trouble arises in this respect because the Lebesgue theory considers only absolute integrability. Roughly speaking, starting from the definition (2.4) for expectations of simple functions it is possible to define $\mathbf{E}(u)$ for general Baire functions by obvious approximations and passages to the limit. The number $\mathbf{E}(u)$ so defined is the Lebesgue-Stieltjes integral of u with respect to F. (The term expectation is preferable when the underlying function F remains fixed so that no ambiguity arises.) We state here without proof[9] the basic fact of the Lebesgue theory; its nature and scope will be analyzed in the following sections. [A constructive definition of $\mathbf{E}(u)$ is given in section 4.]

Main theorem. *Let F be a countably additive interval functions in $\mathcal{R}^r$ with $F\overline{\{-\infty, \infty\}} = 1$. There exists a unique Lebesgue-Stieltjes integral* $\mathbf{E}(u)$ *on the class of Baire functions such that:*

If $u \geq 0$ then $\mathbf{E}(u)$ is a non-negative number or ∞. Otherwise $\mathbf{E}(u)$ exists iff either $\mathbf{E}(u^+)$ or $\mathbf{E}(u^-)$ is finite; in this case $\mathbf{E}(u) = \mathbf{E}(u^+) - \mathbf{E}(u^-)$. A function u is called integrable if $\mathbf{E}(u)$ is finite. then

(i) *If u is a step function, $\mathbf{E}(u)$ is given by* (2.4).

(ii) *Conditions* (2.5)–(2.7) *hold for all integrable functions.*

(iii) (*Monotone convergence principle.*) *Let $u_1 \leq u_2 \leq \cdots \to u$ where u_n is integrable. Then $\mathbf{E}(u_n) \to \mathbf{E}(u)$.*

The change of variables $v_n = u_{n+1} - u_n$ leads to a restatement of the last principle in terms of series:

If v_n is integrable and $v_n \geq 0$, then

$$\sum \mathbf{E}(v_n) = \mathbf{E}(\sum v_n) \tag{2.8}$$

in the sense that both sides are meaningful (finite) or neither is. It follows in particular that if $v \geq u \geq 0$ and $\mathbf{E}(u) = \infty$ then also $\mathbf{E}(v) = \infty$.

What happens if in (iii) the condition of monotonicity is dropped? The answer depends on an important lemma of wide applicability.

Fatou's lemma. *If $u_n \geq 0$ and u_n is integrable, then*

$$\mathbf{E}(\liminf u_n) \leq \liminf \mathbf{E}(u_n). \tag{2.9}$$

In particular, if $u_n \to u$ then $\liminf \mathbf{E}(u_n) \geq \mathbf{E}(u)$.

Proof. Put $v_n = u_n \cap u_{n+1} \cap \cdots$. Then $v_n \leq u_n$ and hence

$$\mathbf{E}(v_n) \leq \mathbf{E}(u_n).$$

[9] The method of proof is indicated in section 5. As usual, u^+ and u^- denote the positive and negative parts of u, that is, $u^+ = u \cup 0$ and $-u^- = u \cap 0$. Thus $u = u^+ - u^-$.

But (as we saw in section 1) v_n tends monotonically to $\liminf u_n$, and so $\mathbf{E}(v_n)$ tends to the left side in (2.9) and the lemma is proved. [Note that each side in (2.9) can represent ∞.] ▶

As example (*e*) will show, the condition of positivity cannot be dropped, but it can be replaced by the formally milder condition that there exists an integrable function U such that $u_n \geq U$. (It suffices to replace u_n by $u_n - U$.) Changing u_n into $-u_n$ we see that *if* $u_n < U$ *and* $\mathbf{E}(U) < \infty$, *then*

$$\limsup \mathbf{E}(u_n) \leq \mathbf{E}(\limsup u_n). \tag{2.10}$$

For convergent sequences the extreme members in (2.9) and (2.10) coincide and the two relations together yield the important

Dominated convergence principle. *Let* u_n *be integrable and* $u_n \to u$ *pointwise. If there exists an integrable* U *such that* $|u_n| \leq U$ *for all* n, *then* u *is integrable and* $\mathbf{E}(u_n) \to \mathbf{E}(u)$.

This theorem relates to the only place in the Lebesgue theory where a naïve formal manipulation may lead to a wrong result. The necessity of the condition $|u_n| \leq U$ is illustrated by

Example. (*e*) We take $\overline{0, 1}$ as basic interval and define expectations by the ordinary integral (with respect to length). Let

$$u_n(x) = (n+1)(n+2)x^n(1-x).$$

These functions tend pointwise to zero, but nevertheless $1 = \mathbf{E}(u_n) \to 1$. Replacing u_n by $-u_n$ it is seen that Fatou's inequality (2.9) does not necessarily hold for non-positive functions. ▶

We mention without proof a rule of ordinary calculus applicable more generally.

Fubini's theorem *for repeated integrals. If* $u \geq 0$ *is a Baire function and* F *and* G *are probability distributions then*

$$\int_{-\infty}^{+\infty} F\{dx\} \int_{-\infty}^{+\infty} u(x, y)\, G\{dy\} = \int_{-\infty}^{+\infty} G\{dy\} \int_{-\infty}^{+\infty} u(x, y)\, F\{dx\} \tag{2.11}$$

with the obvious interpretation in case of divergence. Here x and y may be interpreted as points in $\mathcal{R}^m$ and $\mathcal{R}^n$, and the theorem includes the assertion that the two inner integrals are Baire functions. (This theorem applies to arbitrary product spaces and a better version is given in section 6.)

Mean approximation theorem. *To each integrable* u *and* $\epsilon > 0$ *it is possible to find a step function* v *such that* $\mathbf{E}(|u-v|) < \epsilon$.

Instead of step functions one may use approximation by continuous functions, or by functions with arbitrarily many derivatives and vanishing outside some finite interval. [Compare the approximation theorem of example VIII,3(a).]

Note on Notations. The notation $\mathbf{E}(u)$ emphasizes the dependence on u and is practical in contexts where the interval function F is fixed. When F varies or the dependence on F is to be emphasized, the integral notation (2.1) is preferable. It applies also to integrals extended over a subset A, for the integral of u extended over A is (by definition) the same as the integral of the product $\mathbf{1}_A u$ extended over the whole space. We write

$$\int_A u(x)\, F\{dx\} = \mathbf{E}(\mathbf{1}_A u)$$

(assuming, of course, that the indicator $\mathbf{1}_A$ is a Baire function). The two sides mean exactly the same thing, the left side emphasizing the dependence on F. When $A = \overline{a, b}$ is an interval the notation $\int_a^b$ is sometimes preferred, but to render it unambiguous it is necessary to indicate whether the endpoints belong to the interval. This may be done by writing $a+$ or $a-$. ►

In accordance with the program outlined at the beginning of this section we now define the probability of a set A to equal $\mathbf{E}(\mathbf{1}_A)$ whenever $\mathbf{1}_A$ is a Baire function; for other sets no probabilities are defined. The consequences of this definition will now be discussed in the more general context of arbitrary sample spaces.

3. σ-ALGEBRAS. MEASURABILITY

In discrete sample spaces it was possible to assign probabilities to all subsets of the sample space, but in general this is neither possible nor desirable. In the preceding chapters we have considered the special case of Cartesian spaces $\mathfrak{R}^r$ and started by assigning probabilities to all intervals. It was shown in the preceding section that such an assignment of probabilities can be extended in a natural way to a larger class $\mathfrak{A}$ of sets. The principal properties of this class are:

(i) If a set A is in $\mathfrak{A}$ so is its complement $A' = \mathfrak{S} - A$.

(ii) If $\{A_n\}$ is any countable collection of sets in $\mathfrak{A}$, then also their union $\bigcup A_n$ and intersection $\bigcap A_n$ belong to $\mathfrak{A}$.

In short, $\mathfrak{A}$ is a system closed under complementation and the formation of countable unions and intersections. As was shown in section 1 this implies that also the upper and lower limit of any sequence $\{A_n\}$ of sets in $\mathfrak{A}$ again belongs to $\mathfrak{A}$. In other words, none of the familiar operations on sets in $\mathfrak{A}$

will lead us to sets outside $\mathfrak{A}$, and therefore no need will arise to consider other sets. This situation is typical inasmuch as in general probabilities will be assigned only to a class of sets with the properties (i) and (ii). We therefore introduce the following definition which applies to arbitrary spaces.

Definition 1. *A σ-algebra*[10] *is a family* $\mathfrak{A}$ *of subsets of a given set* $\mathfrak{S}$ *enjoying the properties* (i) *and* (ii).

Given any family $\mathfrak{F}$ *of sets in* $\mathfrak{S}$, *the smallest σ-algebra containing all sets in* $\mathfrak{F}$ *is called the σ-algebra generated by* $\mathfrak{F}$.

In particular, the sets generated by the intervals of $\mathfrak{R}^r$ *are called the Borel sets of* $\mathfrak{R}^r$.

That a *smallest* σ-algebra containing $\mathfrak{F}$ exists is seen by the argument used in the definition of Baire functions in section 1. Note that, $\mathfrak{S}$ being the union of any set A and its complement, *every σ-algebra contains the space* $\mathfrak{S}$.

Examples. The *largest* σ-algebra consists of *all* subsets of $\mathfrak{S}$. This algebra served us well in discrete spaces, but is too large to be useful in general. The other extreme is represented by the trivial algebra containing only the whole space and the empty set. For a non-trivial example consider the sets on the line $\mathfrak{R}^1$ with the property that if $x \in A$ then all points $x \pm 1, x \pm 2, \ldots$ belong to A (periodic sets). Obviously the family of such sets forms a σ-algebra. ►

Our experience so far shows that a principal object of probability theory is random variables, that is, certain functions in sample space. With a random variable $\mathbf{X}$ we wish to associate a distribution function, and for that purpose it is necessary that the event $\{\mathbf{X} \leq t\}$ has a probability assigned to it. This consideration leads us to

Definition 2. *Let* $\mathfrak{A}$ *be an arbitrary σ-algebra of sets in* $\mathfrak{S}$. *A real-valued function* u *on* $\mathfrak{S}$ *is called* $\mathfrak{A}$*-measurable*[11] *if for each* t *the set of all points* x *where* $u(x) \leq t$ *belongs to* $\mathfrak{A}$.

The set where $u(x) < t$ is the union of the countable sequence of sets where $u(x) \leq t - n^{-1}$, and therefore it belongs to $\mathfrak{A}$. Since $\mathfrak{A}$ is closed under complementation it follows that in the above definition the sign $\leq$ may be replaced by $<$, $>$, or $\geq$.

[10] An algebra of sets is defined similarly on replacing the work "countable" in (ii) by finite. A σ-algebra is often called "Borel algebra," but this leads to a confusion with the last part of the definition. (In it intervals may be replaced by open sets, and then this definition applies to arbitrary topological spaces.)

[11] This term is a bad misnomer since no measure is yet defined.

It follows from the definition that *the $\mathfrak{A}$-measurable functions form a closed family* in the sense introduced in section 1.

The following simple lemma is frequently useful.

Lemma 1. *A function u is $\mathfrak{A}$-measurable iff it is the uniform limit of a sequence of simple functions, that is of functions assuming only countably many values, each on a set in $\mathfrak{A}$.*

Proof. By the very definition each simple function is $\mathfrak{A}$-measurable, and because of the closure property of $\mathfrak{A}$-measurable functions every limit of simple functions is again $\mathfrak{A}$-measurable.

Conversely, let u be $\mathfrak{A}$-measurable. For fixed $\epsilon > 0$ define the set A_n as the set of all points x at which $(n-1)\epsilon < u(x) \leq n\epsilon$. Here the integer n runs from $-\infty$ to ∞. The sets A_n are mutually exclusive and their union is the whole space $\mathfrak{S}$. On the set A_n we define $\underline{\sigma}_\epsilon(x) = (n-1)\epsilon$ and $\bar{\sigma}_\epsilon(x) = n\epsilon$. In this way we obtain two functions $\underline{\sigma}_\epsilon$ and $\bar{\sigma}_\epsilon$ defined on $\mathfrak{S}$ and such that

$$(3.1) \qquad \underline{\sigma}_\epsilon \leq u \leq \bar{\sigma}_\epsilon, \qquad \bar{\sigma}_\epsilon - \underline{\sigma}_\epsilon = \epsilon$$

at all points. Obviously u is the uniform limit of $\underline{\sigma}_\epsilon$ and $\bar{\sigma}_\epsilon$ as $\epsilon \to 0$.

Lemma 2. *In $\mathfrak{R}^r$ the class of Baire functions is identical with the class of functions measurable with respect to the σ-algebra $\mathfrak{A}$ of Borel sets.*

Proof. (*a*) It is obvious that every continuous function is Borel measurable. Now these functions form a closed class, while the Baire functions form the *smallest* class containing all continuous functions. Accordingly, every Baire function is Borel measurable.

(*b*) The preceding lemma shows that for the converse it suffices to show that every *simple* Borel-measurable function is a Baire function. This amounts to the assertion that for every Borel set A the indicator $\mathbf{1}_A$ is a Baire function. Now Borel sets may be defined by saying that A is a Borel set if and only if its indicator $\mathbf{1}_A$ belongs to the *smallest* closed class containing all indicators of intervals. Since Baire functions form a closed class containing all indicators of intervals[12] it follows that $\mathbf{1}_A$ is a Baire function for every Borel set A. ►

We apply this result to the special case of the Cartesian space $\mathfrak{R}^r$. In section 2 we started from a completely additive interval function and defined $\mathbf{P}\{A\} = \mathbf{E}(\mathbf{1}_A)$ for every set A whose indicator $\mathbf{1}_A$ is a Baire function. The present setup shows that *under this procedure, the probability* $\mathbf{P}\{A\}$ *is defined iff A is a Borel set.*

[12] To see this for an open interval I, let v be a continuous function vanishing outside I and such that $0 < v(x) \leq 1$ for $x \in I$. Then $\sqrt[n]{v} \to \mathbf{1}_I$.

Approximation of Borel sets by intervals. In view of the last remark, probabilities in $\mathcal{R}^r$ are as a rule defined on the σ-algebra of Borel sets, and it is therefore interesting that *any Borel set* A *can be approximated by a set* B *consisting of finitely many intervals* in the following sense: To each $\epsilon > 0$ there exists a set C such that $\mathbf{P}\{C\} < \epsilon$ and such that outside C the sets A and B are identical (that is, a point in the complement C' belongs either to both A and B, or to neither. One may take for C the union of $A - AB$ and $B - AB$).

Proof. By the mean approximation theorem of section 2 there exists a step function $v \geq 0$ such that $\mathbf{E}(|\mathbf{1}_A - v|) < \frac{1}{2}\epsilon$. Let B be the set of those points x at which $v(x) > \frac{1}{2}$. Since v is a step function, B consists of finitely many intervals. It is easily verified that

$$\mathbf{E}|\mathbf{1}_A(x) - \mathbf{1}_B(x)| \leq 2\,\mathbf{E}|\mathbf{1}_A(x) - v(x)| < \epsilon$$

for all x. But $|\mathbf{1}_A - \mathbf{1}_B|$ is the indicator of the set C consisting of all points that belong to either A or B but not to both. The last inequality states that $\mathbf{P}\{C\} < \epsilon$, and this completes the proof. ▶

4. PROBABILITY SPACES. RANDOM VARIABLES

We are now in a position to describe the general setup used in probability. Whatever the sample space $\mathfrak{S}$ probabilities will be assigned only to the sets of an appropriate σ-algebra $\mathfrak{A}$. We therefore start with

Definition 1. *A probability measure* $\mathbf{P}$ *on a* σ*-algebra* $\mathfrak{A}$ *of sets in* $\mathfrak{S}$ *is a function assigning a value* $\mathbf{P}\{A\} \geq 0$ *to each set* A *in* $\mathfrak{A}$ *such that* $\mathbf{P}\{\mathfrak{S}\} = 1$ *and that for every countable collection of non-overlapping sets* A_n *in* $\mathfrak{A}$

$$\mathbf{P}\{\cup A_n\} = \sum \mathbf{P}\{A_k\}. \tag{4.1}$$

This property is called *complete additivity* and a probability measure may be described as a completely additive non-negative set function on $\mathfrak{A}$ subject to the norming[13] $\mathbf{P}\{\mathfrak{S}\} = 1$.

In individual cases it is necessary to choose an appropriate σ-algebra and construct a probability measure on it. The procedure varies from case to case, and it is impossible to describe a general method. Often it is possible to adapt the approach used in section 2 to construct a probability measure on the Borel sets of $\mathcal{R}^r$. A typical example is provided by sequences of independent random variables (section 6). The starting point for any

[13] The condition $\mathbf{P}\{\mathfrak{S}\} = 1$ serves norming purposes only and nothing essential changes if it is replaced by $\mathbf{P}\{\mathfrak{S}\} < \infty$. One speaks in this case of a finite measure space. In probability theory the case $\mathbf{P}\{\mathfrak{S}\} < 1$ occurs in various connections and in this case we speak of a *defective probability measure*. Even the condition $\mathbf{P}\{\mathfrak{S}\} < \infty$ may be weakened by requiring only that $\mathfrak{S}$ be the union of countably many parts $\mathfrak{S}_n$ such that $\mathbf{P}\{\mathfrak{S}_n\} < \infty$. (Length and area are typical examples.) One speaks then of *σ-finite measures*.

probabilistic problem is a sample space in which a σ-algebra with an appropriate probability measure has been selected. This leads us to

Definition 2. *A probability space is a triple* $(\mathfrak{S}, \mathfrak{A}, \mathbf{P})$ *of a sample space* $\mathfrak{S}$, *a* σ-*algebra* $\mathfrak{A}$ *of sets in it, and a probability measure* $\mathbf{P}$ *on* $\mathfrak{A}$.

To be sure, not every imaginable probability space is an interesting object, but the definition embodies all that is required for the formal setting of a theory following the pattern of the first volume, and it would be sterile to discuss in advance the types of probability spaces that may turn up in practice.

Random variables are functions on the sample space, but for purposes of probability theory we can use only functions for which a distribution function can be defined. Definition 2 of section 3 was introduced to cope with this situation, and leads to

Definition 3. *A random variable* $\mathbf{X}$ *is a real function which is measurable with respect to the underlying* σ-*algebra* $\mathfrak{A}$. *The function* F *defined by* $F(\tau) = \mathbf{P}\{\mathbf{X} \leq \tau\}$ *us called the distribution function of* $\mathbf{X}$.

The elimination of functions that are not random variables is possible because, as we shall presently see, all usual operations, such as taking sums or other functions, passages to the limit, etc., can be performed within the class of random variables without ever leaving it. Before rendering this point more precise let us remark that a random variable $\mathbf{X}$ maps the sample space $\mathfrak{S}$ into the real line $\mathfrak{R}^1$ in such a way that the set in $\mathfrak{S}$ in which $a < \mathbf{X} \leq b$ is mapped into the interval $\overline{a, b}$, with corresponding probability $F(b) - F(a)$. In this way every interval I in $\mathfrak{R}^1$ receives a probability $F\{I\}$. Instead of an interval I we may take an arbitrary Borel set Γ on $\mathfrak{R}^1$ and consider the set A of those points in $\mathfrak{S}$ at which $\mathbf{X}$ assumes a value in Γ. In symbols: $A = \{\mathbf{X} \in \Gamma\}$. It is clear that the collection of all such sets forms a σ-algebra $\mathfrak{A}_1$ which may be identical with $\mathfrak{A}$, but is usually smaller. We say that $\mathfrak{A}_1$ *is the* σ-*algebra generated by the random variable* $\mathbf{X}$. It may be characterized as the *smallest* σ-algebra in $\mathfrak{S}$ with respect to which $\mathbf{X}$ is measurable. The random variable $\mathbf{X}$ maps each set of $\mathfrak{A}_1$ into a Borel set Γ of $\mathfrak{R}^1$, and hence the relation $F\{\Gamma\} = \mathbf{P}\{A\}$ defines uniquely a probability measure on the σ-algebra of Borel sets on $\mathfrak{R}^1$. For an interval $I = \overline{a, b}$ we have $F\{I\} = F(b) - F(a)$ and so F is identical with the unique probability measure in $\mathfrak{R}^1$ associated with the distribution function F by the procedure described in section 2.

This discussion shows that as long as we are concerned with only one particular random variable $\mathbf{X}$ we may forget about the original sample space and pretend that the probability space is the line $\mathfrak{R}^1$ with the σ-algebra of Borel sets on it and the measure induced by the distribution function F. We saw that in $\mathfrak{R}^1$ the class of Baire functions coincides with the Borel measureable functions. Taking $\mathfrak{R}^1$ as sample space this means that the class of random variables coincides with the class of Borel measurable functions. Interpreted in the original sample space this means that the family of Baire functions of the random variable $\mathbf{X}$ coincides with the family of all functions that are measurable with respect to the σ-algebra $\mathfrak{A}_1$ generated by $\mathbf{X}$. Since $\mathfrak{A}_1 \subset \mathfrak{A}$ this implies that any Baire function of $\mathbf{X}$ is again a random variable.

This argument carries over without change to finite collections of random variables. Thus an r-tuple $(\mathbf{X}_1, \ldots, \mathbf{X}_r)$ maps $\mathfrak{S}$ into $\mathfrak{R}^r$ so that to an open interval in $\mathfrak{R}^r$ there corresponds the set in $\mathfrak{S}$ at which r relations of the form $a_k < \mathbf{X}_k < b_k$ are satisfied. This set is $\mathfrak{A}$-measurable because it is the intersection of r such sets. As in the case of one single variable we may now define *the σ-algebra $\mathfrak{A}_1$ generated by* $\mathbf{X}_1, \ldots, \mathbf{X}_r$ as the smallest σ-algebra of sets in $\mathfrak{S}$ with respect to which the r variables are measurable. We have then the basic

Theorem. *Any Baire function of finitely many random variables is again a random variable.*

A random variable $\mathbf{U}$ *is a Baire function of* $\mathbf{X}_1, \ldots, \mathbf{X}_r$ *if it is measurable with respect to the σ-algebra generated by* $\mathbf{X}_1, \ldots, \mathbf{X}_r$.

Examples. (*a*) On the line $\mathfrak{R}^1$ with $\mathbf{X}$ as coordinate variable, the function $\mathbf{X}^2$ generates the σ-algebra of Borel sets that are symmetric with respect to the origin (in the sense that if $x \in A$ then also $-x \in A$).

(*b*) Consider $\mathfrak{R}^3$ with $\mathbf{X}_1$, $\mathbf{X}_2$, $\mathbf{X}_3$ as coordinate variables, and the σ-algebra of Borel sets. The pair $(\mathbf{X}_1, \mathbf{X}_2)$ generates the family of all cylindrical sets with generators parallel to the third axis and whose basis are Borel sets of the $(\mathbf{X}_1, \mathbf{X}_2)$ plane. ►

Expectations

In section 2 we started from an interval function in $\mathfrak{R}^r$ and used it to construct a probability space. There we found it convenient first to define expectations (integrals) of functions and then to define the probability of a Borel set A equal to the expectation $\mathbf{E}(\mathbf{1}_A)$ of its indicator. If one starts from a probability space the procedure must be reversed: the probabilities are given and it is necessary to define the expectations of random variables in terms of the given probabilities. Fortunately the procedure is extremely simple.

As in the preceding section we say that a random variable $\mathbf{U}$ is *simple* if it assumes only countably many values $a_1, a_2, \ldots$ each on a set A_j belonging to the basic σ-algebra $\mathfrak{A}$. To such variables the discrete theory of volume

1 applies and we define the expectation of $\mathbf{U}$ by

$$\mathbf{E}(\mathbf{U}) = \sum a_k \mathbf{P}\{A_k\} \tag{4.2}$$

provided the series converges absolutely; otherwise we say that $\mathbf{U}$ has no expectation.

Given an arbitrary random variable and an arbitrary $\epsilon > 0$ we defined in (3.1) two *simple* random variable $\underline{\sigma}_\epsilon$ and $\bar{\sigma}_\epsilon$ such that $\bar{\sigma}_\epsilon = \underline{\sigma}_\epsilon + \epsilon$ and $\underline{\sigma}_\epsilon \leq \mathbf{U} \leq \bar{\sigma}_\epsilon$. With any reasonable definition of $\mathbf{E}(\mathbf{U})$ we must have

$$\mathbf{E}(\underline{\sigma}_\epsilon) \leq \mathbf{E}(\mathbf{U}) \leq \mathbf{E}(\bar{\sigma}_\epsilon) \tag{4.3}$$

whenever the variables $\underline{\sigma}_\epsilon$ and $\bar{\sigma}_\epsilon$ have expectations. Since these functions differ only by ϵ the same is true of their expectations, or else neither expectation exists. In the latter case we say that $\mathbf{U}$ has no expectation, whereas in the former case $\mathbf{E}(\mathbf{U})$ is uniquely defined by (4.3) letting $\epsilon \to 0$. In brief, since every random variable $\mathbf{U}$ is the uniform limit of a sequence of simple random variables σ_n the expectation of $\mathbf{U}$ can be defined as the limit of $\mathbf{E}(\sigma_n)$. For example, in terms of $\bar{\sigma}_\epsilon$ we have

$$\mathbf{E}(\mathbf{U}) = \lim_{\epsilon \to 0} \sum_{-\infty}^{\infty} n\,\epsilon\; \mathbf{P}\{(n-1)\epsilon < \mathbf{U} \leq n\epsilon\} \tag{4.4}$$

provided the series converges absolutely (for some, and therefore all $\epsilon > 0$). Now the probabilities occurring in (4.4) coincide with the probabilities attributed by the distribution function F of $\mathbf{U}$ to the intervals $\overline{(n-1)\epsilon, n\epsilon}$. It follows that with this change of notations our definition of $\mathbf{E}(\mathbf{U})$ reduces to that given in section 2 for

$$\mathbf{E}(\mathbf{U}) = \int_{-\infty}^{+\infty} t\, F\{dt\}. \tag{4.5}$$

Accordingly, $\mathbf{E}(\mathbf{U})$ *may be defined consistently either in the original probability space or in terms of its distribution function.* (The same remark was made in **1**; IX for discrete variables). For this reason it is superfluous to emphasize that in arbitrary probability spaces expectations share the basic properties of expectations in $\mathcal{R}^r$ discussed in section 2.

5. THE EXTENSION THEOREM

The usual starting point in the construction of probability spaces is that probabilities are assigned a priori to a restricted class of sets, and the domain of definition must be suitably extended. For example, in dealing with unending sequences of trials and recurrent events in volume **1** we were given the probabilities of all events depending on finitely many trials, but this domain of definition has to be enlarged to include events such as ruin, recurrence, and ultimate extinction. Again, the construction of measures in

$\mathfrak{R}^r$ in section 2 proceeded from an assignment of probabilities $F\{I\}$ to intervals, and this domain of definition was extended to the class of all Borel sets. The possibility of such an extension is due to a theorem of much wider applicability, and many constructions of probability spaces depend on it The procedure is as follows.

The additivity of F permits us to define without ambiguity

$$F\{A\} = \sum F\{I_k\} \tag{5.1}$$

for every set A which is the union of *finitely many non-overlapping intervals* I_k. Now these sets form an *algebra* $\mathfrak{A}_0$ (that is, unions, intersections, and complements of *finitely* many sets in $\mathfrak{A}_0$ belong again to $\mathfrak{A}_0$). From here on the nature of the underlying space $\mathfrak{R}^r$ plays no role, and we may consider an arbitrary algebra $\mathfrak{A}_0$ of sets in an arbitrary space $\mathfrak{S}$. There exists always a smallest algebra $\mathfrak{A}$ of sets containing $\mathfrak{A}_0$ which is closed also under *countable* unions and intersections. In other words, there exists a smallest σ-algebra $\mathfrak{A}$ containing $\mathfrak{A}_0$ (see definition 1 of section 3). In the construction of measures in $\mathfrak{R}^r$ the σ-algebra $\mathfrak{A}$ coincided with the σ-algebra of all Borel sets. The extension of the domain of definition of probabilities from $\mathfrak{A}_0$ to $\mathfrak{A}$ is based on the general

Extension theorem. *Let* $\mathfrak{A}_0$ *be an algebra of sets in some space* $\mathfrak{S}$. *Let* F *be a set function defined on* $\mathfrak{A}_0$ *such that* $F\{A\} \geq 0$ *for every set* $A \in \mathfrak{A}_0$, *that* $F\{\mathfrak{S}\} = 1$, *and that the addition rule* (5.1) *holds for any partition of* A *into countably many non-overlapping sets* $I_k \in \mathfrak{A}_0$.

There exists then a unique extension of F *to a countably additive set function* (*that is, to a probability measure*) *on the smallest* σ*-algebra* $\mathfrak{A}$ *containing* $\mathfrak{A}_0$.

A typical application will be given in the next section. Here we give a more general and more flexible version of the extension theorem which is more in line with the development in sections 2 and 3. We started from the expectation (2.4) for step functions (that is, functions assuming only finitely many values, each on an interval). The domain of definition of this expectation was then extended from the restricted class of step functions to a wider class including all bounded Baire functions. This extension leads directly to the Lebesgue-Stieltjes integral, and the measure of a set A is obtained as the expectation of its indicator $\mathbf{1}_A$. The corresponding abstract setup is as follows.

Instead of the algebra $\mathfrak{A}_0$ of sets we consider a class $\mathfrak{B}_0$ of functions closed under linear combinations and the operations $\cap$ and $\cup$. In other words, we suppose that if u_1 and u_2 are in $\mathfrak{B}_0$ so are the functions[14]

$$\alpha_1 u_1 + \alpha_2 u_2, \qquad u_1 \cap u_2, \qquad u_1 \cup u_2. \tag{5.2}$$

[14] Our postulates amount to requiring that $\mathfrak{B}_0$ be a linear lattice.

This implies in particular that every function u of $\mathfrak{B}_0$ can be written in the form $u = u^+ - u^-$ as the difference of two non-negative functions, namely $u^+ = u \cup 0$ and $u^- = u \cap 0$. By a *linear functional on* $\mathfrak{B}_0$ is meant an assignment of values $\mathbf{E}(u)$ to all functions of $\mathfrak{B}_0$ satisfying the addition rule

$$\mathbf{E}(\alpha_1 u_1 + \alpha_2 u_2) = \alpha_1 \mathbf{E}(u_1) + \alpha_2 \mathbf{E}(u_2). \tag{5.3}$$

The functional is *positive if* $u \geq 0$ implies $\mathbf{E}(u) \geq 0$. The *norm* of $\mathbf{E}$ is the least upper bound of $\mathbf{E}(|u|)$ for all functions $u \in \mathfrak{B}_0$ such that $|u| \leq 1$. If the constant function 1 belongs to $\mathfrak{B}_0$ the norm of $\mathbf{E}$ equals $\mathbf{E}(1)$. Finally, we say that $\mathbf{E}$ is *countably additive on* $\mathfrak{B}_0$ if

$$\mathbf{E}\Big(\sum_1^\infty u_k\Big) = \sum_1^\infty \mathbf{E}(u_k) \tag{5.4}$$

whenever $\sum u_k$ happens to be in $\mathfrak{B}_0$. An *equivalent* condition is: if $\{v_n\}$ is a sequence of functions in $\mathfrak{B}_0$ converging *monotonically* to zero, then[15]

$$\mathbf{E}(v_n) \to 0. \tag{5.5}$$

Given the class $\mathfrak{B}_0$ of functions there exists a *smallest class* $\mathfrak{B}$ *containing* $\mathfrak{B}_0$ *and closed under pointwise passages to the limit.* [It is automatically closed under the operations (5.2).] An alternative formulation of the extension theorem is as follows.[16] *Every positive countably additive linear functional of norm* 1 *on* $\mathfrak{B}_0$ *can be uniquely extended to a positive countably additive linear functional of norm* 1 *on all bounded (and many unbounded) functions of* $\mathfrak{B}$.

As an example for the applicability of this theorem we prove the following important result.

F. Riesz representation theorem.[17] *Let* $\mathbf{E}$ *be a positive linear functional of norm* 1 *on the class of continuous functions on* $\mathfrak{R}^r$ *vanishing at infinity.*[18]

---

[15] To prove the equivalence of (5.4) and (5.5) it suffices to consider the case $u_k \geq 0$, $v_k \geq 0$. Then (5.4) follows from (5.5) with $v_n = u_{n+1} + u_{n+2} + \cdots$ and (5.5) follows from (5.4) on putting $u_k = v_k - v_{k+1}$ (that is, $\sum u_k = v_1$).

[16] The basic idea of the proof (going back to Lebesgue) is simple and ingenious. It is not difficult to see that if two sequences $\{u_n\}$ and $\{u_n'\}$ of functions in $\mathfrak{B}_0$ converge *monotonically* to the same limit u then $\mathbf{E}(u_n)$ and $\mathbf{E}(u_n')$ tend to the same limit. For such monotone limits u we can therefore define $\mathbf{E}(u) = \lim \mathbf{E}(u_n)$. Consider now the class $\mathfrak{B}_1$ of functions u such that to each $\epsilon > 0$ there exist two functions $\underline{u}$ and $\bar{u}$ which are either in $\mathfrak{B}_0$ or are monotone limits of sequences $\mathfrak{B}_0$ and such that $\underline{u} < u < \bar{u}$ and $\mathbf{E}(\bar{u}) - \mathbf{E}(\underline{u}) < \epsilon$. The class $\mathfrak{B}_1$ is closed under limits and for functions in $\mathfrak{B}_1$ the definition of $\mathbf{E}(u)$ is obvious since we must have $\mathbf{E}(\underline{u}) \leq \mathbf{E}(u) \leq \mathbf{E}(\bar{u})$.

The *tour de force* in this argument is that the class $\mathfrak{B}_1$ is usually greater than $\mathfrak{B}$ and the simple proof is made possible by proving more than is required. (For a comparison between $\mathfrak{B}$ and $\mathfrak{B}_1$ see section 7.)

[17] Valid for arbitrary locally compact spaces. For an alternative proof see V,1.

[18] u vanishes at infinity if for given $\epsilon > 0$ there exists a sphere (compact set) outside which $|u(x)| < \epsilon$.

There exists a measure **P** *on the σ-algebra of Borel sets with* $\mathbf{P}\{\mathfrak{R}^r\} = 1$ *such that* $\mathbf{E}(u)$ *coincides with the integral of* u *with respect to* **P**.

In other words, our integrals represent the most general positive linear functionals.

Proof. The crucial point is that if a sequence $\{v_n\}$ of continuous functions vanishing at infinity converges monotonically to zero, the convergence is automatically *uniform*. Assume $v_n \geq 0$ and put $\|v_n\| = \max v_n(x)$. Then $\mathbf{E}(v_n) \leq \|v_n\|$, and so the countable additivity condition (5.5) is satisfied. By the extension theorem **E** can be extended to all bounded Baire functions and putting $\mathbf{P}\{A\} = \mathbf{E}(\mathbf{1}_A\}$ we get a measure on the σ-algebra of Borel sets. Given the measures $\mathbf{P}\{A\}$ we saw that the Lebesgue-Stieltjes integral is uniquely characterized by the double inequality (4.3) and this shows that for u continuous and vanishing at infinity this integral coincides with the given functional $\mathbf{E}(u)$. ►

6. PRODUCT SPACES. SEQUENCES OF INDEPENDENT VARIABLES

The notion of combinatorial product spaces (**1**; V,4) is basic for probability theory and is used every time one speaks of repeated trials. Describing a point in the plane $\mathfrak{R}^2$ by two coordinates means that $\mathfrak{R}^2$ is taken as the combinatorial product of its two axes. Denote the two coordinate variables by **X** and **Y**. Considered as functions in the plane they are Baire functions, and if a probability measure **P** is defined on the σ-algebra of Borel sets in $\mathfrak{R}^2$ the two distribution functions $\mathbf{P}\{\mathbf{X} \leq x\}$ and $\mathbf{P}\{\mathbf{Y} \leq y\}$ exist. They induce probability measures on the two axes called the marginal distributions (or projections). In this description the plane appears as the primary notion, but frequently the inverse procedure is more natural. For example, when we speak of two independent random variables with given distributions, the two marginal distributions are the primary notion and probabilities in the plane are derived from it by "the product rule." The procedure is not more complicated in the general setup than for the plane.

Consider then two arbitrary probability spaces, that is, we are given two sample spaces $\mathfrak{S}^{(1)}$ and $\mathfrak{S}^{(2)}$, two σ-algebras $\mathfrak{A}^{(1)}$ and $\mathfrak{A}^{(2)}$ of sets in $\mathfrak{S}^{(1)}$ and $\mathfrak{S}^{(2)}$, respectively, and probability measures $\mathbf{P}^{(1)}$ and $\mathbf{P}^{(2)}$ defined on them. The combinatorial product $(\mathfrak{S}^{(1)}, \mathfrak{S}^{(2)})$ is the set of all ordered pairs $(x^{(1)}, x^{(2)})$ where $x^{(i)}$ is a point in $\mathfrak{S}^{(i)}$. Among the sets in this product space we consider the "rectangles," that is, the combinatorial products $(A^{(1)}, A^{(2)})$ of sets $A^{(i)} \in \mathfrak{A}^{(i)}$. With sets of this form we wish to associate probabilities by the product rule

$$\mathbf{P}\{(A^{(1)}, A^{(2)}\} = \mathbf{P}^{(1)}\{A^{(1)}\}\, \mathbf{P}^{(2)}\{A^{(2)}\}. \tag{6.1}$$

Now sets which are unions of finitely many non overlapping rectangles form an algebra $\mathfrak{U}_0$, and (6.1) defines in a unique way a countably additive function on it. Accordingly, by the extension theorem *there exists a unique probability measure* **P** *defined on the smallest σ-algebra containing all rectangles and such that the probabilities of rectangles are given by the product rule* (6.1). *This smallest σ-algebra containing all rectangles will be denoted by* $\mathfrak{U}^{(1)} \times \mathfrak{U}^{(2)}$, *and the measure will be called product measure.*

Of course, other probability measures can be defined on the product space, for example in terms of conditional probabilities. Under any circumstances the underlying σ-algebra $\mathfrak{U}$ of sets will be at least as large as $\mathfrak{U}^{(1)} \times \mathfrak{U}^{(2)}$, and it is rarely necessary to go beyond this algebra. The following discussion of random variables is valid whenever the underlying algebra $\mathfrak{U}$ is given gy $\mathfrak{U} = \mathfrak{U}^{(1)} \times \mathfrak{U}^{(2)}$.

The notion of random variable (measurable function) is relative to the underlying σ-algebra and with our setup for product spaces we must distinguish between random variables in the product space and those on $\mathfrak{S}^{(1)}$ and $\mathfrak{S}^{(2)}$. The relationship between these three classes is fortunately extremely simple. If u and v are random variables on $\mathfrak{S}^{(1)}$ and $\mathfrak{S}^{(2)}$ we consider in the product space the function w which at the point $(x^{(1)}, x^{(2)})$ takes on the value

$$w(x^{(1)}, x^{(2)}) = u(x^{(1)}) \cdot v(x^{(2)}). \tag{6.2}$$

We show that *the class of random variables in the product space* $(\mathfrak{S}^{(1)}, \mathfrak{S}^{(2)})$ *is the smallest class of finite-valued functions closed under pointwise passages to the limit and containing all linear combinations of functions of the form* (6.2).

To begin with, it is clear that each factor on the right in (6.2) is a random variable even when considered as a function on the product space. It follows that w is a random variable, and hence the class of random variables in $(\mathfrak{S}^{(1)}, \mathfrak{S}^{(2)})$ is at least as extensive as claimed. On the other hand, the random variables form the smallest class of functions that is closed under passages to the limit and contains all linear combinations of indicators of rectangles. Such indicators are of the form (6.2) and therefore the class of random variables cannot be larger than claimed.

The special case of the product of two spaces $\mathfrak{R}^m$ and $\mathfrak{R}^n$ with probability measures F and G occurred indirectly in connection with Fubini's theorem (2.11) concerning repeated integrals. We can now state the more general theorem, which is *not* restricted to $\mathfrak{R}^r$.

Fubini's theorem for product measures. *For arbitrary non-negative Baire functions* u, *the integral of* u *with respect to the product measure equals the repeated integrals in* (2.11).

(It is understood that the integrals may diverge. The theorem is obvious for simple functions and follows in general by the approximation procedure employed repeatedly.) The generalization to product spaces with three or more factors is too obvious to require comment.

We turn to the problem of *infinite sequences of random variables*, which we encountered in volume **1** in connection with unlimited sequences of Bernoulli trials, random walks, recurrent events, etc., and again in chapter III in connection with normal stochastic processes. Nothing need be said when infinitely many random variables are defined on a given probability space. For example, the real line with the normal distribution is a probability space and $\{\sin nx\}$ is an infinite sequence of random variables on it. We are here concerned with the situation when the probabilities are to be defined in terms of the given random variables. More precisely, our problem is as follows.

Let $\mathfrak{R}^\infty$ denote the space whose points are infinite sequences of real numbers $(x_1, x_2, \ldots)$, (that is, $\mathfrak{R}^\infty$ is a denumerable combinatorial product or real lines). We denote the nth coordinate variable by $\mathbf{X}_n$ (that is, $\mathbf{X}_n$ is the function in $\mathfrak{R}^\infty$ which at the point $x = (x, x_2, \ldots)$ assumes the value x_n). We suppose that we are given the probability distributions for $\mathbf{X}_1$, $(\mathbf{X}_1, \mathbf{X}_2)$, $(\mathbf{X}_1, \mathbf{X}_2, \mathbf{X}_3), \ldots$ and wish to define appropriate probabilities in $\mathfrak{R}^\infty$. Needless to say, the given distributions must be mutually consistent in the sense that the distributions of $(\mathbf{X}_1, \ldots, \mathbf{X}_n)$ appear as marginal distributions for $(\mathbf{X}_1, \ldots, \mathbf{X}_{n+1})$, and so on.

Let us now formalize the intuitive notion of an "event determined by the outcome of finitely many trials." We agree to say that *a* set A *in* $\mathfrak{R}^\infty$ *depends only on the first* r *coordinates iff there exists a Borel set* A_r *in* $\mathfrak{R}^r$ *such that* $x = (x_1, x_2, \ldots)$ *belongs to* A *iff* $(x_1, \ldots, x_r)$ *belongs to* A_r. The standard situation in probability is that the probabilities for such sets are prescribed, and we face the problem of extending this domain of definition. We state without proof the basic theorem derived (in slightly greater generality) by A. Kolmogorov in his now classical axiomatic foundation of probability theory (1933). It anticipated and stimulated the development of modern measure theory.

Theorem 1. *A consistent system of probability distributions for* $\mathbf{X}_1$, $(\mathbf{X}_1, \mathbf{X}_2)$, $(\mathbf{X}_1, \mathbf{X}_2, \mathbf{X}_3), \ldots$ *admits of a unique extension to a probability measure on* $\mathfrak{U}$, *the smallest* σ*-algebra of sets in* $\mathfrak{R}^\infty$ *containing all sets depending only on finitely many coordinates.*[19]

The important point is that all probabilities are defined by successive passages to the limit starting with finite-dimensional sets. *Every set* A

---

[19] The theorem applies more generally to products of locally compact spaces; for example, the variables $\mathbf{X}_n$ may be interpreted as vector variables (points in $\mathfrak{R}^r$).

in $\mathfrak{U}$ *can be approximated by finite-dimensional sets* in the following sense. Given $\epsilon > 0$ there exists an n and a set A_n depending only on the first n coordinates and such that

$$\text{(6.3)} \qquad \mathbf{P}\{A - A \cap A_n\} < \epsilon, \qquad \mathbf{P}\{A_n - A \cap A_n\} < \epsilon.$$

In other words, the set of those points that belong to either A or A_n but not to both has probability $< 2\epsilon$. It follows that the sets A_n can be chosen such that

$$\text{(6.4)} \qquad \mathbf{P}\{A_n\} \to \mathbf{P}\{A\}.$$

Theorem 1 enables us to speak of *an infinite sequence of mutually independent random variables with arbitrarily prescribed distributions.* Such sequences did in fact occur in volume **1**, but we had to be careful to define the probabilities in question by specific passages to the limit, whereas theorem 1 provides the desirable freedom of motion. This point is well illustrated by the following two important theorems due, respectively, to A. Kolmogorov (1933) and to E. Hewitt and L. J. Savage (1955). They are typical for probabilistic arguments and play a central role in many contexts.

Theorem 2. (*Zero-or-one law for tail events.*) *Suppose that the variables* $\mathbf{X}_k$ *are mutually independent and that for each* n *the event* A *is independent of*[20] $\mathbf{X}_1, \ldots, \mathbf{X}_n$. *Then either* $\mathbf{P}\{A\} = 0$ *or* $\mathbf{P}\{A\} = 1$.

Proof. In principle the variables $\mathbf{X}_k$ can be defined in an arbitrary probability space, but they map this space into the product space $\mathcal{R}^\infty$ in which they serve as coordinate variables. There is therefore no loss of generality in departing from the setup described in this section. With the notations used in (6.3) the sets A and A_n are independent and so the first inequality implies $\mathbf{P}\{A\} - \mathbf{P}\{A\}\mathbf{P}\{A_n\} < \epsilon$. Therefore $\mathbf{P}\{A\} = \mathbf{P}^2\{A\}$. ▶

Example. (*a*) The series $\Sigma\mathbf{X}_n$ converges with probability zero or one. Similarly, the set of those points where $\limsup \mathbf{X}_n = \infty$ has either probability zero or one. ▶

Theorem 3. (*Zero-or-one law for symmetric events.*) *Suppose that the variables* $\mathbf{X}_k$ *are mutually independent and have a common distribution. If the set* A *is invariant under finite permutations of the coordinates*[21] *then either* $\mathbf{P}\{A\} = 0$ *or* $\mathbf{P}\{A\} = 1$.

---

[20] More precisely, A is independent of every event defined in terms of $\mathbf{X}_1, \ldots, \mathbf{X}_n$. In other words, the indicator of A is a random variable independent of $\mathbf{X}_1, \ldots, \mathbf{X}_n$.

[21] More precisely, if $(a_1, a_2, \ldots)$ is a point of A and n_1 and n_2 are two arbitrary integers it is supposed that A contains also the point obtained by exchanging a_{n_1} and a_{n_2} while leaving all other coordinates fixed. This condition extends automatically to permutations involving k coordinates.

Proof. As in the last proof we use the $\mathbf{X}_k$ as coordinate variables and refer to the sets A_n occurring in (6.3). Let B_n be the set obtained from A_n by reversing the first $2n$ coordinates and leaving the others fixed. By hypothesis then (6.3) remains valid also when A_n is replaced by B_n. It follows that the set of points belonging to either A or $A_n \cap B_n$ but not to both has probability $< 4\epsilon$, and therefore

$$\mathbf{P}\{A_n \cap B_n\} \to \mathbf{P}\{A\}. \tag{6.5}$$

Furthermore A_n depends only on the first n coordinates and hence B_n depends only on the coordinates number $n + 1, \ldots, 2n$. Thus A_n and B_n are independent and from (6.5) we conclude again that $\mathbf{P}\{A\} = \mathbf{P}^2\{A\}$. ▶

Example. (*b*) Put $\mathbf{S}_n = \mathbf{X}_1 + \cdots + \mathbf{X}_n$ and let A be the event $\{\mathbf{S}_n \in I \text{ i. o.}\}$ where I is an arbitrary interval on the line. Then A is invariant under finite permutations. [For the notation see example 1(*a*).] ▶

7. NULL SETS. COMPLETION

Usually a set of probability zero is negligible and two random variables differing only on such a null set are "practically the same." More formally they are called *equivalent*. This means that all probability relations remain unchanged if the definition of a random variable is changed on a null set, and hence we can permit a random variable not to be defined on a null set. A typical example is the epoch of the first occurrence of a recurrent event: with unit probability it is a number, but with probability zero it remains undefined (or is called ∞). Thus we are frequently dealing with classes of equivalent random variables rather than with individual variables, but it is usually simplest to choose a convenient representative rather than to speak of equivalence classes.

Null sets give rise to the only point where our probabilistic setup goes against intuition. The situation is the same in all probability spaces, but it suffices to describe it on the line. With our setup, probabilities are defined only for Borel sets, and in general a Borel set contains many subsets that are not Borel sets. Consequently, a null set may contain sets for which no probability is defined, contrary to the natural expectation that every subset of a null set should be a null set. The discrepancy has no serious effects and it is easily remedied. In fact, suppose we introduce the *postulate: if* $A \subset B$ *and* $\mathbf{P}\{B\} = 0$, *then* $\mathbf{P}\{A\} = 0$. It compels us to enlarge the σ-algebra $\mathfrak{U}$ of Borel sets (at least) to the smallest σ-algebra $\mathfrak{U}_1$ containing all sets of $\mathfrak{U}$ and all subsets of null sets. A direct description is as follows. A set A belongs to $\mathfrak{U}_1$ iff it differs only by a null set[22] from some Borel set A^0.

[22] More precisely, it is required that both $A - A \cap A^0$ and $A^0 - A \cap A^0$ be contained in a null set.

The domain of definition can be extended from $\mathfrak{U}$ to $\mathfrak{U}_1$ simply by putting $\mathbf{P}\{A\} = \mathbf{P}\{A^0\}$. It is almost trivial that this definition is unique and leads to a completely additive measure on $\mathfrak{U}_1$. By this device we have obtained a probability space satisfying our postulate and in which the probabilities of Borel sets remain unchanged.

The construction so described is called the *Lebesgue completion* (of the given probability space). This completion is natural in problems concerned with a unique basic probability distribution. For this reason the length of intervals on $\mathfrak{R}^1$ is usually completed to a Lebesgue measure which is not restricted to Borel sets. But the completion would invite trouble when one deals with families of distributions (for example with infinite sequences of Bernoulli trials with unspecified probability p). In fact, $\mathfrak{U}_1$ depends on the underlying distribution, and so a random variable with respect to $\mathfrak{U}_1$ may stop being a random variable when the probabilities are changed.

Example. Let $a_1, a_2, \ldots$ be a sequence of points on $\mathfrak{R}^1$ carrying probabilities $p_1, p_2, \ldots$ where $\Sigma\, p_k = 1$. The complement of $\{a_j\}$ has probability zero and so $\mathfrak{U}_1$ *contains all sets* of $\mathfrak{R}^1$. Every bounded function u is now a random variable with expectation $\sum p_k u(a_k)$ but it would be dangerous to deal with "arbitrary functions" when the underlying distribution is not discrete. ▶

CHAPTER V

Probability Distributions in $\mathfrak{R}^r$

This chapter develops the notion of probability distribution in the r-dimensional space $\mathfrak{R}^r$. Conceptually the notion is based on the integration theory outlined in the last chapter, but in fact no sophistication is required to follow the development because the notions and formulas are intuitively close to those familiar from volume **1** and from the first three chapters.

The novel feature of the theory is that (in contrast to discrete sample spaces) not *every* set carries a probability and not *every* function serves as random variable. Fortunately this theoretical complication is not noticeable in practice because we can start from intervals and continuous functions, respectively, and restrict our attention to sets and functions that can be derived from them by elementary operations and (possibly infinitely many) passages to the limit. This delimits the classes of Borel sets and Baire functions. Readers interested in facts rather than logical connections need not worry about the precise definitions (given in chapter IV). Rather they should rely on their intuition and assume that all sets and functions are "nice." The theorems are so simple[1] that elementary calculus should suffice for an understanding. The exposition is rigorous under the convention that *the words set and function serve as abbreviations for Borel set and Baire function.*

An initial reading should be restricted to sections 1–4 and 9. Sections 5–8 contain tools and inequalities to which one may refer when occasion arises. The last sections develop the theory of conditional distributions and expectations more fully than required for the present volume where the results are used only incidentally for martingales in VI,11 and VII,9.

[1] It should be understood that this simplicity cannot be achieved by any theory restricted to the use of continuous functions or any other class of "nice" functions. For example, in II,(8.3) we defined a density φ by an infinite series. To establish conditions for φ to be nice would be tedious and pointless, but the formula is obvious in simple cases and the use of Baire functions amounts to a substitute for a vague "goes through generally."—Incidentally, the few occasions where the restriction to Baire functions is not trivial will be pointed out. (The theory of convex functions in 8.b is an example.)

1. DISTRIBUTIONS AND EXPECTATIONS

Even the most innocuous use of the term random variable may contain an indirect reference to a complicated probability space or a complex conceptual experiment. For example, the theoretical model may involve the positions and velocities of 10^{28} particles, but we concentrate our attention on the temperature and energy. These two random variables map the original sample space into the plane $\mathcal{R}^2$, carrying with them their probability distributions. In effect we are dealing with a problem in two dimensions and the original sample space looms dimly in the background. The finite-dimensional Cartesian spaces $\mathcal{R}^r$ therefore represent the most important sample spaces, and we turn to a systematic study of the appropriate probability distributions.

Let us begin with the line $\mathcal{R}^1$. The intervals defined by $a < x < b$ and $a \leq x \leq b$ will be denoted by $\overline{a, b}$ and $\overline{\vert a, b \vert}$. (We do not exclude the limiting case of a closed interval reducing to a single point. Half-open intervals are denoted by $\overline{a, b\,\vert}$ and $\overline{a, b}$. In one dimension all random variables are functions of the coordinate variable $\mathbf{X}$ (that is, the function which at the point x assumes the value x). All probabilities are therefore expressible in terms of the distribution function

$$F(x) = \mathbf{P}\{\mathbf{X} \leq x\}, \qquad -\infty < x < \infty. \tag{1.1}$$

In particular, $I = \overline{a, b\,\vert}$ carries the probability $\mathbf{P}\{I\} = F(b) - F(a)$. The flexible standard notation $\mathbf{P}\{\ \}$ is impractical when we are dealing with varying distributions. A new letter would be uneconomical, and the notation $\mathbf{P}_F\{\ \}$ to indicate the dependence on F is too clumsy. It is by far the simplest to *use the same letter* F *both for the point function* (1.1) *and the corresponding interval function*, and we shall write $F\{I\}$ instead of $\mathbf{P}\{I\}$. In other words, the use of braces $\{\ \}$ will indicate that the argument in $F\{A\}$ is an interval or set, and that F appears as a function of intervals (or measure). When parentheses are used the argument in $F(a)$ is a point. The relationship between the point function $F(\ )$ and the interval function $F\{\ \}$ is indicated by

$$F(x) = F\{\overline{-\infty, x\,\vert}\}, \qquad F\{\overline{a, b\,\vert}\} = F(b) - F(a). \tag{1.2}$$

Actually the notion of the point function $F(x)$ is redundant and serves merely for the convenience of analytical and graphical representation. The primary notion is the assignment of probabilities to *intervals*. The point function $F(\ )$ is called the *distribution function* of the interval function $F\{\ \}$. The symbols $F(\ )$ and $F\{\ \}$ refer to the same thing and no confusion can arise by references to "the probability distribution

F." One should get used to thinking in terms of *interval* functions or measures and using the distribution function only for graphical descriptions.[2]

Definition. *A point function* F *on the line is a distribution function if*
(i) F *is non-decreasing, that is,* $a < b$ *implies* $F(a) \leq F(b)$
(ii) F *is right continuous,*[3] *that is,* $F(a) = F(a+)$
(iii) $F(-\infty) = 0$ *and* $F(\infty) < \infty$.
F *is a probability distribution function if it is a distribution function and* $F(\infty) = 1$. *Furthermore,* F *is defective if* $F(\infty) < 1$.

We proceed to show that every distribution function induces an assignment of probabilities to all sets on the line. The first step consists in assigning probabilities to intervals. Since F is monotone, a left limit $F(a-)$ exists for each point a. We define an interval function $F\{I\}$ by

$$(1.3)\qquad \begin{aligned} F\{\overline{\vert a, b \vert}\} &= F(b) - F(a-), \qquad & F\{\overline{a, b}\} &= F(b-) - F(a) \\ F\{\overline{a, b \vert}\} &= F(b) - F(a), \qquad & F\{\overline{\vert a, b}\} &= F(b-) - F(a-). \end{aligned}$$

For the interval $\overline{\vert a, a \vert}$ reducing to the single point $F\{\overline{\vert a, a \vert}\} = F(a) - F(a-)$, which is the jump of F at the point a. (It will be seen presently that F is continuous "almost everywhere.")

To show that the assignment of values (1.3) to intervals satisfies the requirements of probability theory we prove a simple lemma (which readers may accept as being intuitively obvious).

Lemma 1. (*Countable additivity.*) *If an interval* I *is the union of countably many non-overlapping intervals* $I_1, I_2, \ldots,$ *then*

$$(1.4)\qquad F\{I\} = \sum F\{I_k\}.$$

Proof. The assertion is trivial in the special case $I = \overline{a, b\vert}$ and $I_1 = \overline{a, a_1\vert}$, $I_2 = \overline{a_1, a_2\vert}, \ldots, I_n = \overline{a_{n-1}, b\vert}$. The most general *finite* partition of $I = \overline{a, b}$ is obtained from this by redistributing the endpoints a_k from one subinterval to another, and so the addition rule (1.4) holds for finite partitions.

In considering the case of infinitely many intervals I_k it suffices to assume

[2] Pedantic care in the use of notations seems advisable for an introductory book, but it is hoped that readers will *not* indulge in this sort of consistency and will find the courage to write $F(I)$ and $F(x)$ indiscriminately. No confusion will result and it is (fortunately) quite customary in the best mathematics to use the same symbol (in particular 1 and =) on the same page in several meanings.

[3] As usual we denote by $f(a+)$ the limit, if it exists, of $f(x)$ as $x \to a$ in such a way that $x > a$, and by $f(\infty)$ the limit of $f(x)$ as $x \to \infty$. Similarly for $f(a-)$ and $f(-\infty)$. This notation carries over to higher dimensions.

I closed. In consequence of the right continuity of the given distribution function F it is possible to find an open interval $I_k^{\#}$ containing I_k and such that $0 \leq F\{I_k^{\#}\} - F\{I_k\} \leq \epsilon \cdot 2^{-k}$ for preassigned $\epsilon > 0$. Now there exists a finite collection $I_{k_1}^{\#}, \ldots, I_{k_n}^{\#}$ covering I and hence

$$(1.5)\quad F\{I\} \leq F\{I_{k_1}^{\#}\} + \cdots + F\{I_{k_n}^{\#}\} \leq F\{I_1\} + \cdots + F\{I_n\} + \epsilon.$$

Thus

$$(1.6)\qquad F\{I\} \leq \sum F\{I_k\}.$$

But the reversed inequality is also true since to each n there exists a *finite* partition of I containing $I_1, \ldots, I_n$. This concludes the proof. ►

As explained in IV,2 it is now possible to define

$$(1.7)\qquad F\{A\} = \sum F\{A_k\}$$

for every set A consisting of finitely or denumerably many disjoint intervals A_k. Intuition leads one to expect that every set can be approximated by such unions of intervals, and measure theory justifies this feeling.[4] Using the natural approximations and passages to the limit it is possible to extend the definition of F to all sets in such a way that the countable additivity property (1.7) is preserved. This extension is unique, and the resulting assignment is called a probability distribution or measure.

Note on terminology. In the literature the term distribution is used loosely in various meanings, and so it is appropriate here to establish the usage to which we shall adhere.

A *probability distribution*, or *probability measure*, is an assignment of numbers $F\{A\} \geq 0$ to sets subject to condition (1.7) of countable additivity and the norming $F\{\overline{-\infty, \infty}\} = 1$. More general measures (or mass distributions) are defined by dropping the norming condition; the Lebesgue measure (or ordinary length) is the most notable example.

As will be recalled from the theory of recurrent events in volume **1**, we have sometimes to deal with measures attributing to the line a total mass $p = F\{\overline{-\infty, \infty}\} < 1$. Such a measure will be called *defective probability* measure with *defect* $1 - p$. For stylistic clarity and emphasis we shall occasionally speak of *proper* probability distributions, but the adjective proper is redundant.

The argument of a measure $m\{A\}$ is a set and is indicated by braces. With every bounded measure m there is associated its *distribution function*, that is, a point function defined by $m(x) = m\{\overline{-\infty, x}|\}$. It will be denoted by the same letter with the argument in parentheses. The dual use of the

[4] The convention that the words set and function serve as abbreviations for *Borel set* and *Baire function* should be borne in mind.

same letter can cause no confusion, and by the same token the term distribution may stand as abbreviation both for a probability distribution and its distribution function.

In **1**; IX a *random variable* was defined as a real function on the sample space, and we continue this usage. When the line serves as sample space every real function becomes a random variable. The coordinate variable $\mathbf{X}$ is basic, and all other random variables can be expressed as functions of it. The distribution function of the random variable u is defined by $\mathbf{P}\{u(\mathbf{X}) \leq x\}$ and can be expressed in terms of the distribution F of the coordinate variable $\mathbf{X}$. For example, $\mathbf{X}^3$ has the distribution function given by $F(\sqrt[3]{x})$.

A function u is called *simple* if it assumes only countably many values $a_1, a_2, \ldots$. If A_n denotes the set on which u equals a_n we define the *expectation* $\mathbf{E}(u)$ by

$$\mathbf{E}(u) = \sum a_k F\{A_k\} \tag{1.8}$$

provided the series converges absolutely. In the contrary case u is said not to be integrable with respect to F. Thus u has an expectation iff $\mathbf{E}(|u|)$ exists. Starting from the definition (1.8) we define the expectation for any arbitrary bounded function u as follows. Choose $\epsilon > 0$, and denote by A_n the set of those points x at which $(n-1)\epsilon < u(x) \leq n\epsilon$. With any reasonable definition of $\mathbf{E}(u)$ we must have

$$\sum (n-1)\epsilon \cdot F\{A_n\} \leq \mathbf{E}(u) \leq \sum n\epsilon \cdot F\{A_n\}. \tag{1.9}$$

(The extreme members represent the expectations of two approximating simple functions $\underline{\sigma}$ and $\bar{\sigma}$ such that $\underline{\sigma} \leq u \leq \bar{\sigma}$ and $\bar{\sigma} - \underline{\sigma} = \epsilon$.) Because of the assumed boundedness of u the two series in (1.9) contain only finitely many non-zero terms, and their difference equals $\epsilon \sum F\{A_n\} = \epsilon$. Replacing ϵ by $\frac{1}{2}\epsilon$ will increase the first term in (1.9) and decrease the last. It is therefore not difficult to see that as $\epsilon \to 0$ the two extreme members in (1.9) tend to the same limit, and this limit defines $\mathbf{E}(u)$. For unbounded u the same procedure applies provided the two series in (1.9) converge *absolutely*; otherwise $\mathbf{E}(u)$ remains undefined.

The expectation defined in this simple way is called the *Lebesgue-Stieltjes integral of* u with respect to F. When it is desirable to emphasize the dependence of the expectation on F the integral notation is preferable and we write alternatively

$$\mathbf{E}(u) = \int_{-\infty}^{+\infty} u(x)\, F\{dx\} \tag{1.10}$$

with x appearing as dummy variable. Except on rare occasions we shall be concerned only with piecewise continuous or monotone integrands such that the sets A_n will reduce to unions of finitely many intervals. The sums in (1.9) are then simple rearrangements of the upper and lower sums used in the elementary definition of ordinary integrals. The general Lebesgue-Stieltjes integral shares the basic properties of the ordinary integral and has the additional advantage that formal operations and passages to the limit require less care. Our use of expectations will be limited to situations so simple that no general theory will be required to follow the individual steps. The reader interested in the theoretical background and the basic facts is referred to chapter IV.

Examples. (*a*) Let F be a discrete distribution attributing weights $p_1, p_2, \ldots$ to the points $a_1, a_2, \ldots$. Then clearly $\mathbf{E}(u) = \Sigma\, u(a_k)p_k$ whenever the series converges absolutely. This is in agreement with the definition in **1**; IX.

(*b*) For a distribution defined by a continuous density

$$\mathbf{E}(u) = \int_{-\infty}^{+\infty} u(x)f(x)\, dx \tag{1.11}$$

provided the integral converges absolutely. For the general notion of density see section 3. ▶

The *generalization to higher dimensions* can be described in a few words. In $\mathfrak{R}^2$ a point x is a pair of real numbers, $x = (x_1, x_2)$. Inequalities are to be interpreted coordinate-wise;[5] thus $a < b$ means $a_1 < b_1$ and $a_2 < b_2$ (or "a lies southwest of b"). This induces only a partial ordering, that is, two points a and b need not stand in either of the two relations $a < b$ or $a \geq b$. We reserve the word *interval* for the sets defined by the four possible types of double inequalities $a < x < b$, etc. They are rectangles parallel to the axes which may degenerate into segments or points.

The only novel feature is that the two-dimensional interval $\overline{a, c}$ with $a < b < c$ is not the union of $\overline{a, b}\rvert$ and $\overline{b, c}\rvert$. Corresponding to an interval function assigning the value $F\{I\}$ to the interval I we may introduce its distribution function defined as before by $F\{x\} = F\{\overline{-\infty, x}\}$, but an expression of $F\{\overline{a, b}\rvert\}$ in terms of this distribution function involves all four vertices of the interval. In fact, considering the two infinite strips parallel to the x_2-axis and with the sides of the rectangle $\overline{a, b}\rvert$ as bases one sees immediately that $F\{\overline{a, b}\rvert\}$ is given by the so-called *mixed difference*

$$F\{\overline{a, b}\rvert\} = F(b_1, b_2) - F(a_1, b_2) - F(b_1, a_2) + F(a_1, a_2). \tag{1.12}$$

[5] This notation was introduced in III,5.

For a distribution function the right side is non-negative. This implies that $F(x_1, x_2)$ depends monotonically on x_1 and x_2, but such monotonicity does not guarantee the positivity of (1.12). (See problem 4.)

The limited value of the use of distribution functions in higher dimensions is apparent: were it not for the analogy with $\mathcal{R}^1$ all considerations would probably be restricted to interval functions. Formally the definition of distribution functions in $\mathcal{R}^1$ carries over to $\mathcal{R}^2$ if the condition of monotonicity (i) is replaced by the condition that for $a \leq b$ the mixed difference in (1.12) be non-negative. Such a distribution function induces an interval function as in (1.3) except that again the mixed differences take over the role of the simple differences in $\mathcal{R}^1$. Lemma 1 and its proof remain valid.[6]

A simple, but conceptually important, property of expectations is sometimes taken for granted. Any function $u(\mathbf{X}) = u(\mathbf{X}_1, \mathbf{X}_2)$ of the two coordinate variables is a random variable and as such it has a distribution function G. The expectation $\mathbf{E}(u(\mathbf{X}))$ is now defined in two ways; namely, as the integral of $u(x_1, x_2)$ with respect to the given probability in the plane, but also by

$$\mathbf{E}(u) = \int_{-\infty}^{\infty} y\, G\{dy\} \tag{1.13}$$

in terms of the distribution function G of u. The two definitions are equivalent by the very definition of the former integral by the approximating sums IV,(4.3).[7] The point is that the expectation of a random variable $\mathbf{Z}$ (if it exists) has an intrinsic meaning although $\mathbf{Z}$ may be considered as a function either on the original probability space $\mathfrak{S}$ or on a space obtained by an appropriate mapping of $\mathfrak{S}$; in particular, $\mathbf{Z}$ itself maps $\mathfrak{S}$ on the line where it becomes the coordinate variable.

From this point on there is no difference between the setups in $\mathcal{R}^1$ and $\mathcal{R}^2$. In particular, the definition of expectations is independent of the number of dimensions.

To summarize formally, *any distribution function induces a probability measure on the σ-algebra of Borel sets in $\mathcal{R}^r$, and thus defines a probability space.* Restated more informally, we have shown that the probabilistic setup of discrete sample spaces carries over without formal changes just as in the case of densities, and we have justified the probabilistic terminology employed in the first three chapters. If we speak of r random variables

[6] The proof utilized the fact that in a *finite* partition of a one-dimensional interval the subintervals appear in a natural order from left to right. An equally neat arrangement characterizes the *checkerboard partitions* of a two-dimensional interval $\overline{a, b}$, that is, partitions into mn subintervals obtained by subdividing separately the two sides of $\overline{a, b}$ and drawing parallels to the axes through all points of the subdivisions. The proof of the finite additivity requires no change for such checkerboard partitions, and to an arbitrary partition there corresponds a checkerboard refinement. The passage from finite to denumerable partitions is independent of the number of dimensions.

[7] A special case is covered by theorem 1 in **1**; IX,2.

$\mathbf{X}_1, \ldots, \mathbf{X}_r$ it is understood that they are defined in the same probability space so that a joint probability distribution of $(\mathbf{X}_1, \ldots, \mathbf{X}_r)$ exists. We are then free to interpret the $\mathbf{X}_k$ as coordinate variables io the sample space $\mathcal{R}^r$.

It is hardly necessary to explain the continued use of terms such as *marginal distribution* (see III,1 and **1**; IX,1), or *independent variables.* The basic facts concerning such variables are the same as in the discrete case, namely:

(i) Saying that $\mathbf{X}$ and $\mathbf{Y}$ are independent random variables with (one-dimensional) distributions F and G means that the joint distribution function of $(\mathbf{X}, \mathbf{Y})$ is given by the products $F(x_1)\, G(x_2)$. This statement may refer to two variables in a given probability space or may be an abbreviation for the statement that we introduce a plane with $\mathbf{X}$ and $\mathbf{Y}$ as coordinate variables and *define* probabilities by the product rule. This remark applies equally to pairs or triples of random variables, etc.

(ii) If the m-tuple $(\mathbf{X}_1, \ldots, \mathbf{X}_m)$ is independent of the n-tuple $(\mathbf{Y}_1, \ldots, \mathbf{Y}_n)$ then $u(\mathbf{X}_1, \ldots, \mathbf{X}_m)$ and $v(\mathbf{Y}_1, \ldots, \mathbf{Y}_n)$ are independent (for any pair of functions u and v).

(iii) If $\mathbf{X}$ and $\mathbf{Y}$ are independent, then $\mathbf{E}(\mathbf{XY}) = \mathbf{E}(\mathbf{X})\, \mathbf{E}(\mathbf{Y})$ whenever the expectations of $\mathbf{X}$ and $\mathbf{Y}$ exist (that is, if the integrals converge absolutely).

The following simple result is frequently useful.

Lemma 2. *A probability distribution F is uniquely determined by the knowledge of $\mathbf{E}(u)$ for every continuous function u vanishing outside some finite interval.*

Proof. Let I be a finite open interval and v a continuous function that is positive in I and zero outside I. Then $\sqrt[n]{v(x)} \to 1$ at each point $x \in I$, and hence $\mathbf{E}(\sqrt[n]{v}) \to F\{I\}$. Thus the knowledge of the expectations of our continuous functions uniquely determines the values $F\{I\}$ for all open intervals, and these uniquely determine F. ►

Note I.[8] **The F. Riesz representation theorem.** In the preceding lemma the expectations were defined in terms of a given probability distribution. Often (for example, in the moment problem of VII,3) we start from a given functional, that is, from an assignment of values $\mathbf{E}(u)$ to certain functions. We inquire whether there exists a probability distribution F such that

$$\mathbf{E}(u) = \int_{-\infty}^{+\infty} u(x)\, F\{dx\}. \tag{1.14}$$

It turns out that three evidently necessary conditions are also sufficient.

[8] This note treats a topic of conceptual interest but will not be used in the sequel. For an alternative approach see IV,5.

Theorem. *Suppose that to each continuous function u vanishing outside a finite interval there corresponds a number* $\mathbf{E}(u)$ *with the following properties;* (i) *The functional is linear, that is, for all linear combinations*

$$\mathbf{E}(c_1u_1 + c_2u_2) = c_1\mathbf{E}(u_1) + c_2\mathbf{E}(u_2);$$

(ii) *it is positive, that is,* $u \geq 0$ *implies* $\mathbf{E}(u) \geq 0$; (iii) *it has norm* 1, *that is,* $0 \leq u \leq 1$ *implies* $\mathbf{E}(u) \leq 1$, *but for each* $\epsilon > 0$ *there exists* u *such that* $0 \leq u \leq 1$ *and*

$$\mathbf{E}(u) > 1 - \epsilon.$$

Then there exists a unique probability distribution F for which (1.14) *is true.*

Proof. For arbitrary t and $h > 0$ denote by $z_{t,h}$ the continuous function of x that equals 1 when $x \leq t$, vanishes for $x \geq t + h$, and is linear in the intermediate interval $t \leq x \leq t + h$. This function does not vanish at infinity, but we can define $\mathbf{E}(z_{t,h})$ by simple approximations. Choose a function $|u_n| \leq 1$ such that $\mathbf{E}(u_n)$ is defined and $u_n(x) = z_{t,h}(x)$ for $|x| < n$. If $m > n$ the difference $u_m - u_n$ vanishes identically within the interval $\overline{-n, n}$ and from the fact that $\mathbf{E}$ has norm 1 one concludes easily that $\mathbf{E}(u_n - u_m) \to 0$. It follows that $\mathbf{E}(u_n)$ converges to a finite limit, and this limit is obviously independent of the particular choice of the approximating u_n. It is therefore legitimate to define $\mathbf{E}(z_{t+h}) = \lim \mathbf{E}(u_n)$. It is easily seen that even within this extended domain of definition the functional $\mathbf{E}$ enjoys the three properties postulated in the theorem.

Now put $F_h(t) = \mathbf{E}(z_{t,h})$. For fixed h this is a monotone function going from 0 to 1. It is continuous, because when $0 < \delta < h$ the difference $z_{t+\delta,h} - z_{t,h}$ has a triangular graph with height δ/h, and hence F_h has difference ratios bounded by $1/h$. As $h \to 0$ the functions F_h decreases monotonically to a limit which we denote by F. We show that *F is a probability distribution.* Obviously F is monotone and $F(-\infty) = 0$. Furthermore $F(t) \geq F_h(t-h)$ which implies that $F(\infty) = 1$. It remains to show that F is continuous from the right. For given t and $\epsilon > 0$ choose h so small that $F(t) > F_h(t) - \epsilon$. Because of the continuity of F_h we have then for δ sufficiently small

$$F(t) > F_h(t) - \epsilon > F_h(t+\delta) - 2\epsilon \geq F(t+\delta) - 2\epsilon$$

which proves the right-continuity.

Let u be a continuous function vanishing outside a finite interval $\overline{a, b}$. Choose $a = a_0 < a_1 < \cdots < a_n = b$ such that within each subinterval $\overline{a_{k-1}, a_k}$ the oscillation of u is less than ϵ. If h is smaller that the smallest among these intervals, then

$$(1.15) \qquad u_h = \sum_{k=1}^{n} u(a_k)[z_{a_k,h} - z_{a_{k-1},h}]$$

is a piecewise linear function with vertices at the points a_k and $a_k + h$. Since $u(a_k) = u_h(a_k)$ it follows that $|u - u_h| \leq 2\epsilon$, and hence $|\mathbf{E}(u) - \mathbf{E}(u_h)| \leq 2\epsilon$. But as $h \to 0$

$$(1.16) \qquad \mathbf{E}(u_h) \to \sum_{k=1}^{n} u(a_k)F\{\overline{a_{k-1}, a_k}\}$$

and this sum differs from the integral in (1.14) by less than ϵ. Thus the two sides in (1.14) differ by less than 3ϵ, and hence (1.14) is true. ►

Note II. On independence and correlation. Statistical correlation theory goes back to a time when a formalization of the theory was impossible and the notion of stochastic independence was necessarily tinged with mystery. It was understood that the independence of two bounded random variables with zero expectation implies $\mathbf{E}(\mathbf{XY}) = 0$, but this condition was at first thought also to be sufficient for the independence of $\mathbf{X}$ and $\mathbf{Y}$. The discovery that this was not so led to a long search for conditions under which the vanishing of correlations would imply stochastic independence. As frequently happens, the history of the problem and the luster of partial results easily obscured the fact that the solution is extremely simple by modern methods. The following theorem contains various results proved in the literature by laborious methods.

Theorem. *The random variables* $\mathbf{X}$ *and* $\mathbf{Y}$ *are independent iff*

$$\mathbf{E}(u(\mathbf{X}) \cdot v(\mathbf{Y})) = \mathbf{E}(u(\mathbf{X})) \cdot \mathbf{E}(v(\mathbf{Y})) \tag{1.17}$$

for all continuous functions u *and* v *vanishing outside a finite interval.*

Proof. The necessity of the condition is obvious. To prove the sufficiency it suffices to show that for every bounded continuous function $\mathbf{E}(w)$ agrees with the expectation of w with respect to a pair of *independent* variables distributed as $\mathbf{X}$ and $\mathbf{Y}$. Now (1.17) states this to be the case whenever w is of the form $w(\mathbf{X}, \mathbf{Y}) = u(\mathbf{X})\, v(\mathbf{Y})$. Every bounded continuous function w can be uniformly approximated[9] by linear combinations of the form $\Sigma\, c_k u_k(\mathbf{X})\, v_k(\mathbf{Y})$, and by passing to the limit we see the assertion to be true for arbitrary bounded continuous w. ►

2. PRELIMINARIES

This section is devoted largely to the introduction of a terminology for familiar or obvious things concerning distribution functions in $\mathcal{R}^1$.

Just as in the case of discrete variables we define the kth *moment* of a random variable $\mathbf{X}$ by $\mathbf{E}(\mathbf{X}^k)$, provided the integral exists. By this we mean that the integral

$$\mathbf{E}(\mathbf{X}^k) = \int_{-\infty}^{\infty} x^k\, F\{dx\} \tag{2.1}$$

converges absolutely, and so $\mathbf{E}(\mathbf{X}^k)$ exists iff $\mathbf{E}(|\mathbf{X}|^k) < \infty$. The last quantity is called the kth *absolute moment* of X (and is defined also for non-integral $k > 0$). Since $|x|^a \leq |x|^b + 1$ when $0 < a < b$, the existence of an absolute moment of order b implies the existence of all absolute moments of orders $a < b$.

If $\mathbf{X}$ has an expectation m, the second moment of $\mathbf{X} - m$ is called the *variance* of $\mathbf{X}$:

$$\text{Var}\,(\mathbf{X}) = \mathbf{E}((\mathbf{X}-m)^2) = \mathbf{E}(\mathbf{X}^2) - m^2. \tag{2.2}$$

Its properties and significance are the same as in the discrete case. In particular, if $\mathbf{X}$ *and* $\mathbf{Y}$ *are independent*

$$\text{Var}\,(\mathbf{X}+\mathbf{Y}) = \text{Var}\,(\mathbf{X}) + \text{Var}\,(\mathbf{Y}) \tag{2.3}$$

whenever the variances on the right exist.

---

[9] See problem 10 in VIII,10.

[Two variables satisfying (2.3) are said to be *uncorrelated.* It was shown in **1**; IX,8 that two dependent variables may be uncorrelated.]

It will be recalled how often we have replaced a random variable $\mathbf{X}$ by the "reduced variable" $\mathbf{X}^* = (\mathbf{X}-m)/\sigma$ where $m = \mathbf{E}(\mathbf{X})$ and $\sigma^2 = \text{Var}\,(\mathbf{X})$. The physicist would say that $\mathbf{X}^*$ is "expressed in dimensionless units." More generally a change from $\mathbf{X}$ to $(\mathbf{X}-\beta)/\alpha$ with $\alpha > 0$ amounts to a change of the origin and the unit of measurement. The distribution function of the new variable is given by $F(\alpha x+\beta)$, and in many situations we are actually dealing with the whole class of distributions of this form rather than with an individual representative. For convenience of expression we introduce therefore

Definition 1. *Two distributions* F_1 *and* F_2 *in* $\mathfrak{R}^1$ *are said to be of the same type*[10] *if* $F_2(x) = F_1(\alpha x+\beta)$ *with* $\alpha > 0$. *We refer to* α *as scale factor,* β *as centering* (*or location*) *constant.*

This definition permits the use of clauses such as "F is centered to zero expectation" or "centering does not affect the variance."

A *median* ξ of a distribution F is defined as a number such that $F(\xi) \geq \frac{1}{2}$ but $F(\xi-) \leq \frac{1}{2}$. It is not necessarily defined uniquely; if $F(x) = \frac{1}{2}$ for all x of an interval $\overline{a, b}$ then every such x is a median. It is possible to center a distribution so that 0 becomes a median.

Except for the median these notions carry over to higher dimensions or vector variables of the form $\mathbf{X} = (\mathbf{X}_1, \ldots, \mathbf{X}_n)$; the appropriate vector notation was introduced in III,5, and requires no modification. The expectation of $\mathbf{X}$ is now a vector, the variance a matrix.

The first things one notices looking at the graph of a distribution function are the discontinuities and the intervals of constancy. It is frequently necessary to say that a point is not in an interval of constancy. We introduce the following convenient terminology applicable in all dimensions.

Definition 2. *A point* x *is an atom if it carries a positive mass. It is a point of increase of* F *iff* $F\{I\} > 0$ *for every open interval* I *containing* x.

The distribution F *is concentrated on the set* A *if the complement* A' *has probability* $F\{A'\} = 0$.

The distribution F *is atomic if it is concentrated on the set of its atoms.*

Example. Order the rationals in $\overline{0, 1}$ in a sequence $r_1, r_2, \ldots$ with increasing denominators. Let F attribute probability 2^{-k} to r_k. Then F is purely atomic. Note, however, that every point of the closed interval $\overline{|0, 1|}$ is a point of increase of F. ►

---

[10] The notion was introduced by Khintchine who used the German term *Klasse*, but in English "a class of functions" has an established meaning.

Because of the countable additivity (1.7) the sum of the weights of the atoms cannot exceed unity and so at most one atom carries a weight $>\frac{1}{2}$, at most two atoms carry weights $>\frac{1}{3}$, etc. It is therefore possible to arrange the atoms in a simple sequence $a_1, a_2, \ldots$ such that the corresponding weights decrease: $p_1 \geq p_2 \geq \cdots$. In other words, *there exist at most denumerably many atoms.*

A distribution without atoms is called *continuous.* If there are atoms, denote their weights by $p_1, p_2, \ldots$ and let $p = \Sigma\, p_k > 0$ be their sum. Put

$$F_a(x) = p^{-1} \sum_{a_k \leq x} p_k, \tag{2.4}$$

the summation extending over all atoms in the interval $\overline{-\infty, x}|$. Obviously F_a is again a distribution function, and it is called the *atomic component* of F. If $p = 1$ the distribution F is atomic. Otherwise let $q = 1 - p$. It is easily seen that $[F - pF_a]/q = F_c$ is a *continuous* distribution, and so

$$F = pF_a + qF_c \tag{2.5}$$

is a linear combination of two distribution functions of which F_a is atomic, F_c continuous. If F is atomic (2.5) is true with $p = 1$ and F_c arbitrary; in the absence of atoms (2.5) holds with $p = 0$. We have thus the

Jordan decomposition theorem. *Every probability distribution is a mixture of the form* (2.5) *of an atomic and a continuous distribution; here* $p \geq 0$, $q \geq 0$, $p + q = 1$.

Among the atomic distributions there is a class which sometimes encumbers simple formulation by trite exceptions. Its members differ only by an arbitrary scale factor from distributions of integral-valued random variables, but they occur so often that they deserve a name for reference.

Definition 3. *A distribution* F *in* $\mathcal{R}^1$ *is arithmetic*[11] *if it is concentrated on a set of points of the form* $0, \pm\lambda, \pm 2\lambda, \ldots$. *The largest* λ *with this property is called the span of* F.

3. DENSITIES

The first two chapters were devoted to probability distributions in $\mathcal{R}^1$ such that

$$F\{A\} = \int_A \varphi(x)\, dx \tag{3.1}$$

[11] The term *lattice distribution* is, perhaps, more usual but its usage varies: according to some authors a lattice distribution may be concentrated on a set of points a, $a \pm \lambda$, $\alpha \pm 2\lambda, \ldots$ with a arbitrary. (The binomial distribution with atoms at ± 1 is arithmetic with span 1 in our terminology, but a lattice distribution with span 2 according to the alternative definition.)

for all intervals (and therefore all sets). The distributions of chapter III are of the same form, the integration being with respect to the Lebesgue measure (area or volume) in $\mathcal{R}^r$. If the density φ in (3.1) is concentrated on the interval $\overline{0, 1}$ then (3.1) takes on the form

$$F\{A\} = \int_A \varphi(x)\, U\{dx\} \tag{3.2}$$

where U stands for the uniform distribution in $\overline{0, 1}$. The last formula makes sense for an arbitrary probability distribution U, and whenever $F\{\overline{-\infty, \infty}\} = 1$ it defines a new probability distribution F. In this case we shall say that φ *is the density of* F *with respect to* U.

In (3.1) the measure U is infinite whereas in (3.2) we have $U\{\overline{-\infty, \infty}\} = 1$. The difference is not essential since the integral in (3.1) can be broken up into integrals of the form (3.2) extended over finite intervals. We shall use (3.2) only when U is either a probability distribution or the Lebesgue measure as in (3.1) but the following definition is general.

Definition. *The distribution* F *is absolutely continuous with respect to the measure* U *if it is of the form* (3.2). *In this case* φ *is called a density*[12] *of* F *with respect to* U.

The special case (3.1) where U is the Lebesgue measure is of course the most important and we say in this case that φ is an "*ordinary*" *density*.

We now introduce the *abbreviation*

$$F\{dx\} = \varphi(x)\, U\{dx\}. \tag{3.3}$$

This is merely a shorthand notation to indicate the validity of (3.2) for all sets and no meaning must be attached to the symbol dx. With this notation we would abbreviate (3.1) to $F\{dx\} = \varphi(x)\, dx$ and if U has an ordinary density u then (3.2) is the same as $F\{dx\} = \varphi(x)\, u(x)\, dx$.

Examples. (*a*) Let U be a probability distribution in $\mathcal{R}^1$ with second moment m_2. Then

$$F\{dx\} = \frac{1}{m_2}\, x^2\, U\{dx\}$$

is a new probability distribution. In particular, if U is the uniform distribution in $\overline{0, 1}$ then $F(x) = x^3$ for $0 < x < 1$, and if U has density e^{-x} $(x > 0)$ then F is the gamma distribution with ordinary density $\frac{1}{2}x^2e^{-x}$.

(*b*) Let U be atomic, attaching weights $p_1, p_2, \ldots$ to the atoms $a_1, a_2, \ldots$ (where $\sum p_k = 1$). A distribution F has a density φ with respect to U iff

[12] In measure theory φ is called a Radon-Nikodym derivative of F with respect to U.

it is purely atomic and its atoms are among $a_1, a_2, \ldots$. If F attributes weight q_j to a_j the density φ is given by $\varphi(a_k) = q_k/p_k$. The value of φ at other points plays no role and it is best to leave φ undefined except at the atoms. ▶

In theory the integrand φ in (3.2) is not uniquely determined, for if N is a set such that $U\{N\} = 0$ then φ may be redefined on N in an arbitrary manner without affecting (3.2). However, this is the only indeterminacy and *a density is uniquely determined up to values on a null set.*[13] In practice a unique choice is usually dictated by continuity conditions, and for this reason one speaks usually of "the" density although "a" density would be more correct.

For any bounded function v the relation (3.3) implies obviously[14]

$$v(x)\,F\{dx\} = v(x)\varphi(x)\;I\{dx\}. \tag{3.4}$$

In particular, if φ is bounded away from 0 we can choose $v = \varphi^{-1}$ to obtain the *inversion formula* for (3.2):

$$U\{dx\} = \frac{1}{\varphi(x)}\,F\{dx\}. \tag{3.5}$$

A useful criterion for absolute continuity is contained in a basic theorem of measure theory which we accept without proof.

Randon-Nikodym theorem. [15] *F is absolutely continuous with respect to U iff*

$$F\{A\} = 0 \quad \textit{whenever} \quad U\{A\} = 0. \tag{3.6}$$

[13] In fact, if both φ and φ_1 are densities of F with respect to U consider the set A of all points x such that $\varphi(x) > \varphi_1(x) + \epsilon$. From

$$F\{A\} = \int_A \varphi(x)\,U\{dx\} = \int_A \varphi_1(x)\,U\{dx\}$$

it follows that $U\{A\} = 0$, and since this holds for every $\epsilon > 0$ we see that $\varphi(x) = \varphi_1(x)$ except on a set N such that $U\{N\} = 0$.

[14] Readers who feel uneasy about the new integrals should notice that in the case of continuous densities (3.4) reduces to the familiar substitution rule for integrals. The following proof in the general case uses a standard argument applicable in more general situations. Formula (3.4) is trivial when v is simple, that is, assumes only finitely many values. For every bounded v there exist two simple functions of this nature such that $\underline{v} < v \leq \bar{v}$ and $\bar{v} - \underline{v} < \epsilon$, and so the validity of (3.4) for all simple functions implies its truth in general.

[15] Often called Lebesgue-Nikodym theorem. The relation (3.6) may be taken as a *definition* of absolute continuity, in which case the theorem asserts the existence of a density.

This expression may be rephrased by the statement that U-null sets are also F-null sets. We give an important corollary although it will not be used explicitly in this book.

Criterion. *F is absolutely continuous with respect to U iff to each $\epsilon > 0$ there corresponds a $\delta > 0$ such that for any collection of non-overlapping intervals $I_1, \ldots, I_n$*

$$\sum_1^n U\{I_k\} < \delta \quad \textit{implies} \quad \sum_1^n F\{I_k\} < \epsilon. \tag{3.7}$$

An important special case arises when

$$F\{I\} \leq a \cdot U\{I\} \tag{3.8}$$

for all intervals. Then (3.7) is trivially true with $\delta = \epsilon/a$, and it is easily seen that in this case F has a density φ with respect to U such that $\varphi \leq a$.

*3a. Singular Distributions

The condition (3.6) of the Radon-Nikodym theorem leads one to the study of the extreme counterpart of absolutely continuous distributions.

Definition. *The probability distribution F is singular with respect to U if it is concentrated on a set N such that $U\{N\} = 0$.*

The Lebesgue measure $U\{dx\} = dx$ plays a special role and the word "singular" without further qualification refers to it. Every atomic distribution is singular with respect to dx, but the *Cantor distribution* of example I,11(d) shows that there exist *continuous* distributions in $\mathcal{R}^1$ that are singular with respect to dx. Such distributions are not tractable by the methods of calculus and explicit representations are in practice impossible. For analytic purposes one is therefore forced to choose a framework which leads to absolutely continuous or atomic distributions. Conceptually, however, singular distributions play an important role and many statistical tests depend on their existence. This situation is obscured by the cliché that "in practice" singular distributions do not occur.

Examples. (c) *Bernoulli trials.* It was shown in example I,11(c) that the sample space of sequences $SS \cdots F \cdots$ can be mapped onto the unit interval by the simple device of replacing the symbols S and F by 1 and 0, respectively. The unit interval then becomes the sample space, and the outcome of an infinite sequence of trials is represented by the random variable $\mathbf{Y} = \sum 2^{-k}\mathbf{X}_k$ where the $\mathbf{X}_k$ are independent variables assuming the values 1 and 0 with probabilities p and q. Denote the distribution of $\mathbf{Y}$ by F_p. For symmetric trials $F_{\frac{1}{2}}$ is the *uniform* distribution and the model becomes attractive because of its simplicity. In fact, the equivalence

* Although conceptually of great importance, singular distributions appear in this book only incidentally.

of symmetric Bernoulli trials with "a random choice of a point in $\overline{0, 1}$" has been utilized since the beginnings of probability theory. Now by the law of large numbers the distribution F_p is concentrated on the set N_p consisting of points in whose dyadic expansion the frequency of the digit 1 tends to p. When $p \neq \alpha$ the set N_α has probability zero and hence *the distributions* F_p *are singular with respect to each other;* for $p \neq \frac{1}{2}$ the distribution F_p is singular with respect to the uniform distribution dx. An explicit representation of F_p is impractical and, accordingly, the model is not in common use when $p \neq \frac{1}{2}$. Two points deserve attention.

First, consider what would happen if the special value $p = \frac{1}{3}$ presented a particular interest or occurred frequently in applications. We would replace the dyadic representation of numbers by triadic expansions and introduce a new scale such that now $F_{\frac{1}{3}}$ would coincide with the uniform distribution. "In practice" we would again deal only with absolutely continuous distributions, but the reason for this lies in our choice of tools rather than in the nature of things.

Second, whether a coin is, or is not, biased can be tested statistically and practical certainty can be reached after finitely many trials. This is possible only because what is likely under the hypothesis $p = \frac{1}{2}$ is extremely unlikely under the hypothesis $p = \frac{1}{3}$. A little reflection along these lines reveals that the possibility of a decision after finitely many trials is due to the fact that F_p is singular with respect to $F_{\frac{1}{2}}$ (provided $p \neq \frac{1}{2}$). The existence of singular distributions is therefore essential to statistical practice.

(*d*) *Random directions.* The notion of a unit vector in $\mathcal{R}^2$ with random direction was introduced in I,10. The distribution of such a vector is concentrated on the unit circle and is therefore singular with respect to the Lebesgue measure (area) in the plane. One might object that in this case the circle should serve as sample space, but practical problems sometimes render this choice impossible. [See example 4(*e*).] ▶

Lebesgue decomposition theorem. *Every probability distribution* F *is a mixture of the form*

$$F = p \cdot F_s + q \cdot F_{ac} \tag{3.9}$$

(*where* $p \geq 0$, $q \geq 0$, $p+q = 1$) *of two probability distributions such that* F_s *is singular and* F_{ac} *absolutely continuous with respect to a given measure* U.

The Jordan decomposition (2.5) applies to F_s and hence F can be written as a mixture of three probability distributions of which the first is atomic, the second absolutely continuous with respect to $U\{dx\}$, the third continuous but singular.

Proof. To simplify the language a set N with $U\{N\} = 0$ will be called nullset. Let p be the least upper bound (the sup) of $F\{N\}$ for all nullsets N. To each n there exists a nullset N_n such that $F\{N_n\} > p - \frac{1}{n}$. Then $F\{A\} \leq \frac{1}{n}$ for any nullset A in the complement N'_n. For the union $N = \bigcup N_n$ this implies $U\{N\} = 0$ and $F\{N\} = p$, and hence no nullset in the complement N' can carry positive probability.

If $p = 1$ it follows that F is singular, whereas $p = 0$ means that F is absolutely continuous. When $0 < p < 1$ the assertion holds with the two probability distributions defined by

$$p \cdot F_s\{A\} = F\{AN\}, \qquad q \cdot F_{ac}\{A\} = F\{AN'\}. \tag{3.10}$$ ▶

4. CONVOLUTIONS

It is difficult to exaggerate the importance of convolutions in many branches of mathematics. We shall have to deal with convolution in two ways: as an operation between distributions and as an operation between a distribution and a continuous function.

For definiteness we refer explicitly to distributions in $\mathfrak{R}^1$, but with the vector notation of section 1 the formulas are independent of the number of dimensions. The definition of *convolutions on a circle* follows the pattern described in II,8 and requires no comment. (More general convolutions can be defined on arbitrary groups.)

Let F be a probability distribution and φ a bounded point function. (In our applications φ will be either continuous or a distribution function.) A new function u is then defined by

$$u(x) = \int_{-\infty}^{+\infty} \varphi(x-y)\, F\{dy\}. \tag{4.1}$$

If F has a density f (with respect to dx) this reduces to

$$u(x) = \int_{-\infty}^{+\infty} \varphi(x-y) f(y)\, dy. \tag{4.2}$$

Definition 1. *The convolution of a function φ with a probability distribution F is the function defined by* (4.1). *It will be denoted by $u = F \bigstar \varphi$. When F has a density f we write alternatively $u = f * \varphi$.*

Note that the order of the terms is important: the symbol $\varphi \bigstar F$ is in general meaningless. On the other hand, (4.2) makes sense for arbitrary integrable f and φ (also if f is not non-negative), and the symbol $*$ is used in this generalized sense. Needless to say, the boundedness of φ was assumed only for simplicity and is not necessary.

Examples. (*a*) When F is the uniform distribution in $\overline{0, a}$ then

$$u(x) = a^{-1}\int_{x-a}^{x} \varphi(s)\, ds. \tag{4.3}$$

It follows that u is continuous; if φ is continuous u has a continuous derivative, etc. Generally speaking u will behave better than φ, and so the convolution serves as *smoothing operator*.

(*b*) The convolution formulas for the exponential and the uniform distributions [I,(3.6) and I,(9.1)] are special cases. For examples in $\mathcal{R}^2$ see III,(1.22) and problems 15–17 of chapter III. ▶

Theorem 1. *If* φ *is bounded and continuous, so is* $u = F \star \varphi$; *if* φ *is a probability distribution function, so is* u.

Proof. If φ is bounded and continuous then $u(x+h) \to u(x)$ by the dominated convergence principle. For the same reason right-continuity of φ implies right-continuity of u. Finally, if φ goes monotonically from 0 to 1 the same is obviously true of u. ▶

The next theorem gives an interpretation of $F \star \varphi$ when φ is a distribution function.

Theorem 2. *Let* $\mathbf{X}$ *and* $\mathbf{Y}$ *be independent random variables with distributions* F *and* G. *Then*

$$\mathbf{P}\{\mathbf{X} + \mathbf{Y} \leq t\} = \int_{-\infty}^{+\infty} G(t-x)\, F\{dx\}. \tag{4.4}$$

Proof.[16] Choose $\epsilon > 0$ and denote by I_n the interval $n\epsilon < x \leq (n+1)\epsilon$; here $n = 0, \pm 1, \ldots$. The event $\{\mathbf{X} + \mathbf{Y} \leq t\}$ occurs if $\mathbf{X} \in I_{n-1}$, $\mathbf{Y} \leq t - n\epsilon\}$ for some n. The latter events are mutually exclusive, and as $\mathbf{X}$ and $\mathbf{Y}$ are independent we have therefore

$$\mathbf{P}\{\mathbf{X} + \mathbf{Y} \leq t\} \leq \sum G(t-n\epsilon) \cdot F\{I_n\}. \tag{4.5}$$

On the right we have the integral of the step function G_ϵ assuming in I_n the value $G(t-n\epsilon)$. Since $G_\epsilon(y) \leq G\,(t+\epsilon-y)$ we have

$$\mathbf{P}\{\mathbf{X} + \mathbf{Y} \leq t\} \leq \int_{-\infty}^{+\infty} G(t+\epsilon-x)\, F\{dx\}. \tag{4.6}$$

The same argument leads to the reversed inequality with ϵ replaced by $-\epsilon$. Letting $\epsilon \to 0$ we get (4.4). ▶

[16] (4.4) is a special case of Fubini's theorem IV,(2.11). The converse of theorem 2 is false: we saw in II,4(*e*), and in problem 1 of III,9, that in exceptional cases formula (4.4) may hold for a pair of dependent variables $\mathbf{X}$, $\mathbf{Y}$.

Example. (*c*) Let F and G be concentrated on the integers $0, 1, 2, \ldots$ and denote the weights of k by p_k and q_k. The integral in (4.4) then reduces to the sum $\sum G(t-k)p_k$. This is a function vanishing for $t < 0$ and constant in each interval $n-1 < t < n$. The jump at $t = n$ equals

$$\sum_{k=0}^{n} q_{n-k}p_k = q_n p_0 + q_{n-1}p_1 + \cdots + q_0 p_n \tag{4.7}$$

in agreement with the convolution formula **1**; XI,(2.1) for integral-valued random variables. ▶

Each of the preceding theorems shows that for two distribution functions the convolution operation $F \bigstar G$ yields a new distribution function U. The commutativity of the addition $\mathbf{X} + \mathbf{Y}$ implies that $F \bigstar G = G \bigstar F$. A perfect system might introduce a new symbol for such convolutions among distribution functions, but this would hardly be helpful.[17] Of course, one should think of U as an interval function or measure: for each interval $I = \overline{a, b}$ obviously

$$U\{I\} = \int_{-\infty}^{+\infty} G\{I-y\}\, F\{dy\} \tag{4.8}$$

where, as usual, $I - y$ denotes the interval $\overline{a-y, b-y}$. (This formula automatically carries over to arbitrary sets.) Because of the commutativity the roles of F and G in (4.8) may be interchanged.

Consider now three distributions F_1, F_2, F_3. The associative law of addition for random variables implies that $(F_1 \bigstar F_2) \bigstar F_3 = F_1 \bigstar (F_2 \bigstar F_3)$ so that we can dispense with the parentheses and write $F_1 \bigstar F_2 \bigstar F_3$. We summarize this in theorems 3 and 4.

Theorem 3. *Among distributions the convolution operation* $\bigstar$ *is commutative and associative.*

---

[17] In other words, the symbol $A \star B$ is used when the integration is with respect to the measure A. This convolution is a point function or measure according as B is a point function [as in (4.1)] or a measure [as in (4.6)]. The asterisk $*$ is used for an operation between two functions, the integration being with respect to Lebesgue measure. In our context this type of convolution is restricted almost exclusively to probability densities.

A more general definition of a convolution between two functions may be defined by

$$f * g(x) = \int_{-\infty}^{+\infty} f(x-y)g(y)\, m\{dy\}$$

where m stands for an arbitrary measure. Sums of the form (4.7) represent the special case when m is concentrated on the positive integers and attributes unit weight to each. In this sense the use of the asterisk for the convolutions between sequences in **1**; XI,2 is consistent with our present usage.

Theorem 4. *If G is continuous (= free of atoms), so is $U = F \star G$. If G has the ordinary density φ, then U has the ordinary density u given by* (4.1).

Proof The first assertion is contained in theorem 1. If φ is the density of G then an integration of (4.1) over the interval I leads to (4.8), and so u is indeed the density of the distribution U defined by (4.8). ▶

It follows in particular that if F and G have densities f and g, then the convolution $F \star G$ has a density $h = f * g$ given by

$$h(x) = \int_{-\infty}^{+\infty} f(x-y)\, g(y)\, dy. \tag{4.9}$$

In general h will have much better smoothness properties than either f or g. (See problem 14.)

Sums $\mathbf{S}_n = \mathbf{X}_1 + \cdots + \mathbf{X}_n$ of n mutually independent random variables with a common distribution F occur so frequently that a special notation is in order. The distribution of $\mathbf{S}_n$ is the *n-fold convolution of F with itself.* It will be denoted by $F^{n\star}$. Thus

$$F^{1\star} = F, \qquad F^{(n+1)\star} = F^{n\star} \star F \tag{4.10}$$

A sum with no terms is conventionally interpreted as 0, and for consistency we define $F^{0\star}$ as the atomic distribution concentrated at the origin. Then (4.9) holds also for $n = 0$.

If F has a density f then $F^{n\star}$ has the density $f * f * \cdots * f$ (n times). *We denote it by f^{n*}.* These notations are consistent with the notation introduced in I,2.

Note. The following examples show that *the convolution of two singular distributions can have a continuous density.* They show also that an effective calculation of convolutions need not be based on the defining formula.

Examples. (*d*) *The uniform distribution in $\overline{0, 1}$ is the convolution of two Cantor-type singular distributions.* In fact, let $\mathbf{X}_1, \mathbf{X}_2, \ldots$ be mutually independent random variables assuming the values 0 and 1 with probability $\frac{1}{2}$. We saw in example I,11(*c*) that the variable $\mathbf{X} = \sum 2^{-k}\mathbf{X}_k$ has a uniform distribution. Denote the contributions of the even and odd terms by $\mathbf{U}$ and $\mathbf{V}$, respectively. Obviously $\mathbf{U}$ and $\mathbf{V}$ are independent and $\mathbf{X} = \mathbf{U} + \mathbf{V}$. The uniform distribution is therefore the convolution of the distributions of $\mathbf{U}$ and $\mathbf{V}$. But obviously $\mathbf{U}$ has the same distribution as $2\mathbf{V}$, and the variable $\mathbf{V}$ differs only notationally from the variable $\frac{1}{3}\mathbf{Y}$ of example I,11(*d*). In other words, the distributions of $\mathbf{U}$ and $\mathbf{V}$ differ only by scale factors from the Cantor distribution of that example.

(*e*) *Random vectors in $\mathcal{R}^2$.* The distribution of a unit vector with random direction (see I,10) is concentrated on the unit circle and therefore singular with respect to the Lebesgue measure in the plane. Nevertheless, the resultant of two independent vectors *has a length $\mathbf{L}$ which is a random variable with the density* $\dfrac{2}{\pi}\dfrac{1}{\sqrt{4-r^2}}$ concentrated on $\overline{0, 2}$. In fact, by the law of the cosines $\mathbf{L} = \sqrt{2 - 2\cos\omega} = |2 \sin \frac{1}{2}\omega|$ where ω is the angle between the two vectors. As $\frac{1}{2}\omega$ is distributed uniformly in $\overline{0, \pi}$ we have

$$\mathbf{P}\{\mathbf{L} \le r\} = \mathbf{P}\{|2 \sin \tfrac{1}{2}\omega| \le r\} = \frac{2}{\pi} \arcsin \tfrac{1}{2}r, \qquad 0 < r < 2 \tag{4.11}$$

which proves the assertion. (See problem 12.) ▶

4a. Concerning the Points of Increase

It is necessary here to interrupt the exposition in order to record some elementary facts concerning the points of increase of $F \star G$. The first lemma is intuitively obvious, whereas the second is of a technical nature. It will be used only in renewal theory, and hence indirectly in the theory of random walks.

Lemma 1. *If a and b are points of increase for the distributions F and G, then $a + b$ is a point of increase for $F \star G$. If a and b are atoms, the same is true of $a + b$. Furthermore, all atoms of $F \star G$ are of this form.*

Proof. If $\mathbf{X}$ and $\mathbf{Y}$ are independent then

$$\mathbf{P}\{|\mathbf{X}+\mathbf{Y}-a-b| < \epsilon\} \ge \mathbf{P}\{|\mathbf{X}-a| < \tfrac{1}{2}\epsilon\} \cdot \mathbf{P}\{|\mathbf{Y}-b| < \tfrac{1}{2}\epsilon\}.$$

The right side is positive for every $\epsilon > 0$ if a and b are points of increase, and so $a + b$ is again a point of increase.

Denote by F_a and G_a the atomic components of F and G in the Jordan decomposition (2.5). The atomic component of $F \star G$ is obviously identical with the convolution $F_a \star G_a$, and hence all atoms of $F \star G$ are of the form $a + b$, where a and b are atoms of F and G, respectively. ▶

The intrinsic simplicity of the next lemma suffers by the special role played on one hand by arithmetic distributions, on the other hand by distributions of *positive* variables.

Lemma 2. *Let F be a distribution in $\mathcal{R}^1$ and Σ the set formed by the points of increase of $F, F^{2\star}, F^{3\star}, \ldots$.*

(a) If F is not concentrated on a half-axis then Σ is dense in $\overline{-\infty, \infty}$ for F not arithmetic, and $\Sigma = \{0, \pm\lambda, \pm 2\lambda, \ldots\}$ for F arithmetic with span λ.

(b) Let F be concentrated on $\overline{0, \infty}$ but not at the origin. If F is not arithmetic then Σ is "asymptotically dense at ∞" in the sense that for given $\epsilon > 0$ and x sufficiently large the interval $\overline{x, x+\epsilon}$ contains points of Σ. If F is arithmetic with span λ then Σ contains all points $n\lambda$ for n sufficiently large.

Proof. Let $0 < a < b$ be two points in the set Σ and put $h = b - a$. We distinguish two cases:

(i) For each $\epsilon > 0$ it is possible to choose a, b such that $h < \epsilon$.

(ii) There exists a $\delta > 0$ such that $h \ge \delta$ for all possible choices.

Let I_n denote the interval $na < x \le nb$. If $n(b - a) > a$ this interval contains $\overline{na, (n+1)a}$ as proper subinterval, and hence every point $x > x_0 = a^2/(b - a)$ belongs to at least one among the intervals $I_1, I_2, \ldots$. By lemma 1 the $n + 1$ points $na + kh$, $k = 0, \ldots, n$, belong to Σ, and they partition I_n into n subintervals of length h. Thus every point $x > x_0$ is at a distance $\le h/2$ from a point of Σ.

In the situation of case (i) this implies that Σ is asymptotically dense at $+\infty$. If then F is concentrated on $\overline{0, \infty}$ there is nothing to be proved. Otherwise let $-c < 0$ be a point of increase of F. For arbitrary y and n sufficiently large the interval

$$nc + y < x < nc + y + \epsilon$$

contains a point s of Σ. Since $s - nc$ again belongs to Σ it follows that every interval of length ϵ contains some points of Σ, and thus Σ is everywhere dense.

In the situation of case (ii) we may suppose that a and b were chosen such that $h < 2\delta$. It follows then that the points $na + kh$ exhaust *all* points of Σ within I_n. Since $(n+1)a$ is among these points this means that all points of Σ within I_n are multiples of h. Now let c be an arbitrary (positive or negative) point of increase of F. For n sufficiently large the interval I_n contains a point of the form $kh + c$, and as this belongs to Σ it follows that c is a multiple of h. Thus in case (ii) the distribution F is arithmetic. ▶

A special case of this theorem commands interest. Every number $x > 0$ can be represented uniquely in the form $x = m + \xi$ as the sum of an integer m and a number $0 \leq \xi < 1$. This ξ is called the fractional part of x. Consider now a distribution F concentrated on the two points -1 and $\alpha > 0$. The set Σ contains all points of the form $n\alpha - m$ and hence the fractional parts of $\alpha, 2\alpha, \ldots$. This F is arithmetic if $\alpha = p/q$ where p and q are positive integers without common divisors, and in this case the span of F equals $1/q$. We have thus the following corollary (to be sharpened in the equidistribution theorem 3 of VIII, 7).

Corollary. *If* $\alpha > 0$ *is an irrational number the set formed by the fractional parts of* $\alpha, 2\alpha, 3\alpha, \ldots$ *is dense in* $\overline{0, 1}$.

5. SYMMETRIZATION

If the random variable $\mathbf{X}$ has the distribution F we shall denote the distribution of $-\mathbf{X}$ by ${}^-F$. At points of continuity we have

$$(5.1) \qquad {}^-F(x) = 1 - F(-x)$$

and this defines ${}^-F$ uniquely. The distribution F is called *symmetric* if ${}^-F = F$. [When a density f exists this means that $f(-x) = f(x)$.]

Let $\mathbf{X}_1$ and $\mathbf{X}_2$ be independent with the common distribution F. Then $\mathbf{X}_1 - \mathbf{X}_2$ has the symmetric distribution ${}^0F$ given by

$$(5.2) \qquad {}^0F = F \star {}^-F.$$

Using the symmetry property ${}^0F(x) = 1 - {}^0F(-x)$ it is readily seen that

$$(5.3) \qquad {}^0F(x) = \int_{-\infty}^{+\infty} F(x+y)\, F\{dy\}.$$

We shall say that ${}^0F$ is obtained by *symmetrization* of F.

Examples. (*a*) Symmetrization of the exponential leads to the bilateral exponential [II,4(*a*)]; the uniform distribution on $\overline{0, 1}$ leads to the triangular distribution τ_2 of II,(4.1).

(*b*) The distribution with atoms of weight $\frac{1}{2}$ at ± 1 is symmetric, but not the result of a symmetrization procedure.

(*c*) Let F be atomic, attributing weights $p_0, p_1, \ldots$ to $0, 1, \ldots$. The symmetrized distribution ${}^0F$ is atomic and the points $\pm n$ carry the weight

$$q_n = \sum_{k=0}^{\infty} p_k p_{k+n}, \qquad q_{-n} = q_n \tag{5.4}$$

When F is the *Poisson* distribution we get for $n \geq 0$

$$q_n = e^{-2\alpha} \sum_{k=0}^{\infty} \frac{\alpha^{n+2k}}{k!(n+k)!} = e^{-2\alpha} I_n(2\alpha) \tag{5.5}$$

where I_n is the *Bessel function* defined in II,(7.1). (See problem 9.) ▶

Many messy arguments can be avoided by symmetrization. In this connection it is important that the tails of F and ${}^0F$ are of comparable magnitude, a statement made more precise by the following inequalities. Their meaning appears clearer when expressed in terms of random variables rather than the distribution itself.

Lemma 1. *Symmetrization inequalities. If* $\mathbf{X}_1$ *and* $\mathbf{X}_2$ *are independent and identically distributed, then for* $t > 0$

$$\mathbf{P}\{|\mathbf{X}_1 - \mathbf{X}_2| > t\} \leq 2\mathbf{P}\{|\mathbf{X}_1| > \tfrac{1}{2}t\}. \tag{5.6}$$

If $a \geq 0$ *is chosen so that* $\mathbf{P}\{\mathbf{X}_i \leq a\} \geq p$ *and also* $\mathbf{P}\{\mathbf{X}_i \geq -a\} \geq p$, *then*

$$\mathbf{P}\{|\mathbf{X}_1 - \mathbf{X}_2| > t\} \geq p\,\mathbf{P}\{|\mathbf{X}_1| > t + a\}. \tag{5.7}$$

In particular, if 0 *is a median for* $\mathbf{X}_j$

$$\mathbf{P}\{|\mathbf{X}_1 - \mathbf{X}_2| > t\} \geq \tfrac{1}{2}\mathbf{P}\{|\mathbf{X}_1| > t\}. \tag{5.8}$$

Proof. The event on the left in (5.6) cannot occur unless either $|\mathbf{X}_1| > \frac{1}{2}t$ or $|\mathbf{X}_2| > \frac{1}{2}t$ and hence (5.6) is true. The event on the left in (5.7) occurs if $\mathbf{X}_1 > t + a$, $\mathbf{X}_2 \leq a$, and also if $\mathbf{X}_1 < -t - a$ and $\mathbf{X}_2 \geq -a$. This implies (5.7). ▶

Symmetrization is frequently used for the estimation of sums of independent random variables. In this connection the following inequality is particularly useful.

Lemma 2. *If* $\mathbf{X}_2, \ldots, \mathbf{X}_n$ *are independent and have symmetric distributions then* $\mathbf{S}_n = \mathbf{X}_1 + \cdots + \mathbf{X}_n$ *has a symmetric distribution and*

$$\mathbf{P}\{|\mathbf{X}_1 + \cdots + \mathbf{X}_n| > t\} \geq \tfrac{1}{2}\mathbf{P}\{\operatorname{Max} |\mathbf{X}_j| > t\}. \tag{5.9}$$

If the $\mathbf{X}_j$ *have a common distribution* F *then*

$$\mathbf{P}\{|\mathbf{X}_1 + \cdots + \mathbf{X}_n| \geq t\} \geq \tfrac{1}{2}(1 - e^{-n[1 - F(t) + F(-t)]}). \tag{5.10}$$

Proof. Let the random variable $\mathbf{M}$ equal the first term among $\mathbf{X}_1, \ldots, \mathbf{X}_n$ that is greatest in absolute value and put $\mathbf{T} = \mathbf{S}_n - \mathbf{M}$. The pair $(\mathbf{M}, \mathbf{T})$ is symmetrically distributed in the sense that the four combinations $(\pm\mathbf{M}, \pm\mathbf{T})$ have the same distribution. Clearly

$$(5.11) \qquad \mathbf{P}\{\mathbf{M} > t\} \leq \mathbf{P}\{\mathbf{M} > t, \mathbf{T} \geq 0\} + \mathbf{P}\{\mathbf{M} > t, \mathbf{T} \leq 0\}.$$

The two terms on the right have equal probabilities, and so

$$(5.12) \quad \mathbf{P}\{\mathbf{S} > t\} = \mathbf{P}\{\mathbf{M} + \mathbf{T} > t\} \geq \mathbf{P}\{\mathbf{M} > t, \mathbf{T} \geq 0\} \geq \tfrac{1}{2}\mathbf{P}\{\mathbf{M} > t\}$$

which is the same as (5.9).

(*b*) To prove (5.10) note that at points of continuity

$$(5.13) \qquad \mathbf{P}\{\text{Max}\, |\mathbf{X}_j| \leq t\} = (F(t) - F(-t))^n \leq e^{-n[1-F(t)+F(-t)]}$$

This implies (5.10) because $1 - x < e^{-x}$ when $0 < x < 1$. ▶

6. INTEGRATION BY PARTS. EXISTENCE OF MOMENTS

The familiar formula for integration by parts can be used also for arbitrary expectations in $\mathfrak{R}^1$. *If* u *is bounded and has a continuous derivative* u', *then*

$$(6.1) \qquad \int_a^{b+} u(x)\, F\{dx\} = u(b)\, F(b) - u(a)\, F(a) - \int_a^b u'(x)\, F(x)\, dx.$$

Proof. A simple rearrangement reduces (6.1) to the form

$$(6.2) \qquad \int_a^{b+} [u(b) - u(x)]\, F\{dx\} - \int_a^b u'(x)[F'(x) - F(a)]\, dx = 0.$$

Suppose $|u'| < M$ and partition $\overline{a, b}$ into congruent intervals I_k of length h. It is easily seen that the contribution of I_k to the left side in (6.2) is in absolute value less than $2MhF\{I_k\}$. Summing over k we find that the left side is in magnitude $< 2Mh$, which can be made as small as we please. Thus the left side in (6.2) is indeed zero. ▶

As an application we derive a frequently used formula [generalizing **1**; XI,(1.8)].

Lemma 1. *For any* $\alpha > 0$

$$(6.3) \qquad \int_0^\infty x^\alpha\, F\{dx\} = \alpha \int_0^\infty x^{\alpha-1}[1 - F(x)]\, dx$$

in the sense that if one side converges so does the other.

Proof. Because of the infinite interval of integration (6.1) does not apply directly, but for every $b < \infty$ we have after a trivial rearrangement

$$(6.4) \qquad \int_0^{b+} x^\alpha\, F\{dx\} = -b^\alpha[1 - F(b)] + \alpha \int_0^b x^{\alpha-1}[1 - F(x)]\, dx.$$

Suppose first that the integral on the left converges as $b \to \infty$. The contribution of $\overline{b, \infty}$ to the infinite integral is $\geq b^a[1-F(b)]$, and this quantity therefore tends to zero. In this case the passage to the limit $b \to \infty$ leads from (6.4) to (6.3). On the other hand, the integral on the left is smaller than the integral on the right and hence the convergence of the second entails the convergence of the former, and hence (6.3). ►

An analogue to (6.3) holds for the left tail. Combining the two formulas we get

Lemma 2. *The distribution F possesses an absolute moment of order $\alpha > 0$ iff $|x|^{\alpha-1}[1 - F(x) + F(-x)]$ is integrable over $\overline{0, \infty}$.*

As an application we prove

Lemma 3. *Let $\mathbf{X}$ and $\mathbf{Y}$ be independent random variables, and $\mathbf{S} = \mathbf{X}+\mathbf{Y}$. Then $\mathbf{E}(|\mathbf{S}|^\alpha)$ exists iff both $\mathbf{E}(|\mathbf{X}|^\alpha)$ and $\mathbf{E}(|\mathbf{Y}|^\alpha)$ exist.*

Proof. Since the variables $\mathbf{X}$ and $\mathbf{X}-c$ possess exactly the same moments there is no loss of generality in assuming that 0 is a median for both $\mathbf{X}$ and $\mathbf{Y}$. But then $\mathbf{P}\{|\mathbf{S}| > t\} \geq \frac{1}{2}\mathbf{P}\{|\mathbf{X}| > t\}$, and by the last lemma $\mathbf{E}(|\mathbf{S}|^\alpha) < \infty$ implies $\mathbf{E}(|\mathbf{X}|^\alpha) < \infty$. This proves the "only if" part of the assertion. The "if" part follows from the inequality $|\mathbf{S}|^\alpha \leq 2^\alpha(|\mathbf{X}|^\alpha + |\mathbf{Y}|^\alpha)$ which is valid, because at no point can $|\mathbf{S}|$ exceed the larger of $2|\mathbf{X}|$ and $2|\mathbf{Y}|$.

7. CHEBYSHEV'S INEQUALITY

Chebyshev's inequality is among the most frequently used tools in probability. Both the inequality and its proof are the same as in the discrete case (**1**; IX,6) and we repeat it mainly for reference. Interesting applications will be given in VII,1.

Chebyshev's inequality. *If $\mathbf{E}(\mathbf{X}^2)$ exists*

$$\mathbf{P}\{|\mathbf{X}| \geq t\} \leq t^{-2}\mathbf{E}(\mathbf{X}^2) \qquad t > 0. \tag{7.1}$$

In particular, if $\mathbf{E}(\mathbf{X}) = m$ and $\mathrm{Var}(\mathbf{X}) = \sigma^2$,

$$\mathbf{P}\{|\mathbf{X} - m| \geq t\} \leq \sigma^2/t^2. \tag{7.2}$$

Proof. If F stands for the distribution of $\mathbf{X}$,

$$\mathbf{E}(\mathbf{X}^2) \geq \int_{|x|\geq t} x^2\, F\{dx\} \geq t^2 \int_{|x|\geq t} F\{dx\}$$

which is the same as (7.1). ►

The usefulness of Chebyshev's inequality depends (not on sharp numerical estimates but) on its simplicity and the fact that it is specially adapted to sums of random variables. Many generalizations are possible, but they do not share these desirable properties. (Most of them are so simple that it is better to derive them as occasion arises. For example, a useful combination of Chebyshev's inequality with truncation procedures is described in VII,7.)

A fairly general method for deriving non-trivial inequalities may be described as follows. If $u \geq 0$ everywhere and $u(x) > a > 0$ for all x in an interval I then

$$F\{I\} \leq a^{-1}\mathbf{E}(u(\mathbf{X})). \tag{7.3}$$

On the other hand, if $u \leq 0$ outside I and $u \leq 1$ in I we get the reversed inequality $F\{I\} \geq \mathbf{E}(u(\mathbf{X}))$. Choosing for u polynomials we obtain inequalities depending only on the moments of F.

Examples. (*a*) Let $u(x) = (x+c)^2$ with $c > 0$. Then $u(x) \geq 0$ for all x and $u(x) \geq (t+c)^2$ for $x \geq t > 0$. Therefore

$$\mathbf{P}\{\mathbf{X} > t\} \leq \frac{1}{(t+c)^2}\mathbf{E}((\mathbf{X}+c)^2). \tag{7.4}$$

If $\mathbf{E}(\mathbf{X}) = 0$ and $\mathbf{E}(\mathbf{X}^2) = \sigma^2$ the right side assumes its minimum for $c = \sigma^2/t$ and hence

$$\mathbf{P}\{\mathbf{X} > t\} \leq \frac{\sigma^2}{\sigma^2 + t^2}, \qquad t > 0. \tag{7.5}$$

This interesting inequality was discovered independently by many authors.

(*b*) Let $\mathbf{X}$ be positive (that is, $F(0) = 0$) and $\mathbf{E}(\mathbf{X}) = 1$, $\mathbf{E}(\mathbf{X}^2) = b$. The polynomial $u(x) = h^{-2}(x-a)(a+2h-x)$ is positive only for $a < x < a + 2h$, and $u(x) \leq 1$ everywhere. When $0 < a < 1$ it is readily seen that $\mathbf{E}(u(\mathbf{X})) \geq [2h(1-a) - b]h^{-2}$. Choosing $h = b(1-a)^{-1}$ we get by the remark preceding these examples

$$\mathbf{P}\{\mathbf{X} > a\} \geq (1-a)^2 b^{-1}. \tag{7.6}$$

(*c*) If $\mathbf{E}(\mathbf{X}^2) = 1$ and $\mathbf{E}(\mathbf{X}^4) = M$, the last inequality applied to $\mathbf{X}^2$ shows that

$$\mathbf{P}\{|\mathbf{X}| > t\} \geq (1-t^2)^2 M^{-1} \qquad \text{if } 0 < t < 1. \tag{7.7}$$ ►

For Kolmogorov's generalization of Chebyshev's inequality see section 8(*e*).

8. FURTHER INEQUALITIES. CONVEX FUNCTIONS

The inequalities collected in this section are of widespread use and are by no means typical for probability. Most common is Schwarz' inequality. The others are given mainly because of their use in stochastic processes and statistics. (This section is meant for reference rather than for reading.)

(a) Schwarz' Inequality

In its probabilistic version this inequality states that for two arbitrary random variables φ and ψ defined on the same space

$$(\mathbf{E}(\varphi\psi))^2 \leq \mathbf{E}(\varphi^2)\,\mathbf{E}(\psi^2) \tag{8.1}$$

whenever these expectations exist. Furthermore, the equality sign holds only if a linear combination $a\varphi + b\psi$ is zero with probability one. More generally, if F is an arbitrary measure on the set A then

$$\left(\int_A \varphi(x)\,\psi(x)\,F\{dx\}\right)^2 \leq \int_A \varphi^2(x)\,F\{dx\}\cdot\int_A \psi^2(x)\,F\{dx\} \tag{8.2}$$

for arbitrary functions for which the integrals on the right exist. Taking for F the purely atomic measure attaching unit weight to integers we get *Schwarz' inequality for sums* in the form

$$\left(\sum \varphi_i\psi_i\right)^2 \leq \sum \varphi_i^2 \sum \psi_i^2. \tag{8.3}$$

In view of the importance of (8.1) we give two proofs pointing to different generalizations. The same proofs apply to (8.2) and (8.3).

First proof. We may assume $\mathbf{E}(\psi^2) > 0$. Then

$$\mathbf{E}(\varphi+t\psi)^2 = \mathbf{E}(\varphi^2) + 2t\,\mathbf{E}(\varphi\psi) + t^2\,\mathbf{E}(\psi^2) \tag{8.4}$$

is a quadratic polynomial in t which, being non-negative, has either two complex roots or a double root λ. The standard solution for quadratic equations shows in the first case that (8.1) holds with strict inequality. In the second case $\mathbf{E}(\varphi+t\psi)^2 = 0$ and so $\varphi + t\psi = 0$ except on a set of probability zero.

Second proof. As we are free to replace φ and ψ by constant multiples $a\varphi$ and $b\psi$ it suffices to consider the case $\mathbf{E}(\varphi^2) = \mathbf{E}(\psi^2) = 1$. Then (8.1) follows trivially taking expectations in the inequality $2\,|\varphi\psi| \leq \varphi^2 + \psi^2$. ►

(b) Convex Functions. Jensen's inequality

Let u be a function defined on an open interval I, and $P = (\xi, u(\xi))$ a point on its graph. A line L passing through P is said to *support* u *at* ξ if the graph of u lies entirely above or on L. (This excludes vertical lines.) In analytical terms it is required that

$$u(x) \geq u(\xi) + \lambda\cdot(x-\xi) \tag{8.5}$$

for all x in I, where λ is the slope of L. The function u *is called convex in* I *if a supporting line exists at each point* x of I. (The function u is *concave*, if $-u$ is convex.)

We proceed to show that this definition implies the various properties intuitively associated with convexity as exemplified by convex polygonal lines.

Let F be an arbitrary probability distribution concentrated on I and suppose that the expectation $\mathbf{E}(\mathbf{X})$ exists. Choosing $\xi = \mathbf{E}(\mathbf{X})$ and taking expectations in (8.5) we get

$$\mathbf{E}(u) \geq u(\mathbf{E}(\mathbf{X})) \tag{8.6}$$

whenever the expectation on the left exists. This statement is known as *Jensen's inequality.*

By far the most important is the case where F is concentrated at two points x_1 and x_2 and attributes weights $1-t$ and t to them. Then (8.6) takes on the form

$$(1-t)\,u(x_1) + t\,u(x_2) \geq u((1-t)x_1 + tx_2). \tag{8.7}$$

This inequality admits of a simple geometric interpretation which we state in the following.

Theorem 1. *The function u is convex iff all its chords lie above or on the graph of u.*

Proof. (i) *Necessity.* Let u be convex and consider the chord over an arbitrary interval $\overline{x_1, x_2}$. As t runs from 0 to 1 the point $(1-t)x_1 + tx_2$ runs through the interval $\overline{x_1, x_2}$ and the left side in (8.7) is the ordinate of the corresponding point on the chord. Thus (8.7) states that the points of the chord lie above or on the graph.

(ii) *Sufficiency.* Assume that u has the stated property and consider the triangle formed by three points P_1, P_2, P_3 on the graph of u with abscissas $x_1 < x_2 < x_3$. Then P_2 lies below the chord P_1P_3, and among the three sides of the triangle P_1P_2 has the smallest slope, P_2P_3 the largest. Outside the interval $\overline{x_2, x_3}$ the graph of u therefore lies above the line P_2P_3. Now consider x_3 as a variable and let $x_3 \to x_2+$. The slope of P_2P_3 decreases monotonically but is bounded from below by the slope of P_1P_2. Thus the lines P_2P_3 tend to a line L through P_2. Outside $\overline{x_2, x_3}$ the graph of u is above the line P_2P_3, and hence the whole graph lies above or on L. Thus L supports u at x_2, and as x_2 is arbitrary, this proves the convexity of u. ►

Being the limit of chords, the line L is a right tangent. In the limiting process the abscissa x_3 of P_3 tends to x_2, and P_3 to a point on L. Thus $P_3 \to P_2$. The same argument applies for an approach from the left, and we conclude that the graph of u is continuous and possesses right and left tangents at each point. Furthermore, these tangents are supporting lines and their slopes vary monotonically. Since a monotone function has at most denumerably many discontinuities we have proved

Theorem 2. *A convex function possesses right and left derivatives at all points, and these are non-decreasing functions. They are the same except possibly at countably many points.*

Obviously this theorem again expresses necessary and sufficient conditions for convexity. In particular, if a second derivative exists, u is convex iff $u'' \geq 0$.

Usually (8.7) is taken as definition of convexity. For $t = \frac{1}{2}$ we get the inequality

$$u\left(\frac{x_1+x_2}{2}\right) \leq \frac{u(x_1)+u(x_2)}{2} \tag{8.8}$$

stating that the *midpoint* of the chord lies above or on the graph of u. If u is continuous this property guarantees that the graph can never cross a chord and hence that u is convex. It can be shown more generally that *any Baire function*[18] *satisfying* (8.8) *is convex.*

(c) Moment Inequalities

We prove that for any random variable $\mathbf{X}$

$$u(t) = \log \mathbf{E}(|\mathbf{X}|^t), \qquad t \geq 0, \tag{8.9}$$

is a convex function of t in every interval in which the integral exists. In fact, by Schwarz' inequality (8.1)

$$\mathbf{E}^2(|\mathbf{X}|^t) \leq \mathbf{E}(|\mathbf{X}|^{t+h})\,\mathbf{E}(|\mathbf{X}|^{t-h}), \qquad 0 \leq h \leq t, \tag{8.10}$$

provided the integrals converge. Putting $x_1 = t - h$ and $x_2 = t + h$ we see that (8.8) holds and so u is convex as asserted.

Since $u(0) \leq 0$ the slope $t^{-1}\,u(t)$ of the line joining the origin to $(t, u(t))$ varies monotonically and hence $(\mathbf{E}(|\mathbf{X}|^t))^{1/t}$ *is a non-decreasing function of* $t > 0$.

(d) Hölder's Inequality

Let $p > 1, q > 1$ *and* $p^{-1} + q^{-1} = 1$. *Then for* $\varphi \geq 0$, $\psi \geq 0$

$$\mathbf{E}(\varphi\psi) \leq (\mathbf{E}(\varphi^p))^{1/p}\,(\mathbf{E}(\psi^q))^{1/q} \tag{8.11}$$

whenever the integrals exist.

(Schwarz' inequality (8.1) is the special case $p = q = \frac{1}{2}$, and (8.2) and (8.3) generalize similarly.)

Proof. For $x > 0$ the function $u = \log x$ is concave, that is, it satisfies (8.7) with the inequality reversed. Taking antilogarithms we get for $x_1, x_2 > 0$

$$x_1^{1-t}x_2^t \leq (1 - t)x_1 + tx_2 \tag{8.12}$$

As in the second proof of Schwarz' inequality it suffices to consider integrands normed by $\mathbf{E}(\varphi^p) = \mathbf{E}(\psi^q) = 1$. Let $t = q^{-1}$ and $1 - t = p^{-1}$. The assertion $\mathbf{E}(\varphi\psi) \leq 1$ then follows directly taking expectations in (8.12) with $x_1 = \varphi^p$ and $x_2 = \psi^q$. ►

[18] *Every* u satisfying (8.8) is either convex, or else its oscillations in *every* interval range from $-\infty$ to ∞. See G. H. Hardy, J. E. Littlewood, and G. Plóya, *Inequalities*, Cambridge, England, 1934, in particular p. 91.

(e) Kolmogorov's Inequality

Let $\mathbf{X}_1, \ldots, \mathbf{X}_n$ *be independent random variables with finite variances and* $\mathbf{E}(\mathbf{X}_k) = 0$. *Then for any* $x > 0$

$$\mathbf{P}\{\max [|\mathbf{S}_1|, \ldots, |\mathbf{S}_n|] > x\} \leq x^{-2}\, \mathbf{E}(\mathbf{S}_n^2) \tag{8.13}$$

This important strengthening of Chebyshev's inequality was derived for discrete variables in **1**; IX, 7. The proof carries over without change, but we rephrase it in a form which will make it evident that Kolmogorov's inequality applies more generally to submartingales. We shall return to this point in VII, 9.

Proof. Put $x^2 = t$. For fixed t and $j = 1, 2, \ldots, n$ denote by A_j the event that $\mathbf{S}_j^2 > t$, but $\mathbf{S}_\nu^2 \leq t$ for all subscripts $\nu < j$. In words, A_j is the event that j is the *smallest* among the subscripts k for which $\mathbf{S}_k^2 > t$. Of course, such an index j need not exist, and the union of the events A_j is precisely the event occurring on the left side of Kolmogorov's inequality. Since the events A_j are mutually exclusive this inequality may be restated in the form

$$\sum_{j=1}^{n} \mathbf{P}\{A_j\} \leq t^{-1}\, \mathbf{E}(\mathbf{S}_n^2) \tag{8.14}$$

Denote by $\mathbf{1}_{A_j}$ the indicator of the event A_j, that is, $\mathbf{1}_{A_j}$ is a random variable which equals 1 on A_j and equals 0 on the compliment of A_j. Then $\sum \mathbf{1}_{A_j} \leq 1$ and so

$$\mathbf{E}(\mathbf{S}_n^2 \mathbf{1}) \geq \sum_{j=1}^{n} \mathbf{E}(\mathbf{S}_n^2 \mathbf{1}_{A_j}). \tag{8.15}$$

We shall show that

$$\mathbf{E}(\mathbf{S}_n^2 \mathbf{1}_{A_j}) \geq \mathbf{E}(\mathbf{S}_j^2 \mathbf{1}_{A_j}). \tag{8.16}$$

Since $\mathbf{S}_j^2 > t$ whenever A_j occurs, the right side is $\geq t\mathbf{P}\{A_j\}$, and so (8.15) reduces to the assertion (8.14).

To prove (8.16) we note that $\mathbf{S}_n = \mathbf{S}_j + (\mathbf{S}_n - \mathbf{S}_j)$ and hence

$$\mathbf{E}(\mathbf{S}_n^2 \mathbf{1}_{A_j}) \geq \mathbf{E}(\mathbf{S}_j^2 \mathbf{1}_{A_j}) + 2\mathbf{E}((\mathbf{S}_n - \mathbf{S}_j)\mathbf{S}_j \mathbf{1}_{A_j}). \tag{8.17}$$

The second term on the right vanishes because the variables $\mathbf{S}_n - \mathbf{S}_j = \mathbf{X}_{j+1} + \cdots + \mathbf{X}_n$ and $\mathbf{S}_j \mathbf{1}_{A_j}$ are independent and so the multiplication rule applies to their expectations. Thus (8.17) reduces to the assertion (8.16). ►

9. SIMPLE CONDITIONAL DISTRIBUTIONS. MIXTURES

In III,2 we introduced a "conditional density of a random variable $\mathbf{Y}$ for a given value of another variable $\mathbf{X}$" in the case where the joint distribution of $\mathbf{X}$ and $\mathbf{Y}$ has a continuous density. Without any attempt at generality we proceed to define an analogous concept for a wider class of distributions. (A systematic theory is developed in sections 10 and 11*a*.)

For any pair of intervals A and B on the line put

$$Q(A, B) = \mathbf{P}\{\mathbf{X} \in A, \mathbf{Y} \in B\}. \tag{9.1}$$

With this notation the *marginal distribution for* $\mathbf{X}$ is given by

$$\mu\{A\} = Q(A, \mathfrak{R}^1). \tag{9.2}$$

If $\mu\{A\} > 0$ the conditional probability of the event $\{\mathbf{Y} \in B\}$ given $\{\mathbf{X} \in A\}$ is

$$\mathbf{P}\{\mathbf{Y} \in B \mid \mathbf{X} \in A\} = \frac{Q(A, B)}{\mu\{A\}}. \tag{9.3}$$

(If $\mu\{A\} = 0$ this conditional probability is not defined.) We use this formula when A is the interval $A_h = \overline{x, x + h}$ and let $h \to 0+$. Under appropriate regularity conditions the limit

$$q(x, B) = \lim_{h \to 0} \frac{Q(A_h, B)}{\mu\{A_h\}} \tag{9.4}$$

will exist for all choices of x and B. Following the procedure and reasoning used in III,2 we write in this case

$$q(x, B) = \mathbf{P}\{\mathbf{Y} \in B \mid \mathbf{X} = x\} \tag{9.5}$$

and call q "*the conditional probability of the event* $\{\mathbf{Y} \in B\}$ *given that* $\mathbf{X} = x$." This constitutes an extension of the notion of conditional probabilities to situations in which the "hypothesis" has zero probability. No difficulties arise when q is insufficiently regular, but we shall not analyze the appropriate regularity conditions because a general procedure will be discussed in the next section. This naïve approach usually suffices in individual cases, and the form of the conditional distribution can frequently be derived by intuitive reasoning.

Examples. (*a*) Suppose that the pair $\mathbf{X}, \mathbf{Y}$ has a joint density given by $f(x, y)$. For simplicity we assume that f is continuous and strictly positive. Then

$$q(x, B) = \frac{1}{f_1(x)} \int_B f(x, y)\, dy$$

where $f_1(x) = q(x, \overline{-\infty, \infty})$ is the marginal density of $\mathbf{X}$. In other words, for fixed x the set function q has a density given by $f(x, y)/f_1(x)$.

(*b*) Let $\mathbf{X}$ and $\mathbf{Y}$ be independent random variables with distributions F and G, respectively. For simplicity we assume that $\mathbf{X} > 0$ [that is,

$F(0) = 0$]. Consider the product $\mathbf{Z} = \mathbf{XY}$. Then

$$\mathbf{P}\{\mathbf{Z} \leq t \mid \mathbf{X} = x\} = G(t/x) \tag{9.6}$$

and the distribution function U of $\mathbf{Z}$ is obtained by integrating (9.6) with respect to F. [See II,(3.1). The assertion is a special case of formula (9.8) below.] In particular, when $\mathbf{X}$ is distributed uniformly over $\overline{0, 1}$

$$U(t) = \int_0^1 G(t/x)\, dx. \tag{9.7}$$

This formula can be used as a convenient starting point for the theory of unimodal distributions.[19]

For a further example see problems 18–19. ►

The following theorem (due to L. Shepp) is a probabilistic version of a formal criterion found by A. Khintchine.

Theorem. *U is unimodal iff it is of the form* (9.7), *that is, iff it is the distribution of the product* $\mathbf{Z} = \mathbf{XY}$ *of two independent variables such that* $\mathbf{X}$ *is distributed uniformly in* $\overline{0, 1}$.

Proof. Choose $h > 0$ and denote by U_h the distribution function whose graph is the polygonal line agreeing with U at the points $0, \pm h, \ldots$. [In other words, $U_h(nh) = U(nh)$ and U_h is linear in the interval between nh and $(n+1)h$.] It is obvious from the definition that U is unimodal iff all U_h are unimodal. Now U_h has a density u_h which is a step function, and every step function with discontinuities at the points nh can be written in the form

$$\sum p_n \cdot \frac{1}{|n|\, h} f\left(\frac{x}{nh}\right) \tag{*}$$

where $f(x) = 1$ for $0 < x < 1$ and $f(x) = 0$ elsewhere. The function (*) is monotone in $\overline{0, \infty}$ and in $\overline{-\infty, 0}$ iff $p_n \geq 0$ for all n, and it is a density if $\sum p_n = 1$. But in this case (*) is the density of the product $\mathbf{Z}_h = \mathbf{XY}_h$ of two independent variables such that $\mathbf{X}$ is distributed uniformly in $\overline{0, 1}$ and $\mathbf{P}\{\mathbf{Y}_h = nh\} = p_n$. We have thus proved that U_h is unimodal iff it is of the form (9.7) with G replaced by an arithmetic distribution G_h concentrated on the points $0, \pm h, \ldots$. Letting $h \to 0$ we get the theorem by monotone convergence.

(See problems 25–26 and problem 10 in XV,9.) ►

Under appropriate regularity conditions $q(x, B)$ will for fixed x represent a probability distribution in B and for fixed B a continuous function in x. Then

$$Q(A, B) = \int_A q(x, B)\, \mu\{dx\}. \tag{9.8}$$

[19] A distribution function U is called *unimodal* with the mode at the origin iff the graph of U is *convex* in $\overline{-\infty, 0}$ and *concave* in $\overline{0, \infty}$ [see 8(*b*)]. The origin may be a point of discontinuity, but apart from this unimodality requires that there exist *a density u which is monotone in $\overline{-\infty, 0}$ and in $\overline{0, \infty}$.* (Intervals of constancy are not excluded.)

In fact, the right side obviously represents a probability distribution in the plane, and the differentiation described in (9.4) leads to $q(x, B)$ Formula (9.8) shows how a given distribution in $\mathfrak{R}^2$ can be expressed in terms of a conditional and a marginal distribution. In the terminology of II,5 it represents the given distribution as a *mixture* of the family of distributions $q(x, B)$ depending on the parameter x with μ serving as the distribution of the randomized parameter.

In practice the procedure is frequently reversed. One *starts* from a "*stochastic kernel*" q, that is a function $q(x, B)$ of a point x and a set B such that for fixed x it is a probability distribution and for fixed B a Baire function. Given an arbitrary probability distribution μ the integral in (9.8) defines probabilities for plane sets of the form (A, B) and hence a probability distribution in the plane. Usually (9.8) is expressed in terms of point functions. Consider a family of distribution functions $G(\theta, y)$ depending on a parameter θ, and a probability distribution μ. A new distribution function is then defined by

$$U(y) = \int_{-\infty}^{+\infty} G(x, y)\, \mu\{dx\}. \tag{9.9}$$

[This formula represents the special case of (9.8) when $A = \overline{-\infty, \infty}$ and $q(x, \overline{-\infty, y}\,]) = G(x, y)$.] Such mixtures occur in **1**; V and are discussed in II,5. In the next section it will be shown that q can always be interpreted as a conditional probability distribution.

Examples. (*c*) If F_1 and F_2 are distributions $pF_1 + (1-p)F_2$ is a mixture $(0 < p < 1)$ and represents a special case of (9.9) when μ is concentrated on two atoms.

(*d*) *Random sums.* Let $\mathbf{X}_1, \mathbf{X}_2, \ldots$ be independent random variables with a common distribution F. Let $\mathbf{N}$ be a random variable independent of the $\mathbf{X}_j$ and assuming the values $0, 1, \ldots$ with positive probabilities $p_0, p_1, \ldots$. We are interested in the random variable $\mathbf{S}_\mathbf{N} = \mathbf{X}_1 + \cdots + \mathbf{X}_\mathbf{N}$. The conditional distribution of $\mathbf{S}_\mathbf{N}$ given that $\mathbf{N} = n$ is $F^{n\star}$, and so the distribution of $\mathbf{S}_\mathbf{N}$ is given by

$$U = \sum_{n=0}^{\infty} p_n F^{n\star}, \tag{9.10}$$

which is a special case of (9.9). In this case each hypothesis $\mathbf{N} = n$ carries a positive probability p_n and so we have conditional probability distributions in the strict sense. Other examples are found in II,5–7. (See problems 21 and 24.)

*10. CONDITIONAL DISTRIBUTIONS

It would be pointless to investigate the precise conditions under which conditional probabilities q can be defined by the differentiation process in (9.4). The main properties of conditional probabilities are embodied in the relation (9.8) expressing probabilities of sets in terms of conditional probabilities, and it is simplest to use (9.8) as *definition* of conditional probabilities. It does not determine q uniquely because if for each set B we have $q(x, B) = \bar{q}(x, B)$ except on a set of μ-measure zero, then (9.8) will remain true with q replaced by $\bar{q}$. This indeterminacy is unavoidable, however. For example, if μ is concentrated on an interval I no natural definition of q is possible for x outside I. By the very nature of things we are really dealing with the whole class of equivalent conditional probabilities and should refer to *a* rather than *the* conditional probability distribution q. In individual cases there usually exists a natural choice dictated by regularity requirements.

For definiteness we consider only events specified by conditions of the form $\mathbf{X} \in A$ and $\mathbf{Y} \in B$, where $\mathbf{X}$ and $\mathbf{Y}$ are given random variables and A, B are Borel sets on the line. Let us begin by examining the different meanings that may be attached to the phrase "conditional probability of the event $\{\mathbf{Y} \in B\}$ for given $\mathbf{X}$." The given value of $\mathbf{X}$ may be either a fixed number or indeterminate. With the second interpretation we have a function of $\mathbf{X}$, that is, a random variable. It will be denoted by $\mathbf{P}\{B \mid \mathbf{X}\}$ or $q(\mathbf{X}, B)$, etc. For the value at a fixed point x we write for emphasis $\mathbf{P}\{\mathbf{Y} \in B \mid \mathbf{X} = x\}$ or $q(x, B)$.

Definition 1. *Let the set* B *be fixed. By* $\mathbf{P}\{\mathbf{Y} \in B \mid \mathbf{X}\}$ *(in words, "a conditional probability of the event* $\{\mathbf{Y} \in B\}$ *for given* $\mathbf{X}$*") is meant a function* $q(\mathbf{X}, B)$ *such that for every set* A *in* $\mathfrak{R}^1$

$$\mathbf{P}\{\mathbf{X} \in A, \mathbf{Y} \in B\} = \int_A q(x, B)\, \mu\{dx\} \tag{10.1}$$

where μ *is the marginal distribution of* $\mathbf{X}$.

When x happens to be an atom the hypothesis $\mathbf{X} = x$ has positive probability and $\mathbf{P}\{\mathbf{Y} \in B \mid \mathbf{X} = x\}$ is already defined by (9.3) with A consisting of the single point x. But in this case (10.1) reduces to (9.3) and our definitions and notations are consistent.

We show that a conditional probability $\mathbf{P}\{\mathbf{Y} \in B \mid \mathbf{X}\}$ *always exists.* In fact, clearly

$$\mathbf{P}\{\mathbf{X} \in A, \mathbf{Y} \in B\} \leq \mu(A). \tag{10.2}$$

* This section should be omitted at first reading.

Considered for fixed B as a function of A the left side defines a finite measure, and (10.2) implies that this measure is absolutely continuous with respect to μ (see the Radon-Nikodym theorem in section 3). This means that our measure is defined by a density q, and so (10.1) is true.

So far the set B was fixed, but the notation $q(x, B)$ was chosen with a view to vary B. In other words, we wish to consider q as a function of two variables, a point x and a set B on the line. It is desired that for fixed x the set function q be a probability measure, which requires that $q(x, \mathcal{R}^1) = 1$ and that for any sequence of non-overlapping sets $B_1, B_2, \ldots$ with union B

$$q(x, B) = \sum q(x, B_k). \tag{10.3}$$

Now if the terms on the right represent conditional probabilities for B_k this sum yields *a* conditional probability for B, but there is an additional consistency requirement that (10.3) be true for our choice of q and *all* x. [Note that definition 1 does not exclude the absurd choice $q(x, B) = 17$ at an individual point x.] It is not difficult to see that it is possible to choose $q(x, B)$ so as to satisfy these conditions.[20] This means that there *exists a conditional probability distribution of* $\mathbf{Y}$ *for given* $\mathbf{X}$ *in the sense* of the following.

Definition 2. *By a conditional probability distribution of* $\mathbf{Y}$ *for given* $\mathbf{X}$ *is meant a function* q *of two variables, a point* x *and a set* B, *such that*

(i) *for a fixed set* B

$$q(\mathbf{X}, B) = \mathbf{P}\{\mathbf{Y} \in B \mid \mathbf{X}\} \tag{10.4}$$

is a conditional probability of the event $\{\mathbf{X} \in B\}$ *for given* $\mathbf{X}$.

(ii) q *is for each* x *a probability distribution.*

In effect a conditional probability distribution is a family of ordinary probability distributions and so the whole theory carries over without

[20] It is easiest to choose *directly* only the values $q(x, B)$ when B is an interval in a dyadic subdivision of $\mathcal{R}^1$. For example, let $B_1 = \overline{0, \infty}$ and $B_2 = \overline{-\infty, 0}$. Choose for $q(x, B_1)$ any conditional probability for B_1 such that $0 \leq q(x, B_1) \leq 1$. Then $q(x, B_2) = 1 - q(x, B_1)$ is automatically a legitimate choice. Partition B_1 into B_{11} and B_{12} and choose $q(x, B_{11})$ subject to $0 \leq q(x, B_{11}) \leq q(x, B_1)$. Put $q(x, B_{12}) = q(x, B_1) - q(x, B_{11})$ and proceed in like manner refining the subdivision indefinitely. The additivity requirement (10.3) then defines $q(x, B)$ for all open sets B and hence for all Borel sets.

This construction depends only on the existence of a so-called *net*, namely a partition of the space into finitely many non-overlapping sets each of which is partitioned in like manner and each point of the space is the unique limit of a contracting sequence of sets appearing in the successive partitions. The assertion is therefore true in $\mathcal{R}^r$ and in many other spaces.

change. Thus *when* q *is given*[21] the following definition introduces a new notation rather than a new concept.

Definition 3. *A conditional expectation* $\mathbf{E}(\mathbf{Y} \mid \mathbf{X})$ *is a function of* $\mathbf{X}$ *assuming at* x *the value*

$$\mathbf{E}(\mathbf{Y} \mid x) = \int_{-\infty}^{+\infty} y\, q(x, dy) \tag{10.5}$$

provided the integral converges (except possibly on an x-set of probability zero).

$\mathbf{E}(\mathbf{Y} \mid \mathbf{X})$ is a function of $\mathbf{X}$, that is a random variable. For clarity it is occasionally preferable to denote its value at an individual point x by $\mathbf{E}(\mathbf{Y} \mid \mathbf{X} = x)$. From the very definition we get

$$\mathbf{E}(\mathbf{Y}) = \int_{-\infty}^{+\infty} \mathbf{E}(\mathbf{Y} \mid x)\, \mu\{dx\} \quad or \quad \mathbf{E}(\mathbf{Y}) = \mathbf{E}(\mathbf{E}(\mathbf{Y} \mid \mathbf{X})). \tag{10.6}$$

*11. CONDITIONAL EXPECTATIONS

We have now defined a conditional expectation $\mathbf{E}(\mathbf{Y} \mid \mathbf{X})$ in terms of a conditional distribution, and this is quite satisfactory as long as one deals only with one fixed pair of random variables $\mathbf{X}, \mathbf{Y}$. However, when one deals with whole families of random variables the non-uniqueness of the individual conditional probabilities leads to serious difficulties, and it is therefore fortunate that it is in practice possible to dispense with this unwieldy theory. Indeed, it turns out that a surprisingly simple and flexible theory of conditional expectation can be developed without any reference to conditional distributions. To understand this theory it is best to begin with a closer scrutiny of the identity (10.5).

Let A be a Borel set *on the line* and denote by $\mathbf{1}_A(\mathbf{X})$ the random variable that equals one whenever $\mathbf{X} \in A$ and zero otherwise. We integrate the two sides in (10.5) with respect to the marginal distribution μ of $\mathbf{X}$, taking the set A as domain of integration. The result may be written in the form

$$\mathbf{E}(\mathbf{Y}\mathbf{1}_A(\mathbf{X})) = \int_A \mathbf{E}(\mathbf{Y} \mid x)\, \mu\{dx\} = \int_{-\infty}^{+\infty} \mathbf{1}_A(x)\, \mathbf{E}(\mathbf{Y} \mid x)\, \mu(dx). \tag{11.1}$$

The variable $\mathbf{X}$ maps the sample space $\mathfrak{S}$ on a real line, and the last integral refers only to functions and measures on this line. The random variable $\mathbf{Y}\mathbf{1}_A(\mathbf{X})$, however, is defined in the original sample space, and therefore a better notation is indicated. Obviously $\mathbf{1}_A(\mathbf{X})$ is the indicator of a set B in $\mathfrak{S}$, namely the set of all those points in $\mathfrak{S}$ at which $\mathbf{X}$

[21] For a more flexible general definition see section 11.

* The theory of this section will be used only in connection with martingales in VI,12 and VII,9.

assumes a value in A. As we saw in IV, 3, the sets B that in this manner correspond to arbitrary Borel sets A on the line form a σ-algebra of sets in $\mathfrak{S}$ which is called *the algebra generated by* X. Thus (11.1) states that $\mathbf{U} = \mathbf{E}(\mathbf{Y} \mid \mathbf{X})$ is a function of $\mathbf{X}$ that satisfies the identity

$$\mathbf{E}(\mathbf{Y}\mathbf{1}_B) = \mathbf{E}(\mathbf{U}\mathbf{1}_B) \tag{11.2}$$

for every set B in the σ-algebra generated by $\mathbf{X}$. We shall see that this relation may be used as a *definition* of conditional expectations, and it is therefore important to understand it properly. A simple example will explain its nature.

Examples. (*a*) We take the plane with coordinate variables $\mathbf{X}$ and $\mathbf{Y}$ as sample space and suppose for simplicity that the probabilities are defined by a strictly positive continuous density $f(x, y)$. The random variable $\mathbf{X}$ assumes a constant value along any line parallel to the y-axis. If A is a set on the x-axis, the corresponding plane set B consists of all such lines passing through a point of A. The left side of (11.2) is the ordinary integral of $y\,f(x, y)$ over this set, and this can be written as an iterated integral. Thus

$$\mathbf{E}(\mathbf{Y}\mathbf{1}_B) = \int_A dx \int_{-\infty}^{+\infty} y\,f(x, y)\,dy. \tag{11.3}$$

The right side of (11.2) is the ordinary integral of a function $\mathbf{U}(x)f_1(x)$, where f_1 is the marginal density of $\mathbf{X}$. Thus in this case (11.2) states that

$$\mathbf{U}(x) = \frac{1}{f_1(x)} \int_{-\infty}^{+\infty} y\,f(x, y)\,dy, \tag{11.4}$$

in accordance with the definition (10.5) of conditional expectation and in accordance with intuition.

(*b*) (*Continuation.*) We show now that (11.2) defines a conditional expectation $\mathbf{U}$ even when no densities exist and the probability distribution in the plane is arbitrary. Given a Borel set A on the x-axis, the left side in (11.2) defines a number $\mu_1\{A\}$. Obviously μ_1 is a measure on the Borel sets of the x-axis. Another such measure is given by the marginal distribution μ of X, which is defined by $\mu\{A\} = \mathbf{E}(\mathbf{1}_B)$. It is therefore obvious that if $\mu\{A\} = 0$ then also $\mu_1\{A\} = 0$. In other words, μ_1 is absolutely continuous with respect to μ_1 and by the Radon-Nikodym theorem of section 3 there exists a function U such that

$$\mu_1\{A\} = \int_A \mathbf{U}(x)\,\mu(dx). \tag{11.5}$$

This differs only notationally from (11.2). Of course, (11.5) remains valid if $\mathbf{U}$ is changed on a set of μ-measure 0, but this non-uniqueness is inherent in the notion of conditional expectation. ►

This example shows that (11.2) may be used to define a conditional expectation $\mathbf{U}(\mathbf{X}) = \mathbf{E}(\mathbf{Y} \mid \mathbf{X})$ for an arbitrary pair of random variables $\mathbf{X}, \mathbf{Y}$ in an arbitrary probability space [provided, of course, that $\mathbf{E}(\mathbf{Y})$ exists]. But this approach leads much further. For example, to define a conditional expectation $\mathbf{E}(\mathbf{Y} \mid \mathbf{X}_1, \mathbf{X}_2)$ with respect to a pair $\mathbf{X}_1, \mathbf{X}_2$ of random variables we can use (11.2) unchanged except that B will now be an arbitrary set in the σ-algebra $\mathfrak{B}$ generated by $\mathbf{X}_1$ and $\mathbf{X}_2$ (see IV,3). Of course, $\mathbf{U}$ will be a function of $\mathbf{X}_1, \mathbf{X}_2$, but we saw in IV,4 that the class of Baire functions of the pair $(\mathbf{X}_1, \mathbf{X}_2)$ coincides with the class of all $\mathfrak{B}$-measurable functions. Thus we may cover all imaginable cases by the following definition first proposed by Doob.

Definition. *Let* $(\mathfrak{S}, \mathfrak{A}, \mathbf{P})$ *be a probability space, and* $\mathfrak{B}$ *a* σ*-algebra of sets in* $\mathfrak{A}$ *(that is,* $\mathfrak{B} \subset \mathfrak{A}$*). Let* $\mathbf{Y}$ *be a random variable with expectation.*

A random variable $\mathbf{U}$ *is called conditional expectation of* $\mathbf{Y}$ *with respect to* $\mathfrak{B}$ *if it is* $\mathfrak{B}$*-measureable and* (11.2) *holds for all sets* B *of* $\mathfrak{B}$*. In this case we write* $\mathbf{U} = \mathbf{E}(\mathbf{Y} \mid \mathfrak{B})$.

In the particular case that $\mathfrak{B}$ *is the* σ*-algebra generated by the random variables* $\mathbf{X}_1, \ldots, \mathbf{X}_r$ *the variable* $\mathbf{U}$ *reduces to a Baire function of* $\mathbf{X}_1, \ldots, \mathbf{X}_r$ *and will be denoted by* $\mathbf{E}(\mathbf{Y} \mid \mathbf{X}_1, \ldots, \mathbf{X}_r)$.

The *existence* of $\mathbf{E}(\mathbf{Y} \mid \mathfrak{B})$ is established by the method indicated in example (*b*) using an abstract Radon-Nikodym theorem.

To see the main properties of the conditional expectation $\mathbf{U} = \mathbf{E}(\mathbf{Y} \mid \mathfrak{B})$ note that (11.2) holds trivially when $\mathbf{1}_B$ is replaced by a linear combination of indicators of sets B_j in $\mathfrak{B}$. But we saw in IV,3 that every $\mathfrak{B}$-measurable function can be uniformly approximated by such linear combinations. Passing to the limit we see that (11.2) implies that more generally $\mathbf{E}(\mathbf{YZ}) = \mathbf{E}(\mathbf{UZ})$ for any $\mathfrak{B}$-measurable function $\mathbf{Z}$. Replacing $\mathbf{Z}$ by $\mathbf{Z}\mathbf{1}_B$ and comparing with the definition (11.2) we see that

$$\mathbf{E}(\mathbf{YZ} \mid \mathfrak{B}) = \mathbf{Z}\,\mathbf{E}(\mathbf{Y} \mid \mathfrak{B}) \tag{11.6}$$

for any $\mathfrak{B}$*-measurable function* $\mathbf{Z}$. This is a relation of great importance.

Finally, consider a σ-algebra $\mathfrak{B}_0 \subset \mathfrak{B}$ and let $\mathbf{U}_0 = \mathbf{E}(\mathbf{Y} \mid \mathfrak{B}_0)$. For a set B in $\mathfrak{B}_0$ we can interpret (11.2) relative to $\mathfrak{B}_0$ as well as relative to $\mathfrak{B}$, and thus we find that for B in $\mathfrak{B}_0$

$$\mathbf{E}(\mathbf{Y}\mathbf{1}_B) = \mathbf{E}(\mathbf{U}\mathbf{1}_B) = \mathbf{E}(\mathbf{U}_0\mathbf{1}_B).$$

Thus by the very definition $\mathbf{U}_0 = \mathbf{E}(\mathbf{U} \mid \mathfrak{B}_0)$, and so

$$\mathbf{E}(\mathbf{Y} \mid \mathfrak{B}_0) = \mathbf{E}(\mathbf{E}(\mathbf{Y} \mid \mathfrak{B})\mathfrak{B}_0) \qquad \text{if} \quad \mathfrak{B}_0 \subset \mathfrak{B}. \tag{11.7}$$

For example, $\mathfrak{B}$ may be the algebra generated by the two variables $\mathbf{X}_1, \mathbf{X}_2$

while $\mathfrak{B}_0$ stands for the algebra generated by $\mathbf{X}_1$ alone. Then (11.7) reduces to $\mathbf{E}(\mathbf{Y} \mid \mathbf{X}_1) = \mathbf{E}(\mathbf{E}(\mathbf{Y} \mid \mathbf{X}_1, \mathbf{X}_2) \mid \mathbf{X}_1)$.

Finally we note that (11.2) implies that for the constant function 1 the conditional expectation equals 1, no matter how $\mathfrak{B}$ is chosen. Thus (11.6) implies that $\mathbf{E}(\mathbf{Z} \mid \mathfrak{B}) = \mathbf{Z}$ *for all* $\mathfrak{B}$*-measurable variables* $\mathbf{Z}$

It is hardly necessary to say that the basic properties of expectation carry over to conditional expectation.

12. PROBLEMS FOR SOLUTION

1. Let $\mathbf{X}$ and $\mathbf{Y}$ be independent variables with distribution functions F and G. Find the distribution functions of[22] (*a*) $\mathbf{X} \cup \mathbf{Y}$, (*b*) $\mathbf{X} \cap \mathbf{Y}$, (*c*) $2\mathbf{X} \cup \mathbf{Y}$, (*d*) $\mathbf{X}^3 \cup \mathbf{Y}$.

2. *Mixtures.* Let $\mathbf{X}, \mathbf{Y}, \mathbf{Z}$ be independent; $\mathbf{X}$ and $\mathbf{Y}$ have distributions F and G, while $\mathbf{P}\{\mathbf{Z} = 1\} = p$, $\mathbf{P}\{\mathbf{Z} = 0\} = q$ $(p + q = 1)$. Find the distribution functions of (*a*) $\mathbf{ZX} + (1 - \mathbf{Z})\mathbf{Y}$, (*b*) $\mathbf{ZX} + (1 - \mathbf{Z})(\mathbf{X} \cup \mathbf{Y})$, (*c*) $\mathbf{ZX} + (1 - \mathbf{Z})(\mathbf{X} \cap \mathbf{Y})$.

3. If F is a continuous distribution function show that

$$\int_{-\infty}^{+\infty} F(x)F\{dx\} = \int_0^1 y\,dy = \tfrac{1}{2}$$

(*a*) from the very definition of the first integral (partitioning $\overline{-\infty, \infty}$ into subintervals) and (*b*) from the interpretation of the left side as $\mathbf{E}(F(\mathbf{X}))$ where $F(\mathbf{X})$ has a uniform distribution. More generally, putting $G(x) = F^n(x)$,

$$\int_{-\infty}^{+\infty} F^k(x)G\{dx\} = \frac{n}{n+k}\,.$$

4. Let $F(x, y)$ stand for a probability distribution in the plane. Put $U(x, y) = 0$ when $x < 0$ and $y < 0$, and $U(x, y) = F(x, y)$ at all other points. Show that U is monotone in each variable but is not a probability distribution. [*Hint:* Consider the mixed differences.]

5. *Prescribed marginal distributions.*[23] Let F and G be distribution functions in $\mathfrak{R}^1$ and

$$U(x, y) = F(x)G(y)[1 + \alpha(1 - F(x))(1 - G(y))]$$

where $|\alpha| \leq 1$. Prove that U is a distribution function in $\mathfrak{R}^2$ with marginal distributions F, G and that U has a density iff F and G have densities.

Hint: If $w(x, y) = u(x)v(y)$, the mixed differences of w [defined in (1.12)] are of the form $\Delta u\, \Delta v$. Note also that $\Delta(F^2) \leq 2\,\Delta F$.

---

[22] If a and b are numbers, $a \cup b = \max(a, b)$ denotes the larger of the two, $a \cap b = \min(a, b)$ the smaller. For functions $f \cup g$ denotes the function which at the point x assumes the value $f(x) \cap g(x)$ (see IV,1). Thus $\mathbf{X} \cup \mathbf{Y}$ and $\mathbf{X} \cap \mathbf{Y}$ are random variables.

[23] This problem contains a new example for a *non-normal distribution with normal marginal distributions* (see problems 2 and 3 in III,9). It is due to E. J. Gumbel.

6. Within the unit square put $U(x, y) = x$ if $x \leq y$ and $U(x, y) = y$ if $x \geq y$. Show that U is a distribution function concentrated at the bisector (hence singular).

7. *Fréchet's maximal distribution with given marginal distributions.* Let F and G be distribution functions in $\mathcal{R}^1$ and $U(x, y) = F(x) \cap G(y)$. Prove: (*a*) U is a distribution function with marginal distributions F and G. (*b*) If V is any other distribution function with this property, then $V \leq U$. (*c*) U is concentrated on the curve defined by $F(x) = G(y)$, and hence singular. (Problem 5 contains a special case.)

8. Denote by U the uniform distribution in $\overline{-h, 0}$ and by T the triangular distribution in $\overline{-h, h}$ [see II, 4(*b*)]. Then $F\bigstar U$ and $F\bigstar T$ have the *densities*

$$h^{-1}[F(x + h) - F(x)] \quad \text{and} \quad h^{-2}\int_0^h [F(x + y) - F(x - y)]\, dy.$$

9. The independent variables $\mathbf{X}$ and $\mathbf{Y}$ have Poisson distributions with expectations pt and qt. If I_k is the Bessel function defined in II, (7.1) show that

$$\mathbf{P}\{\mathbf{X} - \mathbf{Y} = k\} = e^{-t}\sqrt{(p/q)^k}\, I_{|k|}(2t\sqrt{pq}).$$

10. For a distribution function F such that

$$\varphi(\alpha) = \int_{-\infty}^{+\infty} e^{\alpha x} F\{dx\}$$

exists for $-a < \alpha < a$ we define a new distribution $F\#$ by $\varphi(\alpha)F\#\{dx\} = e^{\alpha x}F\{dx\}$. Let F_1 and F_2 be two distributions with this property and $F = F_1 \bigstar F_2$. Prove that (with obvious notations) $\varphi(\alpha) = \varphi_1(\alpha)\, \varphi_2(\alpha)$ and $F\# = F_1^\# \bigstar F_2^\#$.

11. Let F have atoms $a_1, a_2, \ldots$ with weights $p_1, p_2, \ldots$. Denote by p the maximum of $p_1, p_2, \ldots$. Using lemma 1 of section 4*a* prove

(*a*) The atoms of $F\bigstar F$ have weights strictly less than p except if F is concentrated at finitely many atoms of equal weight.

(*b*) For the symmetrized distribution ${}^0F$ the origin is an atom of weight $p' = \sum p_\nu^2$. The weights of the other atoms are strictly less than p'.

12. *Random vectors in* $\mathcal{R}^3$. Let $\mathbf{L}$ be the resultant of two independent unit vectors with random directions (that is, the endpoints are distributed uniformly over the unit sphere). Show that $\mathbf{P}\{\mathbf{L} \leq t\} = t^2/2$ for $0 < t < 2$. [See example 4(*e*).]

13. Let the $\mathbf{X}_k$ be mutually independent variables assuming the values 0 and 1 with probability $\frac{1}{2}$ each. In example 4(*d*) it was shown that $\mathbf{X} = \sum 2^{-k}\mathbf{X}_k$ is uniformly distributed over $\overline{0, 1}$. Show that $\sum 2^{-3k}\mathbf{X}_{3k}$ has a singular distribution.

14. (*a*) If F has a density f such that f^2 is integrable, then the density f_2 of $F\bigstar F$ is bounded.

(*b*) Using the mean approximation theorem of IV,2 show that *if f is bounded then f_2 is continuous.*

[If f is unbounded near a single point it can happen that f^{n*} is unbounded for every n. See example XI,3(a).]

15. Using Schwarz' inequality show that if $\mathbf{X}$ is a positive variable then $\mathbf{E}(\mathbf{X}^{-p}) \geq (\mathbf{E}(\mathbf{X}^p))^{-1}$ for all $p > 0$.

16. Let $\mathbf{X}$ and $\mathbf{Y}$ have densities f and g such that $f(x) \geq g(x)$ for $x < a$ and $f(x) \leq g(x)$ for $x > a$. Prove that $\mathbf{E}(\mathbf{X}) \leq \mathbf{E}(\mathbf{Y})$. Furthermore, if $f(x) = g(x) = 0$ for $x < 0$ then $\mathbf{E}(\mathbf{X}^k) \leq \mathbf{E}(\mathbf{Y}^k)$ for all k.

17. Let $\mathbf{X}_1, \mathbf{X}_2, \ldots$ be mutually independent with the common distribution F. Let $\mathbf{N}$ be a positive integral-valued random variable with generating function $P(s)$. If $\mathbf{N}$ is independent of the $\mathbf{X}_j$ then $\max[\mathbf{X}_1, \ldots, \mathbf{X}_\mathbf{N}]$ has the distribution $P(F)$.

18. Let $\mathbf{X}_1, \ldots, \mathbf{X}_n$ be mutually independent with a continuous distribution F. Let $\mathbf{X} = \max[\mathbf{X}_1, \ldots, \mathbf{X}_n]$ and $\mathbf{Y} = \min[\mathbf{X}_1, \ldots, \mathbf{X}_n]$. Then

$$\mathbf{P}\{\mathbf{X} \leq x, \mathbf{Y} > y\} = (F(x) - F(y))^n \quad \text{for} \quad y < x$$

and

$$\mathbf{P}\{\mathbf{Y} > y \mid \mathbf{X} = x\} = [(F(x)-F(y))/F(x)]^{n-1}.$$

19. Using the same notations one has for each fixed $k \leq n$

$$\mathbf{P}\{\mathbf{X}_k \leq x \mid \mathbf{X} = t\} = \begin{cases} \dfrac{n-1}{n}\dfrac{F(x)}{F(t)} & \text{for} \quad x < t \\ 1 & \text{for} \quad x \geq t. \end{cases}$$

Derive this (a) by an intuitive argument considering the event $\{\mathbf{X}_k = \mathbf{X}\}$, and ($b$) formally from (9.4).

20. *Continuation.* Prove that

$$\mathbf{E}(\mathbf{X}_k \mid \mathbf{X} = t) = \frac{n-1}{n}\frac{1}{F(t)}\int_{-\infty}^{t} yF\{dy\} + \frac{t}{n}.$$

21. *Random sums.* In example 9(c) let $\mathbf{X}_k$ equal 1 and -1 with probabilities p and $q = 1 - p$. If $\mathbf{N}$ is a Poisson variable with expectation t the distribution of $S_\mathbf{N}$ is identical with the distribution occurring in problem 9.

22. *Mixtures.* Let the distribution G in (9.9) have expectation $m(x)$ and variance $\sigma^2(x)$. Prove that the mixture U has expectation and variance

$$a = \int_{-\infty}^{+\infty} m(x)\mu\{dx\}, \qquad b = \int_{-\infty}^{+\infty} \sigma^2(x)\mu\{dx\} + \int_{-\infty}^{\infty} (m^2(x) - a^2)\mu\{dx\}.$$

23. With obvious notations $\mathbf{E}(\mathbf{E}(\mathbf{Y} \mid \mathbf{X})) = \mathbf{E}(\mathbf{Y})$ but

$$\text{Var}(\mathbf{Y}) = \mathbf{E}(\text{Var}(\mathbf{Y} \mid \mathbf{X})) + \text{Var}(\mathbf{E}(\mathbf{Y} \mid \mathbf{X})).$$

Problem 22 is a special case.

24. *Random sums.* In example 9(c), $\mathbf{E}(\mathbf{S}_\mathbf{N}) = \mathbf{E}(\mathbf{N})\mathbf{E}(\mathbf{X})$,

$$\text{Var}(\mathbf{S}_\mathbf{N}) = \mathbf{E}(\mathbf{N})\,\text{Var}(\mathbf{X}) + (\mathbf{E}(\mathbf{X}))^2\,\text{Var}(\mathbf{N}).$$

Prove this directly and show that it is contained in the last two problems.

Note. The following problems refer to *convolutions of unimodal distributions* defined in footnote 19 of section 9. It has been conjectured that the convolution of two such distributions is again unimodal. One counterexample is due to K. L. Chung, and problem 25 contains another. Problem 26 shows the conjecture to be valid for *symmetric*[24] distributions. This result is due to A. Wintner.

---

[24] For the difficulties arising in the unsymmetric case see I. A. Ibragimov, Theory of Probability and Its Applications, vol. 1 (1956) pp. 225–260. [Translations.]

25. Let $u(x) = 1$ for $0 < x < 1$ and $u(x) = 0$ elsewhere. Put

$$v(x) = \frac{\epsilon}{a}\, u\left(\frac{x}{a}\right) + \frac{1-\epsilon}{b}\, u\left(\frac{x}{b}\right)$$

where $0 < a < b$. If ϵ and a are small and b large, then $w = v \star v$ is not unimodal although v is.

Hint to avoid calculations: The convolution of two uniform densities is the triangular density and hence $w(a) > \epsilon^2 a^{-1}$ and $w(b) > \epsilon^2 b^{-1}$ and the integral of w from b to $2b$ is $> \frac{1}{2}(1-\epsilon)^2$. It follows that w must have a minimum between a and b.

26. Let F be a uniform distribution and G unimodal. If both F and G are symmetric show by simple differentiation that the convolution $F \bigstar G$ is unimodal. Conclude (without further calculations) that the statement remains true when F is any mixture of symmetric uniform distributions, and hence that *the convolution of symmetric unimodal distributions is unimodal.*

CHAPTER VI

A Survey of Some Important Distributions and Processes

This chapter is the product of the deplorable need to avoid repetition and cross references between chapters intended for independent reading. For example, the theory of stable distributions will be developed independently by semi-group methods (IX), by Fourier analysis (XVII), and—at least partly—by Laplace transforms (XIII). Giving the definitions and examples at a neutral place is economical and makes it possible to scrutinize some basic relations without regard to purity of methods.

The miscellaneous topics covered in this chapter are not necessarily logically connected: the queuing process has little to do with martingale theory or stable distributions. The chapter is not intended for consecutive reading; the individual sections should be taken up as occasion arises or when their turn comes up. Sections 6–9 are somewhat interrelated, but independent of the rest. They treat some important material not covered elsewhere in the book.

1. STABLE DISTRIBUTIONS IN $\mathcal{R}^1$

Stable distributions play a constantly increasing role as a natural generalization of the normal distribution. For their description it is convenient to introduce the short-hand notation

$$\mathbf{U} \stackrel{d}{=} \mathbf{V} \tag{1.1}$$

to indicate that the random variables $\mathbf{U}$ and $\mathbf{V}$ have the same distribution. Thus $\mathbf{U} \stackrel{d}{=} a\mathbf{V} + b$ means that the distributions of $\mathbf{U}$ and $\mathbf{V}$ differ only by location and scale parameters. (See definition 1 in V,2.) Throughout this section $\mathbf{X}, \mathbf{X}_1, \mathbf{X}_2, \ldots$ denote mutually independent random variables with a common distribution R and $\mathbf{S}_n = \mathbf{X}_1 + \cdots + \mathbf{X}_n$.

Definition 1. *The distribution* R *is stable* (*in the broad sense*) *if for each* n *there exist constants* $c_n > 0$, γ_n *such that*[1]

$$\mathbf{S}_n \stackrel{d}{=} c_n\mathbf{X} + \gamma_n \tag{1.2}$$

and R *is not concentrated at one point.* R *is stable in the strict sense if* (1.2) *holds with* $\gamma_n = 0$.

Examples will be found in section 2. An elementary derivation of some basic properties of stable distributions is so instructive that we proceed with it at the cost of some repetition. The systematic theory developed in chapters IX and XVII does not depend on the following discussion.

Theorem 1. *The norming constants are of the form* $c_n = n^{1/\alpha}$ *with* $0 < \alpha \leq 2$. *The constant* α *will be called the characteristic exponent of* R.

Proof. The argument is greatly simplified by symmetrization. If R is stable so is the distribution ${}^0R$ of $\mathbf{X}_1 - \mathbf{X}_2$ and the norming constants c_n are the same. It suffices therefore to prove the assertion for a *symmetric* stable R.

We start from the simple remark that $\mathbf{S}_{m+n}$ is the sum of the *independent* variables $\mathbf{S}_m$ and $\mathbf{S}_{m+n} - \mathbf{S}_m$ distributed, respectively, as $c_m\mathbf{X}$ and $c_n\mathbf{X}$ Thus for symmetric stable distributions

$$c_{m+n}\mathbf{X} \stackrel{d}{=} c_m\mathbf{X}_1 + c_n\mathbf{X}_2. \tag{1.3}$$

Similarly, the sum $\mathbf{S}_{rk}$ can be broken up into r independent blocks of k terms each, whence $c_{rk} = c_r c_k$ for all r and k. For subscripts of the form $n = r^\nu$ we conclude by induction that

$$\text{if} \quad n = r^\nu \qquad \text{then} \quad c_n = c_r^\nu. \tag{1.4}$$

Next put $v = m + n$ and note that because of the symmetry of the variables in (1.3) we have for $t > 0$

$$\mathbf{P}\{\mathbf{X} > t\} \geq \tfrac{1}{2}\mathbf{P}\{\mathbf{X}_2 > tc_v/c_n\}. \tag{1.5}$$

It follows that for $v > n$ the ratios c_n/c_v remain bounded.

To any integer r there exists a unique α such that $c_r = r^{1/\alpha}$. To prove that $c_n = n^{1/\alpha}$ it suffices to show that if $c_\rho = \rho^{1/\beta}$ then $\beta = \alpha$. Now by (1.4)

$$\text{if} \quad n = r^j \qquad \text{then} \quad c_n = n^{1/\alpha}$$
$$\text{if} \quad v = \rho^k \qquad \text{then} \quad c_v = v^{1/\beta}.$$

But for each $v = \rho^k$ there exists an $n = r^j$ such that $n < v \leq rn$. Then

$$c_v = v^{1/\beta} \leq (rn)^{1/\beta} = r^{1/\beta}c_n^{\alpha/\beta}.$$

[1] For an alternative form see problem 1.

Since the ratios c_n/c_ν remain bounded this implies that $\beta \leq \alpha$. Interchanging the roles of r and ρ we find similarly that $\beta \geq \alpha$ and hence $\beta = \alpha$.

To prove that $\alpha \leq 2$ we remark that the normal distribution is stable with $\alpha = 2$. For it (1.3) reduces to the addition rule for variances, and the latter implies that any stable distribution with finite variances necessarily corresponds to $\alpha = 2$. To conclude the proof it suffices therefore to show that any stable distribution with $\alpha > 2$ would have a finite variance.

For symmetric distributions (1.2) holds with $\gamma_n = 0$, and hence we can choose a t such that $\mathbf{P}\{|\mathbf{S}_n| > tc_n\} < \frac{1}{4}$ for all n. For reasons of symmetry this implies that $n[1-R(tc_n)]$ remains bounded [see V,(5.10)]. It follows that $x^\alpha[1 - R(x)] < M$ for all $x > t$ and an appropriate constant M. Thus the contribution of the interval $2^{k-1} < x \leq 2^k$ to the integral for $\mathbf{E}(\mathbf{X}^2)$ is bounded by $M2^{(2-\alpha)k}$, and for $\alpha > 2$ this would be the general term of a convergent series. ▶

The theory of stable distributions simplifies greatly by the gratifying fact that the centering constants γ_n may be disregarded in practice. This is so because we are free to center the distribution R in an arbitrary manner, that is, we may replace $R(x)$ by $R(x+b)$. The next theorem shows that, except when $\alpha = 1$, we can use this freedom to eliminate γ_n from (1.2).

Theorem 2. *If R is stable with an exponent $\alpha \neq 1$ the centering constant b may be chosen so that $R(x + b)$ is strictly stable.*

Proof. $\mathbf{S}_{mn}$ is the sum of m independent variables each distributed as $c_n\mathbf{X} + \gamma_n$. Accordingly

$$\mathbf{S}_{mn} \stackrel{d}{=} c_n\mathbf{S}_m + m\gamma_n \stackrel{d}{=} c_nc_m\mathbf{X} + c_n\gamma_m + m\gamma_n. \tag{1.6}$$

Since m and n play the same role this means that we have identically

$$(c_n-n)\gamma_m = (c_m-m)\gamma_n. \tag{1.7}$$

When $\alpha = 1$ this statement is empty,[2] but when $\alpha \neq 1$ it implies that $\gamma_n = b(c_n-n)$ for all n. From (1.2) one sees finally that the sum $\mathbf{S}'_n$ of n variables distributed as $\mathbf{X}' - b$ satisfies the condition $\mathbf{S}'_n \stackrel{d}{=} c_n\mathbf{X}'$. ▶

The relation (1.3) was derived from (1.2) under the sole assumption that $\gamma_n = 0$ and holds therefore for all strictly stable distributions. It implies that

$$s^{1/\alpha}\mathbf{X}_1 + t^{1/\alpha}\mathbf{X}_2 \stackrel{d}{=} (s+t)^{1/\alpha}\mathbf{X} \tag{1.8}$$

whenever the ratio s/t is rational A simple continuity argument[3] leads to

[2] For the case $\alpha = 1$ see problem 4.

[3] Concerning the continuity of stable distributions see problem 2.

Theorem 3. *If* R *is strictly stable with exponent* α *then* (1.8) *holds for all* $s > 0$ *and* $t > 0$.

For the normal distribution (1.8) merely restates the addition rule for the variances. In general (1.8) implies that *all linear combinations* $a_1\mathbf{X}_2 + a_2\mathbf{X}_2$ *belong to the same type.*

The importance of the normal distribution $\mathfrak{N}$ is due largely to the central limit theorem. Let $\mathbf{X}_1, \ldots, \mathbf{X}_n$ be mutually independent variables with a common distribution F having zero expectation and unit variance. Put $\mathbf{S}_n = \mathbf{X}_1 + \cdots + \mathbf{X}_n$. The central limit theorem[4] asserts that the distribution of $\mathbf{S}_n n^{-\frac{1}{2}}$ tends to $\mathfrak{N}$. For distributions without variance similar limit theorems may be formulated, but the norming constants must be chosen differently. The interesting point is that all stable distributions and no others occur as such limits. The following terminology will facilitate the discussion of this problem.

Definition 2. *The distribution* F *of the independent random variables* $\mathbf{X}_k$ *belongs to the domain of attraction of a distribution* R *if there exist norming constants* $a_n > 0$, b_n *such that the distribution of* $a_n^{-1}(\mathbf{S}_n - b_n)$ *tends to* R.

Our last statement can now be reformulated to the effect that *a distribution* R *possesses a domain of attraction iff it is stable.* Indeed, by the very definition each stable R belongs to its own domain of attraction. That no other distribution appears as limit becomes plausible by the argument used in theorem 1.

Our results have important and surprising consequences. Consider, for example, a stable distribution satisfying (1.8) with $\alpha < 1$. The *average* $(\mathbf{X}_1 + \cdots + \mathbf{X}_n)/n$ has the same distribution as $\mathbf{X}_1 n^{-1+1/\alpha}$, and the last factor tends to ∞. Roughly speaking we can say that the average of n variables is likely to be considerably larger than any given component $\mathbf{X}_k$. This is possible only if *the maximal term* $\mathbf{M}_n = \max\,[\mathbf{X}_1, \ldots, \mathbf{X}_n]$ is likely to grow exceedingly large and to receive a preponderating influence on the sum $\mathbf{S}_n$. A closer analysis bears out this conclusion. In the case of positive variables the expectation of the ratio $\mathbf{S}_n/\mathbf{M}_n$ tends to $(1-\alpha)^{-1}$, and this is true also for any sequence $\{\mathbf{X}_n\}$ whose distribution belongs to the domain of attraction of our stable distribution. (See problem 26 of XIII,11.)

Note on history. The general theory of stable distributions was initiated[5] by P. Lévy (1924), who found the Fourier transforms of all strictly stable distributions. (The others

[4] The central limit theorem proves that the normal distribution is the only stable distribution with variance.

[5] The Fourier transforms of *symmetric* stable distributions were mentioned by Cauchy, but it was not clear that they really corresponded to probability distributions. This point was settled by G. Polya for the case $\alpha < 1$. The Holtsmark distribution of example 2(*c*) was known to astronomers, but not to mathematicians.

were originally called quasi-stable. As we have seen, they play a role only when $\alpha = 1$, and this case was analyzed jointly by P. Lévy and H. Khintchine.) A new and simpler approach to the whole theory was made possible by the discovery of infinitely divisible distributions. This new approach (still based on Fourier analysis) is also due to P. Lévy (1937). The interest in the theory was stimulated by W. Doblin's masterful analysis of the domains of attraction (1939). His criteria were the first to involve regularly varying functions. The modern theory still carries the imprint of this pioneer work although many authors have contributed improvements and new results. Chapter XVIII contains a streamlined treatment of the theory by the now classical Fourier methods, while chapter IX presents the same theory by a direct approach which is more in line with modern methods in Markov processes. Great simplifications and a unification of many criteria were made possible by the systematic exploitation of J. Karamata's theory of regularly varying functions. An improved version of this theory is presented in VIII,8–9.

2. EXAMPLES

(*a*) *The normal distribution* centered to zero expectation is strictly stable with $c_n = \sqrt{n}$.

(*b*) *The Cauchy distribution* with arbitrary location parameters has density

$$\frac{1}{\pi}\frac{c}{c^2 + (x-\gamma)^2}.$$

The convolution property II,(4.6) shows that it is stable with $\alpha = 1$.

(*c*) *Stable distribution with* $\alpha = \frac{1}{2}$. The distribution

$$F(x) = 2[1 - \mathfrak{N}(1/\sqrt{x})], \qquad x > 0 \tag{2.1}$$

with density

$$f(x) = \frac{1}{\sqrt{2\pi x^3}}\, e^{-1/2(x)}, \qquad x > 0 \tag{2.2}$$

[and $f(x) = 0$ for $x < 0$] is strictly stable with norming constants $c_n = n^2$.

This can be shown to be elementary integrations, but it is preferable to take the assertion as a consequence of the fact that F has a domain of attraction. Indeed, in a symmetric random walk let $\mathbf{S}_r$ be the epoch of the rth return to the origin. Obviously $\mathbf{S}_r$ is the sum of r independent identically distributed random variables (the waiting times between successive returns). Now it was shown at the end of **1**; III,(7.7) that

$$\mathbf{P}\{\mathbf{S}_r \le r^2 t\} \to F(t) \qquad r \to \infty. \tag{2.3}$$

Thus F has a domain of attraction and is therefore stable. [Continued in example (*e*).]

(*d*) *The gravitational field of stars* (*Holtsmark distribution*). In astronomical terms the problem is to calculate the x-component of the gravitational force exercised by the stellar system at a randomly chosen point O. The underlying idea is that the stellar system appears as a "random aggregate" of points with "randomly varying masses." These notions could be made precise in terms of Poisson distributions, etc., but fortunately no subtleties are required for the problem at hand.

Let us agree to treat the density of the stellar system as a free parameter and to let $\mathbf{X}_\lambda$ stand for the x-component of the gravitational force of a stellar system with density λ. We seek the conceivable types of such distributions. Now the intuitive notion of a "random aggregate of stars" presupposes that two independent aggregates with densities s and t may be combined into a single aggregate of density $s + t$. Probabilistically this amounts to the postulate that the sum of two independent variables distributed as $\mathbf{X}_s$ and $\mathbf{X}_t$ should have the same distribution as $\mathbf{X}_{s+t}$. We indicate this symbolically by

$$\mathbf{X}_s + \mathbf{X}_t \stackrel{\mathrm{d}}{=} \mathbf{X}_{s+t}. \tag{2.4}$$

Considering that a change of density from 1 to λ amounts to a change of the unit of length from 1 to $1/\sqrt[3]{\lambda}$ and that the gravitational force varies inversely with the square of the distance we see that $\mathbf{X}_t$ must have the same distribution as $t^{\frac{2}{3}}\mathbf{X}_1$. This means that the distributions of $\mathbf{X}_t$ differ only by a scale parameter and (2.4) reduces to (1.8) with $\alpha = \frac{3}{2}$. In other words, $\mathbf{X}_\lambda$ *has a symmetric stable distribution with exponent* $\frac{3}{2}$. It will turn out that (up to the trivial scale parameter) there exists exactly one such distribution, and so we have solved our problem without appeal to deeper theory. The astronomer Holtsmark obtained an equivalent answer by other methods (see problem 7) and, remarkably, before P. Lévy's work.

(*e*) *First-passage times in Brownian motion.* We start from the notion of a one-dimensional diffusion process, that is, we suppose that the increments $\mathbf{X}(s+t) - \mathbf{X}(s)$ for non-overlapping time intervals are independent and have a symmetric normal distribution with variance t. We assume as known that the paths depend *continuously* on time. If $\mathbf{X}(0) = 0$ there exists an epoch $\mathbf{T}_a$ *at which the particle reaches the position* $a > 0$ *for the first time.* To derive the distribution function $F_a(t) = \mathbf{P}\{\mathbf{T}_a \leq t\}$ we observe that the notion of an additive process presupposes a complete lack of after-effect (the strong Markov property). This means that the increment $\mathbf{X}(t+\mathbf{T}_a) - a$ of the abscissa between epochs $\mathbf{T}_a$ and $\mathbf{T}_a + t$ is independent of the process before $\mathbf{T}_a$. Now to reach a position $a+b > a$ the particle must first reach a, and we conclude that the residual waiting time $\mathbf{T}_{a+b} - \mathbf{T}_a$ before reaching $a + b$ is independent of $\mathbf{T}_a$ and has the same distribution as $\mathbf{T}_b$. In other words, $F_a \star F_b = F_{a+b}$. But the transition probabilities depend only on the ratio x^2/t and therefore $\mathbf{T}_a$ must have the same distribution as $a^2\mathbf{T}_1$. This means that the distributions F_a differ only by a scale parameter and hence they are *stable with exponent* $\alpha = \frac{1}{2}$.

This argument, based on dimensional analysis, proves the stability of the first-passage distribution but does not lead to an explicit form. To show that F *coincides with the distribution of example* (*c*) we use a reasoning based on symmetry (the so-called reflection principle). Because of the assumed continuity of paths the event $\{\mathbf{X}(t) > a\}$ can occur only if the level a has been crossed at some epoch $\mathbf{T}_a < t$. Given that $\mathbf{T}_a = \tau < t$ we have $\mathbf{X}(\tau) = a$, and for reasons of symmetry the probability that $\mathbf{X}(t) - \mathbf{X}(\tau) > 0$ is $\frac{1}{2}$. We conclude that

$$\mathbf{P}\{\mathbf{T}_a < t\} = 2\mathbf{P}\{\mathbf{X}(t) > a\} = 2[1 - \mathfrak{N}(a/\sqrt{t})] \tag{2.5}$$

which is equivalent to (2.1).

(*f*) *Hitting points in two-dimensional Brownian motion.* A two-dimensional Brownian motion is formed by a pair $(\mathbf{X}(t), \mathbf{Y}(t))$ of independent one-dimensional Brownian motions. We are interested in the point $(a, \mathbf{Z}_a)$ at which the path first reaches the line $x = a > 0$. As in the preceding example we note that the path can reach the line $x = a+b > a$ only after crossing the line $x = a$; taking $(a, \mathbf{Z}_a)$ as new origin we conclude that $\mathbf{Z}_{a+b}$ has the same distribution as the sum of two independent variables distributed as $\mathbf{Z}_a$ and $\mathbf{Z}_b$. Now an obvious similarity consideration shows that $\mathbf{Z}_a$ has the same distribution as $a\mathbf{Z}_1$ and we conclude that $\mathbf{Z}_a$ has a symmetric stable distribution with exponent $\alpha = 1$. Only the Cauchy distribution fits this description, and so *the hitting point* $\mathbf{Z}_a$ *has a Cauchy distribution.*

This instructive dimensional analysis does not determine the scale parameter. For an explicit calculation note that $\mathbf{Z}_a = \mathbf{Y}(\mathbf{T}_a)$ where $\mathbf{T}_a$ is the epoch when the line $x = a$ is first reached. Its distribution is given in (2.5) while $\mathbf{Y}(t)$ has normal density with variance t. It follows that $\mathbf{Z}_a$ has a density given by[6]

$$\int_0^\infty \frac{e^{-\frac{1}{2}x^2/t}}{t^{\frac{1}{2}}\sqrt{2\pi}} \cdot \frac{ae^{-\frac{1}{2}a^2/t}}{t^{\frac{3}{2}}\sqrt{2\pi}} \cdot dt = \frac{a}{\pi(a^2+x^2)}\,. \tag{2.6}$$

(We have here an example for the subordination of processes to which we shall return in X,7.)

(*g*) *Stable distributions in economics.* Arguments related to the dimensional analysis in the last two examples have been used by B. Mandelbrot to show that various economic processes (in particular income distributions) should be subject to stable (or "Lévy-Pareto") distributions. So far the strength of this interesting theory, which has attracted attention among economists, resides in the theoretical argument rather than observations. [For the apparent fit of the tails of the distribution to many empirical phenomena from city size to word frequency see II,4(*h*).]

[6] The substitution $y = \frac{1}{2}(x^2+a^2)/t$ reduces the integrand to e^{-y}.

(*h*) *Products.* There exist many curious relations between stable distributions of different exponents. The most interesting may be stated in the form of the following proposition. Let $\mathbf{X}$ and $\mathbf{Y}$ be independent strictly stable variables with characteristic exponents α and β respectively. Assume $\mathbf{Y}$ to be a positive variable (whence $\beta < 1$). The *product* $\mathbf{X}\mathbf{Y}^{1/\alpha}$ *has a stable distribution with exponent* $\alpha\beta$. In particular, the product of a normal variable and the square root of the stable variable of example (*c*) is a Cauchy variable.

The assertion follows as a simple corollary to a theorem concerning subordinated processes[7] [example X,7(*c*)]. Furthermore, it is easily verified by Fourier analysis (problem 9 of XVII,12) and, for positive variables, also by Laplace transforms [XIII,7(*e*) and problem 10 of XIII,11].

3. INFINITELY DIVISIBLE DISTRIBUTIONS IN $\mathcal{R}^1$

Definition 1. *A distribution* F *is infinitely divisible if for every* n *there exists a distribution* F_n *such that* $F = F_n^{n\star}$.

In other words,[8] F *is infinitely divisible iff for each* n *it can be represented as the distribution of the sum* $\mathbf{S}_n = \mathbf{X}_{1,n} + \cdots + \mathbf{X}_{n,n}$ *of* n *independent random variables with a common distribution* F_n.

This definition is valid in any number of dimensions, but for the present we shall limit our attention to one-dimensional distributions. It should be noted that infinite divisibility is a property of the *type*, that is, together with F all distributions differing from F only by location parameters are infinitely divisible. *Stable* distributions are infinitely divisible and distinguished by the fact that F_n differs from F only by location parameters.

Examples. (*a*) On account of the convolution property II,(2,3) all *gamma* distributions (including the *exponential*) are infinitely divisible. That the same is true of their discrete counterpart, the "*negative binomial*" (including the *geometric*) distributions was shown in **1**; XII,2(*e*).

(*b*) The Poisson and the compound Poisson distributions are infinitely divisible. It will turn out that all infinitely divisible distributions are *limits* of compound Poisson distributions.

---

[7] For a direct verification requiring a minimum of calculations find the distribution of $\mathbf{Z} = \mathbf{X}_1\sqrt[\alpha]{\mathbf{Y}_1} + \mathbf{X}_2\sqrt[\alpha]{\mathbf{Y}_2}$ by first calculating the conditional distribution of $\mathbf{Z}$ given that $\mathbf{Y}_1 = y_1$ and $\mathbf{Y}_2 = y_2$. The distribution of $\mathbf{Z}$ is a function of $y_1 + y_2$ and the change of variables $u = y_1 + y_2$, $v = y_1 - y_2$ shows that it differs only by a scale factor from that of the two summands. The same calculation works for sums of n similar terms.

[8] It should be understood that the random variables $\mathbf{X}_{k,n}$ serve merely to render notations simpler and more intuitive. For fixed n the variables $\mathbf{X}_{1,n}, \ldots, \mathbf{X}_{n,n}$ are supposed to be mutually independent, but the variables $\mathbf{X}_{j,m}$ and $\mathbf{X}_{k,n}$ with $m \neq n$ need not be defined on the same probability space. (In other words, a joint distribution for $\mathbf{X}_{k,m}$ and $\mathbf{X}_{k,n}$ need not exist.) This remark applies to triangular arrays in general.

(*c*) The distribution II,(7.13) connected with *Bessel* functions is infinitely divisible but this is by no means obvious. See example XIII,7(*d*).

(*d*) *A distribution F carried by a finite interval is not infinitely divisible* except if it is concentrated at one point. Indeed, if $|\mathbf{S}_n| < a$ with probability one then $|\mathbf{X}_{k,n}| < an^{-1}$ and so $\text{Var}(\mathbf{X}_{k,n}) < a^2n^{-2}$. The variance of F is therefore $<a^2n^{-1}$ and hence zero. ▶

Returning to definition 1 let us consider what happens if we drop the requirement that the $\mathbf{X}_{k,n}$ have the same distribution and require only that for each n there exist n distributions $F_{1,n}, \ldots, F_{n,n}$ such that

$$F = F_{1,n} \bigstar \cdots \bigstar F_{n,n}. \tag{3.1}$$

Such generality leads to a new phenomenon best illustrated by examples.

Examples. (*e*) If F is infinitely divisible and U arbitrary, $G = U \bigstar F$ can be written in the form (3.1) with $G_{1,n} = U$ and all other $G_{k,n}$ equal to F_{n-1}. Here the first component plays an entirely different role from all other components.

(*f*) Consider a convergent series $\mathbf{X} = \sum \mathbf{X}_k$ of mutually independent random variables. The distribution F of $\mathbf{X}$ is the convolution of the distributions of $\mathbf{X}_1, \mathbf{X}_2, \ldots, \mathbf{X}_{n-1}$ and the remainder $(\mathbf{X}_n + \mathbf{X}_{n+1} + \cdots)$ and so F is of the form (3.1). Such distributions will be studied under the name of *infinite convolutions*. Example I,11(*c*) shows the uniform distribution to be among them. ▶

The distinguishing feature of these examples is that the contribution of an individual component $\mathbf{X}_{1,n}$ to $\mathbf{S}_n$ is essential, whereas in the case of equally distributed components the contribution of each tends to zero. We wish to connect infinitely divisible distributions to the typical limit theorems involving "many small components." It is then necessary to supplement our scheme by the requirement that the individual components $\mathbf{X}_{k,n}$ become asymptotically negligible in the sense that for each $\epsilon > 0$,

$$\mathbf{P}\{|\mathbf{X}_{k,n}| > \epsilon\} < \epsilon \qquad (k = 1, \ldots, n) \tag{3.2}$$

for n sufficiently large. In the terminology of VIII,2 this means that the $\mathbf{X}_{k,n}$ *tend in probability to zero uniformly in* $k = 1, \ldots, n$. Systems of variables of this type appear so often that it is convenient to give them a name.

Definition 2. *By a triangular array is meant a double sequence of random variables* $\mathbf{X}_{k,n}$ $(k = 1, 2, \ldots, n;\ n = 1, 2, \ldots)$ *such that the variables* $\mathbf{X}_{1,n}, \ldots, \mathbf{X}_{n,n}$ *of the* n*th row are mutually independent.*

The array is a null array (or has asymptotically negligible components) if (3.2) *holds.*

More generally one may consider arrays with r_n variables in the nth row with $r_n \to \infty$. The gain in generality is slight. (See problem 10.)

Example. (*g*) Let $\{\mathbf{X}_j\}$ be a sequence of identically distributed independent random variables and $\mathbf{S}_n = \mathbf{X}_1 + \cdots + \mathbf{X}_n$. The normalized sequence $\mathbf{S}_n a_n^{-1}$ represent the nth row sum of a triangular array in which $\mathbf{X}_{k,n} = \mathbf{X}_k a_n^{-1}$. This array is a null array if $a_n \to \infty$. An array of a different sort was considered in the derivation of the Poisson distribution in **1**; VI,6. ▶

In chapters IX and XVII we shall prove the remarkable fact that a *limit distribution of the row sums* $\mathbf{S}_n$ *of a triangular null array (if it exists) is infinitely divisible.* As long as the asymptotic negligibility condition (3.2) holds it does not matter whether or not the components $\mathbf{X}_{k,n}$ have a common distribution, and in (3.1) we may replace the equality sign by a limit: the class of infinitely divisible distributions coincides with the class of limit distributions of the row sums of triangular null arrays.

Examples for applications. (*h*) *The shot effect in vacuum tubes.* Variants and generalizations of the following stochastic process occur in physics and in communication engineering.

We propose to analyze the fluctuations in electrical currents due to the chance fluctuations of the numbers of electrons arriving at an anode. It is assumed that the arrivals form a Poisson process, and that an arriving electron produces a current whose intensity x time unit later equals $I(x)$. The intensity of the current at epoch t is then formally a random variable

$$\mathbf{X}(t) = \sum_{k=1}^{\infty} I(t-\mathbf{T}_k), \tag{3.3}$$

where the $\mathbf{T}_k$ represent the epochs of past electron arrivals. (In other words, the variables $t-\mathbf{T}_1, \mathbf{T}_2-\mathbf{T}_1, \mathbf{T}_3-\mathbf{T}_2, \ldots$ are mutually independent and have a common exponential distribution.)

A direct analysis of the sum (3.3) by the methods of stochastic processes is not difficult, but the simple-minded approach by triangular arrays may serve as an aid to intuition. Partition the interval $\overline{-\infty, t}$ into small subintervals with endpoints $t_k = t - kh$ (where $k = 0, 1, \ldots$). By the very definition of the Poisson process the contribution of the interval $\overline{t_k, t_{k-1}}$ to the sum in (3.3) is comparable to a binomial random variable assuming the value 0 with probability $1 - \alpha h$ and $I(t-t_k)$ with probability αh. The expectation of this variable is $\alpha h\, I(kh)$, its variance $\alpha h(1-\alpha h)\, I^2(kh)$. We take $h = 1/\sqrt{n}$ and construct the triangular array in which $\mathbf{X}_{k,n}$ is the contribution of the interval $\overline{t_k, t_{k-1}}$. The row sums have then expectation $\alpha h \sum I(kh)$ and variance $\alpha h(1-\alpha h) \sum I^2(kh)$. If any meaning can be attached to the series (3.3) the distributions of the row sums must tend to the distribution of $\mathbf{X}(t)$ and so we must have

$$(3.4) \qquad \mathbf{E}(\mathbf{X}(t)) = \alpha \int_0^\infty I(s)\, ds, \qquad \mathrm{Var}\,(\mathbf{X}(t)) = \alpha \int_0^\infty I^2(s)\, ds.$$

These conclusions are easily confirmed by the theory of triangular arrays. The relations (3.4) are known as *Campbell's theorem.* At present it does not appear deep, but it was proved in 1909 decades ahead of a systematic theory. At that time it appeared remarkable and various proofs have been given for it. (Cf. problems 22 in VIII,10 and 5 in XVII,12.)

(*i*) *Busy trunklines.* A variant of the preceding example may illustrate the types of possible generalizations. Consider a telephone exchange with infinitely many trunklines. The incoming calls form a Poisson process, and an arriving call is directed to a free trunkline. The ensuing holding times have a common distribution F; as usual, they are assumed independent of the arrival process and of each other. The number of busy lines at epoch t is a random variable $\mathbf{X}(t)$ whose distribution can be derived by the method of triangular arrays. As in the preceding example we partition $\overline{0, t}$ into n intervals of length $h = t/n$ and denote by $\mathbf{X}_{k,n}$ the number of conversations that originated between $n - kh$ and $n - (k-1)h$ and are still going on at epoch t. When n is large the variable $\mathbf{X}_{k,n}$ assumes in practice only the values 0 and 1, the latter with probability $\alpha h[1-F(kh)]$. The expectation of $\mathbf{S}_n$ is then the sum of these probabilities, and passing to the limit we conclude that *the number of busy lines has expectation*

$$(3.5) \qquad \mathbf{E}(\mathbf{X}(t)) = \alpha \int_0^\infty [1-F(s)]\, ds.$$

Note that the integral equals the expectation of the holding times. ▶

Historical note. The notion of infinite divisibility goes back to B. de Finetti (1929). The Fourier transforms of infinitely divisible distributions with finite variance were found by A. Kolmogorov (1932), and those of the general infinitely divisible distributions by P. Lévy (1934), who also treated the problem from the point of view of stochastic processes. All subsequent investigations were strongly influenced by his pioneer work. The first purely analytical derivations of the general formula were given in 1937 independently by Feller and Khintchine. These authors proved also that the limit distributions of null arrays are infinitely divisible.

4. PROCESSES WITH INDEPENDENT INCREMENTS

Infinitely divisible distributions are intimately connected with stochastic processes with independent increments. By this we mean a *family of random variables* $\mathbf{X}(t)$ *depending on the continuous time parameter* t *and such that the increments* $\mathbf{X}(t_{k+1}) - \mathbf{X}(t_k)$ *are mutually independent for any finite set*

$t_1 < t_2 < \cdots < t_n$. At this juncture we require no theory of stochastic processes; we argue simply that *if* certain phenomena can be described probabilistically, the theory will lead to infinitely divisible distributions. In this sense we have considered special processes with independent increments in **1**; XVII,1 and in example III,8(*a*). We limit our attention to numerical variables $\mathbf{X}(t)$ although the theory carries over to vector variables.

The process has *stationary increments* if the distribution of $\mathbf{X}(s+t) - \mathbf{X}(s)$ depends only on the length t of the interval but not on s.

Let us partition the interval $\overline{s, s+t}$ by $n+1$ equidistant points $s = t_0 < t_1 < \cdots < t_n = s + t$ and put $\mathbf{X}_{k,n} = \mathbf{X}(t_k) - \mathbf{X}(t_{k-1})$. The variable $\mathbf{X}(s+t) - \mathbf{X}(s)$ of a process with stationary independent increments is the sum of the n independent variables $\mathbf{X}_{k,n}$ with a common distribution and hence $\mathbf{X}(s+t) - \mathbf{X}(s)$ *has an infinitely divisible distribution.* We shall see that the *converse* is also true. In fact, a one-parametric family of probability distributions Q_t defined for $t > 0$ can serve as the distribution of $\mathbf{X}(s+t) - \mathbf{X}(s)$ in a process with stationary independent increments iff

$$Q_{s+t} = Q_s \star Q_t \qquad s, t > 0. \tag{4.1}$$

A family of distributions satisfying (4.1) is said to form a semi-group (see IX,2). Every infinitely divisible distribution can be taken as element Q_t (with $t > 0$ arbitrary) of such a semi-group.

Before passing to the non-stationary case let us consider typical examples.

Examples. (*a*) *The compound Poisson process.* With an arbitrary probability distribution F and $\alpha > 0$

$$Q_t = e^{-\alpha t} \sum_{k=0}^{\infty} \frac{(\alpha t)^k}{k!} F^{k\star} \tag{4.2}$$

defines a compound Poisson distribution, and it is easily verified that (4.1) holds. Suppose now that Q_t represents the distribution of $\mathbf{X}(t) - \mathbf{X}(0)$ in a stochastic process with stationary independent increments. When F is concentrated at the point 1 this process reduces to an ordinary *Poisson process* and (4.2) to

$$\mathbf{P}\{\mathbf{X}(t) - \mathbf{X}(0) = n\} = e^{-\alpha t} \frac{(\alpha t)^n}{n!}. \tag{4.3}$$

The general model (4.2) may be interpreted in terms of this special Poisson process as follows. Let $\mathbf{Y}_1, \mathbf{Y}_2, \ldots$ be independent variables with the common distribution F, and let $\mathbf{N}(t)$ be the variable of a pure Poisson process with $\mathbf{P}\{\mathbf{N}(t) = n\} = e^{-\alpha t}(\alpha t)^n/n!$, and independent of the $\mathbf{Y}_k$. Then (4.2) represents the distribution of the random sum $\mathbf{Y}_1 + \cdots + \mathbf{Y}_{\mathbf{N}(t)}$. In other words, with the nth jump of the Poisson process there is associated an effect $\mathbf{Y}_n$, and $\mathbf{X}(t) - \mathbf{X}(0)$ represents the sum of the effects occurring during $\overline{0, t}$. The *randomized random* walk studied in II,7 is a compound Poisson process with $\mathbf{Y}_k$ assuming the values ± 1 only. Empirical applications are illustrated at the end of this section.

(*b*) *Brownian motion* or the Wiener-Bachelier process. Here $\mathbf{X}(0) = 0$ (the process starts at the origin) and the increments $\mathbf{X}(t+s) - \mathbf{X}(s)$ have a normal distribution with zero expectation and variance t. Wiener and Lévy have shown that the sample functions of this process are *continuous* with probability one, and this property characterizes the normal distribution among all infinitely divisible distributions.

(*c*) *Stable processes.* The relation (1.8) for a strictly stable distribution merely paraphrases (4.1) with $Q_t(x) = R(r^{-1/\alpha}x)$. Thus this distribution defines transition probabilities in a process with stationary independent increments; for $\alpha = 2$ it reduces to Brownian motion. ►

The main theorem of the theory (see chapters IX and XVII) states that the *most general solution of* (4.1)—*and hence the most general infinitely divisible distribution—may be represented as a limit of an appropriate sequence of compound Poisson distributions.* This result is surprising in view of the great formal difference between examples (*a*) and (*b*).

Even in *non-stationary* processes with independent increments the distribution of $\mathbf{X}(t+s) - \mathbf{X}(t)$ appears as the distribution of the row sums of our triangular array $\{\mathbf{X}_{k,n}\}$, but a slight continuity condition must be imposed on the process to assure that (3.2) holds. Example (*e*) will explain this necessity. Under a slight restriction *only infinitely divisible distributions appear as distributions of* $\mathbf{X}(t+s) - \mathbf{X}(t)$.

Examples. (*d*) *Operational time.* A simple change of the time scale will frequently reduce a general process to a more tractable stationary process. Given any continuous increasing function φ we may switch from the variable $\mathbf{X}(t)$ to $\mathbf{Y}(t) = \mathbf{X}(\varphi(t))$. The property of independent increments is obviously preserved, and with an appropriate choice of φ the new process may also have stationary increments. In practice the choice is usually dictated by the nature of things. For example, at a *telephone exchange* nobody would compare an hour at night with the busy hour of the day, while it is natural to measure time in variable units such that the expected number of calls per unit remains constant. Again, in a growing *insurance* business claims will occur at an accelerated rate but this departure from stationarity is removed by the simple expedient of introducing an operational time measuring the frequency of claims.

(*e*) *Empirical applications.* An unending variety of practical problems can be reduced to compound Poisson processes. Here are a few typical examples. (i) The accumulated *damage* due to automobile accidents, fire, lightning, etc. For applications to *collective risk* theory see example 5(*a*).

(ii) The total catch by a fishery boat in search of schools of fish (J. Neyman). (iii) The content of water *reservoirs* due to rainfall and demand. Other *storage facilities* are treated in like manner. (iv) A *stone at the bottom of a river* lies at rest for such long periods that its successive displacements are practically instantaneous. The total displacement within time $\overline{0, t}$ may be treated as a compound Poisson process. (First treated by different methods by Albert Einstein Jr. and G. Polya.) (v) *Telephone calls*, or customers, arriving at a server require service. Under appropriate conditions the total service time caused by arrivals within the time interval $\overline{0, t}$ represents a compound Poisson process. The remarkable feature of this process is that the value of $\mathbf{X}(t)$ *is not observable at epoch* t because it depends on service times that still lie in the future. (vi) For energy changes of physical particles due to *collisions* see example X, 1(*b*). ►

*5. RUIN PROBLEMS IN COMPOUND POISSON PROCESSES

Let $\mathbf{X}(t)$ be the variable of a compound Poisson process, that is, the increment $\mathbf{X}(t+s) - \mathbf{X}(s)$ over any time interval of duration t has the probability distribution Q_t of (4.2). Let $c > 0$ and $z > 0$ be fixed. By *ruin* we mean the event

$$\{\mathbf{X}(t) > z + ct\}. \tag{5.1}$$

We regard c as a constant and $z > 0$ as a free parameter, *and we denote by* $R(z)$ *the probability that no ruin will ever occur.* We shall argue formally that if the problem makes sense $R(z)$ must be a non-increasing solution of the functional equation (5.2). First a few examples may indicate the variety of practical situations to which our problem is applicable.

Examples. (*a*) *Collective risk theory.*[9] Here $\mathbf{X}(t)$ stands for the accumulated amount of claims within the time interval $\overline{0, t}$ against an insurance company. It is assumed that the occurrence of claims is subject to a Poisson process and that the individual claims have the distribution F. In principle these "claims" may be positive or negative. (For example, a death may free the company of an obligation and increase the reserves.) In practice a growing company will measure time in operational units proportional to the

* This section treats a special topic. It is of great practical interest, but will not be referred to in the present book except for examples where it will be treated by new methods.

[9] A huge literature is devoted to this theory (inaugurated by F. Lundberg). For a relatively recent survey see H. Cramér, *On some questions connected with mathematical risk*, Univ. Calif. Publications in Statistics, vol. 2, no. 5 (1954) pp. 99–125. Cramér's asymptotic estimates (obtained by deep Wiener–Hopf techniques) are obtained in an elementary manner in examples XI,7(*a*) and XII,5(*d*).

total incoming premiums [see example 4(*d*)]. It may then be assumed that in the absence of claims the reserves increase at a constant rate c. If z stands for the initial reserves at epoch 0, the company's total reserve at epoch t is represented by the random variable $z + ct - \mathbf{X}(t)$, and "ruin" stands for a negative reserve, that is, failure.

(*b*) *Strong facilities.* An idealized water reservioir is being filled by rivers and rainfall at a constant rate c. At random intervals the reservoir is tapped by amounts $\mathbf{X}_1, \mathbf{X}_2, \ldots$. The compound Poisson model applies and if z stands for the initial content at epoch 0 then $z + ct - \mathbf{X}(t)$ represents the content at epoch t provided that no ruin occurs before t. For the huge literature on related problems see the monographs listed at the end of the book.

(*c*) *Scheduling of patients.*[10] We agree to treat the times devoted by a doctor to his patients as independent random variables with an exponential distribution and mean duration α^{-1}. As long as treatments continue without interruption the departures of treated patients are subject to an *ordinary Poisson process.* Let $\mathbf{X}(t)$ stand for the number of such departures within $\overline{0, t}$. Suppose that z patients are waiting at epoch 0 (the beginning of the office hours) and that thereafter new patients arrive at epochs c^{-1}, $2c^{-1}$, $3c^{-1}, \ldots$. The doctor will not be idle as long as $\mathbf{X}(t) \leq z + ct$. ▶

The following formal argument leads to an equation determining the probability of ruin R. Suppose that the *first* jump of the sample function occurs at epoch τ and has magnitude x. For no ruin ever to occur it is necessary that $x \leq z + c\tau$ and that for all $t > \tau$ the increments $\mathbf{X}(t) - x$ be $\leq z - x + ct$. Such increments being independent of the past the latter event has probability $R(z-x+c\tau)$. Summing over all possible τ and x we get

$$R(z) = \int_0^\infty \alpha e^{-\alpha\tau}\, d\tau \int_{-\infty}^{z+c\tau} R(z+c\tau-x)\, F\{dx\}. \tag{5.2}$$

This is the desired equation, but it can be simplified. The change of variable $s = z + c\tau$ leads to

$$R(z) = \frac{\alpha}{c}\int_z^\infty e^{-(\alpha/c)(s-z)}\, ds \int_{-\infty}^{s} R(s-x)\, F\{dx\}. \tag{5.3}$$

Consequently R is differentiable, and a simple differentiation leads to the final *integro-differential equation*

$$R'(z) = \frac{\alpha}{c} R(z) - \frac{\alpha}{c}\int_{-\infty}^{z} R(z-x)F\{dx\}. \tag{5.4}$$

[10] R. Pyke, *The supremum and infimum of the Poisson process*, Ann. Math. Statist., vol. 30 (1959) pp. 568–576. Pyke treats only the pure Poisson process but obtains more precise results (by different methods).

Note that by definition $R(s) = 0$ for $s < 0$ so that the integral on the right is the *convolution* $F \bigstar R$. We shall return to (5.4) in examples 9(*d*); XI,7(*a*); and XII,5(*d*).

6. RENEWAL PROCESSES

The basic notions of renewal theory were introduced in **1**; XIII in connection with recurrent events. It will be seen that the introduction of a continuous time parameter depends on notational, rather than conceptual, changes. The salient feature of recurrent events is that the successive waiting times $\mathbf{T}_k$ are mutually independent random variables with a common distribution F; the epoch of the nth occurrence is given by the sum

$$\mathbf{S}_n = \mathbf{T}_1 + \cdots + \mathbf{T}_n. \tag{6.1}$$

By convention $\mathbf{S}_0 = 0$ and 0 counts as occurrence number zero.

Even in stochastic processes depending on a continuous time parameter it is frequently possible to discover one or more sequences of epochs of the form (6.1). In such cases surprisingly sharp results are obtainable by simple methods. Analytically we are concerned merely with sums of independent positive variables, and the only excuse for introducing the term "renewal process" is its frequent occurrence in connection with *other* processes and the tacit implication that the powerful tool of the renewal equation is used.[11]

Definition 1. *A sequence of random variables* $\mathbf{S}_n$ *constitutes a renewal process if it is of the form* (6.1) *where the* $\mathbf{T}_k$ *are mutually independent variables with a common distribution* F *such that*[12] $F(0) = 0$.

The variables being positive there is no danger in writing $\mu = \mathbf{E}(\mathbf{T}_k)$ even if the integral diverges (in which case we write $\mu = \infty$). The expectation μ will be called *mean recurrence time*. As usual in similar situations, it is irrelevant for our present analysis whether the variables $\mathbf{T}_k$ occur in some stochastic process or whether the sequence $\{\mathbf{T}_j\}$ itself defines our probability space.

In most (but not all) applications the $\mathbf{T}_j$ can be interpreted as "waiting times" and the $\mathbf{S}_n$ are then referred to as *renewal* (*or regeneration*) epochs.

It seems intuitively obvious that for a fixed finite interval $I = \overline{a, b}$ the number of renewal epochs $\mathbf{S}_n$ falling within I is finite with probability

[11] For a more sophisticated generalization of the recurrent events see J. F. C. Kingman, *The stochastic theory of regenerative events*, Zeitschrift Wahrscheinlichkeitstheorie, vol. 2 (1964) pp. 180–224.

[12] An atom of weight $p < 1$ at the origin would have no serious effect.

one and hence is a well-defined random variable **N**. If the event $\{\mathbf{S}_n \in I\}$ is called "success" then **N** is the total number of successes in infinitely many trials and its expectation equals

$$U\{I\} = \sum_{n=0}^{\infty} \mathbf{P}\{\mathbf{S}_n \in I\} = \sum_{n=0}^{\infty} F^{n\star}\{I\}. \tag{6.2}$$

For the study of this measure we introduce, as usual, its distribution function defined by

$$U(x) = \sum_{n=0}^{\infty} F^{n\star}(x) \tag{6.3}$$

It is understood that $U(x) = 0$ for $x < 0$, but U has an atom of unit weight at the origin.

In the discrete case studied in **1**; XIII the measure U was concentrated at the integers: u_k stood for the probability that one among the $\mathbf{S}_n$ equals k. Since this event can occur only once, u_k can also be interpreted as the expected number of n for which $\mathbf{S}_n = k$. In the present situation $U\{I\}$ must be interpreted as an expectation rather than as a probability because the event $\{\mathbf{S}_n \in I\}$ can occur for many n.

It is necessary to prove that $U(x) < \infty$. From the definition of convolutions for distributions concentrated on $\overline{0, \infty}$ it is clear that $F^{n\star}(x) \leq F^n(x)$, and hence the series in (6.3) converges at least geometrically at each point where $F(x) < 1$. There remains the case of distributions concentrated on a finite interval, but then there exists an integer r such that $F^{r\star}(x) < 1$. The terms with $n = r, 2r, 3r, \ldots$ form a convergent subseries, and this implies the convergence of the whole series in (6.3) because its terms depend monotonically on n.

As in the discrete case the renewal measure U is intimately connected with the *renewal equation*

$$Z = z + F \star Z \tag{6.4}$$

Spelled out it reads

$$Z(x) = z(x) + \int_0^x Z(x-y)\, F\{dy\}, \qquad x > 0, \tag{6.5}$$

where the interval of integration is considered closed. Actually the limits of integration may be replaced by $-\infty$ and ∞ provided it is understood that $z(x) = Z(x) = 0$ for $x < 0$. We shall adhere to this convention.

The basic fact concerning the renewal equation is contained in

Theorem 1. *If* z *is bounded and vanishes for* $x < 0$ *the convolution* $Z = U \star z$ *defined by*

$$Z(x) = \int_0^x z(x-y)\, U\{dy\} \tag{6.6}$$

represents a solution of the renewal equation (6.5). *There exists no other solution vanishing on* $\overline{-\infty, 0}$ *and bounded on finite intervals.*

Proof. We know already that the series (6.3) defining U converges for all x. Taking the convolution with z it is seen that Z is bounded on finite intervals and satisfies the renewal equation. The difference of two such solutions would satisfy $V = F \bigstar V$, and hence also $V = F^{n\star} \bigstar V$ for all n. But $F^{n\star}(x) \to 0$ for all x, and if V is bounded this implies that $V(x) = 0$ for all x. ▶

We shall return to the renewal equation in XI,1 where we shall study the asymptotic properties of U and Z. (For a generalized version of the renewal equation see section 10.)

It should be noticed that U *itself satisfies*

$$U(x) = 1 + \int_0^x U(x-y)\, F\{dy\}, \qquad x > 0, \tag{6.7}$$

which is the special case of the renewal equation with $z = 1$. This can be seen directly by a probabilistic reasoning known as "*renewal argument*" which is of frequent use. Since 0 counts as a renewal epoch the expected number of renewal epochs in the closed interval $\overline{0, x}$ is one plus the expected number in the half-open interval $\overline{0, x}$. This interval contains renewal epochs only if $\mathbf{T}_1 \leq x$; given that $\mathbf{T}_1 = y \leq x$, the expected number of renewal epochs in $\overline{0, x}$ equals $U(x-y)$. Summing over y we get (6.7).

Two simple generalizations of the renewal process are useful. First, by analogy with transient recurrent events we may permit *defective* distributions. The defect $q = 1 - F(\infty)$ is then interpreted as probability of *termination.* Abstractly speaking, the real line is enlarged by a point Ω called "death," and $\mathbf{T}_k$ is either a positive number or Ω. For ease of reference we introduce the informal

Definition 2.[13] *A terminating or transient renewal process is an ordinary renewal process except that* F *is defective. The defect* $q = 1 - F(\infty)$ *is interpreted as probability of termination.*

For consistency, 0 is counted as renewal epoch number zero. The probability that the process effectively survives the renewal epoch number n equals $(1-q)^n$ and tends to 0 as $n \to \infty$. Thus *with probability one a terminating process terminates at a finite time.* The total mass of $F^{n\star}$ is $(1-q)^n$ and so *the expected number of renewal epochs is* $U(\infty) = q^{-1} < \infty$. This is, so to speak, the expected number of generations attained by the

---

[13] See example 7(*f*) for an illustration and problem 4 for a generalization.

process. The probability that $\mathbf{S}_n \leq x$ and the process dies with this nth renewal epoch is $qF^{n\star}(x)$. We have thus the

Theorem 2. *In a terminating renewal process, qU is the proper probability distribution of the duration of the process (age at time of death).*

The second generalization corresponds to delayed recurrent events and consists in permitting the *initial* waiting time to have a different distribution. In such cases we begin the numbering of the $\mathbf{T}_j$ with $j = 0$ so that now $\mathbf{S}_0 = \mathbf{T}_0 \neq 0$.

Definition 3. *A sequence $\mathbf{S}_0, \mathbf{S}_1, \ldots$ forms a delayed renewal process if it is of the form* (6.1) *where the $\mathbf{T}_k$ are mutually independent strictly positive (proper or defective) variables and $\mathbf{T}_1, \mathbf{T}_2, \ldots$ (but not $\mathbf{T}_0$) have a common distribution.*

7. EXAMPLES AND PROBLEMS

Examples such as self-renewing aggregates, counters, and population growth carry over from the discrete case in an obvious manner. A special problem, however, will lead to interesting questions to be treated later on.

Example. (*a*) *An inspection paradox.* In the theory of self-renewing aggregates a piece of equipment, say an electric battery, is installed and serves until it breaks down. Upon failure it is instantly replaced by a like battery and the process continues without interruption. The epochs of renewal form a renewal process in which $\mathbf{T}_k$ is the lifetime of the kth battery.

Suppose now that the actual lifetimes are to be tested by inspection: we take a sample of batteries in operation at epoch $t > 0$ and observe their lifetimes. Since F is the distribution of the lifetimes for *all* batteries one expects that this applies also to the inspected specimen. But this is not so. In fact, for an exponential distribution F the situation differs only verbally from the waiting time paradox in I,4 where the lifetime of the inspected item has an entirely different distribution. *The fact that the item was inspected at epoch t changes its lifetime distribution and doubles its expected duration.* We shall see in XI,(4.6) that this situation is typical of all renewal processes. The practical implications are serious. We see that an apparently unbiased inspection plan may lead to false conclusions because *what we actually observe need not be typical of the population* as a whole. Once noticed the phenomenon is readily understood (see I,4), but it reveals nevertheless possible pitfalls and the necessary interplay between theory and practice. Incidentally, no trouble arises if one decides to test the first item installed after epoch t. ▶

This is a good occasion to introduce three random variables of interest in renewal theory. In the preceding example all three refer to the item in operation at epoch $t > 0$ and may be described by the self-explanatory terms: residual lifetime, spent lifetime, and total lifetime. The formal definition is as follows.

To given $t > 0$ there corresponds a unique (chance-dependent) subscript $\mathbf{N}_t$ such that $\mathbf{S}_{\mathbf{N}_t} \leq t < \mathbf{S}_{\mathbf{N}_t+1}$. Then:

(*a*) *The residual waiting time* is $\mathbf{S}_{\mathbf{N}_t+1} - t$, the time from t to the next renewal epoch.

(*b*) *The spent waiting time* is $t - \mathbf{S}_{\mathbf{N}_t}$, the time elapsed since the last renewal epoch.

(*c*) Their sum $\mathbf{S}_{\mathbf{N}_t+1} - \mathbf{S}_{\mathbf{N}_t} = \mathbf{T}_{\mathbf{N}_t+1}$ is the length of the recurrence interval covering the epoch t.

The terminology is not unique and varies with the context. For example, in random walk our residual waiting time is called *point of first entry* or *hitting point* for the interval $\overline{t, \infty}$. In the preceding example the word lifetime was used for waiting time. We shall investigate the three variables in XI,4 and XIV,3.

The *Poisson process* was defined as a renewal process with an exponential distribution for the recurrence times $\mathbf{T}_j$. In many server and counter problems it is natural to assume that the incoming traffic forms a Poisson process. In certain other processes the interarrival times are constant. To combine these two cases it has become fashionable in queuing theory to admit general renewal processes with arbitrary interarrival times.[14]

We turn to problems of a fairly general character connected with renewal processes. The distribution underlying the process is again denoted by F.

We begin with what could be described roughly as the "*waiting time* $\mathbf{W}$ *for a large gap*." Here a renewal process with recurrence times $\mathbf{T}_j$ is stopped at the first occurrence of a time interval of duration ξ free of renewal epochs, whereupon the process stops. We derive a renewal equation for the *distribution* V *of the waiting time* $\mathbf{W}$. As the latter necessarily exceeds ξ we have $V(t) = 0$ for $t < \xi$. For $t \geq \xi$ consider the mutually exclusive possibilities that $\mathbf{T}_1 > \xi$ or $\mathbf{T}_1 = y \leq \xi$. In the first case the waiting time $\mathbf{W}$ equals ξ. In the second case the process starts from scratch and, given that $\mathbf{T}_1 = y$, the (conditional) probability of $\{\mathbf{W} \leq t\}$ is $V(t-y)$. Summing over all possibilities we get

$$V(t) = 1 - F(\xi) + \int_0^{\xi+} V(t-y)\, F\{dy\}, \qquad t \geq \xi, \tag{7.1}$$

and, of course, $V(t) = 0$ for $t < \xi$. This equation reduces to the standard

[14] The generality is somewhat deceptive because it is hard to find practical examples besides the bus running without schedule along a circular route. The illusion of generality detracts from the sad fact that a non-Poissonian input is usually also non-Markovian.

renewal equation

$$V = z + G \star V \tag{7.2}$$

with the defective distribution G defined by

$$G(x) = F(x) \quad \text{if} \quad x \leq \xi; \qquad G(x) = F(\xi) \quad \text{if} \quad x \geq \xi \tag{7.3}$$

and

$$z(x) = 0 \quad \text{if} \quad x < \xi; \qquad z(x) = 1 - F(\xi) \quad \text{if} \quad x \geq \xi. \tag{7.4}$$

The most important special case is that of *gaps in a Poisson* process where $F(t) = 1 - e^{-ct}$ and the solution V is related to the covering theorems of I,9 [see problem 15 and example XIV,2(*a*). For a different approach see problem 16.]

Examples *for empirical applications.* (*b*) *Crossing a stream of traffic.*[15] Cars move in a single lane at constant speed, the successive passages forming a sample from a Poisson process (or some other renewal process). A pedestrian arriving at the curb—or a car arriving at an intersection—will start crossing as soon as he observes that no car will pass during the next ξ seconds, namely the time required for his crossing. Denote by $\mathbf{W}$ the time required to effect the crossing, that is, the waiting time at the curb plus ξ. The distribution V of $\mathbf{W}$ satisfies (7.1) with $F(t) = 1 - e^{-ct}$. [Continued in examples XI,7(*b*) and XIV,2(*a*).]

(*c*) *Type* II *Geiger counters.* Arriving particles constitute a Poisson process and each *arriving* particle (whether registered or not) locks the counter for a fixed time ξ. If a particle is registered, the counter remains "dead" until the occurrence of an interval of duration ξ without new arrivals. Our theory now applies to the distribution V of the *duration of the dead period.* (See **1**; XIII,11, problem 14.)

(*d*) *Maximal observed recurrence time.* In a primary renewal process denote by $\mathbf{Z}_t$ the maximum of $\mathbf{T}_j$ observed[16] up to epoch t. The event $\{\mathbf{Z}_t \leq \xi\}$ occurs iff up to epoch t no time interval of duration ξ was free of renewal epochs, and so in our notations $\mathbf{P}\{\mathbf{Z}_t > \xi\} = V(t)$. ▶

A great many renewal processes occurring in applications may be described as *alternating or two stage processes.* Depending on the context the two stages may be called active or passive, free or dead, excited or normal. Active and passive periods alternate; their durations are independent random variables, each type being subject to a common distribution.

---

[15] For the older literature and variants (treated by different methods) see J. C. Tanner, *The delay to pedestrians crossing a road*, Biometrika, vol. 38 (1951) pp. 383–392.

[16] More precisely, if n is the (chance-dependent) index for which $\mathbf{S}_{n-1} \leq t < \mathbf{S}_n$ then $\mathbf{Z}_t = \max\,[\mathbf{T}_1, \ldots, \mathbf{T}_{n-1}, \xi]$. Variables of this nature were studied systematically by A. Lamperti.

Examples. (*e*) *Failures followed by delays.* The simplest example is given by actual replacements of a piece of equipment if each failure is followed by a delay (to be interpreted as time for discovery or repair). The successive service times $\mathbf{T}_1, \mathbf{T}_2, \ldots$ alternate with the successive dead periods $\mathbf{Y}_1, \mathbf{Y}_2, \ldots$ and we get a proper renewal process with recurrence times $\mathbf{T}_j + \mathbf{Y}_j$. The same process may be viewed as *delayed* renewal process with the first renewal epoch at $\mathbf{T}_1$, and recurrence times $\mathbf{Y}_j + \mathbf{T}_{j+1}$.

(*f*) *Lost calls.* Consider a single telephone trunkline such that the incoming calls form a Poisson process with interarrival distribution $G(t) = 1 - e^{-ct}$ while the durations of the ensuing conversations are independent random variables with the common distribution F. The trunkline is free or dead, and calls arriving during dead periods are lost and have no influence on the process. We have here a two stage process in which the distribution of the recurrence times is $F \bigstar G$. (See problem 17 as well as problems 3–4 in XIV,10.)

(*g*) *Last come first served.* Sometimes the distributions of the alternating waiting times are not known *a priori* but must be calculated from other data. As an example consider a data processing machine in which new information arrives in accordance with a Poisson process so that the free periods have an exponential distribution. The time required to process the new information arriving at any epoch has a probability distribution G.

Busy and free periods alternate, but the duration of busy periods depends on the manner in which information arriving during a busy period is treated. In certain situations only the latest information is of interest; a new arrival is then processed immediately and all previous information is discarded. The distribution V of the duration of the busy periods must be calculated from a renewal equation (see problem 18).

(*h*) *Geiger counters.* In type I counters each *registration* is followed by a dead period of fixed duration ξ and arrivals within the dead period have no effect. The process is the same as described in example (*e*), the $\mathbf{T}_j$ having an exponential distribution, the $\mathbf{Y}_j$ being equal to ξ. In type II counters also the unregistered arrivals produce locking and the situation is the same except that the distributions of the $\mathbf{Y}_j$ *depend on the primary process* and must be calculated from the renewal equation (7.1) [example (*c*)]. ▶

8. RANDOM WALKS

Let $\mathbf{X}_1, \mathbf{X}_2, \ldots$ be mutually independent random variables with a common distribution F and, as usual,

$$\mathbf{S}_0 = 0, \qquad \mathbf{S}_n = \mathbf{X}_1 + \cdots + \mathbf{X}_n. \tag{8.1}$$

We say that $\mathbf{S}_n$ is the position, at epoch n, of a particle performing a general random walk. No new theoretical concepts are introduced,[17] but merely a terminology for a short and intuitive description of the process $\{\mathbf{S}_n\}$. For example, if I is any interval (or other set), the event $\{\mathbf{S}_n \in I\}$ is called a *visit* to I, and the study of the successive visits to a given interval I reveals important characteristics of the fluctuations of $\mathbf{S}_1, \mathbf{S}_2, \ldots$. The index n will be interpreted as time parameter and we shall speak of the "epoch n." In this section we describe some striking features of random walks in terms of the successive record values. The usefulness of the results will be shown by the applications in section 9. A second (independent) approach is outlined in section 10.

Imbedded Renewal Processes

A record value occurs at epoch $n > 0$ if

$$\mathbf{S}_n > \mathbf{S}_j \qquad j = 0, 1, \ldots, n-1. \tag{8.2}$$

Such indices may not exist for a given sample path; if they do exist they form a finite or infinite ordered sequence. It is therefore legitimate to speak of the first, second, ..., occurrence of (8.2). Their epochs are again random variables, but possibly defective. With these preparations we are now in a position to introduce the important random variables on which much of the analysis of random walks will be based.

Definition. *The* kth (*ascending*) *ladder index is the epoch of the* kth *occurrence of* (8.2). *The* kth *ladder height is the value of* $\mathbf{S}_n$ *at the* kth *ladder epoch.* (*Both random variables are possibly defective.*)

The descending ladder variables are defined in like manner with the inequality in (8.2) *reversed.*[18]

The term *ascending* will be treated as redundant and used only for emphasis or clarity.

In the graph of a sample path $(\mathbf{S}_0, \mathbf{S}_1, \ldots)$ the ladder points appear as the points where the graph reaches an unprecedented height (record value). Figure 1 represents a random walk $\{\mathbf{S}_n\}$ drifting to $-\infty$ with the last positive term at $n = 31$. The 5 ascending and 18 descending ladder points are indicated by ● and ○, respectively. For a random walk with Cauchy variables see figure 2. (page 204)

---

[17] Sample spaces of infinite random walks were considered also in volume **1**, but there we had to be careful to justify notions such as "probability of ruin" by the obvious limiting processes. Now these obvious passages to the limit are justified by measure theory. (See IV,6.)

[18] Replacing the defining strict inequalities by $\geq$ and $\leq$ one gets the *weak* ladder indices. This troublesome distinction is unnecessary when the underlying distribution is continuous. In figure 1 weak ladder points are indicated by the letter w.

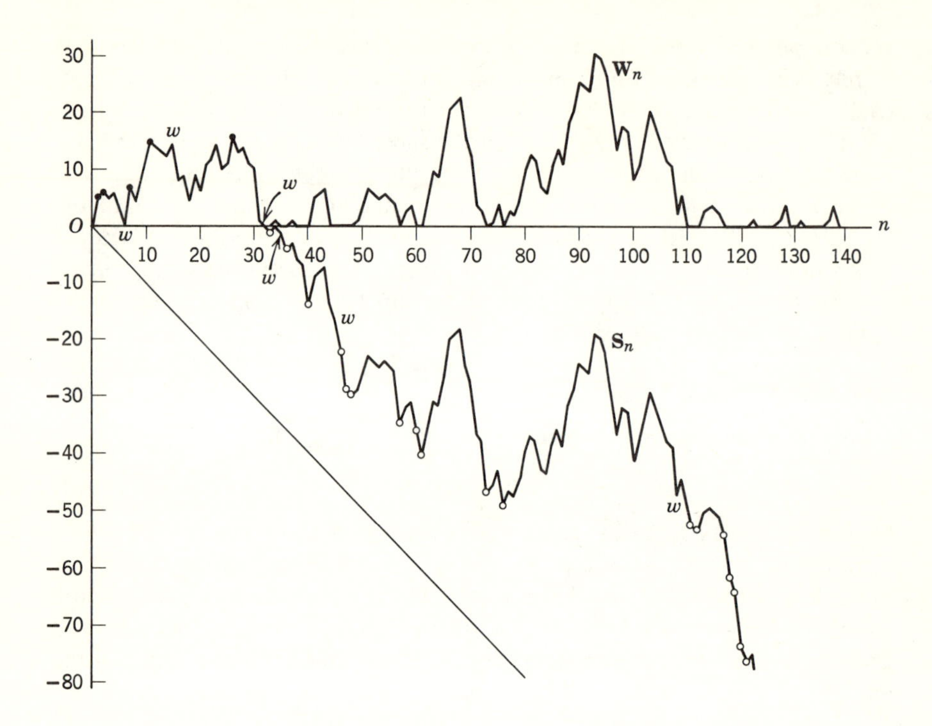

Figure 1. *Random Walk and the Associated Queuing Process.* The variables $\mathbf{X}_n$ of the random walk $\{\mathbf{S}_n\}$ have expectation -1 and variance 16. Ascending and descending ladder points are indicated by ● and ○, respectively. The seventh ladder point is (26, 16) and represents with high probability the maximum of the entire random walk.

[The letter w indicates where a record value is assumed for a second or third time; these are the *weak* ladder points defined by (8.2) when the strict inequality is replaced by $\geq$.]

Throughout the graph $\mathbf{S}_n$ *exceeds its expected value* $-n$. In fact, $n = 135$ *is the first index such that* $\mathbf{S}_n \leq -n$ (namely $\mathbf{S}_{135} = -137$). This accords with the fact that the expectation of such n is infinite.

The variables $\mathbf{X}_n$ are of the form $\mathbf{X}_n = \mathscr{B}_n - \mathscr{A}_n$, where the variables $\mathscr{B}_n$ and $\mathscr{A}_n$ are mutually independent and uniformly distributed over 1, 3, 5, 7, 9 and 2, 4, 6, 8, 10, respectively. In example 9(a) the variable $\mathbf{W}_n$ represents the total waiting time of the nth customer if the interarrival times assume the values 2, 4, 6, 8, 10 with equal probabilities while the service times equal 1, 3, 5, 7, or 9, each with probability $\frac{1}{5}$. The distribution of $\mathbf{X}_n$ attributes probability $(5 - k)/25$ to the points $\pm 2k - 1$, where $k = 0, 1, 2, 3, 4$.

Example. (a) In the "ordinary" random walk F has the atoms 1 and -1 with weights p and q. The ascending ladder variables are defective if $q > p$, the defect p/q [see **1**; XI,(3.9)]. The kth ladder height necessarily equals k and for this reason volume **1** mentions only ladder epochs. The kth ladder index is the epoch of the *first visit* to the point k. Its distribution was found in **1**; XI,4(d) and in the special case $p = \frac{1}{2}$ already in theorem 2 of **1**; III,4.

The *first* ladder index $\mathscr{T}_1$ is the epoch of the first entry into $\overline{0, \infty}$, and the *first* ladder height $\mathscr{H}_1$ equals $\mathbf{S}_{\mathscr{T}_1}$. The continuation of the random walk beyond epoch $\mathscr{T}_1$ is a probabilistic replica of the entire random walk. Given that $\mathscr{T}_1 = n$, the occurrence of a second ladder index a an epoch $k > n$ depends only on $\mathbf{X}_{n+1}, \ldots, \mathbf{X}_k$, and hence the number of trials between the first ladder index and the second is a random variable $\mathscr{T}_2$ which is independent of $\mathscr{T}_1$ and has the same distribution. In this way it is seen more generally that the kth *ladder index and the* kth *ladder height may be written in the form*

$$\mathscr{T}_1 + \cdots + \mathscr{T}_k, \qquad \mathscr{H}_1 + \cdots + \mathscr{H}_k$$

where the $\mathscr{T}_j$ and $\mathscr{H}_j$ are mutually independent random variables distributed, respectively, as $\mathscr{T}_1$ and $\mathscr{H}_1$. In other words, the ladder indices and heights form (possibly terminating) *renewal processes.*

For terminating processes it is intuitively obvious that $\mathbf{S}_n$ drifts to $-\infty$, and with probability one $\mathbf{S}_n$ reaches a finite maximum. The next section will show that the ladder variables provide a powerful tool for the analysis of a class of processes of considerable practical interest.

Example. (b) *Explicit expressions.* Let F have the density defined by

$$\frac{abe^{ax}}{a+b} \quad \text{if} \quad x < 0; \qquad \frac{abe^{-bx}}{a+b} \quad \text{if} \quad x > 0. \tag{8.3}$$

This random walk has the rare distinction that all pertinent distributions can be calculated explicitly. It is of great interest in queuing theory because f is the convolution of two exponential densities concentrated on $\overline{0, \infty}$ and $\overline{-\infty, 0}$, respectively. This means that $\mathbf{X}_j$ may be written as the difference $\mathbf{X}_j = \mathscr{B}_j - \mathscr{A}_j$ of two *positive exponentially distributed random variables.* Without loss of generality we assume $a \leq b$.

The *ascending ladder height $\mathscr{H}_1$ has the density* ae^{-bx}; this variable is defective and its defect equals $(b-a)/b$. The *ascending ladder epoch* $\mathscr{T}_1$ has the generating function $b^{-1}p(s)$ where

$$2p(s) = a + b - \sqrt{(a+b)^2 - 4abs}. \tag{8.4}$$

The defect is again $(b-a)/b$.

The descending ladder height $\mathscr{H}_1^-$ has density ae^{ax} for $x < 0$, the descending ladder epoch $\mathscr{T}_1^-$ has the generating function $a^{-1}p(s)$. In the special case $a = b$ it reduces to $1 - \sqrt{1-s}$, and this generating function is familiar from ordinary random walks (or coin tossing). [For proofs and other results see XII,4–5 and XVIII,3. See also example 4(e).]

9. THE QUEUING PROCESS

An incredibly voluminous literature[19] has been devoted to a variety of problems connected with servers, storage facilities, waiting times, etc. Much progress has been made towards a unification, but the abundance of small variants obscures the view so that it is difficult to see the forest for the trees. The power of new and general methods is still underrated. We begin by a formal introduction of a stochastic process defined by a recursive scheme that at first sight appears artificial. Examples will illustrate the wide applicability of the scheme; later on we shall see that sharp results can be obtained by surprisingly simple methods. (See XII,5.)

Definition 1. *Let* $\mathbf{X}_1, \mathbf{X}_2, \ldots$ *be mutually independent random variables with a common (proper) distribution* F. *The induced queuing process is the sequence of random variables* $\mathbf{W}_0, \mathbf{W}_1, \ldots$ *defined recursively by* $\mathbf{W}_0 = 0$ *and*

$$(9.1) \qquad \mathbf{W}_{n+1} = \begin{matrix} \mathbf{W}_n + \mathbf{X}_{n+1} & \text{if} \quad \mathbf{W}_n + \mathbf{X}_{n+1} \geq 0 \\ 0 & \text{if} \quad \mathbf{W}_n + \mathbf{X}_{n+1} \leq 0 \end{matrix}$$

In short, $\mathbf{W}_{n+1} = (\mathbf{W}_n + \mathbf{X}_{n+1}) \cup 0$.

For an illustration see figure 1.

Examples. (*a*) *The one-server queue.* Suppose that "customers" arrive at a "server" the arrivals forming a proper renewal process with *interarrival times*[20] $\mathscr{A}_1, \mathscr{A}_2, \ldots$ (the epochs of arrivals are $0, \mathscr{A}_1, \mathscr{A}_1 + \mathscr{A}_2, \ldots$ and the customers are labeled $0, 1, 2, \ldots$). With the nth customer there is associated a *service time* $\mathscr{B}_n$, and we assume that the $\mathscr{B}_n$ are independent of the arrivals and of each other and subject to a common distribution. The server is either "free" or "busy"; it is free at the initial epoch 0. The

---

[19] For references consult the specialized books listed in the bibliography. It would be difficult to give a brief outline of the development of the subject with a proper assignment of credits. The most meritorious papers responsible for new methods are now rendered obsolete by the progress which they initiated. [D. V. Lindley's integral equation of queuing theory (1952) is an example.] Other papers are noteworthy by their treatment of (sometimes very intricate) special problems, but they find no place in a skeleton survey of the general theory. On the whole, the prodigal literature on the several subjects emphasizes examples and variants at the expense of general methods. An assignment of priorities is made difficult also by the many duplications. [For example, the solution of a certain integral equation occurs in a Stockholm thesis of 1939 where it is credited to unpublished lectures by Feller in 1934. This solution is now known under several names.] For the history see two survey papers by D. G. Kendall of independent interest: *Some problems in the theory of queues*, and *Some problems in the theory of dams*, J. Roy. Statist. Soc. Series B vol. 13 (1951) pp. 151–185, and vol. 19 (1957) pp. 207–233.

[20] Normally the interarrival times will be constant or exponentially distributed but it is fashionable to permit arbitrary renewal processes; see footnote 14 to section 7.

sequel is regulated by the following rule. If a customer arrives at an epoch where the server is free, his service commences without delay. Otherwise he joins a waiting line (queue) and the server continues uninterruptedly to serve customers in the order of their arrival[21] until the waiting line disappears and the server becomes "free." By *queue length* we mean the number of customers present including the customer being served. The *waiting time* $\mathbf{W}_n$ of the nth customer is the time from his arrival to the epoch where his service *commences*; the total time spent by the customer at the server is $\mathbf{W}_n + \mathscr{B}_n$. (For example, if the first few service times are 4, 4, 1, 3, ... and the interarrival times are 2, 3, 2, 3, ..., customers number 1, 2, ... *join* queues of length 1, 1, 2, 1, ..., respectively, and have waiting times 2, 3, 2, 2, ...).

To avoid trite ambiguities such as when a customer arrives at the epoch of another's departure we shall assume that the distributions A and B of the variables $\mathscr{A}_n$ and $\mathscr{B}_n$ are continuous. Then the queue length at any epoch is well defined.

We proceed to devise a scheme for calculating the waiting times $\mathbf{W}_n$ recursively. By definition customer number 0 arrives at epoch 0 at a free server and so his waiting time is $\mathbf{W}_0 = 0$. Suppose now that the nth customer arrives at epoch t and that we know his waiting time $\mathbf{W}_n$. His service time commences at epoch $t + \mathbf{W}_n$ and terminates at epoch $t + \mathbf{W}_n + \mathscr{B}_n$. The *next* customer arrives at time $t + \mathscr{A}_{n+1}$. He finds the server free if $\mathbf{W}_n + \mathscr{B}_n < \mathscr{A}_{n+1}$ and has a waiting time $\mathbf{W}_{n+1} = \mathbf{W}_n + \mathscr{B}_n - \mathscr{A}_{n+1}$ if this quantity is ≥ 0. In other words, *the sequence* $\{\mathbf{W}_n\}$ *of waiting times coincides with the queuing process induced by the independent random variables*

$$\mathbf{X}_n = \mathscr{B}_{n-1} - \mathscr{A}_n, \qquad n = 1, 2, \ldots \tag{9.2}$$

(*b*) *Storage and inventories.* For an intuitive description we use water *reservoirs* (and dams), but the model applies equally to other storage facilities or inventories. The content depends on the input and the output. The input is due to supplies by rivers and rainfall, the output is regulated by demand except that this demand can be satisfied only when the reservoir is not empty.

Consider now the water contents[22] $0, \mathbf{W}_1, \mathbf{W}_2, \ldots$ at selected epochs $0, \tau_1, \tau_2, \ldots$. Denote by $\mathbf{X}_n$ the actual supply minus the theoretical (ideal)

[21] This "queue discipline" is totally irrelevant to queue length, duration of busy periods, and similar problems. Only the individual customer feels the effect of the several disciplines, among which "first come first served," "first come last served" and "random choice" are the extremes. The whole picture would change if departures were permitted.

[22] For simplicity we start with an empty reservoir. An adjustment to arbitrary initial conditions causes no difficulties [see example (*c*)].

demand during $\overline{\tau_{n-1}, \tau_n}$ and let us pretend that all changes are instantaneous and concentrated at the epochs $\tau_1, \tau_2, \ldots$. We start with $\mathbf{W}_0 = 0$ at epoch 0. In general the change $\mathbf{W}_{n+1} - \mathbf{W}_n$ should equal $\mathbf{X}_{n+1}$ except when the demand exceeds the contents. For this reason the $\mathbf{W}_n$ must satisfy (9.1) and so the *successive contents are subject to the queuing process induced by* $\{\mathbf{X}_k\}$ provided the theoretical net changes $\mathbf{X}_k$ are independent random variables with a common distribution.

The problem (for the mathematician if not for the user) is to find conditions under which the $\mathbf{X}_k$ will appear as independent variables with a common distribution F and to find plausible forms for F. Usually the τ_k will be equidistant or else a sample from a Poisson process, but it suffices for our purposes to assume that the τ_k form a *renewal process* with interarrival times $\mathscr{A}_1, \mathscr{A}_2, \ldots$. The most frequently used models fall into one of the following two categories:

(i) The input is at a constant rate c, the demand $\mathscr{B}_n$ arbitrary. Then $\mathbf{X}_n = c\mathscr{A}_n - \mathscr{B}_n$. We must suppose this $\mathbf{X}_n$ to be independent of the "past" $\mathbf{X}_1, \ldots, \mathbf{X}_{n-1}$. (The usual assumption that $\mathscr{A}_n$ and $\mathscr{B}_n$ be independent is superfluous: there is no reason why the demand $\mathscr{B}_n$ should not be correlated with the duration $\mathscr{A}_n$.)

(ii) The output is at a constant rate, the input arbitrary. The description is the same with the roles of $\mathscr{A}_n$ and $\mathscr{B}_n$ reversed.

(*c*) *Queues for a shuttle train.*[23] A shuttle train with r places for passengers leaves a station every hour on the hour. Prospective passengers appear at the station and wait in line. At each departure the first r passengers in line board the train, and the others remain in the waiting line. We suppose that the number of passengers arriving between successive departures are independent random variables $\mathscr{A}_1, \mathscr{A}_2, \ldots$ with a common distribution. Let $\mathbf{W}_n$ be the number of passengers in line just after the nth departure, and assume for simplicity $\mathbf{W}_0 = 0$. Then $\mathbf{W}_{n+1} = \mathbf{W}_n + \mathscr{A}_{n+1} - r$ if this quantity is positive, and $\mathbf{W}_{n+1} = 0$ otherwise. Thus $\mathbf{W}_n$ *is the variable of a queuing process* (9.1) *generated by the* random walk with variables $\mathbf{X}_n = \mathscr{A}_n - r$. ►

We turn to a description of the queuing process $\{\mathbf{W}_n\}$ in terms of the random walk generated by the variables $\mathbf{X}_k$. As in section 8 we put $\mathbf{S}_0 = 0$, $\mathbf{S}_n = \mathbf{X}_1 + \cdots + \mathbf{X}_n$ and adhere to the notation for the ladder variables. For ease of description we use the terminology appropriate for the server of example (*a*).

[23] P. E. Boudreau, J. S. Griffin Jr., and Mark Kac, *An elementary queuing problem*, Amer. Math. Monthly, vol. 69 (1962) pp. 713–724. The purpose of this paper is didactic, that is, it is written for outsiders without knowledge of the subject. Although a different mode of description is used, the calculations are covered by those in example XII,4(*c*).

Define ν as the subscript for which $\mathbf{S}_1 \geq 0, \mathbf{S}_2 \geq 0, \ldots, \mathbf{S}_{\nu-1} \geq 0$, but $\mathbf{S}_\nu < 0$. In this situation customers number $1, 2, \ldots, \nu-1$ had positive waiting times $\mathbf{W}_1 = \mathbf{S}_1, \ldots, \mathbf{W}_{\nu-1} = \mathbf{S}_{\nu-1}$, and customer number ν was the first to find the server free (the first lucky customer). At the epoch of his arrival the process starts from scratch as a replica of the whole process. Now ν is simply the index of the first negative sum, that is, ν is the first descending ladder index, and we denote it consistently by $\mathcal{T}_1^-$. We have thus reached the *first conclusion: The descending ladder indices correspond to the lucky customers who find the server free.* Put differently, the epochs of arrival of the lucky customers constitute a renewal process with recurrence times distributed as $\mathcal{T}_1^-$.

In practical cases the variable $\mathcal{T}_1^-$ must not be defective, for its defect p would equal the probability that a customer never finds the server free and with probability one there would be a last lucky customer followed by an unending queue. It will turn out that $\mathcal{T}_1^-$ is proper whenever $\mathbf{E}(\mathcal{B}_k) < \mathbf{E}(\mathcal{A}_k)$.

Suppose now that customer number $\nu - 1$ arrives at epoch τ. His waiting time was $\mathbf{W}_{\nu-1} = \mathbf{S}_{\nu-1}$ and so the epoch of his departure is $\tau + \mathbf{W}_{\nu-1} + \mathcal{B}_{\nu-1}$. The first lucky customer (number ν) arrives at epoch $\tau + \mathcal{A}_\nu$ when the server was free for

$$\mathcal{A}_\nu - \mathbf{W}_{\nu-1} - \mathcal{B}_{\nu-1} = -\mathbf{S}_{\nu-1} - \mathbf{X}_\nu = -\mathbf{S}_\nu$$

time units. But by definition $\mathbf{S}_\nu$ is the first descending ladder height $\mathcal{H}_1^-$. As the process starts from scratch we have reached the *second conclusion: The durations of the free periods are independent random variables with the same distribution as* $-\mathcal{H}_1^-$ (the recurrence time for the descending ladder heights). In other words, customer number $\mathcal{T}_1^- + \cdots + \mathcal{T}_r^-$ is the rth customer who finds the server free. At the epoch of his arrival the server has been free for $-\mathcal{H}_r^-$ time units.

It should now be clear that between successive ladder epochs *the segments of the graph for the queuing process* $\{\mathbf{W}_n\}$ *are congruent to those for the random walk* but displayed vertically so as to start at a point of the time axis (figure 1). To describe this analytically denote for the moment by $[n]$ the *last* descending ladder index $\leq n$; in other words, $[n]$ is a (random) index such that $[n] \leq n$ and

$$\mathbf{S}_{[n]} \leq \mathbf{S}_j \qquad j = 0, 1, \ldots, n. \tag{9.3}$$

This defines $[n]$ uniquely with probability 1 (the distribution of $\mathbf{X}_i$ being continuous). Clearly

$$\mathbf{W}_n = \mathbf{S}_n - \mathbf{S}_{[n]}. \tag{9.4}$$

This relation leads to the most important conclusion if we look at the variables $\mathbf{X}_1, \ldots, \mathbf{X}_n$ in *reverse order.* Put for abbreviation $\mathbf{X}_1' = \mathbf{X}_n, \ldots, \mathbf{X}_n' = \mathbf{X}_1$. The partial sums of these variables are

$$\mathbf{S}_k' = \mathbf{X}_1' + \cdots + \mathbf{X}_k' = \mathbf{S}_n - \mathbf{S}_{n-k},$$

and (9.4) shows that the maximal term of the sequence $0, \mathbf{S}_1', \ldots, \mathbf{S}_n'$ has subscript $n - [n]$ and equals $\mathbf{W}_n$. But the distribution of $(\mathbf{X}_1', \ldots, \mathbf{X}_n')$ is identical with that of $(\mathbf{X}_1, \ldots, \mathbf{X}_n)$. We have thus the basic

Theorem.[24] *The distribution of the queuing variable* $\mathbf{W}_n$ *is identical with the distribution of the random variable*

$$\mathbf{M}_n = \max\,[0, \mathbf{S}_1, \ldots, \mathbf{S}_n] \tag{9.5}$$

in the underlying random walk $\{\mathbf{X}_k\}$.

The consequences of this theorem will be discussed in chapter XII. Here we show that it permits us to reduce certain ruin problems to queuing processes despite the dissimilarity of the appearance.

Example. (*d*) *Ruin problems.* In section 5 ruin was defined as the event that $\mathbf{X}(t) > z + ct$ for some t where $\mathbf{X}(t)$ is the variable of a compound Poisson process with distribution (4.2). Denote the epochs of the successive jumps in this process by $\tau_1, \tau_2, \ldots$. If ruin occurs at all it occurs also at some epoch τ_k and it suffices therefore to consider the probability that $\mathbf{S}_n = \mathbf{X}(\tau_n) - c\tau_n > z$ for some n. But by the definition of a compound Poisson process $\mathbf{X}(\tau_n)$ is the sum of n independent variables $\mathbf{Y}_k$ with the common distribution F, while τ_n is the sum of n independent exponentially distributed variables $\mathscr{A}_k$. Accordingly we are in effect dealing with the random walk generated by the variables $\mathbf{X}_k = \mathbf{Y}_k - c\mathscr{A}_k$ whose probability *density* is given by the convolution

$$\frac{\alpha}{c}\int_x^\infty e^{\alpha(x-y)/c}\,F\{dy\}. \tag{9.6}$$

Ruin occurs iff in the random walk the event $\{\mathbf{S}_n \geq z\}$ *takes place for some n.* To find the probability of ruin amounts therefore to finding the distributions of the variables $\mathbf{W}_n$ in the associated queuing process.

(*e*) *A numerical illustration.* The most important queuing process arises when *the interarrival and service times are exponentially distributed* with expectations $1/a$ and $1/b$, respectively, where $a < b$. From the characteristics of this process described in example 8(*b*), one can conclude that the *waiting time of the* nth *customer has a limit distribution* W *with an atom of*

[24] Apparently first noticed by F. Pollaczek in 1952 and exploited (in a different context) by F. Spitzer, *The Wiener–Hopf equation whose kernel is a probability density*, Duke Math. J., vol. 24 (1957) pp. 327–344. For Spitzer's proof see problem 21.

weight $1 - a/b$ *at the origin and density* $\frac{b-a}{b}ae^{-(b-a)x}$ for $x > 0$. The expectation equals $\frac{a}{b(b-a)}$. The *free periods* of the counter have the same density as the first descending ladder height, that is, ae^{-at}. In this case the free periods and the interarrival times have the same distribution (but this is not so in other queuing processes).

The number $\mathbf{N}$ *of the first customer to find the counter empty has the generating function* $p(s)/a$ with p defined in (8.4). Consider now the *busy period* commencing at epoch 0, that is, the time interval to the first epoch when the server becomes free. This period being initiated by customer number 0, *the random variable* $\mathbf{N}$ *also equals the number of customers during the initial busy period.* An easy calculation shows that its expectation equals $b/(b-a)$ its variance $ab(a+b)/(b-a)^3$.

Finally, let $\mathbf{T}$ be the duration of the busy period. Its density is given explicitly by XIV, (6.16) with $cp = a$ and $cq = b$. This formula involving a Bessel function does not lend itself to easy calculations, but the moments of $\mathbf{T}$ can be calculated from its Laplace transform derived by different methods in examples XIV,4(a) and XIV,6(b). The result is

$$\mathbf{E}(\mathbf{T}) = \frac{1}{(b-a)} \quad \text{and} \quad \mathrm{Var}\,(\mathbf{T}) = (a+b)\frac{1}{(b-a)^3}.$$

In the queuing process busy periods alternate with free periods, and their expectations are $1(b-a)$ and $1/a$, respectively. Thus $(b-a)/a$ is a measure of the *fraction of the time during which the server is idle.* More precisely: if $\mathbf{U}(t)$ is the idle time up to epoch t, then $t^{-1}\mathbf{E}\mathbf{U}(t) \to (b-a)/a$.

Table 1

		$b = 1$					
		$a = 0.5$	$a = 0.6$	$a = 0.7$	$a = 0.8$	$a = 0.9$	$a = 0.95$
Waiting time (steady state)	Expectation	1	1.5	2.3	4	9	19
	Variance	3	5.3	10	24	99	399
Busy period	Expectation	2	2.5	3.3	5	10	399
	Variance	12	25	63	225	1900	16,000
No. of customers per busy period	Expectation	2	2.5	3.3	5	10	399
	Variance	6	15	44	200	1700	15,200

In the table the expected service time is taken as unit, and so a represents the *expected number of customers arriving during one service time.* The table shows the huge variances of the busy periods. It follows that *fantastic fluctuations of the busy period must be expected.* One sees that the customary reliance on expectations is very dangerous in practical applications. For a busy period with variance 225 the fact that the expectation is 5 has little practical significance.

The multidimensional analogue to our queuing process is more intricate. The foundations for its theory were laid by J. Kiefer and J. Wolfowitz [*On the theory of queues with many servers*, Trans. Amer. Math. Soc., vol. 78 (1955) pp. 1–18].

10. PERSISTENT AND TRANSIENT RANDOM WALKS

We proceed to a classification of random walks which is independent of section 8 and closely related to the renewal theory of section 6. Given a distribution function F on the line we introduce formally an interval function defined by

$$U\{I\} = \sum_{k=0}^{\infty} F^{k\star}\{I\}. \tag{10.1}$$

The series is the same as in (6.2), but when F is not concentrated on a half-line the series may diverge even when I is a finite interval. It will be shown that the convergence or divergence of (10.1) has a deep significance. The basic facts are simple, but the formulations suffer from the unfortunate necessity of a special treatment for arithmetic distributions.[25]

For abbreviation we let I_h stand for the interval $-h \leq x \leq h$ and $I_h + t$ for $t-h \leq x \leq t+h$.

Theorem 1. (i) *If F is non-arithmetic either $U\{I\} < \infty$ for every finite interval or else $U\{I\} = \infty$ for all intervals.*

(ii) *If F is arithmetic with span λ either $U\{I\} < \infty$ for every finite interval or else $U\{I\} = \infty$ for every interval containing a point of the form $n\lambda$.*

(iii) *If $U\{I_h\} < \infty$ then for all t and $h > 0$*

$$U\{I_h + t\} \leq U\{I_{2h}\}. \tag{10.2}$$

For ease of reference to the two cases we introduce a definition (in which F receives an adjective rightfully belonging to the corresponding random walk).

Definition. *F is transient if $U\{I\} < \infty$ for all finite intervals, and persistent otherwise.*

[25] F is arithmetic if all its points of increase are among the points of the form $0, \pm\lambda, \pm 2\lambda, \ldots$. The largest λ with this property is called the *span* of F. (See V,2.)

Besides its probabilistic significance the theorem has a bearing on the integral equation

$$Z = z + F \star Z \tag{10.3}$$

which is the analogue to the renewal equation (6.4). We use this integral equation as starting point and prove theorem 1 together with

Theorem 2. *Let* z *be continuous, and* $0 \leq z(x) \leq \mu_0$ *for* $|x| < h$ *and* $z(x) = 0$ *outside* I_h. *If* F *is transient then*

$$Z(x) = \int_{-\infty}^{+\infty} z(x-y)\, U\{dy\} \tag{10.4}$$

is a uniformly continuous solution of (10.3) *with*

$$0 \leq Z(x) \leq \mu_0 \cdot U\{I_{2h}\}. \tag{10.5}$$

Z *assumes its maximum at a point in* I_h.

Proof *of the two theorems.* (i) Assume that $U\{I_\alpha\} < \infty$ for some $\alpha > 0$. Choose $h < \frac{1}{2}\alpha$ and let z vanish outside I_h but not identically. We try to solve (10.3) by successive approximations putting $Z_0 = z$ and, recursively,

$$Z_n(x) = z(x) + \int_{-\infty}^{+\infty} Z_{n-1}(x-y)\, F\{dy\}. \tag{10.6}$$

With U_n defined by

$$U_n\{I\} = F^{0\star}\{I\} + \cdots + F^{n\star}\{I\} \tag{10.7}$$

we have obviously

$$Z_n(x) = \int_{-\infty}^{+\infty} z(x-y) U_n\{dy\}, \tag{10.8}$$

(the integration extending in effect over an interval of length $\leq 2h$). The function Z_n so defined is continuous, and we prove by induction that it assumes its maximum μ_n at a point ξ_n such that $z(\xi_n) > 0$. This is trivially true for $Z_0 = z$. If it is true for Z_{n-1} one sees from (10.6) that $z(x) = 0$ implies $Z_n(x) \leq \mu_{n-1}$ whereas $\mu_n \geq Z_n(\xi_{n-1}) > Z_{n-1}(\xi_{n-1}) = \mu_{n-1}$.

It follows that the interval $I_h + \xi_n$ is contained in I_{2h} and so by (10.8)

$$\mu_n \leq \mu_0 \cdot U\{I_{2h}\} \tag{10.9}$$

which proves that the functions Z_n remain uniformly bounded. Since $Z_0 \leq Z_1 \leq \cdots$ it follows that $Z_n \to Z$ with Z satisfying (10.5).

By monotone convergence it follows from (10.6) and (10.8) that the limit Z satisfies the integral equation (10.3) and is of the form (10.4). The inequality (10.5) holds because of (10.9). The upper bound depends only on the maximum μ_0 of z, and we are free to let $z(x) = \mu_0$ for all x within a proper subinterval I_η of I_h. In this case we get from (10.8)

$$Z_n(x) \geq \mu_0 \, U_n\{I_\eta + x\} \tag{10.10}$$

This inequality holds for all $\eta < h$, and hence also for $\eta = h$ (since I_h is closed). The last two inequalities together prove the truth of (10.2). This implies that $U\{I\} < \infty$ for intervals of length $\leq h$. But every finite interval can be partitioned into finitely many intervals of length $< h$, and therefore $U\{I\} < \infty$ for all finite I. Finally, taking differences in (10.4) it is seen that

$$|Z(x+\delta) - Z(x)| \leq U\{I_{2h}\} \cdot \sup |z(x+\delta) - z(x)|.$$

Accordingly, Z is uniformly continuous and all assertions concerning transient F are proved.

(ii) There remains the case where $U\{I_\alpha\} = \infty$ for *every* $\alpha > 0$. Then (10.10) shows that $Z_n(x) \to \infty$ for all x in a neighborhood of the origin. If t is a point of increase of F it follows from (10.6) that $Z_n(x) \to \infty$ for all x in a neighborhood of t. By induction the same is true of each point of increase of $F^{2\star}, F^{3\star}, \ldots$. Assume F non-arithmetic. If F were concentrated on a half-line we would have $U\{I_\alpha\} < \infty$ (section 6). By lemma 2 of V,4*a* therefore the points of increase of $F^{2\star}, F^{3\star}, \ldots$ are dense on the line and so $Z_n(x) \to \infty$ everywhere. This implies $U_n\{I\} \to \infty$, for all intervals. With the obvious modification this argument applies also to arithmetic distributions, and so the theorems are proved. ▶

In chapter XI we shall return to the renewal equation (10.3), but now we turn to the implications of theorem 1 for random walks. Let $\mathbf{X}_1, \mathbf{X}_2, \ldots$ be independent random variables with the common distribution F, and put $\mathbf{S}_n = \mathbf{X}_1 + \cdots + \mathbf{X}_n$. By "visit to I at epoch $n = 1, 2, \ldots$" is meant the event that $\mathbf{S}_n \in I$.

Theorem 3.[26] *If F is transient the number of visits to a finite interval I is finite with probability one, and the expected number of such visits equals $U\{I\}$.*

If F is persistent and non-arithmetic every interval I is visited infinitely often with probability one. If F is persistent and arithmetic with span λ then every point $n\lambda$ is visited infinitely often with probability one.

Proof. Assume F transient. The probability of a visit to I after epoch n does not exceed the nth remainder of the series in (10.1) and so for n sufficiently large the probability of more than n visits is $< \epsilon$. This proves the first assertion.

[26] The theorem is a consequence of the second zero-or-one law in IV,6. If $\varphi(I+t)$ is the probability of entering $I + t$ infinitely often then for fixed I the function φ can assume only the values 0 or 1. On the other hand, considering the first step in the random walk one sees that $\varphi = F \star \varphi$ and hence $\varphi = \text{const}$ (see XI,9).

Assume now F persistent and non-arithmetic. Denote by $\rho_h(t)$ the probability of a visit to $I_h + t$. It suffices to prove that $\rho_h(t) = 1$ for all $h > 0$ and all t, for this obviously implies the certainty of any number of visits to each interval.

Before proceeding we observe that if $\mathbf{S}_n = x$ we may take x as a new origin to conclude that the probability of a subsequent visit to I_h equals $\rho_h(-x)$. In particular, if x is a point in I_{h+t} the probability of a subsequent visit to I_h is $\leq \rho_{2h}(-t)$.

We begin now by showing that $\rho_h(0) = 1$. For an arbitrary, but fixed, $h > 0$ denote by p_r the probability of at least r visits to I_h. Then $p_1 + p_2 + \cdots$ is the expected number of visits to I_h and hence infinite by the definition of persistency. On the other hand, the preliminary remark makes it clear that $p_{r+1} \leq p_r \cdot \rho_{2h}(0)$. The divergence of $\sum p_r$ therefore requires that $\rho_{2h}(0) = 1$.

Passing to general intervals I_{h+t}, assume first that F is not arithmetic. By lemma 2 of V,4*a* every interval contains a point of increase of $F^{k\star}$ for some k and therefore the probability $\rho_h(t)$ of entering I_{h+t} is positive for all $h > 0$ and all t. But we saw already that even after entering I_{h+t} a return to I_h is certain, and by the preliminary remark this implies that $\rho_{2h}(-t) = 1$. Since h and t are arbitrary this concludes the proof for non-arithmetic distributions. But the same argument applies to arithmetic distributions. ▶

In testing whether the series (10.1) for $U\{I\}$ converges one has usually to rely on limit theorems which provide information only for very large intervals. In such situations it is advisable to rely on the following

Criterion. *If F is transient, then $x^{-1}U\{I_x\}$ remains bounded as $x \to \infty$.*

The assertion is obvious from (10.2) since any interval I_{nh} may be partitioned into n intervals of the form I_{h+t}. As an illustration of the method we prove

Theorem 4. *A distribution with expectation μ is persistent if $\mu = 0$, transient if $\mu \neq 0$.*

Proof. Let $\mu = 0$. By the weak law of large numbers there exists an integer n_ϵ such that $\mathbf{P}\{|\mathbf{S}_n| < \epsilon n\} > \frac{1}{2}$ for all $n > n_\epsilon$. Accordingly $F^{n\star}\{I_a\} > \frac{1}{2}$ for all n such that $n_\epsilon < n < a/\epsilon$. If $a > 2\epsilon n_\epsilon$ there are more than $a/(2\epsilon)$ integers n satisfying this condition, and hence $U\{I_a\} > a/(4\epsilon)$. Since ϵ is arbitrary this implies that the ratio $a^{-1}U\{I_a\}$ is not bounded, and hence F cannot be transient.

If $\mu > 0$ the strong law of large numbers guarantees that with a probability arbitrarily close to one $\mathbf{S}_n$ will be positive for all n sufficiently large.

The probability of entering the negative half-line infinitely often is therefore zero, and thus F is transient. ►

In a persistent process the sequence $\{\mathbf{S}_n\}$ necessarily changes sign infinitely often and so the ascending and descending ladder processes are persistent. It may come as a surprise that the converse is false. *Even in a transient random walk* $\{\mathbf{S}_n\}$ *may change signs infinitely often* (with probability one). In fact, this is the case when F is symmetric. Since a finite interval $\overline{-a,\ a}$ will be visited only finitely often this implies (very roughly speaking) that the changes of signs are due to occasional jumps of fantastic magnitude: $|\mathbf{S}_n|$ is likely to grow over all bounds, but the fantastic inequality $\mathbf{X}_{n+1} < -\mathbf{S}_n - a$ will occur infinitely often however large the constant a.

Figure 2 illustrates the occurrence of large jumps but is not fully representative of the phenomenon because it was necessary to truncate the distribution in order to obtain a finite graph.

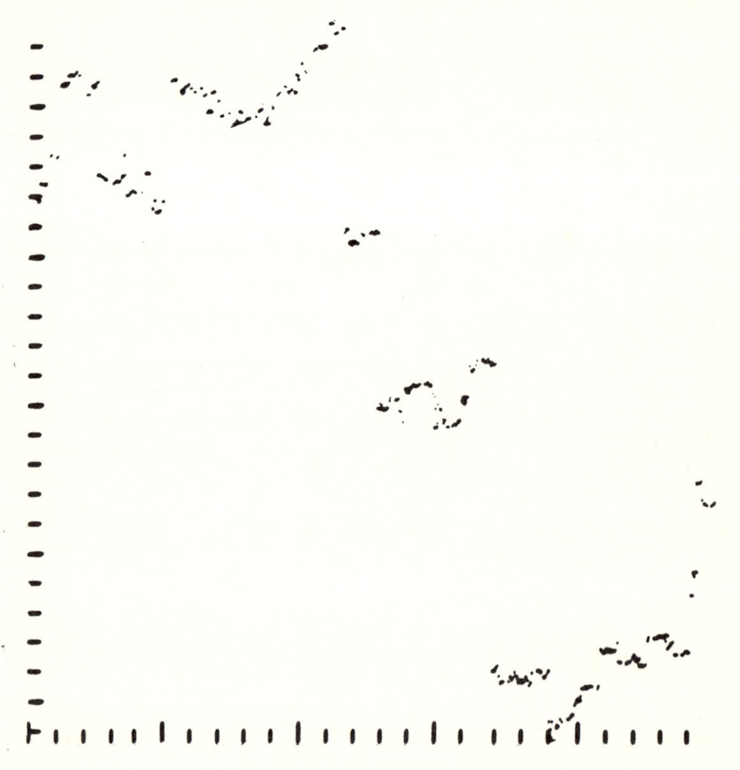

Figure 2. Random Walk Generated by the Cauchy Distribution. (The distribution was truncated so as to eliminate jumps of the magnitude of the graph.)

11. GENERAL MARKOV CHAINS

The generalization of the discrete Markov chains of **1**; XV to Cartesian (and more general) spaces is simple. In the discrete case the transition probabilities were given by a stochastic matrix with elements p_{ij} whose rows were probability distributions. Now we have to consider transitions from a point x to an arbitrary interval or set Γ in $\mathcal{R}^n$; we denote the probability of this transition by $K(x, \Gamma)$. The novel feature is that we must impose some regularity conditions to ensure that the necessary integrations can be performed. Continuity would do for most practical purposes, but nothing is gained by restricting the full generality.

Definition 1. *A stochastic kernel K is a function of two variables, a point and a set, such that $K(x, \Gamma)$ is* (i) *for a fixed x a probability distribution in Γ, and* (ii) *for any interval Γ a Baire function in x.*

It is *not* required that K be defined on the whole space. If x and Γ are restricted to a set Ω we say that *K is concentrated on Ω*. Sometimes it is necessary to admit *defective* distributions and we speak then of *substochastic* kernels. Frequently K will be of the form

$$K(x, \Gamma) = \int_\Gamma k(x, y)\, dy \tag{11.1}$$

and in this case k is called a *stochastic density kernel.* Following the convention of V,(3.3) we indicate (11.1) by the shorthand notation

$$K(x, dy) = k(x, y)\, dy.$$

[Strictly speaking, k represents densities with respect to Lebesgue measure or length; densities with respect to an arbitrary measure m would be denoted by $K(x, dy) = k(x, y)\, m\{dy\}$.]

Before giving a formal definition of Markov chains we can assemble the appropriate analytical apparatus by analogy with the discrete case. The probability of a transition from x to Γ in *two* steps is defined by

$$K^{(2)}(x, \Gamma) = \int_\Omega K(x, dy)\, K(y, \Gamma), \tag{11.2}$$

the integration extending over the whole space or the set Ω on which K is concentrated. Relation (11.2) states that the first step leads from x to some point y and the second from y to Γ. The crucial assumption is that, given the intermediate point y, the past history in no way influences the further transitions. A similar argument holds for the higher transition probabilities $K^{(n)}$. If we put $K^{(1)} = K$ we have for arbitrary positive integers

$$K^{(m+n)}(x, \Gamma) = \int_\Omega K^{(m)}(x, dy) K^{(n)}(y, \Gamma); \tag{11.3}$$

this reduces to (11.2) when $m = n = 1$. Keeping $m = 1$ and letting $n = 1, 2, 3, \ldots$ we get an *inductive definition for* $K^{(n)}$. For consistency we define $K^{(0)}$ to stand for the probability distribution concentrated at the point x (the so-called *Kronecker delta* kernel). Then (11.3) is valid for $m \geq 0$, $n \geq 0$. The operation (11.3) between two kernels occurs frequently also outside probability theory and is known as *composition of kernels*. It is in all respects similar to matrix multiplication.

It is hardly necessary to emphasize that the kernels $K^{(n)}$ are stochastic. If K has a density, the same is true of $K^{(n)}$ and the *composition formula* for densities is

$$k^{(m+n)}(x, z) = \int_\Omega k^{(m)}(x, y)\, k^{(n)}(y, z)\, dy. \tag{11.4}$$

Examples. (*a*) *Convolutions.* If $k(x, y) = f(y-x)$, where f is a probability density, the composition (11.4) reduces to ordinary convolutions. The same is true generally if K is *homogeneous* in the sense that

$$K(x, \Gamma) = K(x+s, \Gamma+s)$$

where $\Gamma + s$ is the set obtained by translating Γ through s. For convolutions on the circle see theorem 3 in VIII,7.

(*b*) *Energy losses under collisions.* In physics successive collisions of a particle are usually treated as a chance process such that if the energy (or mass) before collision equals $x > 0$ the resulting energy (mass) is a random variable $\mathbf{Y}$ such that $\mathbf{P}\{\mathbf{Y} \in \Gamma\} = K(x, \Gamma)$ where K is a stochastic kernel. The standard assumption is that only losses are possible, and that the ratio $\mathbf{Y}/x$ has a distribution function G independent of x; then $\mathbf{P}\{\mathbf{Y} \leq y\} = G(y/x)$ which defines a stochastic kernel.

In a related problem in stellar radiation [example X,2(*b*)] Ambarzumian considered the special case $G(y) = y^\lambda$ for $0 \leq y \leq 1$ where λ is a positive constant. This corresponds to a density kernel $\lambda y^{\lambda-1} x^{-\lambda}$ concentrated on $0 < y < x$ and it is easily verified that the higher densities are given by

$$k^{(n)}(x, y) = \frac{\lambda^n}{(n-1)!} \frac{y^{\lambda-1}}{x^\lambda} \left(\log \frac{x}{y}\right)^{n-1}, \qquad 0 < y < x. \tag{11.5}$$

The particular value $\lambda = 1$ corresponds to a uniform distribution (the *fraction lost* is "randomly distributed") and (11.5) then reduces to I,(8.2). [Continued in example X,1(*a*).]

(*c*) *Random chains.* Consider a chain (or polygonal line) in $\mathfrak{R}^3$ whose links have unit length and where the angles between adjacent links depend on a chance mechanism. Many (frequently rather involved) variants occur in polymer chemistry, but we consider only the case where the successive angles are independent random variables.

By *length* $\mathbf{L}$ of a chain with endpoints A and B we mean the distance between A and B. Addition of a unit link to a chain of length x results in a chain length $\sqrt{x^2 + 1 - 2x \cos \theta}$ where θ is the angle between the new link and the line through A and B. We treat θ as a random variable and consider in particular two distributions that are of special interest in chemistry.

(i) Let θ equal 60° or 120° with probabilities $\frac{1}{2}$ each. Then $\cos \theta = \pm\frac{1}{2}$ and the length of the prolonged chain is subject to the stochastic kernel $K(x, \Gamma)$ attributing probabilities $\frac{1}{2}$ to the two points $\sqrt{x^2 \pm x + 1}$. For fixed x the distribution $K^{(n)}$ is concentrated on 2^n points.

(ii) Let the direction of the new link be chosen "at random," that is, suppose $\cos \theta$ to be uniformly distributed in $\overline{-1, 1}$. (See I,10.) The prolonged chain has a length $\mathbf{L}$ between $x + 1$ and $|x - 1|$. Within this range by the law of the cosines

$$\mathbf{P}\{\mathbf{L} < y\} = \mathbf{P}\{2x \cos \theta > x^2 + 1 + y^2\} = \tfrac{1}{2} - [x^2+1+y^2]/4x.$$

Thus the length is determined by the stochastic density kernel

$$k(x, y) = y/2x \qquad\qquad |x-1| < y < x + 1.$$

The length $\mathbf{L}_{n+1}$ *of a chain with* $n + 1$ *links has density* $k^{(n)}(1, y)$. (See problem 23.)

(*d*) *Discrete Markov chains.* A stochastic matrix (p_{ij}) may be considered as a stochastic density $k(i, j) = p_{ij}$ defined on the set Ω of positive integers and with respect to the measure m attributing unit weight to each integer.

Absolute and Stationary Probabilities

Saying that a sequence $\mathbf{X}_0, \mathbf{X}_1, \ldots$ is subject to the transition probabilities $K^{(n)}$ means that $K^{(n)}(x, \Gamma)$ is the conditional probability of the event $\{\mathbf{X}_{m+n} \in \Gamma\}$ given that $\mathbf{X}_m = x$. If the probability distribution of $\mathbf{X}_0$ is γ_0 *the probability distribution of* $\mathbf{X}_n$ *is given by*

$$\gamma_n(\Gamma) = \int_\Omega \gamma_0\{dx\}\, K^{(n)}(x, \Gamma). \tag{11.6}$$

Definition 2. *The distribution* γ_0 *is a stationary distribution for* K *if* $\gamma_n = \gamma_0$ *for all* n, *that is, if*

$$\gamma_0\{\Gamma\} = \int_\Omega \gamma_0\{dx\}\, K(x, \Gamma). \tag{11.7}$$

The basic facts concerning stationary distributions are the same as in the case of discrete Markov chains. Under mild regularity conditions on K *there exists a unique stationary distribution and it represents the asymptotic distribution of* $\mathbf{X}_n$ *under any initial distribution.* In other words, the influence of the initial state fades away and the system tends to a *steady state* governed by the stationary solution. This is one form of the ergodic theorem. (See VIII,7.)

Examples. (*e*) *The queuing process* $\{\mathbf{W}_n\}$ defined in (9.1) is a Markov process concentrated on the closed interval $\vdash\!\overline{0, \infty}$. The transition probabilities are defined only for $x, y \geq 0$ and there $K(x, \vdash\!\overline{0, y}) = F(y-x)$. The existence of a stationary measure will be proved in VIII,7.

(*f*) Let $\mathbf{X}_1, \mathbf{X}_2, \ldots$ be mutually independent positive variables with a distribution F with a continuous density f concentrated on $\overline{0, \infty}$. Define a sequence of random variables $\mathbf{Y}_k$ recursively by

$$\mathbf{Y}_1 = \mathbf{X}_1, \qquad \mathbf{Y}_{n+1} = |\mathbf{Y}_n - \mathbf{X}_{n+1}|. \tag{11.8}$$

Then $\{\mathbf{Y}_n\}$ is a Markov chain concentrated on $\overline{0, \infty}$ with transition densities

$$k(x, y) = \begin{cases} f(x-y) + f(x+y) & 0 < y < x \\ f(x+y) & y > x > 0. \end{cases} \tag{11.9}$$

The defining equation for a stationary density g is

$$g(y) = \int_0^\infty g(x+y)\,f(x)\,dx + \int_0^\infty g(x)\,f(x+y)\,dx. \tag{11.10}$$

If F has a finite expectation μ then

$$g(y) = \mu^{-1}[1-F(y)] \tag{11.11}$$

is a *stationary probability density*. In fact, a simple integration by parts will show that g satisfies[27] (11.10) and we know from V,(6.3) that g is a probability density. (See problem 22.)

(*g*) *A technical application.*[28] A long transmission line consists of individual pieces of cable whose characteristics are subject to statistical fluctuations. We treat the deviations from the ideal value as independent random variables $\mathbf{Y}_1, \mathbf{Y}_2, \ldots$ and suppose that their effect is additive. Reversing a piece of cable changes the sign of its contribution. Assume that the deviations $\mathbf{Y}_k$ are symmetric and put $\mathbf{X}_k = |\mathbf{Y}_k|$. An efficient construction of a long transmission line now proceeds by the following inductive rule: the $(n+1)$st piece of cable is attached in that position which gives its error a

[27] How does one discover such a thing? Assuming hopefully that g and f have derivatives we may differentiate (11.10) formally. An integration by parts leads to the relation $g'(y) = -g(0)f(y)$ showing that g must be of the form (11.11). Direct verification then proves the validity of (11.11) without differentiability conditions.

[28] Adapted from a discrete model used by H. von Schelling, Elektrische Nachr.-Technik, vol. 20 (1943) pp. 251–259.

sign opposite to the sign of the accumulated error of the preceding n pieces. The accumulated errors then follow the rule (11.8); the stationary density (11.11) is actually a limiting distribution: *the error of a line consisting of* n *pieces is* (*for large* n) *distributed approximately with density* (11.11). On the other hand, if the pieces of cable were combined randomly, the central limit theorem would apply and the variance of the error would increase linearly with n, that is, with the length of the cable. The simple procedure of testing the sign of the error thus permits us to keep it in bounds. ►

In the preceding examples a Markovian sequence $\mathbf{X}_0, \mathbf{X}_1, \ldots$ was defined in terms of an initial distribution γ_0 and the transition probabilities K. The joint distribution of $(\mathbf{X}_0, \mathbf{X}_1, \ldots, \mathbf{X}_n)$ is of the form

$$\gamma_0\{dx_0\}\, K(x_0, dx_1) \cdots K(x_{n-1}, dx_n)$$

discussed in III,8 and **1**; XV,1. We have here a typical example of the advantage of defining absolute probabilities in terms of conditional ones. A more systematic way would be to start from the postulate

$$\mathbf{P}\{\mathbf{X}_{n+1} \in \Gamma \mid \mathbf{X}_0 = x_0, \ldots, \mathbf{X}_n = x_n\} = K(x_n, \Gamma) \tag{11.12}$$

as definition. Here the *Markov property* is expressed by the fact that the right side is independent of $x_0, x_1, \ldots, x_{n-1}$ so that the "past history" has no effect. The disadvantage of this definition is that it would involve us in problems of existence of conditional probabilities, their uniqueness, etc.

(For Markov processes depending on a continuous time parameter see chapter X.)

*12. MARTINGALES

For a first orientation we may consider a stochastic process $\{\mathbf{X}_n\}$ such that the joint distribution of $(\mathbf{X}_1, \ldots, \mathbf{X}_n)$ has a strictly positive continuous density p_n. Conditional densities and expectations are then defined everywhere in the elementary way of III,2. The variables $\mathbf{X}_n$ and $\mathbf{Y}_n$ are supposed to have expectations.

The sequence $\{\mathbf{X}_n\}$ will be called *absolutely fair* if for $n = 1, 2, \ldots$

$$\mathbf{E}(\mathbf{X}_1) = 0, \qquad \mathbf{E}(\mathbf{X}_{n+1} \mid \mathbf{X}_1, \ldots, \mathbf{X}_n) = 0. \tag{12.1}$$

A sequence $\{\mathbf{Y}_n\}$ is a *martingale* if

$$\mathbf{E}(\mathbf{Y}_{n+1} \mid \mathbf{Y}_1, \ldots, \mathbf{Y}_n) = \mathbf{Y}_n, \qquad n = 1, 2, \ldots \tag{12.2}$$

(A more flexible definition will be given presently.)

---

* Martingales are treated because of their great importance, but they are not used as a tool in this book.

The connection between the two types is simple. Given an absolutely fair sequence $\{\mathbf{X}_n\}$ put

$$\mathbf{Y}_n = \mathbf{X}_1 + \cdots + \mathbf{X}_n + c \tag{12.3}$$

where c is a constant. Then

$$\mathbf{E}(\mathbf{Y}_{n+1} \mid \mathbf{X}_1, \ldots, \mathbf{X}_n) = \mathbf{Y}_n. \tag{12.4}$$

The conditioning variables $\mathbf{X}_j$ may be replaced by the $\mathbf{Y}_k$, and so (12.4) is equivalent to (12.2). On the other hand, given a martingale $\{\mathbf{Y}_n\}$ put $\mathbf{X}_1 = \mathbf{Y}_1 - \mathbf{E}(\mathbf{Y}_1)$ and $\mathbf{X}_{n+1} = \mathbf{Y}_n$. Then $\{\mathbf{X}_n\}$ is absolutely fair and (12.3) holds with $c = \mathbf{E}(\mathbf{Y}_1)$. Thus $\{\mathbf{Y}_n\}$ *is a martingale iff it is of the form* (12.3) *with* $\{\mathbf{X}_n\}$ *absolutely fair.*

The concept of martingales is due to P. Lévy, but it was J. L. Doob who realized its unexpected potentialities and developed the theory. It will be shown in VII,9 that under mild boundedness conditions the variables $\mathbf{Y}_k$ of a martingale converge to a limit; this fact is important for the modern theory of stochastic processes.

Examples. (*a*) Classical gambling is concerned with independent variables $\mathbf{X}_n$ with $\mathbf{E}(\mathbf{X}_n) = 0$. Such a game is absolutely fair[29] and the partial sums $\mathbf{S}_n = \mathbf{X}_1 + \cdots + \mathbf{X}_n$ constitute a martingale. Consider now an ordinary coin-tossing game in which the gambler chooses his stakes according to some rule involving the outcome of previous trials. The successive gains cease to be independent random variables but the game remains absolutely fair. The idea of a fair game is that the knowledge of the past should not enable the gambler to improve on his fortunes. Intuitively this means that an absolutely fair game should remain absolutely fair under any system of gambling, that is, under rules of skipping individual trials. We shall see that this is so.

(*b*) *Polya's urn scheme of* [**1**; V,2(*c*)]. An urn contains b black and r red balls. A ball is drawn at random. It is replaced and, moreover, c balls of the color drawn are added. Let $\mathbf{Y}_0 = \dfrac{b}{b+r}$ and let $\mathbf{Y}_n$ be the proportion of black balls attained by the nth drawing. Then $\{\mathbf{Y}_n\}$ is a martingale. In this case the convergence theorem guarantees the existence of a limit distribution [see examples VII,4(*a*) and VII,9(*a*)].

(*c*) *Concordant*[30] *functions.* Let $\{\mathbf{X}_n\}$ be a Markov chain with transition probabilities given by the stochastic kernel K. Nothing is assumed concerning the expectations of $\mathbf{X}_n$. The function u is called concordant with

[29] The practical limitations of this notion are discussed in **1**; X,3. It will be recalled that there exist "fair" games in which with probability $>1 - \epsilon$ the gambler's gain at the nth trial is as large as, say, $n/\log n$.

[30] This term was introduced by G. Hunt.

respect to K if

$$(12.5) \qquad u(x) = \int K(x, dy)\, u(y).$$

Define random variables $\mathbf{Y}_k$ by $\mathbf{Y}_k = u(\mathbf{X}_k)$ and assume that all expectations exist (for example, that u is bounded). The relation (12.5) is the same as $\mathbf{E}(\mathbf{Y}_{k+1} \mid \mathbf{X}_k = x) = u(x)$, and thus $\mathbf{E}(\mathbf{Y}_{k+1} \mid \mathbf{X}_k) = \mathbf{Y}_k$. Since $\{\mathbf{X}_k\}$ is Markovian this implies (12.4), and since $\mathbf{Y}_k$ is a function of $\mathbf{X}_k$, this in turn implies (12.2) (see V,10*a*). Thus $\{\mathbf{Y}_n\}$ *is a martingale.* This result is of great value in the boundary theory for Markov chains because the existence of a limit for $\mathbf{Y}_n$ usually implies the existence of a limit for the given sequence $\{\mathbf{X}_n\}$. [See examples (*f*) and VII,9(*c*).]

(*d*) *Likelihood ratios.* Suppose it is known that in a stochastic process $\mathbf{X}_1, \mathbf{X}_2, \ldots$ the joint densities of $(\mathbf{X}_1, \ldots, \mathbf{X}_n)$ are either p_n or q_n, but we do not know which. To reach a decision statisticians introduce the new random variables

$$(12.6) \qquad \mathbf{Y}_n = \frac{q_n(\mathbf{X}_1, \ldots, \mathbf{X}_n)}{p_n(\mathbf{X}_1, \ldots, \mathbf{X}_n)}.$$

Under sufficient regularity conditions it is plausible that if the true densities are p_n the observed values of $\mathbf{X}_1, \ldots, \mathbf{X}_n$ will on the average cluster around points where p_n is relatively large. If this is so $\mathbf{Y}_n$ is likely to be small or large according as the true density is p_n or q_n. The asymptotic behavior of $\{\mathbf{Y}_n\}$ is therefore of interest in statistical decision theory.

For simplicity we assume that the densities p_n are strictly positive and continuous. If the p_n represent the true densities, then the conditional density of $\mathbf{X}_{n+1}$ for given $\mathbf{X}_1, \ldots, \mathbf{X}_n$ equals the ratio p_{n+1}/p_n, and hence

$$(12.7) \quad \mathbf{E}(\mathbf{Y}_{n+1} \mid \mathbf{X}_1 = x_1, \ldots, \mathbf{X}_n = x_n) = \\ = \int_{-\infty}^{+\infty} \frac{q_{n+1}(x_1, \ldots, x_n, y)}{p_{n+1}(x_1, \ldots, x_n, y)} \cdot \frac{p_{n+1}(x_1, \ldots, x_n, y)}{p_n(x_1, \ldots, x_n)}\, dy.$$

The factors p_{n+1} cancel. The second denominator is independent of y, and the integral of q_{n+1} is given by the marginal density q_{n+1}. Thus (12.7) reduces to q_n/p_n and so (12.4) is true. Accordingly, under the present conditions the likelihood ratios $\mathbf{Y}_n$ form a martingale. ►

The conditioning used in (12.2) is not particularly fortunate because one has frequently to replace the conditioning variables $\mathbf{Y}_1, \ldots, \mathbf{Y}_n$ by some functions of them. [Such was the case in (12.4).] A greater defect is revealed by example (*a*). The underlying process (say coin tossing or roulette) is represented by a sequence of random variables $\mathbf{Z}_n$, and the gambler's gain at the $(n+1)$st trial is some function of $\mathbf{Z}_1, \ldots, \mathbf{Z}_{n+1}$ and, perhaps, other variables. The observable past is represented by $(\mathbf{Z}_1, \ldots, \mathbf{Z}_n)$, which may provide more information than the past gains. For example, if the gambler skips trials number 1, 3, 5, ... the knowledge of his gains up to epoch $2n$ is at best equivalent to the knowledge of $\mathbf{Z}_2, \mathbf{Z}_4, \ldots, \mathbf{Z}_{2n}$. Here the additional knowledge of $\mathbf{Z}_1, \mathbf{Z}_3, \ldots$ could in principle imply an advantage, and absolute fairness in this case must be based on conditioning by $\mathbf{Z}_1, \ldots, \mathbf{Z}_n$. Thus conditioning with respect to several sets of random variables may be necessary, and to take care of all situations it is best to use the conditioning with respect to arbitrary σ-algebras of events.

Consider then a sequence $\{\mathbf{Y}_n\}$ of random variables in an arbitrary probability space and denote by $\mathfrak{A}_n$ the σ-algebra of events generated by $(\mathbf{Y}_1, \ldots, \mathbf{Y}_n)$ (see V,10a). The defining relation (12.2) is now the same as $\mathbf{E}(\mathbf{Y}_{n+1} \mid \mathfrak{A}_n) = \mathbf{Y}_n$. We want to take this as the defining relation but replace the σ-algebra $\mathfrak{A}_n$ by a larger σ-algebra $\mathfrak{B}_n$. In most cases $\mathfrak{B}_n$ will be generated by $\mathbf{Y}_1, \ldots, \mathbf{Y}_n$ and additional random variables depending on the past. The idea is that any random variable depending on the past must be measurable with respect to $\mathfrak{B}_n$, and in this sense $\mathfrak{B}_n$ represents the information contained in the past history of the process. As this information grows richer with time we shall suppose that the $\mathfrak{B}_n$ increase, that is,

$$\mathfrak{B}_1 \subset \mathfrak{B}_2 \subset \cdots. \tag{12.8}$$

Definition 1. *Let* $\mathbf{Y}_1, \mathbf{Y}_2, \ldots$ *be random variables with expectations. Let* $\mathfrak{B}_1, \mathfrak{B}_2, \ldots$ *be σ-algebras of events satisfying* (12.8).

The sequence $\{\mathbf{Y}_n\}$ *is a martingale with respect to* $\{\mathfrak{B}_n\}$ *iff*

$$\mathbf{E}(\mathbf{Y}_{n+1} \mid \mathfrak{B}_n) = \mathbf{Y}_n. \tag{12.9}$$

[Because of the non uniqueness of the conditional expectations, (12.9) should be read "there exists a version of the conditional probability for which (12.9) is true." This remark applies in the sequel.]

Note that (12.9) implies that $\mathbf{Y}_n$ is $\mathfrak{B}_n$ measurable, and this has two important consequences. Since $\mathfrak{B}_n \supset \mathfrak{B}_{n-1}$ the basic identity V,(11.7) for iterated expectations shows that

$$\mathbf{E}(\mathbf{Y}_{n+1} \mid \mathfrak{B}_{n-1}) = \mathbf{Y}_{n-1}.$$

By induction it is seen that the definition (12.9) *entails the stronger relations*

$$\mathbf{E}(\mathbf{Y}_{n+1} \mid \mathfrak{B}_k) = \mathbf{Y}_k\,, \qquad k = 1, 2, \ldots, n. \tag{12.10}$$

It follows in particular that *every subsequence* $\mathbf{Y}_{\nu_1}, \mathbf{Y}_{\nu_2}, \ldots$ *of a martingale is again a martingale.*

Next we note that $\mathfrak{B}_n$ contains the σ-algebra generated by the variables $\mathbf{Y}_1, \ldots, \mathbf{Y}_n$, and the same argument shows that $\{\mathbf{Y}_n\}$ is also a martingale with respect to $\{\mathfrak{A}_n\}$. *Thus* (12.9) *implies* (12.2).

Example. (*e*) Let the σ-algebras $\mathfrak{B}_n$ satisfy (12.8) and let $\mathbf{Y}$ be an arbitrary random variable with expectation. Put $\mathbf{Y}_n = \mathbf{E}(\mathbf{Y} \mid \mathfrak{B}_n)$. Then $\mathbf{Y}_n$ is $\mathfrak{B}_n$-measurable and hence (12.9) is true. Thus $\{\mathbf{Y}_n\}$ *is a martingale.* ►

Returning to example (*a*) it is now easy to prove the impossibility of systems of a fairly general type. Let $\{\mathbf{Y}_n\}$ be a martingale with respect to $\{\mathfrak{B}_n\}$. To describe the gambler's freedom to skip the *n*th trial we introduce a *decision function* $\boldsymbol{\epsilon}_n$; this is a $\mathfrak{B}_{n-1}$ measurable[31] random variable assuming only the values 0 and 1. In the event $\boldsymbol{\epsilon}_n = 0$ the gambler skips the *n*th trial; in the event $\boldsymbol{\epsilon}_n = 1$ he bets and in this case his gain at the *n*th trial is $\mathbf{Y}_n - \mathbf{Y}_{n-1}$. Denoting his accumulated gain up to and including the *n*th trial by $\mathbf{Z}_n$ we have

$$\mathbf{Z}_n = \mathbf{Z}_{n-1} + \boldsymbol{\epsilon}_n[\mathbf{Y}_n - \mathbf{Y}_{n-1}]. \tag{12.11}$$

By induction it is seen that $\mathbf{Z}_n$ has an expectation. Furthermore, $\mathbf{Z}_{n-1}$, $\boldsymbol{\epsilon}_n$, and $\mathbf{Y}_{n-1}$ are $\mathfrak{B}_{n-1}$ measurable and hence [see V,(11.6)]

$$\mathbf{E}(\mathbf{Z}_n \mid \mathfrak{B}_{n-1}) = \mathbf{Z}_{n-1} + \boldsymbol{\epsilon}_n[\mathbf{E}(\mathbf{Y}_n \mid \mathfrak{B}_{n-1}) - \mathbf{Y}_{n-1}]. \tag{12.12}$$

Since $\{\mathbf{Y}_n\}$ is a martingale the expression within brackets vanishes, and so $\{\mathbf{Z}_n\}$ is a martingale. We have thus proved a theorem due to P. R. Halmos implying the

Impossibility of systems. *Every sequence of decision functions* $\boldsymbol{\epsilon}_1, \boldsymbol{\epsilon}_2, \ldots$ *changes the martingale* $\{\mathbf{Y}_n\}$ *into a martingale* $\{\mathbf{Z}_n\}$.

By far the most important special case concerns *optional stopping*. By this is meant a system where the first $\mathbf{N}$ trials are accepted and all succeeding ones skipped; the $\mathbf{N}$th trial is the last. Here $\mathbf{N}$ (*the stopping epoch*) is a random variable such that the event $\{\mathbf{N} > k\}$ is in $\mathfrak{B}_k$. (In the notation of the theorem $\boldsymbol{\epsilon}_k = 1$ for $\mathbf{N} > k - 1$ and $\boldsymbol{\epsilon}_k = 0$ for $\mathbf{N} \leq k - 1$.) We have thus the

Corollary. *Optional stopping does not affect the martingale property.*

Examples. (*f*) A simple random walk on the line starts at the origin; the particle moves with probability p one step to the right, with probability $q = 1 - p$ to the left. If $\mathbf{S}_n$ is the position of the particle at epoch n it is easily seen that $\mathbf{Y}_n = (q/p)^{\mathbf{S}_n}$ constitutes a martingale with $\mathbf{E}(\mathbf{Y}_n) = 1$ and $\mathbf{E}(\mathbf{Y}_0) = 1$. [This is a special case of example (*c*).]

In the ruin problem the random walk is stopped when it first reaches one of the positions $-a$ or b, where a and b are positive integers. In this

[31] This condition guarantees that the decision is made on the basis of past history of observations. No mathematical theory can disprove prescience of the future, we must exclude it from our models.

modified process $-a \leq \mathbf{S}_n \leq b$ and with probability one $\mathbf{S}_n$ is ultimately fixed at b or at $-a$. Denote the corresponding probabilities by x and $1 - x$. Since $\mathbf{S}_n$ is bounded

$$\mathbf{E}(\mathbf{S}_n) \to x \cdot \left(\frac{q}{p}\right)^b + (1-x)\left(\frac{q}{p}\right)^{-a}$$

But $\mathbf{E}(\mathbf{S}_n) = 1$ because the expected value of the martingale remains constant. The right side, therefore, equals 1, and this linear equation determines x. We have thus found the probability x of termination at b derived by different methods in **1**; XIV,2. The formula breaks down when $p = q$, but in this case $\{\mathbf{S}_n\}$ is a martingale and the same argument shows that $x = a/(a+b)$. Although the result is elementary and known, the argument illustrates the possible uses of martingale theory.

(*g*) *On systems.* Consider a sequence of independent random variables $\mathbf{X}_n$ where $\mathbf{X}_n$ assumes the value $\pm 2^n$ with probability $\frac{1}{2}$ each. A gambler tosses a coin to decide whether he takes the nth bet. The probability that his first try occurs at epoch n is 2^{-n} and in this case his gain is $\pm 2^n$. Thus *the gambler's gain at his first try is a random variable without expectation.* The system theorem therefore depends on the fact that we have not changed the time parameter. ►

It is frequently necessary to work with absolute values and inequalities, and for such purposes it is convenient to have a name for processes satisfying (12.9) with the equality replaced by an inequality.

Definition 2. *The sequence* $\{\mathbf{Y}_n\}$ *is a submartingale*[32] *if it satisfies the martingale definition* (12.9) *with the equality sign replaced by* $\geq$.

It follows again immediately that every submartingale satisfies the stronger conditions.

$$\mathbf{E}(\mathbf{Y}_{n+1} \mid \mathfrak{B}_k) \geq \mathbf{Y}_k, \qquad k = 1, \ldots, n. \tag{12.13}$$

Lemma. *If* u *is a convex function and* $\{\mathbf{Y}_n\}$ *a martingale, then* $\{u(\mathbf{Y}_n)\}$ *is a submartingale provided the expectation of* $u(\mathbf{Y}_n)$ *exists. In particular,* $\{|\mathbf{Y}_n|\}$ *is a submartingale.*

The proof is immediate from Jensen's inequality [V,(8.6)] which applies to conditional expectations as well as to ordinary ones. It states that

$$\mathbf{E}(u(\mathbf{Y}_{n+1}) \mid \mathfrak{B}_n) \geq u(\mathbf{E}(\mathbf{Y}_{n+1} \mid \mathfrak{B}_n)), \tag{12.14}$$

and the right side equals $u(\mathbf{Y}_n)$.

---

[32] The older term "lower semi-martingale" is now falling into disuse.

The same proof shows that if $\{\mathbf{Y}_n\}$ is a submartingale and u a convex *non-decreasing* function, then $\{u(\mathbf{Y}_n)\}$ is again a submartingale, provided $u(\mathbf{Y}_n)$ has an expectation.

13. PROBLEMS FOR SOLUTION

1. The definition (1.2) of stable distributions is equivalent to: R is stable iff to arbitrary constants c_1, c_2 there exist constants c and γ such that

$$c_1\mathbf{X}_1 + c_2\mathbf{X}_2 \stackrel{\text{d}}{=} c\mathbf{X} + \gamma.$$

2. *Every stable distribution is continuous.* It suffices to prove it for symmetric R. From (1.3) conclude: If R had an atom of weight p at the point $s > 0$, then it would have an atom of weight $\geq p^2$ at each point of the form $s(c_m + c_n)/c_{m+n}$ (see V,4a). Furthermore, a unique atom of weight $p < 1$ at the origin would induce an atom of weight p^2 at the origin for $R \bigstar R$ whereas stability requires an atom of weight p.

3. For F to be stable it suffices that (1.2) holds for $n = 2$ and 3. (P. Lévy) *Hint:* Products of the form $c_1^jc_2^k$ where $j, k = 0, \pm1, \pm2, \ldots$ are either dense in $\overline{0, \infty}$ or powers of a fixed number c. The latter must be shown to be impossible in the present case.

Note. Curiously enough it does *not* suffice that (1.2) holds for $n = 2$. See example XVII,3(f) and problem 10 of IX,10.

4. For a stable distribution with exponent $\alpha = 1$ the centering constants in the defining relation (1.2) satisfy $\gamma_{mn} = m\gamma_n + n\gamma_m$ [see (1.6)]. The analogue to (1.8) is

$$s(\mathbf{X}_1+\gamma \log s) + t(\mathbf{X}_2+\gamma \log t) \stackrel{\text{d}}{=} (s + t)(\mathbf{X}+\gamma \log (s + t)).$$

5. If F and G are stable with the same exponent α so is $F\bigstar G$. Find the centering constants γ_n for $F\bigstar G$ in terms of the constants for F and G.

6. For a symmetric stable distribution R the symmetrization inequality V,(5.11) implies that $n[1-R(c_nx)]$ remains bounded. Conclude that R *has absolute moments of order* $<\alpha$. [Use V,(6.3).] By symmetrization the last statement carries over to unsymmetric R.

7. *Alternative derivation of the Holtsmark distribution.* Consider a ball of radius r about the origin and n stars (points) placed independently and randomly in it. Let each star have unit mass. Let $\mathbf{X}_1, \ldots, \mathbf{X}_n$ be the x-components of the gravitational force due to the individual stars, and $\mathbf{S}_n = \mathbf{X}_1 + \cdots + \mathbf{X}_n$. Let $r \to \infty$, $n \to \infty$ so that $\frac{4}{3}r^3\pi n^{-1} \to \lambda$. Show that the distribution of $\mathbf{S}_n$ tends to the symmetric stable distribution with characteristic exponent $\frac{3}{2}$.

8. Show that the preceding problem is not essentially modified if the mass of a star is assumed to be a random variable with unit expectation provided the masses of the stars are mutually independent and also independent of the position of the stars.

9. *Holtsmark distribution, four dimensions.* The four-dimensional analogue to the Holtsmark distribution is a symmetric stable distribution with characteristic exponent $\frac{4}{3}$. (In four dimensions the gravitational force varies inversely as the third power of the distance.)

10. A triangular array $\{\mathbf{X}_{k,n}\}$ with r_n components in the nth row can be transformed into an essentially equivalent array with n components in the nth row by adding dummy variables assuming only the value zero and repeating certain rows an appropriate number of times.

11. Let $\{\mathbf{X}_{k,n}\}$ be a triangular null array with a common distribution F_n for $\mathbf{X}_{1,n}, \ldots, \mathbf{X}_{n,n}$. Does $\mathbf{P}\{\max(|\mathbf{X}_{1,n}|, \ldots, |\mathbf{X}_{n,n}|) > \epsilon\}$ tend to zero?

12. Find the density for the renewal function U of (6.3) if F has the density (*a*) $f(x) = e^{-x}$, and (*b*) $f(x) = xe^{-x}$.

13. In a terminating process F has density pce^{-ct}. Find the distributions of the lifetime and of the number of renewal epochs.

14. *Generalized terminating renewal process.* Instead of assuming that with probability q the process terminates instantaneously we let it (with probability q) continue for a random duration with a proper distribution F_0 and *then* stop. In other words, the renewal epochs are of the form $\mathbf{T}_1 + \cdots + \mathbf{T}_n + \mathbf{Y}$ where the *last* variable has a different distribution. Show that the distribution V of the *duration* of the process satisfies the renewal equation

$$V = qF_0 + F \star V \qquad (F(\infty) = 1 - q). \tag{*}$$

15. Show that the waiting time problem for large gaps reduces to a special case of the process described in the last problem. Put (7.1) into the form (*).

16. *Poisson process and covering theorems.* We recall from example III,3(*d*) that if in a Poisson process n renewal epochs occur in $\overline{0, t}$ their (conditional) distribution is *uniform*. The probability $1 - V(t)$ that no gap of length ξ appears follows therefore from the covering theorem 3 in I,9 in the form

$$1 - V(t) = e^{-cx} \sum_{n=1}^{\infty} \frac{(ct)^{n-1}}{(n-1)!} \varphi_n(t). \tag{†}$$

(*a*) Verify that this is indeed a solution of (7.1) when $F(x) = 1 - e^{-cx}$.

(*b*) Given (†) is the unique solution of (7.1), derive the covering theorem from it. (This is an instance of a proof by randomization. See problem 5 of **1**; XII, 6.)

17. *The waiting time for the first lost call* in example 7(*f*) should be interpreted as the total lifetime of the terminating process obtained by stopping the original process when for the first time a call arrives during a busy period.[33] Show that the distribution H of the duration of the busy period of the terminating process is given by $H\{dt\} = e^{-\alpha t} F\{dt\}$, and the recurrence times have the distribution $G \star H$. (See also problems 3–4 in XIV,10.)

18. Let V be the distribution of the busy period in example 7(*g*) ("last come first served"). Show that V satisfies the renewal equation $V(t) = A(t) + B \star V(t)$ where A and B are defective distributions given by $A\{dx\} = e^{-cx} G\{dx\}$, and $B\{dx\} = [1 - G(x)]ce^{-cx}\, dx$. (Show that we are concerned with a generalized terminating process in the sense of problem 14: a renewal process generated by B is followed by an undisturbed dead period, the latter having a distribution proportional to A.)

[33] This simple approach replaces complicated procedures suggested in the literature and leads to simpler explicit results. For explicit solutions and estimates see XI,[6].

19. *Small gaps in a Poisson process.*[34] A "coincidence" is said to occur at epoch $\mathbf{S}_n$ if the distance of the renewal epochs $\mathbf{S}_{n-1}$ and $\mathbf{S}_n$ is $\leq \xi$. Find a renewal equation for the distribution of the waiting time for the first coincidence. From this renewal equation conclude that the distribution is proper.

20.[35] *Generalization.* In the standard renewal process $\mathbf{S}_n = \mathbf{T}_1 + \cdots + \mathbf{T}_n$ find a renewal equation for the distribution of the waiting time to the first occurrence of the event $\{\mathbf{T}_n \leq \mathbf{Y}_n\}$, where the $\mathbf{Y}_k$ are independent of the process and of each other and have common distribution G.

21. Let $a_1, \ldots, a_n$ be a finite numerical sequence with partial sums

$$s_k = a_1 + \cdots + a_k.$$

Define recursively

$$v_1 = a_1 \cup 0, \quad v_2 = (v_1 + a_{n-1}) \cup 0, \ldots, \quad v_n = (v_{n-1} + a_1) \cup 0.$$

Prove by induction that $v_n = \max [0, s_1, \ldots, s_n]$. Show that this implies the theorem of section 9.

22. In example 11(*f*) assume that $f(x) = 1$ for $0 < x < 1$. Prove that $g(y) = 2(1 - y)$ is a stationary density and that $k^{(n)}(x, y) = g(y)$ for $n \geq 2$. If $f(x) = \alpha e^{-\alpha x}$ then $g(x) = f(x)$.

23. Define a stochastic density kernel concentrated on $\overline{0, 1}$ by

$$k(x, y) = \tfrac{1}{2}(1 - x)^{-1} \quad \text{if} \quad 0 < x < y < 1$$

and

$$k(x, y) = \tfrac{1}{2}x^{-1} \quad \text{if} \quad 0 < y < x < 1.$$

Find a stationary density. (It satisfies a simple differential equation.) Interpret probabilistically.

24. A Markov chain on $\overline{0, 1}$ is such that if $\mathbf{X}_n = x$ then $\mathbf{X}_{n+1}$ is uniformly distributed on $1 - x, 1$. Show that a stationary density is given by $2x$. (T. Ugaheri.)

25. A Markov chain on $\overline{0, \infty}$ is defined as follows: If $\mathbf{X}_n = x$, then $\mathbf{X}_{n+1}$ is uniformly distributed over $\overline{0, 2x}$ (here $n = 0, 1, \ldots$). Show by induction that the n-step transitions have density kernels given by

$$k^{(n)}(x, y) = \frac{1}{2^n \times (n-1)!}\left(\log \frac{2^n x}{y}\right)^{n-1} \quad \text{if} \quad 0 < y < 2^n x$$

and $k^{(n)}(x, y) = 0$ elsewhere.

[34] For variants (treated differently) see E. N. Gilbert and H. O. Pollak, *Coincidences in Poisson patterns*, Bell System Technical J., vol. 36 (1957) pp. 1005–1033.

[35] The "large gap" problem admits of a similar generalization with an analogous answer.

CHAPTER VII

Laws of Large Numbers. Applications in Analysis

In the first part of this chapter it is shown that certain famous and deep theorems of analysis can be derived with surprising ease by probabilistic arguments. Sections 7 and 8 treat variants of the laws of large numbers. Section 9 contains a restricted version of the martingale convergence theorem and stands somewhat apart from the remainder.

1. MAIN LEMMA AND NOTATIONS

By way of preparation consider a one-dimensional distribution G with expectation θ and variance σ^2. If $\mathbf{X}_1, \ldots, \mathbf{X}_n$ are independent variables with the distribution G, their arithmetic mean $\mathbf{M}_n = (\mathbf{X}_1 + \cdots + \mathbf{X}_n)n^{-1}$ has expectation θ and variance $\sigma^2 n^{-1}$. For large n this variance is small and $\mathbf{M}_n$ is likely to be close to θ. It follows that for every continuous function $u(\mathbf{M}_n)$ is likely to be close to $u(\theta)$. This remark constitutes the weak law of large numbers. It is slightly generalized in the following lemma, which despite its simplicity will prove a source of valuable information.

For $n = 1, 2, \ldots$ consider a family of distributions $F_{n,\theta}$ with expectation θ and variance $\sigma_n^2(\theta)$; here θ is a parameter varying in a finite or infinite interval. For expectations we use the notation

$$\mathbf{E}_{n,\theta}(u) = \int_{-\infty}^{+\infty} u(x)\, F_{n,\theta}\{dx\}. \tag{1.1}$$

Lemma 1. *Suppose that* u *is bounded and continuous, and that* $\sigma_n^2(\theta) \to 0$ *for each* θ*. Then*

$$\mathbf{E}_{n,\theta}(u) \to u(\theta). \tag{1.2}$$

The convergence is uniform in every closed interval in which $\sigma_n^2(\theta) \to 0$ *uniformly.*

Proof. Obviously

$$|\mathbf{E}_{n,\theta}(u)-u(\theta)| \leq \int_{-\infty}^{+\infty} |u(x)-u(\theta)|\, F_{n,\theta}\{dx\}. \tag{1.3}$$

There exists a δ depending on θ and ϵ such that for $|x-\theta| < \delta$ the integrand is $<\epsilon$. Outside this neighborhood the integrand is less than some constant M, and by Chebyshev's inequality V,(7.2) the probability carried by the region $|x-\theta| > \delta$ is less than $\sigma_n^2(\theta)\delta^{-2}$. Thus the right side will be $<2\epsilon$ as soon as n is so large that $\sigma_n^2(\theta) < \epsilon\delta^2/M$. This bound on n is independent of θ if $\sigma_n^2(\theta) \to 0$ uniformly and if u is uniformly continuous. ►

Examples. (*a*) If $F_{n,\theta}$ is a binomial distribution concentrated on the points k/n $(k = 0, \ldots, n)$, then $\sigma_n^2(\theta) = \theta(1-\theta)n^{-1} \to 0$ and so

$$\sum_{k=0}^{n} u\left(\frac{k}{n}\right)\binom{n}{k}\theta^k(1-\theta)^{n-k} \to u(\theta) \tag{1.4}$$

uniformly in $0 \leq \theta \leq 1$. The implications are discussed in section 2.

(*b*) If $F_{n,\theta}$ is the Poisson distribution attaching probability $e^{-n\theta}(n\theta)^k/k!$ to the point k/n, we have $\sigma_n^2(\theta) = \theta/n$ and so

$$e^{-n\theta}\sum_{k=0}^{\infty} u\left(\frac{k}{n}\right)\frac{(n\theta)^k}{k!} \to u(\theta) \tag{1.5}$$

uniformly in every finite θ-interval. This formula is valid also for non-integral n. (Continued in sections 5 and 6.)

(*c*) Taking for $F_{n,\theta}$ a gamma distribution with expectation θ and variance θ/n we get

$$\frac{1}{(n-1)!}\int_0^{\infty} u(x)\cdot\left(\frac{nx}{\theta}\right)^{n-1} e^{-nx/\theta}\,\frac{n\,dx}{\theta} \to u(\theta) \tag{1.6}$$

uniformly in every finite interval. Again this formula holds for non-integral n provided $(n-1)!$ is replaced by $\Gamma(n)$. It will be shown in section 6 that (1.6) is an inversion formula for Laplace transforms.

(*d*) Statisticians frequently face the situation described at the beginning of this section but consider the expectation θ an unknown parameter to be estimated from observations In statistical language the relation(1.2) then states that $u(\mathbf{M}_n)$ is an *asymptotically unbiased estimator* for the unknown parameter $u(\theta)$. [The estimator would be unbiased if the two sides in (1.2) were equal.] ►

We shall see that each of these examples leads to important results of independent interest, but some preparations are necessary for the further development of the theory.

Notations for Differences

In the next few sections we shall employ the convenient notations of the calculus of finite differences. Given a finite or infinite numerical sequence $a_0, a_1, \ldots$ the *differencing operator* Δ is defined by $\Delta a_i = a_{i+1} - a_i$. It produces a new sequence $\{\Delta a_i\}$, and on applying the operator Δ a second time we get the sequence with elements

$$\Delta^2 a_i = \Delta a_{i+1} - \Delta a_i = a_{i+2} - 2a_{i+1} + a_i.$$

Proceeding in like manner we may define the rth power Δ^r inductively by $\Delta^r = \Delta\Delta^{r-1}$. It is easily verified that

$$\Delta^r a_i = \sum_{j=0}^{r} \binom{r}{j} (-1)^{r-j} a_{i+j}. \tag{1.7}$$

For consistency we define Δ^0 as the identity operator, that is, $\Delta^0 a_i = a_i$. Then (1.7) holds for all $r \geq 0$. Of course, if $a_0, \ldots, a_n$ is a finite sequence the variability of r is limited.

Many tedious calculations can be avoided by noticing once and for all a curious reciprocity relation valid for an arbitrary pair of sequences $\{a_i\}$ and $\{c_i\}$; it enables us to express the differences $\Delta^r a_i$ in terms of $\Delta^r c_i$ and vice versa. To derive it multiply (1.7) by $\binom{\nu}{r} c_r$ and sum over $r = 0, \ldots, \nu$. The coefficient of a_{i+j} is found to equal

$$\sum_{r=j}^{\nu} \binom{\nu}{r}\binom{r}{j}(-1)^{r-j} c_r = (-1)^{\nu-j}\binom{\nu}{j} \sum_{k=0}^{\nu-j} \binom{\nu-j}{k} (-1)^{\nu-j-k} c_{j+k}.$$

(Here the new summation index $k = r - j$ was introduced.) The last sum equals $\Delta^{\nu-j} c_j$, and thus we have found the

General reciprocity formula

$$\sum_{r=0}^{\nu} c_r \binom{\nu}{r} \Delta^r a_i = \sum_{j=0}^{\nu} a_{i+j} \binom{\nu}{j} (-1)^{\nu-j} \Delta^{\nu-j} c_j. \tag{1.8}$$

Examples. (*a*) (*Inversion formula.*) Consider the constant sequence with $a_i = 1$ for all i. Then $\Delta^0 a_i = 1$ but all other differences vanish, and so (1.8) reduces to

$$c_0 = \sum_{j=0}^{\nu} \binom{\nu}{j} (-1)^{\nu-j} \Delta^{\nu-j} c_j. \tag{1.9}$$

If we change the sequence $\{c_j\}$ into $\{c_{k+j}\}$ we see that (1.9) remains valid with c_0 and c_j replaced by c_k and c_{k+j}, respectively, and so (1.9) is an inversion formula expressing the given sequence in terms of its differences Here ν can be chosen arbitrarily.

(*b*) Let $0 < \theta < 1$ be fixed and define $c_r = \theta^r$. Then

$$\Delta^k c_r = \theta^r(1-\theta)^k(-1)^k$$

and so (1.8) takes on the form

$$\sum_{r=0}^{\nu} \theta^r \binom{\nu}{r} \Delta^r a_i = \sum_{j=0}^{\nu} a_{i+j} \binom{\nu}{j} \theta^j (1-\theta)^{\nu-j}. \tag{1.10}$$ ▶

We deal frequently with sequences with terms $a_k = u(x+kh)$ obtained from a function u by fixing a point x and a span $h > 0$. For obvious reasons it is then convenient to replace the difference operator Δ by the *difference ratios* $\underset{h}{\Delta} = h^{-1}\Delta$. Thus

$$\underset{h}{\Delta}\, u(x) = [u(x+h)-u(x)]/h \tag{1.11}$$

and more generally

$$\underset{h}{\Delta^r}\, u(x) = h^{-r}\sum_{j=0}^{r}\binom{r}{j}(-1)^{r-j}u(x+jh). \tag{1.12}$$

In particular, $\underset{h}{\Delta^0}\, u(x) = u(x)$.

2. BERNSTEIN POLYNOMIALS. ABSOLUTELY MONOTONE FUNCTIONS

We return to the important relation (1.4). The left side is a polynomial, called *the Bernstein polynomial of degree* n corresponding to the given function u. To emphasize this dependence we shall denote it by $B_{n,u}$. Thus

$$B_{n,u}(\theta) = \sum_{j=0}^{n} u(jh)\binom{n}{j}\theta^j(1-\theta)^{n-j} \tag{2.1}$$

where for convenience we put $n^{-1} = h$. Comparing with (1.10) one sees that $B_{n,u}$ *may be written in the alternative form*

$$B_{n,u}(\theta) = \sum_{r=0}^{n}\binom{n}{r}(h\theta)^r \underset{h}{\Delta^r}\, u(0), \qquad h = \frac{1}{n}. \tag{2.2}$$

An amazing number of far-reaching conclusions can be drawn from the discovery that the representations (2.1) and (2.2) for the Bernstein polynomial are equivalent. Before proceeding in this direction we restate for the record the result derived in example 1(*a*).

Theorem 1. *If* u *is continuous in the closed interval* $\overline{0,1}$ *the Bernstein polynomials* $B_{n,u}(\theta)$ *tend uniformly to* $u(\theta)$.

In other words, for given $\epsilon > 0$

$$|B_{n,u}(\theta)-u(\theta)| < \epsilon, \qquad 0 \le \theta \le 1, \tag{2.3}$$

for all n sufficiently large. The famous *Weierstrass approximation theorem* asserts the possibility of uniform approximation by some polynomials. The present theorem is sharper inasmuch as it exhibits the approximating polynomials. The above proof is due to S. Bernstein.

As a first application of our dual representation for Bernstein polynomials we derive a characterization of functions that can be represented by a power series with positive coefficients:

$$u(x) = p_0 + p_1x + p_2x^2 + \cdots, \quad p_j \geq 0, \quad 0 \leq x < 1. \tag{2.4}$$

Obviously such a function possesses derivatives of all orders and

$$u^{(n)}(x) \geq 0, \qquad 0 < x < 1. \tag{2.5}$$

For many purposes in analysis it is important that the converse is also true, that is, any function with the property (2.5) admits of a power series representation (2.4). This was first observed by S. Bernstein, but the usual proofs are neither simple nor intuitive. We shall show that the representation (2.2) for Bernstein polynomials leads to a simple proof and to the useful result that the two properties (2.4) and (2.5) are equivalent to a third one, namely

$$\underset{h}{\Delta^k}\, u_c(0) \geq 0, \quad k = 0, \ldots, n-1, \quad h = \frac{1}{n}. \tag{2.6}$$

These results are of interest for probability theory, because if $\{p_r\}$ is a probability distribution on the integers, (2.4) defines its generating function (see **1**; XI), and so we are dealing with the problem of *characterizing probability generating functions.* Among our functions they are distinguished by the obvious norming condition $u(1) = 1$. However, the example $u(x) = (1-x)^{-1}$ shows that functions with the series representation (2.4) need not be bounded.

Theorem 2. *For a continuous function* u *defined for* $0 \leq x < 1$ *the three properties* (2.4), (2.5), *and* (2.6) *are fully equivalent.*

Functions with this property are called absolutely monotone in $\overline{0, 1}$.

Proof. We proceed in two steps and consider first only probability generating functions. In other words, we assume now that u is continuous in the closed interval $\overline{0, 1}$ and that $u(1) = 1$.

Obviously (2.4) implies (2.5). If (2.5) holds then u and all its derivatives are monotone. The monotonicity of u implies that $\underset{h}{\Delta}\, u(x) \geq 0$, while the monotonicity of u' implies that $\underset{h}{\Delta}\, u(x)$ depends monotonically on x and hence $\underset{h}{\Delta^2}\, u(x) \geq 0$. By induction we conclude that (2.5) implies (2.6), and also $\underset{h}{\Delta^n}\, u(0) \geq 0$.

Assume (2.6) for $k = 0, \ldots, n$. In view of (2.2) the polynomial $B_{n,u}$ has non-negative coefficients, and (2.1) shows that $B_{n,u}(1) = 1$. Thus $B_{n,u}$ is a probability generating function, and the continuity theorem of **1**; XI,6 (or below VIII,6) assures us that the limit u of $B_{n,u}$ is itself a probability generating function. (The assumption $u(1) = 1$ guarantees that the coefficients add to unity.) This concludes the proof for bounded functions.

If u is unbounded near 1 we put

$$v(x) = u\left(\frac{m-1}{m}\,x\right)\Big/u\left(\frac{m-1}{m}\right), \qquad 0 \le x \le 1 \tag{2.7}$$

where m is an arbitrary integer. The preceding proof applies to v and shows that each of the properties (2.5) and (2.6) imply the validity of a power series expansion (2.4) at least for $0 < x \le (m-1)/m$. Because of the uniqueness of power series representations and the arbitrariness of m this implies that (2.4) holds for $0 \le x < 1$. ►

3. MOMENT PROBLEMS

In the last theorem we encountered sequences $\{a_k\}$ all of whose differences were positive. In the present section we shall be concerned with the somewhat related class of sequences whose differences alternate in sign, that is, sequences $\{c_k\}$ such that

$$(-1)^r\Delta^r c_k \ge 0, \qquad r = 0, 1, \ldots. \tag{3.1}$$

Such sequences are called *completely monotone.*

Let F be a probability distribution on $\overline{0, 1}$ and denote by $\mathbf{E}(u)$ the integral of u with respect to F. The kth *moment* of F is defined by

$$c_k = \mathbf{E}(\mathbf{X}^k) = \int_0^1 x^k\, F\{dx\} \tag{3.2}$$

it being understood that the interval of integration is closed.

Taking successive differences one finds that

$$(-1)^r\Delta^r c_k = \mathbf{E}(\mathbf{X}^k(1-\mathbf{X})^{r-k}) \tag{3.3}$$

and hence the moment sequence $\{c_k\}$ is completely monotone. Now let u be an arbitrary continuous function in $\overline{0, 1}$ and let us integrate the expression (2.1) for the Bernstein polynomial $B_{n,u}$ with respect to F. In view of (3.3) we get

$$\begin{aligned}\mathbf{E}(B_{n,u}) &= \sum_{j=0}^{n} u(jh)\binom{n}{j}(-1)^{n-j}\Delta^{n-j}c_j \\ &= \sum_{j=0}^{n} u(jh)p_j^{(n)} \qquad (h = n^{-1}),\end{aligned} \tag{3.4}$$

where we put for abbreviation

$$p_j^{(n)} = \binom{n}{j}(-1)^{n-j}\Delta^{n-j}c_j. \tag{3.5}$$

With the special choice $u(x) = 1$ we have $B_{n,u}(x) = 1$ for all x, and hence the $p_j^{(n)}$ add to unity. This means that for each n *the* $p_j^{(n)}$ *define a probability distribution attributing weight* $p_j^{(n)}$ *to the point* $jh = j/n$. (Here $j = 0, \ldots, n$.) We denote this probability distribution by F_n, and the expectations with respect to it by E_n. Then (3.4) reduces to $\mathbf{E}_n(u) = \mathbf{E}(B_{n,u})$. In view of the uniform convergence $B_{n,u} \to u$ this implies that

$$\mathbf{E}_n(u) \to \mathbf{E}(u). \tag{3.6}$$

So far $\{c_k\}$ was the moment sequence corresponding to a given distribution F. But we may start from an *arbitrary completely monotone sequence* $\{c_k\}$ and again define $p_j^{(n)}$ by (3.5). By definition these quantities are non-negative, and we proceed to show that they add to c_0. Indeed, by the basic reciprocity formula (1.8)

$$\sum_{j=0}^{n} u(jh)p_j^{(n)} = \sum_{r=0}^{n} c_r \binom{n}{r} h^r \underset{h}{\Delta^r} u(0). \tag{3.7}$$

For the constant function $u = 1$ the right side reduces to c_0, and this proves the assertion.

We see thus that any completely monotone sequence $\{c_k\}$ subject to the trivial norming condition $c_0 = 1$ defines a probability distribution $\{p_j^{(n)}\}$, and the expectation $\mathbf{E}_n(u)$ of u with respect to it is given by (3.7). It is interesting to see what happens when $n \to \infty$. For simplicity let u be a polynomial of degree N. Since $h = 1/n$ it is not difficult to see that $\underset{h}{\Delta^r} u(0) \to u^{(r)}(0)$. Furthermore $n(n-1) \cdots (n-r+1)h^r \to 1$. The series on the right in (3.7) contains at most $N + 1$ terms, and so we conclude that as $n \to \infty$

$$\mathbf{E}_n(u) \to \sum_{r=0}^{N} \frac{c_r}{r!} u^{(r)}(0) \tag{3.8}$$

for every polynomial of degree N. In particular, when $u(x) = x^r$ we get

$$\mathbf{E}_n(\mathbf{X}^r) \to c_r. \tag{3.9}$$

In other words, *the* rth *moment of the probability distribution* F_n *tends to* c_r. It is therefore plausible that there should exist a probability distribution F whose rth moment coincides with c_r. We formulate this as

Theorem 1. *The moments* c_r *of a probability distribution* F *form a completely monotone sequence with* $c_0 = 1$. *Conversely, an arbitrary completely*

monotone sequence $\{c_r\}$ *subject to the norming* $c_0 = 1$ *coincides with the moment sequence of a unique probability distribution.*

This result is due to F. Hausdorff and was justly celebrated as a deep and powerful result. The systematic use of functional analysis led gradually to simplified proofs, but even the best purely analytical proofs remain relatively intricate. The present approach is new and illustrates how probabilistic reasoning can simplify and render intuitive complicated analytical arguments. We shall not only prove the theorem but give an explicit formula for F.

From (3.8) we know that for any polynomial u the expectations $\mathbf{E}_n(u)$ converge to a finite limit. From the uniform approximation theorem (2.3) it follows that the same is true for any function u continuous in $\overline{0, 1}$. We denote the limit of $\mathbf{E}_n(u)$ by $\mathbf{E}(u)$. The relation (3.6) is then valid under any circumstances, but if we start from an arbitrary completely monotone sequence $\{c_k\}$ we have to prove[1] that there exists a probability distribution F such that the limit $\mathbf{E}(u)$ coincides with the expectation of u with respect to F.

For given $0 \leq t \leq 1$ and $\epsilon > 0$ denote by $u_{t,\epsilon}$ the continuous function on $\overline{0, 1}$ that vanishes for $x \geq t + \epsilon$, equals 1 for $x \leq t$, and is linear between t and $t + \epsilon$. Let $U_\epsilon(t) = \mathbf{E}_\epsilon(u_{t,\epsilon})$. If $t < \tau$ then obviously $u_{t,\epsilon} \leq u_{\tau,\epsilon}$ and the maximum of the difference $u_{\tau,\epsilon} - u_{t,\epsilon}$ is $\leq (\tau - t)/\epsilon$. For fixed $\epsilon > 0$ it follows that $U_\epsilon(t)$ is a continuous non-decreasing function of t. Again, for fixed t the value $U_\epsilon(t)$ can only decrease as $\epsilon \to 0$, and hence $U_\epsilon(t)$ tends to a limit which we denote by $F(t)$. This is a non-decreasing function going from 0 to 1. It is automatically right continuous,[2] but this is of no importance since we could achieve this in any case by changing the definition of F at its jumps.

For the distribution function F_n of the probability distribution (3.5) we have trivially from the definition of $u_{t,\epsilon}$ if $\delta > \epsilon$

$$\mathbf{E}_n(u_{t-\delta,\epsilon}) \leq F_n(t) \leq \mathbf{E}_n(u_{t,\epsilon}). \tag{3.10}$$

As $n \to \infty$ the two extreme members tend to $U_\epsilon(t - \delta)$ and $U_\epsilon(t)$, respectively. If t and $t - \delta$ are points of continuity of F, we let $\epsilon \to 0$ to conclude that all limit points of the sequence $\{F_n(t)\}$ lie between $F(t - \delta)$

[1] We could stop the proof here, because the assertion is contained in either of two results proved elsewhere in the book:

(*a*) In the Riesz representation theorem of V,1 since $\mathbf{E}(u)$ is obviously a positive linear functional of norm 1.

(*b*) In the basic convergence theorem of VIII,1.

The proof in the text (which partly repeats that of V,1) is given to render this chapter self-contained and to lead to the inversion formula.

[2] For a verification see the proof of the Riesz representation theorem in V,1.

and $F(t)$. Finally, letting $\delta \to 0$ we conclude that $F_n(t) \to F(t)$ for each t which is a point of continuity for F. This relation reads explicitly

$$\sum_{j \leq nt} \binom{n}{j} (-1)^{n-j} \Delta^{n-j} c_j \to F(t). \tag{3.11}$$

For $r \geq 1$ an integration by parts permits us to write the rth moment of F_n in the form

$$\mathbf{E}_n(\mathbf{X}^r) = 1 - r \int_0^1 x^{r-1} F_n(x)\, dx. \tag{3.12}$$

We have shown in (3.9) that the left side tends to c_r, and since $F_n \to F$ this proves that the rth moment of F coincides with c_r. This concludes the proof of theorem 1. ▶

Note that if F is an arbitrary probability distribution function on $\overline{0, 1}$ then (3.11) represents an *inversion formula* expressing F in terms of its moments. We restate this in

Theorem 2. *For the probability distribution* F *of theorem* 1 *the limit formula* (3.11) *holds at each point of continuity.*

To avoid misconceptions it should be pointed out that the situation is radically different for distributions that are not concentrated on some finite interval. In fact, in general *a distribution is not uniquely determined by its moments.*

Example. *The log-normal distribution is not determined by its moments.* The positive variable $\mathbf{X}$ is said to have a log-normal distribution if $\log \mathbf{X}$ is normally distributed. With the standard normal distribution the density of $\mathbf{X}$ is defined by

$$f(x) = \frac{1}{\sqrt{2\pi}}\, x^{-1} e^{-\frac{1}{2}(\log x)^2}, \qquad x > 0,$$

and $f(x) = 0$ for $x \leq 0$. For $-1 \leq a \leq 1$ put

$$f_a(x) = f(x)[1 + a \sin(2\pi \log x)]. \tag{3.13}$$

We assert that f_a is a probability density with exactly the same moments of f. Since $f_a \geq 0$ it suffices to show that

$$\int_0^\infty x^k f(x) \sin(2\pi \log x)\, dx = 0, \qquad k = 0, 1, \ldots$$

The substitutions $\log x = t = y + k$ reduce the integral to

$$\frac{1}{\sqrt{2\pi}} \int_{-\infty}^{+\infty} e^{-\frac{1}{2}t^2 + kt} \sin(2\pi t)\, dt = \frac{1}{\sqrt{2\pi}}\, e^{\frac{1}{2}k^2} \int_{-\infty}^{+\infty} e^{-\frac{1}{2}y^2} \sin(2\pi y)\, dy$$

and the last integral vanishes since the integrand is an odd function. (This interesting example is due to C. C. Heyde.) ▶

This negative result should not give rise to undue pessimism, for suitable regularity conditions can remove the source of trouble. The best result is a theorem of Carleman to the effect that a distribution F on $\overline{-\infty, \infty}$ *is uniquely determined by its moments if*

$$\sum \mu_{2n}^{-1/(2n)} = \infty, \tag{3.14}$$

that is, if the series on the left diverges. In this book we shall prove only the weaker statement that F is uniquely determined by its moments whenever the power series $\sum \mu_{2n} t^n/(2n)!$ converges in some interval. (Section 6 and XV,4.) Both criteria put restrictions on the rate of growth of μ_{2n}. Even in the most general situation the knowledge of finitely many moments $\mu_0, \mu_1, \ldots, \mu_n$ leads to useful inequalities for F similar to those derived in V,7 from the knowledge of μ_0 and μ_1.[3]

*4. APPLICATION TO EXCHANGEABLE VARIABLES

We proceed to derive a beautiful result due to B. de Finetti which may serve as a typical example of the ease with which theorem 1 of section 3 leads to surprising results.

Definition. *The random variables* $\mathbf{X}_1, \ldots, \mathbf{X}_n$ *are exchangeable*[4] *if the* $n!$ *permutations* $(\mathbf{X}_{k_1}, \ldots, \mathbf{X}_{k_n})$ *have the same n-dimensional probability distribution. The variables of an infinite sequence* $\{\mathbf{X}_n\}$ *are exchangeable if* $\mathbf{X}_1, \ldots, \mathbf{X}_n$ *are exchangeable for each* n.

As the examples will show, there is an essential difference between finite and infinite sequences. We consider here the special case of an *infinite sequence* $\{\mathbf{X}_n\}$ *of exchangeable variables assuming the values* 0 *and* 1 *only.* The next theorem asserts that *the distribution of such a process* $\{\mathbf{X}_n\}$ *are obtained by randomization of the binomial distribution.* As usual we put $\mathbf{S}_n = \mathbf{X}_1 + \cdots + \mathbf{X}_n$ and interpret the event $\{\mathbf{X}_k = 1\}$ as success.

Theorem. *To every infinite sequence of exchangeable variables* $\mathbf{X}_n$ *assuming only the values* 0 *and* 1 *there corresponds a probability distribution* F *concentrated on* $\overline{0, 1}$ *such that*

$$\mathbf{P}\{\mathbf{X}_1 = 1, \ldots, \mathbf{X}_k = 1, \mathbf{X}_{k+1} = 0, \ldots, \mathbf{X}_n = 0\} = \int_0^1 \theta^k(1-\theta)^{n-k}\, F\{d\theta\} \tag{4.1}$$

$$\mathbf{P}\{\mathbf{S}_n = k\} = \binom{n}{k} \int_0^1 \theta^k(1-\theta)^{n-k}\, F\{d\theta\}. \tag{4.2}$$

---

* Not used in the sequel.

[3] The first sharp results were obtained by Markov and Stieltjes around 1884. The recent literature on the subject is inexhaustible. See, for example, A. Wald, Trans. Amer. Math. Soc., vol. 46 (1939) pp. 280–306; H. L. Royden, Ann. Math. Statist., vol. 24 (1953) pp. 361–376 [gives bounds on $F(x) - F(-x)$]. For a general survey see the monograph by J. A. Shohat and J. D. Tamarkin, *The problem of moments*, New York, 1943 (Math Surveys No. 1). See also S. Karlin and W. Studden (1966).

[4] The term *symmetrically dependent* is also in use.

Proof. For brevity denote the left side in (4.1) by $p_{k,n}$ (with $0 \leq k \leq n$) Put $c_0 = 1$ and for $n = 1, 2, \ldots$

$$c_n = p_{n,n} = \mathbf{P}\{\mathbf{X}_1 = 1, \ldots, \mathbf{X}_n = 1\}. \tag{4.3}$$

Then from the probabilistic meaning

$$p_{n-1,n} = p_{n-1,n-1} - p_{n,n} = -\Delta c_{n-1} \tag{4.4}$$

and hence

$$p_{n-2,n} = p_{n-2,n-1} - p_{n-1,n} = \Delta^2 c_{n-2}. \tag{4.5}$$

Continuing in this way we get for $k \leq n$

$$p_{k,n} = p_{k,n-1} - p_{k+1,n} = (-1)^{n-k}\Delta^{n-k}c_k. \tag{4.6}$$

All these quantities are non-negative and hence the sequence $\{c_n\}$ is completely monotone. It follows that c_r is the rth moment of a probability distribution F, and so (4.1) merely spells out the relation (4.6). The assertion (4.2) is contained in it because there are $\binom{n}{k}$ ways in which k successes can occur in n trials. ▶

Generalizations. It is not difficult to apply the same argument to variables capable of three values, but we have then two free parameters, and instead of (4.2) we get a mixture of trinomial distributions with F a bivariate probability distribution. More generally, the theorem and its proof are readily adapted to random variables assuming only a finite number of values. This fact naturally leads to the conjecture that the most general symmetrically dependent sequence $\{\mathbf{X}_j\}$ is obtained by randomization of a parameter from a sequence of independent variables. Individual cases are not difficult to trust but the general problem presents the inherent difficulty that "parameters" are not well defined and may be chosen in weird ways. A version of the theorem has been proved nevertheless in very great generality.[5]

The theorem makes it possible to apply laws of large numbers and the central limit theorem to exchangeable variables. (See problem 21 in VIII,10.)

The next example shows that in individual cases the theorem may lead to surprising results. The other examples show that the theorem fails for finite sequences.

Examples. (*a*) *In Polya's urn model* of **1**; V,2 an urn contains originally b black and r red balls. After each drawing the ball is returned and c balls of the color drawn are added to the urn. Thus the probability of a

---

[5] E. Hewitt and L. J. Savage, *Symmetric measures on Cartesian products*, Trans. Amer. Math. Soc., vol. 80 (1956) pp. 470–501. A martingale treatment is found in Loève (1963). See also H. Bühlmann, *Austauschbare stochastiche Variabeln und ihre Grenzwertsätze*, Univ. of California Publications in Statistics, vol. 3, No. 1 (1960) pp. 1–36.

black ball in each of the first n drawings equals

$$(4.7)\qquad c_n = \frac{b(b+c)\cdots(b+(n-1)c)}{(b+r)\cdots(b+r+(n-1)c)} = \frac{\Gamma\left(\frac{b}{c}+n\right)\Gamma\left(\frac{b+r}{c}\right)}{\Gamma\left(\frac{b+r}{c}+n\right)\Gamma\left(\frac{b}{c}\right)}.$$

Put $\mathbf{X}_n = 1$ or 0 according as the nth drawing results in black or red. The easy calculation in **1**; V,2 shows that these variables are exchangeable and hence c_n represents the nth moment of a distribution F. The appearance of (4.7) reminds one of the beta integral II,(2.5), and inspection shows that *F is the beta distribution* II,(4.2) *with parameters* $\mu = b/c$ and $\nu = r/c$. Again using the beta integral it is seen that (4.1) agrees with **1**; V,(2.3), and (4.2) with **1**; V,(2.4).

(*b*) Consider the 6 distinguishable distributions of 2 balls in 3 cells and attribute probability $\frac{1}{6}$ to each. Let $\mathbf{X}_i$ equal 1 or 0 according as the cell number i is occupied or empty. The variables are exchangeable but the theorem does not apply. Indeed, from (4.3) we get $c_0 = 1$, $c_1 = \frac{1}{2}$, $c_2 = \frac{1}{6}$, $c_3 = 0$ and here the sequence stops. If it were the beginning of a completely monotone sequence $\{c_n\}$ we would have $c_4 = c_5 = \cdots = 0$. But then $\Delta^4 c_1 = -\frac{1}{6} < 0$ against the rule.

(*c*) Let $\mathbf{X}_1, \ldots, \mathbf{X}_n$ be independent with a common distribution and $\mathbf{S}_n = \mathbf{X}_1 + \cdots + \mathbf{X}_n$. Put $\mathbf{Y}_k = \mathbf{X}_k - n^{-1}\mathbf{S}_n$ for $k = 1, \ldots, n-1$. The variables $(\mathbf{Y}_1, \ldots, \mathbf{Y}_{n-1})$ are exchangeable but their joint distribution is not of the form suggested by de Finetti's theorem. ►

*5. GENERALIZED TAYLOR FORMULA AND SEMI-GROUPS

The preceding three sections dealt with consequences of the limit relation in example 1(*a*) involving the binomial distribution. We now pass to example 1(*b*) involving the Poisson distribution. Since this distribution represents a limiting form of the binomial distribution one may hope that our simple treatment of Bernstein polynomials may be extended to the present situation. The starting point for this treatment was the identity (1.10) in which the binomial distribution appears on the right. If we put $\theta = x/\nu$ and let $\nu \to \infty$ this binomial distribution tends to the Poisson distribution with expectation x, and (1.10) passes into[6]

$$(5.1)\qquad \sum_{r=0}^{\infty} \frac{x^r}{r!}\Delta^r a_i = e^{-x}\sum_{j=0}^{\infty}\frac{x^j}{r!}a_{i+j}.$$

* This section may be omitted at first reading.

[6] For a direct proof of the identity (5.1) it suffices to substitute for $\Delta^r a_i$ its defining expression (1.7). The left side then becomes a double sum, and the right side is obtained by an obvious rearrangement of its terms.

We use this identity with $i = 0$ and $a_j = u(jh)$ where u is an arbitrary bounded continuous function on $\overline{0, \infty}$, and h a positive constant. With $x = h\theta$ the relation (5.1) becomes

$$\sum_{r=0}^{\infty} \frac{\theta^r}{r!} \underset{h}{\Delta^r} u(0) = e^{-\theta/h} \sum_{j=0}^{\infty} u(jh) \frac{(\theta/h)^j}{j!} . \tag{5.2}$$

On the right we recognize the expectation of u with respect to a Poisson distribution, and we know from example 1(b) that it tends to $u(\theta)$. To record this result in a more natural form we replace $u(\theta)$ by $u(\theta + t)$ where $t > 0$ is arbitrary. We have thus proved the

Theorem. *For any bounded continuous function on* $\overline{0, \infty}$

$$\sum_{r=0}^{\infty} \frac{\theta^r}{r!} \underset{h}{\Delta^r} u(t) \to u(t+\theta); \tag{5.3}$$

here[7] $\theta > 0$ *and* $h \to 0+$.

This is a fascinating theorem first proved by E. Hille using much deeper methods. The left side represents the Taylor expansion of u except that the derivatives are replaced by difference ratios. For analytic functions the left side approaches the Taylor series but the theorem applies also to non-differentiable functions. In this sense (5.3) represents a generalization of the Taylor expansion and reveals a new side of its nature.

There is another way of looking at (5.3) which leads to the so-called exponential formula of semi-group theory. (See theorem 2 in X,9.) The left side of (5.3) contains the formal exponential series and it is natural to use it to define an operator $\exp \theta \underset{h}{\Delta}$. The relation (5.3) is then abbreviated to

$$\exp \theta \underset{h}{\Delta} u(t) \to u(t+\theta). \tag{5.4}$$

To write it more consistently in terms of operators we introduce the *translation operator*[8] $T(\theta)$ sending u into the function u_θ defined by $u_\theta(t) = u(t+\theta)$.

---

[7] It will be noticed that the argument remains valid if both θ and h are negative provided u is defined on the whole line.

[8] This is the proof, due to M. Riesz, given (in slightly greater generality) in E. Hille and R. S. Phillips, *Functional analysis and semi-groups*, AMS Colloquium Publications, vol. 31 (1957) p. 314. Understandably the authors did not consider it helpful to refer to the linear interpolation (5.7) as a Poisson *randomization of the semi-group* parameter or to take Chebyshev's inequality for granted. The probabilistic content was noted by D. G. Kendall. It is fully exploited in K. L. Chung, *On the exponential formulas of semi-group theory*, Math Scandinavica, vol. 10 (1962) pp. 153–162.

Then $T(0) = 1$ is the identity operator and

$$(5.5)\qquad \underset{h}{\Delta} = h^{-1}[T(h)-1].$$

In operator language (5.3) now becomes

$$(5.6)\qquad e^{\theta h^{-1}[T(h)-1]} \to T(\theta).$$

The main information conveyed by this formula is that the whole family of operators $T(\theta)$ is determined by the behavior of $T(h)$ for small h.

In retrospect it is now clear that our derivation of (5.6) applies to a much more general class of operators. The right side in (5.2) is simply a linear combination of values of u and may be interpreted as an interpolation formula for u. An analogous interpolatory expression is meaningful for *every* family of operators $\{T(\theta)\}$ defined for $\theta > 0$. Indeed, for fixed θ and $h > 0$ the operator

$$(5.7)\qquad A_h(\theta) = e^{-\theta h^{-1}} \sum_{k=0}^{\infty} \frac{1}{k!}\left(\frac{\theta}{h}\right)^k T(kh)$$

is a weighted linear combination of operators $T(kh)$. The weights are given by the Poisson distribution and are such that as $h \to 0$ a neighborhood of θ preponderates, the complement carrying a weight tending to 0. This makes it at least plausible that with any reasonable notion of convergence and continuity we shall have $A_h(\theta) \to T(\theta)$ *for any continuous family of operators* $T(\theta)$. In particular, if the operators $T(\theta)$ form a *semi-group*, one has $T(kh) = (T(h))^k$ and the interpolatory operator $A_h(\theta)$ is the same as appears on the left in (5.6). It is therefore not surprising that the "exponential formula" (5.6) *is generally valid for continuous semi-groups of bounded operators.* We shall return to the proof in X,9.

6. INVERSION FORMULAS FOR LAPLACE TRANSFORMS

The preceding section and example 1(*b*) were based on a special case of the law of large numbers which may be stated as follows: If $\mathbf{X}$ is a random variable with a Poisson distribution of expectation $\lambda\theta$ then for large λ the probability of the event $|\mathbf{X} - \lambda\theta| > \lambda\epsilon$ is small. For $\mathbf{P}\{\mathbf{X} \leq \lambda x\}$ we get therefore as $\lambda \to \infty$

$$(6.1)\qquad e^{-\lambda\theta} \sum_{k \leq \lambda x} \frac{(\lambda\theta)^k}{k!} \to \begin{matrix} 0 & \textit{if}\ \ \theta > x \\ 1 & \textit{if}\ \ \theta < x. \end{matrix}$$

The expression on the left is a special case of (5.2) when u assumes only the values 0 and 1, and so (6.1) is contained in the theorem of the preceding section. The usefulness of this formula in analysis will now be illustrated by applications to Laplace transforms, a topic treated systematically in chapter XIII.

Let F be a probability distribution concentrated on $\overline{0, \infty}$. The Laplace transform of F is the function φ defined for $\lambda > 0$ by

$$\varphi(\lambda) = \int_0^\infty e^{-\lambda\theta}\, F\{d\theta\}. \tag{6.2}$$

The derivatives $\varphi^{(k)}(\lambda)$ exist and are obtained by formal differentiation:

$$(-1)^k\, \varphi^{(k)}(\lambda) = \int_0^\infty e^{-\lambda\theta}\, \theta^k\, F\{d\theta\}. \tag{6.3}$$

From this identity and (6.1) one sees that at every point of continuity of F

$$\sum_{k \le \lambda x} \frac{(-1)^k}{k!}\, \lambda^k\, \varphi^{(k)}(\lambda) \to F(x). \tag{6.4}$$

This is an *inversion formula* of great use. It shows, in particular, that *a distribution F is uniquely determined by its Laplace transform.*

The same argument leads to a great variety of related inversion formulas applicable under various circumstance. In fact, (1.6) is an inversion formula for Laplace integrals of the form

$$w(\lambda) = \int_0^\infty e^{-\lambda x}\, u(x)\, dx. \tag{6.5}$$

Formal differentiation can be performed as in (6.3), and (1.6) states that *if u is bounded and continuous, then*

$$\frac{(-1)^{n-1}}{(n-1)!}\left(\frac{n}{\theta}\right)^n w^{(n-1)}(n/\theta) \to u(\theta) \tag{6.6}$$

uniformly in every finite interval.

[These inversion formulas hold under much wider conditions, but it seemed undesirable at this juncture to let the ballast of new terminology obscure the simplicity of the argument. An abstract version of (6.6) appears in XIII,9.]

If the distribution F possesses moments $\mu_1, \ldots, \mu_{2n}$ its *Laplace transform satisfies the inequalities*

$$\sum_{k=0}^{2n-1} \frac{(-1)^k \mu_k \lambda^k}{k!} \le \varphi(\lambda) \le \sum_{k=0}^{2n} \frac{(-1)^k \mu_k \lambda^k}{k!} \tag{6.7}$$

which are of frequent use. To verify them we start from the well-known inequalities[9]

$$\sum_{k=0}^{2n-1} \frac{(-1)^k t^k}{k!} < e^{-t} < \sum_{k=0}^{2n} \frac{(-1)^k t^k}{k!}, \qquad t > 0. \tag{6.8}$$

[9] Simple differentiation shows by induction that the difference between any two members in (6.8) is a *monotone* function of t.

Replacing t by λt and integrating with respect to F one gets (6.7). It follows, in particular, that

$$\varphi(\lambda) = \sum_{k=0}^{\infty} \frac{(-1)^k \mu_k \lambda^k}{k!} \tag{6.9}$$

in any interval $0 \leq \lambda < \lambda_0$ *in which the series on the right converges.* It is known from analytic function theory that in this case the series in (6.9) uniquely determines $\varphi(\lambda)$ for *all* $\lambda > 0$, and hence *the moments* $\mu_1, \mu_2, \ldots$ *determine the distribution* F *uniquely whenever the series in* (6.9) *converges in some interval* $|\lambda| < \lambda_0$. This useful criterion holds also for distributions not concentrated on $\overline{0, \infty}$, but the proof depends on the use of characteristic functions (see XV,4).

*7. LAWS OF LARGE NUMBERS FOR IDENTICALLY DISTRIBUTED VARIABLES

Throughout this section we use the notation $\mathbf{S}_n = \mathbf{X}_1 + \cdots + \mathbf{X}_n$. The oldest version of the law of large numbers states that if the $\mathbf{X}_k$ are independent and have a common distribution with expectation μ and finite variance then[10] for fixed $\epsilon > 0$ as $n \to \infty$

$$\mathbf{P}\{|n^{-1}\mathbf{S}_n - \mu| > \epsilon\} \to 0. \tag{7.1}$$

This chapter started from the remark that (7.1) is contained in Chebyshev's inequality. To obtain sharper results we derive a variant of Chebyshev's inequality applicable even when no expectation exists. Define new random variables $\mathbf{X}'_k$ by truncation of $\mathbf{X}_k$ at an arbitrary, but fixed, level $\pm s_n$. Thus

$$\mathbf{X}'_k = \begin{cases} \mathbf{X}_k & \textit{when} \quad |\mathbf{X}_k| \leq s_n \\ 0 & \textit{when} \quad |\mathbf{X}_k| > s_n. \end{cases} \tag{7.2}$$

Put

$$\mathbf{S}'_n = \mathbf{X}'_1 + \cdots + \mathbf{X}'_n, \qquad m'_n = \mathbf{E}(\mathbf{S}'_n) = n\mathbf{E}(\mathbf{X}'_1). \tag{7.3}$$

Then obviously

$$\mathbf{P}\{|\mathbf{S}_n - m'_n| > t\} \leq \mathbf{P}\{|\mathbf{S}'_n - m'_n| > t\} + \mathbf{P}\{\mathbf{S}_n \neq \mathbf{S}'_n\} \tag{7.4}$$

because the event on the left cannot occur unless one of the events on the right occurs.

---

* The topics of this section are related to the oldest probabilistic theory but are of no particular significance in the remainder of this book. They are treated for their historical and methodological interest and because many papers are devoted to partial converses of the law of large numbers.

[10] (7.1) is equivalent to $n^{-1}\mathbf{S}_n - \mu \xrightarrow{\text{P}} 0$, where $\xrightarrow{\text{P}}$ signifies "tends in probability to." (See VIII,2.)

This inequality is valid also for dependent variables with varying distributions, but here we are interested only in identically distributed independent variables. Putting $t = nx$ and applying Chebyshev's inequality to the first term on the right, we get from (7.4) the following

Lemma. *Let the* $\mathbf{X}_k$ *be independent with a common distribution* F. *Then for* $x > 0$

$$\mathbf{P}\left\{\left|\frac{1}{n}\mathbf{S}_n - \mathbf{E}(\mathbf{X}_1')\right| > x\right\} \leq \frac{1}{n^2x^2}\mathbf{E}(\mathbf{X}_1'^2) + nP\{|\mathbf{X}_1| > s_n\}. \tag{7.5}$$

As an application we could derive *Khintchine's law of large numbers* which states that (7.1) holds for all $\epsilon > 0$ whenever the $\mathbf{X}_k$ have finite expectation μ. The proof would be essentially a repetition of the proof for the discrete case given in **1**; X,2. We pass therefore directly to a stronger version which includes a necessary and sufficient condition. For its formulation we put for $t > 0$

$$\tau(t) = [1-F(t)+F(-t)]t \tag{7.6}$$

and

$$\sigma(t) = \frac{1}{t}\int_{-t}^{t} x^2\, F(dx) = -\tau(t) + \frac{2}{t}\int_0^t x\tau(x)\, dx. \tag{7.7}$$

(The identity of these two expressions follows by a simple integration by parts.)

Theorem 1. (*Generalized weak law of large numbers.*) *Let the* $\mathbf{X}_k$ *be independent with a common distribution* F. *In order that there exist constants* μ_n *such that for each* $\epsilon > 0$

$$\mathbf{P}\{|n^{-1}\mathbf{S}_n - \mu_n| > \epsilon\} \to 0 \tag{7.8}$$

it is necessary and sufficient that[11] $\tau(t) \to 0$ *as* $t \to \infty$. *In this case* (7.8) *holds with*

$$\mu_n = \int_{-n}^{n} x\, F\{dx\}. \tag{7.9}$$

Proof. (*a*) *Sufficiency*. Define μ_n by (7.9). We use the truncation (7.2) with $s = n$. Then $\mu_n = \mathbf{E}(\mathbf{X}_1')$ and the preceding lemma the left side of (7.8) is $< \epsilon^{-2}\sigma(n) + \tau(n)$, which tends to 0 whenever $\tau(t) \to 0$. Thus this condition is sufficient.

(*b*) *Necessity*. Assume (7.8). As in V,5 we introduce the variables ${}^0\mathbf{X}_k$ obtained directly by symmetrization of $\mathbf{X}_k$. Their sum ${}^0\mathbf{S}_n$ can be obtained

[11] It follows from (7.7) that $\tau(t) \to 0$ implies $\sigma(t) \to 0$. The converse is also true; see problem 11. For a different proof of theorem 1 see XVII,2*a*.

by symmetrization of $\mathbf{S}_n - n\mu$. Let a be a median of the variables $\mathbf{X}_k$. Using the inequalities V,(5.6), V,(5.10), and V,(5.7) in that order we get

$$2\mathbf{P}\{|\mathbf{S}_n - n\mu| > n\epsilon\} \geq \mathbf{P}\{|{}^0\mathbf{S}_n| > 2n\epsilon\} \geq \tfrac{1}{2}[1 - \exp(-n\mathbf{P}\{|{}^0\mathbf{X}_1| > 2n\epsilon\})]$$
$$\geq \tfrac{1}{2}[1 - \exp(-\tfrac{1}{2}n\mathbf{P}\{|\mathbf{X}_1| > 2n\epsilon + |a|\})].$$

In view of (7.8) the left side tends to 0. It follows that the exponent on the right tends to 0, and this is manifestly impossible unless $\tau(t) \to 0$. ►

The condition $\tau(t) \to 0$ is satisfied whenever F has an expectation μ. The truncated moment μ_n then tends to μ and so in this case (7.8) is equivalent with the classical law of large numbers (7.1). However, *the classical law of large numbers in the form* (7.1) *holds also for certain variables without expectation.* For example, if F is a symmetric distribution such that $t[1 - F(t)] \to 0$ then $\mathbf{P}\{|n^{-1}\mathbf{S}_n| > \epsilon\} \to 0$. But an expectation exists only if $1 - F(t)$ is integrable between 0 and ∞, which is a stronger condition.

(It is interesting to note, that the *strong* law of large numbers holds *only* for variables with expectations. See theorem 4 of section 8).

The empirical meaning of the law of large numbers was discussed in **1**; X with special attention to the classical theory of "fair games." We saw in particular that even when expectations exist a participant in a "fair game" may be strongly on the losing side. On the other hand, the analysis of the St. Petersburg game showed that the classical theory applies also to certain games with infinite expectations except that the "fair entrance fee" will depend on the contemplated number of trials. The following theorem renders this more precise.

We consider independent *positive* variables $\mathbf{X}_k$ with a common distribution F. [Thus $F(0) = 0$.]. The $\mathbf{X}_k$ may be interpreted as possible gains, and a_n as the total entrance fee for n trials. We put

$$\mu(s) = \int_0^s x\, F\{dx\}, \qquad \frac{\mu(s)}{s[1 - F(s)]} = \rho(s). \tag{7.10}$$

Theorem 2. *In order that there exist constants a_n such that*

$$\mathbf{P}\{|a_n^{-1}\mathbf{S}_n - 1| > \epsilon\} \to 0 \tag{7.11}$$

it is necessary and sufficient that[12] $\rho(s) \to \infty$ *as* $s \to \infty$. *In this case there exist numbers* s_n *such that*

$$n\mu(s_n) = s_n \tag{7.12}$$

and (7.11) *holds with* $a_n = n\mu(s_n)$.

Proof. (*a*) *Sufficiency.* Assume $\rho(s) \to \infty$. For large n the function $n\mu(s)/s$ assumes values >1, but it tends to 0 as $s \to \infty$. The function is right continuous, and the limit from the left cannot exceed the limit from the right. If s_n is the lower bound of all s such that $n\mu(s)s^{-1} \leq 1$ it follows that (7.12) holds.

[12] It will be seen in VIII,9 (theorem 2) that $\rho(s) \to \infty$ iff $\mu(s)$ *varies slowly* at infinity. The relation (7.11) is equivalent to $a_n^{-1}\mathbf{S}_n \xrightarrow{\ p\ } 1$ (see VIII,2).

Put $\mu_n = \mu(s_n) = \mathbf{E}(\mathbf{X}_1')$. We use the inequality (7.5) of the lemma with $x = \epsilon\mu_n$ to obtain

$$\mathbf{P}\left\{\left|\frac{\mathbf{S}_n}{n\mu_n} - 1\right| > \epsilon\right\} \leq \frac{1}{\epsilon^2 n\mu_n^2}\mathbf{E}(\mathbf{X}_1'^2) + n[1-F(s_n)]. \tag{7.13}$$

An integration by parts reduces $\mathbf{E}(\mathbf{X}_1'^2)$ to an integral with integrand $x[1 - F(x)]$, and by assumption this function is $o(\mu(x))$. Thus $\mathbf{E}(\mathbf{X}_1'^2) = o(s_n\mu_n)$, and in view of (7.12) this means that the first term on the right in (7.13) tends to 0. Similarly (7.12) and the definition (7.10) of $\rho(s)$ show that $n[1 - F(s_n)] \to 0$. Thus (7.13) reduces to (7.11) with $a_n = n\mu_n$.

(*b*) *Necessity*. We now assume (7.11) and use the truncation (7.2) with $s_n = 2a_n$. Since $\mathbf{E}(\mathbf{X}_1'^2) \leq s_n\mu_n$ we get from the basic inequality (7.5) with $x = \epsilon a_n/n$

$$\mathbf{P}\{\mathbf{S}_n > n\mu_n + \epsilon a_n\} \leq \frac{2}{\epsilon^2}\cdot\frac{n\mu_n}{a_n} + n[1-F(2a_n)]. \tag{7.14}$$

Since we are dealing with positive variables

$$\mathbf{P}\{\mathbf{S}_n < 2a_n\} \leq \mathbf{P}\{\max_{k\leq n} \mathbf{X}_k \leq 2a_n\} = F^n(2a_n). \tag{7.15}$$

By assumption the left side tends to 1, and this implies $n[1 - F(2a_n)] \to 0$ (because $x \leq e^{-(1-x)}$ for $x \leq 1$). If $n\mu_n/a_n$ tended to zero the same would be true of the right side in (7.14) and this inequality would manifestly contradict the assumption (7.11). This argument applies also to subsequences and shows that $n\mu_n/a_n$ remains bounded away from zero; this in turn implies that $\rho(2a_n) \to \infty$.

To show that $\rho(x) \to \infty$ for any approach $x \to \infty$ choose a_n such that $2a_n < x \leq 2a_{n+1}$. Then $\rho(x) \geq (2a_n)a_n/a_{n+1}$, and it is obvious that (7.11) necessitates the boundedness of the ratios a_{n+1}/a_n. ▶

*8. STRONG LAWS

Let $\mathbf{X}_1, \mathbf{X}_2, \ldots$ be mutually independent random variables with a common distribution F and $\mathbf{E}(\mathbf{X}_k) = 0$. As usual we put $\mathbf{S}_n = \mathbf{X}_1 + \cdots + \mathbf{X}_n$. The weak law of large numbers states that for every $\epsilon > 0$

$$\mathbf{P}\{n^{-1}|\mathbf{S}_n| > \epsilon\} \to 0. \tag{8.1}$$

This fact does not eliminate the possibility that $n^{-1}\mathbf{S}_n$ may become arbitrarily large for infinitely many n. For example, in a symmetric random walk the probability that the particle passes through the origin at the nth step tends to 0, and yet it is certain that infinitely many such passages will occur. In practice one is rarely interested in the probability in (8.1) for any particular large value of n. A more interesting question is whether $n^{-1}|\mathbf{S}_n|$ will ultimately become and remain small, that is, whether $n^{-1}|\mathbf{S}_n| < \epsilon$ simultaneously for all $n \geq N$. Accordingly we ask for the probability of the event[13] that $n^{-1}\mathbf{S}_n \to 0$.

---

* This section may be omitted at the first reading.

[13] It follows from the zero-or-one law of IV,6 that this probability equals 0 or 1, but we shall not use this fact.

If this event has probability one we say that $\{\mathbf{X}_k\}$ *obeys the strong law of large numbers.*

The next theorem shows that this is the case whenever $\mathbf{E}(\mathbf{X}_1) = 0$. [That this statement is much stronger than the weak law of large numbers follows from the fact that (8.1) was seen to hold also for certain sequences $\{\mathbf{X}_k\}$ without expectation. By contrast the existence of an expectation is a necessary condition for the strong law. In fact, the converse to the strong law, discussed at the end of this section, shows that in the absence of an expectation the averages $n^{-1}|\mathbf{S}_n|$ are certain infinitely often to exceed any prescribed bound a.]

Theorem 1. (*Strong law of large numbers.*) *Let* $\mathbf{X}_1, \mathbf{X}_2, \ldots$ *be independent identically distributed variables with* $\mathbf{E}(\mathbf{X}) = 0$. *Then* $n^{-1}\mathbf{S}_n \to 0$ *with probability* 1.

The proof depends on truncations and is in effect concerned with sequences with varying distributions. To avoid duplications we therefore postpone the proof and prepare for it by establishing another theorem of wide applicability.

Theorem 2. *Let* $\mathbf{X}_1, \mathbf{X}_2, \ldots$ *be independent random variables with arbitrary distributions. Suppose that* $\mathbf{E}(\mathbf{X}_k) = 0$ *for all* k *and*

$$\sum_{k=1}^{\infty} \mathbf{E}(\mathbf{S}_k^2) < \infty. \tag{8.2}$$

Then the sequence $\{\mathbf{S}_n\}$ *converges with probability one to a finite limit* $\mathbf{S}$.

Proof. We refer to infinite-dimensional sample space defined by the variables $\mathbf{X}_k$. Let $A(\epsilon)$ be the event that the inequality $|\mathbf{S}_n - \mathbf{S}_m| > \epsilon$ holds for some arbitrarily large subscripts n, m. The event that $\{\mathbf{S}_n\}$ does not converge is the monotone limit as $\epsilon \to 0$ of the events $A(\epsilon)$, and so it suffices to prove that $\mathbf{P}\{A(\epsilon)\} = 0$. Let $A_m(\epsilon)$ be the event that $|\mathbf{S}_n - \mathbf{S}_m| > \epsilon$ for some $n > m$. Then $A(\epsilon)$ is the limit as $m \to \infty$ of the decreasing sequence of events $A_m(\epsilon)$, and so it suffices to prove that $\mathbf{P}\{A_m(\epsilon)\} \to 0$. Finally, for $n > m$ let $A_{m,n}(\epsilon)$ be the event that $|\mathbf{S}_k - \mathbf{S}_m| > \epsilon$ for some $m < k < n$. By Kolmogorov's inequality

$$\mathbf{P}\{A_{m,n}(\epsilon)\} \leq \epsilon^{-2} \operatorname{Var}(\mathbf{S}_n - \mathbf{S}_m) = \epsilon^{-2} \sum_{k=m+1}^{n} \mathbf{E}(\mathbf{X}_k^2). \tag{8.3}$$

Letting $n \to \infty$ we conclude that

$$\mathbf{P}\{A_m(\epsilon)\} \leq \epsilon^{-2} \sum_{k=m+1}^{\infty} \mathbf{E}(\mathbf{X}_k^2) \tag{8.4}$$

and the right side tends to 0 as $m \to \infty$. ►

This theorem has many applications. The following is a variant to be used for the proof of the strong law of large numbers.

Theorem 3. *Let* $\mathbf{X}_1, \mathbf{X}_2 \ldots$ *be independent variables with arbitrary distributions. Suppose that* $\mathbf{E}(\mathbf{X}_k) = 0$ *for all* k. *If* $b_1 < b_2 < \cdots \to \infty$ *and if*

$$\sum b_k^{-2}\, \mathbf{E}(\mathbf{X}_k^2) < \infty \tag{8.5}$$

then with probability one the series $\sum b_k^{-1}\mathbf{X}_k$ *converges and*

$$b_n^{-1}\, \mathbf{S}_n \to 0. \tag{8.6}$$

Proof. The first assertion is an immediate consequence of theorem 2 applied to the random variables $b_k^{-1}\mathbf{X}_k$. The following widely used lemma shows that the relation (8.6) takes place at every point at which the series converges, and this completes the proof. ►

Lemma 1. ("*Kronecker's lemma*".) *Let* $\{x_k\}$ *be an arbitrary numerical sequence and* $0 < b_1 < b_2 < \cdots \to \infty$. *If the series* $\sum_1^\infty b_k x_k$ *converges, then*

$$\frac{x_1 + \cdots + x_n}{b_n} \to 0. \tag{8.7}$$

Proof. Denote the remainders of our convergent series by ρ_n. Then for $n = 1, 2, \ldots$

and hence
$$x_n = b_n(\rho_{n-1} - \rho_n),$$

$$\frac{x_1 + \cdots + x_n}{b_n} = -\rho_n + \frac{1}{b_n}\sum_{k=1}^{n-1} \rho_k(b_{k+1} - b_k) + \frac{\rho_0}{b_n}. \tag{8.8}$$

Suppose $|\rho_k| < \epsilon$ for $k \geq r$. Since $b_n \to \infty$ the contribution of the first r terms in the sum tends to zero, while the remaining terms add to at most $\epsilon(b_n - b_r)/b_n < \epsilon$. Thus (8.7) is true. ►

Before returning to the strong law of large numbers we prove another lemma of a purely analytic character.

Lemma 2. *Let the variables* $\mathbf{X}_k$ *have a common distribution* F. *Then for any* $a > 0$

$$\sum \mathbf{P}\{|\mathbf{X}_k| > ak\} < \infty \tag{8.9}$$

if and only if $\mathbf{E}(\mathbf{X}_k)$ *exists.*

Proof. According to lemma 2 of V,6 an expectation $\mathbf{E}(\mathbf{X}_1)$ exists if and only if

$$\int_0^\infty [1 - F(x) + F(-x)]\, dx < \infty. \tag{8.10}$$

The series in (8.9) may be considered as a Riemann sum to the integral, and since the integrand is monotone the two relations (8.9) and (8.10) imply each other. ▶

We are finally in a position to prove the strong law of large numbers.[14]

Proof of theorem 1. We use truncation and define new random variables as follows:

$$\begin{aligned} &\mathbf{X}'_k = \mathbf{X}_k, \quad \mathbf{X}''_k = 0 \qquad \textit{if} \quad |\mathbf{X}_k| \leq k \\ &\mathbf{X}'_k = 0, \quad \mathbf{X}''_k = \mathbf{X}_k \qquad \textit{if} \quad |\mathbf{X}_k| > k. \end{aligned} \tag{8.11}$$

Since a finite expectation exists we conclude from the preceding lemma with $a = 1$ that

$$\sum \mathbf{P}\{\mathbf{X}''_k \neq 0\} < \infty \tag{8.12}$$

and this implies that with probability one only finitely many variables $\mathbf{X}''_k$ will be different from 0. Thus, with obvious notations, $\frac{1}{n}\mathbf{S}''_n \to 0$ with probability one.

Next we shall prove that

$$\sum k^{-2}\mathbf{E}(\mathbf{X}'_k) < \infty. \tag{8.13}$$

By theorem 3 this implies that with probability one

$$n^{-1}[\mathbf{S}'_n - \mathbf{E}(\mathbf{S}'_n)] \to 0. \tag{8.14}$$

But $\mathbf{E}(\mathbf{X}'_k) \to 0$ and hence obviously

$$n^{-1}\mathbf{E}(\mathbf{S}'_n) = n^{-1}\sum_{k=1}^{n}\mathbf{E}(\mathbf{X}'_k) \to 0. \tag{8.15}$$

To conclude the proof it remains only to verify the assertion (8.13). Now

$$\mathbf{E}(\mathbf{X}'^2_k) = \sum_{j=1}^{k}\int_{j-1\leq|x|<j} x^2\, F\{dx\}. \tag{8.16}$$

It follows that

$$\sum_{k=1}^{\infty} k^{-2}\mathbf{E}(\mathbf{X}'^2_k) = \sum_{j=1}^{\infty}\int_{j-1\leq|x|<j} x^2\, F\{dx\}\sum_{k=j}^{\infty}k^{-2}. \tag{8.17}$$

The inner sum is less than $2/j$ and so the right side is less than

$$\sum_{j=1}^{\infty}\int_{j-1\leq|x|<j} |x|\, F\{dx\} = \int_{-\infty}^{+\infty}|x|\, F\{dx\}. \tag{8.18}$$

This accomplishes the proof. ▶

[14] For a more direct proof see problem 12.

We saw that the weak law in the form (7.8) applies also to certain sequences without expectation. This is in striking contrast to the strong law for which the existence of $\mathbf{E}(\mathbf{X}_1)$ is *necessary*. In fact, the next theorem shows that in the absence of a finite expectation the sequence of averages $\mathbf{S}_n/n$ is *unbounded* with probability one.

Theorem 4. (*Converse to the strong law of large numbers.*) *Let* $\mathbf{X}_1, \mathbf{X}_2, \ldots$ *be independent with a common distribution. If* $\mathbf{E}(|\mathbf{X}_1|) = \infty$ *then for any numerical sequence* $\{c_n\}$ *with probability one*

$$\limsup |n^{-1}\mathbf{S}_n - c_n| = \infty \tag{8.19}$$

Proof. Let A_k stand for the event that $|\mathbf{X}_k| > ak$. These events are mutually independent and by lemma 2 the absence of an expectation implies that $\Sigma\, \mathbf{P}\{A_n\}$ diverges. By the second Borel-Cantelli lemma (see **1**; VIII,3) this means that with probability one infinitely many events A_k will occur, and so the sequence $|\mathbf{X}_k|/k$ is unbounded with probability one. But since $\mathbf{X}_k = \mathbf{S}_k - \mathbf{S}_{k-1}$ the boundedness of $|\mathbf{S}_n|/n$ would entail the boundedness of $|\mathbf{X}_k|/k$, and so we conclude that the sequence of averages $\mathbf{S}_n/n$ is unbounded with probability one.

This proves the assertion (8.19) for the special case $c_k = 0$, and the general case may be reduced to it by symmetrization. As in V,5 we denote by ${}^0\mathbf{X}_k$ the symmetrized variables $\mathbf{X}_k$. From the symmetrization inequality V,(5.1) it follows that $\mathbf{E}(|{}^0\mathbf{X}_k|) = \infty$, and so the sequence of averages ${}^0\mathbf{S}_n/n$ is unbounded with probability one. But ${}^0\mathbf{S}_n$ may be obtained by symmetrization of $\mathbf{S}_n - c_n$, and so the probability that $(\mathbf{S}_n - c_n)/n$ remains bounded is zero. ▶

*9. GENERALIZATION TO MARTINGALES

Kolmogorov's inequality of V,8(*e*) provided the main tool for the proofs in section 8. A perusal of these proofs reveals that the assumed independence of the variables was used only to derive certain inequalities among expectations, and hence the main results carry over to martingales and submartingales. Such generalizations are important for many applications and they throw new light on the nature of our theorems.

We recall from VI,12 that a finite or infinite sequence of random variables $\mathbf{U}_r$ constitutes a *submartingale* if for all r

$$\mathbf{E}(\mathbf{U}_r \mid \mathfrak{B}_k) \geq \mathbf{U}_k \qquad \textit{for} \quad k = 1, 2, \ldots, r-1, \tag{9.1}$$

where $\mathfrak{B}_1 \subset \mathfrak{B}_2 \subset \cdots$ is an increasing sequence of σ-algebras of events. When all the inequalities are replaced by equalities then $\{\mathbf{U}_r\}$ is called a *martingale*. [In each case the $r-1$ conditions (9.1) are automatically satisfied if they hold for the particular value $k = r - 1$.] Recall also that if $\{\mathbf{X}_k\}$ is a sequence of independent random variables with $\mathbf{E}(\mathbf{X}_k) = 0$, then the partial sums $\mathbf{S}_n$ form a martingale; furthermore, if the variances exist, $\{\mathbf{S}_n^2\}$ is a submartingale.

Theorem 1. (*Kolmogorov's inequality for positive submartingales.*) *Let* $\mathbf{U}_1, \ldots, \mathbf{U}_n$ *be positive variables. Suppose that the submartingale condition*

* The contents of this section will not be used in the sequel.

(9.1) *is satisfied for* $r \leq n$. *Then for* $t > 0$

$$\mathbf{P}\left\{\max_{k\leq n} \mathbf{U}_k > t\right\} \leq t^{-1}\,\mathbf{E}(\mathbf{U}_n). \tag{9.2}$$

If $\{\mathbf{U}_k\}$ is an arbitrary martingale then the variables $|\mathbf{U}_k|$ form a submartingale. (See the lemma in VI,12.) It follows that theorem 1 entails the important

Corollary. (*Kolmogorov's inequality for martingales.*) *If* $\mathbf{U}_1, \ldots, \mathbf{U}_n$ *constitute a martingale then for* $t > 0$

$$\mathbf{P}\left\{\max_{k\leq n} |\mathbf{U}_k| > t\right\} \leq t^{-1}\,\mathbf{E}(|\mathbf{U}_n|). \tag{9.3}$$

Proof of theorem 1. We repeat literally the proof of Kolmogorov's inequality in V,8(*e*) letting $\mathbf{S}_k^2 = \mathbf{U}_k$. The assumption that the $\mathbf{S}_k$ are sums of independent random variables was used only to establish the inequality V,(8.16) which now reads

$$\mathbf{E}(\mathbf{U}_n \mathbf{1}_{A_j}) \geq \mathbf{E}(\mathbf{U}_j \mathbf{1}_{A_j}). \tag{9.4}$$

Now $\mathbf{1}_{A_j}$ is $\mathfrak{B}_j$-measurable, and therefore

$$\mathbf{E}(\mathbf{U}_n \mathbf{1}_{A_j} \mid \mathfrak{B}_j) = \mathbf{1}_{A_j}\mathbf{E}(\mathbf{U}_n \mid \mathfrak{B}_j) \geq \mathbf{U}_j \mathbf{1}_{A_j} \tag{9.5}$$

[See V,(10.9)]. Taking expectations we get (9.4). ►

We turn to a generalization of the infinite convolution theorem of section 8, although it leads only to a special case of the general martingale convergence theorem. Indeed, Doob has shown that the following theorem remains valid if the condition $\mathbf{E}(\mathbf{S}_n^2) < \infty$ is replaced by the weaker requirement that $\mathbf{E}(|\mathbf{S}_n|)$ remain bounded. The proof of the general theorem is intricate, however, and our version is given because of the great importance of the theorem and the simplicity of the proof. (For a generalization see problem 13.)

Theorem 2. (*Martingale convergence theorem.*) *Let* $\{\mathbf{S}_n\}$ *be an infinite martingale with* $\mathbf{E}(\mathbf{S}_n^2) < C < \infty$ *for all* n. *There exists a random variable* $\mathbf{S}$ *such that* $\mathbf{S}_n \to \mathbf{S}$ *with probability one. Furthermore* $\mathbf{E}(\mathbf{S}_n) = \mathbf{E}(\mathbf{S})$ *for all* n.

Proof. We repeat the proof of theorem 2 in section 8. The assumption that the $\mathbf{S}_k$ are sums of independent variables was used only in (8.3) to prove that

$$\mathbf{E}((\mathbf{S}_n - \mathbf{S}_m)^2) \to 0, \qquad n, m \to 0. \tag{9.6}$$

Now we know from VI,12 that the martingale property of $\{\mathbf{S}_n\}$ implies that

$\mathbf{E}(\mathbf{S}_n \mid \mathbf{S}_m) = \mathbf{S}_m$ for $n > m$. By the fundamental property V,(10.9) of conditional expectations this implies

$$\mathbf{E}(\mathbf{S}_n\mathbf{S}_m \mid \mathbf{S}_m) = \mathbf{S}_m^2, \qquad n > m. \tag{9.7}$$

Taking expectations and recalling the formula V,(10.10) for iterated expectations we conclude that $\mathbf{E}(\mathbf{S}_n\mathbf{S}_m) = \mathbf{E}(\mathbf{S}_m^2)$ and hence

$$\mathbf{E}(\mathbf{S}_n^2 - S_m^2) = \mathbf{E}(\mathbf{S}_n^2) - \mathbf{E}(\mathbf{S}_m^2), \qquad n > m.$$

But by the lemma of VI,12 the variables $\mathbf{S}_n^2$ form a submartingale, and so the sequence $\{\mathbf{E}(\mathbf{S}_n^2)\}$ is monotonically increasing. By assumption it is bounded, and hence it has a finite limit. This implies the truth of (9.6). The martingale properly implies that $\mathbf{E}(\mathbf{S}_n)$ is independent of n, and the identity $\mathbf{E}(\mathbf{S}) = \mathbf{E}(\mathbf{S}_n)$ follows from the boundedness of the sequence $\{\mathbf{E}(\mathbf{S}_n^2)\}$ [example VIII,1(*e*)]. ▶

As an immediate corollary we obtain the following analogue to theorem 3 of section 8.

Theorem 3. *Let* $\{\mathbf{X}_n\}$ *be a sequence of random variables such that*

$$\mathbf{E}(\mathbf{X}_n \mid \mathfrak{B}_{n-1}) = 0 \tag{9.8}$$

for all n. *If* $b_1 < b_2 < \cdots \to \infty$ *and*

$$\sum b_k^{-2}\mathbf{E}(\mathbf{X}_k^2) < \infty, \tag{9.9}$$

then with probability one

$$\frac{\mathbf{X}_1 + \cdots + \mathbf{X}_n}{b_n} \to 0. \tag{9.10}$$

Proof. It is easily seen that the variables

$$\mathbf{U}_n = \sum_{k=1}^{n} b_k^{-1}\mathbf{X}_k \tag{9.11}$$

form a martingale and that $\mathbf{E}(\mathbf{U}_n^2)$ is bounded by the series in (9.9). The preceding theorem therefore guarantees the almost sure convergence of $\{\mathbf{U}_n\}$, and by Kronecker's lemma this implies the assertion (9.10). ▶

Examples. (*a*) *Polya's urn scheme* was treated in examples VI,11(*b*) and above in 4(*a*). If $\mathbf{Y}_n$ is the proportion of black balls at the nth trial it was shown that $\{\mathbf{Y}_n\}$ is a martingale and we see now that a limit $\mathbf{Y} = \lim \mathbf{Y}_n$ exists with probability one. On the other hand, the probability of a black ball at the nth trial is obtained by randomization of the binomial distribution. Thus, if $\mathbf{S}_n$ is the total number of black balls drawn in the first n trials the distribution of $n^{-1}\mathbf{S}_n$ tends to the beta distribution F found in example 4(*a*). It follows that the limit variable $\mathbf{Y}$ has the beta distribution F.

(*b*) *Branching processes.* In the branching process described in **1**; XII,5 the population size $\mathbf{X}_n$ in the nth generation has the expectation $\mathbf{E}(\mathbf{X}_n) = \mu^n$ [see **1**; XII,(4.9)]. Given that the $(n-1)$th generation consisted of ν individuals, the (conditional) expectation of $\mathbf{X}_n$ becomes μ^ν, and this independently of the size of the preceding generations. Thus, if we put $\mathbf{S}_n = \mathbf{X}_n/\mu^n$ the sequence $\{\mathbf{S}_n\}$ forms a martingale. It is not difficult to establish that if $\mathbf{E}(\mathbf{X}_1^2) < \infty$ then $\mathbf{E}(\mathbf{S}_n^2)$ remains bounded (see problem 7 of **1**; XII,6). We thus have the striking result that $\mathbf{S}_n$ converges with probability one to a limit $\mathbf{S}_\infty$. This implies, in particular, that the distribution of $\mathbf{S}_n$ tends to the distribution of $\mathbf{S}_\infty$. These results are due to T. E. Harris.

(*c*) *Harmonic functions.* For clarity we describe a specific example, although the following argument applies to more general Markov chains and concordant functions {example VI,12(*c*)].

Let D denote the unit disk of points $x = (x_1, x_2)$ such that $x_1^2 + x_2^2 \leq 1$. For any point $x \in D$ let C_x be the largest circle centered at x and contained in D. We consider a Markov process $\{\mathbf{Y}_n\}$ in D defined as follows. Given that $\mathbf{Y}_n = x$ the variable $\mathbf{Y}_{n+1}$ is uniformly distributed on the circle C_x; the initial position $\mathbf{Y}_0 = y$ is assumed known. The transition probabilities are given by a stochastic kernel K which for fixed x is concentrated on C_x and reduces there to the uniform distribution. A function u in D is concordant if $u(x)$ equals the average of the value of u on C_x. Consider now a harmonic function u that is continuous on the *closed* disk D. Then $\{u(\mathbf{Y}_n)\}$ is a bounded martingale and hence $\mathbf{Z} = \lim u(\mathbf{Y}_n)$ exists with probability one. Since the coordinate variables x_j are harmonic functions it follows that with probability one $\mathbf{Y}_n$ tends to a limit $\mathbf{Y} \in D$. It is easily seen that the process cannot converge to an interior point of D, and hence with probability one $\mathbf{Y}_n$ *tends to a point* $\mathbf{Y}$ *on the boundary of* D.

An extension of arguments of this sort is used for the study of asymptotic properties of Markov processes, and also to prove general theorems concerning harmonic functions, such as Fatou's theorem concerning the existence almost everywhere of radial boundary values.[15] ▶

10. PROBLEMS FOR SOLUTION

1. If u is bounded and continuous on $\overline{0, \infty}$ then as $n \to \infty$

$$\sum_{k=0}^{\infty} \binom{n+k}{k} \frac{t^k}{(1+t)^{n+k+1}} u\left(\frac{k}{n+1}\right) \to u(t)$$

uniformly in every finite interval.

---

[15] M. Brelot and J. L. Doob, Ann. Inst. Fourier, vol. 13 (1963), pp. 395–415.

Hint: Remember the "negative binomial" distribution of **1**; VI,8. No calculation necessary.

2. If u has a continuous derivative u', the derivative $B'_{n,u}$ of $B_{n,u}$ tends uniformly to u'.

3. *Bernstein polynomials in* $\mathcal{R}^2$. If $u(x, y)$ is continuous in the triangle $x \geq 0$, $y \geq 0$, $x + y \leq 1$, then uniformly

$$\sum u\left(\frac{j}{n}, \frac{k}{n}\right) \frac{n!}{j!k!(n-j-k)!} x^j y^k (1-x-y)^{n-j-k} \to u(x, y).$$

4. A function u continuous in $\overline{0, 1}$ can be uniformly approximated by even polynomials. If $u(0) = 0$ the same is true for odd polynomials.[16]

5. If u is continuous in the interval $\overline{0, \infty}$ and $u(\infty)$ exists, it can be approximated uniformly by linear combinations of e^{-nx}.

6. For the three moment sequences given below find the probabilities $p_k^{(n)}$ of (3.5). Find the corresponding distribution F using the limit relation (3.11).

$$(a)\ \mu_n = p^n \quad (0 < p < 1), \qquad (b)\ \mu_n = \frac{1}{n+1}, \qquad (c)\ \mu_n = \frac{2}{n+2}.$$

7.[17] Let p be a polynomial of degree ν. Show that $\underset{h}{\Delta^n} p$ vanishes identically when $n > \nu$. Conclude that $B_{n,p}$ is a polynomial of degree $\leq \nu$ (despite its formal appearance as polynomial of degree $n > \nu$).

8. When F has a density, (6.4) can be derived by integration from (6.6).

9. *Law of large numbers for stationary sequences.* Let $\{\mathbf{X}_k\}$ $(k = 0, \pm 1, \pm 2, \ldots)$ be a stationary sequence and define the $\mathbf{X}'_k$ by truncation as in (7.2). If $\mathbf{E}(\mathbf{X}_k) = 0$ and $\mathbf{E}(\mathbf{X}'_0 \mathbf{X}'_n) \to 0$ as $n \to \infty$ then

$$\mathbf{P}\{n^{-1} |\mathbf{X}_1 + \cdots + \mathbf{X}_n| > \epsilon\} \to 0.$$

10. Let the $\mathbf{X}_k$ be independent and define the $\mathbf{X}'_k$ by truncation as in (7.2). Let $a_n \to 0$ and suppose that

$$\sum_{k=1}^{n} \mathbf{P}\{|\mathbf{X}_k| > s_n\} \to 0, \qquad a_n^{-2} \sum_{k=1}^{n} \mathbf{E}(\mathbf{X}'^2_k) \to 0.$$

Prove that

$$\mathbf{P}\left\{\left|\mathbf{S}_n - \sum_{k=1}^{n'} \mathbf{E}(\mathbf{X}'_k)\right| > \epsilon a_n\right\} \to 0.$$

11. (To theorem 1 of section 7). Show that $\sigma(t) \to 0$ implies $\tau(t) \to 0$. *Hint:* Prove that $\tau(x) - \frac{1}{2}\tau(2x) < \epsilon$ for x sufficiently large. Apply this inequality successively to $x = t, 2t, 4t, \ldots$ to conclude that $\tau(t) < 2\epsilon$.

12. (*Direct proof of the strong law of large numbers.*) With the notations used in the proof of theorem 1 in section 8 put $\mathbf{Z}_r = \max |\mathbf{S}'_k|$ for $2^r < k \leq 2^{r+1}$.

[16] A famous theorem due to H. Ch. Müntz asserts that uniform approximation is possible in terms of linear combinations of $1, x^{n_1}, x^{n_2}, \ldots$ iff $\sum n_k^{-1}$ diverges.

[17] The use of this result leads to a considerable simplification of the classical solution of the moment problem (for example in the book of Shohat and Tamarkin).

Using Kolmogorov's inequality show that

$$\sum \mathbf{P}(\mathbf{Z}_r > \epsilon 2^r\} < \infty,$$

and that this implies the strong law of large numbers. (This proof avoids reference to theorems 2 and 3 of section 8.)

13. (*Convergence theorem for submartingales.*) Prove that theorem 2 of section 8 applies also to submartingales $\{\mathbf{U}_k\}$ provided $\mathbf{U}_k > 0$ for all k.

14. Generalize the variant of Chebyshev's inequality in the example of V,7(*a*) to martingales.[18]

---

[18] A. W. Marshall, *A one-sided analog of Kolmogorov's inequality*, Ann. Math. Statist., vol. 31 (1960) pp. 483–487.

CHAPTER VIII

The Basic Limit Theorems

The main results of this chapter are found in sections 1, 3, and 6. Sections 4, 5, and 7 may be considered as sources of interesting examples. These are chosen because of their importance in other contexts.

The last two sections are devoted to regularly varying functions in the sense of Karamata. This interesting theory steadily gains in importance, but it is not accessible in textbooks and has not been adapted to distribution functions. A tremendous amount of disconnected calculation in probability can be saved by exploiting the asymptotic relations in section 9. They are of a technical nature in contrast to the simple section 8.

1. CONVERGENCE OF MEASURES

The following theory is independent of the number of dimensions. For convenience of expression the text refers to one-dimensional distributions, but with the conventions of III,5 the formulas apply without change in higher dimensions.

Two examples are typical for the phenomena with which we have to cope.

Examples. (*a*) Consider an arbitrary probability distribution F and put $F_n(x) = F(x-n^{-1})$. At a point x at which F is continuous we have $F_n(x) \to F(x)$, but at points of discontinuity $F_n(x) \to F(x-)$. We shall nevertheless agree to say that the sequence $\{F_n\}$ converges to F.

(*b*) This time we put $F_n(x) = F(x+n)$ where F is a continuous distribution function. Now $F_n(x) \to 1$ for all x: a limit exists, but is not a probability distribution function. Here $F_n\{I\} \to 0$ for every bounded interval, but not when I coincides with the whole line.

(*c*) Let $F_n(x) = F(x+(-1)^n n)$. Then $F_{2n}(x) \to 1$ whereas $F_{2n+1}(x) \to 0$. Accordingly, the distribution functions as such do not converge, but nevertheless $F_n\{I\} \to 0$ for every bounded interval. ▶

Basic Notions and Notations

It will be necessary to distinguish three classes of continuous functions. In one dimension[1] $C(-\infty, \infty)$ is the class of all bounded continuous functions; $C[-\infty, \infty]$ is the subclass of functions with finite limits $u(-\infty)$ and $u(\infty)$; finally, $C_0(-\infty, \infty)$ is the subclass of functions "*vanishing at infinity*," that is, where $u(\pm\infty) = 0$.

We shall say that I is an *interval of continuity* for the probability distribution F if I is open and its endpoints are not atoms.[2] The whole line counts as an interval of continuity. Throughout this section we use the abbreviations

$$\mathbf{E}_n(u) = \int_{-\infty}^{+\infty} u(x)\, F_n\{dx\}, \qquad \mathbf{E}(u) = \int_{-\infty}^{+\infty} u(x)\, F\{dx\}. \tag{1.1}$$

Throughout this section F_n will stand for a proper probability distribution, but F will be permitted to be *defective* (that is, its total mass may be <1; see the definition in V,1). ►

Definition.[3] *The sequence* $\{F_n\}$ *converges to the (possibly defective) distribution* F *if*

$$F_n\{I\} \to F\{I\} \tag{1.2}$$

for every bounded interval of continuity of F. *In this case we write* $F_n \to F$ *or* $F = \lim F_n$.

The convergence is called proper if F *is not defective.*

For stylistic clarity we speak sometimes of *improper* convergence to indicate that the limit F is defective.

For ease of reference we record two simple criteria for proper convergence.

Criterion 1. *The convergence* $F_n \to F$ *is proper iff to each* $\epsilon > 0$ *there correspond numbers* a *and* N *such that* $F_n\{\overline{-a, a}\} > 1 - \epsilon$ *for* $n > N$.

---

[1] For the analogue in higher dimensions note that in one dimension $C[-\infty, \infty]$ is simply the class of continuous functions on the compactified line obtained by adding $\pm\infty$ to $\mathcal{R}^1$. For $C[-\infty, \infty]$ in $\mathcal{R}^2$ both axes are so extended, and this requires the existence of limits $u(x, \pm\infty)$ and $u(\pm\infty, x)$ for each number x. In itself this class is not very interesting, but distribution functions belong to it. For $C_0(-\infty, \infty)$ it is required that $u(x, \pm\infty) = u(\pm\infty, x) = 0$.

[2] In higher dimensions it is required that the boundary of I has probability zero.

[3] For readers interested in general measure theory we remark the following. The definitions and the theorems of this section apply without change to bounded measures in arbitrary locally compact spaces provided "interval of continuity" is replaced by "open set whose boundary has measure zero." To bounded intervals there correspond subsets of compacts sets. Finally, C_0 is the class of conditions functions vanishing at infinity, that is, $u \in C_0$ iff u is continuous and $|u| < \epsilon$ outside some compact set. The other classes play no role in this section.

Proof. Without loss of generality we may suppose that $\overline{-a, a}$ is an interval of continuity for the limit F. The condition is sufficient because it implies that $F\{\overline{-a, a}\} > 1 - \epsilon$, and hence F cannot be defective. Conversely, if F is a probability distribution we may choose a so large that $F\{\overline{-a, a}\} > 1 - \frac{1}{2}\epsilon$. Then $F_n\{\overline{-a, a}\} > 1 - \epsilon$ for n sufficiently large, and hence the condition is necessary. ▶

Criterion 2. *A sequence $\{F_n\}$ of probability distributions converges to a proper probability distribution F iff* (1.2) *holds for every bounded or unbounded interval I which is an interval of continuity for F.*

(This implies that in the case of proper convergence $F_n(x) \to F(x)$ at every point of continuity of F.)

Proof. We may suppose that $F_n \to F$ with F possibly defective. Obviously F is proper iff (1.2) holds for $I = \overline{-\infty, \infty}$, and hence the condition of the criterion is sufficient. Assume then that F is a proper probability distribution. For $x > -a$ the interval $\overline{-\infty, x}$ is the union of $\overline{-\infty, a}$ and $\overline{a, x}$. Using the preceding criterion it is therefore seen that for a and n sufficiently large $F_n\{\overline{-\infty, x}\}$ differs from $F\{\overline{-\infty, x}\}$ by less than 3ϵ. A similar argument applies to $\overline{x, \infty}$ and we conclude that (1.2) holds for all semi-infinite intervals. ▶

We have defined convergence by (1.2), but the next theorem shows that we could have used (1.3) as defining relation.

Theorem 1. (i) *In order that $F_n \to F$ it is necessary and sufficient that*[4]

$$\mathbf{E}_n(u) \to \mathbf{E}(u) \quad \textit{for all} \quad u \in C_0(-\infty, \infty). \tag{1.3}$$

(ii) *If the convergence is proper then $\mathbf{E}_n(u) \to \mathbf{E}(u)$ for all bounded continuous functions.*

Proof. (*a*) It is convenient to begin with the assertion concerning *proper* convergence. Assume then that F is a probability distribution and $F_n \to F$. Let u be a continuous function such that $|u(x)| < M$ for all x. Let A be an interval of continuity for F so large that $F\{A\} > 1 - \epsilon$. For the complement A' we have then $F_n\{A'\} < 2\epsilon$ for all n sufficiently large.

Since u is uniformly continuous in finite intervals it is possible to partition A by intervals $I_1, \ldots, I_n$ so small that within each, u oscillates by less

[4] If $\mathbf{E}_n(u) \to \mathbf{E}(u)$ for a certain class of functions one says that F_n converges to F "weakly with respect to that class." Thus convergence in the sense of definition 1 is equivalent to weak convergence with respect to $C_0(-\infty, \infty)$.

than ϵ. These I_k may be chosen so that they are intervals of continuity for F. Within A we can approximate u by a step function σ assuming a constant value in each I_k and such that $|u(x) - \sigma(x)| < \epsilon$ for all $x \in A$. In the complement A' we put $\sigma(x) = 0$. Then $|u(x)-\sigma(x)| < M$ for $x \in A'$ and

$$(1.4) \qquad |\mathbf{E}(u)-\mathbf{E}(\sigma)| \leq \epsilon F\{A\} + MF\{A'\} \leq \epsilon + M\epsilon.$$

Similarly for n sufficiently large

$$(1.5) \qquad |\mathbf{E}_n(u) - \mathbf{E}_n(\sigma)| \leq \epsilon F_n\{A\} + MF_n\{A'\} \leq \epsilon + 2M\epsilon.$$

Now $\mathbf{E}_n(\sigma)$ is a finite linear combination of values $F_n\{I_k\}$ tending to $F\{I_k\}$. It follows that $\mathbf{E}_n(\sigma) \to \mathbf{E}(\sigma)$ and so for n sufficiently large

$$(1.6) \qquad |\mathbf{E}(\sigma) - \mathbf{E}_n(\sigma)| < \epsilon.$$

Combining the last three inequalities we get

$$(1.7) \quad |\mathbf{E}(u) - \mathbf{E}_n(u)| \leq |\mathbf{E}(u) - \mathbf{E}(\sigma)| + |\mathbf{E}(\sigma) - \mathbf{E}_n(\sigma)| + \\ + |\mathbf{E}_n(\sigma) - \mathbf{E}_n(u)| < 3(M + 1)\epsilon$$

and as ϵ is arbitrary this implies that $\mathbf{E}_n(u) \to \mathbf{E}(u)$.

This argument breaks down in the case of *improper* convergence because then $F_n\{A'\}$ need not be small. However, in this case we consider only functions $u \in C_0(-\infty, \infty)$ and the interval A may be chosen so large that $|u(x)| < \epsilon$ for $x \in A'$. Then $|u(x) - \sigma(x)| < \epsilon$ for *all* x, and the inequalities (1.4)–(1.5) hold in the sharper form with the right side replaced by ϵ. Thus (1.2) implies (1.3).

(*b*) We prove[5] that $\mathbf{E}_n(u) \to \mathbf{E}(u)$ implies $F_n \to F$. Let I be an interval of continuity of F of length L. Denote by I_δ a concentric interval of length $L + \delta$ where δ is chosen so small that $F\{I_\delta\} < F\{I\} + \epsilon$. Let u be a continuous function which within I assumes the constant value 1, which vanishes outside I_δ, and for which $0 \leq u(x) \leq 1$ everywhere. Then $\mathbf{E}_n(u) \geq F_n\{I\}$ and $\mathbf{E}(u) \leq F\{I_\delta\} < F\{I\} + \epsilon$. But for n sufficiently large we have $\mathbf{E}_n(u) < \mathbf{E}(u) + \epsilon$, and so

$$F_n\{I\} \leq \mathbf{E}_n(u) < \mathbf{E}(u) + \epsilon \leq F\{I_\delta\} + \epsilon < F\{I\} + 2\epsilon.$$

Using a similar argument with I_δ replaced by an interval of length $L - \delta$ we get the reversed inequality $F_n\{I\} > F\{I\} - 2\epsilon$, and so $F_n \to F$ as asserted. ►

[5] It would be simpler to apply the proof of theorem 2, but the proof of the text is more intuitive.

It is desirable to have a criterion for convergence which does not presuppose the knowledge of the limit. This is furnished by

Theorem 2. *In order that the sequence* $\{F_n\}$ *of probability distributions converges to a (possibly defective) limit distribution it is necessary and sufficient that for each* $u \in C_0(-\infty, \infty)$ *the sequence of expectations,* $\mathbf{E}_n(u)$, *tends to a finite limit.*

Proof.[6] The necessity is covered by theorem 1. For the proof of sufficiency we anticipate the selection theorem 1 of section 6. (It is elementary, but it is preferable to discuss it together with related topics.)

According to this theorem it is always possible to find a subsequence $\{F_{n_r}\}$ converging to a possibly defective limit Φ. Denote by $\mathbf{E}^*(u)$ the expectation of u with respect to Φ. Let $u \in C_0(-\infty, \infty)$. Then $\mathbf{E}_{n_k}(u) \to \mathbf{E}^*(u)$ by theorem 1. But $\mathbf{E}_{n_k}(u) \to \lim \mathbf{E}_n(u)$ as well. Hence $\mathbf{E}^*(u) = \lim \mathbf{E}_n(u)$ and u was arbitrary in $C_0(-\infty, \infty)$. Another application of theorem 1 gives $F_n \to \Phi$ as required. ►

Examples. (*d*) *Convergence of moments.* If the distributions F_n are concentrated on $\overline{0, 1}$ the definition of u outside this interval is immaterial and in the wording of the theorem it suffices to assume u continuous in $\overline{0, 1}$. Every such function can be approximated uniformly by polynomials (see VII,2) and hence the theorem may be restated as follows. *A sequence of distributions* F_n *concentrated on* $\overline{0, 1}$ *converges to a limit* F *iff for each* k *the sequence of moments* $\mathbf{E}_n(\mathbf{X}^k)$ *converges to a number* μ_k. *In this case* $\mu_k = \mathbf{E}(\mathbf{X}^k)$ *is the* kth *moment of* F *and the convergence is proper because* $\mu_0 = 1$. (See VII,3.)

(*e*) *Convergence of moments* (continued). In general the expectations of F_n need not converge even if $F_n \to F$ properly. For example, if F_n attributes weight n^{-1} to n^2 and weight $1 - n^{-1}$ to the origin, then $\{F_n\}$ converges to the distribution concentrated at the origin, but $\mathbf{E}_n(\mathbf{X}) \to \infty$. We have however, the following useful *criterion. If* $F_n \to F$ *and for some* $\rho > 0$ *the expectations* $\mathbf{E}_n(|\mathbf{X}|^\rho)$ *remain bounded, then* F *is a proper probability distribution.* Indeed, the contribution of the region $|x| \geq a$ to $\mathbf{E}_n(|\mathbf{X}|^\rho)$ is $> a^\rho(1-F_n\{\overline{-a, a}\})$ and this quantity can remain $<M$ only if $1 - F\{\overline{-a, a}\} \leq a^{-\rho}M$. Since a may be chosen arbitrarily large F must

[6] The theorem is trivial if one assumes the Riesz representation theorem (see note I, in V,1). Indeed, $\lim \mathbf{E}_n(u)$ defines a linear functional and according to that theorem the limit is the expectation of u with respect to some F.

be proper. A slight sharpening of this argument shows that *the absolute moments* $\mathbf{E}_n(|\mathbf{X}|^\alpha)$ *of order* $\alpha < \rho$ *converge to* $\mathbf{E}(|\mathbf{X}|^\alpha)$.

(*f*) *Convergence of densities.* If the probability distributions F_n have densities f_n the latter need not converge even if $F_n \to F$ and F has a continuous density. As an example let $f_n(x) = 1 - \cos 2n\pi x$ for $0 < x < 1$ and $f_n(x) = 0$ elsewhere. Here F_n converges to the uniform distribution with density $f(x) = 1$ for $0 < x < 1$, but f_n does not converge to f. On the other hand, *if* $f_n \to f$ *and* f *is a probability density then* $F_n \to F$ where F is the proper distribution with density f. Indeed, Fatou's lemma [IV,(2.9)] implies that $\liminf F_n\{I\} \geq F\{I\}$ for every interval I of continuity. If the inequality sign prevailed for some I it would hold a fortiori for every bigger interval, in particular for $\overline{-\infty, \infty}$. This being impossible (1.2) holds. ►

In dealing with functions u_t such as $\sin tx$ or $v(t+x)$ depending on a parameter t, it is often useful to know that for n sufficiently large the relation $|\mathbf{E}_n(u_t) - \mathbf{E}(u_t)| < \epsilon$ holds simultaneously for all t. We prove that this is so if the family of functions u_t is *equicontinuous*, that is, if to each $\epsilon > 0$ there corresponds a δ independent of t such that $|u_t(x_2) - u_t(x_1)| < \epsilon$ whenever $|x_2 - x_1| < \delta$.

Corollary. *Suppose that* $F_n \to F$ *properly. Let* $\{u_t\}$ *be a family of equicontinuous functions depending on the parameter* t *and such that* $|u_t| < M < \infty$ *for some* M *and all* t. *Then* $\mathbf{E}_n(u_t) \to \mathbf{E}(u_t)$ *uniformly in* t.

Proof. The proof in theorem 2 that $\mathbf{E}_n(u) \to \mathbf{E}(u)$ depended on partitioning the interval A into intervals within each of which u varies by less than ϵ. In the present situation this partition may be chosen independently of t and the assertion becomes obvious. ►

Example. (*g*) Let $u_t(x) = u(tx)$ where u is a differentiable function with $|u'(x)| \leq 1$. By the mean value theorem

$$|u_t(x_2) - u_t(x_1)| \leq |t| \cdot |x_2 - x_1|,$$

and so the family is equicontinuous provided t is restricted to a finite interval $\overline{-a, a}$. Therefore $\mathbf{E}_n(u_t) \to \mathbf{E}(u_t)$ uniformly in every finite t-interval. ►

2. SPECIAL PROPERTIES

According to definition 1 to V,2 two distributions U and V are of the same *type* if they differ only by location and scale parameters, that is, if

$$V(x) = U(Ax+B), \qquad A > 0 \tag{2.1}$$

We now show that convergence is a property of types in the sense that a change of location parameters does not affect the type of the limit distribution. It is this fact which makes it legitimate to speak of an "asymptotically normal sequence" without specifying the appropriate parameters. More precisely we prove

Lemma 1. *Let* U *and* V *be two probability distributions neither of which is concentrated at one point. If for a sequence* $\{F_n\}$ *of probability distributions and constants* $a_n > 0$ *and* $\alpha_n > 0$

$$F_n(a_nx+b_n) \to U(x), \qquad F_n(\alpha_nx+\beta_n) \to V(x) \tag{2.2}$$

at all points of continuity, then

$$\frac{\alpha_n}{a_n} \to A > 0, \qquad \frac{\beta_n - b_n}{a_n} \to B \tag{2.3}$$

and (2.1) *is true. Conversely, if* (2.3) *holds then each of the two relations* (2.2) *implies the other and* (2.1).

Proof. For reasons of symmetry we may assume that the first relation in (2.2) holds. To simplify notations we put $G_n(x) = F_n(a_nx+b_n)$ and also $\rho_n = \alpha_n/a_n$ and $\sigma_n = (\beta_n - b_n)/a_n$. Assume then that $G_n \to U$. If

$$\rho_n \to A, \qquad \sigma_n \to B \tag{2.4}$$

then obviously

$$G_n(\rho_nx+\sigma_n) \to V(x) \tag{2.5}$$

with $V(x) = U(Ax + B)$. We have to prove that (2.5) cannot take place unless (2.4) holds.

Since V is not concentrated at one point there exist at least two values x' and x'' such that the sequences $\{\rho_nx' + \sigma_n\}$ and $\{\rho_nx'' + \sigma_n\}$ remain bounded. This implies the boundedness of the sequences $\{\rho_n\}$ and $\{\sigma_n\}$, and hence it is possible to find a sequence of integers n_k such that $\rho_{n_k} \to A$ and $\sigma_{n_k} \to B$. But then $V(x) = U(Ax + B)$, and hence $A > 0$, for otherwise V cannot be a probability distribution. It follows that the limits A and B are the same for all subsequences, and so (2.4) is true. ►

Example. The lemma breaks down if V is concentrated at one point. Thus if $\rho_n \to \infty$ and $\sigma_n = (-1)^n$ the condition (2.4) is not satisfied but (2.5) holds with V concentrated at the origin. ►

Two types of sequences $\{F_n\}$ of probability distributions occur so frequently that they deserve names. For notational clarity we state the definitions formally in terms of random variables $\mathbf{X}_n$, but the notions really refer only to their distributions F_n. The definitions are therefore meaningful without reference to any probability space.

Definition 1. $\mathbf{X}_n$ *converges in probability to zero if*

$$\mathbf{P}\{|\mathbf{X}_n| > \epsilon\} \to 0 \tag{2.6}$$

for any $\epsilon > 0$. *We indicate this by* $\mathbf{X}_n \to 0$.

By extension $\mathbf{X}_n \xrightarrow{p} \mathbf{X}$ *means the same as* $\mathbf{X}_n - \mathbf{X} \xrightarrow{p} 0$.

Note that (2.6) holds iff the distributions F_n tend to the distribution concentrated at the origin. In general, however, $F_n \to F$ implies nothing about the convergence of $\mathbf{X}_1, \mathbf{X}_2, \ldots$. For example, if the $\mathbf{X}_j$ are independent with a common distribution F then $F_n \to F$ but the sequence $\{\mathbf{X}_n\}$ does not converge in probability.

The following simple lemma is of frequent use but not always mentioned explicitly. (For example, the truncation method used in **1**; X depends implicitly on it.)

Lemma 2. *Denote the distributions of* $\mathbf{X}_n$ *and* $\mathbf{Y}_n$ *by* F_n *and* G_n. *Suppose that* $\mathbf{X}_n - \mathbf{Y}_n \xrightarrow{p} 0$ *and* $G_n \to G$. *Then also* $F_n \to G$.

In particular, if $\mathbf{X}_n \xrightarrow{p} \mathbf{X}$ then $F_n \to F$ where F is the distribution of $\mathbf{X}$.

Proof. If $\mathbf{X}_n \leq x$ then either $\mathbf{Y}_n \leq x + \epsilon$ or $\mathbf{X}_n - \mathbf{Y}_n \leq -\epsilon$. The probability of the latter event tends to 0 and hence $F_n(x) \leq G_n(x+\epsilon) + \epsilon$ for all n sufficiently large. The same argument leads to an analogous inequality in the opposite direction. ▶

Definition 2. *The sequence* $\{\mathbf{X}_n\}$ *is stochastically bounded if for each* $\epsilon > 0$ *there exists an* a *such that for all* n *sufficiently large*

$$\mathbf{P}\{|\mathbf{X}_n| > a\} < \epsilon. \tag{2.7}$$

This notion applies equally to distributions in higher dimensions and vector variables $\mathbf{X}_n$.

A properly convergent sequence is obviously stochastically bounded whereas improper convergence excludes stochastic boundedness. We have therefore the trite but useful *criterion*: *If the distributions* F_n *converge, then the limit* F *is a proper distribution iff* $\{F_n\}$ *is stochastically bounded.*

If $\{\mathbf{X}_n\}$ and $\{\mathbf{Y}_n\}$ are stochastically bounded, so is $\{\mathbf{X}_n+\mathbf{Y}_n\}$. Indeed, the event $|\mathbf{X}_n+\mathbf{Y}_n| > 2a$ cannot occur unless either $\{\mathbf{X}_n| > a$ or $|\mathbf{Y}_n| > a$ and therefore

$$\mathbf{P}\{|\mathbf{X}_n+\mathbf{Y}_n| > 2a\} \leq \mathbf{P}\{|\mathbf{X}_n| > a\} + \mathbf{P}\{|\mathbf{Y}_n| > a\}. \tag{2.8}$$

3. DISTRIBUTIONS AS OPERATORS

The convolution $U = F \bigstar u$ of a point function u and a probability distribution F was defined in V,4. If we define a family of functions u_t by $u_t(x) = u(t-x)$ we can express the value $U(t)$ as an expectation

$$U(t) = \int_{-\infty}^{+\infty} u(t-y)\, F\{dy\} = \mathbf{E}(u_t). \tag{3.1}$$

We use this to derive a criterion for proper convergence. It is based on the class $C[-\infty, \infty]$ of continuous functions with limits $u(\pm\infty)$ because such functions are uniformly continuous.

Theorem 1. *A sequence of probability distributions* F_n *converges properly to a probability distribution* F *iff for each* $u \in C[-\infty, \infty]$ *the convolutions* $U_n = F_n \bigstar u$ *converge uniformly to a limit* U. *In this case* $U = F \bigstar u$.

Proof. The condition is necessary because the uniform continuity of u implies that the family $\{u_t\}$ is equicontinuous and so by the last corollary $U_n \to F \bigstar u$ uniformly. Conversely, the condition of the theorem entails the convergence of the expectations $\mathbf{E}_n(u)$. We saw in section 1 that this implies the convergence $F_n \to F$, but it remains to show that F is proper. For this purpose we use criterion 1 of section 1.

If u increases monotonically from 0 to 1 the same will be true of each U_n. Because of the uniform convergence there exists an N such that $|U_n(x) - U_N(x)| < \epsilon$ for $n > N$ and all x. Choose a so large that $U_N(-a) < \epsilon$. Now U_N is defined by a convolution of the form (3.1); restricting the interval of integration to $-\infty < y \leq -2a$ we see that for $n > N$

$$2\epsilon > U_n(-a) \geq u(a)F_n(-2a).$$

Since u increases to unity it follows that for n and a sufficiently large $F_n(-a)$ will be as small as we please. For reasons of symmetry the same argument applies to $1 - F(a)$, and hence F is proper by virtue of criterion 1.

To illustrate the power of our last result we derive an important theorem of analysis whose proof becomes particularly simple in the present probabilistic setting. (For a typical application see problem 10.)

Example. (*a*) *General approximation theorems.* With an arbitrary probability distribution G we associate the family of distributions G_h differing from it only by a scale parameter: $G_h(x) = G(x/h)$. As $h \to 0$ the distributions G_h tend to the distribution concentrated at the origin, and hence by the preceding theorem $G_h \bigstar u \to u$ *for each* $u \in C[-\infty, \infty]$, *the convergence being uniform.*[7]

[7] For a direct verification note that

$$G_h \bigstar u(t) - u(t) = \int_{-\infty}^{+\infty} [u(t-y) - u(t)]G\{dy/h\}.$$

To given $\epsilon > 0$ there exists a δ such that within each interval of length 2δ the oscillation of u is less than ϵ. The contribution of the interval $|y| \leq \delta$ to the integral is then $<\epsilon$, and the contribution of $|y| \geq \delta$ tends to 0 because G_h attributes to $|y| \geq \delta$ a mass tending to 0 as $h \to 0$.

If G has a density g, the values of $G_h \bigstar u$ are given by

$$G_h \bigstar u(t) = \int_{-\infty}^{+\infty} u(y)\, g\left(\frac{t-y}{h}\right)\frac{1}{h}\, dy. \tag{3.2}$$

When g has a bounded derivative the same is true of G_h and (3.2) may be differentiated under the integral. Taking for g the normal density we get the following ▶

Approximation lemma. *To each* $u \in C[-\infty, \infty]$ *there exists an infinitely differentiable* $v \in C[-\infty, \infty]$ *such that* $|u(x) - v(x)| < \epsilon$ *for all* x. ▶

In the present context it is desirable to replace the clumsy convolution symbol $\bigstar$ by a simpler notation emphasizing that in (3.1) the distribution F serves as an operator sending u into U. This operator will be denoted by the German letter $\mathfrak{F}$ and we agree that $U = \mathfrak{F}u$ means exactly the same as $U = F \bigstar u$. The true advantage of this apparent pedantry will become visible only when other types of operators appear in the same context. It will then be convenient to see at a glance whether a distribution plays its original probabilistic role or serves merely as an analytic operator (even though this fine distinction may lead to schizophrenia among the distributions themselves). With this explanation we introduce the

Notational convention. *With each probability distribution* F *we associate the operator* $\mathfrak{F}$ *from* $C[-\infty, \infty]$ *to itself which associates with the function* u *the transform* $\mathfrak{F}u = F \bigstar u$. *As far as possible distributions and the associated operators will be denoted by corresponding Latin and German letters.*

As usual in operator notation $\mathfrak{F}\mathfrak{G}u$ denotes the result of $\mathfrak{F}$ operating on $\mathfrak{G}u$, and so $\mathfrak{F}\mathfrak{G}$ *denotes the operator associated with the convolution* $F \bigstar G$ of two probability distributions. In particular, $\mathfrak{F}^n$ *is the operator associated with* $F^{n\bigstar}$, the n-fold convolution of F with itself.

Example. (*b*) If H_a denotes the atomic distribution concentrated at a, then $\mathfrak{H}_a$ is the *translation* operator $\mathfrak{H}_a u(x) = u(x-a)$. In particular, $\mathfrak{H}_0$ serves as the *identity* operator: $\mathfrak{H}_0 u = u$.

We now define *the norm* $\|u\|$ of the bounded function u by

$$\|u\| = \sup |u(x)| \tag{3.3}$$

With this notation the statement "u_n converges uniformly to u" simplifies to $\|u_n - u\| \to 0$. Note that the norm satisfies the easily verified *triangle inequality* $\|u + v\| \leq \|u\| + \|v\|$.

An operator T is called *bounded* if there exists a constant a such that $\|Tu\| \leq a \cdot \|u\|$. The smallest number with this property is called the *norm* of T and is denoted by $\|T\|$. With these notations the principal properties of the linear operators associated with distribution functions are:

They are positive, that, is, $u \geq 0$ implies $\mathfrak{F}u \geq 0$. They have norm 1, which implies

$$\|\mathfrak{F}u\| \leq \|u\|. \tag{3.4}$$

Finally, they commute, that is, $\mathfrak{F}\mathfrak{G} = \mathfrak{G}\mathfrak{F}$.

Definition.[8] *If* $\mathfrak{F}_n$ *and* $\mathfrak{F}$ *are operators associated with probability distributions* F_n *and* F *we write* $\mathfrak{F}_n \to \mathfrak{F}$ *iff*

$$\|\mathfrak{F}_n u - \mathfrak{F}u\| \to 0 \tag{3.5}$$

for each $u \in C[-\infty, \infty]$.

In other words, $\mathfrak{F}_n \to \mathfrak{F}$ if $F_n \bigstar u \to F \bigstar u$ uniformly. Theorem 1 may now be restated as follows.

Theorem 1a. $F_n \to F$ *properly iff* $\mathfrak{F}_n \to \mathfrak{F}$.

The next lemma is basic. It has the form of an algebraic inequality and illustrates the suggestive power of the new notation.

Lemma 1. *For operators associated with probability distributions*

$$\|\mathfrak{F}_1\mathfrak{F}_2 u - \mathfrak{G}_1\mathfrak{G}_2 u\| \leq \|\mathfrak{F}_1 u - \mathfrak{G}_1 u\| + \|\mathfrak{F}_2 u - \mathfrak{G}_2 u\|. \tag{3.6}$$

Proof. The operator on the left equals $(\mathfrak{F}_1 - \mathfrak{G}_1)\mathfrak{F}_2 + (\mathfrak{F}_2 - \mathfrak{G}_2)\mathfrak{G}_1$ and (3.6) follows from the triangle inequality and the fact that $\mathfrak{F}_2$ and $\mathfrak{G}_1$ have norms ≤ 1. Notice that this proof applies also to defective probability distributions. ►

An immediate consequence of (3.6) is

Theorem 2. *Let the sequences* $\{F_n\}$ *and* $\{G_n\}$ *of probability distributions converge properly to* F *and* G *respectively. Then*

$$F_n \bigstar G_n \to F \bigstar G. \tag{3.7}$$

(The convergence is proper by the definition of $F \bigstar G$. The theorem is false if F or G is defective. See problem 9.)

As a second application we prove that theorem 1 remains valid if the class of functions u is restricted to the particularly pleasing functions with derivatives of all orders. In this way we obtain the more flexible

---

[8] In Banach space terminology (3.5) is described as *strong convergence*. Note that it does *not* imply $\|\mathfrak{F}_n - \mathfrak{F}\| \to 0$. For example, if F_n is concentrated at $1/n$ and $\mathfrak{F}$ is the identity operator, then $\mathfrak{F}_n u(x) - \mathfrak{F}u(x) = u(x - n^{-1}) - u(x)$ and (3.5) is true but $\|\mathfrak{F}_n - \mathfrak{F}\| = 2$, because there exist functions $|v| \leq 1$ such that $v(0) = 1$ and $v(-n^{-1}) = -1$.

Criterion 3. *Let the F_n be probability distributions. If for each infinitely differentiable*[9] $v \in C[-\infty, \infty]$ *the sequence* $\{\mathfrak{F}_n v\}$ *converges uniformly, then there exists a proper probability distribution F such that $F_n \to F$.*

Proof. It was shown in example (*a*) that to given $u \in C[-\infty, \infty]$ and $\epsilon > 0$ there exists an infinitely differentiable v such that $\|u - v\| < \epsilon$. By the triangle inequality

$$(3.8) \quad \|\mathfrak{F}_n u - \mathfrak{F}_m u\| \leq \|\mathfrak{F}_n u - \mathfrak{F}_n v\| + \|\mathfrak{F}_n v - \mathfrak{F}_m v\| + \|\mathfrak{F}_m v - \mathfrak{F}_m u\|.$$

The first and last terms on the right are $<\epsilon$, and by assumption the middle term is $<\epsilon$ for all n, m sufficiently large. Thus $\{\mathfrak{F}_n u\}$ converges uniformly and $F_n \to F$ by theorem 1. ►

With $\mathbf{E}_n$ and $\mathbf{E}$ defined in (1.1) the same argument yields

Criterion 2. *Let F_n and F be proper probability distributions. If $\mathbf{E}_n(v) \to \mathbf{E}(v)$ for each infinitely differentiable v vanishing at infinity then $F_n \to F$.*

The basic inequality (3.6) extends by induction to convolutions with more than two terms; for ease of reference we record the obvious result in

Lemma 2. *Let $\mathfrak{U} = \mathfrak{F}_1 \cdots \mathfrak{F}_n$ and $\mathfrak{B} = \mathfrak{G}_1 \cdots \mathfrak{G}_n$ where the $\mathfrak{F}_j$ and $\mathfrak{G}_j$ are associated with probability distributions. Then*

$$(3.9) \qquad \|\mathfrak{U}u - \mathfrak{B}u\| \leq \sum_{j=1}^{n} \|\mathfrak{F}_j u - \mathfrak{G}_j u\|.$$

In particular

$$(3.10) \qquad \|\mathfrak{F}^n u - \mathfrak{G}^n u\| \leq n \cdot \|\mathfrak{F}u - \mathfrak{G}u\|.$$

(For applications see problems 14–15.)

4. THE CENTRAL LIMIT THEOREM

The central limit theorem establishes conditions under which sums of independent random variables are asymptotically normally distributed. Its role and meaning has been partly explained in **1**; X,1 and we have applied it on several occasions [last in example VI,11(*g*)]. It occupies a place of honor in probability theory acquired by its age and by the fruitful role which it played in the development of the theory and still plays in applications. It is therefore appropriate to use the central limit theorem as a test case to compare the scope of the various tools at our disposal. For this reason we shall give several proofs. A more systematic treatment (including necessary and sufficient conditions) will be found in chapters IX,

[9] By this is meant that all derivatives exist and belong to $C[-\infty, \infty]$.

XV, and XVI. The present discussion sidetracks us from the development of our main theme; its purpose is to illustrate the advantages of the operator terminology by a striking and significant example. Also, many readers will welcome an easy access to the central limit theorem in its simplest setting. At the cost of some repetitions we begin by a special case.

Theorem 1. (*Identical distributions in* $\mathfrak{R}^1$.) *Let* $\mathbf{X}_1, \mathbf{X}_2, \ldots$ *be mutually independent random variables with a common distribution* F. *Assume*

$$\mathbf{E}(\mathbf{X}_k) = 0, \qquad \text{Var}\,(\mathbf{X}_k) = 1. \tag{4.1}$$

As $n \to \infty$ *the distribution of the normalized sums*

$$\mathbf{S}_n^* = (\mathbf{X}_1 + \cdots + \mathbf{X}_n)/\sqrt{n} \tag{4.2}$$

tends to the normal distribution $\mathfrak{N}$ *with density* $\mathfrak{n}(x) = e^{-\frac{1}{2}x^2}/\sqrt{2\pi}$.

In purely analytical terms: for a distribution F with zero expectation and unit variance

$$F^{n\star}(x\sqrt{n}\,) \to \mathfrak{N}(x). \tag{4.3}$$

For the proof we require the following

Lemma. *If* $\mathfrak{F}_n$ *is the operator associated with* $F_n(x) = F(x\sqrt{n}\,)$ *then for each* $u \in C[-\infty, \infty]$ *with three bounded derivatives*

$$n[\mathfrak{F}_n u - u] \to \tfrac{1}{2}u'' \tag{4.4}$$

uniformly on the line.

Proof. Since $\mathbf{E}(\mathbf{X}_k^2) = 1$ we can define a proper probability distribution $F_n^{\#}$ by

$$F_n^{\#}\{dy\} = ny^2\, F_n\{dy\} = ny^2\, F\{\sqrt{n}\, dy\}. \tag{4.5}$$

The change of variables $\sqrt{n}\, y = s$ shows that $F_n^{\#}$ tends to the distribution concentrated at the origin. In view of (4.1) we have for the difference of the two sides in (4.4)

$$\begin{aligned} &n[\mathfrak{F}_n u(x) - u(x)] - \tfrac{1}{2}u''(x) = \\ &\int_{-\infty}^{+\infty} \left[\frac{u(x-y) - u(x) + yu'(x)}{y^2} - \tfrac{1}{2}u''(x)\right] F_n^{\#}\{dy\}. \end{aligned} \tag{4.6}$$

The Taylor development of the numerator shows that for $|y| < \epsilon$ the integrand is dominated by $\frac{1}{6}|y| \cdot \|u'''\| < \epsilon \cdot \|u'''\|$, and for all y by $\|u''\|$. Since $F_n^{\#}$ tends to concentrate near the origin it follows that for n sufficiently large the quantity is in absolute value less than $\epsilon(\|u''\| + \|u'''\|)$, and so the left side tends uniformly to zero. ►

Proof *of theorem 1.* Denote by $\mathfrak{G}$ and $\mathfrak{G}_n$, respectively, the operators associated with the normal distributions $\mathfrak{N}(x)$ and $\mathfrak{N}(x\sqrt{n})$. Then by the basic inequality (3.10)

$$\|\mathfrak{F}_n^n u - \mathfrak{G}u\| = \|\mathfrak{F}_n^n u - \mathfrak{G}_n^n u\| \leq n\,\|\mathfrak{F}_n u - \mathfrak{G}_n u\| \tag{4.7}$$
$$\leq n\,\|\mathfrak{F}_n u - u\| + n\,\|\mathfrak{G}_n u - u\|.$$

By the preceding lemma the right side tends to zero and hence $\mathfrak{F}_n^n \to \mathfrak{G}$ by criterion 1 of section 3. ▶

Example. (*a*) *Central limit theorem with infinite variances.* It is of methodological interest to note that the proof of theorem 1 applies without change to certain distributions without variance, provided appropriate norming constants are chosen. For example, if the $\mathbf{X}_k$ have a density such that $f(x) = 2\,|x|^{-3}\log|x|$ for $|x| \geq 1$ and $f(x) = 0$ for $|x| \leq 1$, then $(\mathbf{X}_1+\cdots+\mathbf{X}_n)/(\sqrt{2n}\log n)$ has a normal limit distribution. (The proof requires only obvious changes.) Necessary and sufficient conditions for a normal limit are given in IX,7 and XVII,5. ▶

The method of proof is of wide applicability. Problem 16 may serve as a good exercise. Here we use the method to prove the central limit theorem in more general settings. The following theorem refers formally to two dimensions but is valid in $\mathfrak{R}^r$.

Theorem 2. (*Multivariate case*). *Let* $\{\mathbf{X}_n\}$ *stand for a sequence of mutually independent two-dimensional random variables with a common distribution* F. *Suppose that the expectations are zero and that the covariance matrix is given by*

$$C = \begin{pmatrix} \sigma_1^2 & \rho\sigma_1\sigma_2 \\ \rho\sigma_1\sigma_2 & \sigma_2^2 \end{pmatrix}. \tag{4.8}$$

As $n \to \infty$ *the distribution of* $(\mathbf{X}_1+\cdots+\mathbf{X}_n)/\sqrt{n}$ *tends to the bivariate normal distribution with zero expectation and covariance matrix* C.

Proof. The proof requires no essential change if the matrix notation of III,5 is used. Since subscripts are already overtaxed we denote the points of the plane by row vectors $x = (x^{(1)}, x^{(2)})$. Then $u(x)$ denotes a function of the two variables and we denote its partial derivatives by subscripts. Thus $u' = (u_1, u_2)$ is a row vector, and $u'' = (u_{jk})$ is a symmetric two by two matrix. With this notation the Taylor expansion takes on the form

$$u(x-y) = u(x) - yu'(x) + \tfrac{1}{2}yu''(x)y^T + \cdots \tag{4.9}$$

where y^T is the transpose of y, namely the column vector with components $y^{(1)}, y^{(2)}$. In analogy with (4.5) we define a proper probability distribution by

$$F_n^{\#}\{dy\} = nq(y)\,F\{\sqrt{n}\,dy\} \quad \text{where} \quad 2q(y) = \frac{y_1^2}{\sigma_1^2} + \frac{y_2^2}{\sigma_2^2}.$$

As in the last proof $F_n^{\#}$ tends to the probability distribution concentrated at the origin. To (4.6) there corresponds the identity

$$\begin{aligned} &n[\mathfrak{F}_n u(x) - u(x)] - \tfrac{1}{2}m(x) = \\ (4.10) \qquad &= \int_{\mathfrak{R}^2} \frac{u(x-y) - u(x) + yu'(x) - \tfrac{1}{2}yu''(x)y^T}{q(y)} \cdot F_n^{\#}\{dy\} \end{aligned}$$

where[10]

$$(4.11) \qquad m(x) = \mathbf{E}(yu''(x)y^T) = u_{11}(x)\sigma_1^2 + 2u_{12}(x)\rho\sigma_1\rho_2 + u_{22}(x)\sigma_2^2.$$

(Here $\mathbf{E}$ denotes expectations with respect to F.) In view of (4.9) the integrand tends to zero and as in the preceding lemma it is seen that $n[\mathfrak{F}_n u - u] \to m$ uniformly, and the proof of the theorem requires no change. ►

Example. (*b*) *Random walks in d dimensions.* Let $\mathbf{X}_1, \mathbf{X}_2, \ldots$ be independent random vectors with a common distribution that may be described as follows. The $\mathbf{X}_k$ have a random direction in the sense introduced in I,10, and the length $\mathbf{L}$ is a random variable with $\mathbf{E}(\mathbf{L}^2) = 1$. For reasons of symmetry the covariance matrix C is the diagonal matrix with elements $\sigma_j^2 = 1/d$. The distribution of the normalized sum $\mathbf{S}_n/\sqrt{n}$ tends to the normal distribution with covariance matrix C. The distribution of the *squared length of the vector* $\mathbf{S}_n/\sqrt{n}$ therefore tends to the distribution of the sum of squares of independent normal variables. It was shown in II,3 that this limit has the density

$$(4.12) \qquad w_d(r) = \frac{d^{\frac{1}{2}d}}{2^{\frac{1}{2}d-1}\Gamma(\frac{1}{2}d)}\, e^{-\frac{1}{2}dr^2}\, r^{d-1}.$$

This result shows the influence of the number of dimensions and applies, in particular, to the random flight example I,10(*e*).

(*c*) *Random dispersal of populations.* As an empirical application of the foregoing example consider the spread of a population of oak trees in prehistoric times. If new plants were due only to seeds dropped by mature trees, then seedlings would be located near mature trees and the distance of an nth generation tree from its progenitor would be approximately normally

---

[10] Obviously $m(x)$ is the trace (sum of the diagonal elements) of the product Cu''. This is true in all dimensions. [In one dimension $m(x) = \frac{1}{2}\sigma^2 u''(x)$.]

distributed. Under these conditions the area covered by the descendants of a tree would be roughly proportional to the age of the tree. Observations show that the actual development is inconsistent with this hypothesis. Biologists conclude that the actual dispersal was strongly influenced by birds carrying the seeds long distances.[11] ▶

We turn to a generalization of theorem 1 to variable distributions The conditions give the impression that they are introduced artifically with the sole purpose of making the same proof work. Actually it turns out that the conditions are also necessary for the validity of the central limit theorem with the classical norming used in (4.17). (See XV,6.)

Theorem 3. (*Lindeberg*).[12] *Let* $\mathbf{X}_1, \mathbf{X}_2, \ldots$ *be mutually independent one-dimensional random variables with distributions* $F_1, F_2, \ldots$ *such that*

$$\mathbf{E}(\mathbf{X}_k) = 0, \qquad \operatorname{Var}(\mathbf{X}_k) = \sigma_k^2, \tag{4.13}$$

and put

$$s_n^2 = \sigma_1^2 + \cdots + \sigma_n^2. \tag{4.14}$$

Assume that for each $t > 0$

$$s_n^{-2} \sum_{k=1}^{n} \int_{|y| \geq ts_n} y^2 \, F_k\{dy\} \to 0 \tag{4.15}$$

or, what amounts to the same, that

$$s_n^{-2} \sum_{k=1}^{n} \int_{|y| < ts_n} y^2 \, F_k\{dy\} \to 1. \tag{4.16}$$

Then the distribution of the normalized sum

$$\mathbf{S}_n^* = (\mathbf{X}_1 + \cdots + \mathbf{X}_n)/s_n \tag{4.17}$$

tends to the normal distribution $\mathfrak{N}$ *with zero expectation and unit variance.*

The Lindeberg condition (4.15) guarantees that the individual variances σ_k^2 are small as compared to their sum s_n^2 in the sense that for given $\epsilon > 0$

---

[11] J. G. Skellam, Biometrika, vol. 38 (1951) pp. 196–218.

[12] J. W. Lindeberg, Math. Zeit., vol. 15 (1922) pp. 211–235. Special cases and variants had been known before, but Lindeberg gave the first general form containing theorem 1. The necessity of Lindeberg's condition with the classical norming was proved by Feller, *Ibid.*, vol. 40 (1935). (See XV,6.)

Lindeberg's method appeared intricate and was in practice replaced by the method of characteristic functions developed by P. Lévy. That streamlined modern techniques permit presenting Lindeberg's method in a simple and intuitive manner was shown by H. F. Trotter, Archiv. der Mathematik, vol. 9 (1959) pp. 226–234. Proofs of this section utilize Trotter's idea.

and all n sufficiently large

$$\sigma_k < \epsilon s_n \qquad k = 1, \ldots, n. \tag{4.18}$$

In fact, obviously σ_k^2/s_k^2 is less than t^2 plus the left side in (4.15), and taking $t = \frac{1}{2}\epsilon$ we see that (4.15) implies (4.18).

Theorem 3 generalizes to higher dimensions in the way indicated by theorem 2. See also problems 17–20.

Proof. To each distribution F_k we make correspond a normal distribution G_k with zero expectation and the same variance σ_k^2. The distribution $F_k(xs_n)$ of $\mathbf{X}_k/s_n$ now depends on both k and n, and we denote the associated operator by $\mathfrak{F}_{k,n}$. Similarly $\mathfrak{G}_{k,n}$ is associated with the normal distribution $G_k(xs_n)$. By (3.9) it suffices to prove that

$$\sum_{k=1}^{n} \| \mathfrak{F}_{k,n} u - \mathfrak{G}_{k,n} u \| \to 0 \tag{4.19}$$

for every $u \in C[-\infty, \infty]$ with three bounded derivatives. We proceed as in theorem 1, but (4.6) is now replaced by the n relations

$$\begin{aligned} &\mathfrak{F}_{k,n} u(x) - u(x) - \frac{\sigma_k^2}{2s_n^2} u''(x) = \\ &= \int_{-\infty}^{+\infty} \left[\frac{u(x-y) - u(x) + yu'(x)}{y^2} - \frac{1}{2} u''(x) \right] \cdot y^2 F_k\{s_n\, dy\}. \end{aligned} \tag{4.20}$$

Splitting the interval of integration into $|y| \le \epsilon$ and $|y| > \epsilon$ and using the same estimates as in (4.6) we obtain

$$\left\| \mathfrak{F}_{k,n} u - u - \frac{\sigma_k^2}{2s_n^2} u'' \right\| \le \epsilon \, \|u'''\| \, \frac{\sigma_k^2}{s_n^2} + \|u''\| \cdot \int_{|y|<\epsilon} y^2 \, F_k\{s_n\, dy\}. \tag{4.21}$$

The Lindeberg condition (4.15) with $t = \epsilon$ now guarantees that for n sufficiently large

$$\sum_{k=1}^{n} \left\| \mathfrak{F}_{k,n} u - u - \frac{\sigma_k^2}{2s_n^2} u'' \right\| \le \epsilon(\|u'''\| + \|u''\|). \tag{4.22}$$

For our normal distributions G_k the Lindeberg condition (4.15) is satisfied as a simple consequence of (4.18), and therefore the inequality (4.22) remains valid with $\mathfrak{F}_{k,n}$ replaced by $\mathfrak{G}_{k,n}$. Adding these two inequalities we obtain (4.19), and this concludes the proof. ▶

Examples. (*d*) *Uniform distributions.* Let $\mathbf{X}_k$ be uniformly distributed [with density $1/(2a_k)$] between $-a_k$ and a_k. Then $\sigma_k^2 = \frac{1}{3}a_k^2$. It is easily seen that the conditions of the theorem are satisfied if the a_k remain bounded and $a_1^2 + \cdots + a_n^2 \to \infty$; indeed, in this case the sum (4.15) vanishes identically for all n sufficiently large. On the other hand, if $\sum a_k^2 < \infty$ then s_n remains bounded and (4.15) cannot hold: in this case the central limit theorem does *not* apply. (Instead we get an example of an infinite convolution to be studied in section 5.)

A less obvious case where the central limit theorem does *not* hold is $\sigma_k^2 = 2^k$. Then $3s_n^2 = 2^{n+1} - 2 < 2a_n^2$ and obviously the left side of (4.15) is $> \frac{1}{2}$ if, say, $t < \frac{1}{100}$. These examples show that (4.15) serves to insure that the individual $\mathbf{X}_k$ will be asymptotically negligible: the probability that *any* term $\mathbf{X}_k$ will be of the same order of magnitude as the sum $\mathbf{S}_n$ must tend to zero.

(*e*) *Bounded variables.* Assume that the $\mathbf{X}_k$ are uniformly bounded, that is, that all the distributions F_k are carried by some finite interval $\overline{-a, a}$. The Lindeberg condition (4.15) is then satisfied iff $s_n \to \infty$.

(*f*) Let F be a probability distribution with zero expectation and unit variance. Choose a sequence of positive numbers σ_k and put $F_n(x) = F(x/\sigma_n)$ (so that F_k has variance σ_k^2). *The Lindeberg condition is satisfied iff* $s_n \to \infty$ *and* $\sigma_n/s_n \to 0$. Indeed, we know that these conditions are necessary. On the other hand, the left side in (4.15) reduces to

$$s_n^{-2}\sum_{k=1}^{n}\sigma_k^2\int_{|x|<ts_n/\sigma_k} x^2\, F\{dx\}.$$

Under the stated conditions s_n/σ_k tends to ∞ uniformly in $k = 1, \ldots, n$, and so for n sufficiently large all the integrals appearing in the sum will be $<\epsilon$. This means that the sum is $<\epsilon s_n^2$, and so (4.15) is true. ►

It is of methodological interest to observe that the same method of proof works even for certain sequences of random variables *without expectations*, but the norming factors are, of course, different. We shall return to this problem in XV,6 where we shall also further analyze the nature of the Lindeberg condition. (See problems 19–20.)

We conclude this excursion by a version of the central limit theorem for *random sums*. The idea is as follows. If in theorem 1 we replace the fixed number n of terms by a Poisson variable $\mathbf{N}$ with expectation n it is plausible that the distribution of $\mathbf{S_N}$ will still tend to $\mathfrak{N}$. Similar situations arise in statistics and physics when the number of observations is not fixed in advance.

We consider only sums of the form $\mathbf{S_N} = \mathbf{X}_1 + \cdots + \mathbf{X_N}$ where the $\mathbf{X}_j$ and $\mathbf{N}$ are mutually independent random variables. We suppose that the $\mathbf{X}_j$ have a common distribution F with zero expectation and variance 1. Using the notation of section 2 we have

Theorem 4.[13] (*Random sums.*) *Let* $\mathbf{N}_1, \mathbf{N}_2, \ldots$ *be positive integral-valued random variables such that*

$$n^{-1}N_n \xrightarrow{p} 1. \tag{4.23}$$

Then the distribution of $\mathbf{S}_{\mathbf{N}_n}/\sqrt{n}$ *tends to* $\mathfrak{N}$.

The interesting feature is that $\mathbf{S}_{\mathbf{N}_n}/\sqrt{n}$ is *not* normalized to unit variance. In fact, the theorem applies to cases with $\mathbf{E}(\mathbf{N}_n) = \infty$ and even when expectations exist, (4.23) does *not* imply that $n^{-1}\mathbf{E}(\mathbf{N}_n) \to 1$. Normalization to unit variance may be impossible, and when possible it complicates the proof.

Proof. To avoid double subscripts we write $\mathbf{P}\{\mathbf{N}_n = k\} = a_k$ with the understanding that the a_k depend on n. The operator associated with $\mathbf{S}_{\mathbf{N}_n}$ is given by the formal power series $\sum a_k \mathfrak{F}^k$. As in the proof of theorem 1 let $\mathfrak{F}_n$ be the operator associated with $F(x\sqrt{n})$. Since $F_n^{n\star} \to \mathfrak{N}$ it suffices to prove that

$$\sum_{k=1}^{\infty} a_k \mathfrak{F}_n^k\, u - \mathfrak{F}_n^n\, u \to 0 \tag{4.24}$$

uniformly for each $u \in C[-\infty, \infty]$ with three bounded derivatives.

Using the obvious factoring and the basic inequality (3.9) it is seen that

$$\|\mathfrak{F}_n^k u - \mathfrak{F}_n^n u\| \leq \|\mathfrak{F}_n^{|k-n|} u - u\| \leq |k-n| \cdot \|\mathfrak{F}_n u - u\|. \tag{4.25}$$

Because of (4.23) the coefficients a_k with $|k - n| > \epsilon n$ add to less than ϵ provided n is sufficiently large. For such n the norm of the left side in (4.24) is

$$\leq \sum_{k=1}^{\infty} a_k \|\mathfrak{F}_n^k u - \mathfrak{F}_n^n u\| \leq 2\epsilon \cdot \|u\| + 2\epsilon \cdot n\, \|\mathfrak{F}_n u - u\|. \tag{4.26}$$

We saw in the proof of the lemma the right side is $\leq 2\epsilon\, \|u\| + 3\epsilon\, \|u''\|$ for all n sufficiently large, and so (4.24) holds uniformly. ►

*5. INFINITE CONVOLUTIONS

The following theorem is given for its intrinsic interest and because it is a good example for the working of our criteria. Stronger versions are found in VII,8, IX,9, and XVII,10.

---

[13] For generalizations to mutually dependent $\mathbf{X}_j$ see P. Billingsley, *Limit theorems for randomly selected partial sums*, Ann. Math. Statist., vol. 33 (1963) pp. 85–92. When (4.23) is dropped one gets limit theorems of a novel form. See H. E. Robbins, *The asymptotic distribution of the sum of a random number of random variables*, Bull. Amer. Math. Soc. vol. 54 (1948) pp. 1151–1161.

For generalizations of the central limit theorem to other types of dependent variables the reader is referred to the book by Loève. (For exchangeable variables, see problem 21.)

* This section is not used in the sequel.

We denote by $\mathbf{X}_1, \mathbf{X}_2, \ldots$ mutually independent variables with distributions $F_1, F_2, \ldots$. It is assumed that $\mathbf{E}(\mathbf{X}_j) = 0$ and $\sigma_k^2 = \mathbf{E}(\mathbf{X}_k^2)$ exist.

Theorem. *If* $\sigma^2 = \sum \sigma_k^2 < \infty$ *the distributions*[14] G_n *of the partial sums* $\mathbf{X}_1 + \cdots + \mathbf{X}_n$ *tend to a probability distribution* G *with zero expectation and variance* σ^2.

Proof. To establish the existence of a proper limit G it suffices (theorem 1 of section 3) to show that for infinitely differentiable $u \in C[-\infty, \infty]$ the sequence of functions $\mathfrak{F}_1\mathfrak{F}_2 \cdots \mathfrak{F}_n u$ converges uniformly as $n \to \infty$. Now for $n > m$ by the obvious factorization

$$\|\mathfrak{F}_1 \cdots \mathfrak{F}_n u - \mathfrak{F}_1 \cdots \mathfrak{F}_m u\| \le \|\mathfrak{F}_{m+1} \cdots \mathfrak{F}_n u - u\|. \tag{5.1}$$

Since $\mathbf{E}(\mathbf{X}_k) = 0$ we have the identity

$$\mathfrak{F}_k u(x) - u(x) = \int_{-\infty}^{+\infty} [u(x-y) - u(x) + yu'(x)]\, F_k\{dy\}. \tag{5.2}$$

By the second-order Taylor expansion the integrand is in absolute value $\le \|u''\| \cdot y^2$ and therefore $\|\mathfrak{F}_k u - u\| \le \sigma_k^2 \cdot \|u''\|$. By the basic inequality (3.9) the quantity (5.1) is therefore $\le (\sigma_{m+1}^2 + \cdots + \sigma_n^2) \cdot \|u''\|$ and thus there exists a proper distribution G such that $G_n \to G$. Since G_n has variance $\sigma_1^2 + \cdots + \sigma_n^2$ the second moment of G exists and is $\le \sigma^2$. By the criterion of example 1(*e*) this implies that G has zero expectation. Finally, G is the convolution of G_n and the limit distribution of $\mathbf{X}_{n+1} + \cdots + \mathbf{X}_{n+k}$, and hence the variance of G cannot be smaller than that of G_n. This concludes the proof. ▶

Examples. (*a*) In example I,11(*c*) a random choice of a point between 0 and 1 is effected by a succession of coin tossings. In the present terminology this means representing the uniform distribution as an infinite convolution. Example I,11(*d*) shows that the infinite convolution of the corresponding even-numbered terms is a singular distribution. (See XVII,10.)

(*b*) Let the $\mathbf{Y}_k$ be independent with $\mathbf{E}(\mathbf{Y}_k) = 0$ and $\mathbf{E}(\mathbf{Y}_k^2) = 1$. Then the distributions of the partial sums of $\sum b_k \mathbf{Y}_k$ converge if $\sum b_k^2 < \infty$. This fact was exploited in the discussion of normal stochastic processes in III,7.

(*c*) *Application to birth processes.* Let $\mathbf{X}_n$ be a positive variable with density $\lambda_n e^{-\lambda_n t}$. Then $\mathbf{E}(\mathbf{X}_n) = \sqrt{\text{Var}(\mathbf{X}_n)} = \lambda_n^{-1}$ and in case $m = \sum \lambda_n^{-1} < \infty$ our theorem applies to the centered variables $\mathbf{X}_n - \lambda_n^{-1}$. This observation leads to a probabilistic interpretation of the divergent *pure birth process* described in **1**; XVII,3–4. A "particle" moves by successive

[14] It was shown in VII,8 that *the random variables* $\mathbf{S}_n$ *themselves converge to a limit.*

jumps, the sojourn times $\mathbf{X}_1, \mathbf{X}_2, \ldots$ being independent exponentially distributed variables. Here $\mathbf{S}_n = \mathbf{X}_1 + \cdots + \mathbf{X}_n$ represents the epoch of the nth jump. If $\lim \mathbf{E}(\mathbf{S}_n) = m < \infty$, the distribution of $\mathbf{S}_n$ tends to a proper limit G. Then $G(t)$ is the probability that infinitely many jumps will occur before epoch t.

(*d*) For applications to shot noise, trunking problems, etc., see problem 22. ▶

6. SELECTION THEOREMS

A standard method of proving the convergence of a numerical sequence consists in proving first the existence of at least one point of accumulation and then its uniqueness. A similar procedure is applicable to distributions, the analogue to a point of accumulation being provided by the following important theorem usually ascribed to Helly. As for all theorems of this section, it is independent of the number of dimensions. (A special case was used in **1**; XI,6.)

Theorem 1. (i) *Every sequence* $\{F_k\}$ *of probability distributions in* $\mathcal{R}^r$ *possesses a subsequence* $F_{n_1}, F_{n_2}, \ldots$ *that converges (properly or improperly) to a limit* F.

(ii) *In order that all such limits be proper it is necessary and sufficient that* $\{F_n\}$ *be stochastically bounded. (See definition* 2 *of section* 2.)

(iii) *In order that* $F_n \to F$ *it is necessary and sufficient that the limit of every convergent subsequence equals* F.

The proof is based on the following

Lemma. *Let* $a_1, a_2, \ldots$ *be an arbitrary sequence of points. Every sequence* $\{u_n\}$ *of numerical functions contains a subsequence* $u_{n_1}, u_{n_2}, \ldots$ *that converges at all points* a_j *(possibly to* $\pm\infty$).

Proof. We use G. Cantor's "diagonal method." It is possible to find a sequence $\nu_1, \nu_2, \ldots$ such that the sequence of values $u_{\nu_k}(a_1)$ converges. To avoid multiple indices we put $u_k^{(1)} = u_{\nu_k}$ so that $\{u_k^{(1)}\}$ is a subsequence of $\{u_n\}$ and converges at the particular point a_1. Out of this subsequence we extract a further subsequence $u_1^{(2)}, u_2^{(2)}, \ldots$ that converges at the point a_2. Proceeding by induction we construct for each n a sequence $u_1^{(n)}, u_2^{(n)}, \ldots$ converging at a_n and contained in the preceding sequence. Consider now the diagonal sequence $u_1^{(1)}, u_2^{(2)}, u_3^{(3)}, \ldots$. Except for its first $n-1$ terms this sequence is contained in the nth sequence $u_1^{(n)}, u_2^{(n)}, \ldots$ and hence it converges at a_n. This being true for each n, the diagonal sequence $\{u_n^{(n)}\}$ converges at all points $a_1, a_2, \ldots$ and the lemma is proved. ▶

Proof *of theorem* 1. (i) Choose for $\{a_j\}$ a sequence that is everywhere dense, and choose a subsequence $\{F_{n_k}\}$ that converges at each point a_j.

Denote the limits by $G(a_j)$. For any point x not belonging to the set $\{a_j\}$ we define $G(x)$ as the greatest lower bound of all $G(a_j)$ with $a_j > x$. The function G thus defined increases from 0 to 1, but it need not be right-continuous: we can only assert that $G(x)$ lies between the limits $G(x+)$ and $G(x-)$. However, it is possible to redefine G at the points of discontinuity so as to obtain a right-continuous function F which agrees with G at all points of continuity. Let x be such a point. There exist two points $a_i < x < a_j$ such that

$$(6.1) \qquad G(a_j) - G(a_i) < \epsilon \qquad G(a_i) \leq F(x) \leq G(a_j).$$

The F_n being monotone we have $F_{n_k}(a_i) \leq F_{n_k}(x) \leq F_{n_k}(a_j)$. Letting $k \to \infty$ we see from (6.1) that no limit point of the sequence $\{F_{n_k}(x)\}$ can differ from $F(x)$ by more than ϵ, and so $F_{n_k}(x) \to F(x)$ at all points of continuity.

(ii) Next we recall that a convergent sequence of distributions converges properly iff it is stochastically bounded. Given (i) the remaining assertions are therefore almost tautological. ▶

The selection theorem is extremely important. The following famous theorem in number theory may give an idea of its amazing power and may also serve as a reminder that our probabilistic terminology must not be allowed to obscure the much wider scope of the theory developed.

Examples. (*a*) *An equidistribution theorem in number theory.*[15] *Let* α *be an irrational number and* α_n *the fractional part of* $n\alpha$. *Denote by* $N_n(x)$ *the number of terms among* $\alpha_1, \alpha_2, \ldots, \alpha_n$ *that are* $\leq x$. *Then* $n^{-1}N_n(x) \to x$ *for all* $0 < x < 1$.

Proof. We consider distributions and functions on the *circle* of unit length; in other words, additions of coordinates are reduced modulo 1. (The idea was explained in II,8. The convenient tool of distribution functions becomes meaningless on the circle, but distributions in the sense of measures are meaningful.) Let F_n be the atomic distribution concentrated on the n points $\alpha, 2\alpha, \ldots, n\alpha$ and assigning probability $1/n$ to each. By the selection theorem there exists a sequence $n_1, n_2, \ldots$ such that $F_{n_k} \to F$, where F is a *proper* probability distribution (the circle being bounded). Taking convolutions with an arbitrary continuous function u we get

$$(6.2) \qquad \frac{1}{n_k}\,[u(x-\alpha) + u(x-2\alpha) + \cdots + u(x-n_k\alpha)] \to v(x).$$

[15] Usually attributed to H. Weyl although discovered independently by Bohl and by Sierpiński. See G. H. Hardy and E. M. Wright, *Theory of numbers*, Oxford, 1945, pp. 378–381, to appreciate the difficulties of the proof when the theorem is considered in a non-probabilistic setting.

Now it is obvious that replacing x by $x-\alpha$ does not affect the asymptotic behavior of the left side, and hence $v(x) = v(x-\alpha)$ for all x. This in turn implies $v(x) = v(x-k\alpha)$ for $k = 1, 2, \ldots$. By the corollary to lemma 2 in V,4 the points $\alpha, 2\alpha, \ldots$ lie everywhere dense, and hence $v = \text{const.}$ We have thus shown that for each continuous u the convolution $F \bigstar u$ is a constant. It follows that F must attribute the same value to intervals of equal length, and so $F\{I\}$ equals the length of the interval I. The impossibility of other limits proves that the whole sequence $\{F_n\}$ converges to this distribution, and this proves the theorem. We call F *the uniform distribution* on the circle.

(*b*) *Convergence of moments.* Let F_n and F be probability distributions with finite moments of all orders, which we denote by $\mu_k^{(n)}$ and μ_k, respectively. We know from VII,3 that different distribution functions can have the same moment sequence and it is therefore not always possible from the behavior of $\mu_k^{(n)}$ to conclude that $F_n \to F$. However, *if* F *is the only distribution with the moments* $\mu_1, \mu_2, \ldots$ *and if* $\mu_k^{(n)} \to \mu_k$ *for* $k = 1, 2, \ldots$ *then* $F_n \to F$. In fact, the result of example 1(*e*) shows that every convergent subsequence of $\{F_n\}$ converges to F.

(*c*) *Separability.* For brevity call a distribution rational if it is concentrated at finitely many rational points and attributes a rational weight to each. An arbitrary distribution F is the limit of a sequence $\{F_n\}$ of rational distributions, and we may choose F_n with zero expectation since this can be achieved by the addition of an atom and adjustment of the weights by arbitrarily small amounts. But there are only denumerably many rational distributions and they may be ordered into a simple sequence $G_1, G_2, \ldots$. Thus *there exists a sequence* $\{G_n\}$ *of distributions with zero expectations and finite variances such that every distribution* F *is the limit of some subsequence* $\{G_{n_k}\}$. ►

Theorem 1 was formulated in the form most useful for probability but is unnecessarily restrictive. The proof depended on the fact that a sequence $\{F_n\}$ of monotone functions with $F_n(-\infty) = 0$, $F_n(\infty) = 1$ contains a convergent subsequence. Now this remains true also when the condition $F_n(\infty) = 1$ is replaced by the less stringent requirement that the numerical sequence $\{F_n(x)\}$ be bounded for each fixed x. The limit F will then be finite but possibly unbounded; the induced measure will be finite on intervals $\overline{-\infty, x}|$, but possibly infinite for $\overline{-\infty, \infty}$. A similar relaxation is possible for $-\infty$ and we are led to the following generalization of theorem 1, in which the symbol $\mu_n \to \mu$ is used in the obvious sense that the relation holds in finite intervals.

Theorem 2. *Let* $\{\mu_n\}$ *be a sequence of measures such that the numerical sequence* $\mu_n\{\overline{-x, x}\}$ *is bounded for each* x. *There exists a measure* μ *and a sequence* $n_1, n_2, \ldots$ *such that* $\mu_{n_k} \to \mu$.

Variants of the selection theorem hold for many classes of functions Particularly useful is the following theorem, usually called after either Ascoli or Arzelà.

Theorem 3. *Let* $\{u_n\}$ *be an equicontinuous*[16] *sequence of functions* $|u_n| \leq 1$. *There exists a subsequence* $\{u_{n_k}\}$ *converging to a continuous limit* u. *The convergence is uniform in every finite interval.*

Proof. Choose again a dense sequence of points a_j and a subsequence $\{u_{n_k}\}$ converging at each a_j; denote the limit by $u(a_j)$. Then

$$(6.3)\quad |u_{n_r}(x)-n_{n_s}(x)| \leq |u_{n_r}(x)-u_{n_r}(a_j)| + |u_{n_s}(x)-u_{n_s}(a_j)| + |u_{n_r}(a_j)-u_{n_s}(a_j)|.$$

By assumption the last term tends to 0. Because of the assumed equicontinuity there exists for each x a point a_j such that

$$(6.4)\quad |u_n(x) - u_n(a_j)| < \epsilon$$

for all n, and finitely many such a_j suffice for any finite interval I. It follows that the right side in (6.3) will be $<3\epsilon$ for all r and s sufficiently large uniformly in I. Thus $u(x) = \lim u_{n_r}(x)$ exists, and because of (6.4) we have $|u(x) - u(a_j)| \leq \epsilon$ which implies the continuity of u. ▶

*7. ERGODIC THEOREMS FOR MARKOV CHAINS

Let K be a stochastic kernel concentrated on a finite or infinite interval Ω. (By definition 1 of VI,11 this means: K is a function of two variables, a point x and a set Γ, which for fixed Γ reduces to a Baire function of x and for fixed $x \in \Omega$ a probability distribution concentrated on Ω.) In higher dimensions the interval Ω may be replaced by more general regions and the theory requires no change.

It was shown in VI,11 that there exist Markov chains $(\mathbf{X}_0, \mathbf{X}_1, \ldots)$ with transition probabilities K. The distribution γ_0 of the initial variable $\mathbf{X}_0$ may be chosen arbitrarily and the distributions of $\mathbf{X}_1, \mathbf{X}_2, \ldots$ are then

[16] That is, to each $\epsilon > 0$ there corresponds a $\delta > 0$ such that $|x' - x''| < \delta$ implies $|u_n(x') - u_n(x'')| < \epsilon$ for all n.

* This material is treated because of its importance and as a striking example for the use of the selection theorems. It is not used explicitly in the sequel.

given recursively by

$$\gamma_n(\Gamma) = \int_\Omega \gamma_{n-1}(dx)\, K(x, \Gamma). \tag{7.1}$$

In particular, if γ_0 is concentrated at a point x_0 then $\gamma_n(\Gamma) = K^{(n)}(x_0, \Gamma)$ coincides with the transition probability from x_0 to Γ.

Definition 1. *A measure* α *is strictly positive in* Ω *if* $\alpha\{I\} > 0$ *for each open interval* $I \subset \Omega$. *The kernel* K *is strictly positive if* $K(x, I) > 0$ *for each* x *and each open interval in* Ω.

Definition 2. *The kernel is ergodic if there exists a strictly positive probability distribution* α *such that* $\gamma_n \to \alpha$ *independently of the initial probability distribution* γ_0.

This amounts to saying that

$$K^{(n)}(x, I) \to \alpha(I) > 0 \tag{7.2}$$

for each interval of continuity for α. The definition is the same as in the discrete case (**1**; XV); its meaning has been discussed and clarified by examples in VI,11.

The most general stochastic kernels are subject to various pathologies, and we wish to restrict the theory to kernels depending in a continuous manner on x. The simplest way of expressing this is by considering the transformations on continuous functions induced by K. Given a function u which is bounded and continuous in the underlying interval Ω we define $u_0 = u$ and, by induction,

$$u_n(x) = \int_\Omega K(x, dy)\, u_{n-1}(y). \tag{7.3}$$

This transformation on functions is dual to the transformation (7.1) on measures. Note that in both cases throughout this section indices serve to indicate the effect of a transformation induced by K.

The regularity property that we wish to impose on K is, roughly speaking, that u_1 should not be worse than u_0. The following definition expresses exactly our needs but looks formal. The examples will show that it is trivially satisfied in typical situations.

Definition 3. *The kernel* K *is regular if the family of transforms* u_k *is equicontinuous*[17] *whenever* u_0 *is uniformly continuous in* Ω.

---

[17] See footnote[16]. Our "regularity" in analogous to "complete continuity" as used in Hilbert space theory.

Examples. (*a*) *Convolutions.* If F is a probability distribution the convolutions

$$u_n(x) = \int_{\infty}^{+\infty} u_{n-1}(x-y)\, F\{dy\}$$

represent a special case of the transformation (7.3). (With self-explanatory notations in this case $K(x, I) = F\{I-x\}$. This transformation is regular because, u_0 being uniformly continuous, there exists a δ such that $|x' - x''| < \delta$ implies $|u_0(x') - u_0(x'')| < \epsilon$ and by induction this entails $|u_n(x') - u_n(x'')| < \epsilon$ for all n.

(*b*) Let Ω be the unit interval and let K be defined by a density k which is continuous in the closed unit square. Then K is regular. Indeed

$$(7.4) \qquad |u_n(x') - u_n(x'')| \le \int_0^1 |k(x', y) - k(x'', y)| \cdot |u_{n-1}(y)|\, dy.$$

By induction it is seen that if $|u_0| < M$ also $|u_n| < M$ for all n. Because of the uniform continuity of k there exists a δ such that

$$|k(x', y) - k(x'', y)| < \epsilon/M \quad \text{whenever} \quad |x' - x''| < \delta,$$

and then $|u_n(x') - u_n(x'')| < \epsilon$ independently of n. ►

The condition of strict positivity in the following theorems is unnecessarily restrictive. Its main function is to eliminate the nuisance of decomposable and periodic chains with which we had to cope in **1**; XV.

Theorem 1. *Every strictly positive regular kernel K on a bounded closed interval Ω is ergodic.*

This theorem fails when Ω is unbounded, for the limit in (7.2) can be identically zero. A universal criterion may be formulated in terms of stationary measures. We recall that a measure α is called *stationary* for K if $\alpha_1 = \alpha_2 = \cdots = \alpha$, that is, if all its transforms (7.1) are identical.

Theorem 2. *A strictly positive regular kernel K is ergodic iff it possesses a strictly positive stationary probability distribution α.*

Proof *of theorem* 1. Let v_0 be a continuous function and v_1 its transform (7.3). The proof depends on the obvious fact that for a strictly positive kernel K the maximum of the transform v_1 is strictly less than the maximum of v_0 except if v_0 is a constant.

Consider now the sequence of transforms u_n of a continuous function u_0. Since Ω is closed, u_0 is uniformly continuous on Ω and hence the sequence $\{u_n\}$ is equicontinuous. By theorem 3 of section 6 there exists therefore a subsequence $\{u_{n_k}\}$ converging uniformly to a continuous function v_0. Then u_{n_k+1} converges to the transform v_1 of v_0. Now the numerical sequence of the maxima m_n of u_n is monotone, and hence m_n converges to a limit m. Because of the uniform convergence both v_0 and v_1 have the maximum m, and hence $v_0(x) = m$ for all x. This limit being independent of the subsequence $\{u_{n_k}\}$ we conclude that $u_n \to m$ uniformly.

Let γ_0 be an arbitrary probability distribution on Ω and denote by $\mathbf{E}_n$ expectations with respect to its transform γ_n defined in (7.1). A comparison of (7.1) and (7.3) shows that

$$\mathbf{E}_n(u_0) = \mathbf{E}_0(u_n) \to \mathbf{E}_0(m) = m.$$

The convergence of $\mathbf{E}_n(u_0)$ for arbitrary continuous u_0 implies the existence of a probability measure α such that $\gamma_n \to \alpha$. (See theorem 2 of section 1; the convergence is proper since the distributions γ_n are concentrated on a finite interval.) From (7.1) it follows that α is stationary for K. The strict positivity of α is an immediate consequence of the strict positivity of K. ►

Proof *of theorem* 2. Denote by $\mathbf{E}$ expectations with respect to the given stationary distribution α. For an arbitrary $u_0 \in C[-\infty, \infty]$, and its transforms u_k we have on account of the stationarity $\mathbf{E}(u_0) = \mathbf{E}(u_1) = \cdots$. Furthermore, $\mathbf{E}(|u_k|)$ decreases with k and so $\lim \mathbf{E}(|u_k|) = m$ exists.

As in the preceding proof we choose a subsequence such that $u_{n_k} \to v_0$. Then $u_{n_k+1} \to v_1$, where v_1 is the transform of v_0. By bounded convergence this entails $\mathbf{E}(u_{n_k}) \to \mathbf{E}(v_0)$ and $\mathbf{E}(|u_{n_k}|) \to \mathbf{E}(|v_0|)$. Thus

$$\mathbf{E}(v_1) = \mathbf{E}(v_0) = \mathbf{E}(u_0) \quad \text{and} \quad \mathbf{E}(|v_1|) = \mathbf{E}(|v_0|) = m.$$

In view of the strict positivity of K the last equality implies that the continuous function v_0 cannot change signs. When $\mathbf{E}(u_0) = 0$ we have therefore $v_0(x) = 0$ identically. It follows that for arbitrary initial u_0 we have $v_0(x) = \mathbf{E}(u_0)$ for all x. This proves that $u_n(x) \to \mathbf{E}(u_0)$ which is the same as $K^{(n)}(x, \Gamma) \to \alpha(\Gamma)$ at all intervals of continuity. ►

We now apply this theory to *convolutions on the circle* of circumference 1, that is, to transformations of the form

$$(7.5) \qquad u_{n+1}(x) = \int_0^1 u_n(x-y)\, F\{dy\}$$

where F is a probability distribution on the circle and addition is modulo 1. [See II,8 and example 6(*a*).] This transformation may be written in the form (7.3) with $\Omega = \overline{0, 1}$ and $K^{(n)}(x, \Gamma) = F^{n\star}\{x - \Gamma\}$. Theorem 1 applies directly if F is strictly positive, but we prove the following more general analogue to the central limit theorem.

Theorem 3.[18] *Let F be a probability distribution on the circle and suppose that it is not concentrated on the vertices of a regular polygon. Then $F^{n\bigstar}$ tends to the distribution with constant density.*

Proof. It suffices to show that for an arbitrary continuous function u_0 the transforms u_n tend to a constant m (depending on u_0). Indeed, as the second part of the proof of theorem 1 shows, this implies that $F^{n\bigstar}$ converges to a probability distribution α on the circle, and since $\alpha \bigstar u_0$ is constant for every continuous function u_0 it follows that α coincides with the uniform distribution.

To show that $u_n \to m$ we use the first part of the proof of theorem 1 except that we require a new proof for the proposition that the maximum of the transform v_1 of a continuous function v_0 is strictly less than the maximum of v_0 except if v_0 is a constant. To prove the theorem it suffices therefore to establish the following proposition. *If v_0 is a continuous function such that $v_0 \leq m$ and $v_0(x) < m$ for all x of an internal I of length $\lambda > 0$, then there exists an r such that $v_r(x) < m$ for all x.*

Since rotations do not affect the maxima there is no loss of generality in assuming that 0 is a point of increase of F. If b is another point of increase then $0, b, 2b, \ldots, rb$ are points of increase of $F^{r\bigstar}$, and it is possible to choose b and r such that every interval of length λ contains at least one among these points (see lemma 1 and the corollary in V,4*a*). By definition

$$v_r(x) = \int_0^1 v_0(x-y)\, F^{r\bigstar}\{dy\}. \tag{7.6}$$

To every point x it is possible to find a point y of increase of $F^{r\bigstar}$ such that $x - y$ is contained in I. Then $v_0(x-y) < m$, and hence $v_r(x) < m$. Since x is arbitrary this proves the assertion. ►

Note. The proof is easily adapted to show that *if F is concentrated on the vertices of a regular polygon with one vertex at* 0, *then $F^{n\bigstar}$ tends to an atomic distribution with atoms of equal weight.* Convergence need *not* take place if 0 is not among the atoms.

Example. (*c*) Let F be concentrated on the two irrational points a and $a + \frac{1}{2}$. Then $F^{n\bigstar}$ is concentrated on the two points na and $na + \frac{1}{2}$, and convergence is impossible. ►

[18] For the analogue on the open line see problems 23 and 24. For generalizations to variable distributions see P. Lévy, Bull. Soc. Math. France, vol. 67 (1939) pp. 1–41; A. Dvoretzky and J. Wolfowitz, Duke Math. J., vol. 18 (1951) pp. 501–507.

8. REGULAR VARIATION

The notion of regular variation (introduced by J. Karamata in 1930) proved fruitful in many connections, and finds an ever increasing number of applications in probability theory. The reason for this is partly explained in the next lemma, which is basic despite its simplicity. The examples of this section contain interesting probabilistic results, and problem 29–30 contains a basic result concerning stable distributions which follow from the lemma in an elementary way. (See also problem 31.)

We have frequently to deal with monotone functions U obtained from a probability distribution F by integrating $y^p\, F\{dy\}$ over $\overline{0, x}$ or $\overline{x, \infty}$. [See, for example, (4.5), (4.15), (4.16).] The usual changes of parameters lead from such a function U to the family of functions of the form $a_t\, U(tx)$, and we have to investigate their asymptotic behavior as $t \to \infty$. If a limit $\psi(x)$ exists, it suffices to consider norming factors of the form $a_t = \psi(1)/U(t)$ provided $\psi(1) > 0$. The next lemma is therefore wider in scope than appears at first sight. It shows that the class of possible limits is surprisingly restricted.

Lemma 1. *Let U be a positive monotone function on $\overline{0, \infty}$ such that*

$$\frac{U(tx)}{U(t)} \to \psi(x) \le \infty \qquad t \to \infty \tag{8.1}$$

at a dense set A of points. Then

$$\psi(x) = x^\rho \tag{8.2}$$

where $-\infty \le \rho \le \infty$.

The senseless symbol x^∞ is introduced only to avoid exceptions. It is, of course, to be interpreted as ∞ for $x > 1$ and as 0 for $x < 1$. Similarly $x^{-\infty}$ is ∞ or 0 according as $x < 1$ or $x > 1$. (See problem 25.)

Proof. The identity

$$\frac{U(tx_1x_2)}{U(t)} = \frac{U(tx_1x_2)}{U(tx_2)} \cdot \frac{U(tx_2)}{U(t)} \tag{8.3}$$

shows that if in (8.1) a finite positive limit exists for $x = x_1$ and $x = x_2$, then also for $x = x_1x_2$, and

$$\psi(x_1x_2) = \psi(x_1)\,\psi(x_2). \tag{8.4}$$

Suppose first that $\psi(x_1) = \infty$ for some point x_1. Then by induction $\psi(x_1^n) = \infty$ and $\psi(x_1^{-n}) = 0$ for all n. Since ψ is monotone this implies that either $\psi(x) = x^\infty$ or $\psi(x) = x^{-\infty}$. It remains to prove the lemma for finite valued ψ. (See problem 25.) Because of the assumed monotonicity we may define ψ everywhere by right-continuity, in which case (8.4) holds at all points x_1, x_2. Now (8.4) differs only notationally from the equation which we have used repeatedly to characterize the exponential distribution. In fact, letting $x = e^\xi$ and $\psi(e^\xi) = u(\xi)$ the relation (8.4) is transformed into $u(\xi_1+\xi_2) = u(\xi_1)\, u(\xi_2)$. We know from **1**; XVII,6 that all solutions that are bounded in finite intervals are of the form $u(\xi) = e^{\rho\xi}$. This, however, is the same as $\psi(x) = x^\rho$. ▶

A function U satisfying the conditions of lemma 1 with a *finite* ρ will be said to vary regularly at infinity, and this definition will be extended to non-monotone functions. If we put

$$U(x) = x^\rho\, L(x) \tag{8.5}$$

the ratio $U(tx)/U(t)$ will approach x^ρ iff

$$\frac{L(tx)}{L(t)} \to 1, \qquad t \to \infty, \tag{8.6}$$

for every $x > 0$. Functions with this property are said to *vary slowly*, and thus the transformation (8.5) reduces regular variation to slow variation. It is convenient to use this fact for a formal definition of regular variation.

Definition. *A positive* (*not necessarily monotone*) *function* L *defined on* $\overline{0, \infty}$ *varies slowly at infinity iff* (8.6) *is true.*

A function U *varies regularly with exponent* ρ $(-\infty < \rho < \infty)$ *iff it is of the form* (8.5) *with* L *slowly varying.*

This definition extends to *regular variation at the origin*: U varies regularly at 0 iff $U(x^{-1})$ varies regularly at ∞. Thus no new theory is required for this notion.

The property of regular variation depends only on the behavior at infinity and it is therefore not necessary that $L(x)$ be positive, or even defined, for *all* $x > 0$.

Example. (*a*) All powers of $|\log x|$ vary slowly both at 0 and at ∞. Similarly, a function approaching a *positive* limit varies slowly.

The function $(1 + x^2)^p$ varies regularly at ∞ with exponent $2p$.

e^x does not vary regularly at infinity, but it satisfies the conditions of lemma 1 with $\rho = \infty$. Finally, $2 + \sin x$ does not satisfy (8.1). ▶

For ease of reference we rephrase lemma 1 in the form of a

Theorem. *A monotone function* U *varies regularly at infinity iff* (8.1) *holds on a dense set and the limit* ψ *is finite and positive in some interval.*[19]

[19] The notion of regular variation may be generalized as follows: Instead of postulating the existence of a limit in (8.1) we require only that every sequence $\{t_n\}$ tending to infinity contains a subsequence such that $U(t_{n_k}x)/U(t_{n_k})$ tends to finite positive limit. We say then that U *varies dominatedly*. See the end of the problem section 10.

This theorem carries over to non-monotone functions except that it must be assumed that convergence takes place at *all* points.

The following lemma should serve to develop a feeling for regular variation. It is an immediate consequence of the general form (9.9) of slowing varying functions.

Lemma 2. *If* L *varies slowly at infinity then*

$$x^{-\epsilon} < L(x) < x^{\epsilon} \tag{8.7}$$

for any fixed $\epsilon > 0$ *and all* x *sufficiently large.*

The passage to the limit in (8.6) *is uniform in finite intervals* $0 < a < x < b$.

We conclude this survey by a frequently used criterion.

Lemma 3. *Suppose that*

$$\frac{\lambda_{n+1}}{\lambda_n} \to 1, \quad \text{and} \quad a_n \to \infty.$$

If U *is a monotone function such that*

$$\lim \lambda_n\, U(a_n x) = \chi(x) \leq \infty \tag{8.8}$$

exists on a dense set, and χ *is finite and positive in some interval, then* U *varies regularly and* $\chi(x) = cx^{\rho}$ *where* $-\infty < \rho < \infty$.

Proof. We may assume that $\chi(1) = 1$ and that (8.8) is true for $x = 1$ (because this can be achieved by a trivial change of scale). For given t define n as the *smallest* integer such that $a_{n+1} > t$. Then $a_n \leq t < a_{n+1}$ and for a non-decreasing U

$$\frac{U(a_n x)}{U(a_{n+1})} \leq \frac{U(tx)}{U(t)} \leq \frac{U(a_{n+1}x)}{U(a_n)}\,; \tag{8.9}$$

for a non-increasing U the reversed inequalities hold. Since $\lambda_n\, U(a_n) \to 1$ the extreme members tend to $\chi(x)$ at each point where (8.8) holds. The assertion is therefore contained in the last theorem. ▶

To illustrate typical applications we derive first a limit theorem due to R. A. Fisher and B. V. Gnedenko, and next a new result.

Example. (*b*) *Distribution of maxima.* Let the variables $\mathbf{X}_k$ be mutually independent and have a common distribution F. Put

$$\mathbf{X}_n^* = \max\,[\mathbf{X}_1, \ldots, \mathbf{X}_n].$$

We ask whether there exist scale factors a_n such that the variables $\mathbf{X}_n^*/a_n$ have a limit distribution G. We exclude two cases on account of their triviality. If F has a largest point of increase ξ then the distribution of $\mathbf{X}_n^*$ trivially tends to the distribution concentrated at ξ. On the other hand, it is always possible to choose scale factors a_n increasing so rapidly that $\mathbf{X}_n^*/a_n$ tends to 0 in probability. The remaining cases are covered by the following *proposition*. ▶

Let $F(x) < 1$ for all x. In order that with appropriate scale factors a_n the distributions G_n of $\mathbf{X}_n^*/a_n$ tend to a distribution G not concentrated at 0 *it is necessary and sufficient that $1 - F$ varies regularly with an exponent $\rho < 0$. In this case,*

$$G(x) = e^{-cx^\rho} \tag{8.10}$$

for $x > 0$ and $G(x) = 0$ for $x < 0$. (Clearly $c > 0$.)

Proof. If a limit distribution G exists we have

$$F^n(a_nx) \to G(x) \tag{8.11}$$

at all points of continuity. Passing to logarithms and remembering that $\log(1-z) \sim -z$ as $z \to 0$ we get

$$n[1-F(a_nx)] \to -\log G(x). \tag{8.12}$$

Since $0 < G(x) < 1$ in some interval the last lemma guarantees the regular variation of $1 - F$. Conversely, if $1 - F$ varies regularly it is possible to determine a_n such that $n[1-F(a_n)] \to 1$, and in this case the left side in (8.12) tends to x^ρ. (See problem 26.) ▶

Example. (*c*) *Convolutions.* From the definition (8.6) it is obvious that the sum of two slowly varying functions is again slowly varying. We now prove the following ▶

Proposition. *If F_1 and F_2 are two distribution functions such that as $x \to \infty$*

$$1 - F_i(x) = x^{-\rho} L_i(x) \tag{8.13}$$

with L_i slowly varying, then the convolution $G = F_1 * F_2$ has a regularly varying tail such that

$$1 - G(x) \sim x^{-\rho}(L_1(x)+L_2(x)). \tag{8.14}$$

Proof. Let $\mathbf{X}_1$ and $\mathbf{X}_2$ be independent random variables with distributions F_1 and F_2. Put $t' = (1 + \delta)t > t$. The event $\mathbf{X}_1 + \mathbf{X}_2 > t$ occurs whenever one of the variables is $> t'$ and the other $> -\delta t$. As $t \to \infty$ the probability of the latter contingency tends to 1, and hence for any $\epsilon > 0$ and t sufficiently large

$$1 - G(t) \geq [(1-F_1(t')) + (1-F_2(t'))](1 - \epsilon). \tag{8.15}$$

On the other hand, if we put $t'' = (1-\delta)t$ with $0 < \delta < \frac{1}{2}$ then the event $\mathbf{X}_1 + \mathbf{X}_2 > t$ cannot occur unless either one of the variables exceeds t'', or else both are $>\delta t$. In view of (8.13) it is clear that the probability of the latter contingency is asymptotically negligible compared with the probability that $\mathbf{X}_i > t''$, and this implies that for t sufficiently large

$$1 - G(t) \leq [(1-F_1(t'')) + (1-F_2(t''))](1 + \epsilon). \tag{8.16}$$

Since δ and ϵ can be chosen arbitarily small the two inequalities (8.15) and (8.16) together entail the assertion (8.14). ▶

By induction on r one gets the interesting

Corollary. *If* $1 - F(x) \sim x^{-\rho}L(x)$ *then* $1 - F^{r\star}(x) \sim rx^{-\rho}L(x)$.

When applicable, this theorem[20] supplements the central limit theorem by providing information concerning the tails. (For applications to stable distributions see problems 29 and 30. For a related theorem concerning the compound Poisson distribution see problem 31.)

*9. ASYMPTOTIC PROPERTIES OF REGULARLY VARYING FUNCTIONS

The purpose of this section is to investigate the relations between the tails and the truncated moments of distributions with regularly varying tails. The main result is that if $1 - F(x)$ and $F(-x)$ vary regularly so do all the truncated moments. This is asserted by theorem 2, which contains more than what we shall need for the theory of stable distributions. It could be proved directly, but it may also be considered a corollary to theorem 1 which embodies Karamata's[21] striking characterization of regular variation. It seems therefore best to give a complete exposition of the theory in particular since the arguments can now be significantly simplified.[22]

We introduce the formal abbreviations

$$Z_p(x) = \int_0^x y^p\, Z(y)\, dy, \qquad Z_p^*(x) = \int_x^\infty y^p\, Z(y)\, dy. \tag{9.1}$$

* This section is used only for the theory of stable distributions, but the use of theorem 2 would simplify many lengthy calculations in the literature.

[20] Special cases were noticed by S. Port.

[21] J. Karamata, *Sur un mode de croissance regulière*, Mathematica (Cluj), vol. 4 (1930) pp. 38–53. Despite frequent references to this paper, no newer exposition seems to exist. For recent generalizations and applications to Tauberian theorems see W. Feller, *One-sided analogues of Karamata's regular variation*, in the Karamata memorial volume (1968) of L'Enseignement Mathématique.

[22] Although new, our proof of theorem 1 uses Karamata's ideas.

It will now be shown that in the case of a regularly varying Z these functions are asymptotically related to Z just as in the simple case $Z(x) = x^a$.

The asymptotic behavior of Z_p at infinity is not affected by the behavior of Z near the origin. Without loss of generality we may therefore assume that Z vanishes identically in some neighborhood of 0 and so the integral defining Z_p will be meaningful for all p.

Lemma. *Let $Z > 0$ vary slowly. The integrals in* (9.1) *converge at ∞ for $p < -1$, diverge for $p > -1$.*

If $p \geq -1$ then Z_p varies regularly with exponent $p + 1$. If $p < -1$ then Z_p^* varies regularly with exponent $p + 1$, and this remains true for $p + 1 = 0$ if Z_{-1}^* exists.

Proof. For given positive x and ϵ choose η such that for $y \geq \eta$

$$(1-\epsilon)\,Z(y) \leq Z(xy) \leq (1+\epsilon)\,Z(y). \tag{9.2}$$

Assume that the integrals in (9.1) converge. From

$$Z_p^*(tx) = x^{p+1}\int_t^\infty y^p\, Z(xy)\, dy \tag{9.3}$$

it follows for $t > \eta$ that

$$(1-\epsilon)x^{p+1}\, Z_p^*(t) \leq Z_p^*(tx) \leq (1+\epsilon)x^{p+1}\, Z_p^*(t).$$

Since ϵ is arbitrary we conclude that as $t \to \infty$

$$\frac{Z_p^*(tx)}{Z_p^*(t)} \to x^{p+1}. \tag{9.4}$$

This proves the regular variation of Z_p^*. Furthermore, since Z_p^* is a decreasing function it follows that $p + 1 \leq 0$. Thus the integrals in (9.1) cannot converge unless $p \leq -1$.

Assume then that these integrals diverge. Then for $t > \eta$

$$Z_p(tx) = Z_p(\eta x) + x^{p+1}\int_\eta^t y^p\, Z(xy)\, dy$$

and hence

$$(1-\epsilon)x^{p+1}\, Z_p(t) \leq Z_p(tx) - Z_p(\eta x) \leq (1+\epsilon)x^{p+1}\, Z_p(t).$$

On dividing by $Z_p(t)$ and letting $t \to \infty$ we conclude as above that $Z_p(tx)/Z_p(t)$ tends to x^{p+1}. In case of divergence therefore Z_p varies regularly, and divergence is possible only when $p \geq -1$. ►

The next theorem shows that regular variation of Z ensures that of Z_p and Z_p^*; the converse is also true except if Z_p or Z_p^* vary slowly. Furthermore we get a useful criterion for the regular variation of these functions.

Parts (*a*) and (*b*) of the theorem treat the functions Z_p^* and Z_p, respectively. They are parallel in all respects, but only part (*a*) is used extensively in probability theory.

Theorem 1. (*a*) *If* Z *varies regularly with exponent* γ *and* Z_p^* *exists, then*

$$\frac{t^{p+1}Z(t)}{Z_p^*(t)} \to \lambda \tag{9.5}$$

where $\lambda = -(p+\gamma+1) \geq 0$.

Conversely, if (9.5) *holds with* $\lambda > 0$, *then* Z *and* Z_p^* *vary regularly with exponents* $\gamma = -\lambda - p - 1$ *and* $-\lambda$, *respectively. If* (9.5) *holds with* $\lambda = 0$ *then* Z_p^* *varies slowly* (*but nothing can be said about* Z).

(*b*) *If* Z *varies regularly with exponent* γ *and if* $p \geq -\gamma - 1$ *then*

$$\frac{t^{p+1}\, Z(t)}{Z_p(t)} \to \lambda \tag{9.6}$$

with $\lambda = p + \gamma + 1$.

Conversely, if (9.6) *holds with* $\lambda > 0$ *then* Z *and* Z_p *vary regularly with exponents* $\lambda - p - 1$ *and* λ, *respectively. If* (9.6) *holds with* $\lambda = 0$ *then* Z_p *varies slowly.*

Proof. The proofs are identical for both parts, and we conduct it for part (*a*). Put

$$\frac{y^p\, Z(y)}{Z_p^*(y)} = \frac{\eta(y)}{y}. \tag{9.7}$$

The numerator on the left is the negative derivative of the denominator, and hence we get for $x > 1$

$$\log \frac{Z_p^*(t)}{Z_p^*(tx)} = \int_t^{tx} \eta(y)\, \frac{dy}{y} = \eta(t) \int_1^x \frac{\eta(ts)}{\eta(t)}\, \frac{ds}{s}. \tag{9.8}$$

Suppose now that Z varies regularly with exponent γ. By the preceding lemma Z_p^* varies regularly with exponent $\lambda = \gamma + p + 1$ and so the two sides in (9.7) vary regularly with exponent -1. Thus η is a slowly varying function. As $t \to \infty$ the last integrand in (9.8) therefore tends to s^{-1}. Unfortunately we do not know that η is bounded, and so we can only assert that the lower limit of the integral is $\geq \log x$ by virtue of Fatou's theorem [see IV,(2.9)]. But because of the regular variation of Z_p^* the left side tends to $\lambda \log x$, and so

$$\limsup \eta(t) \leq \lambda.$$

But this implies the boundedness of η, and hence we may choose a sequence $t_n \to \infty$ such that $\eta(t_n) \to c < \infty$. Because of the slow variation this implies that $\eta(t_n s) \to c$ for all s, and the convergence is bounded. Thus the right side in (9.8) approaches $c \log x$, and hence $c = \lambda$. It follows that the limit c is independent of the sequence $\{t_n\}$, and so $\eta(t) \to \lambda$. This proves that (9.5) is true.

The converse is easier. Suppose $\eta(t) \to \lambda \geq 0$. The two sides in (9.8) then approach $\lambda \log x$, and hence the ratio $Z_p^*(t)/Z_p^*(tx)$ approaches x^λ as asserted. If $\lambda > 0$ this together with (9.5) proves that Z varies regularly with exponent $-\lambda - p - 1$. ▶

Although we shall not use it we mention the following interesting

Corollary. *A function Z varies slowly iff it is of the form*

$$Z(x) = a(x) \exp\left(\int_1^x \frac{\epsilon(y)}{y}\, dy\right) \tag{9.9}$$

where $\epsilon(x) \to 0$ and $a(x) \to c < \infty$ as $x \to \infty$.

Proof. It is easily verified that the right side represents a slowly varying function. Conversely, assume that Z varies slowly. Using (9.6) with $p = \gamma = 0$ we get

$$\frac{Z(t)}{Z_0(t)} = \frac{1 + \epsilon(t)}{t}$$

with $\epsilon(t) \to 0$. On the left the numerator is the derivative of the denominator, and by integration we get

$$Z_0(x) = Z_0(1) \cdot x \exp\left(\int_1^x \frac{\epsilon(t)}{t}\, dt\right)$$

which is equivalent to (9.9) because $Z(x) \sim Z_0(x)x^{-1}$ by (9.6). ▶

We proceed to apply theorem 1 to the truncated moment functions of a probability distribution F. We can consider each tail separately or else combine them by considering $F(x) - F(-x)$ instead of F. It suffices therefore to study distributions F concentrated on $\overline{0, \infty}$. For such a distribution we define the truncated moment functions U_ζ and V_η by

$$U_\zeta(x) = \int_0^x y^\zeta\, F\{dy\}, \qquad V_\eta(x) = \int_x^\infty y^\eta\, F\{dy\}. \tag{9.10}$$

It will be understood that the second integral converges while the first integral tends to ∞ as $x \to \infty$. This requires that $\zeta > 0$ and $-\infty < \eta < \zeta$. In particular, $V_0 = 1 - F$ is the tail of the distribution F.

We prove a generalization of part (*a*) of theorem 1; part (*b*) generalizes in like manner.

Theorem 2.[23] *Suppose that* $U_\zeta(\infty) = \infty$.

(i) *If either* U_ζ *or* V_η *varies regularly then there exists a limit*

$$\lim_{x\to\infty} \frac{t^{\zeta-\eta}V_\eta(t)}{U_\zeta(t)} = c, \qquad 0 \leq c \leq \infty. \tag{9.11}$$

We write this limit uniquely in the form

$$c = \frac{\zeta - \alpha}{\alpha - \eta}, \qquad \eta \leq \alpha \leq \zeta \tag{9.12}$$

with $\alpha = \eta$ *if* $c = \infty$.

(ii) *Conversely, if* (9.11) *is true with* $0 < c < \infty$ *then automatically* $\alpha \geq 0$ *and there exists a slowly varying function* L *such that*

$$U_\zeta(x) \sim (\alpha-\eta)x^{\zeta-\alpha}L(x), \qquad V_\eta(x) \sim (\zeta-\alpha)x^{\eta-\alpha}L(x) \tag{9.13}$$

where the sign $\sim$ *indicates that the ratio of the two sides tends to* 1.

(iii) *The statement remains true when* $c = 0$ *or* $c = \infty$, *provided the sign* $\sim$ *is interpreted in the obvious manner.*

For example, if (9.11) holds with $c = 0$ then $\alpha = \zeta$ and U_ζ varies slowly, but about V_η we know only that $V_\eta(x) = o(x^{\eta-\zeta}L(x))$. In this case V_η need not vary regularly (see problem 31). However, slow variation is the only case in which regular variation of one of the functions U_ζ or V_η does not imply regular variation of the others.

Proof. (i) We write V_η in the form

$$V_\eta(x) = \int_x^\infty y^{\eta-\zeta}\, U_\zeta(dy). \tag{9.14}$$

Integrating by parts between x and $t > x$ we get

$$V_\eta(x) - V_\eta(t) = -x^{\eta-\zeta}\, U_\zeta(x) + t^{\eta-\zeta}\, U_\zeta(t) + (\zeta-\eta)\int_x^t y^{\eta-\zeta-1}\, U_\zeta(y)\, dy.$$

The last two terms on the right are positive and therefore the integral must converge as $t \to \infty$. Because of the monotonicity of U_ζ this implies that $y^{\eta-\zeta}\, U_\zeta(t) \to 0$ and hence

$$V_\eta(x) = -x^{\eta-\zeta}\, U_\zeta(x) + (\zeta-\eta)\int_x^\infty y^{\eta-\zeta-1}\, U_\zeta(y)\, dy \tag{9.15}$$

or

$$\frac{x^{\zeta-\eta}V_\eta(x)}{U_\zeta(x)} = -1 + \frac{\zeta-\eta}{x^{\eta-\zeta}U_\zeta(x)}\int_x^\infty y^{\eta-\zeta-1}\, U_\zeta(y)\, dy. \tag{9.16}$$

[23] For a generalization see problems 34 and 35.

Assume now that U_ζ varies regularly. Since $U_\zeta(\infty) = \infty$ the exponent is necessarily $\leq\zeta$ and we denote it by $\zeta - \alpha$. (Since the integral in (9.16) converges we have necessarily $\alpha \geq \eta$.) The relation (9.5) with $Z = U_\zeta$ and $p = \eta - \zeta - 1$ asserts that the right side in (9.16) tends to

$$-1 + (\zeta-\eta)/(\alpha-\eta) = (\zeta-\alpha)/(\alpha-\eta)$$

if $\lambda \neq 0$ and to ∞ if $\lambda = 0$. We have thus shown that if U_ζ varies regularly with exponent $\zeta - \alpha$, then (9.11) holds with c given by (9.12) and ≥ 0.

Assume then that V_η varies regularly. Its exponent is $\leq\eta$ and we denote it by $\eta - \alpha$. We use the same argument except that (9.15) is replaced by the analogous relation

$$U_\zeta(x) = -x^{\zeta-\eta}\, V_\eta(x) + (\zeta-\eta)\int_0^x y^{\zeta-\eta-1}\, V_\eta(y)\, dy. \tag{9.17}$$

An application of (9.6) with $Z = V_\eta$ and $p = \zeta - \eta - 1$ now shows that (9.11) holds with c given by (9.12) where $\alpha \geq 0$.

(ii) To prove the converse, assume (9.11) and write c in the form (9.12). Suppose first that $0 < c < \infty$. From (9.16) we see then that

$$\frac{x^{\eta-\zeta}\, U_\zeta(x)}{\displaystyle\int_x^\infty y^{\eta-\zeta-1}\, U_\zeta(y)\, dy} \to \frac{\zeta - \eta}{c + 1} = \alpha - \eta. \tag{9.18}$$

From theorem 1(*a*) it follows directly that U_ζ varies regularly with exponent $\zeta - \alpha > 0$, and (9.11) then implies that V_η varies regularly with exponent $\eta - \alpha$. It follows that U_ζ and V_η can be written in the form (9.13) where $\alpha \geq 0$.

If $c = 0$ the same argument shows that U_ζ varies slowly, but (9.11) does not permit the conclusion that V_η varies regularly.

Finally, if (9.11) holds with $c = \infty$ we conclude from (9.18) that

$$\frac{x^{\zeta-\eta}\, V_\eta(x)}{\displaystyle\int_0^x y^{\zeta-\eta-1}\, V_\eta(y)\, dy} \to 0, \tag{9.19}$$

and by theorem 1(*b*) this implies that V_η varies slowly. ▶

10. PROBLEMS FOR SOLUTION

1. *Alternative definition of convergence.* Let F_n and F be probability distributions. Show that $F_n \to F$ (properly) iff for given $\epsilon > 0, h > 0$ and t there exists an $N(\epsilon, h, t)$ such that for $n > N(\epsilon, h, t)$

$$F(t - h) - \epsilon < F_n(t) < F(t + h) + \epsilon. \tag{10.1}$$

2. *Improper convergence.* If F is a defective distribution then (10.1) implies that $F_n \to F$ improperly. The converse is *not* true. Show that *proper* convergence may be *defined* by requiring that (10.1) holds for $n \geq N(\epsilon, h)$, independently of t.

3. Let $\{F_n\}$ converge properly to a limit that is not concentrated at one point. The sequence $\{F_n(a_n x + b_n)\}$ converges to the distribution concentrated at the origin iff $a_n \to \infty$, $b_n = o(a_n)$.

4. *Let* $\mathbf{X}_1, \mathbf{X}_2, \ldots$ *be independent random variables with a common distribution* F *and* $\mathbf{S}_n = \mathbf{X}_1 + \cdots + \mathbf{X}_n$. *Let the variables* $a_n^{-1}\mathbf{S}_n - b_n$ *have a proper limit distribution* U *not concentrated at one point. If* $a_n > 0$ *then*

$$a_n \to \infty, \qquad a_n/a_{n-1} \to 1.$$

[*Hint:* Using theorem 2 of section 3 show that a_{2n}/a_n approaches a finite limit. It suffices to consider symmetric distributions.] (The limit distribution is stable, see VI,1.)

5. Let $\{u_n\}$ be a sequence of bounded monotone functions converging pointwise to a bounded *continuous* limit (which is automatically monotone). Prove that the convergence is uniform. [*Hint:* Partition the axis into subintervals within each of which u varies by less than ϵ.]

6. Let F_n be concentrated at n^{-1} and $u(x) = \sin(x^2)$. Then $F_n \bigstar u \to u$ pointwise, but not uniformly.

7. (*a*) If the joint distribution of $(\mathbf{X}_n, \mathbf{Y}_n)$ converges to that of $(\mathbf{X}, \mathbf{Y})$, then the distribution of $\mathbf{X}_n + \mathbf{Y}_n$ tends to that of $\mathbf{X} + \mathbf{Y}$.

(*b*) Show that theorem 2 of section 3 is a special case.

(*c*) The conclusion does not hold in general if it is only known that the marginal distributions for $\mathbf{X}_n$ and $\mathbf{Y}_n$ converge.

8. Let $F_n \to F$ with F defective. If $u \in C_0(-\infty, \infty)$ then $F_n \bigstar u \to F \bigstar u$ uniformly *in every finite interval.* (This generalizes theorem 1 of section 3.)

9. If $F_n \to F$ improperly it is not necessarily true that $F_n \bigstar F_n \to F \bigstar F$.
Example. Let F_n have atoms of weight $\frac{1}{3}$ at the points $-n$, 0, and n.

10. In the plane every continuous function vanishing at infinity can be approximated uniformly by finite linear combinations $\Sigma\, c_k \varphi_k(x)\psi_k(y)$ with infinitely differentiable φ_k and ψ_k.
[*Hint:* Use example 3(*a*) choosing $G_k(x, y) = \mathfrak{N}_k(x)\mathfrak{N}_k(y)$ where $\mathfrak{N}$ is the normal density.]

Metrics. A function ρ is called a *distance function* for probability distributions if $\rho(F, G)$ is defined for every pair F, G of probability distributions and has the following three properties: $\rho(F, G) \geq 0$ and $\rho(F, G) = 0$ iff $F = G$; next $\rho(F, G) = \rho(G, F)$; and finally, ρ satisfies the triangle inequality

$$\rho(F_1, F_2) \leq \rho(F_1, G) + \rho(F_2, G).$$

11. *P. Lévy metric.* For two proper distributions F and G define $\rho(F, G)$ as the infimum of all $h > 0$ such that

$$F(x - h) - h \leq G(x) \leq F(x + h) + h \tag{10.2}$$

for all x. Verify that ρ is a distance function. Show that $F_n \to F$ properly iff $\rho(F_n, F) \to 0$.

12. *Distance "in variation."* Put $\rho(F, G) = \sup \|\mathfrak{F}u - \mathfrak{G}u\|$ where $u \in C_0$ and $\|u\| = 1$. Show that ρ is a distance function.[24] If F and G are atomic and attribute weights p_k and q_k to the point a_k, then

$$\rho(F, G) = \sum |p_k - q_k|. \tag{10.3}$$

If F and G have densities f and g

$$\rho(F, G) = \int_{-\infty}^{\infty} |f(x) - g(x)|\, dx. \tag{10.4}$$

[*Hint:* It suffices to prove (10.4) for continuous f and g. The general case follows by approximation.]

13. *Continuation.* Show that $\rho(F_n, G) \to 0$ implies proper convergence $F_n \to G$. To see that the converse is false consider the normal distribution functions $\mathfrak{N}(nx)$ and the distribution F_n concentrated at n^{-1}.

14. *Continuation.* If $U = F_1 \bigstar \cdots \bigstar F_n$ and $V = G_1 \bigstar \cdots \bigstar G_n$ show that

$$\rho(U, V) \leq \sum_{k=1}^{n} \rho(F_k, G_k). \tag{10.5}$$

This extends the basic inequality (3.9). [*Hint:* Use (3.9) and a test function u such that $\|\mathfrak{A}u - \mathfrak{B}u\|$ is close to $\rho(U, V)$.]

15. *Approximation by the Poisson distribution.*[25] Let F attribute weight p to the point 1 and $q = 1 - p$ to the point 0. If G is the Poisson distribution with expectation p show that $\rho(F, G) \leq \frac{9}{4}p^2$, where ρ is the distance defined in (10.3). Conclude: If F is the distribution of the number of successes in n Bernoulli trials with probabilities $p_1, \ldots, p_n$ and if G is the Poisson distribution with expectation $p_1 + \cdots + p_n$ then $\rho(F, G) \leq \frac{9}{4}(p_1^2 + \cdots + p_n^2)$.

16. The law of large numbers of VII,7 states that if the $\mathbf{X}_k$ are independent and identically distributed, and if $\mathbf{E}(\mathbf{X}_k) = 0$, then $(\mathbf{X}_1 + \cdots + \mathbf{X}_n)/n \xrightarrow{\text{p}} 0$. Prove this by the method used for theorem 1 in section 4.

17. The Lindeberg condition (4.15) is satisfied if $\alpha_k = \mathbf{E}(|\mathbf{X}_k^{2+\delta}|)$ exists for some $\delta > 0$ and $\alpha_1 + \cdots + \alpha_n = o(s_n^{2+\delta})$ (Liapunov's condition).

18. Let F_k be symmetric and $1 - F_k(x) = \frac{1}{2}x^{-2-1/k}$ for $x > 1$. Show that the Lindeberg condition (4.15) is not satisfied.

19. Let $\mathbf{X}_k = \pm 1$ with probability $\frac{1}{2}(1 - k^{-2})$ and $\mathbf{X}_k = \pm k$ with probability $\frac{1}{2}k^{-2}$. By simple truncation prove that $\mathbf{S}_n/\sqrt{n}$ behaves asymptotically in the same way as if $\mathbf{X}_k = \pm 1$ with probability $\frac{1}{2}$. Thus *the distribution of* $\mathbf{S}_n/\sqrt{n}$ *tends to* $\mathfrak{N}$ *but* $\operatorname{Var}(\mathbf{S}_n/\sqrt{n}) \to 2$.

20. Construct variants of the preceding problem where $\mathbf{E}(\mathbf{X}_k^2) = \infty$ and yet the distribution of $\mathbf{S}_n/\sqrt{n}$ tends to $\mathfrak{N}$.

---

[24] The definition can be extended to differences of arbitrary finite measures and defines the "*norm topology*" for measures. Problem 13 shows that the resulting notion of convergence is not natural for probability theory.

[25] Suggested by inequalities in L. LeCam, *An approximation theorem for the Poisson binomial distribution*, Pacific J. Math., vol. 10 (1960) pp. 1181–1197.

21.[26] *Central limit theorem for exchangeable variables.* For fixed θ let F_θ be a distribution with zero expectation and variance $\sigma^2(\theta)$. A value θ is chosen according to the probability distribution G and one considers mutually independent variables $\mathbf{X}_n$ with the common distribution F_θ. If a^2 is the expectation of σ^2 with respect to G show that the distribution of $\mathbf{S}_n/(a\sqrt{n})$ tends to the distribution

$$\int_{-\infty}^{+\infty} \mathfrak{N}(ax/(\sigma(\theta))G\{d\theta\}.$$

It is not normal unless G is concentrated at one point.

22. *Shot noise in vacuum tubes, etc.* Consider the stochastic process of example VI,3(h) with discretized time parameter. Assuming that at epoch kh an arrival occurs with probability αh show that the intensity of the current in the discrete model is given by an *infinite convolution.* The passage to the limit $h \to 0$ leads to Campbell's theorem VI,(3.4).

Do the same for the busy-trunkline example VI,3(i). Generalize the model to the situation where the after-effect at epoch kh is a random variable assuming the values 1, 2, . . . with probabilities $p_1, p_2, \ldots$.

23. The sequence $\{F^{n\star}\}$ is never stochastically bounded. [*Hint:* It suffices to consider symmetric distributions. Also, one may suppose that F has infinite tails, for otherwise $F^{n\star} \to 0$ by the central limit theorem. Use V, (5.10).]

Note. It will be shown in example XV,3(a) that $F^{n\star} \to 0$.

24. *Continuation.* It is nevertheless possible that for every x

$$\limsup_{n\to\infty} F^{n\star}(x) = 1, \qquad \liminf_{n\to\infty} F^{n\star}(x) = 0.$$

In fact, it is possible to choose two extremely rapidly increasing sequences of integers a_k and n_k such that

$$(-1)^k \frac{1}{2ka_k} \mathbf{S}_{n_k} \xrightarrow{\text{p}} 1.$$

[*Hint:* Consider the distribution $\mathbf{P}\{\mathbf{X} = (-1)^k a_k\} = p_k$. With an appropriate choice of the constants there is an overwhelming probability that about $2k$ among the terms $\mathbf{X}_1, \ldots, \mathbf{X}_{n_k}$ will equal $(-1)^k a_k$ and none will exceed a_k in absolute value. Then for k even $\mathbf{S}_{n_k} > a_k - n_k a_{k-1}$. Show that

$$n_k = (2k)!, \qquad p_k \sim \frac{1}{(2k-1)!}, \qquad a_k \sim (n_k)^k$$

will do.]

25. In the proof of lemma 1 of section 8 it suffices to assume that the set A is dense in some open interval.

26. *Distribution of maxima.* Let $\mathbf{X}_1, \ldots, \mathbf{X}_n$ be independent with the common distribution F and $\mathbf{X}_n^* = \max(\mathbf{X}_1, \ldots, \mathbf{X}_n)$. Let G_n be the distribution of $a_n^{-1}\mathbf{X}_n^*$.

---

[26] J. R. Blum, H. Chernoff, M. Rosenblatt, and H. Teicher, *Central limit theorems for interchangeable processes*, Canadian J. Math., vol. 10 (1958) pp. 222–229.

(*a*) If $F(x) = 1 - e^{-x}$ and $a_n = n$ then G_n tends to the distribution concentrated at the point 0. Show directly that no choice of a_n leads to more discriminating results.

(*b*) If F is the Cauchy distribution with density $\dfrac{1}{\pi(1 + x^2)}$ and $a_n = n/\pi$, then $G_n(x) \to e^{-x^{-1}}$ for $x > 0$.

27. If **X** and **Y** have a common distribution F such that $1 - F(x) \sim x^{-\rho}L(x)$ with L slowly varying, then $\mathbf{P}\{\mathbf{X} > t \mid \mathbf{X} + \mathbf{Y} > t\} \to \frac{1}{2}$ as $t \to \infty$. Roughly speaking, a large value for the sum is likely to be due to the contribution of one of the two variables.[27]

28. Let $v > 0$ and $a > 0$ on $\overline{0, \infty}$ and suppose that

$$\lim_{t \to \infty} [a(t)v(tx) + b(t)x] = z(x)$$

exists and depends continuously on x. For fixed $x_0 > 0$ prove that $\dfrac{v(x_0x)}{x_0x} - \dfrac{v(x)}{x}$ varies regularly. Conclude that either $z(x) = cx^\alpha$ or $z(x) = cx + c_1x \log x$, provided only that v itself does not vary regularly [in which case $z(x) = cx^\alpha + c_1x$].

29. Let G be a symmetric stable distribution, that is, $G^{r\star}(c_rx) = G(x)$ (see VI,1). From the last corollary in section 8 conclude that $1 - G(x) \sim x^{-\alpha}L(x)$ with $\alpha < 2$ unless $r[1 - G(c_rx)] \to 0$ in which case G is the normal distribution.

[*Hint:* The sequence $r[1 - G(c_rx)]$ remains bounded by the symmetrization inequality V,(5.13). The remainder is easy.]

30. Generalize to unsymmetric stable distributions.

31. Let $\{\mathbf{X}_n\}$ be a sequence of mutually independent positive random variables with a common distribution F concentrated on $\overline{0, \infty}$. Let **N** be a Poisson variable. The random sum $\mathbf{S_N} = \mathbf{X}_1 + \cdots + \mathbf{X_N}$ has the compound Poisson distribution

$$U = e^{-c} \sum \frac{c^n}{n!} F^{n\star}.$$

Let L vary slowly at infinity. Prove that

$$\textit{if} \quad 1 - F(x) \sim x^{-\rho} L(x) \qquad \textit{then} \quad 1 - U(x) \sim cx^{-\rho} L(x).$$

[*Hint:* Obviously $\mathbf{P}\{\mathbf{S_N} > x\}$ exceeds the probability that exactly one among the components $\mathbf{X}_j > x$, that is

$$1 - U(x) \geq c[1 - F(t)]e^{-c[1-F(t)]}.$$

On the other hand, for sufficiently large x the event $\mathbf{S_N} > x$ cannot occur unless either one among the components $\mathbf{X}_j > (1 - \epsilon)x$, or at least two exceed $x^{\frac{3}{8}}$, or finally $\mathbf{N} > x^{\frac{1}{8}}$. The probability of the second contingency is $o(1 - F(x))$, while the probability of $\mathbf{N} > \log x$ tends to 0 more rapidly than any power of x.]

32. Let F be atomic with weight proportional to $n^{-1}2^{-2n}$ at the point 2^n. Show that U_2, as defined in (9.10), is slowly varying and $U_2(\infty) = \infty$, but that $1 - F$ does not vary regularly.

[*Hint:* For the last statement it suffices to consider the magnitude of the jumps.]

---

[27] The phenomenon as such seems to have been noticed first by B. Mandelbrot.

Note. The remaining problems refer to a generalization of the notion of regular variation.[28] A convenient starting point is provided by the following

Definition. *A monotone function* u *varies dominatedly at infinity if the ratios* $u(2x)/u(x)$ *remain bounded away from* 0 *and* ∞.

33. Show that a non-decreasing function u varies dominatedly iff there exist constants $A, p,$ and t_0 such that

$$\frac{u(tx)}{u(t)} < Ax^p, \qquad t > t_0, \quad x > 1. \tag{10.5}$$

For non-increasing u the same criterion applies with $x > 1$ replaced by $x < 1$.

34. (*Generalization of theorem* 2 *in section* 9.) Define U_ζ and V_η as in (9.10) (which requires that $-\infty < \eta < \zeta$). Put $R(t) = t^{\zeta-\eta}V_\eta(t)/U_\zeta(t)$.
Show that U_ζ *varies dominatedly iff* $\limsup R(t) < \infty$. Similarly V_η *varies dominatedly iff* $\liminf R(t) > 0$.

35. (*Continuation.*) More precisely: If $R(t) \leq M$ for $t > t_0$, then

$$\frac{U_\zeta(tx)}{U_\zeta(t)} \leq (M+1)x^p, \qquad x > 1, \quad t > t_0 \tag{10.6}$$

with $p = (\zeta - \eta)M/(M+1)$. Conversely, (10.6) with $p < \zeta - \eta$ implies

$$R(t) \leq \frac{M(\zeta - \eta) + p}{\zeta - \eta - p} \tag{10.7}$$

These statements remain true if R is replaced by its reciprocal R^{-1} and at the same time the ratio $U_\zeta(tx)/U_\zeta(t)$ is replaced by $V_\eta(t)/V_\eta(tx)$.

36. Prove the following *criterion*: If there exists a number $s > 1$ such that $\liminf U_\zeta(st)/U_\zeta(t) > 1$ then V_η varies dominatedly. Similarly, if $\liminf V_\eta(t/s)/V_\eta(t) > 1$ then U_ζ varies dominatedly.

---

[28] For further results and details see W. Feller, *One-sided analogues of Karamata's regular variation*, in the Karamata Memorial volume of l'Enseignement Mathématique, vol. 15 (1969), pp. 107–121. See also W. Feller, *On regular variation and local limit theorems*, Proc. Fifth Berkeley Symposium Math. Statistics and Probability, vol. 2, part 1, pp. 373–388 (1965–66).

CHAPTER IX

Infinitely Divisible Distributions and Semi-Groups

The purpose of this chapter is to show that the basic theorems concerning infinitely divisible distributions, processes with independent increments, and stable distributions and their domains of attraction can be derived by a natural extension of the argument used to prove the central limit theorem. *The theory will be developed anew and amplified by methods of Fourier analysis*, and for this reason the present outline is limited to the basic facts. The interest in the chapter is largely methodological, to tie the present topics to the general theory of Markov processes; when applicable, the methods of Fourier analysis lead to sharper results. To provide easy access to important facts some theorems are proved twice. Thus the general structure theorem is first proved for semi-groups of distributions with variances. In this way sections 1–4 present a self-contained exposition of basic facts.

The semi-group operators in this chapter are convolutions. Other semi-groups will be considered independently in the next chapter by new methods.

1. ORIENTATION

The limit theorems of this chapter are a natural extension of the central limit theorem, and the infinitely divisible distributions are closely related to the normal distribution. To see this it is worthwhile to repeat the proof of theorem 1 in VIII,4 in a slightly different setting.

We consider this time an arbitrary triangular array $\{\mathbf{X}_{k,n}\}$ where for each n the n variables[1] $\mathbf{X}_{1,n}, \ldots, \mathbf{X}_{n,n}$ are independent and have a

[1] Triangular arrays were defined in VI,3. It should be borne in mind that we are really dealing with distribution functions $F_{k,n}$; the random variables $\mathbf{X}_{k,n}$ serve merely to simplify notations. Accordingly, the variables of different rows need not be related in any way (and need not be defined on the same probability space).

common distribution F_n. For the row sums we write $\mathbf{S}_n = \mathbf{X}_{1,n} + \cdots + \mathbf{X}_{n,n}$. In chapter VIII we dealt with the special case where $\mathbf{X}_{k,n} = \mathbf{X}_k a_n^{-1}$ and $F_n(x) = F(a_n x)$. There the row sums were denoted by $\mathbf{S}_n^*$.

Throughout this chapter we use the operational notation of VIII,3. Thus $\mathfrak{F}_n$ is the operator associated with F_n and $\mathfrak{F}_n^n$ is associated with the distribution of $\mathbf{S}_n$. Finally, $\|u\|$ denotes the upper bound of the continuous function $|u|$.

Example. (*a*) *Central limit theorem.* Suppose that there exist numbers $\epsilon_n \to 0$ such that

$$|\mathbf{X}_{1,n}| < \epsilon_n, \qquad \mathbf{E}(\mathbf{X}_{1,n}) = 0, \qquad n\mathbf{E}(\mathbf{X}_{1,n}^2) \to 1. \tag{1.1}$$

For a function u with three bounded derivatives we have the identity

$$n[\mathfrak{F}_n u(x) - u(x)] = \int_{-\epsilon_n}^{\epsilon_n} \frac{u(x-y) - u(x) + y\,u'(x)}{y^2} \cdot ny^2\, F_n\{dy\}. \tag{1.2}$$

The finite measure $ny^2\, F_n\{dy\}$ converges by assumption to the probability distribution concentrated at the origin. The fraction under the integral is a continuous function of y and differs from $\frac{1}{2}u''(x)$ by less than $\epsilon_n\, \|u'''\|$. Thus

$$n[\mathfrak{F}_n u - u] \to \tfrac{1}{2}u'' \tag{1.3}$$

uniformly in x.

Suppose now that $\{\mathfrak{G}_n\}$ is a second sequence of operators such that $n[\mathfrak{G}_n u - u]$ tends uniformly to $\frac{1}{2}u''$. Then

$$n(\mathfrak{F}_n u - \mathfrak{G}_n u) \to 0 \tag{1.4}$$

uniformly. By the basic inequality VIII,(3.10) (which will be used constantly in the sequel)

$$\|\mathfrak{F}_n^n u - \mathfrak{G}_n^n u\| \leq n\, \|\mathfrak{F}_n u - \mathfrak{G}_n u\|, \tag{1.5}$$

and the right side tends to zero in consequence of (1.4). As we have seen in the proof of theorem 1 in VIII,4, we may choose for $\mathfrak{G}_n$ the operator associated with the symmetric normal distribution with variance $1/n$. Then $\mathfrak{G}_n^n = \mathfrak{G}_1$ and hence $\mathfrak{F}_n^n \to \mathfrak{G}_1$. We have thus proved that *the distribution of* $\mathbf{S}_n$ *tends to the normal distribution* $\mathfrak{N}$. ►

In scrutinizing the structure of this proof it is seen that the form of the right side in (1.3) played no role. Suppose we had an array such that (uniformly)

$$n[\mathfrak{F}_n u - u] \to \mathfrak{A}u \tag{1.6}$$

where $\mathfrak{A}$ is an arbitrary, but fixed, operator. Our argument permits us to compare any two arrays satisfying (1.6) and to conclude that their row sums behave asymptotically in the same way. If for *one* such array the distributions of $\mathbf{S}_n$ tend to a limit G then the same will be true for *all* our arrays. We shall prove that this is always the case.

Example. (*b*) *Poisson distribution.* Suppose $\mathbf{X}_{1,n}$ equals 1 with probability p_n, and 0 with probability $1 - p_n$. If $np_n \to \alpha$

$$(1.7) \qquad n[\mathfrak{F}_n u(x) - u(x)] = np_n[u(x-1) - u(x)] \to \alpha[u(x-1) - u(x)].$$

This time we take for $\mathfrak{G}_n$ the operator associated with the Poisson distribution with expectation α/n. An easy calculation shows that also $n[\mathfrak{G}_n u - u]$ tends to the right side in (1.7) and we conclude as before that $\mathfrak{F}_n^n u \to \mathfrak{G}_1$. Thus the distribution of $\mathbf{S}_n$ tends to the Poisson distribution with expectation α. [The right side in (1.7) illustrates one possible form for the operator $\mathfrak{A}$ in (1.6). For another example of a simple triangular array see problem 2.] ►

In our two examples we were fortunate in knowing the limit distribution in advance. In general the triangular array as such will serve to define the limit and in this way we shall derive new distribution functions. This procedure was used in **1**; VI to *define* the Poisson distribution as a limit of binomial distributions.

We recall from VI,3 that the limit distributions of the sums $\mathbf{S}_n$ are called *infinitely divisible*. We shall show that such a limit distribution exists whenever a relation of the form (1.6) holds, and that this condition is also necessary. Another approach to the problem depends on the study of the measures $ny^2 F_n\{dy\}$. In both examples a limit measure existed; in example (*a*) it was concentrated at the origin, in (*b*) at the point 1. In general, the relation (1.6) is intimately connected with the existence of a measure Ω such that $ny^2 F_n\{dy\} \to \Omega\{dy\}$, and infinitely divisible distributions will be characterized either by the operator $\mathfrak{A}$ or the measure Ω (which may be unbounded).

A third approach to the problem starts from the solution of the convolution equation

$$(1.8) \qquad Q_s \star Q_t = Q_{s+t}$$

in which Q_t is a probability distribution depending on the parameter $t > 0$.

Example. (*c*) The normal and the Poisson distributions satisfy (1.8) with t proportional to the variance. The gamma distributions of II,(2.2) have the convolution property II,(2.3), which is a special case of (1.8). The same is true of the analogous convolution properties derived for the Cauchy distribution II,(4.5) and the one-sided stable distribution of II,(4.7).

For a triangular array with $F_n = Q_{1/n}$ the relation (1.6) states that as t runs through $\frac{1}{2}, \frac{1}{3}, \ldots$. One should expect that (1.6) will hold for an arbitrary approach $t \to 0+$.

Now (1.8) is the basic equation for processes with *stationary independent increments* (VI,4) and is closely connected with semi-group theory. In this context $\mathfrak{A}$ appears as a "generator." It turns out that this theory provides the easiest access to limit theorems and to infinitely divisible distributions, and hence we begin with it.

2. CONVOLUTION SEMI-GROUPS

For $t > 0$ let Q_t be a probability distribution in $\mathcal{R}^1$ satisfying (1.8) and $\mathfrak{Q}(t)$ the associated operator, that is,

$$\mathfrak{Q}(t)\, u(x) = \int_{-\infty}^{+\infty} u(x-y)\, Q_t\{dy\}. \tag{2.1}$$

Then (1.8) is equivalent to

$$\mathfrak{Q}(s+t) = \mathfrak{Q}(s)\, \mathfrak{Q}(t). \tag{2.2}$$

A family of operators satisfying (2.2) is called a *semi-group*. [It fails to be a group because in general $\mathfrak{Q}(t)$ has no inverse.] The operators of a semi-group may be of an arbitrary nature and it is convenient to have a word to indicate our requirement that $\mathfrak{Q}(t)$ be associated with a probability distribution.

Definition 1. *A convolution semi-group* $\{\mathfrak{Q}(t)\}$ *(where* $t > 0$*) is a family of operators associated with probability distributions and satisfying* (2.2).

We take $C_0[-\infty, \infty]$ as domain of definition. The operators $\mathfrak{Q}(t)$ are transition operators, that is, $0 \leq u \leq 1$ implies $0 \leq \mathfrak{Q}(t)u \leq 1$ and we have $\mathfrak{Q}(t)1 = 1$.

We shall have to deal with operators [such as d^2/dx^2 in (1.3)] which are not defined for all continuous functions. For our present purposes it is fortunately possible to avoid tedious discussions of the precise domain of definition of such operators since we need consider only the class of functions u such that $u \in C[-\infty, \infty]$ and u has derivatives of all orders belonging to $C[-\infty, \infty]$. Such functions are called *infinitely differentiable*,[2] and their class is denoted by C^∞. For the present we consider only operators $\mathfrak{A}$ defined for all $u \in C^\infty$ and such that $\mathfrak{A}u \in C^\infty$, and so all occurring operators may be taken as *operators from* C^∞ *to* C^∞. For operators associated with probability distributions we saw in VIII,3 that $\mathfrak{F}_n \to \mathfrak{F}$ iff $\mathfrak{F}_n u \to \mathfrak{F}u$ for $u \in C^\infty$. We now extend this definition of convergence consistently to arbitrary operators.

---

[2] The class C^∞ is introduced only to avoid a new term. It could be replaced by the class of functions with (say) four bounded derivatives, or (simpler still) by the class of all linear combinations of normal distribution functions with arbitrary expectations and variances.

Definition 2. *Let* $\mathfrak{A}_n$ *and* $\mathfrak{A}$ *be operators from* C^∞ *to* C^∞. *We say that* $\mathfrak{A}_n$ *converges to* $\mathfrak{A}$, *in symbols* $\mathfrak{A}_n \to \mathfrak{A}$, *if*

$$\|\mathfrak{A}_n u - \mathfrak{A}u\| \to 0 \tag{2.3}$$

for each $u \in C^\infty$.

Now (2.3) states that $\mathfrak{A}_n u \to \mathfrak{A}u$ *uniformly*. Conversely, if for each $u \in C^\infty$ the sequence $\{\mathfrak{A}_n u\}$ converges uniformly to a limit $v \in C^\infty$ an operator $\mathfrak{A}$ is defined by $\mathfrak{A}u = v$, and clearly $\mathfrak{A}_n \to \mathfrak{A}$.

Definition 3. *The convolution semi-group* $\{\mathfrak{Q}(t)\}$ *is continuous if*

$$\mathfrak{Q}(h) \to \mathbf{1} \qquad h \to 0+ \tag{2.4}$$

where $\mathbf{1}$ *is the identity operator. In this case we put* $\mathfrak{Q}(0) = \mathbf{1}$.

Since $\|\mathfrak{Q}(t)u\| \leq \|u\|$ we get from the definition (2.2) for $h > 0$

$$\|\mathfrak{Q}(t+h)u - \mathfrak{Q}(t)u\| \leq \|\mathfrak{Q}(h)u - u\|. \tag{2.5}$$

For h sufficiently small the left side will be $< \epsilon$ independently of t, and in this sense a continuous convolution semi-group is *uniformly continuous.*

Definition 4. *An operator* $\mathfrak{A}$ *from* C^∞ *to* C^∞ *is said to generate the convolution semi-group* $\{\mathfrak{Q}(t)\}$ *if as* $h \to 0+$

$$h^{-1}[\mathfrak{Q}(h) - \mathbf{1}] \to \mathfrak{A}. \tag{2.6}$$

We say, equivalently, that $\mathfrak{A}$ *is the generator.*[3]

More explicitly, whenever the limit exists the operator $\mathfrak{A}$ is defined by

$$t^{-1}\int_{-\infty}^{\infty} [u(x-y) - u(x)]\, Q_t\{dy\} \to \mathfrak{A}u(x). \tag{2.7}$$

Obviously a semi-group with a generator is automatically continuous. It will be shown that all continuous convolution semi-groups possess generators, but this is by no means obvious.

Formally (2.6) defines $\mathfrak{A}$ as derivative of $\mathfrak{Q}(t)$ at $t = 0$. Its existence implies differentiability at $t > 0$ since

$$\frac{\mathfrak{Q}(t+h) - \mathfrak{Q}(t)}{h} = \frac{\mathfrak{Q}(h) - \mathbf{1}}{h}\,\mathfrak{Q}(t) \to \mathfrak{A}\mathfrak{Q}(t) \tag{2.8}$$

as $h \to \mathbf{0}+$ and similarly for $h \to 0-$.

The following examples will be used in the sequel.

---

[3] Since we restrict the domain of definition of $\mathfrak{A}$ to C^∞ our terminology departs slightly from canonical usage as developed in E. Hille and R. S. Phillips (1957).

Examples. (*a*) *Compound Poisson semi-groups.* Let

$$Q_t = e^{-\alpha t}\sum_{k=0}^{\infty}\frac{(\alpha t)^k}{k!}F^{k\star} \tag{2.9}$$

be a compound Poisson distribution. Here

$$\mathfrak{Q}(h)u - u = (e^{-\alpha h}-1)u + \alpha h e^{-\alpha h}\left[\mathfrak{F}u+\frac{\alpha h}{2!}\mathfrak{F}^2u+\cdots\right]. \tag{2.10}$$

Dividing by h we see that (2.6) holds with $\mathfrak{A} = \alpha(\mathfrak{F}-\mathbf{1})$. Thus *the compound Poisson semi-group* (2.9) *is generated by* $\alpha(\mathfrak{F}-\mathbf{1})$ *and we shall indicate its elements by the abbreviation* $\mathfrak{Q}(t) = e^{\alpha(\mathfrak{F}-1)t}$.

(*b*) *Translations.* Denote by T_α the distribution concentrated at a and by $\mathfrak{T}(a)$ the associated operator. For fixed $\beta > 0$ the semi-group property $T_{\beta s}\star T_{\beta t} = T_{\beta(t+s)}$ holds and $\mathfrak{T}(\beta t)\,u(x) = u(x-\beta t)$. The graph of $\mathfrak{T}(\beta t)u$ is obtained by a translation from that of u and we speak of a *translation semi-group. The generator is given by* $-\beta\,\dfrac{d}{dx}$. Note that this generator is the limit as $h\to 0$ of the generator $\alpha(\mathfrak{F}-\mathbf{1})$ when $\alpha = \beta/h$ and F is concentrated at h. Now $\alpha(\mathfrak{F}-\mathbf{1})$ is a difference operator, and the passage to the limit was studied in VII,5. It is suggestive to indicate this semi-group by $\mathfrak{T}(t) = \exp\left(-\beta t\,\dfrac{d}{dx}\right)$.

(*c*) *Addition of generators.* Let $\mathfrak{A}_1$ and $\mathfrak{A}_2$ generate the convolution semi-groups $\{\mathfrak{Q}_1(t)\}$ and $\{\mathfrak{Q}_2(t)\}$. Then $\mathfrak{A}_1 + \mathfrak{A}_2$ *generates the convolution semi-group of operators* $\mathfrak{Q}(t) = \mathfrak{Q}_1(t)\,\mathfrak{Q}_2(t)$. [Such $\mathfrak{Q}(t)$ is associated with the convolution of the distributions associated with $\mathfrak{Q}_1(t)$ and $\mathfrak{Q}_2(t)$; see theorem 2 of VIII,3.] The assertion is obvious from the simple rearrangement

$$\frac{\mathfrak{Q}_1(h)\mathfrak{Q}_2(h) - \mathbf{1}}{h} = \frac{\mathfrak{Q}_1(h) - \mathbf{1}}{h} + \mathfrak{Q}_1(h)\cdot\frac{\mathfrak{Q}_2(h) - \mathbf{1}}{h}. \tag{2.11}$$

(*d*) *Translated semi-groups.* As a special case we get the rule: if $\mathfrak{A}$ generates the semi-group of operators $\mathfrak{Q}(t)$ associated with the distributions Q_t, then $\mathfrak{A} - \beta\, d/dx$ generates a semi-group $\{\mathfrak{Q}^\#(t)\}$ such that $Q_t^\#(x) = Q_t(x-\beta t)$.

(*e*) *Normal semi-groups.* Let Q_t stand for the normal distribution with zero expectation and variance ct. As already mentioned, these Q_t determine a semi-group, and we seek its generator as defined by (2.7). By Taylor's formula

$$u(x-y) - u(x) = -y\,u'(x) + \tfrac{1}{2}y^2\,u''(x) - \tfrac{1}{6}y^3\,u'''(x-\theta y). \tag{2.12}$$

The third absolute moment of Q_t is proportional to $t^{\frac{3}{2}}$, and we see thus that for functions with three bounded derivatives the limit in (2.7) exists and equals $\frac{1}{2}cu''(x)$. We express this by saying that $\mathfrak{A} = \frac{1}{2}cd^2/dx^2$. ▶

(For further examples see problems 3–5.)

Note *on the Fokker-Planck equation.* Consider the family of functions defined by $v(t, x) = \mathfrak{Q}(t)f(x)$. The relation (2.8) states that for smooth f

$$\frac{\partial v}{\partial t} = \mathfrak{A}v. \tag{2.13}$$

This is the Fokker–Planck equation of the process, and v is its unique solution satisfying the initial condition $v(0,x) = f(x)$. Equation (2.13) describes the process, and unnecessary complications are introduced by the traditional attempts to replace (2.13) by an equation for the transition probabilities Q_t themselves. Consider, for example, a translated compound Poisson semi-group generated by $\mathfrak{A} = \alpha(\mathfrak{F}-\mathbf{1}) - \beta\frac{d}{dx}$. The Fokker–Planck equation (2.13) holds whenever the initial function $f(x) = v(0,x)$ has a continuous derivative. Its formal analogue for the transition probabilities is given by

$$\frac{\partial Q_t}{\partial t} = -\beta\,\frac{\partial Q_t}{\partial x} - \alpha Q_t + \alpha F \bigstar Q_t \tag{2.14}$$

This equation makes sense only if Q has a density and is therefore not applicable to discrete processes. The usual reliance on (2.14) instead of (2.13) only causes complications.

3. PREPARATORY LEMMAS

In this section we collect a few simple lemmas on which the whole theory depends. Despite its simplicity the following inequality is basic.

Lemma 1. *If* $\mathfrak{A}$ *and* $\mathfrak{A}^\#$ *generate the convolution semi-groups* $\{\mathfrak{Q}(t)\}$ *and* $\{\mathfrak{Q}^\#(t)\}$, *respectively, then for all* $t > 0$

$$\|\mathfrak{Q}(t)u - \mathfrak{Q}^\#(t)u\| \le t\,\|\mathfrak{A}u - \mathfrak{A}^\#u\|. \tag{3.1}$$

Proof. From the semi-group property and the basic inequality (1.5) we get for $r = 1, 2, \ldots$

$$\begin{aligned} \|\mathfrak{Q}(t)u - \mathfrak{Q}^\#(t)u\| &\le r\left\|\mathfrak{Q}\left(\frac{t}{r}\right)u - \mathfrak{Q}^\#\left(\frac{t}{r}\right)u\right\| \\ &= t\left\|\frac{\mathfrak{Q}(t/r) - \mathbf{1}}{t/r}\,u - \frac{\mathfrak{Q}^\#(t/r) - \mathbf{1}}{t/r}\,u\right\|. \end{aligned} \tag{3.2}$$

As $r \to \infty$ the right side tends to the right side in (3.1) and so this inequality is true. ▶

Corollary. *Distinct convolution semi-groups cannot have the same generator.*

Lemma 2. (*Convergence.*) *For each n let $\mathfrak{A}_n$ generate the convolution semi-group $\{\mathfrak{Q}_n(t)\}$.*

If $\mathfrak{A}_n \to \mathfrak{A}$, *then $\mathfrak{A}$ generates a convolution semi-group $\{\mathfrak{Q}(t)\}$, and $\mathfrak{Q}_n(t) \to \mathfrak{Q}(t)$ for each $t > 0$.*

Proof. For each $t > 0$ the sequence $\{\mathfrak{Q}_n(t)u\}$ converges uniformly, since by (3.1)

$$\|\mathfrak{Q}_n(t)u - \mathfrak{Q}_m(t)u\| \leq t\,\|\mathfrak{A}_n u - \mathfrak{A}_m u\|. \tag{3.3}$$

By criterion 1 of VIII,3 there exists therefore an operator $\mathfrak{Q}(t)$ associated with a probability distribution such that $\mathfrak{Q}_n(t) \to \mathfrak{Q}(t)$. Then

$$\mathfrak{Q}_n(s+t) = \mathfrak{Q}_n(s)\,\mathfrak{Q}_n(t) \to \mathfrak{Q}(s)\,\mathfrak{Q}(t) \tag{3.4}$$

(by theorem 2 of VIII,3) and so $\{\mathfrak{Q}(t)\}$ is a convolution semi-group. To show that it is generated by $\mathfrak{A}$ note that

$$\left\|\frac{\mathfrak{Q}(t) - \mathbf{1}}{t}\,u - \mathfrak{A}u\right\| \leq \left\|\frac{\mathfrak{Q}_n(t) - \mathbf{1}}{t}\,u - \mathfrak{A}u\right\| + \frac{\|\mathfrak{Q}(t)u - \mathfrak{Q}_n(t)u\|}{t}.$$

The first term on the right tends to $\|\mathfrak{A}_n u - \mathfrak{A}u\|$ as $t \to 0$. Letting $m \to \infty$ in (3.3) we see that the second term is $< \|\mathfrak{A}u - \mathfrak{A}_n u\|$. For fixed n the upper limit of the left side is therefore $<2\,\|\mathfrak{A}u - \mathfrak{A}_n u\|$ which can be made arbitrarily small by choosing n sufficiently large. ▶

The next lemma makes it at least plausible that every continuous convolution semi-group has a generator.

Lemma 3. *Let $\{\mathfrak{Q}(t)\}$ be a continuous convolution semi-group. If for some sequence $t_1, t_2, \ldots, \to 0$*

$$\frac{\mathfrak{Q}(t_k) - \mathbf{1}}{t_k} \to \mathfrak{A}, \tag{3.5}$$

then $\mathfrak{A}$ generates the semi-group.

Proof. Call the left side $\mathfrak{A}_k$. As was shown in example 2(*a*) this $\mathfrak{A}_k$ generates a compound Poisson semi-group and by the last lemma there exists a semi-group $\{\mathfrak{Q}^{\#}(t)\}$ generated by $\mathfrak{A}$. To show that $\mathfrak{Q}^{\#}(t) = \mathfrak{Q}(t)$ we proceed as in (3.2) to obtain

$$\|\mathfrak{Q}(rt_k)u - \mathfrak{Q}^{\#}(rt_k)u\| \leq rt_k \left\|\mathfrak{A}_k u - \frac{\mathfrak{Q}^{\#}(t_k) - \mathbf{1}}{t_k}\,u\right\|. \tag{3.6}$$

Let $k \to \infty$ and $r \to \infty$ so that $rt_k \to t$. The right side tends to 0 and the left to $\|\mathfrak{Q}(t)u - \mathfrak{Q}^{\#}(t)u\|$ by virtue of (2.5). ▶

These are the lemmas that will be of immediate use. The next is recorded here because it is merely a variant of lemma 2 and the proof is nearly the same. We shall use only the special case $\nu_n = n$ and $t = 1$, which will serve as the connecting link between triangular arrays and convolution semi-groups.

Lemma 4. *For each* n *let* $\mathfrak{F}_n$ *be the operator associated with the probability distribution* F_n. *If*

$$n(\mathfrak{F}_n - \mathbf{1}) \to \mathfrak{A}, \tag{3.7}$$

then $\mathfrak{A}$ *generates a convolution semi-group* $\{\mathfrak{Q}(t)\}$. *If* $n \to \infty$ *and* $\dfrac{\nu_n}{n} \to t$ *then*

$$\mathfrak{F}_n^{\nu_n} \to \mathfrak{Q}(t). \tag{3.8}$$

In particular, $\mathfrak{F}_n^n \to \mathfrak{Q}(1)$. The lemma remains true if n is restricted to a sequence $n_1, n_2, \ldots$.

Proof. The left side in (3.7) generates a compound Poisson semi-group [example 2(*a*)] and so $\mathfrak{A}$ is a generator by lemma 2. By the basic inequality (1.5)

$$\|\mathfrak{F}_n^{\nu_n} u - \mathfrak{Q}(\nu_n/n)u\| \leq \nu_n/n \, \|n[\mathfrak{F}_n u - u] - n[\mathfrak{Q}(1/n)u - u]\| \tag{3.9}$$

and for $u \in C^\infty$ each of the terms within the norm signs tends uniformly to $\mathfrak{A}u$. ▶

4. FINITE VARIANCES

Semi-groups of distributions with finite variances are of special importance and their theory is so simple that it deserves a special treatment. Many readers will not be interested in the more complicated general semi-groups, and for others this section may provide an interesting introductory example.

We consider a convolution semi-group $\{\mathfrak{Q}(t)\}$ and denote the associated probability distributions by Q_t. Suppose that Q_t has a finite variance $\sigma^2(t)$. Because of the semi-group property $\sigma^2(s+t) = \sigma^2(s) + \sigma^2(t)$ and the only positive solution of this equation[4] is of the form $\sigma^2(t) = ct$.

Suppose that Q_t is centered to zero expectation. The second moment then induces a *probability distribution* Ω_t defined by

$$\Omega_t\{dy\} = \frac{1}{ct} y^2 Q_t\{dy\}. \tag{4.1}$$

[4] The equation $\varphi(s+t) = \varphi(s) + \varphi(t)$ is called the Hamel equation. Putting $u(t) = e^{\varphi(t)}$ one gets $u(s+t) = u(s)\,u(t)$ in which form the equation was encountered several times and is treated in **1**; XVII,6. The expectation of Q_t is also a solution of the Hamel equation; it is therefore either of the form mt or exceedingly weird. See section 5*a*.

By the selection theorem there exists a sequence $\{t_n\}$ tending to 0 such that as t runs through it Ω_t tends to a possibly defective distribution Ω.

Since Q_t is centered to zero expectation we have the identity

$$\frac{\mathfrak{Q}(t) - \mathbf{1}}{t}\, u(x) = c\int_{-\infty}^{+\infty} \frac{u(x-y) - u(x) + y\, u'(x)}{y^2}\, \Omega_t\{dy\} \tag{4.2}$$

the integrand being (for fixed x) a continuous function of y assuming at the origin the value $\frac{1}{2}u''(x)$. At infinity the integrand and its derivative vanish. This implies that as t runs through $\{t_n\}$, and consequently $\Omega_t \to \Omega$, the integral in (4.2) tends uniformly to the analogous integral with respect to the limit distribution Ω. According to lemma 3 of section 3 this means that our semi-group has a generator $\mathfrak{A}$ given by

$$\mathfrak{A}u(x) = c\int_{-\infty}^{+\infty} \frac{u(x-y) - u(x) + y\, u'(x)}{y^2}\, \Omega\{dy\}. \tag{4.3}$$

This representation of $\mathfrak{A}$ is *unique* because for functions of the form

$$u(x) = 1 + \frac{x^2}{1 + x^2} f(-x) \tag{4.4}$$

one gets

$$\mathfrak{A}u(0) = c\int_{-\infty}^{+\infty} \frac{f(y)}{1 + y^2}\, \Omega\{dy\}. \tag{4.5}$$

The knowledge of $\mathfrak{A}u$ for all $u \in C^\infty$ therefore uniquely determines the measure $(1+y^2)^{-1}\,\Omega\{dy\}$ and hence Ω itself.

In consequence of this uniqueness the limit distribution Ω is independent of the sequence $\{t_k\}$ and hence $\Omega_t\{dy\} \to \Omega\{dy\}$ for any approach $t \to 0$.

We shall show that Ω is a proper probability distribution and that every operator of the form (4.3) is a generator. The proof depends on two special cases contained in the following

Examples. (*a*) *Normal semi-groups.* When Ω is the probability distribution concentrated at the origin (4.3) reduces to $\mathfrak{A}u(x) = \frac{1}{2}cu''(x)$. We saw in example 2(*e*) that this $\mathfrak{A}$ generates a semi-group of normal distributions with zero expectations and variances ct. It is easy to verify directly that the distributions Ω_t tend to the probability distribution concentrated at the origin.

(*b*) *Compound Poisson semi-groups.* Let F be a probability distribution concentrated on the intervals $|x| > \eta$ having expectation m_1 and variance m_2. The distributions Q_t of the compound Poisson semi-group of example 2(*a*) have expectations $\alpha m_1 t$ and variances $\alpha m_2 t$. The semi-group is generated by $\alpha(\mathfrak{F} - \mathbf{1})$. In accordance with example 2(*d*) the same Q_t but centered to zero expectations form a semi-group generated by

$$\alpha[\mathfrak{F}-\mathbf{1}-m_1 d/dx]$$

that is

$$\mathfrak{A}u(x) = \int_{|y|>\eta} [u(x-y) - u(x) - y\,u'(x)]\,F\{dy\}. \tag{4.6}$$

With the change of notations $\Omega\{dy\} = y^2 F\{dy\}/m_2$ and $\alpha m_2 = c$ this reduces to (4.3). Conversely, if Ω is a probability distribution concentrated on $|x| > 0$, then (4.3) may be rewritten in the form (4.6) and hence such $\mathfrak{A}$ generates a compound Poisson distribution whose distributions Q_t have zero expectation and variances $m_2 t = ct$.

We are now in a position to formulate the basic

Theorem. *Let* Q_t *have zero expectation and variance* ct. *The convolution semi-group* $\{\mathfrak{Q}(t)\}$ *then has a generator* $\mathfrak{A}$ *of the form* (4.3) *where* Ω *is a proper probability distribution. The representation* (4.3) *is unique. Conversely, every operator of this form generates a convolution semi-group of distributions with zero expectation and variance* ct.

Proof. We have shown the existence of a generator of the form (4.3) but have proved only that Ω has a total mass $\omega \leq 1$. It remains to prove that if Ω has a mass ω then the operator $\mathfrak{A}$ of (4.3) generates a semi-group such that Q_t has zero expectation and a variance $\leq c\omega t$.

Let $\mathfrak{A}_\eta$ be the operator obtained from (4.3) by deleting the intervals $0 < |y| \leq \eta$ from the domain of integration. Denote the masses attributed by Ω to the origin and to $|y| > \eta$ respectively by m and ω_η. It follows from the preceding examples that $\mathfrak{A}_\eta$ is the sum of two operators, of which the first generates a normal semi-group with variances cmt, and the second generates a compound Poisson semi-group with variances $c\omega_\eta t$. By the addition rule of example 2(*c*) the operator $\mathfrak{A}_\eta$ itself generates a semi-group with variances $c(m+\omega_\eta)t$. Being the limit of $\mathfrak{A}_\eta$ as $\eta \to 0$ the operator $\mathfrak{A}$ itself is the generator of a semi-group. The variances of the associated distributions lie between the variances $c(m+\omega_\eta)t$ corresponding to $\mathfrak{A}_\eta$, and their limit $c\omega t$. This proves that $\mathfrak{A}$ indeed generates a semi-group with variances $c\omega t$. ▶

5. THE MAIN THEOREMS

In this section $\{\mathfrak{Q}(t)\}$ stands for an arbitrary continuous convolution semi-group, and the associated distribution functions are again denoted by Q_t. By analogy with (4.1) we define a new measure Ω_t by

$$\Omega_t\{dy\} = t^{-1}y^2\,Q_t\{dy\}. \tag{5.1}$$

The novel feature is that, in the absence of a second moment of Q_t, the measure Ω_t need not be finite on the whole line. However, $\Omega_t\{I\}$ is finite for every finite interval I. Furthermore, since $Q_t\{dy\} = ty^{-2}\,\Omega_t\{dy\}$ it follows that y^{-2} is integrable with respect to Ω_t over any domain excluding a neighborhood of the origin. We shall see that as $t \to 0$ the measures Ω_t will converge to a measure Ω with similar properties and this Ω will determine the generator of the semi-group. It is therefore convenient to introduce the

Definition. *A measure Ω on the real line will be called canonical if $\Omega\{I\} < \infty$ for finite intervals I and if the integrals*

$$\psi^+(x) = \int_x^\infty y^{-2}\,\Omega\{dy\}, \qquad \psi^-(-x) = \int_{-\infty}^{-x} y^{-2}\,\Omega\{dy\} \tag{5.2}$$

converge for each $x > 0$.

(For definiteness we take the intervals of integration closed.)

We proceed to show that the theory of the preceding section carries over except that we have to deal with canonical measures rather than with probability distributions and that in the absence of expectations we must resort to an artificial centering. We define *the truncation function τ_s as the continuous monotone function such that*

$$\tau_s(x) = x \quad when \quad |x| \le s, \qquad \tau_s(x) = \pm s \quad when \quad |x| \ge s \tag{5.3}$$

where $s > 0$ is arbitrary, but fixed.

In analogy with (4.2) we now have the identity

$$\frac{\mathfrak{Q}(t) - \mathbf{1}}{t}\,u(x) = \int_{-\infty}^{+\infty} \frac{u(x-y) - u(x) - \tau_s(y)\,u'(x)}{y^2}\,\Omega_t\{dy\} + b_t\,u'(x) \tag{5.4}$$

where

$$b_t = \int_{-\infty}^{+\infty} \tau_s(y) y^{-2}\,\Omega_t\{dy\} = t^{-1}\int_{-\infty}^{+\infty} \tau_s(y)\,Q_t\{dy\}. \tag{5.5}$$

The integrand in (5.4) is again (for fixed x) a bounded continuous function assuming at the origin the value $\frac{1}{2}u''(x)$. It will be noted that the special choice (5.3) for the truncation function is not important: we could choose for τ any bounded continuous function provided it is near the origin twice continuously differentiable with $\tau(0) = \tau''(0) = 0$ and $\tau'(0) = 1$.

We have now a setup similar to the one in the preceding section, and we derive a similar theorem. The integral in (5.4) makes sense with Ω_t replaced by any canonical measure, and we define an operator $\mathfrak{A}^{(\tau)}$ by

$$\mathfrak{A}^{(\tau)}u(x) = \int_{-\infty}^{+\infty} \frac{u(x-y) - u(x) - \tau_s(y)\,u'(x)}{y^2}\,\Omega\{dy\}. \tag{5.6}$$

The superscript τ serves to indicate the dependence on the truncation function τ. A change of τ_s [or of the point s in the definition (5.3)] amounts to adding a term bd/dx to $\mathfrak{A}^{(\tau)}$, and so the family of operators

$$\mathfrak{A} = \mathfrak{A}^{(\tau)} + bd/dx \tag{5.7}$$

is independent of the choice of τ_s.

Theorem 1. *A continuous convolution semi-group* $\{\mathfrak{Q}(t)\}$ *has a generator* $\mathfrak{A}$, *and* $\mathfrak{A}$ *is of the form determined by* (5.6)–(5.7) *where* Ω *is a canonical measure.*

Conversely, every operator $\mathfrak{A}$ *of this form generates a continuous convolution semi-group* $\{\mathfrak{Q}(t)\}$. *The measure* Ω *is unique, and as* $t \to 0$

$$\Omega_t\{I\} \to \Omega\{I\} \tag{5.8}$$

for infinite intervals[5] I *and*

$$t^{-1}[1-\mathfrak{Q}_t(x)] \to \psi^+(x), \qquad t^{-1}\mathfrak{Q}_t(-x) \to \psi^-(-x) \tag{5.9}$$

for $x > 0$.

Proof. We return for the moment to the notations of section 1. For each n we consider the sum $\mathbf{S}_n = \mathbf{X}_{1,n} + \cdots + \mathbf{X}_{n,n}$ of n mutually independent variables with the common distribution $Q_{1/n}$. Then $\mathbf{S}_n$ has the distribution Q_1, and hence $\mathbf{S}_n$ remains stochastically bounded (definition 2 of VIII,2). We now anticipate lemma 1 of section 7 according to which this implies that for each finite interval I the measures $\Omega_{1/n}\{I\}$ remain bounded, and that to each $\epsilon > 0$ there corresponds a number $a > 0$ such that

$$n[1 - Q_{1/n}(a) + Q_{1/n}(-a)] < \epsilon \tag{5.10}$$

for all n. (The lemma is quite simple, but its proof is postponed in order not to interrupt the argument.)

By the selection theorem there exists a measure Ω and a sequence of integers n_k such that as t^{-1} runs through it $\Omega_t\{I\} \to \Omega\{I\}$ for finite intervals. The contribution of I to the integral in (5.4) then converges to the corresponding contribution of I to the integral (5.6) defining $\mathfrak{A}^{(\tau)}$. Remembering that $y^{-2}\,\Omega_t\{dy\} = Q_t\{dy\}$ it is seen from (5.10) that the contribution of $|y| > a$ to these integrals is uniformly small if a is chosen large enough. We conclude that Ω is a canonical measure and that as t^{-1} runs through $\{n_k\}$ the integral in (5.4) converges uniformly to the integral in (5.6). Furthermore, under the present conditions the quantity b_t of (5.5) remains bounded and hence there is no loss of generality in supposing that the sequence $\{n_k\}$ was picked so that as t^{-1} runs through it b_t converges to a number b. Then as t^{-1} runs through $\{n_k\}$

$$t^{-1}[(\mathfrak{Q}t)-1]u(x) \to \mathfrak{A}u(x) \tag{5.11}$$

[5] Here and in the sequel it is understood convergence is required only for intervals and points of continuity.

the convergence being uniform. By lemma 3 of section 3 this means that the semi-group $\{\mathfrak{Q}(t)\}$ is generated by $\mathfrak{A}$, and so (5.11) holds for any approach $t \to 0$.

We have thus shown that a generator $\mathfrak{A}$ exists and can be expressed in the form (5.6)–(5.7) in terms of a canonical measure Ω. As in the preceding section the uniqueness of Ω in this representation follows from the fact that for functions of the form (4.4) the value $\mathfrak{A}u(0)$ is given by (4.5) (with c now absorbed in Ω).

The uniqueness of Ω implies that (5.8) holds for an arbitrary approach $t \to 0$. Furthermore, (5.10) guarantees that the quantities in (5.9) are uniformly small for x sufficiently large. These quantities were defined by (5.2) and the analogous relations with Ω replaced by Ω_t. This shows that (5.9) is a consequence of (5.8).

It remains to show that the measure Ω can be chosen arbitrarily. The proof is the same as in the case of finite variances. As in example 4(*b*) it is seen that if Ω is concentrated on $|y| > \eta > 0$ the operator $\mathfrak{A}$ of (5.6)–(5.7) generates a compound Poisson semi-group with modified centering (but without finite expectations). Thus $\mathfrak{A}$ can again be represented as a limit of generators and is therefore a generator. ►

Example. *Cauchy semi-groups.* The distributions Q_t with density $\pi^{-1}t(t^2+x^2)^{-1}$ form a semi-group. It is easily verified that the limits in (5.9) are given by $\psi^+(x) = \psi^-(-x) = \pi x^{-1}$, and $\pi\Omega$ coincides with the Lebesgue measure or ordinary length. ►

The following theorem embodies various important characterizations of infinitely divisible distributions. The proof of part (v) is postponed to section 7. (For an alternative direct proof see problem 11.) This part admits of further generalizations to triangular arrays with variable distributions. (See section 9. The full theory will be developed in chapter XVII.) For the history of the theory see VI,3.

Theorem 2. *The following classes of probability distributions are identical:*

(i) *Infinitely divisible distributions.*

(ii) *Distributions associated with continuous convolution semi-groups (that is, distributions of increments in processes with stationary independent increments).*

(iii) *Limits of sequences of compound Poisson distributions.*

(iv) *Limits of sequences of infinitely divisible distributions.*

(v) *Limit distributions of row sums in triangular arrays* $\{\mathbf{X}_{k,n}\}$ *where the variables* $\mathbf{X}_{k,n}$ *of the* nth *row have a common distribution.*

Proof. Let $\{\mathfrak{Q}(t)\}$ be a continuous convolution semi-group. It was shown in example 2(*a*) that for fixed $h > 0$ the operator $\mathfrak{A}_h = [\mathfrak{Q}(h)-1]/h$ generates a compound Poisson semi-group of operators $\mathfrak{Q}_h(t)$. As $h \to 0$

the generators $\mathfrak{A}_h$ converge to the generator $\mathfrak{A}$, and hence $\mathfrak{Q}_h(t) \to \mathfrak{Q}(t)$ by lemma 2 of section 3. Thus Q_t is the limit of compound Poisson distributions, and so the class (ii) is contained in (iii). The class (iii) is trivially contained in (iv).

For each n let $G^{(n)}$ be an infinitely divisible distribution. By definition $G^{(n)}$ is the distribution of the sum of n independent identically distributed random variables, and in this way the sequence $\{G^{(n)}\}$ gives rise to a triangular array as described under (v). Thus the class (iv) is contained in (v). In section 7 it will be shown that (v) is contained in (ii), and so the classes (ii)–(v) are identical. Finally, the class of all infinitely divisible distributions is a subclass of (iv) and contains (ii). ►

According to the last theorem every infinitely divisible distribution F appears as a distribution Q_t of an appropriate convolution semi-group. The value of the parameter t can be fixed arbitrarily by an appropriate change of scale on the t-axis. However, *there exists only one semi-group* $\{\mathfrak{Q}(t)\}$ *to which an infinitely divisible distribution* F *belongs.* This amounts to saying that the representation $F = F_n^{n\star}$ of F as an n-fold convolution is unique. This assertion is plausible, but requires proof. As a matter of fact, in its Fourier theoretic version the uniqueness becomes obvious, whereas in the present context the proof would detract from the main topic without being illuminating. For this reason we desist for once from proving the statement within both frameworks.

Application to stochastic processes. Let $\mathbf{X}(t)$ be the variable of a stochastic process with *stationary independent increments* (VI,4) and let us interpret Q_t as the distribution of the increment $\mathbf{X}(t+s) - \mathbf{X}(s)$. Consider a time interval $\overline{s, s+1}$ of unit length and subdivide it by the points $s = s_0 < s_1 < \cdots < s_n = s + 1$ into subintervals of length n^{-1}. Then $\mathbf{P}\{\mathbf{X}(s_k) - \mathbf{X}(s_{k-1}) > x\} = 1 - Q_{1/n}(x)$ and so $n[1-Q_{1/n}(x)]$ equals the expected number of intervals $\overline{s_{k-1}, s_k}$ with increment $>x$. As $n \to \infty$ this expected number tends to $\psi^+(x)$. For simplicity of discussion suppose that the limits $\mathbf{X}(t+)$ and $\mathbf{X}(t-)$ exist for all t and that $\mathbf{X}(t)$ lies between them. Let $\overline{s_{k-1}, s_k}$ be the interval of our partition containing t. For n sufficiently large the increment $\mathbf{X}(s_k) - \mathbf{X}(s_{k-1})$ will be close to the jump $\mathbf{X}(t+) - \mathbf{X}(t-)$ and it is intuitively clear that the limit $\psi^+(x)$ *represents the expected number of epochs* t *per unit time at which* $\mathbf{X}(t+) - \mathbf{X}(t-) > x$. The argument may be justified rigorously but we shall not enter into details. It follows from this result that the expected number of discontinuities is zero only if $\psi^+(x) = 0$ and $\psi^-(-x) = 0$ for all $x > 0$. In this case Ω is concentrated at the origin, that is, the increments $\mathbf{X}(t+s) - \mathbf{X}(s)$ are normally distributed. For such a process the paths are continuous with probability one (theorem of P. Lévy and N. Wiener) and so *the paths are continuous with probability* 1 *iff the process is normal.*

As a second illustration consider the *compound Poisson process* (2.8). The expected number of jumps per unit time is α, and the probability of a jump exceeding $x > 0$ is $1 - F(x)$. Thus $\alpha[1-F(x)]$ is the expected number of jumps $>x$ in full agreement with our intuitive argument.

*5a. Discontinuous Semi-groups

It is natural to ask whether there exist discontinuous semi-groups. The question is of no practical importance but the answer has some curiousity value: *Every convolution semi-group* $\{\mathfrak{Q}(t)\}$ *differs merely by centering from a continuous semi-group* $\{\mathfrak{Q}^{\#}(t)\}$. In particular, *if the distributions* Q_t *are symmetric the semi-group is necessarily continuous.* In the general case there exists a function φ such that the distributions $Q_t^{\#}$ defined by $Q_t(x+\varphi(t))$ *are associated with a continuous semi-group.* The function φ must obviously satisfy

$$\varphi(t+s) = \varphi(t) + \varphi(s). \tag{5.12}$$

This is the famous Hamel equation whose only continuous solution is of the form ct (see footnote 4 to section 4). In fact, the only Baire function satisfying (5.12) is linear. The other solutions are weird indeed; for example, a non-linear solution assumes in *every* interval arbitrarily large and arbitrarily small values, and it is impossible to represent it analytically by limiting processes. In short, it is fair to ask in what precise sense it "exists."

To return to earth, consider an arbitrary convolution semi-group $\{\mathfrak{Q}(t)\}$ and the triangular array $\{\mathbf{X}_{k,n}\}$ associated with the distributions $Q_{1/n}$. The row sums $\mathbf{S}_n$ have the common distribution Q_1 and hence we can use the last lemma to extract a sequence $n_1, n_2, \ldots$ such that as n runs through it $n[\mathfrak{Q}(1/n)-1] \to \mathfrak{A}^{\#}$ where $\mathfrak{A}^{\#}$ is the generator of a continuous semi-group $\{\mathfrak{Q}^{\#}(t)\}$. We may choose the n_k of the form 2^ν. The inequality (3.2) now shows that $\mathfrak{Q}(t) = \mathfrak{Q}^{\#}(t)$ for all t that are multiples of $1/n_k$ for arbitrarily large k, that is, for all t of the form $t = a2^{-\nu}$ with a and ν integers. Thus *there exists always a continuous semi-group* $\{\mathfrak{Q}^{\#}(t)\}$ *such that* $\mathfrak{Q}(t) = \mathfrak{Q}^{\#}(t)$ *for all* t *of a dense set* Σ.

We are now in a position to prove the initial proposition. Choose $\epsilon_n > 0$ such that $t + \epsilon_n$ is in Σ. Then

$$\mathfrak{Q}^{\#}(t+\epsilon_n) = \mathfrak{Q}(t+\epsilon_n) = \mathfrak{Q}(t)\,\mathfrak{Q}(\epsilon_n). \tag{5.13}$$

As $\epsilon_n \to 0$ the left side tends to $\mathfrak{Q}^{\#}(t)$ and hence it suffices to show that if $\mathfrak{Q}(\epsilon_n) \to \mathfrak{F}$ then the distribution F is concentrated at a single point. Choose points h_n in Σ such that $0 < \epsilon_n < h_n$ and $h_n \to 0$. Then $\mathfrak{Q}^{\#}(h_n) = \mathfrak{Q}(h_n-\epsilon_n)\,\mathfrak{Q}(\epsilon_n)$. The left side tends to the identity operator, and so F can indeed have only one point of increase.

6. EXAMPLE: STABLE SEMI-GROUPS

A semi-group $\{\mathfrak{Q}(t)\}$ is called *stable* if its distributions are of the form

$$Q_t(x) = G(\lambda_t(x-\beta_t)) \tag{6.1}$$

where $\lambda_t > 0$ and β_t are constants depending continuously on t, and G is a fixed distribution. Obviously, G is a stable distribution as defined in VI,1. The theory of stable semi-groups is here developed principally as an illustration for the results of the last section and to put on record the form of their generators. In an indirect way the results of this section are derived independently in section 8.

Because of the assumed continuity of λ_t and β_t the semi-group (if it exists) is continuous. As $t \to 0$ the distribution Q_t tends to the distribution concentrated at the origin, and hence $\lambda_t \to \infty$ and $\beta_t \to 0$. The first relation in (5.14) takes on the form

$$t^{-1}[1 - G(\lambda_t(x-\beta_t))] \to \psi^+(x), \qquad x > 0. \tag{6.2}$$

Since $\beta_t \to 0$ and G is monotone this relation remains valid also when β_t is dropped, and then (6.2) may be rewritten in the form

$$\frac{1 - G(\lambda_t x)}{1 - G(\lambda_t)} \to \psi^+(x). \tag{6.3}$$

(Here we assume that 1 is a point of continuity for ψ^+, which can be achieved by a change of scale.) Now (6.3) is a particular case of the relation VIII,(8.1) defining regular variation. We conclude that either ψ^+ vanishes identically, or else the tail $1 - G$ varies regularly at infinity and

$$\psi^+(x) = c^+x^{-\alpha}, \qquad x > 0, \quad c^+ > 0. \tag{6.4}$$

On the positive half-axis the measure Ω has therefore the density $\alpha c^+ x^{-\alpha-1}$. We conclude that $0 < \alpha < 2$ because Ω attributes finite masses to finite neighborhoods of the origin, and $\psi^+(x) \to 0$ as $x \to \infty$. For similar reasons either ψ^- vanishes identically or else $\psi^-(x) = c^- |x|^{-\alpha}$ for $x < 0$. The exponent α is the same for the two tails, because also the tail sum $1 - G(x) + G(-x)$ varies regularly.

The functions ψ^+ and ψ^- determine the measure Ω up to a possible atom at the origin. We shall see that such an atom cannot exist unless both ψ^+ and ψ^- vanish identically and Ω is concentrated at the origin.

The generator $\mathfrak{A}$ is given by (5.7). In the present case it is convenient to write it in the form

$$\mathfrak{A} = c^+\mathfrak{A}_\alpha^+ + c^-\mathfrak{A}_\alpha^- + b\,d/dx, \tag{6.5}$$

where the operators $\mathfrak{A}^+$ and $\mathfrak{A}^-$ describe the contributions of the two half axes and are defined as follows.

If $0 < \alpha < 1$

$$\mathfrak{A}_\alpha^+ u(x) = \int_0^\infty [u(x-y) - u(x)]y^{-\alpha-1}\,dy. \tag{6.6}$$

If $1 < \alpha < 2$

$$\mathfrak{A}_\alpha^+ u(x) = \int_0^\infty [u(x-y) - u(x) - y\,u'(x)]y^{-\alpha-1}\,dy, \tag{6.7}$$

and finally if $\alpha = 1$

$$\mathfrak{A}_1^+ u(x) = \int_0^\infty [u(x-y) - u(x) - \tau_s(y)\, u'(x)]y^{-2}\, dy. \tag{6.8}$$

$\mathfrak{A}_\alpha^-$ is defined by the analogous integral over $\overline{-\infty, 0}$ with $y^{-\alpha-1}$ replaced by $|y|^{-\alpha-1}$. The centering in (6.6) and (6.7) differs from that in the canonical form (5.6), but the difference is absorbed in the term bd/dx in (6.5). An atom of Ω at the origin would add a term $\gamma d^2/dx^2$ to the generator $\mathfrak{A}$. It was shown in example 4(*a*) that this term by itself generates a semi-group of normal distributions.

Theorem. (*a*) *When* $b = 0$ *and* $0 < \alpha < 1$ *or* $1 < \alpha < 2$ *the operator* (6.5) *generates a strictly stable semi-group of the form*

$$Q_t(x) = G(xt^{-1/\alpha}); \tag{6.9}$$

when $\alpha = 1$ *and* $b = 0$ *it generates a stable semi-group of the form*

$$Q_t(x) = G(xt^{-1} - (c^+ - c^-)\log t). \tag{6.10}$$

(*b*) *A stable semi-group is either generated by* (6.5) *or else it is a semi-group of normal distributions.*

[We saw in 2(*b*) that bd/dx generates a translation semi-group and to obtain the semi-group generated by (6.5) with $b \neq 0$ it suffices in (6.9) and (6.10) to replace x by $x + bt$.]

Proof. (*a*) A change of scale changes the distribution Q_t of a semi-group into distributions defined by $Q_t^\#(x) = Q_t(x/\rho)$. These form a new semi-group $\{\mathfrak{Q}^\#(t)\}$. If we put $v(x) = u(\rho x)$ it is seen from the definition of a convolution that $\mathfrak{Q}^\#(t)\, u(x) = \mathfrak{Q}(t)\, v(x/\rho)$. For the generators this means that to find $\mathfrak{A}^\# u(x)$ we have simply to calculate $\mathfrak{A}v(x)$ and replace x by x/ρ. The substitution $y = z/\rho$ in (6.7) and (6.8) shows that for the corresponding generators $\mathfrak{A}_\alpha^\# = \rho^\alpha \mathfrak{A}_\alpha$. How $\rho^\alpha \mathfrak{A}_\alpha$ is obviously the generator of the semi-group $\{\mathfrak{Q}(\rho^\alpha t)\}$, and from the uniqueness of generators we conclude therefore that $Q_t(x/\rho) = Q_{t\rho^\alpha}(x)$. Letting $G = Q_1$ and $\rho = t^{-1/\alpha}$ we get (6.9).

A similar argument applies in the case $\alpha = 1$, except that when the substitution $y = z/\rho$ is used in (6.8) the centering function gives rise to an additional term of the form $(c^+ - c^-)(\rho \log \rho)\, u'(x)$, and this leads to (6.10).

(*b*) To measure Ω concentrated at the origin there corresponds a normal semi-group. We saw that the generator of any other stable semi-group is of the form $\mathfrak{A}_\alpha + \gamma d^2/dx^2$. As shown in example 2(*c*) the distributions of the corresponding semi-group would be the convolutions of our stable Q_t with normal distributions with variance $2\gamma t$ and it is clear that such a semi-group cannot be stable. ▶

The theorem asserts that the function λ_t occurring in the definition of stable semi-groups is of the form $\lambda_t = t^{-1/\alpha}$. Considering (6.2) and its analogue for the negative half-axis we get the

Corollary. *If* $0 < \alpha < 2$, $c^+ \geq 0$, $c^- \geq 0$ *(but* $c^+ + c^- > 0$*) there exists exactly one stable distribution function* G *such that as* $x \to \infty$

$$x^\alpha[1-G(x)] \to c^+, \qquad x^\alpha G(-x) \to c^-. \tag{6.11}$$

The normal distribution is the only remaining stable distribution [and satisfies (6.11) *with* $\alpha = 2$ *and* $c^+ = c^- = 0$].

The assertion within brackets will be proved in section 8.

7. TRIANGULAR ARRAYS WITH IDENTICAL DISTRIBUTIONS

For each n let $\mathbf{X}_{1,n}, \ldots, \mathbf{X}_{n,n}$ be mutually independent random variables with a common distribution F_n. We are interested in the possible limit distributions of the sums $\mathbf{S}_n = \mathbf{X}_{1,n} + \cdots + \mathbf{X}_{n,n}$, but it is useful to begin by investigating a necessary condition for the existence of a limit distribution, namely the requirement that the sequence $\{\mathbf{S}_n\}$ be *stochastically bounded.* We recall from VIII,2 that $\{\mathbf{S}_n\}$ is said to be stochastically bounded if to each $\epsilon > 0$ there corresponds an a such that $\mathbf{P}\{|\mathbf{S}_n| > a\} < \epsilon$ for all n. Very roughly speaking this means that no probability mass flows out to infinity. Obviously this is a necessary condition for the existence of a proper limit distribution.

We shall rely heavily on truncation. It is most convenient to use once more the truncation function τ_s introduced in (5.3) in order to avoid discontinuities: τ_s is the continuous monotone function such that $\tau_s(x) = x$ when $|x| \leq s$ and $\tau_s(x) = \pm s$ when $|x| \geq s$. With this truncation function we put

$$\mathbf{X}'_{k,n} = \tau_s(\mathbf{X}_{k,n}), \qquad \mathbf{X}_{k,n} = \mathbf{X}_{k,n} = \mathbf{X}'_{k,n} + \mathbf{X}''_{k,n}. \tag{7.1}$$

The new variables depend on the parameter s even though our notation does not emphasize it. The row sums of the triangular arrays $\{\mathbf{X}'_{k,n}\}$ and $\{\mathbf{X}''_{k,n}\}$ will be denoted by $\mathbf{S}'_n$ and $\mathbf{S}''_n$. Thus $\mathbf{S}_n = \mathbf{S}'_n + \mathbf{S}''_n$. The variables $\mathbf{X}'_{k,n}$ are bounded, and for their expectation we write

$$\beta_n = \mathbf{E}(\mathbf{X}'_{k,n}). \tag{7.2}$$

(β_n is, of course, independent of k.) Finally we introduce the analogue to the measures Ω_t of section 5, namely the measure Φ_n defined by

$$\Phi_n\{dx\} = nx^2 F_n\{dx\}. \tag{7.3}$$

$\Phi_n\{I\}$ is finite for finite intervals I, but the whole line may receive an infinite mass.

It is plausible that $\{\mathbf{S}_n\}$ cannot remain stochastically bounded unless the individual components become small in the sense that

$$\mathbf{P}\{|\mathbf{X}_{k,n}| > \epsilon\} \to 0, \qquad n \to \infty, \tag{7.4}$$

for every $\epsilon > 0$. (The left side is independent of k.) Arrays with this property are called *null-arrays*. We shall see that only null-arrays can have stochastically bounded row sums, but for the time being we introduce (7.4) as a starting assumption.

The "necessary" part of the following lemma was used in the proof of theorem 1 in section 5; there the condition (7.4) was fulfilled because the semi-group was continuous.

Lemma. (*Compactness.*) *In order that the row sums* $\mathbf{S}_n$ *of a null-array* $\{\mathbf{X}_{k,n}\}$ *remain stochastically bounded it is necessary and sufficient*

(i) *that* $\Phi_n\{I\}$ *remains bounded for every finite interval* I, *and*

(ii) *that for large* x *the tail sums*

$$T_n(x) = n[1 - F_n(x) + F_n(-x)] \tag{7.5}$$

are uniformly small.

In other words, it is required that to each $\epsilon > 0$ there corresponds a t such that $T_n(x) < \epsilon$ for $x > t$. (Note that T_n is a decreasing function.)

Proof. In the special case of symmetric distributions F_n the necessity of condition (ii) is apparent from the inequality

$$\mathbf{P}\{|\mathbf{S}_n| > a\} \geq \tfrac{1}{2}(1 - \exp(-T_n(a))) \tag{7.6}$$

[see V,(5.10)]. For arbitrary F_n we apply the familiar symmetrization. Together with $\mathbf{S}_n$ the symmetrized variables ${}^0\mathbf{S}_n$ also remain stochastically bounded, and therefore condition (ii) applies to the tails ${}^0T_n$ of the symmetrized distributions. But for a null-array it is clear that for each $\delta > 0$ ultimately ${}^0T_n(a) \geq \tfrac{1}{2}T_n(a + \delta)$, and so condition (ii) is necessary in all cases.

Assuming that condition (ii) is satisfied, the truncation point s can be chosen so large that $T_n(s) < 1$ for all n. The number of terms among $\mathbf{X}''_{1,n}, \ldots, \mathbf{X}''_{n,n}$ that are different from 0 is then a binomial random variable with expectation and variance less than 1. It is therefore possible to pick numbers N and c such that with probability arbitrarily close to 1 fewer than N among the variables $\mathbf{X}''_{k,n}$ will be different from 0 and all of them $\leq c$. This means that the sums $\mathbf{S}''_n$ remain stochastically bounded, and under these circumstances $\{\mathbf{S}_n\}$ is stochastically bounded iff $\{\mathbf{S}'_n\}$ is.

It remains to show that condition (i) is necessary and sufficient for the stochastic boundedness of $\{\mathbf{S}'_n\}$.

Put $\sigma_n^2 = \text{Var}(\mathbf{S}_n)$. If $\sigma_n \to \infty$ the central limit theorem of example 1(*a*) applied to the variables $(\mathbf{X}'_{k,n} - \beta_n)/\sigma_n$ shows that for large n the distribution of $\mathbf{S}'_n$ will be approximately normal with variance $\sigma_n^2 \to \infty$ and hence $\mathbf{P}\{\mathbf{S}'_n \in I\} \to 0$ for any finite interval I. The same argument applies to subsequences and proves that $\{\mathbf{S}'_n\}$ cannot remain stochastically bounded unless $\text{Var}(\mathbf{S}'_n)$ remains bounded. But in this case Chebyshev's inequality shows that $\{\mathbf{S}'_n - n\beta_n\}$ is stochastically bounded, and the same will be true of $\mathbf{S}'_n$ iff $\{n\beta_n\}$ is bounded. Now $\beta_n \to 0$ because $\{\mathbf{X}_{k,n}\}$ is a null-array, and hence the boundedness of $n\beta_n$ implies that $\text{Var}(\mathbf{S}'_n) \sim \mathbf{E}(\mathbf{S}'^2_n)$.

We have thus shown that the boundedness of $\mathbf{E}(\mathbf{S}'^2_n)$ is a necessary condition for the stochastic boundedness of $\{\mathbf{S}'_n\}$, and by Chebyshev's inequality this condition is also sufficient. But

$$\mathbf{E}(\mathbf{S}'^2_n) = \Phi_n\{\overline{-s, s}\} + s^2\, T_n(s) \tag{7.7}$$

and hence under the present circumstances the condition (i) is equivalent to the condition that $\mathbf{E}(\mathbf{S}'^2_n)$ remains bounded. ►

The assumption that $\{\mathbf{X}_{k,n}\}$ is a null-array was used only in connection with the symmetrization and could be omitted for arrays with symmetric distributions F_n. However, the boundedness of $\Phi_n\{I\}$ implies the $\mathbf{E}(\mathbf{X}'^2_{k,n}) = O(n^{-1})$, and one concludes easily from the lemma that an array with stochastically bounded row sums and symmetric distributions is necessarily a null-array. By symmetrization it follows that in the general case there exist numbers μ_n (for example, the medians of $\mathbf{X}_{k,n}$) such that $\{\mathbf{X}_{k,n} - \mu_n\}$ is a null-array. In other words, an appropriate centering will produce a null-array, and in this sense only null-arrays are of interest.

Example. Let the $\mathbf{X}_{k,n}$ have normal distributions with expectation β_n and variance n^{-1}. Then $\mathbf{S}_n - n\beta_n$ has the standard normal distribution but since the β_n are arbitrary, $\{\mathbf{S}_n\}$ need not be stochastically bounded. This illustrates the importance of centering. ►

For theoretical purposes it would be possible to center the array so that $\beta_n = 0$ for all n, but the resulting criterion would be difficult to use in concrete situations. With arbitrary centerings the criteria involve non-linear terms and become unwieldy. We shall cover this case in full generality in XVII,7. Here we shall strike a compromise: we shall require only that

$$\beta_n^2 = o(\mathbf{E}(\mathbf{X}'^2_{k,n})), \qquad n \to \infty. \tag{7.8}$$

This condition seems to be satisfied in all cases occurring in practice. In any case, it is so mild that it is usually easy to satisfy it by an appropriate centering whereas the more stringent requirement that $\beta_n = 0$ may require complicated calculations.

Theorem. *Let* $\{\mathbf{X}_{k,n}\}$ *be a null-array such that* (7.8) *holds.*

In order that there exist centering constants b_n *such that the distributions of* $\mathbf{S}_n - b_n$ *tend to a proper limit it is necessary and sufficient that there exist a canonical measure* Ω *such that*

$$\Phi_n\{I\} \to \Omega\{I\} \tag{7.9}$$

for every finite interval and that for $x > 0$

$$n[1-F_n(x)] \to \psi^+(x), \qquad nF_n(-x) \to \psi^-(-x). \tag{7.10}$$

In this case the distribution of $\mathbf{S}_n - n\beta_n$ *tends to the distribution* Q_1 *associated with the convolution semi-group generated by the operator* $\mathfrak{A}$ *defined by*

$$\mathfrak{A}u(x) = \int_{-\infty}^{+\infty} \frac{u(x-y) - u(x) + \tau_s(y)\,u'(x)}{y^2}\,\Omega\{dy\}. \tag{7.11}$$

Proof. We observe first that with arbitrary b_n the conditions (i) and (ii) of the lemma are necessary for $\{\mathbf{S}_n - b_n\}$ to be stochastically bounded. The proof is the same with the simplification that the relation $\mathbf{E}(\mathbf{S}_n'^2) \sim \mathrm{Var}\,(\mathbf{S}_n')$ is now a consequence of (7.8) whereas before we had to derive it from the boundedness of $\mathbf{S}_n'$.

Assume then the conditions of the lemma satisfied. By the selection theorem there exists a sequence $\{n_k\}$ such that as n runs through it (7.9) holds for finite intervals. For a finite interval $0 < a < x \leq b$ we have

$$n[F_n(b) - F_n(a)] = \int_a^b y^{-2}\,\Phi_n\{dy\} \tag{7.12}$$

and (7.9) entails that also this quantity tends to a limit. Condition (ii) assures us that $n[1 - F_n(b)]$ will be less than an arbitrary $\epsilon > 0$ provided only that b is sufficiently large. It follows that for $0 < a < b \leq \infty$ the integral in (7.12) converges to the analogous integral with respect to Ω. Thus Ω is a canonical measure, and (7.10) is true as n runs through $\{n_k\}$.

We know that the operator $\mathfrak{A}$ of (7.11) defines a semi-group $\{\mathfrak{Q}(t)\}$ of convolution operators. Let $\mathfrak{G}_n$ be the operator induced by the distribution G_n of $\mathbf{X}_{k,n} - \beta_n$, namely $G_n(x) = F_n(x+\beta_n)$. It was shown in section 1 that to show that the distribution of $S_{n_k} - n_k\beta_{n_k}$ tends to the distribution Q_1 associated with $\mathfrak{Q}(1)$ it suffices to show that as n runs through $\{n_k\}$

$$n[\mathfrak{Q}_n - \mathbf{1}] \to \mathfrak{A}. \tag{7.13}$$

Now

$$n[\mathfrak{G}_n - 1]u(x) = n\int_{-\infty}^{+\infty} [u(x+\beta_n-y) - u(x-y)]\,F_n\{dy\}. \tag{7.14}$$

We express $u(x+\beta_n-y)$ using Taylor's formula to second-order terms. Since $\beta_n \to 0$ it follows from (7.8) and the boundedness of $\Phi_n\{I\}$ that $n\beta_n^2 \to 0$. As also u'' is bounded it is seen that

$$(7.15)\quad n[\mathfrak{G}_n - 1]\, u(x) = \int_{-\infty}^{+\infty} [u(x-y) - u(x) + \tau_s(y)\, u'(x)] nF_n\{dy\} + \epsilon_n(x)$$

where ϵ_n is a quantity tending uniformly to zero. The integral may be rewritten in the form (7.11) except that the integration is with respect to Φ_n rather than Ω. As was shown repeatedly, the limit relations (7.9)–(7.10) imply that the integral in (7.15) converges to that in (7.11) and so (7.13) holds as n runs through $\{n_k\}$. Finally, the uniqueness of the semi-group containing Q_1 shows that our limit relations must be true for an arbitrary approach $n \to \infty$, and this concludes the proof. ►

8. DOMAINS OF ATTRACTION

In this section $\mathbf{X}_1, \mathbf{X}_2, \ldots$ are independent variables with a common distribution F. By definition 2 of VI,1 the distribution F *belongs to the domain of attraction of* G if there exist constants $a_n > 0$ and b_n such that the distribution of $a_n^{-1}(\mathbf{X}_1 + \cdots + \mathbf{X}_n) - b_n$ tends to G, where G is a proper distribution *not concentrated at a point*. Despite preliminary results in VI,1 and in section 6 we here develop the theory from scratch. (In XVII,5 the theory will be developed independently and in greater detail.)

Throughout this section we use the notation

$$(8.1)\qquad U(x) = \int_{-x}^{x} y^2\, F\{dy\}, \qquad x > 0.$$

We recall from the theory of regular variation in VIII,8 that a positive function L defined on $\overline{0, \infty}$ *varies slowly* (at ∞) if for $x > 0$

$$(8.2)\qquad \frac{L(sx)}{L(s)} \to 1, \qquad s \to \infty.$$

Theorem 1. *A distribution* F *belongs to the domain of attraction of some distribution* G *iff there exists a slowly varying* L *such that*

$$(8.3)\qquad U(x) \sim x^{2-\alpha} L(x), \qquad x \to \infty,$$

with $0 < \alpha \leq 2$, *and when* $\alpha < 2$

$$(8.4)\qquad \frac{1 - F(x)}{1 - F(x) + F(-x)} \to p, \qquad \frac{F(-x)}{1 - F(x) + F(-x)} \to q.$$

When $\alpha = 2$ *condition* (8.3) *alone is sufficient provided* F *is not concentrated at one point.*[6]

We shall see that (8.3) with $\alpha = 2$ implies convergence to the normal distribution. This covers distributions with finite variance, but also many distributions with unbounded slowly varying U [see example VIII,4(*a*)].

Using theorem 2 of VIII,9 with $\xi = 2$ and $\eta = 0$ it is seen that *the relation* (8.3) *is fully equivalent to*[7]

$$\frac{x^2[1-F(x)+F(-x)]}{U(x)} \to \frac{2-\alpha}{\alpha} \tag{8.5}$$

in the sense that the two relations imply each other.

When $0 < \alpha < 2$ we can rewrite (8.5) in the form

$$1 - F(x) + F(-x) \sim \frac{2-\alpha}{\alpha}\, x^{-\alpha} L(x), \tag{8.6}$$

and conversely (8.6) implies (8.3) and (8.5). This leads us to a reformulation of the theorem which is more intuitive inasmuch as it describes the behavior of the individual tails. (For other alternatives see problem 17.)

Theorem 1a. (*Alternative form*). (i) *A distribution* F *belongs to the domain of attraction of the normal distribution iff* U *varies slowly.*

(ii) *It belongs to some other domain of attraction iff* (8.6) *and* (8.4) *hold for some* $0 < \alpha < 2$.

Proof. We shall apply the theorem of section 7 to the array of variables $\mathbf{X}_{k,n} = \mathbf{X}_k/a_n$ with distributions $F_n(x) = F(a_n x)$. The row sums of the array $\{\mathbf{X}_{k,n}\}$ are given by

$$\mathbf{S}_n = (\mathbf{X}_1 + \cdots + \mathbf{X}_n)/a_n.$$

Obviously $a_n \to \infty$ and hence $\{\mathbf{X}_{k,n}\}$ is a null-array. To show that the condition (7.8) is satisfied we put

$$v(x) = \int_{-x}^{x} y\, F\{dy\} \tag{8.7}$$

---

[6] For distributions with finite variance, U varies slowly except when F is concentrated at the origin. In all other cases (8.3) and (8.4) remain unchanged if $F(x)$ is replaced by $F(x+b)$.

[7] Condition (8.4) requires a similar relation for each tail separately:

$$\frac{x^2[1-F(x)]}{U(x)} \to p\,\frac{2-\alpha}{\alpha}, \qquad \frac{x^2F(-x)}{U(x)} \to q\,\frac{2-\alpha}{\alpha}. \tag{*}$$

When $\alpha = 2$ these relations follow from (8.5), which explains the absence of a second condition when $\alpha = 2$. Theorem 1 could have been formulated more concisely (but more artificially) as follows: F *belongs to some domain of attraction iff* (*) *is true* with $0 < \alpha \le 2$, $p \ge 0$, $q \ge 0$, $p + q = 1$.

and note that (7.8) is certainly true if

$$v^2(x) = o(U(x)). \tag{8.8}$$

Now if $U(x) \to \infty$ as $x \to \infty$ it is clear that $v(x) = o(x\,U(x))$ and $U(x) = o(x^{-2})$, and therefore (8.8) holds. The case of a bounded function U is of no interest since we know that the central limit theorem applies to variables with finite variances. However, even with bounded functions U the relation (8.8) holds provided that the distribution F is centered to zero expectation.

Condition (i) of the last theorem requires[8] that for $x > 0$

$$n\,a_n^{-2}\,U(a_nx) \to \Omega(-x, x\}, \qquad n \to \infty, \tag{8.9}$$

while (ii) reduces to

$$n[1-F(a_nx] \to \psi^+(x), \qquad nF(-a_nx) \to \psi^-(-x) \tag{8.10}$$

[for the notation see (5.2)]. It is easily seen that[9] $a_{n+1}/a_n \to 1$. According to lemma 3 of VIII,8 it follows therefore from (8.9) that U varies regularly, and the limit on the right is proportional to a power of x. Following a custom established by P. Lévy we denote this power by $2 - \alpha$. Thus

$$\Omega\{-x, x\} = Cx^{2-\alpha}, \qquad x > 0. \tag{8.11}$$

The left side being a non-decreasing function of x and bounded near the origin, we have $0 < \alpha \leq 2$. It follows that U is indeed of the form asserted in (8.3).

Again, the same lemma 3 of VIII,8 assures us that the limits in (8.10) are either identically zero, or else proportional to a power of x. Now (8.5) shows that the only possible power is $x^{-\alpha}$; in fact, when $\alpha = 2$ both limits are identically zero, whereas for $\alpha < 2$ the limits are necessarily of the form $Ax^{-\alpha}$ and $Bx^{-\alpha}$ where $A \geq 0$ and $B \geq 0$, but $A + B > 0$. It follows that the conditions of the theorem are necessary.

Assuming (8.3) to be true it is possible to construct a sequence $\{a_n\}$ such that

$$na_n^{-2}\,U(a_n) \to 1. \tag{8.12}$$

For example, one may take for a_n the lower bound of all t such that $nt^{-2}\,U(t) \leq 1$. Then (8.3) guarantees that for $x > 0$

$$na_n^{-2}\,U(a_nx) \to x^{2-\alpha}. \tag{8.13}$$

---

[8] As usual, it is tacitly understood that convergence is required only at points of continuity.

[9] For symmetric distributions this follows from the fact that $(\mathbf{X}_1 + \cdots + \mathbf{X}_n)/a_n$ and $(\mathbf{X}_1 + \cdots + \mathbf{X}_n)/a_{n+1}$ have the same limit distribution. For arbitrary F the assertion follows by symmetrization.

Thus condition (7.9) is satisfied for intervals of the form $I = \overline{\{-x, x\}}$. In case $\alpha = 2$ the limit measure Ω is concentrated at the origin and therefore (7.9) is automatically satisfied for all finite intervals; in this case it follows from (8.5) that also condition (7.10) is satisfied with ψ^+ and ψ^- identically zero. When $\alpha < 2$ the relations (8.3)–(8.5) together imply that as $x \to \infty$

$$(8.14)\quad 1 - F(x) \sim p\,\frac{2-\alpha}{\alpha}\,x^{-\alpha}\,L(x), \qquad F(-x) \sim q\,\frac{2-\alpha}{\alpha}\,x^{-\alpha}\,L(x)$$

provided $p > 0$ and $q > 0$. (In the contrary case the symbol $\sim$ is to be replaced by "little oh" and there is no essential change.) It follows that condition (7.10) holds, and this in turn implies that (7.9) applies to arbitrary intervals at a positive distance from the origin. ►

It is noteworthy that all the results of section 7 are implicitly contained in the present theorem and its proof. The proof leads also to other valuable information. First we have the obvious

Corollary. *If* $\alpha = 2$ *the limit distribution is normal, and otherwise it is a stable distribution satisfying* (6.11). *In either case it is determined up to arbitrary scale parameters.*

We saw also that (8.12) leads to a possible sequence of norming factors a_n. It is easily seen that another sequence $\{a_n'\}$ will do iff the ratios a_n'/a_n tend to a positive limit.

Under the conditions of theorem 1 we have established the existence of a limit distribution for $\mathbf{S}_n - n\beta_n$ where [with v defined in (8.7)]

$$(8.15)\qquad \beta_n = \mathbf{E}(\mathbf{X}_{k,n}) = a_n^{-1}\,v(sa_n) + s[1-F(sa_n)-F(-sa_n)].$$

We now proceed to prove the pleasant fact that the centering constants β_n are really unnecessary except when $\alpha = 1$.

When $\alpha < 1$ we apply theorem 2 of VIII,9 with $\zeta = 1$ and $\eta = 0$ separately to the two half-axes to find that as x

$$(8.16)\qquad v(x) \sim \frac{\alpha}{1-\alpha}\,x[1-F(x)-F(-x)].$$

From this and (8.10) it follows that $n\beta_n$ tends to a finite limit and therefore plays no essential role.

When $\alpha > 1$ the same theorem 2 of VIII,9 with $\zeta = 2$ and $\eta = 1$ shows that F has an expectation, and we naturally center F to zero expectation. The domain of integration in the integral (8.7) for v may then be replaced by $|y| > x$ and it is found that (8.16) holds without change. Thus the distributions of $\mathbf{S}_n$ tend to a limit which is again centered to zero expectation. Similarly, when $\alpha < 1$ the limit distribution is centered so as to be strictly stable. We have thus proved

Theorem 2. *Suppose that F satisfies the conditions of theorem 1. If $\alpha < 1$ then $F^{n\star}(a_n x) \to G(x)$ where G is a strictly stable distribution satisfying* (6.11). *If $\alpha > 1$ the same is true provided F is centered to zero expectation.*

(For the centering when $\alpha = 1$ see XVII,5. Concerning the moments of F see problem 16.)

9. VARIABLE DISTRIBUTIONS. THE THREE-SERIES THEOREM

We turn very briefly to general triangular arrays $\{\mathbf{X}_{k,n}\}$ where the variables[10] $\mathbf{X}_{1,n}, \ldots, \mathbf{X}_{n,n}$ of the nth row are mutually independent, but have arbitrary distributions $F_{k,n}$. To preserve the character of our limit theorems we consider only null-arrays: it is required that for arbitrary $\eta > 0$ and $\epsilon > 0$ and n sufficiently large

$$\mathbf{P}\{|\mathbf{X}_{k,n}| > \eta\} < \epsilon, \qquad k = 1, \ldots, n. \tag{9.1}$$

The theory developed in section 7 carries over with the sole change that expressions like $n \operatorname{Var}(\mathbf{X}'_{k,n})$ are replaced by the corresponding sums. In particular, *only infinitely divisible distributions occur as limit distributions of row sums of null-arrays.* The verification may be left to the reader as a matter of routine.

We proceed to discuss some interesting special cases. The notations are the same as in section 7, but in the following it does not matter which type of truncation is used; it is perhaps simplest to define the truncated variables by $\mathbf{X}'_{k,n} = \mathbf{X}_{k,n}$ when $|\mathbf{X}_{k,n}| < s$ and $\mathbf{X}'_{k,n} = 0$ otherwise. Here the truncation level s is arbitrary.

The first theorem is a variant of the compactness lemma and is equivalent to it.

Theorem 1. (*Law of large numbers.*) *Let $\mathbf{S}_n$ stand for the row sums of a null-array. In order that there exist constants b_n such that*[11] $\mathbf{S}_n - b_n \xrightarrow{\mathrm{P}} 0$ *it is necessary and sufficient that for each $\eta > 0$ and each truncation level s*

$$\sum_{k=1}^{n} \mathbf{P}\{|\mathbf{X}_{k,n}| > \eta\} \to 0, \qquad \sum_{k=1}^{n} \operatorname{Var}(\mathbf{X}'_{k,n}) \to 0 \tag{9.2}$$

In this case one may take $b_n = \sum_k \mathbf{E}(\mathbf{X}'_{k,n})$.

[10] Concerning the number of variables in the nth row see problem 10 in VII,13.

[11] We recall from VIII,2 that $\mathbf{Z}_n$ converges in probability to 0 if $\mathbf{P}\{|\mathbf{Z}_n| > \epsilon\} \to 0$ for every $\epsilon > 0$.

As an application we prove the following theorem which was already discussed in VIII,5.

Theorem 2. (*Infinite convolutions.*) *Let* $\mathbf{Y}_1, \mathbf{Y}_2, \ldots$ *be independent random variables with distributions* $G_1, G_2, \ldots$. *In order that the distributions* $G_1 \bigstar G_2 \cdots \bigstar G_n$ *of the sums* $\mathbf{T}_n = \mathbf{Y}_1 + \cdots + \mathbf{Y}_n$ *tend to a proper limit distribution* G *it is necessary and sufficient that for each* $s > 0$

$$\sum \mathbf{P}\{|\mathbf{Y}_k| > s\} < \infty, \qquad \sum \mathrm{Var}\,(\mathbf{Y}'_k) < \infty \tag{9.3}$$

and

$$\sum_{k=1}^{n} \mathbf{E}(\mathbf{Y}'_k) \to m. \tag{9.4}$$

Proof. For a given increasing sequence of integers $\nu_1, \nu_2, \ldots$ and $k = 1, \ldots, n$ put $\mathbf{X}_{k,n} = \mathbf{Y}_{\nu_n+k}$. The distributions $G_1 \bigstar \cdots \bigstar G_n$ converge iff *all* triangular arrays of this type obey the law of large numbers with centering constants $b_n = 0$. From theorem 1 it is clear that the conditions (9.3) and (9.4) are necessary and sufficient for this. ▶

Theorem 2 may be reformulated more strikingly as follows.

Theorem 3. (*Kolmogorov's "three-series theorem".*) *The series* $\sum \mathbf{Y}_k$ *converges with probability one if* (9.3) *and* (9.4) *hold, and with probability zero otherwise.*

Proof. Assume (9.3) and (9.4). By theorem 2 of VII,8 the second condition in (9.3) guarantees that $\sum [\mathbf{Y}'_k - \mathbf{E}(\mathbf{Y}'_k)]$ converges with probability one, and then (9.4) implies the same for $\sum \mathbf{Y}'_k$. By the Borel-Cantelli lemma (see **1**; VIII,3) the first condition in (9.3) entails that with probability one only finitely many $\mathbf{Y}_k$ differ from $\mathbf{Y}'_k$, and so $\sum \mathbf{Y}_k$ converges with probability one.

To prove the necessity of our conditions recall from IV,6 that the probability of convergence is either zero or one. In the latter case the distribution of the partial sums must converge, and so (9.3) and (9.4) hold. ▶

Processes with Non-stationary Increments

The semi-group theory developed in this chapter is the tool particularly adapted to processes with stationary independent increments. Without the condition of stationarity the increment $\mathbf{X}(t) - \mathbf{X}(\tau)$ will have a distribution depending on the two parameters t and τ, and we have to deal with a two-parametric family of operators $\mathfrak{Q}(\tau, t)$, $0 < \tau < t$ satisfying the convolution equation

$$\mathfrak{Q}(\tau, s)\mathfrak{Q}(s, t) = \mathfrak{Q}(\tau, t), \qquad \tau < s < t \tag{9.5}$$

Are the distributions associated with such operators infinitely divisible? We can partition $\overline{\tau, t}$ into n intervals $\overline{t_{k-1}, t_k}$ and consider the variables $\mathbf{X}(t_k) - \mathbf{X}(t_{k-1})$, but to apply the theory of triangular arrays we require the condition (9.1) amounting to a uniform continuity of the distributions in their dependence on the two time parameters. But $\mathfrak{Q}(\tau, t)$ need not depend continuously on t. In fact, the partial sums of a sequence $\mathbf{X}_1, \mathbf{X}_2, \ldots$ of independent random variables represent a process with independent increments where all changes occur at integer-valued epochs and so the process is basically discontinuous. In a certain sense, however, this is the only type of essential discontinuity. The qualification "essential" is necessary, for it was shown in section 5*a* that even with ordinary semi-groups artificial centering can produce mischief which, though inconsequential, requires caution in formulations. For simplicity we stick therefore to symmetric distributions and prove

Lemma. *If the distributions associated with* $\mathfrak{Q}(\tau, t)$ *are symmetric, a one-sided limit* $\mathfrak{Q}(\tau, t-)$ *exists for each* t.

Proof. Let $\tau < t_1 < t_2 < \cdots$ and $t_n \to t$. The sequence of distributions associated with $\mathfrak{Q}(\tau, t_n)$ is stochastically bounded and so there exists a convergent subsequence. Dropping double subscripts we may suppose that $\mathfrak{Q}(\tau, t_n) \to \mathfrak{U}$ where $\mathfrak{U}$ is associated with a proper distribution U. It follows easily that $\mathfrak{Q}(t_n, t_{n+1}) \to 1$ and this implies $\mathfrak{Q}(t_n, s_n) \to 1$ for any sequence of epochs such that $t_n < s_n < t_{n+1}$. In view of (9.5) this means that $\mathfrak{Q}(\tau, s_n) \to \mathfrak{U}$, and so the limit $\mathfrak{U}$ is independent of the sequence $\{t_n\}$, and the lemma is proved. ▶

Following Paul Lévy, we say that a *fixed discontinuity* occurs at t if the two limits $\mathfrak{Q}(\tau, t+)$ and $\mathfrak{Q}(\tau, -1)$ are different. It follows readily from theorem 2 that *the set of fixed discontinuities is countable*. Using symmetrization it follows also in the general case that, except for at most denumerably many epochs, discontinuities are due only to centering (and are removable by an adequate centering). The contribution $\mathfrak{Q}_d(\tau, t)$ of all fixed discontinuities to $\mathfrak{Q}(\tau, t)$ is an infinite convolution and it is possible to *decompose the process into a discrete and a continuous part*. For the triangular arrays arising from continuous processes it is not difficult to see (using theorem 2) that the uniformity condition (9.1) is automatically satisfied and we reach the conclusion that *the distributions associated with continuous processes are infinitely divisible*. P. Lévy has shown that the sample function of such processes are well-behaved in the sense that with probability one right and left limits exist at every epoch t.

10. PROBLEMS FOR SOLUTION

1. In Example 1(*a*) show that $\sum \mathbf{X}_{2,n}^k \xrightarrow{\text{p}} 1$ as $n \to \infty$. (*Hint:* Use variances.)

2. In an ordinary symmetric random walk let $\mathbf{T}$ be the epoch of the first passage through $+1$. In other words, $\mathbf{T}$ is a random variable such that

$$\mathbf{P}\{\mathbf{T} = 2r - 1\} = \frac{1}{2r}\binom{2r}{r} 2^{-2r}$$

Consider a triangular array in which $\mathbf{X}_{k,n}$ has the same distribution as $\mathbf{T}/n^2$. Using the elementary methods of section 1 show by direct calculation that

$$(*) \qquad n[\mathfrak{F}_n u(x) - u(x)] \to \frac{1}{\sqrt{2\pi}} \int_0^\infty \frac{u(x-y) - u(x)}{\sqrt{y^3}}\, dy.$$

Conclude that the distribution of the row sums tends to the stable distribution F_1 defined in II,(4.7) with the convolution property II,(4.9). Interpret the result in terms of a random walk in which the steps are $\pm 1/n$ and the times between successive steps $1/n^2$.

[*Hint*: The series defining $\mathfrak{F}_n u(x)$ can be approximated by an integral.]

3. Consider the gamma distributions of II,(2.2) for the parameter value $\alpha = 1$. Show that the convolution property II,(2.3) implies that they form a semi-group with a generator given by

$$\mathfrak{U}u(x) = \int_0^\infty \frac{u(x-y) - u(x)}{y} e^{-y} dy.$$

Discuss the absence of a centering term.

4. The one-sided stable distributions of II,(4.7) enjoy the convolution property II,(4.9) and therefore form a semi-group. Show that the generator is given by the right side of (*) in problem 2.

5. Let the distributions Q_t of a semi-group be concentrated on the integers and denote the weight of k by $q_k(t)$. Show that

$$\mathfrak{U}u(x) = -q'(0)\, u(x) + \sum_{k \neq 0} q'(0)\, u(x - k).$$

Compare with the canonical form (5.9). Interpret in this light the generating functions for infinitely divisible distributions obtained in **1**; XII,2.

6. Generalize the notions of section 2 to semi-groups with defective distributions. Show that if $\mathfrak{U}$ generates $\{\mathfrak{Q}(t)\}$ then $\mathfrak{U} - c\mathbf{1}$ generates $\{e^{-ct}\mathfrak{Q}(t)\}$.

7. The notation $e^{t(\mathfrak{F}-\mathbf{1})}$ for the compound Poisson semi-group tempts one to write in general $\mathfrak{Q}(t) = e^{t\mathfrak{U}}$. For the normal semi-group this leads to the formal operational equation

$$\exp\left(\frac{1}{2}\, t\, \frac{d^2}{dx^2}\right) u(x) = \sum \frac{1}{n!} \left(\frac{t}{2}\right)^n u^{(2n)}(x).$$

Show that it is valid whenever the Taylor series of u converges for all x and the series on the right converges for $t > 0$. (*Hint:* start from the Taylor series for the convolution of u and a normal distribution. Use the moments of the normal distribution.)

8. The distributions of a semi-group have finite expectations iff $\dfrac{1}{1 + |x|}$ is integrable with respect to the measure Ω appearing in the generator.

9. Show directly that if $n[\mathfrak{F}_n - \mathbf{1}] \to \mathfrak{U}$, the operator $\mathfrak{U}$ is necessarily of the form of a generator. [Use the method of section 4 considering functions of the form (4.4) but do not use semi-group theory. The intention is to derive the general form of a generator without first proving its existence.]

10. Let F_k attach probabilities $\frac{1}{2}$ to the two points $\pm \mu^k$. Then

$$\sum_{h=-\infty}^{+\infty} 2^{-k}(\mathfrak{F}_k - \mathbf{1})$$

generates a semi-group such that $Q_{2t}(x) = Q_t(x\mu)$, but Q_t is not stable. (P. Lévy.)

11. (*A direct proof that the limit distributions of row sums of triangular arrays are infinitely divisible.*) Let $\{\mathbf{X}_{k,n}\}$ be a triangular array with identically distributed variables. The stochastic boundedness of $\mathbf{S}_n$ implies the same for the partial sums $\mathbf{X}_{1,n} + \cdots + \mathbf{X}_{m,n}$, where m stands for the largest integer $\leq n/r$. Using the

selection theorem show that this implies that the limit distribution G of $\mathbf{S}_n$ is the r-fold convolution of a distribution G^r.

12. For any distribution F and smooth u

$$\|(\mathfrak{F} - \mathbf{1})u\| \leq 100(\|u\| + \|u''\|) \int_{-\infty}^{+\infty} \frac{x^2}{1 + x^2} F\{dx\} + \|u'\|.$$

13. To the triangular array $\{\mathbf{X}_{k,n}\}$ with distributions F_n there corresponds another array $\{\mathbf{X}^{\#}_{k,n}\}$ with compound Poisson distributions

$$\mathfrak{F}^{\#}_n = e^{\mathfrak{F}_n - 1}.$$

Show that $n[\mathfrak{F}_n - \mathfrak{F}^{\#}_n] \to 0$ whenever $\{\mathbf{S}_n\}$ is stochastically bounded. This shows that *the row sums* $\mathbf{S}_n$ *and* $\mathbf{S}^{\#}_n$ *are asymptotically equivalent.* Since the distribution of $\mathbf{S}^{\#}_n$ is associated with $e^{[n\mathfrak{F}_n - 1]}$ this yields a second method for deriving the main theorems of section 7. This method can be used also for arrays with variable distributions. (*Hint:* Use problem 12.)

14. With the notations of section 5 put $\mathbf{M}_n = \max [\mathbf{X}_{1,n}, \ldots, \mathbf{X}_{n,n}]$. If $\mathbf{S}_n$ has a limit distribution show that $\psi^+(x) = -\lim \log \mathbf{P}\{\mathbf{M}_n < x\}$.

15. If $\mathbf{S}_n$ has a limit distribution so do the row sums of the array formed by the squares $\mathbf{X}^2_{k,n}$.

16. (*Domains of attraction.*) Let F belong to the domain of attraction of a stable distribution with index α. Using theorem 2 of VIII,9 show that F possesses absolute moments of all orders $<\alpha$. If $\alpha < 2$ no moments of order $>\alpha$ exist. The last statement is false when $\alpha = 2$.

17. (*Continuation.*) In section 8 the theory was based on the truncated second moment function, but this was done only for reasons of tradition. Theorem 2 of VIII,9 permits us to replace y^2 in (8.1) by $|y|^p$ with other exponents ρ, and for each ρ to replace (8.3) and (8.5) by equivalent relations.

18. Let $\mathbf{X}_1, \mathbf{X}_2, \ldots$ be independent variables with a common distribution F. If $1 - F(x) + F(-x)$ varies slowly, deduce from the compactness lemma that a sequence $\mathbf{S}_{n_k}/a_k + b_k$ can have no proper limit distribution G except G concentrated at a point. (This may be expressed by saying that F *belongs to no domain of partial attraction.* See XVII,9.) *Hint:* Use symmetrization.

CHAPTER X

Markov Processes and Semi-Groups

This chapter starts out with an elementary survey of the most common types of Markov processes—or rather, of the basic equations governing their transition probabilities. From this we pass to Bochner's notion of subordination of processes and to the treatment of Markov processes by semi-groups. The so-called exponential formula of semi-group theory is the connecting link between these topics. The existence of generators will be proved only in chapter XIII by the theory of resolvents. In theory the present exposition might have covered the processes and semi-groups of the preceding chapter as a special case, but the methods and uses are so different that the following theory is self-contained and *independent of chapter IX.* The results will be amplified in chapter XIII, but *the theory of Markov processes is not used* for the remaining topics in this book.

This chapter is largely in the nature of a survey, and no attempt is made at either generality or completeness.[1] Specifically, we shall not discuss properties of the sample functions, and throughout this chapter the existence of the processes will be taken for granted. Our interest centers entirely on the analytical properties of the transition probabilities and of the defining operators.

The theory of the induced semi-groups of transformations will be treated in fair generality in sections 8–9. In the earlier sections the basic space is an interval on the line or the whole line although parts of the theory apply more generally. To avoid special symbols it is therefore agreed that when no limits are indicated *integrals are taken over a fixed set* Ω *serving as the basic space.*

[1] The semi-group treatment of Markov processes is described in greater detail in Dynkin (1965) and Loève (1963). Yosida (1966) contains a succinct introduction to the analytic theory of semi-groups and their applications to diffusion and to ergodic theory.

1. THE PSEUDO-POISSON TYPE

Throughout this chapter we limit our discussion to Markov processes with *stationary transition probabilities* Q_t defined by

$$Q_t(x, \Gamma) = \mathbf{P}\{\mathbf{X}(t+\tau) \in \Gamma \mid \mathbf{X}(\tau) = x\} \tag{1.1}$$

and supposed to be independent of τ. (See VI,11.)

A simple generalization of the compound Poisson process leads to an important class of such processes from which all others can be derived by approximation. The theory of semi-groups hinges on an analytical counterpart to this situation (section 10).

Let $\mathbf{N}(t)$ denote the variable of an ordinary Poisson process. In VI,4 the compound Poisson process was introduced by considering the random sums $\mathbf{S}_{\mathbf{N}(t)}$ where $\mathbf{S}_0, \mathbf{S}_1, \ldots$ are the partial sums of a sequence of independent identically distributed random variables. The pseudo-Poisson process is defined in like manner except that now $\mathbf{S}_0, \mathbf{S}_1, \ldots$ are the variables of Markov chain with transition probabilities given by a stochastic kernel K (see VI,11). The variables $\mathbf{X}(t) = \mathbf{S}_{\mathbf{N}(t)}$ define a new stochastic process which can be described formally as follows.

Between the jumps of the Poisson process the typical sample path remains constant. A transition from x to Γ can occur in 0, 1, 2, . . . steps, and hence

$$Q_t(x, \Gamma) = e^{-\alpha t} \sum_{n=0}^{\infty} \frac{(\alpha t)^n}{n!} K^{(n)}(x, \Gamma), \qquad t > 0. \tag{1.2}$$

This generalizes the compound Poisson distribution VI,(4.2) and reduces to it in the special case when Ω is the whole line and $\mathbf{S}_n$ is the sum of n independent random variables with a common distribution F.

The composition rule

$$Q_{t+\tau}(x, \Gamma) = \int Q_t(x, dy)\, Q_\tau(y, \Gamma) \tag{1.3}$$

$(t, \tau > 0)$ analogous to VI,(4.1) is easily verified analytically.[2] It is called the *Chapman-Kolmogorov* equation and states that a transition from x at epoch 0 to Γ at epoch $t + \tau$ occurs via a point y at epoch τ and that the subsequent change is independent of the past.[3] [See **1**; XVII,9 and also VI,(11.3).]

Examples. (*a*) *Particles under collision.* Let a particle travel at uniform speed through homogeneous matter occasionally scoring a collision. Each

[2] Approximating Q_t and Q_τ by their partial sums with n terms shows that the right side in (1.3) is $\leq$ the left side but $\geq$ the nth partial sum of $Q_{t+\tau}$.

[3] It is sometimes claimed that (1.3) is a law either of nature or of compound probabilities, but it is *not* true for non-Markovian processes. See **1**; XV,13.

collision produces a change of energy regulated by a stochastic kernel K. The transition probabilities for the energy $\mathbf{X}(t)$ are of the form (1.2) if the number of collisions obeys a Poisson process. This will be the case under the now familiar assumptions concerning homogeneity of space and lack of memory.

It is usually assumed that the *fraction of energy* lost at each collision is independent of the initial amount, which means that $K(x, dy) = V\{dy/x\}$ where V is a probability distribution concentrated on $\overline{0, 1}$. For later applications we consider the special case where $V(x) = x^{\lambda}$. Then K has a density given by

$$k(x, y) = \lambda x^{-\lambda} y^{\lambda-1}, \qquad 0 < y < x. \tag{1.4}$$

For $\lambda = 1$ this implies that the fraction of energy lost is *uniformly distributed*.[4] The iterated kernels $k^{(n)}$ were calculated in VI,(11.5). Substituting into (1.2) it is seen that Q_t has an atom of weight $e^{-\alpha t}$ at the origin (accounting for the event of no collision) and for $0 < y < x$ the density

$$q_t(x, y) = e^{-\alpha t}\sqrt{\lambda \alpha t}\,\frac{y^{\lambda-1}}{x^{\lambda}\sqrt{\log(x/y)}}\, I_1(2\sqrt{\alpha t\lambda \log(x/y)}) \tag{1.5}$$

where I_1 is the Bessel function defined in II,(7.1). [See examples 2(*a*) and 2(*b*).]

(*b*) *The energy loss of fast particles by ionization.*[5] An instructive variant of the last example is obtained by considering the extreme case of a particle whose energy may be considered infinitely large. The energy losses at successive collisions are then independent random variables with a common distribution V concentrated on $\overline{0, \infty}$. If $\mathbf{X}(t)$ is the total energy loss within the time interval $\overline{0, t}$ then $\mathbf{X}(t)$ is the variable of a *compound Poisson process*. Its transition probabilities are given by (1.2) with $K^{(n)}$ replaced by the convolutions $V^{n\star}$.

(*c*) *Changes in direction.* Instead of the energy of a particle we may consider the direction in which it travels and derive a model analogous to example (*a*). The main difference is that a direction in $\mathfrak{R}^3$ is determined by two variables, and so the density kernel k now depends on four real variables.

[4] This assumption is used by W. Heitler and L. Janossy, *Absorption of meson producing nucleons*, Proc. Physical Soc., Series A, vol. 62 (1949) pp. 374–385, where the Fokker-Planck equation (1.8) is derived (but not solved).

[5] Title of a paper by L. Landau, J. Physics, USSR, vol. 8 (1944) pp. 201–205. Landau uses a different terminology, but his assumptions are identical with ours and he derives the forward equation (1.6).

(*d*) *The randomized random walk* of example II,7(*b*) represents a pseudo-Poisson process restricted to the integers. For a fixed integer x the kernel K attributes weight $\frac{1}{2}$ to the two points $x \pm 1$. ▶

From (1.2) one gets easily

$$\frac{\partial Q_t(x, \Gamma)}{\partial t} = -\alpha Q_t(x, \Gamma) + \alpha \int Q_t(x, dz)\, K(z, \Gamma). \tag{1.6}$$

This is *Kolmogorov's forward equation* which will be discussed in a more general setting in section 3 where it will be shown in the next section that (1.2) is its *only* solution satisfying the obvious probabilistic requirements. The equation (1.6) takes on a more familiar form when K has a density k. At the point x the distribution Q_t has an atom of weight $e^{-\alpha t}$, which is the probability of no change; except for this atom Q_t has a density q_t satisfying the equation

$$\frac{\partial q_t(x, \xi)}{\partial t} = -\alpha q_t(x, \xi) + \alpha \int q_t(x, z)\, k(z, \xi)\, dz. \tag{1.6a}$$

If μ_0 is the probability distribution at epoch 0 the distribution at epoch t is given by

$$\mu_t\{\Gamma\} = \int \mu_0\{dx\}\, Q_t(x, \Gamma) \tag{1.7}$$

and (1.6) implies

$$\frac{\partial \mu_t\{\Gamma\}}{\partial t} = -\alpha \mu_t\{\Gamma\} + \alpha \int \mu_t\{dz\}\, K(z, \Gamma). \tag{1.8}$$

This version of (1.6) is known to physicists as the *Fokker-Planck* (or continuity) equation. Its nature will be analyzed in section 3. When K and the initial distribution μ_0 have densities, then μ_t also has a density m_t, and the Fokker-Planck equation reduces to

$$\frac{\partial m_t(\xi)}{\partial t} = -\alpha m_t(\xi) + \alpha \int m_t(z)\, k(z, \xi)\, dz. \tag{1.8a}$$

2. A VARIANT: LINEAR INCREMENTS

A simple variant of our process occurs in physics, queuing theory, and other applications. The assumptions concerning the jumps remain the same but *between jumps* $\mathbf{X}(t)$ *varies linearly at a rate* c. This means that $\mathbf{X}(t) - ct$ is the variable of the described pseudo-Poisson process; if Q_t stands for the transition probabilities of the new process, then $Q_t(x, \Gamma + ct)$ must satisfy (1.6). The resulting equation for Q_t is of an unfamiliar form, but if differentiable densities exist they satisfy familiar equations. For m_t we have to replace ξ in (1.8a) by $\xi + ct$. With the change of variables $y = \xi + ct$ we get the Fokker-Planck equation.[6]

$$\frac{\partial m_t(y)}{\partial t} = -c\,\frac{\partial m_t(y)}{\partial y} - \alpha m_t(y) + \alpha \int m_t(z)\, k(z, y)\, dz. \tag{2.1}$$

The analogue to (1.6a) is obtained similarly by adding the term $-c\, \partial q_t/\partial y$ to the right side.

In connection with semi-group theory we shall cast the Fokker-Planck equation in a more flexible form quite independent of the unnatural differentiability conditions. [See example 10(b).]

Examples. (a) *Particles under collision.* In the physical literature example 1(a) occurs usually in a modified form where it is assumed that between collisions energy is dissipated at a constant rate due to absorption or friction. The model of (2.1) fits this situation if the energy loss is proportional with m_t standing for the probability density of the energy at epoch t.

In other situations physicists assume that between collisions energy is dissipated at a rate proportional to the instantaneous energy. In this case the logarithm of the energy decreases at a constant rate and an equation of the form (2.1) now governs the probability density for the logarithm of the energy.

(b) *Stellar radiation.*[7] In this model the variable t stands for *distance* and $\mathbf{X}(t)$ for the intensity of a light ray traveling through space. It is assumed that (within the equatorial plane) each element of volume radiates at a constant rate and hence $\mathbf{X}(t)$ increases linearly. But the space also contains absorbing dark clouds which we treat as a Poissonian ensemble of points. On meeting a cloud each ray experiences a chance-determined loss and we have the exact situation that led us to (2.1). It is plausible (and it

[6] Many special cases of the Fokker-Planck equation (1.8) have been discovered independently, and much fuss has been made about the generalization (2.1). The general notion of Fokker-Planck equations was developed by Kolmogorov in his celebrated paper, *Über die analytischen Methoden in der Wahrscheinlichkeitsrechnung*, Math. Ann., vol. 104 (1931) pp. 415–458. In it Kolmogorov mentions the possibility of adding an arbitrary diffusion term

$$\gamma \frac{\partial^2 m_t}{\partial y^2} - c\,\frac{\partial m_t}{\partial y}$$

to the right side, and (2.1) is merely a special case of this. Even the first existence theorems covered the general equation in the non-stationary case. [Feller, Math. Ann., vol. 113 (1936).]

[7] The physical assumptions are taken from V. A. Ambarzumian, *On the brightness fluctuations in the Milky Way*, Doklady Akad. Nauk SSSR, vol. 44 (1944) pp. 223–226, where a version of (2.1) is derived by an indirect approach.

can be proved) that the density m_t of $\mathbf{X}(t)$ approaches a *steady state density* m which is independent of t and satisfies (2.1) with the left side replaced by 0.

Ambarzumian assumes specifically that the loss of intensity at an individual passage through a cloud is regulated by the transition kernel (1.4). In this case an explicit solution is available in (1.5) but it is of minor interest. More important (and easily verified) is the fact that (2.1) has a time-independent (or steady state) solution, namely the gamma density

$$m(y) = \left(\frac{\alpha}{c}\right)^{\lambda+1} \frac{1}{\Gamma(\lambda+1)}\, y^{\lambda} e^{-(\alpha/c)y}, \qquad y > 0. \tag{2.2}$$

This result shows that pertinent information can be derived from (2.1) even without finding explicit solutions. For example, it is readily verified by direct integration that the steady state solution has expectation $c[\alpha(1-\mu)]^{-1}$ where μ is the expectation of the absorption distribution V.

(*c*) *The ruin problems* of VI,5 represent the special case where k is a convolution kernel. The variable of these processes is obtained by adding $-ct$ to the variable of a compound Poisson process. Analogous ruin problems can be formulated for arbitrary pseudo-Poissonian processes, and they lead to (2.1). ▶

3. JUMP PROCESSES

In the pseudo-Poisson process the waiting time to the next jump has a fixed exponential distribution with expectation $1/\alpha$. A natural generalization consists in permitting this distribution to depend on the present value of the path function $\mathbf{X}(t)$. [In example 1(*a*) this amounts to assuming that the probability of scoring a hit depends on the energy of the particle.] The Markovian character of the process requires that the distribution be exponential, but its expectation can depend on the present value of $\mathbf{X}(t)$. Accordingly, we start from the following

Basic postulates. *Given that* $\mathbf{X}(t) = x$, *the waiting time to the next jump has an exponential distribution with expectation* $1/\alpha(x)$ *and is independent of the past history. The probability that the following jump leads to a point in* Γ *equals* $K(x, \Gamma)$.

In analytical terms these postulates lead to an integral equation for the transition probabilities $Q_t(x, \Gamma)$ of the process (assuming that such a process does in fact exist). Consider a fixed point x and a fixed set Γ *not containing* x. The event $\{\mathbf{X}(t) \in \Gamma\}$ cannot occur unless the first jump from x has occurred at some epoch $s < t$. Given this, the conditional probability of $\{\mathbf{X}(t) \in \Gamma\}$ is obtained by integrating $K(x, dy)Q_{t-s}(y, \Gamma)$ over the set Ω of all possible y. Now the epoch of the first jump is a random variable with exponential density $\alpha(x)e^{-\alpha(x)s}$. Integrating with respect to it we get the probability of $\{\mathbf{X}(t) \in \Gamma\}$ in the form

$$Q_t(x, \Gamma) = \alpha(x)\int_0^t e^{-\alpha(x)s}\,ds \int_\Omega K(x, dy)\, Q_{t-s}(y, \Gamma). \tag{3.1a}$$

For a set Γ containing x we must add the probability that no jump occurs before t and thus we get

$$Q_t(x, \Gamma) = e^{-\alpha(x)t} + \alpha(x)\int_0^t e^{-\alpha(x)s}\,ds \int_\Omega K(x, dy)\, Q_{t-s}(y, \Gamma). \tag{3.1b}$$

These two equations, valid for $x \notin \Gamma$ and $x \in \Gamma$, respectively, are the analytic equivalent of the basic postulates. They simplify by the change of the variable of integration to $\tau = t - s$. Differentiation with respect to t then reduces the two equations to the same form, and the pair is replaced by the single integro-differential equation

$$\frac{\partial Q_t(x, \Gamma)}{\partial t} = -\alpha(x)\, Q_t(x, \Gamma) + \alpha(x)\int_\Omega K(x, dy)\, Q_t(y, \Gamma). \tag{3.2}$$

This is *Kolmogorov's backward equation*, which serves as point of departure for the analytical development because it avoids the annoyance of distinguishing between two cases.

The backward equation (3.2) admits of a simple intuitive interpretation which may serve to reformulate the basic postulates in more practical terms. In terms of difference ratios (3.2) is equivalent to

$$Q_{t+h}(x, \Gamma) = [1-\alpha(x)h]\, Q_t(x, \Gamma) + \alpha(x)h\int_\Omega K(x, dy)\, Q_t(y, \Gamma) + o(h). \tag{3.3}$$

For an intuitive interpretation of this relation consider the change within the time interval $\overline{0, t+h}$ as the result of the change within the initial short interval $\overline{0, h}$ and the subsequent interval $\overline{h, t+h}$ of duration t. Evidently then (3.3) states that if $\mathbf{X}(0) = x$, the probability of one jump within $\overline{0, h}$ is $\alpha(x)h + o(h)$; and the probability of more than one jump is $o(h)$; finally, if a jump does occur within $\overline{0, h}$, the conditional probabilities of the possible transitions are given by $K(x, dy)$. These three postulates lead to (3.3) and hence to (3.2). In essence they repeat the basic postulates.[8]

From a probabilistic point of view the backward equation is somewhat artificial inasmuch as in it the terminal state Γ plays the role of a parameter,

[8] The differentiability of Q_t with respect to t and the fact that the probability of more than one jump is $o(h)$ are now stated as new postulates, whereas they are implied by the original more sophisticated formulation.

and (3.2) describes the dependence of $Q_t(x, \Gamma)$ on the initial position x. Offhand it would seem more natural to derive an equation for Q_{t+h} by splitting up the interval $\overline{0, t+h}$ into a long initial interval $\overline{0, t}$ and the short terminal interval $\overline{t, t+h}$. Instead of (3.3) we get then *formally*

$$(3.4)\quad Q_{t+h}(x, \Gamma) = \int_\Gamma Q_t(x, dz)[1-\alpha(z)h] + \int_\Omega Q_t(x, dz)\,\alpha(z)h\, K(z, \Gamma) + o(h)$$

and hence

$$(3.5)\qquad \frac{\partial Q_t(x, \Gamma)}{\partial t} = -\int_\Gamma Q_t(x, dz)\,\alpha(z) + \int_\Omega Q_t(x, dz)\,\alpha(z)\, K(z, \Gamma).$$

This is *Kolmogorov's forward equation* (in special cases known to physicists as the *continuity or Fokker-Planck equation*). It reduces to (1.6) when α is independent of z.

The formal character of the derivation was emphasized because the forward equation is really not implied by our basic postulates. This is because the term $o(h)$ in (3.3) depends on z and since z appears as variable of integration in (3.4), the term $o(h)$ should have appeared under the integral sign. But then the problem arises as to whether the integrals in (3.5) converge and whether the passage to the limit $h \to 0$ is legitimate. [No such problems occurred in connection with the backward equation because the initial value x was fixed and the integral in (3.2) exists in consequence of the boundedness of Q_t.]

It is possible to justify the forward equation by adding to our basic postulates an appropriate condition on the error term in (3.3), but such a derivation would lose its intuitive appeal and, besides, it seems impossible to formulate conditions which will cover all typical cases occurring in practice. Once unbounded functions α are admitted, the existence of the integrals in (3.5) is in doubt and the equation cannot be justified a priori. On the other hand, the backward equation is a necessary consequence of the basic assumptions, and it is therefore best to use it as a starting point and to investigate the extent to which the forward equation can be derived from it.

A solution of the backward equation is easily constructed using successive approximations with a simple probabilistic significance. Denote by $Q_t^{(n)}(x, \Gamma)$ the probability of a transition from $\mathbf{X}(0) = x$ to $\mathbf{X}(t) \in \Gamma$ *with at most* n *jumps*. A transition without jumps is possible only if $x \in \Gamma$, and since the sojourn time at x has an exponential distribution we have

$$(3.6)\qquad Q_t^{(0)}(x, \Gamma) = e^{-\alpha(x)t} K^{(0)}(x, \Gamma)$$

(where $K^{(0)}(x, \Gamma)$ equals 1 or 0 according as x is, or is not, contained in Γ). Suppose next that the first jump occurs at epochs $s < t$ and leads from x to y. Summing over all possible s and y we get [as in (3.1)] the recursion formula

$$(3.7)\quad Q_t^{(n+1)}(x, \Gamma) = Q_t^{(0)}(x, \Gamma) + \int_0^t e^{-\alpha(x)s}\alpha(x)\, ds \int K(x, dy)\, Q_{t-s}^{(n)}(y, \Gamma)$$

valid for $n = 0, 1, \ldots$. Obviously $Q_t^{(0)} \leq Q_t^{(1)}$ and hence by induction $Q_t^{(1)} \leq Q_t^{(2)} \leq \cdots$.

It follows that for every pair x, Γ the limit

$$(3.8)\qquad Q_t^{(\infty)}(x, \Gamma) = \lim_{n\to\infty} Q_t^{(n)}(x, \Gamma)$$

exists, but conceivably it could be infinite. We show that actually

$$(3.9)\qquad Q_t^{(\infty)}(x, \Omega) \leq 1$$

[which implies $Q_t^{(\infty)}(x, \Gamma) \leq 1$ for all sets in Ω]. It suffices to prove that

$$(3.10)\qquad Q_t^{(n)}(x, \Omega) \leq 1$$

for all n. This is trivially true for $n = 0$ and we proceed by induction: Assuming (3.10) for some fixed n we get from (3.7) (recalling that K is stochastic)

$$(3.11)\qquad Q_t^{(n+1)}(x, \Omega) \leq e^{-\alpha(x)t} + \int_0^t e^{-\alpha(x)s}\, \alpha(x)\, ds = 1,$$

and hence (3.10) is true for all n.

From (3.7) it follows by monotone convergence that $Q_t^{(\infty)}$ satisfies the backward equations in the original integral version (3.1) [and hence also in the integro-differential version (3.2)]. For any other positive solution Q_t of (3.1) it is clear that $Q_t \geq Q_t^{(0)}$; comparing (3.1) with (3.7) we conclude that $Q_t \geq Q_t^{(n)}$ for all n, and hence $Q_t \geq Q_t^{(\infty)}$. For this reason $Q_t^{(\infty)}$ is called the *minimal solution* of the backward equations; (3.9) shows that $Q_t^{(\infty)}$ is stochastic or substochastic.

It follows from (3.8) that $Q_t^{(\infty)}(x, \Gamma)$ is the probability of a passage from x to Γ in finitely many steps. Accordingly, with a substochastic solution the defect $1 - Q_t^{(\infty)}(x, \Omega)$ represents the probability that, from x as starting point, infinitely many jumps will occur within time t. We know from **1**; XVII,4 and example VIII,5(c) that this phenomenon occurs in certain pure birth processes, and hence substochastic solutions exist. But they are the exception rather than the rule. In particular, if the coefficient $\alpha(x)$ is bounded the minimal solution is strictly stochastic, that is,

$$(3.12)\qquad Q_t^{(\infty)}(x, \Omega) = 1, \qquad t > 0.$$

Indeed, if $\alpha(x) < a < \infty$ for all x we show by induction that

$$(3.13)\qquad Q_t^{(n)}(x, \Omega) \geq 1 - (1-e^{-at})^n$$

for all n and $t > 0$. This is trivially true for $n = 0$. Assume (3.12) for some n, and note that the right side is a decreasing function, say $f(t)$. Consider (3.7) with $\Gamma = \Omega$. In consequence of (3.13) the inner integral is $\geq f(t)$ which does not depend on the variable of integration. Integrating $\alpha(x)e^{-\alpha(x)s}$ it is then seen that (3.13) holds with n replaced by $n+1$.

Finally we note that in the strictly stochastic case (3.12) the solution $Q_t^{(\infty)}$ is unique. In fact, because of the minimal character of $Q_t^{(\infty)}$ any other acceptable solution would satisfy

$$(3.14)\qquad \begin{aligned} 1 \geq Q_t(x, \Omega) &= Q_t(x, \Gamma) + Q_t(x, \Omega - \Gamma) \\ &\geq Q_t^{(\infty)}(x, \Gamma) + Q_t^{(\infty)}(x, \Omega - \Gamma) = Q_t^{(\infty)}(x, \Omega) = 1 \end{aligned}$$

which is impossible unless the equality sign prevails in both places. We have thus proved the

Theorem. *The backward equations admit of a minimal solution* $Q_t^{(\infty)}$ *defined by* (3.8) *and corresponding to a process in which transitions from* x *to* Γ *occur only with finitely many jumps. It is stochastic or substochastic.*

In the substochastic case the defect $1 - Q_t^{(\infty)}(x, \Omega)$ *accounts for the probability of the event that infinitely many jumps occur within time* t, *in which case the minimal process terminates.*

In the strictly stochastic case (3.12) *the minimal solution is the unique probabilistic solution of the backward equation. This case arises whenever the coefficient* $\alpha(x)$ *is bounded.*

The discovery of defective solutions came as a shocking surprise during the early stages of the theory in the 1930's, but it has given impetus to research leading to a unified theory of Markov processes. Processes in which transitions from x to Γ are possible after the occurrence of infinitely many jumps are the analogue to diffusion processes with boundary conditions and therefore not as pathological as they appeared at the beginning. The possibility of infinitely many jumps also explains the difficulties in deriving the forward equations directly.[9] The backward equations were derived from the assumption that, given the present state x, the *next* jump occurs after an exponentially distributed waiting time with expectation $1/\alpha(x)$. The forward equations depend on the state just prior to epoch t, and therefore depend on the whole space Ω. In particular, it is not easy to express directly the requirement that there exist a *last* jump prior to epoch t.

However, it will be shown in the Appendix that *the minimal solution* $Q_t^{(\infty)}$ *automatically satisfies the forward equations and is minimal also for the latter.* It follows, in particular, that if α is bounded, $Q_t^{(\infty)}$ represents the unique

[9] Compare the analogous discussion in **1**; XVII,9.

solution of the forward equation. When $Q_t^{(\infty)}$ is substochastic there exist various processes involving transitions through infinitely many jumps and satisfying the backward equations. The transition probabilities Q_t of such processes may, but need not, satisfy the forward equations. This surprising fact shows that the forward equations may be satisfied in situations where the derivation from (3.4) breaks down.

In conclusion we note again that (in contrast to the process in section 2 and to diffusion processes involving derivatives with respect to x) the pure jump process does not depend on the nature of the underlying space: our formulas apply to any set Ω on which a stochastic kernel K is defined.

Example. *Denumerable sample spaces.* If the random variables $\mathbf{X}(t)$ are positive and integral-valued the underlying sample space consists of the integers $1, 2, \ldots$. It suffices now to know the transition probabilities $P_{ik}(t)$ from one integer to another; all other transition probabilities are obtained by summation over k. The theory of Markovian processes on the integers was outlined in **1**; XVII,9 where, however, also non-stationary transition probabilities were considered. To restrict that theory to the stationary case the coefficients c_i and probabilities p_{ik} must be assumed independent of t. The assumptions are then identical with the present ones and the two systems of Kolmogorov equations derived in **1**; XVII,9 are easily seen to be the special cases of (3.2) and (3.5) [replacing $\alpha(i)$ by c_i and $K(i, j)$ by p_{ij}]. The divergent birth process of **1**; XVII,4 is an example of a process with infinitely many jumps within a finite time interval. We shall return to this process in XIV,7 to present the possibility of a passage from i to j involving infinitely many jumps. ►

Appendix.[10] *The minimal solution for the forward equation.* The construction of the minimal solution $Q_t^{(\infty)}$ for the backward equation can be adapted to the forward equation. We indicate briefly how this can be done and how one can verify that the two solutions are in fact identical. Details will be left to the reader.

Let

$$K_t^{\#}(x, \Gamma) = \int_\Gamma e^{-\alpha(y)t} K(x, dy). \tag{3.15}$$

To construct a solution of the forward equation we define $Q_t^{(0)}$ by (3.6) and put

$$Q_t^{(n+1)}(x, \Gamma) = Q_t^{(0)}(x, \Gamma) + \iint_0^t Q_{t-s}^{(n)}(x, dy)\, \alpha(y)\, K_s^{\#}(y, \Gamma). \tag{3.16}$$

[10] In XIV,7 the theory is developed by means of Laplace transforms. (For simplicity only countable spaces are treated, but the argument applies generally without essential change.) The direct method of the text is less elegant, but has the advantage that it applies also to non-stationary processes with transition probabilities depending on the time parameter. In this general form the theory was developed by Feller, Trans. Amer. Math. Soc. vol. 48 (1940) pp. 488–515 [*erratum* vol. 58, p. 474].

This defines the probabilities for transitions in at most $n + 1$ steps in terms of the *last* jump, just as (3.7) refers to the first jump. Repeating the proof of the last theorem it is seen that $Q_t^{(\infty)} = \lim Q_t^{(n)}$ is the minimal solution of the forward equations.

Although we used the same letters, the two recursion formulas (3.7) and (3.16) are independent, and it is by no means clear that the resulting kernels are identical. To show that this is so we put $P_t^{(n)} = Q_t^{(n)} - Q_t^{(n-1)}$, which corresponds to transitions in exactly n steps. Then (3.16) reduces to

$$P_t^{(n+1)}(x, \Gamma) = \iint_0^t P_{t-s}^{(n)}(x, dy)\, \alpha(y)\, K_s^{\#}(y, \Gamma). \tag{3.17}$$

We indicate this by the shorthand notation $P_t^{(n+1)} = P_t^{(n)}\mathfrak{A}$. For the recursion formula (3.7) we write similarly $P_t^{(n+1)} = \mathfrak{B}P_t^{(n)}$. The starting $P_t^{(0)}$ is the same in either case [defined by (3.6)]. We now prove that the two recursion formulas lead to the same result. More precisely: the $P_t^{(n)}$ defined by $P_t^{(n+1)} = P_t^{(n)}\mathfrak{A}$ satisfy also $P_t^{(n+1)} = \mathfrak{B}P_t^{(n)}$. We proceed by induction. Assume the assertion to be true for all $n \leq r$. Then

$$P_t^{(r+1)} = P_t^{(r)}\mathfrak{A} = (\mathfrak{B}P_t^{(r-1)})\mathfrak{A} = \mathfrak{B}(P_t^{(r-1)}\mathfrak{A}) = \mathfrak{B}P_t^{(r)}$$

and thus the induction hypothesis holds also for $n = r + 1$.

We have thus proved that *the minimal solution is common to the backward and forward equations.*

4. DIFFUSION PROCESSES IN $\mathcal{R}^1$

Having considered processes in which all changes occur by jumps we turn to the other extreme where the sample functions are (with probability one) continuous. Their theory is parallel to that developed in the last section, but the basic equations require more sophisticated analysis. We shall therefore be satisfied with a derivation of the backward equation and with a brief summary concerning the minimal solution and other problems. The prototype for diffusion processes is the Brownian motion (or Wiener process). This is the process with independent normally distributed increments. Its transition probabilities have densities $q_t(x, y)$ given by the normal density with expectation x and variance at, where $a > 0$ is a constant. These densities satisfy the standard diffusion equation,

$$\frac{\partial q_t(x, y)}{\partial t} = \frac{1}{2} a \frac{\partial^2 q_t(x, y)}{\partial x^2}. \tag{4.1}$$

It will now be shown that other transition probabilities are governed by related partial differential equations. The object of this derivation is merely to give an idea concerning the types of processes and the problems involved and thus to serve as a first introduction; for this reason we shall not strive at generality or completeness.

From the nature of the normal distribution it is evident that in Brownian motion the increments during a short time interval of duration t have the following properties: (i) for fixed $\delta > 0$ the probability of a displacement exceeding δ is $o(t)$; (ii) the expected value of the displacement is zero; (iii) its variance is at. We retain the first condition, but adapt the others to an inhomogeneous medium; that is, we let a depend on x and permit a non-zero mean displacement. Under such circumstances the expectation and the variance of the displacement will not be strictly proportional to t and we can postulate only that given $\mathbf{X}(\tau) = x$ the displacement $\mathbf{X}(t+\tau) - \mathbf{X}(\tau)$ has an expectation $b(x)t + o(t)$ and variance $a(x) + o(t)$. Moments do not necessarily exist, but in view of the first condition it is natural to consider truncated moments. These considerations lead us to the following

Postulates[11] *for the transition probabilities* Q_t. *For every* $\delta > 0$ *as* $t \to 0$

$$t^{-1}\int_{|y-x|\geq\delta} Q_t(x, dy) \to 0 \tag{4.2}$$

$$t^{-1}\int_{|y-x|<\delta} (y-x)Q_t(x, dy) \to b(x) \tag{4.3}$$

$$t^{-1}\int_{|y-x|<\delta} (y-x)^2Q_t(x, dy) \to a(x). \tag{4.4}$$

Note that if (4.2) holds for *all* $\delta > 0$, the asymptotic behavior of the quantities in (4.3) and (4.4) is independent of δ; it is then permissible in the last two relations to replace δ by 1.

The first condition makes large displacements improbable and was introduced in 1936 in the hope that it is necessary and sufficient for the continuity of the sample functions.[12] It was named in honor of Lindeberg because of its similarity to his condition in the central limit theorem. It can be shown that under mild regularity conditions on the transition probabilities the existence of the limits in (4.3) and (4.4) is really a consequence of (4.2). We shall not discuss such details because we are not at this juncture interested in developing a systematic theory.[13] Our modest aim is to explain the nature

[11] The original derivation of (4.1) from probabilistic assumptions is due to Einstein. The first systematic derivation of the backward equation (4.6) and forward equation (5.2) was given in Kolmogorov's famous paper of 1931 (see section 2). The improved postulates of the text are due to Feller (1936), who gave the first existence proof and investigated the relation between the two equations.

[12] This conjecture was verified by D. Ray.

[13] Modern semi-group theory enabled the author to derive the most general backward equation (generator) for Markov processes satisfying a Lindeberg type condition. The classical differential operators are replaced by a modernized version, in which a "natural scale" takes over the role of the coefficient b, and a "speed measure" the role of a. The study of such processes was the object of fruitful research by E. B. Dynkin and his school on one hand, by K. Ito and H. P. McKean on the other. The whole theory is developed in the books by these authors quoted in the bibliography.

and the empirical meaning of the diffusion equations in the simplest situation. For this purpose we show how certain differential equations can be derived formally from (4.2)–(4.4), but we shall not discuss under what conditions there exist solutions to these equations.[14] The coefficients a and b may be therefore assumed as bounded continuous functions and $a(x) > 0$.

We take as our basic space a finite or infinite interval I on the line and continue the convention that when no limits are indicated, the integration is over the interval I. To simplify writing and to prepare for the applications of semi-group theory we introduce the transformations.

$$u(t, x) = \int Q_t(x, dy)\, u_0(y) \tag{4.5}$$

changing (for fixed t) a bounded continuous "initial function" u_0 into a function[15] with values $u(t, x)$.

Clearly the knowledge of the left side in (4.5) for all initial u_0 uniquely determines Q_t. It will now be shown that under mild regularity conditions u must satisfy the *backward equation*

$$\frac{\partial u}{\partial t} = \frac{1}{2} a \frac{\partial^2 u}{\partial x^2} + b \frac{\partial u}{\partial x}, \tag{4.6}$$

generalizing the standard diffusion equation (4.1). We seek a function u satisfying it and such that $u(t, x) \to u_0(x)$ as $t \to 0$. In case of uniqueness this solution is necessarily of the form (4.5) and Q_t is called the Green function of the equation. Cases of non-uniqueness will be discussed in the next section.

To derive the backward equation (4.6) we start from the identity

$$u(s+t, x) = \int Q_s(x, dy)\, u(t, y), \qquad s, t > 0 \tag{4.7}$$

which is an immediate consequence of the Chapman-Kolmogorov equation (1.3). From it we get for $h > 0$

$$\frac{u(t+h, x) - u(t, x)}{h} = \frac{1}{h} \int Q_h(x, dy)[u(t, y) - u(t, x)]. \tag{4.8}$$

We now suppose that the transition probabilities Q_t are sufficiently regular to ensure that in (4.5) the transform u has two bounded continuous derivatives with respect to x, at least when u_0 is infinitely differentiable. To given

[14] For the treatment of diffusion equations by Laplace transforms see XIV,5.

[15] In terms of the stochastic process, $u(t, x)$ is the conditional expectation of $u_0(\mathbf{X}(t))$ on the hypothesis that $\mathbf{X}(0) = x$.

$\epsilon > 0$ and fixed x there corresponds then by Taylor's formula a $\delta > 0$ such that

(4.9)

$$\left| u(t, y) - u(t, x) - (y-x)\frac{\partial u(t, x)}{\partial x} - \tfrac{1}{2}(y-x)^2 \frac{\partial^2 u(t, x)}{\partial x^2} \right| < \epsilon\, |y-x|^2$$

for all $|y - x| \leq \delta$. With this δ consider in (4.8) separately the contributions of the domains $|y - x| > \delta$ and $|y - x| \leq \delta$. The former tends to 0 in consequence of (4.2) and of the boundedness of u. Owing to the conditions (4.3) and (4.4) it is clear from (4.9) that for sufficiently small h the contribution of $|y - x| \leq \delta$ differs from the right side in (4.6) by less than $\epsilon \cdot a(x)$. Since ϵ is arbitrary, this means that as $h \to 0$ the right side in (4.8) tends to that of (4.6). Accordingly, at least a right-sided derivative $\partial u/\partial t$ exists and is given by (4.6). The principal result of the theory may be summarized roughly as follows. *If the transition probabilities of a Markov process satisfy the continuity condition* (4.2) *the process is determined by the two coefficients* b *and* a. This sounds theoretical, but in practical situations the coefficients b and a are given a priori from their empirical meaning and the nature of the process.

To explain the meaning of b and a consider the increment $\mathbf{X}(t+\tau) - \mathbf{X}(\tau)$ over a short time interval assuming that $\mathbf{X}(\tau) = x$. If the moments in (4.3) and (4.4) were complete, this increment would have the conditional expectation $b(x)t + o(t)$ and the conditional variance $a(x)t - b^2(x)t^2 + o(t) = a(x)t + o(t)$. Thus $b(x)$ is a measure for the local average rate of displacement (which may be zero for reasons of symmetry), and $a(x)$ for the variance. For want of a better word we shall refer to b as the *infinitesimal velocity* (or drift) and a as the *infinitesimal variance.*

The following examples illustrate the way in which these coefficients are determined in concrete situations.

Examples. (*a*) *Brownian motion.* If the x-axis is assumed homogeneous and symmetric, $a(x)$ must be independent of x and $b(x)$ must vanish. We are thus led to the classical diffusion equation (4.1).

(*b*) *The Ornstein-Uhlenbeck process* is obtained by subjecting the particles of a Brownian motion to an elastic force. Analytically this means a drift towards the origin of a magnitude proportional to the distance, that is, $b(x) = \rho x$. As this does not affect the infinitesimal variance, $a(x)$ remains a constant, say 1. The backward equation takes on the form

$$\text{(4.10)} \qquad \frac{u(t, x)}{\partial t} = \frac{1}{2}\frac{\partial^2 u(t, x)}{\partial x^2} - \rho x \frac{\partial u(t, x)}{\partial x}.$$

It is fortunately easy to solve this equation. Indeed, the change of variables

$$v(t, x) = u(t, xe^{\rho t})$$

reduces it to

$$e^{2\rho t}\frac{\partial v}{\partial t} = \frac{1}{2}\frac{\partial^2 v}{\partial x^2} \tag{4.11}$$

and the further change of variables

$$\tau = \frac{1 - e^{-2\rho t}}{2\rho} \tag{4.12}$$

changes (4.11) into the standard diffusion equation (4.1). It follows that *the transition densities* $q_t(x, y)$ *of the Ornstein-Uhlenbeck process coincide with the normal density centered at* $xe^{-\rho t}$ *and with variance* τ *given by* (4.12).

It was shown in example III,8(e) that the Ornstein-Uhlenbeck process determined by (4.10) and an initial normal distribution is the only *normal Markovian process* with stationary transition probabilities. (Brownian motion is included as the special case $\rho = 0$.)

(c) *Diffusion in genetics.* Consider a population with distinct generations and a constant size N. (A cornfield represents a typical example.) There are $2N$ genes and each belongs to one of two genotypes. We denote by $\mathbf{X}_n$ the *proportion* of genes of type A. If selection advantages and mutations are disregarded, the genes in the $(n+1)$st generation may be taken as a random sample of size $2N$ of the genes in the nth generation. The $\mathbf{X}_n$ process is then Markovian, $0 \leq \mathbf{X}_n \leq 1$, and given that $\mathbf{X}_n = x$ the distribution of $2N\mathbf{X}_{n+1}$ is binomial with mean $2Nx$ and variance $2Nx(1-x)$. The change per generation has expectation 0 and variance proportional to $x(1-x)$.

Suppose now that we look over a tremendous number of generations and introduce a time scale on which the development appears continuous. In this approximation we deal with a Markov process whose transition probabilities satisfy our basic conditions with $b(x) = 0$ and $a(x)$ proportional to $x(1 - x)$. The proportionality factor depends on the unit of time scale and may be normalized to 1. Then (4.6) takes on the form

$$\frac{\partial u(t, x)}{\partial t} = x(1 - x)\frac{\partial^2 u(t, x)}{\partial x^2} \tag{4.13}$$

and this time the process is restricted to the finite interval $\overline{0, 1}$. Selection and mutation pressures would cause a drift and lead to an equation (4.13) with a first-order term added. The resulting model is mathematically equivalent to the models developed by R. A. Fisher and S. Wright although their arguments were of a different nature. The genetical implications are somewhat dubious because of the assumption of constant population size, the effect of which is not generally appreciated. The correct description[15] depends on an equation in two space variables (gene frequency and population size).

(*d*) *Population growth.* We wish to describe the growth of a large population in which the individuals are stochastically independent and the reproduction rate does not depend on the population size. For a very large population the process is approximately continuous, that is, governed by a diffusion equation. The independence of the individuals implies that the infinitesimal velocity and variance must be proportional to the population size. Thus the process is governed by the backward equation (4.6) with $a = \alpha x$ and $b = \beta x$. The constants α and β depend on the choice of the units of time and population size, and with appropriate units of measurement it is possible to achieve that $\alpha = 1$ and $\beta = 1, -1$, or 0 (depending on the net rate of growth).

In **1**; XVII,(5.7) the *same* population growth is described by a discrete model. Given $\mathbf{X}(\tau) = n$ it was assumed that the probabilities of the contingencies $\mathbf{X}(t+\tau) = n+1$, $n-1$, and n differ from λnt, μnt, and $1 - (\lambda+\mu)nt$, respectively, by terms $o(t^2)$, and so the infinitesimal velocity and variance are $(\lambda-\mu)n$ and $(\lambda+\mu)n$. The diffusion process is obtained by a simple passage to the limit, and it can be shown that its transition probabilities represent the limit of the transition probabilities for the discrete model.

Similar approximations of discrete processes by diffusion processes are often practical; the passage from ordinary random walks to diffusion processes described in **1**; XIV,6 provides a typical example. [Continued in example 5(*a*).] ▶

5. THE FORWARD EQUATION. BOUNDARY CONDITIONS

In this section we assume for simplicity that the transition probabilities Q_t have probability densities q_t given by a stochastic density kernel $q_t(x, y)$.

The transformation (4.5) and the ensuing backward equation (4.6) describe the transition probabilities in their dependence on the initial point x. From a probabilistic point of view it appears more natural to keep the initial point x fixed and to consider $q_t(x, y)$ as a function of the terminal point y. From this point of view the transformation (4.5) should be replaced by

$$v(s, y) = \int v_0(x)\, q_s(x, y)\, dx \tag{5.1}$$

[15] W. Feller, Proc. Second Berkeley Symposium on Math. Statist. and Probability, 1951, pp. 227–246.

Here v_0 is an arbitrary probability density. From the stochastic character of q_s it follows that for arbitrary fixed $s > 0$ the transform v is again a probability density. In other words, whereas the transformation (4.5) operated on continuous functions, the new transformation changes probability densities into new densities.

In the preceding section we were able by probabilistic arguments to show that the transform (4.5) satisfies the backward equation (4.6). Even though the new transformation is more natural from a probabilistic point of view, a similar direct derivation of the forward equation is impossible. However, the general theory of adjoint partial differential equations make it plausible that (under sufficient regularity conditions) v should satisfy the equation[17]

$$\frac{\partial v(s, y)}{\partial s} = \frac{1}{2}\frac{\partial^2}{\partial y^2}[a(y)\, v(s, y)] - \frac{\partial}{\partial y}[b(y)\, v(s, y)]. \tag{5.2}$$

[17] Here is an informal sketch of the derivation of (5.2). From the Chapman-Kolomogorov equation (1.3) for the transition probabilities it follows that

$$\int v(s, y)\, u(t, y)\, dy$$

depends only on the sum $s + t$. Accordingly,

$$\int \frac{\partial v(s, y)}{\partial s} u(t, y)\, dy = \int v(s, y)\, \frac{\partial u(t, y)}{\partial t}\, dy. \tag{*}$$

We now express $\partial u/\partial t$ in accordance with the backward equation (4.6) and apply the obvious integrations by parts to the resulting integral. If $R(s, y)$ stands for the right side in (5.2) we conclude that (*) equals

$$\int R(s, y)\, u(t, y)\, dy$$

plus a quantity depending only on the values of u, v and their derivatives at the boundaries (or at infinity). Under appropriate conditions these boundary terms may be neglected, and in this case the passage to the limit $t \to 0$ leads to the identity

$$\int \left[\frac{\partial v(s, y)}{\partial s} - R(s, y)\right] u_0(y)\, dy.$$

If this is to be valid for arbitrary u_0 the expression within brackets must vanish, that is (5.2) must hold.

This argument is justified in most situations of practical interest and accordingly the forward equation (5.2) is generally valid. However, in the so-called return processes the boundary terms which we have neglected actually play a role. The transition probabilities of such processes therefore satisfy the backward equation (4.6), but *not* (5.2); the correct forward equation is in this case an equation of a different form.

It is also noteworthy that (5.2) is meaningless unless a and b are differentiable whereas no such restriction applies to the backward equation. The true forward equation can be written down also when a and b are not differentiable, but it involves the generalized differential operators mentioned in footnote 13 of section 4.

In probability theory this equation is known as the *forward* or *Fokker-Planck* equation.

Before proceeding let us illustrate the kind of information that can be derived from (5.2) more easily than from the backward equation.

Example. (*a*) *Population growth.* Example 4(*d*) leads to the forward equation

$$\frac{\partial v(s, y)}{\partial s} = \alpha \frac{\partial^2 y\, v(s, y)}{\partial y^2} - \beta \frac{\partial y\, v(s, y)}{\partial y}. \tag{5.3}$$

It can be proved that for a given initial density v_0 there exists a unique solution. Although explicit formulas are hard to come by, much relevant information can be obtained directly from the equation. For example, to calculate the expected population size $M(s)$ multiply (5.3) by y and integrate with respect to y from 0 to ∞. On the left we get the derivative $M'(s)$. Using integration by parts and assuming that v vanishes at infinity faster than $1/y^2$ it is seen that the right side equals $\beta M(s)$. Thus

$$M'(s) = \beta M(s)$$

and hence $M(s)$ is proportional to $e^{\beta s}$. Similar formal manipulations show that the variance is proportional to $2\alpha\beta^{-1}e^{\beta s}(e^{\beta s}-1)$. [Compare the analogous result in the discrete case, formulas (5.10) and (10.9) of **1**; XVII.] Admittedly the manipulations require justification, but the result has at least heuristic value and could not be obtained from the backward equation without explicit calculation of q_t. ▶

The connection between the forward and backward equations is similar to that described in the case of jump processes in section 3. We give a brief summary without proof.

Consider the backward equation (4.6) in an open interval $\overline{x_1, x_2}$ which may be finite or infinite. We assume, of course, $a > 0$ and that the coefficients a and b are sufficiently regular for (5.2) to make sense. Under these conditions there exists a *unique minimal solution* Q_t such that (4.5) yields a solution of the backward equation (4.6). The catch is that for fixed t and x the kernel $Q_t(x, \Gamma)$ may represent a *defective distribution.* Under any circumstances Q_t possesses densities and the function v of (5.1) satisfies the forward equation. In fact, this solution is again minimal in the obvious sense made precise in section 3. To this extent the forward equation is a consequence of the backward equation. However, these equations determine the process uniquely only when the minimal solution is not defective. In all other cases the nature of the process is determined by additional boundary conditions.

The nature of boundary conditions is best understood by analogy with the simple random walk on $\overline{0, \infty}$ discussed in **1**; XIV. Various conventions can be in effect when the origin is reached for the first time. In the *ruin problem* the process stops; in this case the origin is said to act as an *absorbing barrier*. On the other hand, when the origin acts as *reflecting barrier*, the particle is returned instantaneously to the position 1 and the process continues forever. The point is that boundary conditions appear iff a boundary point can be reached. The event "the boundary point x_2 has been reached before epoch t" is well defined in diffusion processes because of the continuity of the path functions. It is closely related to the event "infinitely many jumps have occurred before t" in jump processes.

In some diffusion processes with probability one no boundary point is ever reached. Such is the Brownian motion [example 4(*a*)]. Then the minimal solution stands for a proper probability distribution and no other solutions exist. In all other situations the minimal solution regulates the process until a boundary is reached. It corresponds to absorbing barriers, that is, it describes a process that stops when a boundary point is reached. This is the most important type of process not only because all other processes are extensions of it, but even more because all *first-passage* probabilities can be calculated by imposing artificial absorbing barriers. The method is generally applicable but will be explained by the simplest example. (It was used implicitly in random walks and elsewhere, for example in problem 18 of **1**; XVII,10.)

In the following examples we limit our attention to the simple equation (5.4). More general diffusion equations will be treated by the method of Laplace transforms in XIV,5.

Examples. (*b*) *One absorbing barrier. First-passage times.* Consider Brownian motion on $\overline{0, \infty}$ with an absorbing barrier at the origin. More precisely, a Brownian motion starting at the point $x > 0$ at epoch 0 is stopped at the epoch of the first arrival at the origin. Because of symmetry both the backward and the forward equation take on the form of the classical diffusion equation

$$\frac{\partial u}{\partial t} = \frac{1}{2}\frac{\partial^2 u}{\partial x^2}. \tag{5.4}$$

The appropriate *boundary condition* is $q_t(0, y) = 0$ for all t, just as in the case of random walks. (The assertion can be justified either by the passage to the limit in **1**; XIV,6 or from the minimal character of the solution.)

For a given u_0 we seek a solution of (5.4) defined for $t \geq 0$, $x \geq 0$ and such that $u(0, x) = u_0(x)$ and $u(t, 0) = 0$. Its construction depends on *the method of images* due to Lord Kelvin.[18] We extend u_0 to the left half-line by $u_0(-x) = -u_0(x)$ and solve (5.4) with this initial condition in

[18] See problem 15–18 in **1**; XIV,9 for the same method applied to difference equations.

$\overline{-\infty, \infty}$. For reasons of symmetry the solution satisfies the condition $u(t, 0) = 0$, and restricting x again to $\overline{0, \infty}$ we have the desired solution. It is given by the integral of $u_0(y)\, q_t(x, y)$ over $\overline{0, \infty}$ where

$$q_t(x, y) = \frac{1}{\sqrt{2\pi t}}\left[\exp\left(-\frac{(y-x)^2}{2t}\right) - \exp\left(-\frac{(y+x)^2}{2t}\right)\right]. \tag{5.5}$$

Thus q_t *represents the transition densities of our process* $(t > 0, x > 0, y > 0)$. It is easily seen that q_t is, for fixed y, a solution (5.4) satisfying the boundary condition $q_t(0, y) = 0$. [For a more systematic derivation see example XIV,5(a).]

Integrating over y one gets the total probability mass at epoch t

$$\int_0^\infty q_t(x, y)\, dy = 2\mathfrak{N}(x/\sqrt{t}) - 1, \tag{5.6}$$

where $\mathfrak{N}$ stands for the standard normal distribution. In other words, (5.6) *is the probability that a path starting from* $x > 0$ *does not reach the origin before epoch* t. In this sense (5.6) represents the distribution of *first-passage times in a free Brownian motion.* Note that (5.6) may be characterized as the solution of the differential equation (5.4) defined for $x > 0$ and satisfying the initial condition $u(0, x) = 1$ together with the boundary condition $u(t, 0) = 0$.

[One recognizes in (5.6) the stable distribution with exponent $\alpha = \frac{1}{2}$; the same result was found in VI,2 by a passage to the limit from random walks.]

(c) *Two absorbing barriers.* Consider now a Brownian motion impeded by two absorbing barriers at 0 and $a > 0$. This means that for fixed $0 < y < a$ the transition densities q_t should satisfy the differential equation (5.4) together with the boundary conditions $q_t(0, y) = q_t(a, y) = 0$.

It is easily verified that the solution is given by[19]

$$q_t(x, y) = \frac{1}{\sqrt{2\pi t}} \sum_{k=-\infty}^{+\infty} \left\{\exp\left(-\frac{(y-x+2ka)^2}{2t}\right) - \exp\left(-\frac{(y+x+2ka)^2}{2t}\right)\right\} \tag{5.7}$$

where $0 < x, y < a$. Indeed, the series is manifestly convergent, and the obvious cancellation of terms shows that $q_t(0, y) = q_t(a, y) = 0$ for all $t > 0$ and $0 < y < a$. [The Laplace transform of (5.7) is given in XIV,(5.17).]

[19] The construction depends on successive approximations by repeated reflections. In (5.5) we have a solution of the differential equation satisfying the boundary condition at 0, but not a. A reflection at a leads to a four-term solution satisfying the boundary condition at a, but not at 0. Alternating reflections at 0 and a lead in the limit to (5.7). The analogous solution for random walks is given in **1**; XIV,(9.1), and (5.7) could be derived from it by the passage to the limit described in **1**; XIV,6.

Integrating (5.7) over $0 < y < a$ we get the total probability mass at epoch t in the form

$$(5.8)\quad \lambda_a(t, x) = \sum_{k=-\infty}^{\infty} \left\{ \mathfrak{N}\left(\frac{2ka + a - x}{\sqrt{t}}\right) - \mathfrak{N}\left(\frac{2ka - x}{\sqrt{t}}\right) - \mathfrak{N}\left(\frac{2ka + a + x}{\sqrt{t}}\right) + \mathfrak{N}\left(\frac{2ka + x}{\sqrt{t}}\right) \right\}.$$

This is the probability that a *particle starting at* x *will not be absorbed before epoch* t.

The function λ_a is a solution of the diffusion equation (5.4) tending to 1 as $t \to 0$ and satisfying the boundary conditions $\lambda_a(t, 0) = \lambda_a(t, a) = 0$. This solution can be obtained also by a routine application of the method of Fourier series in the form[20]

$$(5.9)\quad \lambda_a(t, x) = \frac{4}{\pi} \sum_{n=0}^{\infty} \frac{1}{2n + 1} \cdot \exp\left(-\frac{(2n+1)^2\pi^2}{2a^2} t\right) \cdot \sin \frac{(2n+1)\pi x}{a}.$$

We have thus obtained two very different representations[21] for the same function λ_a. This is fortunate because the series in (5.8) converges reasonably only when t is small, whereas (5.9) is applicable for large t.

For an alternative interpretation of λ_a consider the position $\mathbf{X}(t)$ of a particle in free Brownian motion starting at the origin. To say that during the time interval $\overline{0, t}$ the particle remained within $\overline{-\frac{1}{2}a, \frac{1}{2}a}$ amounts to saying that in a process with absorbing barriers at $\pm\frac{1}{2}a$ and starting at 0 no absorption took place before epoch t. Thus $\lambda_a(t, \frac{1}{2}a)$ *equals the probability that in an unrestricted Brownian motion starting at the origin* $|\mathbf{X}(s)| < \frac{1}{2}a$ *for all* s *in the interval* $0 < s < t$.

(*d*) *Application to limit theorems and Kolmogorov-Smirnov tests.* Let $\mathbf{Y}_1, \mathbf{Y}_2, \ldots$ be independent random variables with a common distribution and suppose that $\mathbf{E}(\mathbf{Y}_j) = 0$ and $\mathbf{E}(\mathbf{Y}_j^2) = 1$. Put $\mathbf{S}_n = \mathbf{Y}_1 + \cdots + \mathbf{Y}_n$ and $\mathbf{T}_n = \max\,[|\mathbf{S}_1|, \ldots, |\mathbf{S}_n|]$. In view of the central limit theorem it is plausible that the asymptotic behavior of $\mathbf{T}_n$ will be nearly the same as in the case where the $\mathbf{Y}_j$ are normal variables, and in the latter case the normed sum $\mathbf{S}_k/\sqrt{n}$ is comparable to the variable of a Brownian motion

---

[20] The analogous formula for random walks is derived in **1**; XIV,5 where, however, the boundary conditions are $\lambda_a(t, 0) = 1$ and $\lambda_a(t, a) = 0$.

[21] The identity between (5.8) and (5.9) serves as a standard example for the Poisson summation formula [see XIX,(5.10)]. It has acquired historical luster, having been discovered originally in connection with Jacobi's theory of transformations of theta functions. See Satz 277 in E. Landau, *Verteilung der Primzahlen*, 1909.

at epoch k/n $(k = 0, 1, \ldots, n)$. The probability that this Brownian motion remains constrained to the interval $(-\frac{1}{2}a, \frac{1}{2}a)$ was shown to equal $\lambda_a(1, \frac{1}{2}a)$. Our plausibility argument would therefore lead us to the conjecture that as $n \to \infty$

$$\mathbf{P}\{\mathbf{T}_n < z\} \to L(z) \tag{5.10}$$

where $L(z) = \lambda_{2z}(1, z)$ is obtained from (5.8) and (5.9):

$$\begin{aligned} L(z) &= 2 \sum_{k=-\infty}^{\infty} \{\mathfrak{N}((4k+1)z) - \mathfrak{N}((4k-1)z)\} = \\ &= \frac{4}{\pi} \sum_{n=0}^{\infty} \frac{(-1)^n}{2n+1} \exp\left(-\frac{(2n+1)^2\pi^2}{8z^2}\right). \end{aligned} \tag{5.11}$$

This conjecture was proved in 1946 by P. Erdös and M. Kac, and the underlying idea has since become known as *invariance principle*. It states, roughly speaking, that the asymptotic distribution of certain functions of random variables is insensitive to changes of the distributions of these variables and may be obtained by considering an appropriate approximating stochastic process. This method has been perfected by M. Donsker, P. Billingsley, Yu. V. Prohorov, and others, and has become a powerful tool for proving limit theorems.

For similar reasons the distribution (5.11) plays a prominent part also in the vast literature on non-parametric tests of the type discussed in I,12.[22]

(*e*) *Reflecting barriers.* By analogy with the ordinary random walk we define a reflecting barrier at the origin by the boundary condition $\dfrac{\partial q_t(0, y)}{\partial y} = 0$ for a reflecting barrier at the origin is imposed by analogy with random walks. It is readily verified that the solution for the interval $\overline{0, \infty}$ is given by (5.5) with the minus sign replaced by plus. The formal derivation by the method of images is the same, except that one puts $u_0(-x) = u_0(x)$. The solution for $\overline{0, a}$ with reflecting barriers at both 0 and a is obtained similarly by changing the minus sign to a plus in (5.7). (An alternative expression obtained by Fourier expansions or the Poisson summation formula is given in problem 11 of XIX,9.)

It should be noted that in the case of reflecting barriers q_t is a proper probability density. ►

[22] The topic is relatively new, and yet the starting point of the much used identity (5.11) seems already to have fallen into oblivion. [A. Renyi, *On the distribution function L(z).* Selected Translations in Math. Statist. and Probability, vol. 4 (1963) pp. 219–224. Renyi's supposedly new proof depends on the classical argument involving theta functions, thus obscuring the simple probabilistic meaning of (5.11).]

6. DIFFUSION IN HIGHER DIMENSIONS

It is easy to generalize the foregoing theory to two dimensions. To avoid the nuisance of subscripts we denote the coordinate variables by $(\mathbf{X}(t), \mathbf{Y}(t))$ and the values of the transition *densities* by $q_t(x, y; \xi, \eta)$; here x, y is the initial point and q_t is a density in (ξ, η). The postulates are as in section 4 except that the infinitesimal velocity $b(x)$ is replaced by a vector, and the variance $a(x)$ by a covariance matrix. Instead of (4.6) we get for the *backward diffusion equation*

$$\frac{\partial u}{\partial t} = a_{11}\frac{\partial^2 u}{\partial x^2} + 2a_{12}\frac{\partial^2 u}{\partial x\,\partial y} + a_{22}\frac{\partial^2 u}{\partial y^2} + b_1\frac{\partial u}{\partial x} + b_2\frac{\partial u}{\partial y}, \tag{6.1}$$

the coefficients depending on x and y. In the case of two-dimensional Brownian motion we require rotational symmetry, and up to an irrelevant norming constant we must have

$$\frac{\partial u}{\partial t} = \frac{1}{2}\left[\frac{\partial^2 u}{\partial x^2} + \frac{\partial^2 u}{\partial y^2}\right]. \tag{6.2}$$

The corresponding transition densities are normal with variance t, centered at (x, y). The obvious factoring of this density shows that $\mathbf{X}(t)$ and $\mathbf{Y}(t)$ are stochastically independent.

The most interesting variable in this process is the distance $\mathbf{R}(t)$ from the origin $(\mathbf{R}^2 = \mathbf{X}^2 + \mathbf{Y}^2)$. It is intuitively obvious that $\mathbf{R}(t)$ is the variable of a one-dimensional diffusion process and it is interesting to compare the various ways of getting at the diffusion equation for this process. In polar coordinates our normal transition densities for (6.2) take on the form

$$\frac{\rho}{2\pi t}\exp\left(-\frac{\rho^2 + r^2 - 2\rho r\cos(\theta-\alpha)}{2t}\right) \tag{6.3}$$

(with $x = r\cos\alpha$, etc.). Given the position r, α at epoch 0, the marginal density of $\mathbf{R}(t)$ is obtained by integrating (6.3) with respect to θ. The parameter α drops out and we get[23] for *the transition densities of the* $\mathbf{R}(t)$ *process*

$$w_t(r, \rho) = \frac{1}{t}\exp\left(-\frac{r^2+\rho^2}{2t}\right)I_0\left(\frac{r\rho}{t}\right) \tag{6.4}$$

where I_0 is the Bessel function defined in II,(7.1). Here r stands for the initial position at epoch 0. From the derivation it is clear that for fixed ρ the

[23] The integral is well known. For a routine verification expand $e^{\cos\theta}$ into a power series in $\cos\theta$.

transition probabilities w_t satisfy (6.2) in polar coordinates; that is to say

$$\frac{\partial w_t}{\partial t} = \frac{1}{2}\left(\frac{\partial^2 w_t}{\partial r^2} + \frac{1}{r}\frac{\partial w_t}{\partial r}\right). \tag{6.5}$$

This is the *backward equation for the* $\mathbf{R}(t)$ *process* and is obtained from (6.2) simply by requiring rotational symmetry.

Equation (6.5) shows that the $\mathbf{R}(t)$ process has an infinitesimal velocity $1/(2r)$. The existence of a drift away from the origin can be understood if one considers a plane Brownian motion starting at the point r of the x-axis. For reasons of symmetry its *abscissa* at epoch $h > 0$ is equally likely to be $>r$ or $<r$. In the first case certainly $\mathbf{R}(h) > r$, but this relation can occur also in the second case. Thus the relation $\mathbf{R}(h) > r$ has probability $>\frac{1}{2}$, and on the average $\mathbf{R}$ is bound to increase.

The same derivation of transition probabilities applies to three dimensions with one essential simplification: the Jacobian ρ in (6.3) is now replaced by $\rho^2 \sin \theta$, and an elementary integration is possible. Instead of (6.4) we get for the *transition densities of the* $\mathbf{R}(t)$ *process in three dimensions*

$$w_t(r, \rho) = \frac{1}{\sqrt{2\pi t}}\frac{\rho}{r}\left[\exp\left(-\frac{(\rho-r)^2}{2t}\right) - \exp\left(-\frac{(\rho+r)^2}{2t}\right)\right]. \tag{6.6}$$

(Again r stands for the initial position at epoch 0.)

7. SUBORDINATED PROCESSES

From a Markov process $\{\mathbf{X}(t)\}$ with stationary transition probabilities $Q_t(x, \Gamma)$ it is possible to derive a variety of new processes by introducing what may be called a *randomized operational time*. Suppose that to each $t > 0$ there corresponds a random variable $\mathbf{T}(t)$ with distribution U_t. A new stochastic kernel P_t may then be defined by

$$P_t(x, \Gamma) = \int_{0-}^{\infty} Q_s(x, \Gamma)\, U_t\{ds\}. \tag{7.7}$$

This represents the distribution of $\mathbf{X}(\mathbf{T}(t))$ given that $\mathbf{X}(0) = 0$.

Example. (*a*) If $\mathbf{T}(t)$ has a Poisson distribution with expectation αt

$$P_t(x, \Gamma) = \sum_{n=0}^{\infty} e^{-\alpha t}\frac{(\alpha t)^n}{n!}\, Q_n(x, \Gamma). \tag{7.2}$$

These P_t are the transition probabilities of a pseudo-Poisson process. It will be shown in section 9 that the randomization by Poisson distributions leads to the so-called exponential formula which is basic for the theory of Markov semi-groups. We shall now see that similar results are obtained with a variety of other distributions, each of which leads to an analogue of the exponential formula. ►

The variables $\mathbf{X}(\mathbf{T}(t))$ form a new stochastic process which need not be Markovian. For the process to be Markovian it is obviously necessary that the P_t satisfy the Chapman-Kolmogorov equation

$$P_{s+t}(x, \Gamma) = \int_{-\infty}^{+\infty} P_s(x, dy)\, P_t(y, \Gamma). \tag{7.3}$$

This means that the distribution of $\mathbf{X}(\mathbf{T}(t+s))$ is obtained by integration of $P_t(y, \Gamma)$ with respect to the distribution of $\mathbf{X}(\mathbf{T}(s))$ and so

$$P_t(y, \Gamma) = \mathbf{P}\{\mathbf{X}(\mathbf{T}(t+s)) \in \Gamma \mid \mathbf{X}(\mathbf{T}(s)) = y\} \tag{7.4}$$

by the definition of conditional probabilities. A similar calculation of higher order transition probabilities shows that (7.3) suffices to ensure the Markovian character of the derived process $\{\mathbf{X}(\mathbf{T}(t))\}$.

We wish now to find the distributions U_t that lead to solutions P_t of (7.3). A direct attack on this problem leads to considerable difficulties, but these can be avoided by first considering the simple special case where the variables $\mathbf{T}(t)$ are restricted to the multiples of a fixed number $h > 0$. For the distribution of $\mathbf{T}(t)$ we write

$$\mathbf{P}\{\mathbf{T}(t) = nh\} = a_n(t). \tag{7.5}$$

Given that $\mathbf{X}(0) = x$ the variable $\mathbf{X}(\mathbf{T}(t))$ has the distribution

$$P_t(x, \Gamma) = \sum_{k=0}^{\infty} a_k(t)\, Q_{kh}(x, \Gamma). \tag{7.6}$$

Since the kernels $\{Q_t\}$ satisfy the Chapman-Kolmogorov equation we have

$$\int_{-\infty}^{+\infty} P_s(x, dy)\, P_t(y, \Gamma) = \sum_{j,k} a_j(s)\, a_k(t) \cdot Q_{(j+k)h}(x, \Gamma), \tag{7.7}$$

and it is seen that the kernels P_t satisfy the Chapman-Kolmogorov equation (7.3) iff

$$a_0(s)\, a_n(t) + a_1(s)\, a_{n-1}(t) + \cdots + a_n(s)\, a_0(t) = a_n(s+t) \tag{7.8}$$

for all $s > 0$ and $t > 0$. This relation holds if $\{\mathbf{T}(t)\}$ *is a process with stationary independent increments*, and the most general solution of (7.8) was found in **1**; XII,2.

This result leads to the conjecture that in general (7.3) will be satisfied whenever the $\mathbf{T}(t)$ are the variables of a process with stationary independent increments, that is, whenever the distributions U_t satisfy[24]

$$U_{t+s}(x) = \int_{0-}^{x} U_s(x-y)\, U_t\{dy\}, \tag{7.9}$$

We verify this conjecture by a passage to the limit.[25] We represent U_t as the limit of a sequence of arithmetic distributions of the type just considered: $U_t^{(\nu)}$ is concentrated on the multiples of a number h_ν, and the weights that it attaches to the points nh_ν satisfy a relation of the form (7.8). For each ν we get thus a kernel $P_t^{(\nu)}$ corresponding to (7.6) that satisfies the Chapman-Kolmogorov equation (7.3). To show that also the kernel P_t of (7.1) satisfies this equation it suffices therefore to show that $P_t^{(\nu)} \to P_t$, or, what amounts to the same, that

$$\int_{-\infty}^{+\infty} P_t^{(\nu)}(x, dy) f(y) \to \int_{-\infty}^{+\infty} P_t(x, dy) f(y) \tag{7.10}$$

for every continuous function $f^{-\infty}$ vanishing at infinity. If we put

$$F(t, x) = \int_{-\infty}^{+\infty} Q_t(x, dy)\, f(y) \tag{7.11}$$

(7.10) may be rewritten in the form

$$\int_0^\infty F(s, x)\, U_t^{(\nu)}\{ds\} \to \int_0^\infty F(s, x)\, U_t\{ds\}. \tag{7.12}$$

This relation holds certainly if F is continuous, and this imposes only an extremely mild regularity condition on the Q_t. We have thus proved the following basic result:

Let $\{\mathbf{X}(t)\}$ *be a Markov process with continuous transition probabilities* Q_t *and* $\{\mathbf{T}(t)\}$ *a process with non-negative independent increments. Then* $\{\mathbf{X}(\mathbf{T}(t))\}$ *is a Markovian process with transition probabilities* P_t *given by* (7.7). This process is said to be *subordinate*[26] *to* $\{\mathbf{X}(t)\}$ using the operational time $\mathbf{T}(t)$. The process $\{\mathbf{T}(t)\}$ is called the *directing process.*

The most interesting special case arises when also the $\mathbf{X}(t)$ process has independent increments. In this case the transition probabilities depend only on the differences $\Gamma - x$ and may be replaced by the equivalent distribution

[24] The most general solution of (7.9) will be found by means of Laplace transforms in XIII,7. It can be obtained also from the general theory of infinitely divisible distributions.

[25] A direct verification requires analytic skill. Our procedure shows once again that a naive approach is sometimes most powerful.

[26] The notion of subordinated semi-groups was introduced by S. Bochner in 1949. For a high-level systematic approach see E. Nelson, *A functional calculus using singular Laplace integrals*, Trans. Amer. Math. Soc., 88 (1958), pp. 400–413.

functions. Then (7.7) takes on the simpler form

$$(7.13) \qquad P_t(x) = \int_0^\infty Q_s(x)\, U_t\{ds\}.$$

All our examples are of this type.

Examples. (*b*) *The Cauchy process is subordinated to Brownian motion.* Let $\{\mathbf{X}(t)\}$ be the Brownian motion (Wiener process) with transition *densities* given by $q_t(x) = (2\pi t)^{-\frac{1}{2}} e^{-\frac{1}{2}x^2/t}$. For $\{\mathbf{T}(t)\}$ we take the stable process with exponent $\frac{1}{2}$ with transition densities given by

$$u_t(x) = \frac{t}{\sqrt{2\pi}\,\sqrt{x^3}}\, e^{-\frac{1}{2}t^2/x}.$$

The distribution (7.13) has then a density given by

$$(7.14) \qquad p_t(x) = \frac{t}{2\pi} \int_{-\infty}^{+\infty} s^{-2} e^{-(x^2+t^2)/(2s)}\, ds = \frac{t}{\pi(t^2+x^2)}$$

and thus our subordination procedure leads to a Cauchy process.

This result may be interpreted in terms of two independent Brownian motions $\mathbf{X}(t)$ and $\mathbf{Y}(t)$ as follows.

It was shown in example VI,2(*e*) that U_t may be interpreted as the distribution of the waiting time for the epoch at which the $\mathbf{Y}(s)$-process for the first time attains the value $t > 0$. Accordingly, *a Cauchy process* $\mathbf{Z}(t)$ *may be realized by considering the value of the* $\mathbf{X}$*-process at the epoch* $\mathbf{T}(t)$ *when* $\mathbf{Y}(s)$ *first attains the value* t. [For another connection of the Cauchy process with hitting times in Brownian motion see example VI,2(*f*).]

(*c*) *Stable processes.* The last example generalizes easily to arbitrary strictly stable processes $\{\mathbf{X}(t)\}$ and $\{\mathbf{T}(t)\}$ with exponents α and β, respectively. Here $\alpha \leq 2$, but since $\mathbf{T}(t)$ must be positive we have necessarily $\beta < 1$. The transition probabilities Q_t and U_t are of the form $Q_t(x) = Q(xt^{-1/\alpha})$ and $U_t(x) = U(xt^{-1/\beta})$ where Q and U are fixed stable distributions. We show that *the subordinated process* $\mathbf{X}(\mathbf{T}(t))$ *is stable with exponent* $\alpha\beta$. This assertion is equivalent to the relation $P_{\lambda t}(x) = P_t(x^{-1/\alpha\beta})$. In view of the given form of Q_s and U_t this relation follows trivially from (7.13) by the substitution $s = y\lambda^{1/\beta}$.

[Our result is essentially equivalent to the product formula derived in example VI,2(*h*). When $\mathbf{X}(t) > 0$ the formula can be restated in terms of Laplace transforms as in XIII,7(*e*). For the Fourier version see problem 9 in XVII,12.]

(*d*) *Compound Poisson process directed by gamma process.* Let Q_t be the compound Poisson distribution generated by the probability distribution F and let U_t have the gamma density $e^{-x}x^{t-1}/\Gamma(t)$. Then (7.13) takes on the form

$$P_t = \sum_{n=0}^{\infty} a_n(t)\, F^{n\star} \tag{7.15}$$

where

$$a_n(t) = \int_0^{\infty} e^{-s}\frac{s^n}{n!}\cdot e^{-s}\frac{s^{t-1}}{\Gamma(t)}\, ds = \frac{\Gamma(n+t)}{n!\,\Gamma(t)}\cdot 2^{-n-t} \tag{7.16}$$

It is easily verified that the probabilities $a_n(t)$ have the infinitely divisible generating function $\sum a_n(t)\zeta^n = (2-\zeta)^{-t}$.

(*e*) *Gamma process directed by the Poisson process.* Let us now consider the same distributions but with reversed roles. The operational time is then integral-valued and 0 has weight e^{-t}. It follows that the resulting distribution has an atom of weight e^{-t} at the origin. The continuous part has the density

$$\sum_{n=1}^{\infty} e^{-x}\frac{x^{n-1}}{(n-1)!}\cdot e^{-t}\frac{t^n}{n!} = e^{-t-x}\sqrt{\frac{t}{x}}\, I_1(2\sqrt{xt}), \tag{7.17}$$

where I_1 is the Bessel function of II,(7.1). It follows that this distribution is infinitely divisible, but a direct verification is not easy. ►

8. MARKOV PROCESSES AND SEMI-GROUPS

Chapter VIII revealed the advantages of treating probability distributions as operators on continuous functions. The advantages of the operator approach to stochastic kernels are even greater, and the theory of semi-groups leads to a unified theory of Markov processes not attainable by other methods. Given a stochastic kernel K in $\mathfrak{R}^1$ and a bounded continuous function u the relation

$$U(x) = \int_{-\infty}^{+\infty} K(x, dy)\, u(y) \tag{8.1}$$

defines a new function. Little generality is lost in assuming that the transform U is again continuous and we could proceed to study properties of the kernel K in terms of the induced transformation $u \to U$ on continuous functions. There are two main reasons for a more general setup. First, transformations of the form (8.1) make sense in arbitrary spaces, and it would be exceedingly uneconomical to develop a theory which does not cover the simplest and most important special case, namely processes with a denumerable state space [where (8.1) reduces to a matrix transformation]. Second, even in a theory restricted to continuous functions on the line various types of boundary conditions compel one to introduce special classes of continuous functions. On the other hand, the greater generality is bought at no expense. Readers so inclined are urged to ignore the generality and refer all theorems to one (or several) of the following typical situations. (i) The underlying space Σ

is the real line and $\mathscr{L}$ the class of bounded continuous functions vanishing at infinity. (ii) the space Σ is a finite closed interval I in $\mathfrak{R}^1$ or $\mathfrak{R}^2$ and $\mathscr{L}$ the class of continuous functions on it. (iii) Σ consists of the integers and $\mathscr{L}$ of bounded sequences. In this case it is best to think of sequences as column vectors and of transformations as matrices.

As in chapter VIII the *norm* of a bounded real function u is defined by $\|u\| = \sup |u(x)|$. A sequence of functions u_n converges uniformly to u iff $\|u_n - u\| \to 0$.

From now on $\mathscr{L}$ *will denote a family of real functions on some set* Σ *with the following properties:* (i) If u_1 and u_2 belong to $\mathscr{L}$ then every linear combination $c_1u_1 + c_2u_2 \in \mathscr{L}$. (ii) If $u_n \in \mathscr{L}$ and $\|u_n - u\| \to 0$ then $u \in \mathscr{L}$. (iii) If $u \in \mathscr{L}$ then also u^+ and u^- belong to $\mathscr{L}$ (where $u = u^+ - u^-$ is the usual decomposition of u into its positive and negative parts.) In other words, $\mathscr{L}$ is closed under linear combinations, uniform limits, and absolute values. The first two properties make $\mathscr{L}$ a *Banach* space, the last a lattice.

The following definitions are standard. A linear transformation T is an *endomorphism* on $\mathscr{L}$ if each $u \in \mathscr{L}$ has an image $Tu \in \mathscr{L}$ such that $\|Tu\| \leq m\,\|u\|$ where m is a constant independent of u. The smallest constant with this property is called the *norm* $\|T\|$ of T. The transformation T is *positive* if $u \geq 0$ implies $Tu \geq 0$. In this case $-Tu^- \leq Tu \leq Tu^+$. A *contraction* is a positive operator T with $\|T\| \leq 1$. If the constant function 1 belongs to $\mathscr{L}$ and T is a positive operator such that $T1 = 1$, then T is called a *transition operator*. (It is automatically a contraction.)

Given two endomorphisms S and T on $\mathscr{L}$, their *product* ST is the endomorphism mapping u into $S(Tu)$. Obviously $\|ST\| \leq \|S\| \cdot \|T\|$. In general $ST \neq TS$, in contrast to the particular class of convolution operators of VIII,3 which commuted with each other.

We are seriously interested only in transformations of the form (8.1) where K is a stochastic, or at least substochastic, kernel. Operators of this form are contractions or transition operators and they also enjoy the

Monotone convergence property: *if* $u_n \geq 0$ *and* $u_n \uparrow u$ *(with* u_n *and* u *in* $\mathscr{L}$*) then* $Tu_n \to Tu$ *pointwise.*

In practically all situations contractions with this property are of the form (8.1). Two examples will illustrate this point.

Examples. (*a*) Let Σ stand for the real line and $\mathscr{L} = C$ for the family of all bounded continuous functions on it. Let $C_0 \subset \mathscr{L}$ the subclass of functions vanishing at $\pm\infty$. If T is a contraction on $\mathscr{L}$ then for $u \in C_0$ the value $Tu(x)$ of Tu at a fixed x is a positive linear functional on $\mathscr{L}$. By the F. Riesz representation theorem there exists a possibly defective probability distribution F such that $Tu(x)$ is the expectation of u with respect to F. Since F depends on x we write $K(x, \Gamma)$ for $F(\Gamma)$. Then for $u \in C_0$

$$Tu(x) = \int_{-\infty}^{+\infty} K(x, dy)\, u(y), \tag{8.2}$$

and when T has the monotone convergence property this relation automatically extends to all bounded continuous functions.

For fixed x, as a function of Γ, the kernel K is a measure. If Γ is an open interval and $\{u_n\}$ an increasing sequence of continuous functions such that $u_n(x) \to 1$ if $x \in \Gamma$ and $u_n(x) \to 0$ otherwise, then $K(x, \Gamma) = \lim Tu_n(x)$ by the basic properties of integrals. Since Tu_n is continuous it follows that for fixed Γ the kernel K is a Baire function of x and therefore K has all the properties required of stochastic or substochastic kernels. The same situation prevails when the line is replaced by an interval, or $\mathcal{R}^n$.

(*b*) Let Σ be the set of integers, and $\mathscr{L}$ the set of numerical sequences $u = \{x_n\}$ with $\|u\| = \sup |x_n|$. If p_{ij} stands for a stochastic or substochastic matrix we define a transformation T such that the ith component of Tu is given by

$$(Tu)_i = \sum p_{ik} u_k, \tag{8.3}$$

Evidently T is a contraction operator enjoying the monotone convergence property; if the matrix is strictly stochastic, then T is a transition operator. ►

These examples are typical and it is actually difficult to find contractions not induced by a stochastic kernel. Anyhow, we are justified to proceed with the general theory of contractions with the assurance that applications to probabilistically significant problems will be obvious. (In fact, we shall never have to go beyond the scope of these examples.)

The transition probabilities of a Markov process form a one-parameter family of kernels satisfying the Chapman-Kolmogorov equation

$$Q_{s+t}(x, \Gamma) = \int Q_s(x, dy)\, Q_t(y, \Gamma) \tag{8.4}$$

($s > 0, t > 0$), the integration extending over the underlying space. Each individual kernel induces a transition operator $\mathfrak{Q}(t)$ defined by

$$\mathfrak{Q}(t)\, u(x) = \int Q_t(x, dy)\, u(y). \tag{8.5}$$

Obviously then (8.4) is equivalent with

$$\mathfrak{Q}(s+t) = \mathfrak{Q}(s)\, \mathfrak{Q}(t), \qquad s > 0, t > 0. \tag{8.6}$$

A family of endomorphisms with this property is a *semi-group*. Clearly $\mathfrak{Q}(s)\,\mathfrak{Q}(t) = \mathfrak{Q}(t)\,\mathfrak{Q}(s)$, that is, the *elements of a semi-group commute with each other*.

A sequence of endomorphisms T_n *on* $\mathscr{L}$ *is said to converge*[27] *to the endomorphism* T *iff* $\|T_n u - Tu\| \to 0$ *for each* $u \in \mathscr{L}$. *In this case we write* $T_n \to T$.

From now on we concentrate on semi-groups of contraction operators and impose a regularity condition on them. Denote again by $\mathbf{1}$ the identity operator, $\mathbf{1}u = u$.

Definition. *A semi-group of contraction operators* $\mathfrak{Q}\,(t)$ *will be called continuous*[28] *if* $\mathfrak{Q}(0) = \mathbf{1}$ *and* $\mathfrak{Q}(h) \to \mathbf{1}$ *as* $h \to 0+$.

If $0 \le t' < t''$ we have

$$\|\mathfrak{Q}(t'')u - \mathfrak{Q}(t')u\| \le \|\mathfrak{Q}(t''-t')u - u\|. \tag{8.7}$$

For continuous semi-groups there exists a $\delta > 0$ such that the right side is $<\epsilon$ for $t'' - t' < \delta$. Thus not only is it true that $\mathfrak{Q}(t) \to \mathfrak{Q}(t_0)$ as $t \to t_0$, but (8.7) shows that $\mathfrak{Q}(t)u$ is a uniformly continuous function of t for each fixed u.[29]

The transformation (8.1) is, of course, the same as (4.5) and served as starting point for the derivation of the backward equation for diffusion processes. Now a family of Markovian transition probabilities induces also a semi-group of transformations of *measures* such that the measure μ is transformed into a measure $T(t)\mu$ attributing to the set Γ the mass

$$T(t)\mu(\Gamma) = \int \mu\{dx\}\, Q_t(x, \Gamma). \tag{8.8}$$

When the Q_t have a density kernel q_t this transformation is the same as (5.1) and was used for the forward equation. Probability theory being concerned primarily with measures, rather than functions, the question arises, why we do not start from the semi-group $\{T(t)\}$ rather than $\mathfrak{Q}(t)$? The answer is interesting and throws new light on the intricate relationship between the backward and forward equations.

The reason is that (as evidenced by the above examples) with the usual setup the continuous semi-groups of contractions on the function space $\mathscr{L}$ come from transition probabilities: studying our semi-groups $\mathfrak{Q}(t)$ is in practice the same as studying Markovian

[27] This mode of convergence was introduced in VIII,3 and is called *strong*. It does *not* imply that $\|T_n - T\| \to 0$ (which type of convergence is called uniform). A weaker type of convergence is defined by the requirement that $T_n u(x) \to Tu(x)$ for each x, but not necessarily uniformly. See problem 6 in VIII,10.

[28] We use this word as abbreviation for the standard term "strongly continuous at the origin."

[29] There exist semi-groups such that $\mathfrak{Q}(h)$ tends to an operator $T \ne \mathbf{1}$, but they are pathological. For an example define an endomorphism T by

$$Tu(x) = \tfrac{1}{2}u(0)[1+\cos x] + \tfrac{1}{2}u(\pi)[1-\cos x]$$

and put $\mathfrak{Q}(t) = T$ for all $t \ge 0$.

transition probabilities. For semi-groups of measures this is *not* true. There exist analytically very reasonable contraction semi-groups that are not induced by Markov processes. To get an example consider any Markovian semi-group of the form (8.8) on the line assuming only that an absolutely continuous μ is transformed into an absolutely continuous $T(t)\mu$ [for example, let $T(t)$ be the convolution with a normal distribution with variance t]. If $\mu = \mu_c + \mu_s$ is the decomposition of μ into its absolutely continuous and singular parts define, a new semi-group $\{S(t)\}$ by

$$S(t)\mu = T(t)\mu_c + \mu_s. \tag{8.9}$$

This semi-group is continuous and $S(0) = \mathbf{1}$, but it is not difficult to see that it is not connected with any system of transition probabilities and that it is probabilistically meaningless.

9. THE "EXPONENTIAL FORMULA" OF SEMI-GROUP THEORY

The pseudo-Poisson processes of section 1 are by far the simplest Markov processes, and it will now be shown that practically all Markov processes represent limiting forms of pesudo-Poisson processes.[30] An abstract version of the theorem plays a fundamental role in semi-group theory, and we shall now see that it is really a consequence of the law of large numbers.

If T is the operator induced by the stochastic kernel K, the operator $\mathfrak{Q}(t)$ induced by the pseudo-Poisson distribution (1.2) takes on the form

$$\mathfrak{Q}(t) = e^{-\alpha t} \sum_{n=0}^{\infty} \frac{(\alpha t)^n}{n!} T^n, \tag{9.1}$$

the series being defined as the limit of the partial sums. These operators form a semi-group by virtue of the Chapman-Kolmogorov equation (1.3). It is better, however, to start afresh and to prove the assertion for arbitrary contractions T.

Theorem 1. *If T is a contraction on $\mathscr{L}$, the operators* (9.1) *form a continuous semi-group of contractions. If T is a transition operator so is $\mathfrak{Q}(t)$.*

Proof. Obviously $\mathfrak{Q}(t)$ is positive and $\|\mathfrak{Q}(t)\| \leq e^{-\alpha t + \alpha t\|T\|} \leq 1$. The semi-group property is easily verified from the formal product of the series for $\mathfrak{Q}(s)$ and $\mathfrak{Q}(t)$ (see footnote 1 to section 1). The relation $\mathfrak{Q}(h) \to \mathbf{1}$ is obvious from (9.1). ►

We shall abbreviate (9.1) to

$$\mathfrak{Q}(t) = e^{\alpha t(T-1)}. \tag{9.2}$$

[30] The special case where the $\mathfrak{Q}(t)$ are convolution operators is treated in chapter IX.

Semi-groups of contractions of this form will be called pseudo-Poissonian[31] *and we shall say that* $\{\mathfrak{Q}(t)\}$ *is generated by* $\alpha(T-\mathbf{1})$.

Consider now an arbitrary continuous semi-group of contractions $\mathfrak{Q}(t)$. It behaves in many respects just as a real-valued continuous function and the approximation theory developed in chapter VII using the law of large numbers carries over without serious change. We show in particular that the procedure of example VII,1(*b*) leads to an important formula of general semi-group theory.

For fixed $h > 0$ we define the operators

$$\mathfrak{Q}_h(t) = e^{-t/h} \sum_{n=0}^{\infty} \frac{(t/h)^n}{n!} \mathfrak{Q}(nh) \tag{9.3}$$

which could be described as obtained by randomization of the parameter t in $\mathfrak{Q}(t)$. Comparing with (9.1) is it seen that the $\mathfrak{Q}_h(t)$ form a pseudo-Poissonian semi-group generated by $[\mathfrak{Q}(h)-\mathbf{1}]/h$. We now prove that

$$\mathfrak{Q}_h(t) \to \mathfrak{Q}(t), \qquad h \to 0. \tag{9.4}$$

Because of the importance of this result we formulate it as

Theorem 2. *Every continuous semi-group of contractions* $\mathfrak{Q}(t)$ *is the limit* (9.4) *of the pseudo-Poisson semi-group* $\{\mathfrak{Q}_h(t)\}$ *generated by the endomorphism* $h^{-1}[\mathfrak{Q}(h)-\mathbf{1}]$.

Proof. The starting point is the identity

$$\mathfrak{Q}_h(t)u - \mathfrak{Q}(t)u = e^{-th} \sum_{n=0}^{\infty} \frac{(th^{-1})^n}{n!} [\mathfrak{Q}(nh)u - \mathfrak{Q}(t)u]. \tag{9.5}$$

Choose δ such that $\|\mathfrak{Q}(s)u - u\| < \epsilon$ for $0 < s < \delta$. In view of (8.7) we have then

$$\|\mathfrak{Q}(nh)u - \mathfrak{Q}(t)u\| \leq \epsilon \qquad \text{for } |nh-t| < \eta t. \tag{9.6}$$

The Poisson distribution appearing in (9.5) has expectation and variance equal to t/h. The contribution of the terms with $|nh - t| \geq \delta$ can be estimated using Chebyshev's inequality, and we find that

$$\|\mathfrak{Q}_h(t)u - \mathfrak{Q}(t)u\| < \epsilon + 2\,\|u\|\, th\delta^{-2}.$$

It follows that (9.4) holds uniformly in finite t-intervals. ►

Equivalent variants of this section are obtained by letting other infinitely divisible distributions take over the role of the Poisson distribution. We know from **1**; XII,2 that the

[31] There exist contraction semi-groups of the form e^{tS} where S is an *endomorphism not* of the form $\alpha(T-\mathbf{1})$. Such are the semi-groups associated with the solutions of the jump processes of section 3 if $\alpha(x)$ remains bounded.

generating function of an infinitely divisible distribution $\{u_n(t)\}$ concentrated on the integers $n \geq 0$ is of the form

$$\sum_0^\infty u_n(t)\zeta^n = \exp(t\alpha[p(\zeta)-1]) \tag{9.7}$$

where

$$p(\zeta) = p_0 + p_1\zeta + \cdots, \qquad p_i \geq 0, \Sigma p_i = 1. \tag{9.8}$$

Suppose that the distribution $\{u_n(t)\}$ has expectation bt and a finite variance ct. Replacing in (9.3) the Poisson distribution by $\{u_n(t/b)\}$ leads to the operator

$$\mathfrak{Q}_h(t) = \sum u_n\left(\frac{t}{bh}\right)\mathfrak{Q}(nh) = \sum u_n\left(\frac{t}{bh}\right)\mathfrak{Q}^n(h). \tag{9.9}$$

As in the preceding proof a simple application of the law of large numbers shows that $\mathfrak{Q}_h(t) \to \mathfrak{Q}(t)$ as $h \to 0$. In this way we get a substitute "exponential formula" in which $\{u_n(t)\}$ takes over the role of the Poisson distribution.[32]

To see the probabilistic content and the possible generalizations of this argument, denote by $\mathbf{X}(t)$ the variables of the Markov process with the semi-group $\{\mathfrak{Q}(t)\}$, and by $\mathbf{T}(t)$ the variables of the process with independent increments subject to $\{u_n(t)\}$. The operators (9.9) correspond to the transition probabilities for the variables $\mathbf{X}\left(h\mathbf{T}\left(\frac{t}{bh}\right)\right)$. In other words, we have introduced a particular subordinated process; the law of large numbers for the $\mathbf{T}$-process makes it plausible that as $h \to 0$ the distributions of the new process tend to those of the initial Markov process. This approximation procedure is by no means restricted to integral valued variables $\mathbf{T}(t)$. Indeed, we may take for $\{\mathbf{T}(t)\}$ an arbitrary process with positive independent increments such that $\mathbf{E}(\mathbf{T}(t)) = bt$ and that variances exist.

The given Markov process $\{\mathbf{X}(t)\}$ thus appears as the limit of the subordinated Markov processes with variables $\mathbf{X}\left(h\mathbf{T}\left(\frac{t}{bh}\right)\right)$.

The point is that the approximating semi-groups may be of a much simpler structure than the original one. In fact, *the semi-group of the operators* $\mathfrak{Q}_h(t)$ *of* (9.9) *is of the simple pseudo-Poisson type.* To see this put

$$\mathfrak{Q}^{\#} = p(\mathfrak{Q}(h)) = \sum_{n=0}^{\infty} p_n\mathfrak{Q}^n(h) = \sum_{n=0}^{\infty} p_n\mathfrak{Q}(nh). \tag{9.10}$$

This is a mixture of transition operators and therefore itself a transition operator. A comparison of (9.7) and (9.9) now shows that formally

$$\mathfrak{Q}_h(t) = \exp\frac{\alpha}{bh}(\mathfrak{Q}^{\#} - 1) \tag{9.11}$$

which is indeed of the form (9.2). It is not difficult to justify (9.11) by elementary methods, but we shall see that it is really only a special case of a formula for the generators of subordinated semi-groups (see example XIII,9(*b*)).

---

[32] This was pointed out by K. L. Chung (see VII,5).

10. GENERATORS. THE BACKWARD EQUATION

Consider a pseudo-Poisson semi-group $\{\mathfrak{Q}(t)\}$ of contractions generated by the operator $\mathfrak{A} = \alpha(T-\mathbf{1})$. This operator being an endomorphism, $\mathfrak{A}u = v$ is defined for all $u \in \mathscr{L}$, and

$$\frac{\mathfrak{Q}(h) - \mathbf{1}}{h} u \to v, \qquad h \to 0+. \tag{10.1}$$

It would be pleasant if the same were true of all semi-groups, but this is too much to expect. For example, for the semi-group associated with Brownian motion the diffusion equation (4.1) implies that for twice continuously differentiable u the left side in (10.1) tends to $\frac{1}{2}u''$, but no limit exists when u is not differentiable. The diffusion equation nevertheless determines the process uniquely because a semi-group is determined by its action on twice differentiable functions. We must therefore not expect that (10.1) will hold for *all* functions u, but for all practical purposes it will suffice if it holds for sufficiently many functions. With this in mind we introduce the

Definition. *If for some elements u, v in $\mathscr{L}$ the relation* (10.1) *holds* (*in the sense of uniform convergence*) *we put $v = \mathfrak{A}u$. The operator so defined is called the generator*[33] *of the semi-group* $\{\mathfrak{Q}(t)\}$.

Premultiplying (10.1) by $\mathfrak{Q}(t)$ we see that it implies

$$\frac{\mathfrak{Q}(t+h) - \mathfrak{Q}(t)}{h} u \to \mathfrak{Q}(t)v. \tag{10.2}$$

Thus, if $\mathfrak{A}u$ exists then all functions $\mathfrak{Q}(t)u$ are in the domain of the $\mathfrak{A}$ and

$$\frac{\mathfrak{Q}(t+h) - \mathfrak{Q}(t)}{h} u \to \mathfrak{Q}(t)\, \mathfrak{U}u = \mathfrak{U}\mathfrak{Q}(t)u. \tag{10.3}$$

This relation is essentially the same as the backward equation for Markov processes. In fact, with the notations of section 4 we should put

$$u(t, x) = \mathfrak{Q}(t)\, u_0(x),$$

where u_0 is the initial function. Then (10.3) becomes

$$\frac{\partial u(t, x)}{\partial t} = \mathfrak{U}u(t, x). \tag{10.4}$$

[33] The treatment of convolution semi-groups in chapter IX restricts the consideration to infinitely differentiable functions with the result that all generators are defined on the same domain. No such convenient device is applicable for general semi-groups.

This is the familiar backward equation, but it must be interpreted properly. The transition probabilities of the diffusion processes in section 4 are so smooth that the backward equation is satisfied for *all* continuous initial functions u_0. This is not necessarily so in general.

Examples. (*a*) *Translations.* Let $\mathscr{L}$ consist of the continuous functions on the line vanishing at infinity and put $\mathfrak{Q}(t)\,u(x) = u(x+t)$. Obviously (10.1) holds iff u possesses a continuous derivative u' vanishing at infinity, and in this case $\mathfrak{A}u = u'$.

Formally the backward equation (10.4) reduces to

$$\frac{\partial u}{\partial t} = \frac{\partial u}{\partial x}. \tag{10.5}$$

The formal solution reducing for $t = 0$ to a given initial u_0 would be given by $u(t, x) = u_0(t+x)$. But this is a true solution only if u_0 is differentiable.

(*b*) As in section 2 consider a pseudo-Poisson process with variables $\mathbf{X}(t)$ and another process defined by $\mathbf{X}^{\#}(t) = \mathbf{X}(t) - ct$. The corresponding semi-groups are in the obvious relationship that the value of $\mathfrak{Q}^{\#}(t)u$ at x equals the value of $\mathfrak{Q}(t)u$ at $x + ct$. For the generators this implies

$$\mathfrak{A}^{\#} = \mathfrak{A} - c\frac{d}{dx} \tag{10.6}$$

and so the domain of $\mathfrak{A}^{\#}$ is restricted to differentiable functions. The backward equation is satisfied whenever the initial function u_0 has a continuous derivative but not for arbitrary functions. In particular, the transition probabilities themselves need not satisfy the backward equation. This explains the difficulties of the old-fashioned theories [discussed in connection with IX,(2.14)] and also why we had to introduce unnatural regularity assumptions to derive the forward equation (2.1). ▶

The usefulness of the notion of generator is due to the fact that *for each continuous semi-group of contractions the generator defines the semi-group uniquely*. A simple proof of this theorem will be given in XIII,9.

This theorem enables us to handle backward equations without unnecessary restrictions and greatly simplifies their derivation. Thus the most general form of diffusion operators alluded to in footnote 12 of section 4 could not have been derived without the a priori knowledge that a generator does in fact exist.

CHAPTER XI

Renewal Theory

Renewal processes were introduced in VI,6 and illustrated in VI,7. We now begin with the general theory of the so-called renewal equation, which occurs frequently in various connections. A striking example for the applicability of the general renewal theorem is supplied by the limit theorem of section 8. Sections 6 and 7 contain an improved and generalized version of some asymptotic estimates originally derived laboriously by deep analytic methods. This illustrates the economy of thought and tools to be achieved by a general theoretical approach to hard individual problems. For a treatment of renewal problems by Laplace transforms see XIV,1–3.

Many papers and much ingenuity have been spent on the elusive problem of freeing the renewal theorem of the condition that the variables be positive. In view of this impressive history a new and greatly simplified proof of the general theorem is incorporated in section 9.

1. THE RENEWAL THEOREM

Let F be a distribution concentrated[1] on $\overline{0, \infty}$, that is, we suppose $F(0) = 0$. We do not require the existence of an expectation, but because of the assumed positivity we can safely write

$$\mu = \int_0^\infty y\, F\{dy\} = \int_0^\infty [1-F(y)]\, dy \tag{1.1}$$

where $\mu \leq \infty$. When $\mu = \infty$ we agree to interpret the symbol μ^{-1} as 0.

In this section we investigate the asymptotic behavior as $x \to \infty$ of the function

$$U = \sum_{n=0}^{\infty} F^{n\star}. \tag{1.2}$$

[1] No essential changes occur if one permits an atom of weight $p < 1$ at the origin. (See problem 1.)

It will be seen presently that this problem is intimately connected with the asymptotic behavior of the solution Z of the *renewal equation*

$$Z(x) = z(x) + \int_0^x Z(x-y)\, F\{dy\}, \qquad x > 0. \tag{1.3}$$

For definiteness we take the interval of integration closed, but in the present context it will be understood that z and Z vanish on the negative half-axis; the limits of integration may then be replaced by $-\infty$ and ∞, and the renewal equation may be written in the form of the convolution equation

$$Z = z + F \bigstar Z. \tag{1.4}$$

(A similar remark applies to all convolutions in the sequel.)

The probabilistic meaning of U and probabilistic applications of the renewal equation were discussed at some length in VI,6–7. For the present we shall therefore proceed purely analytically. However, it should be borne in mind that in a renewal process $U(x)$ equals the expected number of renewal epochs in $\overline{0, x}$, the origin counting as a renewal epoch. Accordingly, U should be interpreted as a measure concentrated on $\overline{0, \infty}$, the interval $I = \overline{a, b}$ carrying the mass $U\{I\} = U(b) - U(a)$. The origin is an atom of unit weight contributed by the zeroth term in the series (1.2).

The following lemma merely restates theorem 1 of VI,6 but a new proof is given to render the present section self-contained.

Lemma. $U(x) < \infty$ *for all* x. *If* z *is bounded the function* Z *defined by*

$$Z(x) = \int_0^x z(x-y)\, U\{dy\}, \qquad x > 0 \tag{1.5}$$

is the unique solution of the renewal equation (1.3) *that is bounded on finite intervals.*

[With the convention that $z(x) = Z(x) = 0$ for $x < 0$ we may write (1.5) in the form $Z = U \bigstar z$.]

Proof. Put $U_n = F^{0\bigstar} + \cdots + F^{n\bigstar}$ and choose positive numbers τ and η such that $1 - F(\tau) > \eta$. Then

$$\int_0^x [1 - F(x-y)]\, U_n\{dy\} = 1 - F^{(n+1)\bigstar}(x), \qquad x > 0 \tag{1.6}$$

and hence $\eta[U_n(x) - U_n(x-\tau)] < 1$. Letting $n \to \infty$ we conclude that $U\{I\} \leq \eta^{-1}$ for every interval I of length $<\tau$. Since an arbitrary interval of length a is the union of at most $1 + a/\tau$ intervals of length τ it follows that

$$U(x) - U(x-a) \le C_a \tag{1.7}$$

where $C_a = (a+\tau)/(\tau\eta)$. Thus $U\{I\}$ is uniformly bounded for all intervals I of a given length.

Now $Z_n = U_n \bigstar z$ satisfies $Z_{n+1} = z + F \bigstar Z_n$. Letting $n \to \infty$ one sees that the integral in (1.5) makes sense and that Z is a solution of (1.3).

To prove its uniqueness note that the difference of two solutions would satisfy $V = F \bigstar V$, and therefore also

$$V(x) = \int_0^x V(x-y) F^{r\bigstar}\{dy\}, \qquad x > 0 \tag{1.8}$$

for $r = 1, 2, \ldots$. But $F^{r\bigstar}(x) \to 0$ as $r \to \infty$ and since V is supposed bounded in $\overline{0, x}$ this implies $V(x) = 0$ for all $x > 0$. ►

The formulation of the renewal theorem is encumbered by the special role played by distributions concentrated on the multiples of a number λ. According to definition 3 of V,2, such a distribution is called arithmetic, and the largest λ such that F is concentrated on $\lambda, 2\lambda, \ldots$ is called the span of F. In this case the measure U is purely atomic, and we denote by u_n the weight of $n\lambda$. The renewal theorem of **1**; XIII,11 states that $u_n \to \lambda/\mu$. The following theorem[2] generalizes this result to arbitrary distributions concentrated on $\overline{0, \infty}$. The case of arithmetic F is repeated for completeness. (We recall the convention that $\mu^{-1} = 0$ if $\mu = \infty$.)

Renewal theorem (*first form*). *If F is not arithmetic*

$$U(t) - U(t-h) \to h/\mu, \qquad t \to \infty \tag{1.9}$$

for every $h > 0$. If F is arithmetic the same is true when h is a multiple of the span λ.

Before proving the theorem we reformulate it in terms of the asymptotic behavior of the solutions (1.5) of the renewal equation. Since the given function z may be decomposed into its positive and negative parts we may suppose that $z \ge 0$. For definiteness we suppose at first that the distribution F is non-arithmetic and has an expectation $\mu < \infty$.

---

[2] The discrete case was proved in 1949 by P. Erdös, W. Feller and H. Pollard. Their proof was immediately generalized by D. Blackwell. The present proof is new. For a generalization to distributions not concentrated on $\overline{0, \infty}$ see section 9. A far reaching generalization in another direction is contained in Y. S. Chow and H. E. Robbins, *A renewal theorem for random variables which are dependent or non-identically distributed.* Ann. Math. Statist., vol. 34 (1963), pp. 390–401.

If $z(x) = 1$ for $0 \leq a \leq x < b < \infty$ and $z(x) = 0$ for all other x we get from (1.5)

$$Z(t) = U(t - a) - U(t - b) \to (b - a)/\mu, \qquad t \to \infty. \tag{1.10}$$

This result generalizes immediately to finite step functions: Let $I_1, \ldots, I_r$ be non-overlapping intervals on the positive half-axis of lengths $L_1, \ldots, L_r$. If z assumes the value a_k in I_k and vanishes outside the union of the I_k, then clearly

$$Z(t) \to \mu^{-1} \sum_{k=1}^{r} a_k L_k = \mu^{-1} \int_0^\infty z(x)\, dx. \tag{1.11}$$

Now the classical Riemann integral of a function z is defined in terms of approximating finite step functions, and it is therefore plausible that the limit relation (1.11) should hold whenever z is Riemann integrable. To make this point clear we recall the definition of the Riemann integral of z over a finite interval $0 \leq x \leq a$. It suffices to consider partitions into subintervals of equal length $h = a/n$. Let $\underline{m}_k$ be the largest, and $\bar{m}_k$ the smallest number such that

$$\underline{m}_k \leq z(x) \leq \bar{m}_k \quad \textit{for} \quad (k-1)h \leq x < kh. \tag{1.12}$$

The obvious dependence of $\underline{m}_k$ and $\bar{m}_k$ on h should be kept in mind. The *lower* and *upper Riemann sums* for the given span h are defined by

$$\underline{\sigma} = h \sum \underline{m}_k, \qquad \bar{\sigma} = h \sum \bar{m}_k. \tag{1.13}$$

As $h \to 0$ both $\underline{\sigma}$ and $\bar{\sigma}$ approach finite limits. If $\bar{\sigma} - \underline{\sigma} \to 0$ these limits are the same, and the Riemann integral of z is defined by this common limit. Every bounded function that is continuous except for jumps is integrable in this sense.

When it comes to integrals over $\overline{0, \infty}$ the classical definition introduces an avoidable complication. To make the class of integrable functions as extensive as possible the integral over $\overline{0, \infty}$ is conventionally defined as the limit of integrals over $\overline{0, a}$. A continuous non-negative function z is integrable in this sense iff the area between its graph and the x-axis is finite. Unfortunately this does not preclude the effective oscillation of $z(x)$ and ∞ as $x \to \infty$. (See example *a*.) It is obviously not reasonable to assume that the solution Z will tend to a finite limit if the given function z oscillates in a wild manner. In other words, the sophisticated standard definition makes too many functions integrable, and for our purposes it is preferable to proceed in the naive manner by extending the original definition also to infinite intervals. For want of an established term we speak of a *direct* integration in contrast to the indirect procedure involving a passage to the limit from finite intervals.

Definition. *A function* $z \geq 0$ *is called directly Riemann integrable if the upper and lower Riemann sums defined in* (1.12)–(1.13) *are finite and tend to the same limit as* $h \to 0$.

This definition makes no distinction between finite and infinite intervals. It is easily seen that z is directly integrable over $\overline{0, \infty}$ if it is integrable over every finite interval $\overline{0, a}$ and if $\bar{\sigma} < \infty$ for some h. (Then automatically $\bar{\sigma} < \infty$ for all h.) It is this last property that excludes wild oscillations.

We may restate the definition in terms of approximating step functions. For fixed $h > 0$ put $z_k(x) = 1$ when $(k-1)h \leq x < kh$ and $z_k(x) = 0$ elsewhere. Then

$$\underline{z} = \sum \underline{m}_k z_k \quad \text{and} \quad \bar{z} = \sum \bar{m}_k z_k \tag{1.14}$$

are two finite step functions and $\underline{z} \leq z \leq \bar{z}$. The integral of z is the common limit as $h \to 0$ of the integrals of these step functions. Denote by Z_k the solution of the renewal equation corresponding to z_k. The solutions corresponding to $\underline{z}$ and $\bar{z}$ are then given by

$$\underline{Z} = \sum Z_k \underline{m}_k \quad \text{and} \quad \bar{Z} = \sum \bar{m}_k Z_k. \tag{1.15}$$

By the renewal theorem $Z_k(x) \to h/\mu$ for each fixed k. Furthermore, (1.7) assures us that $Z_k(x) \leq C_h$ for all k and x. The remainders of the series in (1.15) therefore tend uniformly to 0 and we conclude that

$$\underline{Z}(x) \to \underline{\sigma}/\mu, \qquad \bar{Z}(x) \to \bar{\sigma}/\mu \qquad (x \to \infty). \tag{1.16}$$

But $\underline{Z} \leq Z \leq \bar{Z}$ and hence all limit values of $Z(x)$ lie between $\underline{\sigma}/\mu$ and $\bar{\sigma}/\mu$. If z is directly Riemann integrable it follows that

$$Z(x) \to \mu^{-1} \int_0^\infty z(y)\, dy, \qquad x \to \infty. \tag{1.17}$$

So far we have assumed that F is non-arithmetic and $\mu < \infty$. The argument applies without change when $\mu = \infty$ if μ^{-1} is interpreted as 0.

If F is arithmetic with span λ the solution Z of (1.5) is of the form

$$Z(x) = \sum z(x - k\lambda) u_k \tag{1.18}$$

where $u_k \to \lambda/\mu$. One concludes easily that for fixed x

$$Z(x + n\lambda) \to \lambda\mu^{-1} \sum_{j=1}^{\infty} z(x + j\lambda), \qquad n \to \infty. \tag{1.19}$$

provided the series converges, which is certainly the case if z is directly integrable.

We have derived (1.17) and (1.19) from the renewal theorem, but these relations contain the renewal theorem as a special case when z reduces to the indicator of an interval $\overline{0, h}$. We have thus proved the following

Renewal theorem. (*Alternative form*).[3] *If* z *is directly Riemann integrable the solution* Z *of the renewal equation satisfies* (1.17) *if* F *is non-arithmetic, and* (1.19) *if* F *is arithmetic with span* λ.

One may ask whether the condition of direct integrability may be dropped at least for continuous functions z tending to 0 at infinity. The following examples show that this is not so. Example 3(*b*) will show similarly that the renewal theorem may fail for an unbounded function z even when it vanishes outside a finite interval. Improper Riemann integrals are therefore not usable in renewal theory and direct integrability appears as the natural basis.

Examples. (*a*) *A continuous function* z *may be unbounded and yet Riemann integrable over* $\overline{0, \infty}$. To see this let $z(n) = a_n$ for $n = 1, 2, \ldots$ and let z vanish identically outside the union of the intervals $|x - n| < h_n < \frac{1}{2}$; between n and $n \pm h_n$ let z vary linearly with x. The graph of z then consists of a sequence of triangles of areas $a_n h_n$, and hence z is Riemann integrable iff $\sum a_n h_n < \infty$. This does not preclude that $a_n \to \infty$.

(*b*) To explore the role of direct integrability in the renewal theorem it suffices to consider arithmetic distributions F. Thus we may suppose that the measure U is concentrated on the integers and that the weight u_n carried by the point n tends to the limit $\mu^{-1} > 0$. For any positive integer n we have then

$$Z(n) = u_n\, z(0) + u_{n-1}\, z(1) + \cdots + u_0 z(n).$$

Now choose for z the function of the preceding example with $a_n = 1$; then $Z(n) \sim n\mu^{-1}$, and so Z is not even bounded. The same is obviously true if a_n tends to 0 sufficiently slowly, and thus we get an example of *a continuous integrable function* z *such that* $z(x) \to 0$, *but* $Z(x)$ *does not remain bounded.* ▶

[3] It is hoped that this form and the preceding discussion will end the sorry confusion now prevailing in the literature. The most widely used reference is the report by W. L. Smith, *Renewal theory and its ramifications*, in the J. Roy. Stat. Soc. (Series B), vol. 20 (1958), pp. 243–302. Since its appearance Smith's "key renewal theorem" has in practice replaced all previously used versions (which were not always correct). The key theorem proves (1.17) under the superfluous assumption that z be monotone. Smith's proof (of 1954) is based on Wiener's deep Tauberian theorems, and the 1958 report gives the impression that this tortuous procedure is simpler than a direct reduction to the first form of the renewal theorem. Also, the condition that z be bounded was inadvertently omitted in the report. [Concerning its necessity see example 3(*b*).]

2. PROOF OF THE RENEWAL THEOREM

For arithmetic distributions F the renewal theorem was proved in **1**; XIII,11 and we suppose therefore F non-arithmetic. For the proof we require two lemmas. (The first reappears in a stronger form in the corollary in section 9.)

Lemma 1. *Let ζ be a bounded uniformly continuous function such that $\zeta(x) \leq \zeta(0)$ for $-\infty < x < \infty$. If*

$$\zeta(x) = \int_0^\infty \zeta(x-y)\, F\{dy\} \tag{2.1}$$

then $\zeta(x) = \zeta(0)$ identically.

Proof. Taking convolutions with F we conclude from (2.1) by induction that

$$\zeta(x) = \int_0^\infty \zeta(x-y)\, F^{r\star}\{dy\}, \qquad r = 1, 2, \ldots. \tag{2.2}$$

The integrand is $\leq \zeta(0)$, and for $x = 0$ the equality is therefore possible only if $\zeta(-y) = \zeta(0)$ for every y that is a point of increase of $F^{r\star}$. By lemma 2 of V,4*a* the set Σ formed by such points is asymptotically dense at infinity, and in view of the uniform continuity of ζ this implies that $\zeta(-y) \to \zeta(0)$ as $y \to \infty$. Now as $r \to \infty$ the mass of $F^{r\star}$ tends to be concentrated at ∞. For large r the integral in (2.2) therefore depends essentially only on large values of y, and for such values $\zeta(x-y)$ is close to $\zeta(0)$. Letting $r \to \infty$ we conclude therefore from (2.2) that $\zeta(x) = \zeta(0)$ as asserted. ▶

Lemma 2. *Let z be a continuous function vanishing outside $\overline{0, h}$. The corresponding solution Z of the renewal equation is uniformly continuous and for every a*

$$Z(x+a) - Z(x) \to 0, \qquad x \to \infty. \tag{2.3}$$

Proof. The differences $z(x+\delta) - z(x)$ vanish outside an interval of length $h + 2\delta$ and therefore by (1.5) and (1.7)

$$|Z(x+\delta) - Z(x)| \leq C_{h+2\delta} \max |z(x+\delta) - z(x)|. \tag{2.4}$$

This shows that if z is uniformly continuous the same is true of Z.

Suppose now that z has a continuous derivative z'. Then z' exists and satisfies the renewal equation

$$Z'(x) = z'(x) + \int_0^x Z'(x-y)\, F\{dy\}. \tag{2.5}$$

Thus Z' is bounded and uniformly continuous. Let

$$\limsup Z'(x) = \eta, \tag{2.6}$$

and choose a sequence such that $Z'(t_n) \to \eta$. The family of functions ζ_n defined by

$$\zeta_n(x) = Z'(t_n+x) \tag{2.7}$$

is equicontinuous and

$$\zeta_n(x) = z'(t_n+x) + \int_0^{x+t} \zeta_n(x-y)\, F\{dy\}. \tag{2.8}$$

Hence there exists a subsequence such that ζ_{n_r} converges to a limit ζ. It follows from (2.8) that this limit satisfies the conditions of lemma 1 and therefore $\zeta'(x) = \zeta'(0) = \eta$ for all x.

Thus $Z'(t_{n_r} + x) \to \eta$ or

$$Z(t_{n_r} + a) - Z(t_{n_r}) \to \eta a. \tag{2.9}$$

This being true for every a and Z being bounded it follows that $\eta = 0$. The same argument applies to the lower limit and proves that $Z'(x) \to 0$.

We have thus proved the lemma for continuously differentiable z. But an arbitrary continuous z can be approximated by a continuously differentiable function z_1 vanishing outside $\overline{0, h}$. Let Z_1 be the corresponding solution of the renewal equation. Then

$$|z - z_1| < \epsilon \quad \text{implies} \quad |Z - Z_1| < C_h\epsilon,$$

and thus $|Z(x+a) - Z(x)| < (2C_h+1)\epsilon$ for all x sufficiently large. Thus (2.3) holds for arbitrary continuous z. ►

The conclusion of the proof is now easy. If I is the interval $\alpha \leq x \leq \beta$ we denote by $I + t$ the interval $\alpha + t \leq x \leq \beta + t$. We know from (1.9) that $U\{I + t\}$ remains bounded for every finite interval I. By the selection theorem 2 of VIII,6 there exists therefore a sequence $t_k \to \infty$ and a measure V such that

$$U\{t_k + dy\} \to V\{dy\}. \tag{2.10}$$

The measure V is finite on finite intervals, but is *not* concentrated on $\overline{0, \infty}$.

Now let z be a continuous function vanishing outside the finite interval $\overline{0, a}$. For the corresponding solution Z of the renewal equation we have then

$$Z(t_k + x) = \int_0^a z(-s)\, U\{t_k + x + ds\} \to \int_0^a z(-s)\, V\{x + ds\}. \tag{2.11}$$

From the preceding lemma it follows that the family of measures $V\{x + ds\}$

is independent of x, and hence $V\{I\}$ must be proportional to the length of I. Thus (2.10) may be put in the form

$$U(t_k) - U(t_k - h) \to \gamma h. \tag{2.12}$$

This is the same as the assertion (1.9) of the renewal theorem except that the factor η^{-1} is replaced by the unknown γ and that t is restricted to the sequence $\{t_k\}$. However, our derivation of the alternative form of the renewal theorem remains valid and thus

$$Z(t_k) \to \gamma \int_0^\infty z(y)\, dy \tag{2.13}$$

whenever z is directly integrable.

The function $z = 1 - F$ is monotone, and its integral equals μ. The corresponding solution Z reduces to the constant 1. If $\mu < \infty$ the function z is directly integrable, and (2.13) states that $\gamma\mu = 1$. When $\mu = \infty$ we truncate z and conclude from (2.13) that γ^{-1} exceeds the integral of z over an arbitrary interval $\overline{0, a}$. Thus $\mu = \infty$ implies $\gamma = 0$. Hence the limit in (2.12) is independent of the sequence $\{t_k\}$, and (2.12) reduces to the assertion (1.9) of the renewal theorem. ▶

*3. REFINEMENTS

In this section we show how regularity properties of the distribution F may lead to sharper forms of the renewal theorem. The results are not exciting in themselves, but they are useful in many applications.

Theorem 1. *If F is non-arithmetic with expectation μ and variance σ^2, then*

$$0 \le U(t) - \frac{t}{\mu} \to \frac{\sigma^2 + \mu^2}{2\mu^2}. \tag{3.1}$$

The renewal theorem itself states only that $U(t) \sim t/\mu$, and the estimate (3.1) is much sharper. It is applicable even when no variance exists with the right side replaced by ∞. [The analogue for arithmetic distributions is given by **1**; XIII,(12.2).]

Proof. Put

$$Z(t) = U(t) - t/\mu. \tag{3.2}$$

It is easily verified that this is the solution of the renewal equation corresponding to

$$z(t) = \frac{1}{\mu} \int_t^\infty [1 - F(y)]\, dy. \tag{3.3}$$

* This section should be omitted at the first reading.

Integrating by parts we get

$$\int_0^\infty z(t)\,dt = \frac{1}{2\mu}\int_0^\infty y^2\,F\{dy\} = \frac{\sigma^2+\mu^2}{2\mu}. \tag{3.4}$$

Being monotone z is directly integrable, and the alternative form of the renewal theorem asserts that (3.1) is true. ▶

Next we turn to smoothness properties of the renewal function U. If F has a density f the renewal equation for U takes on the form

$$U(x) = 1 + \int_0^x U(x-y)\,f(y)\,dy. \tag{3.5}$$

If f is continuous a formal differentiation would indicate that U should have a derivative u satisfying the equation

$$u(x) = f(x) + \int_0^x u(x-y)\,f(y)\,dy. \tag{3.6}$$

This is a renewal equation of the standard type, and we know that it has a unique solution whenever f is bounded (not necessarily continuous). It is easily verified that the function U defined by

$$U(t) = 1 + \int_0^t u(y)\,dy, \qquad t > 0. \tag{3.7}$$

satisfies (3.5) and hence the solution u of (3.6) is indeed a density for U. As a corollary to the alternative form of the renewal theorem we get thus

Theorem 2. *If F has a directly integrable density f, then U has a density u such that $u(t) \to \mu^{-1}$.*

Densities that are not directly integrable will hardly occur in practice but certain conclusions are possible even for them. In fact, consider the density

$$f_2(t) = \int_0^t f(t-y)\,f(y)\,dy \tag{3.8}$$

of $F \star F$. In general f_2 will behave much better than f. For example, if $f < M$ we get for reasons of symmetry

$$f_2(t) < 2M[1 - F(\tfrac{1}{2}t)]. \tag{3.9}$$

If $\mu < \infty$ the right side is a monotone integrable function and this implies that f_2 is directly integrable. Now $u - f$ is the solution of the renewal equation with $z = f_2$, and we have thus

Theorem 2a. *If F has a bounded density f and a finite expectation μ, then*

$$u(t) - f(t) \to \mu^{-1}. \tag{3.10}$$

This result is curious because it shows that if the oscillations of f are wild, u will oscillate in a manner to compensate them. (For related results see problems 7–8.)

The condition that f be bounded is essential. We illustrate this by an example which also throws new light on the condition of direct integrability in the renewal theorem.

Examples. (*a*) Let G be the probability distribution concentrated on $\overline{0, 1}$ and defined by

(3.11) $$G(x) = \frac{1}{\log (e/x)}, \qquad 0 < x \leq 1.$$

It has a density that is continuous in the open interval, but since $x^{-1}G(x) \to \infty$ as $x \to 0$ the density is unbounded near the origin. The sum of n independent random variables with the distribution G is certainly $<x$ if each component is $<x/n$, and hence

(3.12) $$G^{n\star}(x) \geq (G(x/n))^n.$$

It follows that for each n the density of $G^{n\star}$ is unbounded near the origin.

Now put $F(x) = G(x - 1)$. Then $F^{n\star}(x) = G^{n\star}(x - n)$, and hence $F^{n\star}$ has a density which vanishes for $x < n$, is continuous for $x > n$, but unbounded near n. The density u of the renewal function $U = \sum F^{n\star}$ is therefore unbounded in the neighborhood of every integer $n > 0$.

(*b*) The density u of the preceding example satisfies (3.6) which agrees with the standard renewal equation (1.3) with $z = f$. This is an integrable function vanishing for $x > 2$ and continuous except at the point 1. The fact that the solution $Z = u$ is unbounded near every integer shows that *the renewal theorem breaks down if z is not properly Riemann integrable* (bounded), even when z is concentrated on a finite interval. ▶

4. PERSISTENT RENEWAL PROCESSES

The renewal theorem will now be used to derive various limit theorems for the renewal processes introduced in VI,6. We are concerned with a sequence of mutually independent random variables $\mathbf{T}_1, \mathbf{T}_2, \ldots,$ the *interarrival times*, with a common distribution F. In this section we assume that F is a proper distribution and $F(0) = 0$. In addition to the $\mathbf{T}_k$ there may be defined a non-negative variable $\mathbf{S}_0$ with a proper distribution F_0. We put

(4.1) $$\mathbf{S}_n = \mathbf{S}_0 + \mathbf{T}_1 + \cdots + \mathbf{T}_n.$$

The variables $\mathbf{S}_n$ are called *renewal epochs*. The renewal process $\{\mathbf{S}_n\}$ is called *pure* if $\mathbf{S}_0 = 0$ and *delayed* otherwise.

We adhere to the notation $U = \sum F^{n\star}$ introduced in (1.2). The *expected number of renewal epochs in* $\overline{0, t}$ equals

(4.2) $$V(t) = \sum_{n=0}^{\infty} \mathbf{P}\{\mathbf{S}_n \leq t\} = F_0 \star U$$

For $h > 0$ we have therefore[4]

(4.3) $$V(t+h) - V(t) = \int_0^{t+h} [U(t+h-y) - U(t-y)]\, F_0\{dy\}.$$

[4] We recall from section 1 that the intervals of integration are taken closed; the limits of integration may therefore be replaced by $-\infty$ and ∞.

If F is not arithmetic the integrand tends to $\mu^{-1}h$ as $t \to \infty$, and thus the basic theorem extends also to delayed processes: if F is non-arithmetic *the expected number of renewal epochs within* $\overline{t, t+h}$ *tends to* $\mu^{-1}h$. This statement contains two equidistribution theorems; first, the renewal rate tends to a constant, and second, this constant rate is independent of the initial distribution. In this sense we have an analogue to the ergodic theorems for Markov chains in **1**; XV.

If $\mu < \infty$ it follows that $V(t) \sim \mu^{-1}t$ as $t \to \infty$. It is natural to ask whether F_0 can be chosen as to get the identity $V(t) = \mu^{-1}t$, meaning a constant renewal rate. Now V satisfies the renewal equation

$$V = F_0 + F \star V \tag{4.4}$$

and thus $V(t) = \mu^{-1}t$ iff

$$F_0(t) = \frac{t}{\mu} - \frac{1}{\mu}\int_0^t (t-y)\, F\{dy\}. \tag{4.5}$$

Integration by parts shows this to be the same as

$$F_0(t) = \frac{1}{\mu}\int_0^t [1 - F(y)]\, dy. \tag{4.6}$$

This F_0 is a probability distribution and so the answer is affirmative: *with the initial distribution* (4.6) *the renewal rate is constant,* $V(t) = \mu^{-1}t$.

The distribution (4.6) appears also as the limit distribution of the residual waiting times, or hitting probabilities. To given $t > 0$ there corresponds a chance-dependent subscript $\mathbf{N}_t$ such that

$$\mathbf{S}_{\mathbf{N}_t} \leq t < \mathbf{S}_{\mathbf{N}_t+1}. \tag{4.7}$$

In the terminology introduced in VI,7 the variable $\mathbf{S}_{\mathbf{N}_t+1} - t$ is called *residual waiting time* at epoch t. We denote by $H(t, \xi)$ the probability that it is $\leq \xi$. In other words, $H(t, \xi)$ is the probability that the *first* renewal epoch following epoch t lies within $\overline{t, t+\xi}$, or that the level t be overshot by an amount $\leq \xi$. This event occurs if some renewal epoch $\mathbf{S}_n$ equals $x \leq t$ and the following interarrival time lies between $t - x$ and $t - x + \xi$. In the case of a pure renewal process we get, summing over x and n,

$$H(t, \xi) = \int_0^t U\{dx\}[F(t-x+\xi) - F(t-x)]. \tag{4.8}$$

This integral contains ξ as a free parameter but is of the standard form $U \star z$ with $z(t) = F(t+\xi) - F(t)$, which function is directly integrable.[5]

[5] For $n\xi < x < (n+1)\xi$ we have $z(x) \leq F((n+2)\xi) - F(n\xi)$, and the series with these terms is obviously convergent.

Assume F non-arithmetic. Since

$$(4.9)\qquad \int_0^\infty z(t)\,dt = \int_0^\infty([1-F(t)]-[1-F(t+\xi)])\,dt = \int_0^\xi (1-F(s))\,ds$$

we have the *limit theorem*

$$(4.10)\qquad \lim_{t\to\infty} H(t,\xi) = \mu^{-1}\int_0^\xi [1-F(s)]\,ds.$$

(It is easily verified that this is true also for the delayed process regardless of the initial distribution F_0.) This limit theorem is remarkable in several respects. As the following discussion shows, it is closely connected with the inspection paradox of VI,7 and the waiting time paradox in I,4.

When $\mu < \infty$ the limit distribution (4.10) agrees with (4.6) and thus if $\mu < \infty$ *the residual waiting time has a proper limit distribution which coincides with the distribution attaining a uniform renewal rate.* In this pattern we recognize one more the tendency towards a "*steady state.*"

The limit distribution of (4.10) has a finite expectation only if F has a variance. This indicates that, roughly speaking, *the entrance probabilities behave worse than* F. Indeed, when $\mu = \infty$ we have

$$(4.11)\qquad H(t,\xi) \to 0$$

for all ξ: the probability tends to 1 that *the level* t *will be overshot by an arbitrarily large amount* ξ. (For the case of regularly varying tails more precise information is derived in XIV,3.)

Examples. (*a*) *Superposition of renewal processes.* Given n renewal processes, a new process can be formed by combining all their renewal epochs into one sequence. In general the new process is *not* a renewal process, but it is easy to calculate the waiting time W for the first renewal following epoch 0. We shall show that under fairly general conditions *the distribution of* W *is approximately exponential* and so the combined process is close to a Poisson process. This result explains why many processes (such as the incoming traffic at a telephone exchange) are of the Poisson type.

Consider n mutually independent renewal processes induced by the distributions of their interarrival times by $F_1, \ldots, F_n$ with expectations $\mu_1, \ldots, \mu_n$. Put

$$(4.12)\qquad \frac{1}{\mu_1} + \cdots + \frac{1}{\mu_n} = \frac{1}{\alpha}.$$

We require, roughly speaking, that the renewal epochs of each individual renewal process are extremely rare so that the cumulative effect is due to many small causes. To express this we assume that for fixed k and y the probabilities $F_k(y)$ are small and μ_k large—an assumption that becomes meaningful in the form of a limit theorem.

Consider the "steady state" situation where the processes have been going on for a long time. For the waiting time $\mathbf{W}_k$ to the nearest renewal epoch in the kth process we have then approximately

$$\text{(4.13)} \qquad \mathbf{P}\{\mathbf{W}_k \leq t\} \approx \frac{1}{\mu_k} \int_0^t (1 - F_k(y))\, dy \approx \frac{t}{\mu_k}\,.$$

[The last approximation is justified by the smallness of $F_k(y)$.] The waiting time $\mathbf{W}$ in the cumulative process is the smallest among the waiting times $\mathbf{W}_k$ and hence

$$\text{(4.14)} \qquad \mathbf{P}\{\mathbf{W} > t\} \approx \left(1 - \frac{t}{\mu_1}\right) \cdots \left(1 - \frac{t}{\mu_n}\right) \approx e^{-t\alpha}.$$

This estimate is easily made precise, and under the indicated conditions the exponential distribution emerges as the limit distribution as $n \to \infty$.

(*b*) *Hitting probabilities in random walks.* For a sequence of independent random variables $\mathbf{X}_1, \mathbf{X}_2, \ldots$ let

$$\mathbf{Y}_n = \mathbf{X}_1 + \cdots + \mathbf{X}_n.$$

For positive $\mathbf{X}_k$ the random walk $\{\mathbf{Y}_n\}$ reduces to a renewal process, but we consider arbitrary $\mathbf{X}_k$. Assume the random walk to be persistent so that for each $t > 0$ with certainty $\mathbf{Y}_n > t$ for some n. If $\mathbf{N}$ is the smallest index for which this is true $\mathbf{Y}_\mathbf{N}$ is called the point of *first entry* into $\overline{t, \infty}$. The variable $\mathbf{Y}_\mathbf{N} - t$ is the amount by which the t level is overshot at the first entry and corresponds to the residual waiting time in renewal processes. We put again $\mathbf{P}\{\mathbf{Y}_\mathbf{N} \leq t + \xi\} = H(t, \xi)$, and show how the limit theorem for residual waiting times applies to this distribution.

Define $\mathbf{S}_1$ as the point of first entry into $\overline{0, \infty}$ and, by induction, $\mathbf{S}_{n+1}$ as the point of first entry into $\overline{\mathbf{S}_n, \infty}$. The sequence $\mathbf{S}_1, \mathbf{S}_2, \ldots$ coincides with the *ladder heights* introduced in VI,8 and forms a renewal process: the differences $\mathbf{S}_{n+1} - \mathbf{S}_n$ are evidently mutually independent and have the same distribution as $\mathbf{S}_0$. Thus $\mathbf{Y}_\mathbf{N} - t$ is actually the residual waiting time in the renewal process $\{\mathbf{S}_n\}$, and so (4.10) applies. ►

By the method used to derive (4.10) it can be shown that *the spent waiting time* $t - \mathbf{S}_{\mathbf{N}_t}$ *has the same limit distribution. For the length* $\mathbf{L}_t = \mathbf{S}_{\mathbf{N}_t+1} - \mathbf{S}_{\mathbf{N}_t}$ *of the interarrival time containing the epoch* t *we get*

$$\text{(4.15)} \qquad \mathbf{P}\{\mathbf{L}_t < \xi\} = \int_{t-\xi}^{t} U\{dx\}[F(\xi) - F(t-x)]$$

and hence

$$\text{(4.16)} \qquad \lim_{t\to\infty} \mathbf{P}\{\mathbf{L}_t \leq \xi\} = \mu^{-1} \int_0^\xi [F(\xi) - F(y)]\, dy = \mu^{-1} \int_0^\xi x\, F(dx).$$

The curious implications of this formula were discussed in connection with the inspection paradox in VI,7 and the waiting time paradox in I,4.

It is easily seen that the three families of random variables $t - \mathbf{S}_{\mathbf{N}_t}$, $\mathbf{S}_{\mathbf{N}_{t+1}} - t$, and $\mathbf{L}$ form Markov processes with stationary transition probabilities. Our three limit theorems therefore represent examples for ergodic theorems for Markov processes. (See also XIV,3.)

5. THE NUMBER $\mathbf{N}_t$ OF RENEWAL EPOCHS

For simplicity we consider a pure renewal process so that the rth renewal epoch is the sum of r independent variables with f as distribution. The origin counts as renewal epoch. We denote by $\mathbf{N}_t$ the number of renewal epochs within $\overline{0, t}$. The event $\{\mathbf{N}_t > r\}$ occurs iff the rth renewal epoch falls within $\overline{0, t}$, and hence

$$\mathbf{P}\{\mathbf{N}_t > r\} = F^{r\star}(t). \tag{5.1}$$

Obviously $\mathbf{N}_t \geq 1$. It follows that

$$\mathbf{E}(\mathbf{N}_t) = \sum_{r=0}^{\infty} \mathbf{P}\{\mathbf{N}_t > r\} = U(t). \tag{5.2}$$

(For higher moments see problem 13.)

The variable $\mathbf{N}_t$ occurs also in sequential sampling. Suppose that a sampling $\{\mathbf{T}_n\}$ is to continue until the sum of the observations for the first time exceeds t. Then $\mathbf{N}_t$ represents the total number of trials. Many tedious calculations might have been saved by the use of the estimate (3.1) provided by the refined renewal theorem.

If F has expectation μ and variance σ^2 the asymptotic behavior of the distribution of $\mathbf{N}_t$ is determined by the fact that $F^{r\star}$ is asymptotically normally distributed. The necessary calculations can be found in **1**; XIII,6 and do not depend on the arithmetic character of F. We have therefore the general

Central limit theorem for $\mathbf{N}_t$. *If F has expectation μ and variance σ^2 then for large t the number $\mathbf{N}_t$ of renewal epochs is approximately normally distributed with expectation $t\mu^{-1}$ and variance $t\sigma^2\mu^{-3}$.*

Example. (*a*) *Type* I *counters.* The incoming particles constitute a Poisson process. A particle reaching the counter when it is free is registered but locks the counter for a *fixed* duration ξ. Particles reaching the counter during a locked period have no effect whatever. For simplicity we start the process at an epoch when a new particle reaches a free counter. We have then *two* renewal processes. The primary process—the incoming traffic—is a

Poisson process, that is, its interarrival times have an exponential distribution $1 - e^{-ct}$ with expectation c^{-1} and variance c^{-2}. The successive registrations form a secondary renewal process in which the interarrival times represent the sum of ξ plus an exponential random variable. The waiting time between registrations has therefore expectation $\xi + c^{-1}$ and variance c^{-2}. Thus *the number of registrations within the time interval* $\overline{0, t}$ *is approximately normally distributed with expectation* $tc(1+c\xi)^{-1}$ *and variance* $tc(1+c\xi)^{-3}$.

The discrepancy between these quantities shows that the registrations are not Poisson distributed. In the early days it was not understood that the registration process differs essentially from the primary process, and the observations led some physicists to the erroneous conclusion that cosmic ray showers do not conform to the Poisson pattern of "perfect randomness." ▶

A limit distribution for $\mathbf{N}_t$ exists iff F belongs to some domain of attraction. These domains of attraction are characterized in IX,8 and XVII,5 and it follows that $\mathbf{N}_t$ *has a proper limit distribution iff*

$$1 - F(x) \sim x^{-\alpha} L(x), \qquad x \to \infty \tag{5.3}$$

where L is slowly varying and $0 < \alpha < 2$. The limit distribution for $\mathbf{N}_t$ is easily obtained and reveals the paradoxical properties of fluctuations. The behavior is radically different for $\alpha < 1$ and $\alpha > 1$.

Consider the case $0 < \alpha < 1$. If a_r is chosen so that

$$r[1 - F(a_r)] \to \frac{2 - \alpha}{\alpha} \tag{5.4}$$

then $F^{r\star}(a_r x) \to G_\alpha(x)$ where G_α is the one-sided stable distribution satisfying the condition $x^\alpha[1 - G_\alpha(x)] \to (2 - \alpha)/\alpha$ as $x \to \infty$. (Cf. IX,6 and XVII,5 as well as XIII,6.) Let r and t increase in such a manner that $t \sim a_r x$. On account of the slow variation of L we get then from (5.3) and (5.4)

$$r \sim \frac{2 - \alpha}{\alpha} \frac{x^{-\alpha}}{1 - F(t)} \tag{5.5}$$

whence from (5.1)

$$\mathbf{P}\left\{[1 - F(t)]\mathbf{N}_t \geq \frac{2 - \alpha}{\alpha} x^{-\alpha}\right\} \to G_\alpha(x). \tag{5.6}$$

This is an analogue to the central limit theorem. The special case $\alpha = \frac{1}{2}$ is covered in **1**; XIII,6. The surprising feature is conveyed by the norming factor $1 - F(t)$ in (5.6). Very roughly $1 - F(t)$ is of the order of magnitude $t^{-\alpha}$ and so the probable order of magnitude of $\mathbf{N}_t$ is of the order t^α; the density of the renewal epochs must decrease radically (which agrees with the asymptotic behavior of the hitting probabilities).

When $1 < \alpha < 2$ the distribution F has an expectation $\mu < \infty$ and the same type of calculation shows that

$$\mathbf{P}\left\{\mathbf{N}_t \geq \frac{t - \lambda(t)x}{\mu}\right\} \to G_\alpha(x) \tag{5.7}$$

where $\lambda(t)$ satisfies

$$t[1 - F(\lambda(t))] \to \frac{2 - \alpha}{\alpha}\mu. \tag{5.8}$$

In this case the expected number of renewal epochs increases linearly, but the norming $\lambda(t)$ indicates that the fluctuations about the expectation are extremely violent.

6. TERMINATING (TRANSIENT) PROCESSES

The general theory of renewal processes with a defective distribution F reduces almost to a triviality. The corresponding renewal equation, however, frequently appears under diverse disguises with accidental features obscuring the general background. A clear understanding of the basic facts will avoid cumbersome argument in individual applications. In particular, the asymptotic estimate of theorem 2 will yield results previously derived by special adaptations of the famous Wiener-Hopf techniques.

To avoid notational confusion we replace the underlying distribution F by L. Accordingly, in this section L *stands for a defective distribution with* $L(0) = 0$ *and* $L(\infty) = L_\infty < 1$. It serves as distribution of the (defective) interarrival times $\mathbf{T}_k$, the defect $1 - L_\infty$ representing the probability of a termination. The origin of the time axis counts as renewal epoch number zero, and $\mathbf{S}_n = \mathbf{T}_1 + \cdots + \mathbf{T}_n$ is the nth renewal epoch; it is a defective variable with distribution $L^{n\star}$ whose total mass equals $L^{n\star}(\infty) = L_\infty^n$. The defect $1 - L_\infty^n$ is the probability of extinction *before the* nth renewal epoch. We put again

$$U = \sum_{n=0}^{\infty} L^{n\star}. \tag{6.1}$$

As in the persistent process, $U(t)$ equals the *expected number of* renewal epochs within $\overline{0, t}$; this time, however, the expected number of renewal epochs ever occurring is finite, namely

$$U(\infty) = \frac{1}{1 - L_\infty}. \tag{6.2}$$

The probability that the nth renewal epoch $\mathbf{S}_n$ is the last and $\leq x$ equals $(1-L_\infty)\, L^{n\star}(x)$. We have thus

Theorem 1. *A transient renewal process commencing at the origin terminates with probability one. The epoch of termination* $\mathbf{M}$ (*that is, the maximum attained by the sequence* $0, \mathbf{S}_1, \mathbf{S}_2, \ldots$) *has the proper distribution*

$$\mathbf{P}\{\mathbf{M} \leq x\} = (1-L_\infty)\, U(x). \tag{6.3}$$

The probability that the nth renewal epoch is the last equals $(1-L_\infty)L_\infty^n$, and so the number of renewal epochs has a geometric distribution.

It is possible to couch these results in terms of the (defective) *renewal equation*

$$(6.4) \qquad Z(t) = z(t) + \int_0^t Z(t-y)\, L\{dy\},$$

but with a defective L the theory is trite. Assuming again that $z(x) = 0$ for $x \le 0$ the unique solution is given by

$$(6.5) \qquad Z(t) = \int_0^t z(t-y)\, U\{dy\}$$

and evidently

$$(6.6) \qquad Z(t) \to \frac{z(\infty)}{1 - L_\infty}$$

whenever $z(t) \to z(\infty)$ as $t \to \infty$.

Examples. (*a*) The event $\{\mathbf{M} \le t\}$ occurs if the process terminates with $\mathbf{S}_0$, or else if $\mathbf{T}_1$ assumes some positive value $y \le t$ and the residual process attains an age $\le t - y$. Thus $Z(t) = \mathbf{P}\{\mathbf{M} \le t\}$ *satisfies the renewal equation*

$$(6.7) \qquad Z(t) = 1 - L_\infty + \int_0^t Z(t-y)\, L\{dy\},$$

This is equivalent to (6.3).

(*b*) *Calculation of moments.* The last equation represents the proper distribution Z as the sum of two defective distributions, namely a convolution and the distribution with a single atom at the origin. To calculate the expectation of Z put

$$(6.8) \qquad E_L = \int_0^\infty x\, L\{dx\}$$

and similarly for other distributions, whether defective or not. Since L is defective the convolution in (6.7) has expectation $L_\infty \cdot E_Z + E_L$. Thus $E_Z = E_L/(1 - L_\infty)$. For the more general equation (6.4) we get in like manner.

$$(6.9) \qquad E_Z = \frac{E_z + E_L}{1 - L_\infty}.$$

Higher moments can be calculated by the same method. ▶

Asymptotic estimates

In applications $z(t)$ usually tends to a limit $z(\infty)$, and in this case $Z(t)$ tends to the limit $Z(\infty)$ given by (6.6). It is frequently important to obtain asymptotic estimates for the difference $Z(\infty) - Z(t)$. This can be achieved by a method of wide applicability in the theory of random walks [see the associated random walks in example XII,4(b)]. It depends on the (usually harmless) assumption that there exists a number κ such that

$$\int_0^\infty e^{\kappa y}\, L\{dy\} = 1. \tag{6.10}$$

This root κ is obviously unique and, the distribution L being defective, $\kappa > 0$. We now define a *proper* probability distribution $L^\#$ by

$$L^\#\{dy\} = e^{\kappa y}\, L\{dy\} \tag{6.11}$$

and associate with each function f a new function $f^\#$ defined by

$$f^\#(x) = e^{\kappa x} f(x).$$

A glance at (6.4) shows that the renewal equation

$$Z^\#(t) = z^\#(t) + \int_0^t Z^\#(t-y)\, L^\#\{dy\} \tag{6.12}$$

holds. Now if $Z^\#(t) \to a \neq 0$ then $Z(t) \sim ae^{-\kappa t}$. Accordingly, if $z^\#$ is directly integrable [in which case $z(\infty) = 0$] the renewal theorem implies that

$$e^{-\kappa t} Z(t) \to \frac{1}{\mu^\#} \int_0^\infty e^{\kappa x} z(x)\, dx \tag{6.13}$$

where

$$\mu^\# = \int_0^\infty e^{\kappa y} y\, L\{dy\} \tag{6.14}$$

In (6.13) one has a good estimate for $Z(t)$ for large t.

With a slight modification this procedure applies also when $z(\infty) \neq 0$. Put

$$z_1(t) = z(\infty) - z(t) + z(\infty) \frac{L_\infty - L(t)}{1 - L_\infty}.$$

It is easily verified that the difference $Z(\infty) - Z(t)$ satisfies the standard renewal equation (6.4) with z replaced by z_1. A simple integration by parts shows that the integral of $z_1^\#(x) = z_1(x)e^{\kappa x}$ is given by the right side in (6.15). Applying (6.13) to Z_1 we get therefore

Theorem 2. *If* (6.10) *holds then the solution of the renewal equation satisfies*

$$\mu^\# e^{\kappa t}[Z(\infty) - Z(t)] \to \frac{z(\infty)}{\kappa} + \int_0^\infty e^{\kappa x}[z(\infty) - z(x)]\, dx \tag{6.15}$$

provided $\mu \neq \infty$ *and* z_1 *is directly integrable.*

For the particular case (6.7) we get

$$\mathbf{P}\{\mathbf{M} > t\} \sim \frac{1 - L_\infty}{\kappa\mu} e^{-\kappa t}. \tag{6.16}$$

The next section will show the surprising power of this estimate. In particular, example (*b*) will show that our simple method sometimes leads quickly to results that used to require deep and laborious methods.

The case $L_\infty > 1$. If $L_\infty > 1$ there exists a constant $\kappa < 0$ such that (6.10) is true and the transformation described reduces the integral equation (6.4) to (6.12). The renewal theorem therefore leads to precise estimates of the asymptotic behavior of $Z(t)e^{\kappa t}$ [The discrete case is covered in theorem 1 of **1**; XIII,10. For applications in demography see example **1**; XII,10(*e*).]

7. DIVERSE APPLICATIONS

As has been pointed out already the theory of the last section may be applied to problems which are conceptually not directly related to renewal processes. In this section we give two independent examples.

(*a*) *Cramér's estimates for ruin.* It was shown in VI,5 that the ruin problem in compound Poisson processes and problems connected with storage facilities, scheduling of patients, etc., depends on *a probability distribution* R, *concentrated on* $\overline{0, \infty}$, *and satisfying the integro-differential equation*

$$R'(z) = \frac{\alpha}{c} R(z) - \frac{\alpha}{c} \int_0^z R(z-x)\, F\{dx\} \tag{7.1}$$

where F is a proper distribution.[6] Integrating (7.1) over $\overline{0, t}$ and performing the obvious integration by parts one gets

$$R(t) - R(0) = \frac{\alpha}{c} \int_0^t R(t-x)[1 - F(x)]\, dx. \tag{7.2}$$

Here $R(0)$ is an unknown constant, but otherwise (7.2) is a renewal equation with a *defective* distribution L with density $\alpha/c\ [1 - F(x)]$. Denoting the expectation of F by μ the mass of L equals $L_\infty = \alpha\mu/c$. [The process is meaningful only if $L_\infty < 1$ for otherwise $R(t) = 0$ for all t.] Note that (7.2) is a special case of (6.4) and that $R(\infty) = 1$. Recalling (6.6) we conclude that

$$R(0) = 1 - \alpha\mu/c, \tag{7.3}$$

[6] This is the special case of VI,(5.4) when F is concentrated on $\overline{0, \infty}$. It will be treated by Laplace transforms in XIV,2(*b*). The general situation will be taken up in XII,5(*d*).

and with this value the integral equation (7.2) reduces to the form (6.7) for the distribution of the lifetime $\mathbf{M}$ of a terminating process with interarrival time distribution L. From (6.16) it follows that *if there exists a constant* κ *such that*

$$\frac{\alpha}{c}\int_0^\infty e^{\kappa x}[1 - F(x)]\,dx = 1 \tag{7.4}$$

and

$$\mu^{\#} = \frac{\alpha}{c}\int_0^\infty e^{\kappa x}x[1 - F(x)]\,dx < \infty \tag{7.5}$$

then as $t \to \infty$

$$1 - R(t) \sim \frac{1}{\kappa\mu^{\#}}\left(1 - \frac{\alpha\mu}{c}\right)e^{-\kappa t}. \tag{7.6}$$

This is Cramér's famous estimate originally derived by deep complex variable methods. The moments of R may be calculated as indicated in example 6(b).

(b) *Gaps in Poisson processes.* In VI,7 we derived a renewal equation for the distribution V of the waiting time for the first gap of length $\geq \xi$ in a renewal process. When the latter is a Poisson process the interarrival times have an exponential distribution, and the renewal equation VI,(7.1) is of the standard form $V = z + V \star L$ with

$$\begin{aligned} L(x) &= 1 - e^{-cx}, \qquad z(x) = 0 \qquad &\textit{for} \quad x < \xi \\ L(x) &= 1 - e^{-c\xi}, \qquad z(x) = e^{-c\xi} \qquad &\textit{for} \quad x \geq \xi. \end{aligned} \tag{7.7}$$

Since $z(\infty) = 1 - L_\infty$ the solution V is a proper distribution as required by the problem.

The *moments* of our waiting time $\mathbf{W}$ are easily calculated by the method described in example 6(b). We get

$$\mathbf{E}(\mathbf{W}) = \frac{e^{c\xi} - 1}{c}, \qquad \operatorname{Var}(\mathbf{W}) = \frac{e^{2c\xi} - 1 - 2c\xi\, e^{c\xi}}{c^2}. \tag{7.8}$$

If we interpret $\mathbf{W}$ as the *waiting time for a pedestrian to cross a stream of traffic* these formulas reveal the effect of an increasing traffic rate. The average number of cars during a crossing time is $c\xi$. Taking $c\xi = 1, 2$ we get $\mathbf{E}(\mathbf{W}) \approx 1.72\xi$ and $\mathbf{E}(\mathbf{W}) \approx 3.2\xi$, respectively. The variance increases from about ξ^2 to $6\xi^2$. [For explicit solutions and connection with covering theorems see example XIV,2(a).] The asymptotic estimate (6.14) applies. If $c\xi > 1$ the determining equation (6.10) reduces to

$$ce^{(\kappa - c)\xi} = \kappa, \qquad 0 < \kappa < c \tag{7.9}$$

and by a routine calculation we get from (6.14)

$$1 - V(t) \sim \frac{1 - \kappa/c}{1 - \kappa\xi}\, e^{-\kappa t}. \tag{7.10}$$

►

8. EXISTENCE OF LIMITS IN STOCHASTIC PROCESSES

Perhaps the most striking proof of the power of the renewal theorem is that it enables us without effort to derive the existence of a "steady state" in a huge class of stochastic processes. About the process itself we need assume only that the probabilities in question are well defined; otherwise the theorem is purely analytic.[7]

Consider a stochastic process with denumerably many states $E_0, E_1, \ldots$ and denote by $P_k(t)$ the probability of E_k at epoch $t > 0$. The following theorem depends on the existence of "recurrent events," that is, of epochs at which the process starts from scratch. More precisely, we assume that with probability one there exists an epoch $\mathbf{S}_1$ such that the continuation of the process beyond $\mathbf{S}_1$ is a probabilistic replica of the whole process commencing at epoch 0. This implies the existence of further epochs $\mathbf{S}_2, \mathbf{S}_3, \ldots$ with the same property. The sequence $\{\mathbf{S}_n\}$ forms a persistent renewal process, and we assume that the mean recurrence time $\mu = \mathbf{E}(\mathbf{S}_1)$ is finite. We denote by $P_k(t)$ the conditional probability of the state E_k at epoch $t + s$ given that $\mathbf{S}_1 = s$. It is assumed that these probabilities are independent of s. Under these conditions we prove the important

Theorem

$$\lim_{t\to\infty} P_k(t) = p_k \tag{8.1}$$

exists with $p_k \geq 0$ *and* $\sum p_k = 1$.

Proof. Let $q_k(t)$ be the probability of the joint event that $\mathbf{S}_1 > t$ and that at epoch t the system is in state E_k. Then

$$\sum_{k=0}^{\infty} q_k(t) = 1 - F(t) \tag{8.2}$$

where F is the distribution of the recurrence times $\mathbf{S}_{n+1} - \mathbf{S}_n$. By hypothesis

$$P_k(t) = q_k(t) + \int_0^t P_k(t-y)\, F\{dy\}. \tag{8.3}$$

The function q_k is directly integrable since it is dominated by the monotone integrable function $1 - F$. Therefore

$$\lim_{t\to\infty} P_k(t) = \frac{1}{\mu}\int_0^{\infty} q_k(t)\, dt \tag{8.4}$$

by the second renewal theorem. Integration of (8.2) shows that these limits add to unity, and the theorem is proved. ►

---

[7] For more sophisticated results see V. E. Beneš, *A "renewal" limit theorem for general stochastic processes*, Ann. Math. Statist., vol. 33 (1962) pp. 98–113, or his book (1963).

It is noteworthy that the *existence* of the limit (8.1) has been established without indication of a way to compute them.

Note. If F_0 is a proper distribution, then (8.1) implies that

$$\int_0^t P_k(t-y)\, F_0\{dy\} \to p_k \qquad as \quad t \to \infty. \tag{8.5}$$

Thus the theorem also covers the case of a *delayed* renewal process $\{\mathbf{S}_n\}$ in which $\mathbf{S}_0$ has distribution F_0.

Examples. (*a*) *Queuing theory.* Consider an installation (telephone exchange, post office, or part of a computer) consisting of one or more "servers," and let the state E_k signify that there are k "customers" in the installation. In most models the process starts from scratch whenever an arriving customer finds the system in state E_0; in this case our limit theorem holds iff such an epoch occurs with probability one and the expectations are finite.

(*b*) *Two-stage renewal process.* Suppose that there are two possible states E_1, E_2. Initially the system is in E_1. The successive sojourn times in E_1 are random variables $\mathbf{X}_j$ with a common distribution F_1. They alternate with sojourn times $\mathbf{Y}_j$ in E_2, having a common distribution F_2. Assuming, as usual, independence of all the variables we have an imbedded renewal process with interarrival distribution $F = F_1 \star F_2$. Suppose $\mathbf{E}(\mathbf{X}_j) = \mu_1 < \infty$ and $\mathbf{E}(\mathbf{Y}_j) = \mu_2 < \infty$. Clearly $q_1(t) = 1 - F_1(t)$ and therefore as $t \to \infty$ the probabilities of E_k tend to the limits

$$P_1(t) \to \frac{\mu_1}{\mu_1 + \mu_2}, \qquad P_2(t) \to \frac{\mu_2}{\mu_1 + \mu_2}. \tag{8.6}$$

This argument generalizes easily to multi-stage systems.

(*c*) The differential equations of **1**; XVII correspond to stochastic processes in which the successive returns to *any* state form a renewal process of the required type. Our theorem therefore guarantees the existence of limit probabilities. Their explicit form can be determined easily from the differential equations with the derivatives replaced by zero. [See, for example, **1**; XVII,(7.3). We shall return to this point more systematically in XIV,9. The same argument applies to the semi-Markov process described in problem 14 of XIV,10.] ▶

*9. RENEWAL THEORY ON THE WHOLE LINE

In this section the renewal theory will be generalized to distributions that are not concentrated on a half-line. To avoid trivialities we assume

* Not used in the sequel.

that $F\{\overline{-\infty, 0}\} > 0$ and $F\{\overline{0, \infty}\} > 0$ and that F *is non-arithmetic.* The modifications necessary for arithmetic distributions will be obvious by analogy with section 1.

We recall from VI,10 that the distribution F is *transient* iff

$$U\{I\} = \sum_{n=0}^{\infty} F^{n\star}\{I\} \tag{9.1}$$

is finite for all finite intervals. Otherwise $U\{I\} = \infty$ for *every* interval and F is called persistent. For transient distributions the question imposes itself: do the renewal theorems of section 1 carry over? This problem has intrigued many mathematicians, perhaps less because of its intrinsic importance than because of its unsuspected difficulties. Thus the renewal theorem was generalized step by step to various special classes of transient distributions by Blackwell, Chung, Chung and Pollard, Chung and Wolfowitz, Karlin, and Smith, but the general theorem was proved only in 1961 by Feller and Orey using probabilistic and Fourier analytic tools. The following proof is considerably simpler and more elementary. In fact, when F has a finite expectation the proof given in section 2 carries over without change. (For renewal theory in the plane see problem 20.)

For the following it must be recalled that a distribution with an expectation $\mu \neq 0$ is transient (theorem 4 of VI,10). As usual, $I + t$ denotes the interval obtained by translating I through t.

General renewal theorem. *If* F *has an expectation* $\mu > 0$ *then for every finite interval* I *of length* $h > 0$

$$U\{I + t\} \to \frac{h}{\mu} \qquad t \to \infty \tag{9.2}$$

$$U\{I + t\} \to 0 \qquad t \to -\infty. \tag{9.3}$$

(*b*) *If* F *is transient and without expectation then* $U\{I + t\} \to 0$ *as* $t \to \pm\infty$ *for every finite interval* I.

From now on it is understood that F is transient and that z is a continuous function vanishing outside the finite interval $-h < x < h$ where $z \geq 0$.

Before proceeding to the proof we recall a few facts proved in VI,10.

The convolution $Z = U \star z$ is well defined by

$$Z(x) = \int_{-\infty}^{+\infty} z(x-y)\, U\{dy\} \tag{9.4}$$

because the effective domain of integration is finite. According to theorem 2 of VI,10 this Z is a continuous function satisfying the *renewal equation*

$$Z = z + F \star Z, \tag{9.5}$$

and it assumes its maximum at a point ξ at which $z(\xi) > 0$.

Put $U_n = F^{0\star} + \cdots + F^{n\star}$. Every non-negative solution Z of (9.5) satisfies $Z \geq z = U_0 \star z$, and hence by induction $Z \geq U_n \star z$. It follows that the solution (9.4) is *minimal* in the sense that $Z_1 \geq Z$ for any other non-negative solution Z_1. Since $Z_1 = Z + \text{const.}$ is again a solution, it follows that

$$\liminf Z(x) = 0, \qquad x \to \pm\infty. \tag{9.6}$$

Lemma 1. *For every constant* a

$$Z(x + a) - Z(x) \to 0 \qquad x \to \pm\infty. \tag{9.7}$$

Proof. The proof is identical with the proof of lemma 2 in section 2. There we used the fact that a bounded uniformly continuous solution of the convolution equation

$$\zeta = F \star \zeta \tag{9.8}$$

attaining its maximum at $x = 0$ reduces to a constant. This remains true also for distributions not concentrated on $\overline{0, \infty}$, and the proof is actually simpler because now the set $\sum$ formed by the points of increase of F, $F^{2\star}, \ldots$ is everywhere dense. ▶

Although we shall not use it explicitly, we mention the following interesting

Corollary.[8] *Every bounded continuous solution of* (9.8) *reduces to a constant.*

Proof. If ξ is uniformly continuous the proof of lemma 1 applies without change. Now if G is an arbitrary probability distribution then $\xi_1 = G \star \xi$ is again a solution of (9.8). We may choose G such that ξ_1 has a bounded derivative and is therefore uniformly continuous. In particular, the convolution of ξ with an arbitrary normal distribution reduces to a constant. Letting the variance of G tend to zero we see that ξ itself reduces to a constant. ▶

Proof *of the renewal theorem when expectations exist.* When $0 < \mu < \infty$ the proof in section 2 applies with one trite change. In the final relation (2.13) we used the trial function $z = 1 - F$ for which the solution Z reduced to the constant 1. Now we use instead

$$z = F^{0\star} - F. \tag{9.9}$$

For it $U_n \star z = F^{0\star} - F^{(n+1)\star}$, and since $\mu > 0$ it is clear that $Z = F^{0\star}$.

It should be noticed that this proof applies also if $\mu = +\infty$ in the obvious sense that the integral of $x\, F\{dx\}$ diverges over $\overline{0, \infty}$, but converges over $\overline{-\infty, 0}$. ▶

---

[8] This corollary holds for distributions on arbitrary groups. See G. Choquet and J. Deny, C. R. Acad. Sci. Paris, vol. 250 (1960) pp. 799–801.

When no expectation exists the proof requires a more delicate analysis. The next lemma shows that it suffices to prove the assertion for one tail.

Lemma 2. *Suppose that no expectation exists and that*

$$U\{I - t\} \to 0, \qquad t \to +\infty. \tag{9.10}$$

Then also

$$U\{I + t\} \to 0, \qquad t \to +\infty. \tag{9.11}$$

Proof. We use the result of example 4(*b*) concerning hitting probabilities in the random walk governed by F. Denote by $H(t, \xi)$ the probability that the first entry into $\overline{t, \infty}$ takes place between t and $t + \xi$. Relative to $t + x$ the interval $I + t$ occupies the same position as $I - x$ relative to t and hence

$$U\{I + t\} = \int_0^\infty H(t, d\xi)\, U\{I - \xi\}. \tag{9.12}$$

Considering the first step in the random walk one sees that

$$1 - H(0, \xi) \geq 1 - F(\xi), \tag{9.13}$$

We know already that the assertion is true if $\mu < \infty$ or $\mu = -\infty$, that is, if the right side is integrable over $\overline{0, \infty}$. Otherwise H has an infinite expectation, and hence $H(t, \xi) \to 0$ as $t \to \infty$ for every ξ. For large t therefore only large values of ξ play an effective role and for them $U\{I - \xi\}$ is small. Thus (9.11) is an immediate consequence of (9.10) and (9.12). ►

Lemma 3. *Suppose* $Z(x) \leq m$ *and choose* $p > 0$ *such that* $p' = 1 - pm > 0$. *To given* $\epsilon > 0$ *there exists an* s_ϵ *such that for* $s > s_\epsilon$ *either*

$$Z(s) < \epsilon \tag{9.14}$$

or else

$$Z(s + x) \geq p\, Z(s)\, Z(x) \quad \textit{for all} \quad x. \tag{9.15}$$

Proof. Because of the uniform continuity of Z and lemma 1 we can choose s_ϵ such that

$$Z(s + x) - Z(x) > -\epsilon p' \quad \textit{for} \quad s > s_\epsilon \quad \textit{and} \quad |x| < h. \tag{9.16}$$

Put

$$V_s(x) = Z(s + x) - p\, Z(s)\, Z(x), \qquad v_s(x) = z(s + x) - p\, Z(s)\, z(x). \tag{9.17}$$

V_s satisfies the renewal equation $V_s = v_s + F \bigstar V_s$ and from the remark preceding lemma 1 it follows that if V_s assumes negative values, then it assumes a minimum at a point ξ where $v(\xi) < 0$, and hence $|\xi| < h$. In view of (9.16) we have then if $s > s_\epsilon$

$$V_s(\xi) > -\epsilon p' + Z(s)[1 - p\,Z(\xi)] \geq p'[Z(s) - \epsilon]. \tag{9.18}$$

Accordingly, either (9.14) holds, or else V_s assumes no negative values in which case (9.15) is true. ▶

Lemma 4. *Let*

$$\limsup Z(x) = \eta \qquad x \to \pm\infty. \tag{9.19}$$

Then also

$$\limsup\,[Z(x) + Z(-x)] = \eta \qquad x \to \infty. \tag{9.20}$$

Proof. Choose a such that $Z(a) < \delta$; this is possible in consequence of (9.6). By the preceding lemma we have for sufficiently large s either

$$p\,Z(s)\,Z(a-s) \leq Z(a) < \delta, \tag{9.21}$$

or else $Z(s) < \epsilon$. Since ϵ is arbitrary, the inequality (9.21) will hold in any case for all s sufficiently large. In view of lemma 1 this implies[9]

$$Z(s)\,Z(-s) \to 0. \tag{9.22}$$

Thus for large x either $Z(x)$ or $Z(-x)$ is small, and since $Z \geq 0$ it is clear that (9.19) implies (9.20). ▶

Proof *of the theorem.* Assume $\eta > 0$ because otherwise there is nothing to be proved. Consider the convolutions of Z and z with the uniform distribution $\overline{0, t}$, namely

$$W_t(x) = \frac{1}{t}\int_{x-t}^{x} Z(y)\,dy, \qquad w_t(x) = \frac{1}{t}\int_{x-t}^{x} z(y)\,dy. \tag{9.23}$$

Our next goal is to show that as $t \to \infty$ one of the relations

$$W_t(t) = \frac{1}{t}\int_0^t Z(y)\,dy \to \eta \quad \text{or} \quad W_t(0) = \frac{1}{t}\int_{-t}^{0} Z(y)\,dy \to \eta \tag{9.24}$$

must take place.

Because of (9.7) the upper bounds for $Z(x)$ and $W_t(x)$ (with t fixed) are the same, and hence the maximum of W_t is $\geq \eta$. On the other hand, $\mathbf{W}_t$ satisfies the renewal equation (9.5) with z replaced by w_t. As noted before,

[9] It is easily seen that (9.22) is equivalent to $U\{I+t\}\,U\{I-t\} \to 0$. If $\rho\{I\}$ stands for the probability that the random walk $\{\mathbf{S}_n\}$ governed by F enters I, then (9.22) is also equivalent to $\rho\{I+t\}\,\rho\{I-t\} \to 0$. If this were false the probability of coming near the origin *after a visit* to $I + t$ would not tend to 0, and F could not be transient.

this implies that the maximum of W_t is attained at a point where w_t is positive, that is, between $-h$ and $t+h$. Now for $\frac{1}{2}t \leq x < t$

$$(9.25) \qquad W_t(x) = \frac{1}{t}\int_{t-x}^{x} Z(y)\,dy + \frac{1}{t}\int_0^{t-x} [Z(y) + Z(-y)]\,dy.$$

The combined length of the two intervals of integration is x and so it follows from (9.20) that for t sufficiently large $W_t(x) < \eta(x/t) + \epsilon$. Thus if $W_t(x) \geq \eta$ the point x must be close to t and $W_t(t)$ close to η. If the maximum of W_t is attained at a point $x \leq \frac{1}{2}t$ a similar argument shows that x must be close to 0 and $W_t(0)$ close to η.

We have now proved that for large t either $W_t(t)$ or $W_t(0)$ is close to η. But a glance at (9.23) shows that in view of (9.20)

$$(9.26) \quad \limsup\, [W_t(t) + W_t(0)] = \limsup\, t^{-1}\int_0^t [Z(y) + Z(-y)]\,dy \leq \eta.$$

Because of the continuity of the two functions therefore either $W_t(t) \to \eta$ and $W_t(0) \to 0$, or else these relations hold with the limits interchanged.

For reasons of symmetry we may assume that $W_t(t) \to \eta$, that is

$$(9.27) \quad W_t(t) = t^{-1}\int_0^t Z(y)\,dy = t^{-1}\int_0^{t/2} [Z(\tfrac{1}{2}t+y) + Z(\tfrac{1}{2}t - y)]\,dy \to \eta.$$

It follows that for arbitrarily large t there exist values of x such that both $Z(x)$ and $Z(t-x)$ are close to η. By lemma 3 this implies that for large t the values of $Z(t)$ are bounded away from 0, and therefore $Z(-t) \to 0$ in consequence of (9.22). Thus $U\{I - t\} \to 0$ as $t \to \infty$, and in view of lemma 2 this accomplishes the proof. ►

10. PROBLEMS FOR SOLUTION

(See also problems 12–20 in VI,13.)

1. Dropping the assumption $F(0) = 0$ amounts to replacing F by the distribution $F^{\#} = pH_0 + qF$ where H_0 is concentrated at the origin and $p + q = 1$. Then U is replaced by $U^{\#} = U/q$. Show that this is a probabilistically obvious consequence of the definition and verify the assertion formally (*a*) by calculating the convolutions, (*b*) from the renewal equation.

2. If F is the uniform distribution in $\overline{0, 1}$ show that

$$U(t) = \sum_{k=0}^{n} (-1)^k e^{t-k} \frac{(t-k)^k}{k!} \qquad \text{for} \quad n \leq t \leq n+1.$$

This formula is frequently rediscovered in queuing theory, but it reveals little about the nature of U. The asymptotic formula $0 \leq U(t) - 2t \to \frac{2}{3}$ is much more interesting. It is an immediate consequence of (3.1).

3. Suppose that $z \geq 0$ and that $|z'|$ is integrable over $\overline{0, \infty}$. Show that z is directly integrable.

Note: In the next three problems it is understood that Z and Z_1 are the solutions of the standard renewal equation (1.3) corresponding to z and z_1.

4. If $z \to \infty$ and $z_1 \sim z$ as $x \to \infty$ show that $Z_1 \sim Z$.

5. If z_1 is the integral of z and $z_1(0) = 0$, then Z_1 is the integral of Z. Conclude that if $z = x^{n-1}$ then $Z \sim x^n/(n\mu)$ provided $\mu < \infty$.

6. (*Generalization.*) If $z_1 = G \bigstar z$ (where G is a measure finite on finite intervals) then also $Z_1 = G \bigstar Z$. For $G(x) = x^a$ with $a \geq 0$ conclude:

$$\textit{if} \quad z(x) \sim x^{a-1} \quad \textit{then} \quad Z(x) \sim x^a/(a\mu).$$

7. (*To theorem* 2 *of section* 3). Denote by f_r the density of $F^{r\bigstar}$ and put $v = u - f - \cdots - f_r$. Show that if f_r is bounded, then $v(x) \to 1/\mu$. In particular, if $f \to 0$ then $u \to 1/\mu$.

8. If $F^{2\bigstar}$ has a directly integrable density then $V = U - 1 - F$ has a density v tending to $1/\mu$.

9. From (4.4) show that $Z(t) = V(t) - V(t-h)$ satisfies the standard renewal equation with $z(t) = F_0(t) - F_0(t-h)$. Derive the result $V(t) - V(t-h) \to h/\mu$ directly from the renewal theorem.

10. *Joint distribution for the residual and spent waiting times.* With the notation (4.7) prove that as $t \to \infty$

$$\mathbf{P}\{t - \mathbf{S}_{\mathbf{N}_t} > x, \mathbf{S}_{\mathbf{N}_t+1} - t > y\} \to \mu \frac{1}{\mu}\int_{x+y}^{\infty} [1 - F(s)]\, ds.$$

(*Hint:* Derive a renewal equation for the left side.)

11. *Steady-state properties.* Consider a delayed renewal process with initial distribution F_0 given by (4.6). The probability $H(t, \xi)$ that a renewal epoch $\mathbf{S}_n$ occurs between t and $t + \xi$ satisfies the renewal equation

$$H(t, \xi) = F_0(t+\xi) - F_0(t) + F_0 \bigstar H(t, \xi).$$

Conclude without calculations that $H(t, \xi) = F_0(t)$ identically.

12. *Maximal observed lifetime.* In the standard persistent renewal process let $V(t, \xi)$ be the probability that the maximal interarrival time observed up to epoch t had a duration $> \xi$. Show that

$$V(t, \xi) = 1 - F(\xi) + \int_0^{\xi} V(t-y, \xi) F\{dy\}.$$

Discuss the character of the solution.

13. For the number $\mathbf{N}_t$ of renewal epochs defined in section 5 show that

$$\mathbf{E}(\mathbf{N}_t^2) = \sum_{k=0}^{\infty} (2k+1) F^{k\bigstar}(t) = 2U \bigstar U(t) - U(t).$$

Using an integration by parts conclude from this and the renewal theorem that

$$\mathbf{E}(\mathbf{N}_t^2) = \frac{2}{\mu}\int_0^t U(x)\, dx + \frac{\sigma^2}{\mu^3} t + o(t)$$

and hence $\operatorname{Var}(\mathbf{N}_t) \sim (\sigma^2/\mu^3)t$ in accordance with the estimate in the central limit theorem. (*Note:* This method applies also to arithmetic distributions and is preferable to the derivation of the same result outlined in problem 23 of **1**; XIII,12.)

14. If F is proper and a a constant, reduce the integro-differential equation

$$Z' = aZ - aZ \star F \quad \text{to} \quad Z(t) = Z(0) + a\int_0^t Z(t-x)[1 - F(x)]\,dx.$$

15. *Generalized type* II *counters.* The incoming particles constitute a Poisson process. The jth arriving particle locks the counter for a duration $\mathbf{T}_j$ and *annuls the aftereffect* (if any) of its predecessors. The $\mathbf{T}_j$ are independent of each other and of the Poisson process and have the common distribution G. If $\mathbf{Y}$ is the duration of a locked interval and $Z(t) = \mathbf{P}\{\mathbf{Y} > t\}$, show that $\mathbf{Y}$ is a proper variable and

$$Z(t) = [1 - G(t)]e^{-\alpha t} + \int_0^t Z(t-x) \cdot [1 - G(x)]\alpha e^{-\alpha x}\,dx.$$

Show that this renewal process is terminating if and only if G has an expectation $\mu < \alpha^{-1}$. Discuss the applicability of the asymptotic estimates of section 6.

16. *Effect of a traffic island.* [Example 7(*b*).] A two-way traffic moves in two independent lanes, representing Poisson processes with equal densities. The expected time required to effect a crossing is 2ξ, and formulas (7.10) apply with this change. A traffic island, however, has the effect that the total crossing time is the sum of two independent variables with expectations and variances given in (7.10). Discuss the practical effect.

17. Arrivals at a counter constitute a persistent renewal process with distribution F. After each registration the counter is locked for a fixed duration ξ during which all arrivals are without effect. Show that the distribution of the time from the end of a locked period to the next arrival is given by

$$\int_0^\xi [F(\xi+t-y) - F(\xi-y)]\,U\{dy\}.$$

If F is exponential so is this distribution.

18. *Non-linear renewal.* A particle has an exponential lifetime at the expiration of which it has probability p_k to produce k independent replicas acting in the same manner $(k = 0, 1, \ldots)$. The probability $F(t)$ that the whole process stops before epoch t satisfies the equation

$$F(t) = p_0(1 - e^{-\alpha t}) + \sum_{\kappa=1}^{\infty} p_k \int_0^t \alpha e^{-\alpha(t-x)} F^k(x)\,dx.$$

(No general method for handling such equations is known.)

19. Let F be an arbitrary distribution in $\mathcal{R}^1$ with expectation $\mu > 0$ and finite second moment m_2. Show that

$$\sum_{n=0}^{\infty} F^{n\star}(x) - \frac{x_+}{\mu} \to \frac{m_2}{2\mu^2}$$

where, as usual, x_+ denotes the positive part of x. *Hint:* If $Z(t)$ stands for the left side then Z satisfies renewal equation with

$$z(x) = \begin{cases} \dfrac{1}{\mu}\displaystyle\int_{-\infty}^{t} F(x)\,dx, & t < 0 \\[2ex] \dfrac{1}{\mu}\displaystyle\int_{t}^{\infty} (1-F(x))\,dx, & t > 0 \end{cases}$$

20. *Renewal theorem in* $\mathfrak{R}^2$. Let the distribution of the pair $(\mathbf{X}, \mathbf{Y})$ be concentrated on the positive quadrant. Let I be the interval $0 \leq x, y \leq 1$. For an arbitrary vector $\mathbf{a}$ denote by $I + \mathbf{a}$ the interval obtained by translating I through $\mathbf{a}$. Lemma 1 of section 9 generalizes as follows. For any fixed vectors $\mathbf{a}$ and $\mathbf{b}$

$$U\{I + \mathbf{a} + t\mathbf{b}\} - U\{I + t\mathbf{b}\} \to 0$$

as $t \to \infty$.

(*a*) Taking this for granted show that the renewal theorem for the marginal distributions implies that $U\{I + t\mathbf{b}\} \to 0$.

(*b*) Show that the proof of lemma carries over trivially.[10]

---

[10] A more appropriate formulation of renewal problems in the plane has been introduced recently by P. J. Bickel and J. A. Yahav, [*Renewal theory in the plane*, Ann. Math. Statist., vol. **36** (1965) pp. 946–955]. They consider the expected number of visits to the region between circles of radii r and $r + a$, and let $r \to \infty$.

CHAPTER XII

Random Walks in $\mathcal{R}^1$

This chapter treats random-walk problems with emphasis on combinatorial methods and the systematic use of ladder variables. Some of the results will be derived anew and supplemented in chapter XVIII by Fourier methods. (Other aspects of random walks were covered in VI,10.) In the main our attention will be restricted to two central topics. First, it will be shown that the curious results derived in **1**; III for fluctuations in coin tossing have a much wider validity and that essentially the same methods are applicable. The second topic is connected with first passages and ruin problems. It has become fashionable to relate such topics to the famous Wiener-Hopf theory, but the connections are not as close as they are usually made to appear. They will be discussed in sections 3*a* and XVIII,4.

E. Sparre Andersen's discovery in 1949 of the power of combinatorial methods in fluctuation theory put the whole theory of random walks into a new light. Since then progress has been extremely rapid, stimulated also by the unexpected discovery of the close connection between random walks and queuing problems.[1]

The literature is vast and bewildering. The theory presented in the following pages is so elementary and simple that the newcomer would never suspect how difficult the problems used to be before their natural setting was understood. For example, the elementary asymptotic estimates in section 5 cover a variety of practical results obtained previously by deep methods and sometimes with great ingenuity.

Sections 6–8 are nearly independent of the first part. It is hardly necessary to say that our treatment is one-sided and neglects interesting aspects of random walks such as connections with potential theory and group theory.[2]

---

[1] The first such connection seems to have been pointed out by D. V. Lindley in 1952. He derived an integral equation which would now be considered of the Wiener-Hopf type.

[2] For other aspects see Spitzer's book (1964), although it is limited to arithmetic distributions. For combinatorial methods applicable to higher dimensions see C. Hobby and R. Pyke, *Combinatorial results in multidimensional fluctuation theory*, Ann. Math. Statist., vol. 34 (1963) pp. 402–404.

1. BASIC CONCEPTS AND NOTATIONS

Throughout this chapter $\mathbf{X}_1, \mathbf{X}_2, \ldots$ are independent random variables with a common distribution F not concentrated on a half-axis. [For distributions with $F(0) = 0$ or $F(0) = 1$ the topic is covered by renewal theory.] The *induced random walk* is the sequence of random variables

$$\mathbf{S}_0 = 0, \qquad \mathbf{S}_n = \mathbf{X}_1 + \cdots + \mathbf{X}_n. \tag{1.1}$$

Sometimes we consider a section $(\mathbf{X}_{j+1}, \ldots, \mathbf{X}_k)$ of the given sequence $\{\mathbf{X}_j\}$; its partial sums $0, \mathbf{S}_{j+1}-\mathbf{S}_j, \ldots, \mathbf{S}_k-\mathbf{S}_j$ will be called a *section of the random walk.* The subscripts are treated in the usual manner as a time parameter. Thus an epoch n is said to divide the whole random walk into a *preceding* and a *residual section.* Because $\mathbf{S}_0 = 0$ the random walk is said to start at the origin. By adding a constant a to all terms we obtain a random walk starting at a. Thus $\mathbf{S}_n, \mathbf{S}_{n+1}, \ldots$ is a random walk induced by F and starting at $\mathbf{S}_n$.

Orientation. Looking at the graph of a random walk one notices as a striking feature the points where $\mathbf{S}_n$ reaches a record value, that is, where $\mathbf{S}_n$ exceeds all previously attained values $\mathbf{S}_0, \ldots, \mathbf{S}_{n-1}$. These are the *ladder points* according to the terminology introduced in VI,8. (See fig. 1 in that section.) The theoretical importance of ladder points derives from the fact that the sections between them are probabilistic replicas of each other, and therefore important conclusions concerning the random walk can be derived from a study of the first ladder point.

In volume **1** we have studied repeatedly random walks in which the $\mathbf{X}_k$ assume the values $+1$ and -1 with probabilities p and q respectively. In such walks each record value exceeds the preceding one by $+1$, and the successive ladder points represent simply the first passages through $1, 2, \ldots$. In the present terminology we would say that the ladder heights are known in advance, and only the waiting times between successive ladder points require study. These are independent random variables with the same distribution as the first passage time through $+1$. The generating function of this distribution was found in **1**; XI,(3.6) and is given by

$$[1 - \sqrt{1 - 4pqs^2}\,]/(2qs) \tag{1.2}$$

where $\sqrt{}$ denotes the positive root [see also **1**; XIV,4; for explicit formulas see **1**; XI,3(*d*) and **1**; XIV,5]. When $p < q$ the first passage times are defective random variables since the probability that a positive value will ever be attained equals p/q.

The same record value may be repeated several times before a new record value is reached. Points of such relative maxima are called *weak* ladder points. [In the simple binomial random walk the first weak ladder point is either $(1, 1)$ or else it is of the form $(2r, 0)$.]

After these preliminary remarks we proceed to a formal introduction of ladder variables repeating in part what was said in VI,8. The definition depends on an inequality, and there exist therefore four types of ladder variables corresponding to the four possibilities $<, \leq, >, \geq$. This leads to a twofold classification to be described by the self-explanatory terms *ascending* and *descending*, *strict* and *weak*. The ascending and descending variables are related by the familiar symmetry between plus and minus, or maxima and minima. The distinction between strict and weak variables, however, puts a burden on description and notation. The simplest way out is to consider only continuous distributions F, for then the strict and weak variables are the same with probability one. Beginners are advised to proceed in this way and not to distinguish between strict and weak ladder variables, but this distinction is unavoidable for the general theory on one hand, and for examples such as the coin-tossing game on the other.

To introduce the necessary notations and conventions we consider the ascending strict ladder variables. We shall then show that the theory of weak ladder variables follows as a simple corollary of the theory of strict variables. Descending ladder variables require no new theory. We shall therefore take the ascending strict ladder variables as typical, and when no danger of confusion arises, we shall drop the qualifications "ascending" and "strict."

Ascending strict ladder variables. Consider the sequence of points $(n, \mathbf{S}_n)$ for $n = 1, 2, \ldots$ (the origin is excluded). The *first strict ascending ladder point* $(\mathscr{T}_1, \mathscr{H}_1)$ *is the first term in this sequence for which* $\mathbf{S}_n > 0$. In other words, $\mathscr{T}_1$ is the epoch of the first entry into the (strictly) positive half-axis defined by

$$\{\mathscr{T}_1 = n\} = \{\mathbf{S}_1 \leq 0, \ldots, \mathbf{S}_{n-1} \leq 0, \mathbf{S}_n > 0\}, \tag{1.3}$$

and $\mathscr{H}_1 = \mathbf{S}_{\mathscr{T}_1}$. The variable $\mathscr{T}_1$ is called first *ladder epoch*, $\mathscr{H}_1$ the first *ladder height*. These variables remain *undefined* if the event (1.3) does not take place, and hence both variables are possibly defective.[3]

For the joint distribution of $(\mathscr{T}_1, \mathscr{H}_1)$ we write

$$\mathbf{P}\{\mathscr{T}_1 = n, \mathscr{H}_1 \leq x\} = H_n(x). \tag{1.4}$$

The marginal distributions are given by

$$\mathbf{P}\{\mathscr{T}_1 = n\} = H_n(\infty), \qquad n = 1, 2, \ldots \tag{1.5}$$

$$\mathbf{P}\{\mathscr{H}_1 \leq x\} = \sum_{n=1}^{\infty} H_n(x) = H(x). \tag{1.6}$$

The two variables have the same defect, namely $1 - H(\infty) \geq 0$.

[3] Problems 3–6 provide illustrative exercises accessible without general theory.

The section of the random walk following the first ladder epoch is a probabilistic replica of the whole random walk. Its first ladder point is the *second* point of the whole random walk with the property that

$$\mathbf{S}_n > \mathbf{S}_0, \ldots, \mathbf{S}_n > \mathbf{S}_{n-1}; \tag{1.7}$$

it will be called the second ladder point of the entire random walk. It is of the form $(\mathscr{T}_1 + \mathscr{T}_2, \mathscr{H}_1 + \mathscr{H}_2)$ where the pairs $(\mathscr{T}_1, \mathscr{H}_1)$ and $(\mathscr{T}_2, \mathscr{H}_2)$ are independent and identically distributed. (See also VI,8.) Proceeding in this way we define the third, fourth, ... ladder points of our random walk. Thus *a point* $(n, \mathbf{S}_n)$ *is an ascending ladder point if it satisfies* (1.7). The rth ladder point (if it exists) is of the form $(\mathscr{T}_1 + \cdots + \mathscr{T}_r, \mathscr{H}_1 + \cdots + \mathscr{H}_r)$ where the pairs $(\mathscr{T}_k, \mathscr{H}_k)$ are mutually independent and have the common distribution (1.4). (See fig. 1 in VI,8.)

For economy of notation no new letters will be introduced for the sums $\mathscr{T}_1 + \cdots + \mathscr{T}_r$ and $\mathscr{H}_1 + \cdots + \mathscr{H}_r$. They form (possibly terminating) *renewal processes with "interarrival times"* $\mathscr{T}_k$ and $\mathscr{H}_k$. In the random walk, of course, only $\mathscr{T}_k$ is of the nature of a time variable. The ladder points themselves form a two-dimensional renewal process.

We shall denote by

$$\psi = \sum_{n=0}^{\infty} H^{n\star} \tag{1.8}$$

the renewal measure for the ladder height process. (Here $H^{0\star} = \psi_0$.) Its improper distribution function given by $\psi(x) = \psi\{\overline{-\infty, x}\}$ vanishes when $x < 0$, while for x positive $\psi(x)$ *equals one plus the expected number of ladder points in the strip* $\overline{0, x}$ (no limitation on time). We know from VI,6 and XI,1 that $\psi(x) < \infty$ for all x and in the case of defective ladder variables

$$\psi(\infty) = \sum_{n=0}^{\infty} H^n(\infty) = \frac{1}{1 - H(\infty)}. \tag{1.9}$$

Finally we introduce the notation ψ_0 for the atomic distribution with unit mass at the origin; thus for any interval I

$$\psi_0\{I\} = 1 \quad \textit{if} \quad x \in I, \qquad \psi_0\{I\} = 0 \quad \textit{otherwise}. \tag{1.10}$$

Ascending weak ladder variables. The point $(n, \mathbf{S}_n)$ is a weak (ascending) ladder point iff $\mathbf{S}_n \geq \mathbf{S}_k$ for $k = 0, 1, \ldots, n$. The theories of strict and weak ladder variables run parallel, and we shall systematically use the same letters, indicating weak variables by bars: thus $\overline{\mathscr{T}}_1$ is the smallest index n such that $\mathbf{S}_1 < 0, \ldots, \mathbf{S}_{n-1} < 0$, but $\mathbf{S}_n \geq 0$. As was mentioned before, the tedious distinction between strict and weak variables becomes unnecessary when the distribution F is continuous. Even in the general situation it is easy to express the distribution $\bar{H}$ of weak ladder heights in terms of the distribution H, and this will enable us to confine our attention to the single distribution defined in (1.6).

The first weak ladder point is identical with the first strict ladder point except if the random walk returns to the origin having passed only through negative values; in this case $\overline{\mathscr{H}}_1 = 0$ and we put $\zeta = \mathbf{P}\{\overline{\mathscr{H}}_1 = 0\}$. Thus

$$\zeta = \sum_{n=1}^{\infty} \mathbf{P}\{\mathbf{S}_1 < 0, \ldots, \mathbf{S}_{n-1} < 0, \mathbf{S}_n = 0\}. \tag{1.11}$$

(The event cannot occur if $\mathbf{X}_1 > 0$ and hence $0 \leq \zeta < 1$.) With probability $1 - \zeta$ the first strict ladder point coincides with the first weak ladder point and hence

$$\bar{H} = \zeta\psi_0 + (1-\zeta)H. \tag{1.12}$$

In words, the distribution of the first weak ladder height is a mixture of the distribution H and the atomic distribution concentrated at the origin.

Example. In the simple binomial random walk the first weak ladder height equals 1 iff the first step leads to $+1$. If the first step leads to -1 the (conditional) probability of a return to 0 equals 1 if $p \geq q$, and p/q otherwise. In the first case $\zeta = q$, in the second $\zeta = p$. The possible ladder heights are 1 and 0, and they have probabilities p and q if $p \leq q$, while both probabilities equal p when $p < q$. In the latter case the ladder height is a defective variable. ▶

The probability that prior to the first entry into $\overline{0, \infty}$ the random walk returns to the origin exactly k times equals $\zeta^k(1-\zeta)$. The expected number of such returns is $1/(1-\zeta)$ and this is also the expected multiplicity of each weak ladder height prior to the appearance of the next strict ladder points. Therefore

$$\bar{\psi} = \frac{1}{1-\zeta}\,\psi. \tag{1.13}$$

(See problem 7.) The simplicity of these relations enables us to avoid explicit use of the distribution $\bar{H}$.

Descending ladder variables. The *strict and weak descending ladder variables are defined by symmetry*, that is, by changing $>$ into $<$. On the rare occasions where a special notation will be required we shall denote descending order by the superscript minus. Thus the first strict descending ladder point is $(\mathscr{T}_1^-, \mathscr{H}_1^-)$, and so on.

It will be seen presently that the probabilities $\mathbf{P}\{\overline{\mathscr{H}}_1 = 0\}$ and $\mathbf{P}\{\overline{\mathscr{H}}_1 = 0\}$

are identical because

$$(1.14)\quad \mathbf{P}\{\mathbf{S}_1 > 0, \ldots, \mathbf{S}_{n-1} > 0, \mathbf{S}_n = 0\} = \mathbf{P}\{\mathbf{S}_1 < 0, \ldots, \mathbf{S}_{n-1} < 0, \mathbf{S}_n = 0\}.$$

It follows that the analogue to (1.12) and (1.13) for the descending ladder variables depend on the same quantity ζ.

2. DUALITY. TYPES OF RANDOM WALKS

The amazing properties of the fluctuations in coin tossing were derived in **1**; III by simple combinatorial arguments depending on taking the variables $(\mathbf{X}_1, \ldots, \mathbf{X}_n)$ in reverse order. The same device will now lead to important results of great generality.

For fixed n we introduce n new variables by $\mathbf{X}_1^* = \mathbf{X}_n, \ldots, \mathbf{X}_n^* = \mathbf{X}_1$. Their partial sums are given by $\mathbf{S}_k^* = \mathbf{S}_n - \mathbf{S}_{n-k}$ where $k = 0, \ldots, n$. The joint distributions of $(\mathbf{S}_0, \ldots, \mathbf{S}_n)$ and $(\mathbf{S}_0^*, \ldots, \mathbf{S}_n^*)$ being the same, the correspondence $\mathbf{X}_k \to \mathbf{X}_k^*$ maps any event A defined by $(\mathbf{S}_0, \ldots, \mathbf{S}_n)$ into an event A^* of equal probability. The mapping is easy to visualize because the graphs of $(0, \mathbf{S}_1, \ldots, \mathbf{S}_n)$ and $(0, \mathbf{S}_1^*, \ldots, \mathbf{S}_n^*)$ are rotations of each other through 180 degrees.

Example. (*a*) If $\mathbf{S}_1 < 0, \ldots, \mathbf{S}_{n-1} < 0$ but $\mathbf{S}_n = 0$, then

$$\mathbf{S}_1^* > 0, \ldots, \mathbf{S}_{n-1}^* > 0 \quad \text{and} \quad \mathbf{S}_n^* = 0.$$

This proves the validity of the relation (1.14) used in the preceding section. ▸

We now apply the reversal procedure to the event $\mathbf{S}_n > \mathbf{S}_0, \ldots, \mathbf{S}_n > \mathbf{S}_{n-1}$ defining a (ascending strict) ladder point. The dual relations are $\mathbf{S}_n^* > \mathbf{S}_{n-k}^*$ for $k = 1, \ldots, n$. But $\mathbf{S}_n^* > \mathbf{S}_{n-k}^*$ is the same as $\mathbf{S}_k > 0$, and hence we have for every finite interval $I \subset \overline{0, \infty}$

$$(2.1)\quad \mathbf{P}\{\mathbf{S}_n > \mathbf{S}_j \;\; \textit{for} \;\; j = 0, \ldots, n-1 \;\; \textit{and} \;\; \mathbf{S}_n \in I\} = \\ = \mathbf{P}\{\mathbf{S}_j > 0 \;\; \textit{for} \;\; j = 1, \ldots, n \;\; \textit{and} \;\; \mathbf{S}_n \in I\}.$$

The left side is the probability that there exists a ladder point with abscissa n and ordinate in I. The right side is the probability of the event that a visit to I at epoch n takes place without prior visit to the closed half-line $\overline{-\infty, 0}$.

Consider then the result of summing (2.1) over all n. On the left we get $\psi\{I\}$ by the definition (1.8) of the renewal measure ψ. On the right we get the expected number of visits to the interval I prior to the first entry into $\overline{-\infty, 0}$. It is finite because $\psi\{I\} < \infty$. We have thus proved the basic

Duality lemma. *The renewal measure* ψ *admits of two interpretations. For every finite interval* $I \subset \overline{0, \infty}$ *the value* $\psi\{I\}$ *equals*

(*a*) *the expected number or ladder points in* I;

(*b*) *the expected number of visits* $\mathbf{S}_n \in I$ *such that* $\mathbf{S}_k > 0$ *for* $k = 1, 2, \ldots, n$.

This simple lemma will enable us to prove in an elementary way theorems that would otherwise require deep analytic methods. In its analytic formulation the lemma does not seem exciting, but it has immediate consequences that are most surprising and contrary to naive intuitions.

Example. (*b*) *Simple random walk*. In the random walk of the example in section 1 there exists a ladder point with ordinate k iff the event $\{\mathbf{S}_n = k\}$ occurs for some n, and we saw that the probability for this is 1 or $(p/q)^k$ according as $p \geq q$ or $p \leq q$. By the duality lemma this means that *in a symmetric random walk the expected number of visits to* $k \geq 1$ *prior to the first return to the origin equals* 1 for all k. The fantastic nature of this result appears clearer in the coin-tossing terminology. The assertion is that on the average *Peter's accumulated gain passes once through every value* k, *however large, before reaching the zero level for the first time.* This statement usually arouses incredulity, but it can be verified by direct calculation (problem 2). (Our old result that the waiting time for the first return to 0 has infinite expectation follows by summation over k.) ▶

In the symmetric binomial random walk (coin tossing) each of the values ± 1 is attained with probability one, but the expected waiting time for each of these events is infinite. The next theorem shows that this is not a peculiarity of the coin tossing game since a similar statement is true for all random walks in which both positive and negative values are assumed with probability one.

Theorem 1. *There exist only two types of random walks.*

(i) *The oscillating type. Both the ascending and the descending renewal processes are persistent,* $\mathbf{S}_n$ *oscillates with probability* 1 *between* $-\infty$ *and* ∞, *and*

$$\mathbf{E}(\mathscr{T}_1) = \infty, \qquad \mathbf{E}(\mathscr{T}_1^-) = \infty. \tag{2.2}$$

(ii) *Drift to* $-\infty$, (*say*). *The ascending renewal process is terminating, the descending one proper. With probability one* $\mathbf{S}_n$ *drifts to* $-\infty$ *and reaches a finite maximum* $\mathbf{M} \geq 0$. *The relations* (2.5) *and* (2.7) *are true.*

[Walks of type (ii) are obviously transient, but type (i) includes both persistent and transient walks. See end of VI,10.]

Proof. The identity (2.1) holds also when the strict inequalities are replaced by weak ones. For $I = \overline{0, \infty}$ it reduces to

$$\text{(2.3)} \quad \mathbf{P}\{\mathbf{S}_n \geq \mathbf{S}_k \quad for \quad 0 \leq k \leq n\} = \mathbf{P}\{\mathbf{S}_k \geq 0 \quad for \quad 0 \leq k \leq n\} = 1 - \mathbf{P}\{\mathcal{T}_1^- \leq n\}.$$

The left side equals the probability that $(n, \mathbf{S}_n)$ be a weak ascending ladder point. These probabilities add to $\bar{\psi}(\infty) \leq \infty$, and in view of (1.13) we have therefore

$$\text{(2.4)} \qquad \frac{1}{1-\xi}\,\psi(\infty) = \sum_{n=0}^{\infty}[1 - \mathbf{P}\{\mathcal{T}_1^- \leq n\}].$$

When the descending ladder process is defective the terms of the series are bounded away from zero and the series diverges. In this case $\psi(\infty) = \infty$ which means that the ascending process is persistent. We have thus an analytic proof for the intuitively obvious fact that it is impossible that both the ascending and the descending ladder processes terminate.

If $\mathcal{T}_1^-$ is proper (2.4) reduces to

$$\text{(2.5)} \qquad \mathbf{E}(\mathcal{T}_1^-) = \frac{1}{1-\zeta}\,\psi(\infty) = \frac{1}{(1-\zeta)(1-H(\infty))}$$

with the obvious interpretation when $H(\infty) = 1$. It follows that $\mathbf{E}(\mathcal{T}_1^-) < \infty$ iff $H(\infty) < 1$, that is, iff the ascending variable $\mathcal{T}_1$ is defective. Thus either one of these variables is defective, or else (2.2) holds.

If $\mathbf{E}(\mathcal{T}_1^-) < \infty$ the ascending renewal process is terminating. With probability one there occurs a last ladder point, and so

$$\text{(2.6)} \qquad \mathbf{M} = \max\{\mathbf{S}_0, \mathbf{S}_1, \ldots\}.$$

is finite. Given that the nth ladder point occurred, the probability that it is the last equals $1 - H(\infty)$, and so [see XI,(6.3)]

$$\text{(2.7)} \qquad \mathbf{P}\{\mathbf{M} \leq x\} = [1 - H(\infty)]\sum_{n=0}^{\infty} H^{n\star}(x) = [1 - H(\infty)]\,\psi(x). \qquad \blacktriangle$$

In the following theorem we agree to write $\mathbf{E}(\mathbf{X}) = +\infty$ if the defining integral diverges only at $+\infty$ or, what amounts to the same, if $\mathbf{P}\{\mathbf{X} < t\}$ is integrable over $\overline{-\infty, 0}$.

Theorem 2. (i) *If* $\mathbf{E}(\mathbf{X}_1) = 0$, *then* $\mathcal{H}_1$ *and* $\mathcal{T}_1$ *are proper,*[4] *and* $\mathbf{E}(\mathcal{T}_1) = \infty$.

---

[4] Theorem 4 of VI,10 contains the stronger result that the random walk is persistent whenever $\mathbf{E}(\mathbf{X}_1) = 0$.

(ii) *If* $\mathbf{E}(\mathbf{X}_1)$ *is finite and positive, then* $\mathscr{H}_1$ *and* $\mathscr{T}_1$ *are proper, have finite expectations, and*

$$\mathbf{E}(\mathscr{H}_1) = \mathbf{E}(\mathscr{T}_1)\,\mathbf{E}(\mathbf{X}_1) \tag{2.8}$$

The random walk drifts to $+\infty$.

(iii) *If* $\mathbf{E}(\mathbf{X}_1) = +\infty$ *then* $\mathbf{E}(\mathscr{H}_1) = \infty$ *and the random walk drifts to* $+\infty$.

(iv) *Otherwise either the random walk drifts to* $-\infty$ *(in which case* $\mathscr{T}_1$ *and* $\mathscr{H}_1$ *are defective), or else* $\mathbf{E}(\mathscr{H}_1) = \infty$.

The identity (2.8) was discovered by A. Wald in a more general setting to be discussed in XVIII,2. The following proof is based on the strong law of large numbers of VII,8. Purely analytic proofs will be given in due course. (See theorem 3 of section 7 and problems 9–11 as well as XVIII,4.)

Proof. If n coincides with the kth ladder epoch we have the identity

$$\frac{\mathbf{S}_n}{n} = \frac{(\mathscr{H}_1+\cdots+\mathscr{H}_k)/k}{(\mathscr{T}_1+\cdots+\mathscr{T}_k)/k}. \tag{2.9}$$

We now observe that the strong law of large numbers applies also if $\mathbf{E}(\mathbf{X}_1) = +\infty$ as can be seen by the obvious truncation.

(i) Let $\mathbf{E}(\mathbf{X}_1) = 0$. As $k \to \infty$ the left side in (2.9) tends to 0. It follows that the denominator tends to infinity. This implies that $\mathscr{T}_1$ is proper and $\mathbf{E}(\mathscr{T}_1) = \infty$.

(ii) If $0 < \mathbf{E}(\mathbf{X}_1) < \infty$ the strong law of large numbers implies that the random walk drifts to ∞. In view of (2.5) this means that $\mathscr{T}_1$ is proper and $\mathbf{E}(\mathscr{T}_1) < \infty$. Numerator and denominator in (2.9) therefore tend to finite limits, and (2.8) now follows from the converse to the law of large numbers (theorem 4 in VII,8).

(iii) If $\mathbf{E}(\mathbf{X}_1) = +\infty$ the same argument shows that $\mathbf{E}(\mathscr{H}_1) = \infty$.

(iv) In the remaining cases we show that if $\mathscr{H}_1$ is proper and $\mathbf{E}(\mathscr{H}_1) < \infty$ the random walk drifts to $-\infty$. Considering the first step in the random walk it is clear that for $x > 0$

$$\mathbf{P}\{\mathscr{H}_1 > x\} \geq \mathbf{P}\{\mathbf{X}_1 > x\}. \tag{2.10}$$

If $\mathscr{H}_1$ is improper the random walk drifts to $-\infty$. If it is proper the integral of the left side extended over $\overline{0, \infty}$ equals $\mathbf{E}(\mathscr{H}_1)$. If $\mathbf{E}(\mathscr{H}_1) < \infty$ it follows that $\mathbf{E}(\mathbf{X}_1)$ is finite or $-\infty$. The case $\mathbf{E}(\mathbf{X}_1) \geq 0$ has been taken care of, and if $\mathbf{E}(\mathbf{X}_1) < 0$ (or $-\infty$) the random walk drifts to $-\infty$. ►

It follows from (2.10) and the analogous inequality for $x < 0$ that if both $\mathscr{H}_1$ and $\mathscr{H}_1^-$ are proper and have finite expectations, then $\mathbf{P}\{|\mathbf{X}_1| > x\}$ is integrable and hence $\mathbf{E}(\mathbf{X}_1)$ exists (lemma 2 of V,6). With $\mathbf{E}(\mathbf{X}_1) \neq 0$ one of the ladder variables would be defective, and hence we have the

Corollary. *If both* $\mathscr{H}_1$ *and* $\mathscr{H}_1^-$ *are proper and have finite expectations, then* $\mathbf{E}(\mathbf{X}_1) = 0$.

The converse is not true. However, if $\mathbf{E}(\mathbf{X}_1) = 0$ and $\mathbf{E}(\mathbf{X}_1^2) < \infty$ then $\mathscr{H}_1$ and $\mathscr{H}_1^-$ have finite expectations. (See problem 10. A more precise result is contained in theorem 1 of XVIII,5 where $\mathbf{S}_\mathbf{N}$ and $\mathbf{S}_{\bar{\mathbf{N}}}$ are the first entry variables with distributions $\mathscr{H}_1$ and $\mathscr{H}_1^-$.)

3. DISTRIBUTION OF LADDER HEIGHTS. WIENER-HOPF FACTORIZATION

The calculation of the ladder height distributions H and $\bar{H}$ seems at first to present a formidable problem, and was originally considered in this light. The duality lemma leads to a simple solution, however. The idea is that the first entry into, say, $\overline{-\infty, 0}|$ should be considered together with the section of the random walk prior to this first entry. We are thus led to the study of the modified random walk $\{\mathbf{S}_n\}$ which terminates at the epoch of the first entry into $\overline{-\infty, 0}|$. We denote by ψ_n the defective probability distribution of the position at epoch n in this restricted random walk; that is, for an arbitrary interval I and $n = 1, 2, \ldots$ we put

$$\psi_n\{I\} = \mathbf{P}\{\mathbf{S}_1 > 0, \ldots, \mathbf{S}_n > 0, \mathbf{S}_n \in I\}. \tag{3.1}$$

(Note that this implies $\psi_n\{\overline{-\infty, 0}|\} = 0$.) As before, ψ_0 is the probability distribution concentrated at the origin. Now it was shown in (2.1) that $\psi_n\{I\}$ equals the probability that $(n, \mathbf{S}_n)$ be a ladder point with $\mathbf{S}_n \in I$. Summing over n we get therefore

$$\psi\{I\} = \sum_{n=0}^{\infty} \psi_n\{I\} \tag{3.2}$$

where ψ is the renewal function introduced in (1.8). In other words, for an interval in the open positive half-axis $\psi\{I\}$ is the expected number of (strict ascending) ladder points with ordinate in I. For I in the negative half-axis we now define $\psi\{I\} = 0$. It follows that the series in (3.2) converges for every *bounded* interval I (though not necessarily for $I = \overline{0, \infty}$). It is this unexpected result that renders the following theory so incredibly simple.

Studying the first entry into $\overline{-\infty, 0}|$ means studying the weak descending ladder process, and with the notations of section 1 the point of first entry is $\bar{\mathscr{H}}_1^-$, its distribution $\bar{H}^-$. For typographical convenience, however, we replace $\bar{H}^-$ by ρ and denote by $\rho_n\{I\}$ *the probability that the first entry to* $\overline{-\infty, 0}$ *takes place at epoch* n and within the interval I. Formally for $n = 1, 2, \ldots$

$$\rho_n\{I\} = \mathbf{P}\{\mathbf{S}_1 > 0, \ldots, \mathbf{S}_{n-1} > 0, \mathbf{S}_n \leq 0, \mathbf{S}_n \in I\}. \tag{3.3}$$

(This implies $\rho_n\{\overline{0, \infty}\} = 0$. The term ρ_0 remains undefined.) This time the series

$$\rho\{I\} = \sum_{n=1}^{\infty} \rho_n\{I\} \tag{3.4}$$

obviously converges and represents the possibly defective distribution of the point of the first entry. (In other words, $\rho\{I\} = \overline{\mathscr{H}}_1^-\{I\}$.)

It is easy to derive recurrence relations for ψ_n and ρ_n. Indeed, given the position y of $\mathbf{S}_n$ the (conditional) probability that $\mathbf{S}_{n+1} \in I$ equals $F\{I - y\}$, where $I - y$ is the translate of I through $-y$. Thus

$$\rho_{n+1}\{I\} = \int_{0-}^{\infty} \psi_n\{dy\}\, F\{I - y\} \qquad \textit{if} \quad I \subset \overline{-\infty, 0} \tag{3.5a}$$

$$\psi_{n+1}\{I\} = \int_{0-}^{\infty} \psi_n\{dy\}\, F\{I - y\} \qquad \textit{if} \quad I \subset \overline{0, \infty} \tag{3.5b}$$

(the origin contributing only when $n = 0$). For bounded intervals I the duality lemma assures the convergence of $\sum \psi_n\{I\}$, and $\sum \rho_n\{I\}$ always converges to a number ≤ 1. We have thus series representations for ρ and ψ. It is clear that these sums satisfy

$$\rho\{I\} = \int_{0-}^{\infty} \psi\{dy\}\, F\{I - y\} \qquad \textit{if} \quad I \subset \overline{-\infty, 0} \tag{3.6a}$$

$$\psi\{I\} = \int_{0-}^{\infty} \psi\{dy\}\, F\{I - y\} \qquad \textit{if} \quad I \subset \overline{0, \infty} \tag{3.6b}$$

with the proviso that (3.6*b*) is restricted to bounded intervals I. We shall see that in practice the relations (3.6) are more useful than the theoretical series representations for ρ and ψ. It is sometimes convenient to replace the interval function ρ and ψ by the equivalent point functions

$$\rho(x) = \rho\{\overline{-\infty, x}\} \quad \text{and} \quad \psi(x) = \psi\{\overline{-\infty, x}\}.$$

Clearly (3.6*a*) is equivalent to

$$\rho(x) = \int_{0-}^{\infty} \psi\{dy\}\, F(x-y), \qquad x \leq 0. \tag{3.7a}$$

From (3.6*b*) we get for $x > 0$

$$\psi(x) = 1 + \psi\{\overline{0, x}\} = 1 + \int_{0-}^{\infty} \psi\{dy\}\, [F(x-y) - F(-y)].$$

Taking into account (3.7*a*) we see thus that (3.6*b*) is equivalent to

$$\psi(x) = 1 - \rho(0) + \int_{0-}^{\infty} \psi\{dy\}\, F(x-y), \qquad x \geq 0. \tag{3.7b}$$

To simplify notations we introduce the convolution

$$\psi \bigstar F = \sum_{n=0}^{\infty} \psi_n \bigstar F. \tag{3.8}$$

Since ψ is concentrated on $\overline{0, \infty}$ the value $\psi \bigstar F\{I\}$ equals the sum of the two integrals in (3.6) and is therefore finite. As ψ has a unit atom at the origin we can combine the two relations (3.6) into the single convolution equation

$$\rho + \psi = \psi_0 + \psi \bigstar F. \tag{3.9}$$

In view of the fact that ρ and $\psi - \psi_0$ are concentrated on $\overline{-\infty, 0}$ and $\overline{0, \infty}$, respectively, the relation (3.9) is fully equivalent to the pair (3.6).

We shall use (3.9) as an integral equation determining the unknown measures ρ and ψ. A great many conclusions of theoretical importance can be derived directly from (3.9). We list the most remarkable such theorem under the heading of an example in order to indicate that it will not be used in the sequel and that we embark on a digression.

Examples. (*a*) *Wiener-Hopf type factorization.* It follows from the definition (1.7) of ψ that it satisfies the renewal equation

$$\psi = \psi_0 + \psi \bigstar H. \tag{3.10}$$

Using this relation we show that (3.9) *may be rewritten in the equivalent form*

$$F = H + \rho - H \bigstar \rho. \tag{3.11}$$

Indeed, convolving (3.9) with H we get

$$H \bigstar \rho + \psi - \psi_0 = H - F + \psi \bigstar F.$$

Subtracting this from (3.9) yields (3.11). Conversely, convolving (3.11) with ψ we get

$$\psi \bigstar F = \psi - \psi_0 + \psi \bigstar \rho - (\psi - \psi_0) \bigstar \rho = \psi - \psi_0 + \rho$$

which is the same as (3.9).

The identity (3.11) is remarkable in that it represents an arbitrary probability distribution F in terms of two (possibly defective) distributions H and ρ concentrated on $\overline{0, \infty}$ and $\overline{-\infty, 0}$, respectively. The first interval is open to the second closed, but this asymmetry can be remedied by expressing the first entrance probabilities ρ into $\overline{-\infty, 0}$ by the first entrance probabilities H^- into $\overline{-\infty, 0}$. The relationship between these probabilities is given by the analogue to (1.12) for $x < 0$, namely

$$\rho = \zeta\psi_0 + (1-\zeta)H^-$$

with ζ defined by (1.11) [which relation holds also with the inequalities reversed; see example 2(*a*)]. Substituting into (3.11) we get after a trite rearrangement

$$\psi_0 - F = (1-\zeta)\, [\psi_0 - H] \bigstar [\psi_0 - H^-]. \tag{3.12}$$

Of course, for a continuous distribution F the relations (3.11) and (3.12) are identical.

Various versions of this formula have been discovered independently by different methods and have caused much excitement. For a different variant see problem 19, and for the Fourier analytic equivalent see XVIII,3. The connection with the Wiener-Hopf techniques is discussed in section 3*a*. Wald's identity (2.8) is an easy consequence of (3.11). (See problem 11 as well as XVIII,2.)

(*b*) Explicit expressions for H and H^- are usually difficult to come by. An interesting distribution for which the calculations are particularly simple was found in example VI,8(*b*). If F is the convolution of two exponentials concentrated on $\overline{0, \infty}$ and $\overline{-\infty, 0}$, respectively, then it has a density of the form

$$f(x) = \begin{cases} \dfrac{ab}{a+b}\, e^{ax} & x < 0 \\[2ex] \dfrac{ab}{a+b}\, e^{-bx} & x > 0. \end{cases}$$

We suppose $b \leq a$, so that $\mathbf{E}(\mathbf{X}) \geq 0$. Then H and H^- have densities given by be^{-bx} and be^{ax}. Here $H \bigstar H^- = (b/a)F$, and (3.11) is trivially true. ▶

(For further explicit examples see problem 9.)

We now turn to the consideration of (3.9) as an integral equation for the unknown measures ρ and ψ. It will be shown that in the present context the solution is unique. For brevity we agree to say that a pair (ρ, ψ) is *probabilistically possible* if ρ is a possibly defective probability distribution concentrated on $\overline{-\infty, 0}|$, and $\psi - \psi_0$ a measure concentrated on $\overline{0, \infty}$ such that for each bounded interval I the measures $\psi\{I + t\}$ remain bounded. (The last condition follows from the renewal theorems since $\psi = \sum H^{n\bigstar}$.)

Theorem 1. *The convolution equation* (3.9) [*or, equivalently, the pair* (3.6)] *admits of exactly one probabilistically possible solution* (ρ, ψ).

This implies that ρ is the distribution of the point of first entry into $\overline{-\infty, 0}|$ and $\psi = \sum H^{n\star}$ where H is the distribution of the point of first entry into $\overline{0, \infty}$.

Proof. Let $\rho^\#$ and $\psi^\#$ be two non-negative measures satisfying (3.6), and $\psi^\# \geq \psi_0$. From (3.6*b*) we get by induction that $\psi^\# \geq \psi_0 + \cdots + \psi_n$ for every n, and hence our solution ψ is *minimal* in the sense that for any other solution $\psi^\#$ with a unit atom at the origin $\psi^\#\{I\} \geq \psi\{I\}$ for all intervals. In other words, $\delta = \psi^\# - \psi$ is a *measure*. From (3.6*a*) it is now seen that the same is true of $\gamma = \rho^\# - \rho$. Since both (ρ, ψ) and $(\rho^\#, \psi^\#)$ satisfy (3.9) we have

$$\delta + \gamma = \delta \bigstar F. \tag{3.13}$$

Let I be a fixed finite interval and put $z(t) = \delta\{I + t\}$. Two cases are possible. If ρ is a proper distribution the fact that $\rho^\# \geq \rho$ implies that $\rho^\# = \rho$ and hence $\gamma = 0$. Then z is a bounded solution of the convolution equation $z = F \bigstar z$ and hence by induction

$$z(t) = \int_{-\infty}^{+\infty} z(t-y)\, F^{n\star}\{dy\} \tag{3.14}$$

for all n. Now $z \geq 0$ and $z(t) = 0$ for every t such that $I + t$ is contained in the negative half-axis. For such t it is clear from (3.14) that $z(t - y) = 0$ for every y which is a point of increase for some $F^{n\star}$. By lemma 2 of V,4*a* the set of such y is everywhere dense and we conclude that z vanishes identically.

For a defective ρ we know only that $\gamma \geq 0$, and then (3.13) implies only that $z \leq F \bigstar z$. In this case (3.14) holds with the equality sign replaced by $\leq$. But the random walk then drifts to ∞ and so the mass of $F^{n\star}$ tends to concentrate near ∞. Again, $z(t-y) = 0$ for all sufficiently large y and so z must vanish identically. Thus $\psi^\# = \psi$ as asserted. ►

3a. THE WIENER-HOPF INTEGRAL EQUATION

To explain the connection between the integral equation (3.9) and the standard Wiener-Hopf equation it is best to begin by a probabilistic problem where the latter occurs.

Example. (*c*) *Distribution of maxima.* For simplicity let us assume that the distribution F has a density f and a negative expectation. The random walk $\{\mathbf{S}_n\}$ drifts to $-\infty$ and a finite-valued random variable

$$\mathbf{M} = \max\,[0, \mathbf{S}_1, \mathbf{S}_2, \ldots] \tag{3.15}$$

is defined with probability one. We propose to calculate its probability distribution $M(x) = \mathbf{P}\{\mathbf{M} \leq x\}$ which is by definition concentrated on $\overline{0, \infty}$. The event $\{\mathbf{M} \leq x\}$ occurs iff

$$\mathbf{X}_1 = y \leq x \quad \textit{and} \quad \max\,[0, \mathbf{X}_2, \mathbf{X}_2+\mathbf{X}_3, \ldots] \leq x-y.$$

Summing over all possible y we get

$$M(x) = \int_{-\infty}^{x} M(x-y)f(y)\,dy, \qquad x > 0 \tag{3.16}$$

which is the same as

$$M(x) = \int_{0}^{\infty} M(s)f(x-s)\,ds, \qquad x > 0. \tag{3.17}$$

On the other hand, we know from (2.7) that $M(x) = [1 - H(\infty)]\,\psi(x)$. We saw that ψ satisfies integral equation (3.7*b*) where under the present conditions $\rho(0) = 1$. A simple integration by parts now shows that (3.7*b*) and (3.17) are actually identical. ►

The standard form of the Wiener-Hopf integral equation is represented by (3.17) and our example illustrates the way in which it can occur in probability theory. General references to the Wiener-Hopf techniques are misleading, however, because the restriction to positive functions and measures changes (and simplifies) the nature of the problem.

The ingenious method[5] used by N. Wiener and E. Hopf to treat (3.17) has attracted wide attention and has been adapted to various probabilistic problems, for example, by H. Cramér for asymptotic estimates for probabilities of ruin. The method involves a formidable analytic apparatus and hence the ease with which these estimates are obtained from the present approach is almost disquieting. The deeper reason can be understood as follows. The equation (3.17) represents, at best, only one of the two equations (3.7), and when $\rho(0) < 1$ even less. Taken by itself (3.17) is much more difficult to handle than the pair (3.7). For example, the uniqueness theorem *fails* for (3.17) even if only probability distributions are admitted. In fact, the basic idea of the Wiener-Hopf technique consists in introducing an auxiliary function which in the general theory lacks any particular meaning. This *tour de force* in effect replaces the individual equation (3.17) by a pair equivalent to (3.7) but the uniqueness is lost. We proceeded in the opposite direction, starting from the obvious recursion system (3.5) for the probabilities connected with the two inseparable problems: the

[5] Dating back to 1931. A huge literature followed the first presentation in book form: E. Hopf, *Mathematical problems of radiative equilibrium*, Cambridge tracts, No. 31, 1934.

first entry to $\overline{-\infty, 0}|$ and the random walk restricted to $x > 0$ prior to this first entry. In this way we derived the integral equation (3.9) from the known solution, and the uniqueness of the *probabilistic* solution was easy to establish. The convergence proof, the properties of the solutions, as well as the connection between the distribution M of the maxima and the renewal measure ψ depend on the duality lemma.

The possibility of attacking the Wiener-Hopf equation (3.17) using the duality principle was noticed by F. Spitzer.[6] The usual way of connecting the Wiener-Hopf theory with probabilistic problems starts from formulas related to (9.3) below in their Fourier version to which we shall return in chapter XVIII. There exists now a huge literature relating Wiener-Hopf techniques to probabilistic problems and extending the scope of combinatorial methods. Most of this literature uses Fourier techniques.[7]

4. EXAMPLES

Explicit formulas for the first entry distributions are in general difficult to obtain. By a stroke of good fortune there is a remarkable exception to this rule, discussed in example (*a*). At first sight the distribution F of this example appears artificial, but the type turns up frequently in connection with Poisson processes, queuing theory, ruin problems, etc. Considering the extreme simplicity of our general results it is unbelievable how much ingenuity and analytical skill has been spent (often repeatedly) on individual special cases.

Example (*c*) exhibits (in a rather pedestrian fashion) the complete calculations in the case of an arithmetic F with rational generating functions. The calculations are given because the same method is used for rational Laplace or Fourier transforms. Another example is found in problems 3–6. Example (*b*) deals with a general relationship of independent interest.

We adhere to the notations of the preceding section. Thus H and ρ are the distributions of the point of first entry into $\overline{0, \infty}$ and $\overline{-\infty, 0}|$ respectively. (In other words, H and ρ are the distributions of the first strict ascending and the first weak descending ladder heights.) Finally, $\psi = \sum H^{n\star}$ is the renewal function corresponding to H. Our main tool is the equation (3.7*a*) stating that for $x < 0$ the distribution of the first

[6] *The Wiener-Hopf equation whose kernel is a probability density*, Duke Math. J., vol. 24 (1957) pp. 327–343.

[7] A meaningful short survey of the literature is impossible on account of the unsettled state of affairs and because the methodology of many papers suffers under the influence of accidents of historical developments. Generalizations beyond probability theory are illustrated by G. Baxter, *An operator identity*, Pacific J. Math., vol. 4 (1958) pp. 649–663.

entry into $\overline{-\infty, 0}|$ is given by

$$\rho(x) = \int_{0-}^{\infty} \psi\{dy\}\, F(x-y). \tag{4.1}$$

Examples. (*a*) *Distributions with one exponential tail* occur more frequently than might be expected. For example, in the random walk of example VI,8(*b*) and in the corresponding queuing process VI,9(*e*) *both* tails are exponential. Suppose, by way of introduction, that the *left* tail of F is exponential, that is, $F(x) = qe^{\beta x}$ for $x < 0$. Whatever ψ is, (4.1) shows that $\rho(x) = Ce^{\beta x}$ for $x < 0$ where C is a constant. Having made this discovery we interchange the role of the two half-axes (partly to facilitate reference to our formulas, partly with a view to the most important applications in queuing theory). Assume then that

$$F(x) = 1 - pe^{-\alpha x} \qquad \textit{for} \quad x \geq 0 \tag{4.2}$$

without any conditions imposed for $x < 0$. To avoid unnecessary complications we assume that F *has a finite expectation* μ and that F is continuous. It follows from the preliminary remark that the ladder height distribution H has a density proportional to $e^{-\alpha x}$. We now distinguish two cases.

(i) If $\mu \geq 0$ the distribution H is proper and hence for $x > 0$

$$H(x) = 1 - e^{-\alpha x}, \qquad \psi(x) = 1 + \alpha x. \tag{4.3}$$

[The latter follows trivially from $\psi = \sum H^{n\star}$ or the renewal equation (3.10).] From (4.1) we get

$$\rho(x) = F(x) + \alpha \int_{-\infty}^{x} F(s)\, ds, \qquad x < 0, \tag{4.4}$$

and thus we have explicit expressions for all desired probabilities. An easy calculation shows that

$$\rho(0) = 1 - \alpha\mu. \tag{4.5}$$

This is a special case of (2.8) because $(1 - \rho(0))^{-1} = \mathbf{E}(\mathcal{T}_1)$ by virtue of (2.5).

(ii) If $\mu < 0$, the relations (4.3) and (4.4) still represent a solution of the integral equation (3.9), but because of (4.5) it is probabilistically impossible when $\mu < 0$. For the correct solution we know that H has a density $h(x) = (\alpha - \kappa)e^{-\alpha x}$ where $0 < \kappa < \alpha$ because H is defective. An easy calculation shows that $\psi'(x) = (\alpha - \kappa)e^{-\kappa x}$ for $x > 0$. The unknown constant κ is obtained from the condition that $\rho(0) = 1$. A routine calculation shows that κ must be the unique positive root of the equation (4.6). Given the root of this transcendental equation we have again explicit formulas for H, ρ, and ψ.

The reader will easily verify that the same theory applies when the variables $\mathbf{X}_1, \mathbf{X}_2, \ldots$ of the random walk are integral-valued and the distribution F has a *geometric right tail*, that is, if F attributes to the integer $k > 0$ the weight $q\beta^k$.

(*b*) *Associated random walks.* Suppose that F has an expectation $\mu \neq 0$ and that there exists a number $\kappa \neq 0$ such that

$$\int_{-\infty}^{+\infty} e^{\kappa y} F\{dy\} = 1. \tag{4.6}$$

Given an arbitrary measure γ on the line we associate with it a new measure ${}^a\gamma$ defined by

$${}^a\gamma\{dy\} = e^{\kappa y}\,\gamma\{dy\}. \tag{4.7}$$

The measure ${}^aF$ associated with F is again a proper probability distribution and we say that the random walks generated by ${}^aF$ and F are *associated with each other.*[8] It is easily seen that the n-fold convolution of ${}^aF$ with itself is associated with $F^{n\star}$ so that the notation ${}^aF^{n\star}$ is unambiguous. Furthermore, the recursion formulas (3.5) show that the transforms ${}^a\rho_n$ and ${}^a\psi_n$ have the same probabilistic meaning in the new random walk as ρ_n and ψ_n in the old one. It follows generally that the transforms ${}^a\rho$, ${}^aH$, ${}^a\psi$, etc., *have the obvious meaning for the random walk associated with* ${}^aF$. [This can be seen also directly from the integral equation (3.9).]

The integral

$$\phi(t) = \int_{-\infty}^{+\infty} e^{yt}\, F\{dy\}$$

exists for all t between 0 and κ and in this interval φ may be differentiated indefinitely. The second derivative being positive, φ is a convex function. If $\varphi'(\kappa)$ exists, the fact that $\varphi(0) = \varphi(\kappa)$ implies that $\varphi'(0)$ and $\varphi'(\kappa)$ have opposite signs. The random walks induced by F and ${}^aF$ have therefore drifts in opposite directions. (This remains obviously true even in the exceptional case that ${}^aF$ has no finite expectation.)

We have thus devised a widely applicable method of translating facts about a random walk with $\mu < 0$ into results for a random walk with positive expectation, and vice versa.

If $\mu < 0$ the ladder height distribution H is defective, but ${}^aH$ is a proper distribution. This means that

$$\int_0^{\infty} e^{\kappa y}\, H\{dy\} = 1. \tag{4.8}$$

[8] This notion was used by A. Khintchine, A. Wald, and others but was never fully exploited. The transformation (4.7) was used for renewal theory in XI,6 and (in a form disguised by the use of generating functions) in theorem 1(iii) of **1**; XIII,10 and will be used for Laplace transforms in XIII,(1.8). The equation (4.6) serves also in the Wiener-Hopf theory.

The power of the method of associated random walks derives largely from this remark. Indeed, we know from XI,6 that excellent asymptotic estimates are available for the ascending ladder process if one knows the root of the equation (4.8). These estimates would be illusory if they required a knowledge of H, but we see now that *the roots of the two equations* (4.6) *and* (4.8) *are identical.*

(*c*) *Bounded arithmetic distributions.* Let a and b be positive integers and let F be an arithmetic distribution with span 1 and jumps f_k at $k = -b, \ldots, a$. The measures ψ and ρ are also concentrated on integers and we denote their jumps at k by ψ_k and ρ_k, respectively. The first entry into $\overline{-\infty, 0}$ occurs at an integer $\geq -b$ and so $\rho_k = 0$ for $k < -b$. We introduce the generating functions

$$(4.9) \qquad \Phi(s) = \sum_{k=-b}^{a} f_k s^k, \qquad \Psi(s) = \sum_{k=0}^{\infty} \psi_k s^k, \qquad R(s) = \sum_{k=-b}^{0} \rho_k s^k.$$

They differ from those of **1**; XI in that Φ and R involve also negative powers of s, but it is clear that the basic properties and rules remain unchanged. In particular, $\mu = \Phi'(1)$ is the expectation of F. To a convolution of distributions there corresponds the product of the generating functions, and so the basic integral equation (3.9) is equivalent to $\Psi + R = 1 + \Psi\Phi$ or

$$(4.10) \qquad \Psi(s) = \frac{s^b(R(s)-1)}{s^b(\Phi(s)-1)}.$$

The numerator and denominator are polynomials of degrees b and $a + b$, respectively. The power series on the left is regular for $|s| < 1$, and so all roots of the denominator located within the unit circle must cancel against roots of the numerator. We proceed to show that this requirement uniquely determines R and Ψ.

For concreteness suppose $\mu = 0$. (For $\mu \neq 0$ see problems 12–13.) Then $s = 1$ is a double root of the equation $\Phi(s) = 1$. For $|s| = 1$ we have $|\Phi(s)| \leq \Phi(1) = 1$ where the inequality sign holds only if $s^k = 1$ for every k such that $f_k > 0$. As the distribution F is assumed to have span 1 this is true only if $s = 1$, and hence no other roots of $\Phi(s) = 1$ are located on the unit circle itself. To discover how many roots are located in the interior we consider the polynomial of degree $a + b$ defined by

$$P(s) = s^b[\Phi(s) - q], \qquad q > 1.$$

For $|s| = 1$ we have $|P(s)| \geq q - 1 > 0$ and

$$|P(s) + qs^b| = |\Phi(s)| \leq 1 < q\,|s^b|.$$

By Rouché's theorem[9] this implies that the polynomials $P(s)$ and qs^b have the same number of zeros located inside the unit circle. It follows that P has exactly b roots with $|s| < 1$ and a roots with $|s| > 1$. Now $P(0) = 0$ and $P(1) = 1 - q < 0$ while $P(s) > 0$ for large s, and so P has two real roots $s' < 1 < s''$. As $q \to 1$ the roots of P tend to the roots of $\Phi(s) = 1$. Thus we conclude finally that the denominator in (4.10) has 1 as a double root, and furthermore $b - 1$ roots $s_1, \ldots, s_{b-1}$ with $|s_j| < 1$ and $a - 1$ roots $\sigma_1, \ldots, \sigma_{a-1}$ with $|\sigma_j| > 1$. Then the denominator is of the form

$$(4.11)\quad s^b(\Phi(s)-1) = C(s-1)^2(s-s_1)\cdots(s-s_{b-1})(s-\sigma_1)\cdots(s-\sigma_{a-1}).$$

The roots $s_1, \ldots, s_{b-1}$ must cancel against those of the numerator, and since the coefficients ψ_n remain bounded the same is true of *one* root $s = 1$. This determines Ψ up to a multiplicative constant. But by definition $\Psi(0) = 1$ and hence we have the *desired explicit formula*

$$(4.12)\qquad \Psi(s) = \frac{1}{(1-s)(1-s/\sigma_1)\cdots(1-s/\sigma_{a-1})}.$$

Expansion into partial fractions leads to explicit expressions for ψ_n, the great advantage of this method being that the knowledge of the dominant root leads to reasonable asymptotic estimates (see **1**; XI,4).

For the generating function R of the first entrance probabilities ρ_k we get from (4.10) and (4.12)

$$(4.13)\quad R(s) = 1+C\cdot(-1)^{a-1}\sigma_1\cdots\sigma_{a-1}(1-1/s)(1-s_1/s)\cdots(1-s_{b-1}/s).$$

[The coefficient C is defined in (4.11) and depends only on the given distribution $\{f_k\}$.] Again a partial fraction expansion leads to asymptotic estimates. (Continued in problems 12–15.)

5. APPLICATIONS

It was shown in VI,9 that a basic problem of queuing theory consists in finding the distribution M of

$$(5.1)\qquad \mathbf{M} = \max\,[0, \mathbf{S}_1, \ldots]$$

in a random walk with variables $\mathbf{X}_k$ such that $\mu = \mathbf{E}(\mathbf{X}_k) < 0$. Examples VI,9($a$) to ($c$) show that the same problem turns up in other contexts, for example in connection with ruin problems in compound Poisson

[9] See, for example, E. Hille, *Analytic function theory*, vol. I, section 9.2. (Ginn and Co., 1959.)

processes. In this case, as well as in queuing theory, the underlying distribution is of the form

$$F = A \bigstar B \tag{5.2}$$

where A is concentrated on $\overline{0, \infty}$ and B on $\overline{-\infty, 0}$. We suppose that A and B have finite expectations a and $-b$, so that F has expectation $\mu = a - b$. We suppose also that F is *continuous* so as to avoid the tedious distinction between strict and weak ladder variables.

As in the preceding two sections we denote the ascending and descending ladder height distributions by H and ρ respectively. (For consistency we should write H^- for ρ.) In other words, H and ρ are the distributions of the point of first entry into $\overline{0, \infty}$ and $\overline{-\infty, 0}$ (and also into corresponding closed intervals). It was shown in example 3(*c*) and in (2.7) that *if* $\mu < 0$

$$M(x) = \frac{\psi(x)}{\psi(\infty)} = [1 - H(\infty)] \sum_0^\infty H^{n\bigstar}(x). \tag{5.3}$$

Example 4(*a*) contains an *explicit formula*[10] *for this distribution if one of the tails of* F *is exponential*, that is,

$$F(x) = 1 - pe^{-\alpha x} \qquad \textit{for} \quad x > 0, \tag{5.4}$$

or else $F(x) = qe^{\alpha x}$ for $x < 0$.

By extreme good luck *the condition* (5.4) *holds if* F *is of the form* (5.2) with

$$A(x) = 1 - e^{-\alpha x} \qquad \textit{for} \quad x > 0. \tag{5.5}$$

Then

$$p = \int_{-\infty}^{0} e^{\alpha y}\, B\{dy\}. \tag{5.6}$$

Our simple results are therefore applicable in queuing theory whenever *either* the incoming traffic is Poissonian *or* the service time is exponential. Furthermore, the ruin problem in compound Poisson processes is covered by the present conditions. There exists an immense applied literature treating special problems under various assumptions on the distribution B, sometimes, as in ruin problems, in a disguised form. As it turns out, greater generality and much greater simplicity can be achieved by using only the

---

[10] Another explicit formula is contained in example 4(*c*) for the case of an arithmetic distribution F with finitely many atoms. This explicit formula is too unwieldy to be practical, but an expansion into partial fractions leads to good *asymptotic estimates if the dominant root of the denominator is known*. The same method applies to Fourier transforms whenever the *characteristic function of* F *is rational*. This remark covers many special cases treated in the literature.

condition (5.4) instead of the combination (5.5) and (5.2). We see here a prime example of the economy of thought inherent in a general theory where one's view is not obscured by accidents of special cases.

Examples. (*a*) *The Khintchine-Pollaczek formula.* Suppose that F is of the form (5.2) with A given by (5.5) and $\mu = 1/\alpha - b > 0$. The random walk drifts to ∞ and we have to replace the maximum in (5.1) by the minimum. This means replacing H in (5.3) by the distribution ρ given in (4.4). A simple integration by parts shows that for $x < 0$

$$\rho(x) = \alpha \int_{-\infty}^{x} B(y)\, dy. \tag{5.7}$$

Hence $\rho(0) = \alpha b$ and so for $x < 0$

$$\mathbf{P}\{\min(\mathbf{S}_0, \mathbf{S}_1, \ldots) \leq x\} = (1 - \alpha b) \sum_{0}^{\infty} \rho^{n\star}(x). \tag{5.8}$$

This is the celebrated Khintchine-Pollaczek formula, which has been rediscovered time and again in special situations, invariably using Laplace transforms [which method is inapplicable for the more general distributions of the form (5.4)]. We return to it in problems 10–11 of XVIII,7.

(*b*) *The dual case.* Consider the same distributions as in the last example but with $\mu < 0$. As was shown in the second part of example 4(*a*) in this case

$$\mathbf{P}\{\max(\mathbf{S}_0, \mathbf{S}_1, \ldots) \leq x\} = \frac{\kappa}{\alpha}\, \psi(x) = 1 - \left(1 - \frac{\kappa}{\alpha}\right) e^{-\kappa x} \tag{5.9}$$

where κ is the unique positive root of the "characteristic equation" (4.6). [This result can be obtained also by the method of associated random walks recalling the fact that when $\mu \geq 0$ one has $\psi(x) = 1 + \alpha x$ for $x > 0$.] In queuing theory (5.9) implies that *at a server with exponential servicing times the distribution of the waiting times tends to an exponential limit.*

(*c*) *Asymptotic estimates.* The method of associated random walk described in section 4*b* leads easily to useful estimates for the tail of the distribution

$$M(x) = \mathbf{P}\{\max(\mathbf{S}_0, \mathbf{S}_1, \ldots) \leq x\}. \tag{5.10}$$

The following simple method replaces many complicated calculations used for special problems in the applied literature. It represents a special case of the general theory in XI,6, but for convenience the following exposition is self-contained.

The distribution M is given by (5.3). In it ψ stands for the renewal measure corresponding to the defective distribution H. To the associated proper distribution ${}^{a}H$ there corresponds the renewal measure ${}^{a}\psi$ given by ${}^{a}\psi\{dx\} = e^{\kappa x}\psi\{dx\}$, with κ given by (4.6) or (4.8). Hence (5.3) may be rewritten in the form

$$M\{dx\} = [1-H(\infty)]e^{-\kappa x}\cdot {}^a\psi\{dx\}. \tag{5.11}$$

By the basic renewal theorem of XI,1 the renewal measure ${}^a\psi$ is asymptotically uniformly distributed with a density β^{-1} where

$$\beta = \int_0^\infty xe^{kx}H\{dx\}. \tag{5.12}$$

Integrating (5.11) between t and ∞ we see therefore that as $t\to\infty$

$$1 - M(t) \sim \frac{1-H(\infty)}{\beta\kappa}\, e^{-\kappa t} \tag{5.13}$$

provided only that $\beta < \infty$. [Otherwise $1 - M(t) = o(e^{-\kappa t})$.]

The constant β depends on the distribution H which is usually not known explicitly, but the exponent κ depends only on the given distribution F. At worst therefore (5.13) represents an estimate involving an unknown factor, and even this result is not easily obtained by other methods. The next example illustrates important applications.

(*d*) *Cramér's estimate for probabilities of ruin.* We now apply the preceding result to example (*a*). Here the drift is toward $+\infty$, and hence the roles of the positive and negative half-axes must be interchanged. This means that $\kappa < 0$, and the distribution H is to be replaced by the distribution (5.7) of the first entry into $\overline{-\infty, 0}$. Thus (5.12) takes on the form

$$\beta = \alpha\int_{-\infty}^0 e^{-|\kappa|y}\,|y|\,B(y)\,dy. \tag{5.14}$$

We saw that $\rho(0) = \alpha b$, and so (5.13) is equivalent to the statement that as $x\to\infty$

$$\mathbf{P}\{\min(\mathbf{S}_0, \mathbf{S}_1, \ldots) \le x\} \sim \frac{1-\alpha b}{|\kappa|\,\beta}\, e^{|\kappa|x}. \tag{5.15}$$

This formula has many applications. In queuing theory the left side represents the limit distribution for the waiting time of the nth customer (see the theorem in VI,9). In example VI,9(*d*) it was shown that the basic ruin problem of VI,5 may be reduced to this queuing problem. With different notations the same problem is treated in example XI,7(*a*).[11] It is therefore not surprising that the estimate (5.15) has been derived repeatedly under special circumstances, but the problem becomes simpler in its natural general setting.

[11] To our defective distribution ρ, which is concentrated on $\overline{-\infty, 0}$, there corresponds in XI,7 the defective distribution L with *density* $(\alpha/c)(1 - F(x))$ concentrated on $\overline{0, \infty}$.

From the point of view of the general theory (5.15) is equivalent to a famous estimate for the probability of ruin in the theory of risk due to H. Cramér.[12] ▶

6. A COMBINATORIAL LEMMA

The distribution of ladder epochs depends on a simple combinatorial lemma, and the probabilistic part of the argument will appear clearer if we isolate this lemma.

Let $x_1, \ldots, x_n$ be n numbers, and consider their partial sums

$$s_0 = 0, \ldots, s_n = x_1 + \cdots + x_n.$$

We say that $\nu > 0$ *is a ladder index* if $s_\nu > s_0, \ldots, s_\nu > s_{\nu-1}$, that is, if s_ν exceeds all preceding partial sums. There are n ladder indices if all x_ν are positive, whereas there are none if all x_ν are negative.

Consider the n cyclical reorderings $(x_1, \ldots, x_n)$, $(x_2, \ldots, x_n, x_1), \ldots,$ $(x_n, x_1, \ldots, x_{n-1})$ and number them from 0 to $n - 1$. The partial sums $s_k^{(\nu)}$ in the arrangement number ν are given by

$$(6.1) \qquad s_k^{(\nu)} = \begin{cases} s_{\nu+k} - s_\nu & \text{for } k = 1, \ldots, n-\nu \\ s_n - s_\nu + s_{k-n+\nu} & \text{for } k = n-\nu+1, \ldots, n. \end{cases}$$

Lemma 1. *Suppose* $s_n > 0$. *Denote by* r *the number of cyclical rearrangements in which* n *is a ladder index. Then* $r \geq 1$, *and in each such cyclical arrangement there are exactly* r *ladder indices.*

Examples. For $(-1, -1, -1, 0, 1, 10)$ we have $r = 1$: the given order is the only one in which the last partial sum is maximal. For $(-1, 4, 7, 1)$ we have $r = 3$; the permutations number 0, 2, and 3 yield 3 ladder indices each. ▶

Proof. Choose ν so that s_ν is maximal, and if there are several such indices choose ν as small as possible. In other words,

$$(6.2) \qquad s_\nu > s_1, \ldots, s_\nu > s_{\nu-1}, \qquad s_\nu \geq s_{\nu+1}, \ldots, s_\nu \geq s_n.$$

It is then seen from (6.1) that in the νth permutation the last partial sum is strictly maximal and so n is a ladder index. Thus $r \geq 1$. Without loss of generality we now suppose that n is a ladder index in the original arrangement, that is, $s_n > s_j$ for all j. The quantities in the first line in (6.1) are then $< s_n$, and the second line in (6.1) shows that n is a ladder index also in the νth permutation iff $s_\nu > s_1, \ldots, s_\nu > s_{\nu-1}$, that is, iff ν is a ladder index in the original arrangement. Thus the number of

[12] For a newer derivation by Wiener-Hopf techniques in the complex plane see Cramér's paper cited in VI,5. Our (5.15) is Cramér's (57).

permutations in which n is a ladder index equals the number of ladder indices, and the lemma is proved. ►

Weak ladder indices are defined analogously except that the strict inequality $>$ is replaced by $\geq$. The preceding argument applies to them and leads to

Lemma 2. *If* $s_n \geq 0$, *lemma* 1 *applies also to weak ladder indices.*

7. DISTRIBUTION OF LADDER EPOCHS

In the preceding sections we have focused our attention on the ladder height, but now we turn to the ladder epochs. Let

$$\tau_n = \mathbf{P}\{\mathbf{S}_1 \leq 0, \ldots, \mathbf{S}_{n-1} \leq 0, \mathbf{S}_n > 0\}. \tag{7.1}$$

This is the probability that the first entry into $\overline{0, \infty}$ occurs at the nth step, and so $\{\tau_n\}$ is the (possibly defective) distribution of the first ladder epoch $\mathcal{T}_1$. We introduce its generating function

$$\tau(s) = \sum_{n=1}^{\infty} \tau_n s^n, \qquad 0 \leq s \leq 1. \tag{7.2}$$

The following remarkable theorem shows that the distribution $\{\tau_n\}$ is completely determined by the probabilities $\mathbf{P}\{\mathbf{S}_n > 0\}$ and vice versa. It was discovered by E. Sparre Andersen whose ingenious but extremely complicated proof was gradually simplified by several authors. We derive it as a simple corollary to our combinatorial lemma. [A stronger version is contained in (9.3) and will be treated by Fourier methods in chapter XVIII.]

Theorem 1:

$$\log \frac{1}{1 - \tau(s)} = \sum_{n=1}^{\infty} \frac{s^n}{n} \mathbf{P}\{S_n > 0\}. \tag{7.3}$$

Note: The theorem and its proof remain valid if in (7.1) and (7.3) the signs $>$ and $\leq$ are replaced by $\geq$ and $<$, respectively. In this case $\{\tau_n\}$ stands for the distribution of the first weak ladder epoch.

Proof. For each sample point consider the n cyclical permutations $(\mathbf{X}_\nu, \ldots, \mathbf{X}_n, \mathbf{X}_1, \ldots, \mathbf{X}_{\nu-1})$ and denote the corresponding partial sums by $\mathbf{S}_0^{(\nu)}, \ldots, \mathbf{S}_n^{(\nu)}$. Fix an integer r and define n random variables $\mathbf{Y}^{(\nu)}$ as follows: $\mathbf{Y}^{(\nu)} = 1$ if n is the rth ladder index for $(\mathbf{S}_1^{(\nu)}, \ldots, \mathbf{S}_n^{(\nu)})$ and $\mathbf{Y}^{(\nu)} = 0$ otherwise. To $\nu = 1$ there corresponds the unpermuted sequence $(\mathbf{S}_0, \ldots, \mathbf{S}_n)$ and hence

$$\mathbf{P}\{\mathbf{Y}^{(1)} = 1\} = \tau_n^{(r)} \tag{7.4}$$

where $\{\tau_n^{(r)}\}$ is the distribution of the rth ladder epoch. This epoch is the sum of r independent random variables distributed as $\mathcal{T}_1$, and hence $\tau_n^{(r)}$ is the coefficient of s^n in the rth power $\tau^r(s)$.

For reasons of symmetry the variables $\mathbf{Y}^{(\nu)}$ have a common distribution; since they assume only the values 0 and 1 we conclude from (7.4) that

$$\tau_n^{(r)} = \mathbf{E}(\mathbf{Y}^{(1)}) = \frac{1}{n}\mathbf{E}(\mathbf{Y}^{(1)}+\cdots+\mathbf{Y}^{(n)}). \tag{7.5}$$

By our last lemma the sum $\mathbf{Y}^{(1)} + \cdots + \mathbf{Y}^{(n)}$ can assume only the values 0 or r, and hence

$$\frac{1}{r}\tau_n^{(r)} = \frac{1}{n}\mathbf{P}\{\mathbf{Y}^{(0)} + \cdots + \mathbf{Y}^{(n)} = r\}. \tag{7.6}$$

For fixed n and $r = 0, 1, \ldots$ the events on the right are mutually exclusive and their union is the event $\{\mathbf{S}_n > 0\}$. Summing over r we get therefore

$$\sum_{r=1}^{\infty}\frac{1}{r}\tau_n^{(r)} = \frac{1}{n}\mathbf{P}\{\mathbf{S}_n > 0\} \tag{7.7}$$

On multiplying by s^n and summing over n we obtain

$$\sum_{r=1}^{\infty}\frac{1}{r}\tau(s) = \sum_{n=1}^{\infty}\frac{s^n}{n}\mathbf{P}\{\mathbf{S}_n > 0\} \tag{7.8}$$

which is the same as the assertion (7.3). ►

Corollary. *If F is continuous and symmetric, then*

$$\tau(s) = 1-\sqrt{1-s}. \tag{7.9}$$

Proof. All the probabilities occurring in (7.3) equal and so the right side equals $\log(1/\sqrt{1-s})$. ►

It is of interest to generalize this result assuming only that

$$\mathbf{P}\{\mathbf{S}_n > 0\} \to \tfrac{1}{2}. \tag{7.10}$$

Such is the case whenever the distribution of $\mathbf{S}_n/a_n$ tends to the normal distribution $\mathfrak{N}$. We shall assume a trifle more than (7.10), namely that the series

$$\sum_{n=1}^{\infty}\frac{1}{n}[\mathbf{P}\{\mathbf{S}_n > 0\} - \tfrac{1}{2}] = c \tag{7.11}$$

converges (not necessarily absolutely). It will be shown in XVIII,5 that (7.11) *holds whenever F has zero expectation and a finite variance.*

The following theorem is given not only because of its intrinsic interest, but also as an illustration for the use of refined Tauberian theorems.

Theorem 1a. *If* (7.11) *holds then*

$$\mathbf{P}\{\mathcal{T}_1 > n\} \sim \frac{1}{\sqrt{\pi}} e^{-c} \frac{1}{\sqrt{n}}. \tag{7.12}$$

Thus when F has zero expectation and a finite variance the distribution $\{\tau_n\}$ is very similar to the one encountered in the binomial random walk.

Proof. From (7.3) we see that as $s \to 1$

$$\log \frac{\sqrt{1-s}}{1-\tau(s)} = \sum_{n=1}^{\infty} \frac{s^n}{n} [\mathbf{P}\{\mathbf{S}_n > 0\} - \tfrac{1}{2}] \to c. \tag{7.13}$$

It follows that

$$\frac{1-\tau(s)}{1-s} \sim e^{-c} \frac{1}{\sqrt{1-s}}. \tag{7.14}$$

On the left one recognizes the generating function of the probabilities in (7.12). These decrease monotonically, and hence (7.12) is true by the last part of the Tauberian theorem 5 in XIII,5. ▶

Theorem 2. *The random walk drifts to* $-\infty$ *iff*

$$\sum_{n=1}^{\infty} \frac{1}{n} \mathbf{P}\{\mathbf{S}_n > 0\} < \infty. \tag{7.15}$$

This criterion remains valid[13] *with* $\{\mathbf{S}_n > 0\}$ *replaced by* $\{\mathbf{S}_n \geq 0\}$.

Proof. Drift to $-\infty$ takes place iff the ascending ladder processes are terminating that is, iff the distribution of $\mathcal{T}_1$ is defective. This is the same as $\tau(1) < 1$, and in this case the two sides in (7.3) remain bounded as $s \to 1$. The condition (7.15) is seen to be necessary and sufficient. The same argument applied to weak ladder epochs justifies the concluding assertion. ▶

We know that drift to $-\infty$ takes place if F has an expectation $\mu < 0$, but it is not analytically obvious that $\mu < 0$ implies (7.15). The verification of this fact provides an excellent technical exercise of methodological interest. (See problem 16.)

This theorem has surprising implications.

Examples. (*a*) Let F be a strictly stable distribution with $F(0) = \delta < \frac{1}{2}$. Intuitively one would expect a drift toward ∞, but in fact the random walk is of the oscillating type. Indeed, the series (7.15) reduces to $(1-\delta)\sum n^{-1}$ and diverges. Thus $\mathcal{T}_1$ is proper. But the same argument applies to the negative half-axis and shows that also the descending ladder variable $\mathcal{T}_1^-$ is proper.

(*b*) Let F stand for the symmetric Cauchy distribution and consider the random walk generated by the variables $\mathbf{X}_n' = \mathbf{X}_n + 1$. The median of the

[13] We shall see that $\sum n^{-1}\mathbf{P}\{\mathbf{S}_n = 0\} < \infty$ under any circumstances [see 9(*c*)].

sums $\mathbf{S}'_n = \mathbf{S}_n + n$ lies at n, and intuitively one should expect a strong drift to ∞. Actually the probabilities $\mathbf{P}\{\mathbf{S}'_n > 0\}$ are again independent of n, and as in the preceding example we conclude that the random walk is of the oscillating type. ►

Theorem 3. *The ladder epoch* $\mathcal{T}_1$ *has a finite expectation (and is proper) iff the random walk drifts to* ∞. *In this case*

$$\log \mathbf{E}(\mathcal{T}_1) = \log \sum k\tau_k = \sum_{n=1}^{\infty} \frac{1}{n} \mathbf{P}\{\mathbf{S}_n \leq 0\}. \tag{7.16}$$

(The series diverges in all other cases).

Proof. Subtracting the two sides of (7.3) from $\log (1-s)^{-1}$ we get for $0 < s < 1$

$$\log \frac{1 - \tau(s)}{1 - s} = \sum_{n=1}^{\infty} \frac{s^n}{n} [1 - \mathbf{P}\{S_n > 0\}]. \tag{7.17}$$

As $s \to 1$ the left side converges iff $\mathcal{T}_1$ is proper and has a finite expectation. The right side tends to the right side in (7.16), and by theorem 2 this series converges iff the random walk drifts to ∞. ►

In conclusion we show that the generating function occurring in theorem 1 has an alternative probabilistic interpretation which will lead directly to the amazing arc sine laws.

Theorem 4. *The generating function of the probabilities*

$$p_n = \mathbf{P}\{\mathbf{S}_1 > 0, \mathbf{S}_2 > 0, \ldots, \mathbf{S}_n > 0\} \tag{7.18}$$

is given by

$$p(s) = \frac{1}{1 - \tau(s)} \tag{7.19}$$

that is,

$$\log p(s) = \sum_{n=1}^{\infty} \frac{s^n}{n} \mathbf{P}\{\mathbf{S}_n > 0\}. \tag{7.20}$$

For reasons of symmetry the probabilities

$$q_n = \mathbf{P}\{\mathbf{S}_1 \leq 0, \ldots, \mathbf{S}_n \leq 0\} \tag{7.21}$$

have the generating function q *given by*

$$\log q(s) = \sum_{n=1}^{\infty} \frac{s^n}{n} \mathbf{P}\{\mathbf{S}_n \leq 0\}. \tag{7.22}$$

(Cf. problem 21.)

Proof. We use the duality lemma of section 2. From the theory of recurrent events it is clear that (7.19) is the generating function of the probabilities p_n that n is a ladder epoch, that is

$$p_n = \mathbf{P}\{\mathbf{S}_n > \mathbf{S}_0, \ldots, \mathbf{S}_n > \mathbf{S}_{n-1}\}. \tag{7.23}$$

Reversing the order of the variables $\mathbf{X}_j$ we get the dual interpretation (7.18). [This is really contained in (2.1) when $I = \overline{0, \infty}$.] ►

8. THE ARC SINE LAWS

One of the surprising features of the chance fluctuations in coin tossing finds its expression in the two arc sine laws (**1**; III,4 and 8). One of them implies that the number of positive terms in the sequence $\mathbf{S}_1, \ldots, \mathbf{S}_n$ is more likely to be relatively close to 0 or n than to $n/2$ as one would naïvely expect. The second implies the same for the position of the maximal term. *We show now that these laws are valid for arbitrary symmetric and for many other distributions.* This discovery proves the general relevance and applicability of the discussions of **1**; III.

In the following we have to cope with the nuisance that the maximum may be assumed repeatedly and that partial sums may vanish. These possibilities can be disregarded if F is continuous, for then the probability is zero that any two partial sums are equal. (Readers are advised to consider only this case.) For the general theory we agree to consider the index of the *first* maximum, that is, the index k such that

$$\mathbf{S}_k > \mathbf{S}_0, \ldots, \mathbf{S}_k > \mathbf{S}_{k-1}, \qquad \mathbf{S}_k \geq \mathbf{S}_{k+1}, \ldots, \mathbf{S}_k \geq \mathbf{S}_n. \tag{8.1}$$

Here n is fixed and k runs through the values $0, 1, \ldots, n$. The event (8.1) must occur for some $k \leq n$, and so we may define the (proper) random variable $\mathbf{K}_n$ *as the index of the first maximum*, that is, the index where (8.1) occurs. Here $(\mathbf{S}_0 = 0)$.

The event (8.1) requires the simultaneous realization of the two events $\{\mathbf{S}_k > \mathbf{S}_0, \ldots, \mathbf{S}_k > \mathbf{S}_{k-1}\}$ and $\{\mathbf{S}_{k+1} - \mathbf{S}_k \leq 0, \ldots, \mathbf{S}_n - \mathbf{S}_k \leq 0\}$. The first involves only $\mathbf{X}_1, \ldots, \mathbf{X}_k$, the second only $\mathbf{X}_{k+1}, \ldots, \mathbf{X}_n$, and hence the two events are independent. But these are the events occurring in the last theorem and so we have proved

Lemma 1. *For all* k, n

$$\mathbf{P}\{\mathbf{K}_n = k\} = p_k q_{n-k}. \tag{8.2}$$

Suppose now that $\mathbf{P}\{\mathbf{S}_n > 0\} = \mathbf{P}\{\mathbf{S}_n \leq 0\} = \frac{1}{2}$ for all n. The right sides in (7.15) and (7.17) then reduce to $\frac{1}{2}\log(1-s)^{-1}$, and hence

$$p(s) = q(s) = 1/\sqrt{1-s}.$$

Thus

$$p_k q_{n-k} = \binom{-\frac{1}{2}}{k}\binom{-\frac{1}{2}}{n-k}(-1)^n, \tag{8.3}$$

and this may be written in the more pleasing form

$$p_k q_{n-k} = \binom{2k}{k}\binom{2n-2k}{n-k}\frac{1}{2^{2n}}. \tag{8.4}$$

This expression was used in **1**; III,(4.1) to define the discrete arc sine distribution which was found in **1**; III,4 and 8 to govern various random variables connected with coin tossing; its limiting form was derived in **1**; III,(4.4). Using in particular the arc sine law of **1**; III,8f we can state

Theorem 1. *If F is symmetric and continuous, the probability distribution of $\mathbf{K}_n$ (the index of the first maximum in $\mathbf{S}_0, \mathbf{S}_1, \ldots, \mathbf{S}_n$) is the same as in the coin tossing game. It is given by* (8.3) *or* (8.4). *For fixed $0 < \alpha < 1$ as $n \to \infty$*

$$\mathbf{P}\{\mathbf{K}_n < n\alpha\} \to 2\frac{1}{\pi}\text{ arc sin }\sqrt{\alpha}. \tag{8.5}$$

The limit distribution has the density $1/[\pi\sqrt{\alpha(1-\alpha)}]$ which is unbounded at the endpoints 0 and 1 and has its minimum at the midpoint $\frac{1}{2}$. This shows that the reduced maximum $\mathbf{K}_n/n$ is much more likely to be close to 0 or 1 than to $\frac{1}{2}$. For a fuller discussion see **1**; III,4 and 8. For an alternative form of the theorem see problem 22.

This theorem can be generalized just as theorem 1 of the preceding section.

Theorem 1a.[14] *If the series*

$$\sum_{n=1}^{\infty}\frac{1}{n}\,[\mathbf{P}\{\mathbf{S}_n > 0\} - \tfrac{1}{2}] = c \tag{8.6}$$

converges, then as $n \to \infty$ and $n - k \to \infty$

$$\mathbf{P}\{\mathbf{K}_n = k\} \sim \binom{2k}{k}\binom{2n-2k}{n-k}\frac{1}{2^{2n}}. \tag{8.7}$$

and hence the arc sine law (8.5) *holds.*

It will be shown in XVIII,5 that the series (8.6) converges whenever F has zero expectation and finite variance. The arc sine laws therefore hold for such distributions.

Proof. From (7.20) and the elementary theorem of Abel on power series we conclude that as $s \to 1$

$$\log\,(p(s)\sqrt{1-s}\,) = \sum_{n=1}^{\infty}\frac{s^n}{n}\,[\mathbf{P}\{\mathbf{S}_n > 0\} - \tfrac{1}{2}] \to c, \tag{8.8}$$

[14] This theorem was proved by laborious calculations by Sparre-Andersen. The remark that the Tauberian theorem removes all trouble is due to Spitzer. For a generalization see 9.d.

and so

$$p(s) \sim e^c \cdot (1-s)^{-\frac{1}{2}} \tag{8.9}$$

By the definition (7.18) the p_n decrease monotonically, and hence the last part of the Tauberian theorem 5 of XIII,5 implies that coefficients of the two power series in (8.9) exhibit the same asymptotic behavior. Thus

$$p_n \sim e^c \binom{-\frac{1}{2}}{n} (-1)^n, \qquad n \to \infty \tag{8.10}$$

For q_n we get the same relation with c replaced by $-c$, and hence the assertion (8.7) follows from (8.2). The derivation of the arc sine law depends only on the asymptotic relation (8.7) and not on the identity (8.4). ▶

Theorem 1 and its proof carry over to arbitrary strictly stable distributions. If $\mathbf{P}\{\mathbf{S}_n > 0\} = \delta$ is independent of n we get from (7.20) and (7.22)

$$p(s) = (1-s)^{-\delta}, \qquad q(s) = (1-s)^{\delta-1} \tag{8.11}$$

and hence

$$\mathbf{P}\{\mathbf{K}_n = k\} = p_k q_{n-k} = (-1)^n \binom{-\delta}{k} \binom{\delta-1}{n-k}. \tag{8.12}$$

The limit theorem (8.5) holds with the arc sine distribution on the right replaced by the distribution with density

$$\frac{\sin \pi\delta}{\pi} \frac{1}{x^{1-\delta}(1-x)^{\delta}}, \qquad 0 < x < 1. \tag{8.13}$$

Theorem 1*a* carries over to distributions belonging to the domain of attraction of a stable distribution.

In **1**; III we had to prove the two arc sine laws separately, but the next theorem shows that they are equivalent. Theorem 2 (for continuous distributions) was the point of departure of the investigations by E. Sparre-Andersen introducing the new approach to fluctuation theory. The original proof was exceedingly intricate. Several proofs are now in existence, but the following seems simplest.

Theorem 2. *The number* Π_n *of strictly positive terms among* $\mathbf{S}_1, \ldots, \mathbf{S}_n$ *has the same distribution* (8.2) *as* $\mathbf{K}_n$, *the index of the first maximal term in* $\mathbf{S}_0 = 0, \mathbf{S}_1, \ldots, \mathbf{S}_n$.

(See problem 23.)

This theorem will be reduced to a purely combinatorial lemma. Let $x_1, \ldots, x_n$ be n arbitrary (not necessarily distinct) real numbers and put

$$s_0 = 0, \qquad s_k = x_1 + \cdots + x_k. \tag{8.14}$$

The maximum among $s_0, \ldots, s_n$ may be assumed repeatedly and we must therefore distinguish between the index of the first and the last maximal term.

Consider now the $n!$ permutations $x_{i_1}, \ldots, x_{i_n}$ (some of which may have the same outer appearance). With each we associate the sequence of its $n+1$ partial sums $0, x_{i_1}, \ldots, x_{i_1}+\cdots+x_{i_n}$.

Example. (*a*) Let $x_1 = x_2 = 1$ and $x_3 = x_4 = -1$. Only 6 rearrangements $(x_{i_1}, \ldots, x_{i_4})$ are distinguishable, but each represents four permutations of the subscripts. In the arrangement $(1, 1, -1, -1)$ three partial sums are strictly positive, and the (unique) maximum occurs at the third place. In the arrangement $(-1, -1, 1, 1)$ no partial sum is positive, but the last is zero. The first maximum has index 0, the last index 4. ▶

Theorem 2 will be shown to be a simple consequence of

Lemma 2. *Let r be an integer $0 \le r \le n$. The number A_r of permutations with exactly r strictly positive partial sums is the same as the number B_r of permutations in which the first maximum among these partial sums occurs at the place r.*

(See problem 24.)

Proof.[15] We proceed by induction. The assertion is true for $n = 1$ since $x_1 > 0$ implies $A_1 = B_1 = 1$ and $A_0 = B_0 = 0$ while $x_1 \le 0$ implies $A_1 = B_1 = 0$ and $A_0 = B_0 = 1$. Assume the lemma true when n is replaced by $n - 1 \ge 1$. Denote by $A_r^{(k)}$ and $B_r^{(k)}$ the numbers corresponding to A_r and B_r when the n-tuple $(x_1, \ldots, x_n)$ is replaced by the $(n-1)$-tuple obtained by omitting x_k. The induction hypothesis then states that $A_r^{(k)} = B_r^{(k)}$ for $1 \le k \le n$ and $r = 0, \ldots, n-1$. This is true also for $r = n$ since trivially $A_n^{(k)} = B_n^{(k)} = 0$.

(*a*) Suppose $x_1 + \cdots + x_n \le 0$. The $n!$ permutations of $(x_1, \ldots, x_n)$ are obtained by choosing the element x_k at the last place and permuting the remaining $n-1$ elements. The nth partial sum being ≤ 0, it is clear that the number of positive partial sums and the index of the first maximal term depend only on the first $n-1$ elements. Thus

$$A_r = \sum_{k=1}^{n} A_r^{(k)}, \qquad B_r = \sum_{k=1}^{n} B_r^{(k)}, \tag{8.15}$$

and hence $A_r = B_r$ by the induction hypothesis.

[15] The following proof is due to Mr. A. W. Joseph of Birmingham (England). Its extreme simplicity comes almost as a shock if one remembers that in 1949 Sparre Andersen's discovery of theorem 2 was a sensation greeted with incredulity, and the original proof was of an extraordinary intricacy and complexity. A reduction to the purely combinatorial lemma 2 and an elementary proof of the latter was given by the author. (See the first edition of the present book.) Joseph's proof is not only simpler, but is the first constructive proof establishing a one-to-one correspondence between the two types of permutations. Our discussion of this aspect in lemma 3 exploits an idea of Mr. M. T. L. Bizley of London (England). The author is grateful to Messrs. Joseph and Bizley for permission to use their unpublished results (communicated when the typescript was already at the printers).

(*b*) Assume $x_1 + \cdots + x_n > 0$. The nth partial sum is then positive and the preceding argument shows that now

$$(8.16) \qquad A_r = \sum_{k=1}^{n} A_{r-1}^{(k)}.$$

To obtain an analogous recursion formula for B_r consider the arrangements $(x_k, x_{j_1}, \ldots, x_{j_{n-1}})$ *starting* with x_k. The nth partial sum being positive the maximal terms of the partial sums have positive subscripts. Clearly the first maximum occurs at the place r $(1 \leq r \leq n)$ iff the first maximum of the partial sums for $(x_{j_1}, \ldots, x_{j_{n-1}})$ occurs at the place $r - 1$. Thus

$$(8.17) \qquad B_r = \sum_{k=1}^{n} B_{r-1}^{(k)}$$

A comparison of (8.16) and (8.17) shows again that $A_r = B_r$, and this completes the proof. ▶

We shall presently see that this argument yields further results, but first we return to the

Proof of theorem 1. We proceed as in the proof of theorem 1 in section 7. Consider the $n!$ permutations $(x_{i_1}, \ldots, x_{i_n})$ and number them so that the natural order $(x_1, \ldots, x_n)$ counts as number one. For a fixed integer $0 \leq r \leq n$ define $\mathbf{Y}^{(\nu)} = 1$ if the permutation number ν has exactly r positive partial sums, and $\mathbf{Y}^{(\nu)} = 0$ otherwise. For reasons of symmetry the $n!$ random variables have a common distribution, and hence

$$(8.18) \qquad \mathbf{P}\{\mathbf{\Pi}_n = r\} = \mathbf{P}\{\mathbf{Y}^{(1)} = 1\} = \mathbf{E}(\mathbf{Y}^{(1)}) = \frac{1}{n!} \sum \mathbf{E}(\mathbf{Y}^{(\nu)}).$$

Similarly

$$(8.19) \qquad \mathbf{P}\{\mathbf{K}_n = r\} = \frac{1}{n!} \sum \mathbf{E}(\mathbf{Z}^{(\nu)})$$

where $\mathbf{Z}^{(\nu)} = 1$ if in the permutation number ν the first maximal partial sum has index r, and $\mathbf{Z}^{(\nu)} = 0$ otherwise. By the last lemma the sums $\sum \mathbf{Y}^{(\nu)}$ and $\sum \mathbf{Z}^{(\nu)}$ are identical, and hence the probabilities in (8.18) and (8.19) are the same. ▶

Note *on Sparre Andersen transformations.* It follows from lemma 2 that there exists a transformation such that each n-tuple $(x_1, \ldots, x_n)$ of real numbers is mapped into a rearrangement $(x_{i_1}, \ldots, x_{i_n})$ in such a way that: (i) if exactly r $(0 \leq r \leq n)$ among the partial sums s_k in (8.14) are strictly positive, then a maximum of the partial sums of $(x_{i_1}, \ldots, x_{i_n})$ occurs for the first time with index r, and (i^i) the transformation is invertible (or one-to-one). Such transformations will be called after E. Sparre Andersen even though he was concerned with independent random variables without being aware of the possibility to reduce theorem 2 to the purely combinatorial lemma 2. A perusal of the proof of lemma 2 reveals that it contains implicitly a prescription for a construction of a Sparre Andersen transformation. The procedure is recursive, the first step being given by the rule: if $s_n \leq 0$ leave the n-tuple $(x_1, \ldots, x_n)$ unchanged, but if $s_n > 0$ replace it by the cyclical rearrangement $(x_n, x_1, \ldots, x_{n-1})$. The next step consists in applying the same rule to the $(n-1)$-tuple $(x_1, \ldots, x_{n-1})$. The desired rearrangement $(x_{i_1}, \ldots, x_{i_n})$ is obtained after $n-1$ steps.

Examples. (*b*) Let $(x_1, \ldots, x_6) = (-1, 2, -1, 1, 1, -2)$. No change occurs at the first step while the second leads to $(1, -1, 2, -1, 1, -2)$. As $s_4 = 1$ the third step yields $(1, 1, -1, 2, -1, -2)$, and the fourth step introduces no change because $s_3 = 0$. Since $s_2 = 1$ the final step leads to the arrangement $(1, 1, 2, -1, -1, -2)$. The unique maximum of the partial sums occurs at the third place, and in the original arrangement exactly three partial sums are positive.

(*c*) Suppose that $x_j \leq 0$ for all j. The initial and the final arrangement of the x_j are identical. No partial sum is positive, and $s_0 = 0$ represents a maximum (which is repeated if $x_1 = 0$). ▶

It is preferable to replace the recursive construction by a direct description of the final result. We give it in the following lemma; because of its intrinsic interest a new proof is given which is independent of the preceding lemma. (See also problem 24.)

Lemma 3. *Let $(x_1, \ldots, x_n)$ be an n-tuple of real numbers such that the partial sums $s_{\nu_1}, \ldots, s_{\nu_r}$ are positive and all others negative or zero: here $\nu_1 > \nu_2 > \cdots > \nu_r > 0$. Write down $x_{\nu_1}, \ldots, x_{\nu_r}$ followed by the remaining x_j in their original order. (If all partial sums are ≤ 0 then $r = 0$ and the order remains unchanged.) Among the maxima of the partial sums in the new arrangement the first occurs at the* r*th place, and the transformation thus defined is one-to-one.*

Proof. Denote the new arrangement by $(\xi_1, \ldots, \xi_n)$ and its partial sums by $\sigma_0, \ldots, \sigma_n$. To every subscript $j \leq n$ there corresponds a unique subscript k such that $\xi_j = x_k$. In particular, $\xi_1, \ldots, \xi_r$ agree with $x_{\nu_1}, \ldots, x_{\nu_r}$ in that order.

Consider first a j such that $s_k \leq 0$. It is clear from the construction that $j \geq k$ and the elements $\xi_{j-k+1}, \ldots, \xi_j$ represent a permutation of $x_1, \ldots, x_k$. Thus $\sigma_j = \sigma_{j-k} + s_k \leq \sigma_{j-k}$, and so σ_j cannot be the first maximal partial sum.

If $r = 0$ it follows that the first maximum among the σ_j is assumed for $j = 0$. When $r > 0$ the first maximum occurs at one among the places $0, 1, \ldots, r$, and we show that only r is possible. Indeed, if $j \leq r$ it follows from the construction that the ν_j elements $\xi_j, \ldots, \xi_{j-1+\nu_j}$ coincide with some rearrangement of $(x_1, \ldots, x_{\nu_j})$. Thus $\sigma_{j-1+\nu_j} = \sigma_{j-1} + s_{\nu_j} > \sigma_{j-1}$, and hence no maximum can occur at the place $j-1$.

To complete the proof of the lemma it remains to show that the transformation is one-to-one. As a matter of fact, the inverse to $\xi_1, \ldots, \xi_n$ may be constructed by the following recursive rule. If all $\sigma_j \leq 0$ leave the arrangement unchanged. Otherwise let k be the largest subscript such that $\sigma_k > 0$. Replace $(\xi_1, \ldots, \xi_n)$ by $(\xi_2, \ldots, \xi_k, \xi_1, \xi_{k+1}, \ldots, \xi_n)$, and apply the same procedure to the $(k-1)$-tuple $(\xi_2, \ldots, \xi_k)$. ▶

Note *on exchangeable variables.* It should be noticed that the proof does not depend on the independence of the variables $\mathbf{X}_j$ but only the identity of the joint distribution for each of the $n!$ arrangements $(\mathbf{X}_{i_1}, \ldots, \mathbf{X}_{i_n})$. In other words, *theorem* 2 *remains valid for every n-tuple of exchangeable variables* (VII,4) although naturally the common distribution of $\mathbf{K}_n$ and $\mathbf{\Pi}_n$ will depend on the joint distribution of the $\mathbf{X}_j$. As an interesting example let $\mathbf{X}_1, \mathbf{X}_2, \ldots$ be independent with a common distribution F, and put $\mathbf{Y}_k = \mathbf{X}_k - \mathbf{S}_n/n$ (where $k = 1, \ldots, n$). The variables $\mathbf{Y}_1, \ldots, \mathbf{Y}_n$ are exchangeable and their partial sums are

$$\mathbf{\Sigma}_k = \mathbf{S}_k - k\mathbf{S}_n/n, \qquad k = 1, \ldots, n-1. \tag{8.20}$$

With reference to the graph of $(\mathbf{S}_0, \mathbf{S}_1, \ldots, \mathbf{S}_n)$ we can describe $\mathbf{\Sigma}_k$ as the vertical distance of the vertex $\mathbf{S}_k$ from the chord joining the origin to the endpoint $(n, \mathbf{S}_n)$.

We now suppose that F is continuous (in order to avoid the necessity of distinguishing between the first and the last maximum). With probability 1 there is a unique maximum among the terms $0, \mathbf{\Sigma}_1, \ldots, \mathbf{\Sigma}_{n-1}$. To the cyclical rearrangement $(\mathbf{Y}_2, \ldots, \mathbf{Y}_n, \mathbf{Y}_1)$ there correspond the partial sums $0, \mathbf{\Sigma}_2 - \mathbf{\Sigma}_1, \ldots, \mathbf{\Sigma}_{n-1} - \mathbf{\Sigma}_1, -\mathbf{\Sigma}_1$, and it is clear that the location of the maximum has moved one place ahead in cyclical order. (If the original maximum was at the zero place then $\mathbf{\Sigma}_k < 0$ for $k = 1, \ldots, n-1$, and the new maximum is at the place $n-1$.) In the n cyclical permutations the maximum is therefore assumed exactly once at each place, and its position is uniformly distributed over $0, 1, \ldots, n-1$. We have thus the following theorem due to Sparre-Andersen and related to the theorem 3 of **1**; III,9 in coin tossing.

Theorem 3. *In any random walk with continuous F and for any n the number of vertices among $\mathbf{S}_1, \ldots, \mathbf{S}_{n-1}$ that lie above the chord from $(0, 0)$ to $(n, \mathbf{S}_n)$ is uniformly distributed over $0, 1, \ldots, n-1$.*

(The same is true of the index of the vertex with greatest maximal distance.)

9. MISCELLANEOUS COMPLEMENTS

(a) Joint Distributions

The argument leading to theorem 1 of section 7 requires only notational changes to yield the joint distribution of the ladder variables. Adapting the notation of section 1, let I be an interval in $\overline{0, \infty}$ and denote by $H_n^{(r)}\{I\}$ *the probability that n be the rth ladder epoch and* $\mathbf{S}_n \in I$. Put

$$H\{I, s\} = \sum_{n=1}^{\infty} s^n H_n\{I\}, \qquad 0 \leq s \leq 1. \tag{9.1}$$

It is seen by induction for fixed s

$$H^{r\star}\{I, s\} = \sum_{n=1}^{\infty} s^n H_n^{(r)}\{I\}. \tag{9.2}$$

The argument leading to (7.3) now yields without difficulty the following result due to G. Baxter which reduces to (7.3) when $I = \overline{0, \infty}$.

Theorem. *For* $I \subset \overline{0, \infty}$ *and* $0 \leq s \leq 1$

$$\sum_{r=1}^{\infty} \frac{1}{r} H^{r\star}\{I, s\} = \sum_{n=1}^{\infty} \frac{s^n}{n} \mathbf{P}\{\mathbf{S}_n \in I\}. \tag{9.3}$$

A simpler and more tractable form will be derived in XVIII,3.

(b) A Mortality Interpretation for Generating Functions

The following interpretation may help intuition and simplify formal calculations. For fixed s with $0 < s < 1$ consider the *defective* random walk which at each step has probability $1 - s$ to terminate and otherwise subject to the distribution sF. Now $s^n F^{n\star}\{I\}$ is the probability of a position in I at time n, the defect $1 - s^n$ representing the probability of a prior termination. All considerations carry over without change, except that *all* distributions become defective. In particular, *in our random walk with mortality*, (9.1) *is simply the first ladder height distribution*, and (9.2) the analogue to $H^{r\star}$ of sections 2–3. The generating function $\tau(s)$ now equals the probability that a ladder index will occur.

(c) The Recurrent Event

$$\{\mathbf{S}_1 \leq 0, \ldots, \mathbf{S}_{n-1} \leq 0, \mathbf{S}_n = 0\} \tag{9.4}$$

represents a return to the origin without previous visits to the right half-axis. It was considered in section 1 in the definition of weak ladder variables. Denote by ω_n the probability of the *first* occurrence of the event (9.4) at epoch n, that is,

$$\omega_n = \mathbf{P}\{\mathbf{S}_1 < 0, \ldots, \mathbf{S}_{n-1} < 0, \mathbf{S}_n = 0\}. \tag{9.5}$$

If $\omega(s) = \sum \omega_r s^r$, then ω^r is the generating function for the rth occurrence and so $1/[1-\omega(s)]$ is the generating function for the probabilities (9.4). A simplified version of the proof of (7.3) leads to the basic identity

$$\log \frac{1}{1 - \omega(s)} = \sum_{n=1}^{\infty} \frac{s^n}{n} \mathbf{P}\{\mathbf{S}_n = 0\}. \tag{9.6}$$

Comparing this with (7.3), (7.16), (7.22), etc., one sees how easy it is to pass from weak to strict ladder variables and vice versa. Formula (9.6) confirms also the remark of section 1 that the probabilities of (9.4) remain unchanged if all inequalities are reversed.

(d) Generalization to Arbitrary Intervals

The theory of section 3 generalizes with trite notational changes to the situation in which $\overline{0, \infty}$ is replaced by an arbitrary interval A, and $\overline{-\infty, 0}$ by the complement A'. In particular, the Wiener-Hopf integral equation remains unchanged. The reader is invited to work out the details; they are fully developed in XVIII,1. (See also problem 15.)

10. PROBLEMS FOR SOLUTION

1. In the binomial random walk [example 2(*b*)] let e_k be the expected number of indices $n \geq 0$ such that $\mathbf{S}_n = k, \mathbf{S}_1 \geq 0, \ldots, \mathbf{S}_{n-1} \geq 0$ (visits to k preceding the first negative value). Denote by f the probability of ever reaching -1, that is, $f = 1$ if $q \geq p$ and $f = q/p$ otherwise. Taking the point $(1, 1)$ as new origin prove that $e_0 = 1 + pfe_0$ and $e_k = p(e_{k-1} + fe_k)$ for $k \geq 1$. Conclude that for $k \geq 0$

$$e_k = p^{-1} \quad \text{if} \quad p \geq q, \qquad e_k = (p/q)^k q^{-1} \quad \text{if} \quad p \leq q.$$

2. *Continuation.* For $k \geq 1$ let a_k be the expected number of indices $n \geq 1$ such that $\mathbf{S}_n = k, \mathbf{S}_1 > 0, \ldots, \mathbf{S}_{n-1} > 0$ (visits to k preceding the first return to the origin). Show that $a_k = pe_{k-1}$ and hence

$$a_k = 1 \quad \text{if} \quad p \geq q, \qquad a_k = (p/q)^k \quad \text{if} \quad p \leq q.$$

This gives a direct proof of the paradoxical result of example 2(*b*).

Note. The following problems 3–6 may serve as introduction to the problems of this chapter *and can be solved before studying it.* They present also examples for explicit solutions of the basic integral equations. Furthermore, they illustrate the power and elegance of generating functions [try to solve equation (1) directly!].

3. The variables $\mathbf{X}_k$ of a random walk have a common arithmetic distribution attaching probabilities $f_1, f_2, \ldots$ to the integers $1, 2, \ldots$ and q to -1 (where $q + f_1 + f_2 + \cdots = 1$). Denote by λ_r $(r = 1, 2, \ldots)$ the probability that the first positive term of the sequence $\mathbf{S}_1, \mathbf{S}_2, \ldots$ assumes the value r. (In other words, $\{\lambda_r\}$ is the distribution of the first ladder height.) Show that:

(*a*) The λ_r satisfy the recurrence relations

$$\lambda_r = f_r + q(\lambda_{r+1} + \lambda_1\lambda_r). \tag{1}$$

(*b*) The generating functions satisfy

$$\lambda(s) = 1 - \frac{f(s) + qs^{-1} - 1}{\lambda_1 q + qs^{-1} - 1}, \qquad 0 < s < 1. \tag{2}$$

(*c*) If $\mathbf{E}(\mathbf{X}_k) = \mu = f'(1) - q > 0$, there exists a unique root, $0 < \sigma < 1$, of the equation

$$f(s) + q/s = 1. \tag{3}$$

From the fact that λ must be monotone and <1 in $\overline{0, 1}$ conclude that

$$\lambda(s) = s\frac{\sigma}{q}\frac{f(s) - f(\sigma)}{s - \sigma}. \tag{4}$$

This is equivalent to

$$\lambda_r = \frac{[f_r\sigma + f_{r+1}\sigma^2 + \cdots]}{q}. \tag{5}$$

(*d*) If $\mathbf{E}(\mathbf{X}_k) < 0$ the appropriate solution is obtained letting $\lambda_1 = (1-q)/q$ in (2). Then (4) and (5) hold with $\sigma = 1$.

4. Adapt the preceding problem to weak ladder heights. In other words, instead of λ_r consider the probability that the first non-negative term of $\mathbf{S}_1, \mathbf{S}_2, \ldots$ assumes the value r $(r = 0, 1, \ldots)$. Show that (1) and (4) are replaced by

$$\gamma_r = f_r + \frac{q}{1 - \gamma_0}\gamma_{r+1} \tag{1a}$$

$$\gamma(s) = 1 - \frac{q}{\sigma} + s\frac{f(s) - f(\sigma)}{s - \sigma}. \tag{4a}$$

5. In the random walk of problem 3 (but without using this problem) let x be the probability that $\mathbf{S}_n < 0$ for some n. Show that x satisfies the equation (3) and hence $x = \sigma$.

6. *Continuation.* Show that the probability that $\mathbf{S}_n \leq 0$ for some $n > 0$ is $q + f(\sigma) = 1 - q(\sigma^{-1} - 1)$. Verify that $\lambda'(1) = \mu\sigma[q(1-\sigma)]^{-1}$, which is a special case of relation (2.8) (or Wald's equation).

7. Derive (1.13) by straight calculation from (1.12).

8. *Hitting probabilities.* For $t \geq 0$ and $\xi > 0$ denote by $G(t, \xi)$ the probability that the *first* sum $\mathbf{S}_n$ exceeding t will be $\leq t + \xi$. Prove that G satisfies the integral equation

$$G(t, \xi) = F(t+\xi) - F(t) + \int_{-\infty}^{t+} G(t-y, \xi)\, F\{dy\}.$$

In case of non-uniqueness, G is the minimal solution. The ladder height distribution H is uniquely determined by $H(\xi) = G(0, \xi)$.

9. Let H be a continuous probability distribution concentrated on $\overline{0, \infty}$ and H^- a possibly defective continuous distribution concentrated on $\overline{-\infty, 0}$. Suppose that

$$H + H^- - H \star H^- = F \tag{6}$$

is a probability distribution. From the uniqueness theorem in section 3 it follows that H and H^- are the distributions of the points of first entry $\mathscr{H}$ and $\mathscr{H}^-$ in the random walk generated by F. In this way it is possible to find distributions F admitting of an explicit representation of the form (6). Such is the case if $0 < q \leq 1$ and H and H^- have densities defined by either

(*a*) be^{-bx} *for* $x > 0$, $\quad q$ *for* $-1 < x < 0$

or

(*b*) b^{-1} *for* $0 < x < b$, $\quad q$ *for* $-1 < x < 0$.

In case (*a*) the distribution has a density given by

$$(b - q)e^{-bx} + qe^{-b(x+1)} \quad \text{for} \quad x > 0, \quad \text{and} \quad qe^{-b(x+1)} \quad \text{for} \quad -1 < x < 0.$$

In case (*b*) if $b > 1$ the density of F is given by $qb^{-1}(1 + x)$ for $-1 < x < 0$, by qb^{-1} for $0 < x < b - 1$, and by $qb^{-1}(b - x)$ for $b - 1 < x < b$. In either case F has 0 expectation iff $q = 1$.

10. From (3.11) conclude: If H and H^- are proper and have variances then $\mathbf{E}(\mathbf{X}_1) = 0$ and $\mathbf{E}(\mathbf{X}_1^2) = -2\mathbf{E}(\mathscr{H}_1)\mathbf{E}(\mathscr{H}_1^-)$.

11. *Analytic proof of Wald's relation* (2.8). From (3.11) conclude

$$1 - F(x) = [1 - \rho(0)][1 - H(x)] + \int_{-\infty}^{0+} \rho\{dy\}[H(x-y) - H(x)]$$

$$F(x) = \int_{-\infty}^{x} \rho\{dy\}[1 - H(x-y)]$$

for $x > 0$ and $x < 0$, respectively. Conclude that F has a positive expectation μ iff H has a finite expectation ν and $\rho(0) < 1$. Conclude by integration over $\overline{-\infty, \infty}$ that $\mu = [1 - \rho(0)]\nu$, which is equivalent to (2.8).

12. *To example* 4(*c*). If $\mu > 0$ the denominator has a positive root $s_0 < 1$, exactly $b - 1$ complex roots in $|s| < s_0$, and $a - 1$ complex roots in $|s| > 1$. The situation for $\mu < 0$ is described by changing s into $1/s$.

13. The generating function of the ascending ladder height distribution in example 4(*c*) is given by

$$\chi(s) = 1 - (1 - s)(1 - s/\sigma_1) \cdots (1 - s/\sigma_{a-1}).$$

For descending ladder heights change s/σ_k into s_k/s.

14. *To example* 4(*c*). Suppose that the $\mathbf{X}_j$ assume the values $-2, -1, 0, 1, 2$ each with probability $\frac{1}{5}$. Show that the *ascending* ladder height distribution is given by

$$\lambda_1 = \frac{1 + \sqrt{5}}{3 + \sqrt{5}}, \qquad \lambda_2 = \frac{2}{3 + \sqrt{5}}.$$

For the *weak* heights $\bar{\lambda}_0 = \frac{1}{10}(7 - \sqrt{5})$, $\bar{\lambda}_1 = \frac{1}{10}(1 + \sqrt{5})$, $\bar{\lambda}_2 = \frac{1}{5}$.

15. In example 4(*c*) denote by $\varphi_k^{(n)}$ the probability that the first n steps do not lead out of the interval $\overline{-B, A}$ and that the nth step leads to the position k. (Thus $\varphi_k^{(n)} = 0$ for $k > A$ and $k < -B$. As usual $\varphi_k^{(0)}$ equals 1 when $k = 0$ and 0 otherwise.) Let $\varphi_k = \sum \varphi_k^{(n)}$ be the expected number of visits to k prior to leaving $\overline{-B, A}$. Show that

$$\varphi_k = \sum_{\nu=-B}^{A} \varphi_\nu f_{k-\nu} + \varphi_k^{(0)}, \qquad -B \le k \le A,$$

and that for $k > A$ and $k < -B$

$$\rho_k = \sum_{\nu=-B}^{A} \varphi_\nu f_{k-\nu}$$

is the probability that the *first exit* from the interval $\overline{-B, A}$ leads to the point k. [This problem is important in sequential *analysis*. It illustrates the situation described in 9(*d*).]

16. Theorem 2 of section 7 implies that if $\mu < 0$ then $\sum n^{-1}\mathbf{P}\{\mathbf{S}_n > 0\} < \infty$. Fill in the following *direct proof.* It suffices to show (Chebyshev) that

$$\sum \int_{|y|>n} F\{dy\} < \infty, \qquad \sum \frac{1}{n^2} \int_{-n}^{n} y^2\, F\{dy\} < \infty.$$

The first is obvious. To prove the second relation write the integral as a sum of n integrals over $k - 1 < |y| \leq k$, $k = 1, \ldots, n$. Reverse the order of summation and conclude that the whole series is $<2\mathbf{E}(|\mathbf{X}|)$.

17. For the coin-tossing game show that

$$\sum \frac{s^n}{2} \mathbf{P}\{\mathbf{S}_n = 0\} = \log \frac{2}{1 + \sqrt{1 - s^2}}.$$

Hint: Easy by observing that the left side may be written as the integral on $[(1-x^2)^{-\frac{1}{2}} - 1]x^{-1}$ from 0 to s.

18. Suppose that the random walk is transient, that is $U\{I\} = \sum_0^\infty F^{n\star}\{I\} < \infty$ for every bounded interval. In the notation of section 3 put $\Phi = \sum_0^\infty \rho^{n\star}$. Prove the truth of the renewal equation

$$U = \Phi + U \star H.$$

If ψ^- is the analogue of ψ for the negative half-line then $\psi^- = (1 - \zeta)\Phi$ as in (1.13).

19. Conclude that

$$U = \frac{1}{1 - \zeta} \psi \star \psi^-$$

and show this to be equivalent to the Wiener-Hopf decomposition (3.12).

20. Derive Wald's identity $\mathbf{E}(\mathscr{H}_1) = \mathbf{E}(\mathscr{T}_1)\mathbf{E}(\mathbf{X}_1)$ directly from the renewal equation in problem 18.

21. *To theorem* 4 *of section* 7. The probabilities $p_n^* = \mathbf{P}\{\mathbf{S}_1 \geq 0, \ldots, \mathbf{S}_n \geq 0\}$ *and* $q_n^* = \mathbf{P}\{\mathbf{S}_1 < 0, \ldots, \mathbf{S}_n < 0\}$ have generating functions given by

$$\log p^*(s) = \sum_{n=1}^{\infty} \frac{s^n}{n} \mathbf{P}\{\mathbf{S}_n \geq 0\} \quad \textit{and} \quad \log q_n^* = \sum_{n=1}^{\infty} \frac{s^n}{n} \mathbf{P}\{\mathbf{S}_n < 0\}.$$

22. *On the last maximum.* Instead of the variable $\mathbf{K}_n$ of Section 8 consider the index $\mathbf{K}_n^*$ of the *last* maximum of the partial sums $\mathbf{S}_0, \ldots, \mathbf{S}_n$. With the notations of the preceding problem, prove that

$$\mathbf{P}\{\mathbf{K}_n^* = k\} = p_k^* q_{n-k}^*.$$

23. *Alternative form of theorem* 2, *section* 8. *The number* Π_n^* *of non-negative terms among* $\mathbf{S}_0, \ldots, \mathbf{S}_n$ *has the same distribution as the variable* $\mathbf{K}_n^*$ *of the preceding problem.* Prove this by applying theorem 2 to $(-\mathbf{X}_n - \mathbf{X}_{n-1}, \ldots, -\mathbf{X}_1)$.

24. The combinatorial lemma 2 of section 8 remains valid if the first maximum of $\mathbf{S}_0, \ldots, \mathbf{S}_n$ is replaced by the *last* maximum, and the number of positive partial sums by the number of non-negative terms in $\mathbf{S}_1, \ldots, \mathbf{S}_n$ (excluding $\mathbf{S}_0 = 0$). The proof is the same except for the obvious changes of the inequalities. In like manner lemma 3 carries over.

CHAPTER XIII

Laplace Transforms. Tauberian Theorems. Resolvents

The Laplace transforms are a powerful practical tool, but at the same time their theory is of intrinsic value and opens the door to other theories such as semi-groups. The theorem on completely monotone functions and the basic Tauberian theorem have rightly been considered pearls of hard analysis. (Although the present proofs are simple and elementary, the pioneer work in this direction required originality and power.) Resolvents (sections 9–10) are basic for semi-group theory.

As this chapter must cover diverse needs, a serious effort has been made to keep the various parts as independent of each other as the subject permits, and to make it possible to skip over details. Chapter XIV may serve for collateral reading and to provide examples. The remaining part of this book is entirely independent of the present chapter.

Despite the frequent appearance of regularly varying functions only the quite elementary theorem 1 of VIII,8 is used.

1. DEFINITIONS. THE CONTINUITY THEOREM

Definition 1. *If F is a proper or defective probability distribution concentrated on* $\overline{0, \infty}$, *the Laplace transform φ of F is the function defined for $\lambda \geq 0$ by*

$$\varphi(\lambda) = \int_0^\infty e^{-\lambda x}\, F\{dx\}. \tag{1.1}$$

Here and in the sequel it is understood that the *interval of integration is closed* (and may be replaced by $\overline{-\infty, \infty}$). Whenever we speak of the Laplace transform of a distribution F it is tacitly understood that F is concentrated on $\overline{0, \infty}$. As usual we stretch the language and speak of "the

Laplace transform of the random variable **X**," meaning the transform of its distribution. With the usual notation for expectations we have then

$$\varphi(\lambda) = \mathbf{E}(e^{-\lambda \mathbf{X}}). \tag{1.2}$$

Example. (*a*) Let **X** assume the values 0, 1, ... with probabilities $p_0, p_1, \ldots$. Then $\varphi(\lambda) = \sum p_n e^{-n\lambda}$ whereas the generating function is $P(s) = \sum p_n s^n$. Thus $\varphi(\lambda) = P(e^{-\lambda})$ and the Laplace transform differs from the generating function only by the change of variable $s = e^{-\lambda}$. This explains the close analogy between the properties of Laplace transforms and generating functions.

(*b*) *The gamma distribution* with density $f_\alpha(x) = (x^{\alpha-1}/\Gamma(\alpha))e^{-x}$ has the transform

$$\varphi_\alpha(\lambda) = \frac{1}{\Gamma(\alpha)} \int_0^\infty e^{-(\lambda+1)x}\, x^{\alpha-1}\, dx = \frac{1}{(\lambda+1)^\alpha}, \qquad \alpha > 0. \tag{1.3}$$

The next theorem shows that a distribution is recognizable by its transform; without this the usefulness of Laplace transforms would be limited.

Theorem 1. (*Uniqueness.*) *Distinct probability distributions have distinct Laplace transforms.*

First proof. In VIII,(6.4) we have an explicit inversion formula which permits us to calculate F when its transform is known. This formula will be derived afresh in section 4.

Second proof. Put $y = e^{-x}$. As x goes from 0 to ∞ the variable y goes from 1 to 0. We now define a probability distribution G concentrated on $\overline{0, 1}$ by letting $G(y) = 1 - F(x)$ at points of continuity. Then

$$\varphi(\lambda) = \int_0^\infty e^{-\lambda x}\, F\{dx\} = \int_0^1 y^\lambda\, G\{dy\} \tag{1.4}$$

as is obvious from the fact that the Riemann sums $\sum e^{-\lambda x_k}[F(x_{k+1}) - F(x_k)]$ coincide with the Riemann sums $\sum y_k^\lambda [G(y_k) - G(y_{k+1})]$ when $y_k = e^{-x_k}$. We know from VII,3 that the distribution G is uniquely determined by its moments, and these are given by $\varphi(k)$. Thus the knowledge of $\varphi(1)$, $\varphi(2), \ldots$ determines G, and hence F. This result is stronger than the assertion of the theorem.[1] ▶

The following basic result is a simple consequence of theorem 1.

---

[1] More generally, a completely monotone function is uniquely determined by its values at a sequence $\{a_n\}$ of points such that $\sum a_n^{-1}$ diverges. However, if the series converges there exist two distinct completely monotone functions agreeing at all points a_n. For an elementary proof of this famous theorem see W. Feller, *On Müntz' theorem and completely monotone functions*, Amer. Math. Monthly, vol. 75 (1968), pp. 342–350.

Theorem 2. (*Continuity theorem.*) *For* $n = 1, 2, \ldots$ *let* F_n *be a probability distribution with transform* φ_n.

If $F_n \to F$ *where* F *is a possibly defective distribution with transform* φ *then* $\varphi_n(\lambda) \to \varphi(\lambda)$ *for* $\lambda > 0$.

Conversely, if the sequence $\{\varphi_n(\lambda)\}$ *converges for each* $\lambda > 0$ *to a limit* $\varphi(\lambda)$, *then* φ *is the transform of a possibly defective distribution* F, *and* $F_n \to F$.

The limit F *is not defective iff* $\varphi(\lambda) \to 1$ *as* $\lambda \to 0$.

Proof. The first part is contained in the basic convergence theorem of VIII,1. For the second part we use the selection theorem 1 of VIII,6. Let $\{F_{n_k}\}$ be a subsequence converging to the possibly defective distribution F. By the first part of the theorem the transforms converge to the Laplace transform of F. It follows that F is the unique distribution with Laplace transform φ, and so all convergent subsequences converge to the same limit F. This implies the convergence of F_n to F. The last assertion of the theorem is clear by inspection of (1.1). ►

For clarity of exposition we shall as far as possible reserve the letter F for probability distributions, but instead of (1.1) we may consider more general integrals of the form

$$\omega(\lambda) = \int_0^\infty e^{-\lambda x}\, U\{dx\}. \tag{1.5}$$

where U is a measure attributing a finite mass $U\{I\}$ to the finite interval I, but may attribute an infinite mass to the positive half axis. As usual, we describe this measure conveniently in terms of its improper distribution function defined by $U(x) = U\{\overline{0, x}\}$. In the important special case where U is defined as the integral of a function $u \geq 0$ the integral (1.5) reduces to

$$\omega(\lambda) = \int_0^\infty e^{-\lambda x}\, u(x)\, dx \tag{1.6}$$

Examples. (*c*) If $u(x) = x^a$ with $a > -1$, then $\omega(\lambda) = \Gamma(a+1)/\lambda^{\alpha+1}$ for all $\lambda > 0$.

(*d*) If $u(x) = e^{ax}$ then $\omega(\lambda) = 1/(\lambda - a)$ for $\lambda > a > 0$, but the integral (1.6) diverges for $\lambda \leq a$.

(*e*) If $u(x) = e^{x^2}$ the integral (1.6) diverges everywhere.

(*f*) By differentiation we get from (1.1)

$$-\varphi'(\lambda) = \int_0^\infty e^{-\lambda x} x\, F\{dx\} \tag{1.7}$$

and this is an integral of the form (1.5) with $U\{dx\} = x\, F\{dx\}$. This example illustrates how integrals of the form (1.5) arise naturally in connection with proper probability distributions. ►

We shall be interested principally in measures U derived by simple operations from probability distributions, and the integral in (1.5) will generally converge for all $\lambda > 0$. However, nothing is gained by excluding measures for which convergence takes place only for *some* λ. Now $\omega(a) < \infty$ implies $\omega(\lambda) < \infty$ for all $\lambda > a$, and so the values of λ for which the integral in (1.5) converges fill an interval $\overline{a, \infty}$.

Definition 2. *Let U be a measure concentrated on $\overline{0, \infty}$. If the integral in* (1.5) *converges for $\lambda > a$, then the function ω defined for $\lambda > a$ is called the Laplace transform of U.*

If U has a density u, the Laplace transform (1.6) *of U is also called the ordinary Laplace transform of u.*

The last convention is introduced merely for convenience. To be systematic one should consider more general integrals of the form

$$\int_0^\infty e^{-\lambda x}\, v(x)\, U\{dx\} \tag{1.8}$$

and call them "Laplace transform of v with respect to the measure U." Then (1.6) would be the "transform of u with respect to Lebesgue measure" (or ordinary length). This would have the theoretical advantage that one could consider functions u and v of variable signs. For the purposes of this book it is simplest and least confusing to associate Laplace transforms only with measures, and we shall do so.[2]

If U is a measure such that the integral in (1.5) converges for $\lambda = a$, then for all $\lambda > 0$

$$\omega(\lambda+a) = \int_0^\infty e^{-\lambda x} \cdot e^{-ax}\, U\{dx\} = \int_0^\infty e^{-\lambda x}\, U^{\#}\{dx\} \tag{1.9}$$

is the Laplace transform of the bounded measure $U^{\#}\{dx\} = e^{-ax}\, U\{dx\}$, and $\omega(\lambda+a)/\omega(a)$ is the transform of a *probability* distribution. In this way every theorem concerning transforms of probability distributions automatically generalizes to a wider class of measures. Because the graph of the new transform $\omega(\lambda+a)$ is obtained by translation of the graph of ω we shall refer to this extremely useful method as the *translation principle.* For example, since U is uniquely determined by $U^{\#}$, and $U^{\#}$ by $\omega(\lambda+a)$ for $\lambda > 0$, we can generalize theorem 1 as follows.

Theorem 1a. *A measure U is uniquely determined by the values of its Laplace transform* (1.5) *in some interval $a < \lambda < \infty$.*

[2] The terminology is not well established, and in the literature the term "Laplace transform of F" may refer either to (1.1) or to (2.6). We would describe (2.6) as the "ordinary Laplace transform of the distribution function F," but texts treating principally such transforms would drop the determinative "ordinary." To avoid ambiguities in such cases the transform (1.1) is then called the *Laplace-Stieltjes* transform.

Corollary. *A continuous function* u *is uniquely determined by the values of its ordinary Laplace transform* (1.6) *in some interval* $a < \lambda < \infty$.

Proof. The transform determines uniquely the integral U of u, and two distinct continuous[3] functions cannot have identical integrals. ▶

[An explicit formula for u in terms of ω is given in VII,(6.6).]

The continuity theorem generalizes similarly to sequences of arbitrary measures U_n with Laplace transforms. The fact that U_n has a Laplace transform implies that $U_n\{I\} < \infty$ for finite intervals I. We recall from VIII,1 and VIII,6 that a sequence of such measures is said to converge to a measure U iff $U_n\{I\} \to U\{I\} < \infty$ for every *finite* interval of continuity of U.

Theorem 2a. (*Extended continuity theorem.*) *For* $n = 1, 2, \ldots$ *let* U_n *be a measure with Laplace transform* ω_n. *If* $\omega_n(\lambda) \to \omega(\lambda)$ *for* $\lambda > a$, *then* ω *is the Laplace transform of a measure* U *and* $U_n \to U$.

Conversely, if $U_n \to U$ *ad the sequence* $\{\omega_n(a)\}$ *is bounded, then* $\omega_n(\lambda) \to \omega(\lambda)$ *for* $\lambda > a$.

Proof. (*a*) Assume that $U_n \to U$ and that $\omega_n(a) < A$. If $t > 0$ is a point of continuity of U then

$$\int_0^t e^{-(\lambda+a)x}\, U_n\{dx\} \to \int_0^t e^{-(\lambda+a)x}\, U\{dx\} \tag{1.10}$$

and the left side differs from $\omega_n(\lambda+a)$ by at most

$$\int_t^\infty e^{-(\lambda+a)x}\, U_n\{dx\} < Ae^{-\lambda t} \tag{1.11}$$

which can be made $<\epsilon$ by choosing t sufficiently large. This means that the upper and lower limits of $\omega_n(\lambda+a)$ differ by less than an arbitrary ϵ, and hence for every $\lambda > 0$ the sequence $\{\omega_n(\lambda+a)\}$ converges to a finite limit.

(*b*) Assume then that $\omega_n(\lambda) \to \omega(\lambda)$ for $\lambda > a$. For fixed $\lambda_0 > a$ the function $\omega_n(\lambda+\lambda_0)/\omega_n(\lambda_0)$ is the Laplace transform of the probability distribution $U_n^{\#}\{dx\} = (1/\omega_n(\lambda_0))e^{-\lambda_0 x}\, U_n\{dx\}$. By the continuity theorem therefore $U_n^{\#}$ converges to a possibly defective distribution $U^{\#}$, and this implies that U_n converges to a measure U such that $U\{dx\} = \omega(\lambda_0)e^{\lambda_0 x}\, U^{\#}\{dx\}$. ▶

The following example shows the necessity of the condition that $\{\omega_n(a)\}$ remain bounded.

---

[3] The same argument shows that in general u is determined up to values on an arbitrary set of measure zero.

Example. (*g*) Let U_n attach weight e^{n^2} to the point n, and zero to the complement. Since $U_n\{\overline{0, n}\} = 0$ we have $U_n \to 0$, but $\omega_n(\lambda) = e^{n(n-\lambda)} \to \infty$ for all $\lambda > 0$. ►

One speaks sometimes of the *bilateral transform* of a distribution F with two tails, namely

$$\varphi(\lambda) = \int_{-\infty}^{+\infty} e^{-\lambda x} F\{dx\}, \tag{1.12}$$

but this function need not exist for any $\lambda \neq 0$. If it exists, $\varphi(-\lambda)$ is often called the *moment generating function*, but in reality it is the generating function of the sequence $\{\mu_n/n!\}$ where μ_n is the nth moment.

2. ELEMENTARY PROPERTIES

In this section we list the most frequently used properties of the Laplace transforms; the parallel to generating functions is conspicuous.

(i) Convolutions. Let F and G be probability distributions and U their convolution, that is,

$$U(x) = \int_0^x G(x-y)\, F\{dy\}. \tag{2.1}$$

The corresponding Laplace transforms obey the *multiplication rule*

$$\omega = \varphi\gamma. \tag{2.2}$$

This is equivalent to the assertion that for independent random variables $\mathbf{E}(e^{-\lambda(\mathbf{X}+\mathbf{Y})}) = \mathbf{E}(e^{-\lambda\mathbf{X}})\, \mathbf{E}(e^{-\lambda\mathbf{Y}})$, which is a special case of the multiplication rule for expectations.[4]

If F and G have densities f and g, then U has a density u given by

$$u(x) = \int_0^x g(x-y) f(y)\, dy \tag{2.3}$$

and the multiplication rule (2.2) applies to the "ordinary" Laplace transforms (1.6) of f, g, and u.

We now show that the multiplication rule can be extended as follows. *Let F and G be arbitrary measures with Laplace transforms φ and γ converging for $\lambda > 0$. The convolution U has then a Laplace transform ω given by* (2.2). This implies in particular that the multiplication rule applies to the "ordinary" transforms of any two integrable functions f and g and their convolution (2.3).

[4] The converse is false: two variables may be dependent and yet such that the distribution of their sum is given by the convolution formula. [See II,4(*e*) and problem 1 of III,9.]

To prove the assertion we introduce the finite measures F_n obtained by truncation of F as follows: for $x \leq n$ we put $F_n(x) = F(x)$, but for $x > n$ we let $F_n(x) = F(n)$. Define G_n similarly by truncating G. For $x < n$ the convolution $U_n = F_n * G_n$ does not differ from U, and hence not only $F_n \to F$ and $G_n \to G$, but also $U_n \to U$. For the corresponding Laplace transforms we have $\omega_n = \varphi_n \gamma_n$ and letting $n \to \infty$ we get the assertion $\omega = \varphi\gamma$.

Examples. (*a*) *Gamma distributions.* In example 1(*b*) the familiar convolution rule $f_\alpha * f_\beta = f_{\alpha+\beta}$ is mirrored in the obvious relation $\varphi_\alpha \varphi_\beta = \varphi_{\alpha+\beta}$.

(*b*) *Powers.* To $u_\alpha(x) = x^{\alpha-1}/\Gamma(\alpha)$ there corresponds the ordinary Laplace transform $\omega_\alpha(\lambda) = \lambda^{-\alpha}$. It follows that the convolution (2.3) of u_α and u_β is given by $u_{\alpha+\beta}$. The preceding example follows from this by the translation principle since $\varphi_\alpha(\lambda) = \omega_\alpha(\lambda+1)$.

(*c*) *If* $a > 0$ *then* $e^{-a\lambda}\omega(\lambda)$ *is the Laplace transform of the measure with distribution function* $U(x-a)$. This is obvious from the definition, but may be considered also as a special case of the convolution theorem inasmuch as $e^{-a\lambda}$ is the transform of the distribution concentrated at the point a. ▶

(ii) **Derivatives and moments.** If F is a probability distribution and φ its Laplace transform (1.1), then φ *possesses derivatives of all orders given by*

$$(-1)^n \varphi^{(n)}(\lambda) = \int_0^\infty e^{-\lambda x} x^n \, F\{dx\} \tag{2.4}$$

(as always, $\lambda > 0$). The differentiation under the integral is permissible since the new integrand is bounded and continuous.

It follows in particular that F *possesses a finite* nth *moment iff a finite limit* $\varphi^{(n)}(0)$ *exists.* For a random variable $\mathbf{X}$ we can therefore write

$$\mathbf{E}(\mathbf{X}) = -\varphi'(0), \qquad \mathbf{E}(\mathbf{X}^2) = \varphi''(0) \tag{2.5}$$

with the obvious conventions in case of divergence. The differentiation rule (2.4) remains valid for arbitrary measures F.

(iii) **Integration by parts** leads from (1.1) to

$$\int_0^\infty e^{-\lambda x} \, F(x) \, dx = \frac{\varphi(\lambda)}{\lambda}, \qquad \lambda > 0. \tag{2.6}$$

For probability distributions it is sometimes preferable to rewrite (2.6) in terms of the tail

$$\int_0^\infty e^{-\lambda x}[1 - F(x)] \, dx = \frac{1 - \varphi(\lambda)}{\lambda}. \tag{2.7}$$

This corresponds to formula **1**; XI,(1.6) for generating functions.

(iv) Change of scale. From (1.2) we have $\mathbf{E}(e^{-a\lambda\mathbf{X}}) = \varphi(a\lambda)$ for each fixed $a > 0$, and so $\varphi(a\lambda)$ *is the transform of the distribution* $F\{dx/a\}$ [with distribution function $F(x/a)$]. This relation is in constant use.

Example. (*d*) *Law of large numbers.* Let $\mathbf{X}_1, \mathbf{X}_2, \ldots$ be independent random variables with a common Laplace transform φ. Suppose $\mathbf{E}(\mathbf{X}j) = \mu$. The Laplace transform of the sum $\mathbf{X}_1 + \cdots + \mathbf{X}_n$ is φ^n, and hence the transform of the average $[\mathbf{X}_1 + \cdots + \mathbf{X}_n]/n$ is given by $\varphi^n(\lambda/n)$. Near the origin $\varphi(\lambda) = 1 - \mu\lambda + o(\lambda)$ [see (2.5)] and so as $n \to \infty$

$$\lim \varphi^n\left(\frac{\lambda}{n}\right) = \lim \left(1 - \frac{\mu\lambda}{n}\right)^n = e^{-\mu\lambda}. \tag{2.8}$$

But $e^{-\mu\lambda}$ is the transform of the distribution concentrated at μ, and so the distribution of $[\mathbf{X}_1 + \cdots + \mathbf{X}_n]/n$ tends to this limit. This is the weak law of large numbers in the Khintchine version, which does not require the existence of a variance. True, the proof applies directly only to positive variables, but it illustrates the elegance of Laplace transform methods. ►

3. EXAMPLES

(*a*) *Uniform distribution.* Let F stand for the uniform distribution concentrated on $\overline{0, 1}$. Its Laplace transform is given by $\varphi(\lambda) = (1 - e^{-\lambda})/\lambda$. Using the binomial expansion it is seen that the n-fold convolution $F^{n\star}$ has the transform

$$\varphi^n(\lambda) = \sum_{k=0}^{n} (-1)^k \binom{n}{k} e^{-\lambda k} \lambda^{-n}. \tag{3.1}$$

As λ^{-n} is the transform corresponding to $U(x) = x^n/n!$ example 2(*c*) shows that $e^{-k\lambda}\lambda^{-n}$ corresponds to $(x-k)_+^n/n!$ where x_+ denotes the function that equals 0 for $x \leq 0$ and x for $x \geq 0$. Thus

$$F^{n\star}(x) = \frac{1}{n!} \sum_{k=0}^{n} (-1)^k \binom{n}{k} (x-k)_+^n. \tag{3.2}$$

This formula was derived by direct calculation in I,(9.5) and by a passage to the limit in problem 20 of **1**; XI.

(*b*) *Stable distributions with exponent* $\frac{1}{2}$. The distribution function

$$G(x) = 2[1 - \mathfrak{N}(1/\sqrt{x})], \qquad x > 0 \tag{3.3}$$

(where $\mathfrak{N}$ is the standard normal distribution) has the Laplace transform

$$\gamma(\lambda) = e^{-\sqrt{2\lambda}}. \tag{3.4}$$

This can be verified by elementary calculations, but they are tedious and we prefer to derive (3.4) from the limit theorem 3 in **1**; III,7 in which the distribution G was first encountered. Consider a simple symmetric random walk (coin tossing), and denote by $\mathbf{T}$ the epoch of the first return to the origin. The cited limit theorem states that G is the limit distribution of the normalized sums $(\mathbf{T}_1+\cdots+\mathbf{T}_n)/n^2$, where $\mathbf{T}_1, \mathbf{T}_2, \ldots$ are independent random variables distributed like $\mathbf{T}$. According to **1**; XI,(3.14) the generating function of $\mathbf{T}$ is given by $f(s) = 1 - \sqrt{1-s^2}$, and therefore

$$(3.5) \qquad \gamma(\lambda) = \lim\,[1-\sqrt{1-e^{-2\lambda/n^2}}]^n = \lim\left[1 - \frac{\sqrt{2\lambda}}{n}\right]^n = e^{-\sqrt{2\lambda}}.$$

We have mentioned several times that G *is a stable distribution*, but again the direct computational verification is laborious. Now obviously $\gamma^n(\lambda) = \gamma(n^2\lambda)$ which is the same as $G^{n\star}(x) = G(n^{-2}x)$ and proves the stability without effort.

(*c*) *Power series and mixtures.* Let F be a probability distribution with Laplace transform $\varphi(\lambda)$. We have repeatedly encountered distributions of the form

$$(3.6) \qquad G = \sum_{k=0}^{\infty} p_k F^{k\star}$$

where $\{p_k\}$ is a probability distribution. If $P(s) = \sum p_k s^k$ stands for the generating function of $\{p_k\}$, the Laplace transform of G is obviously given by

$$(3.7) \qquad \gamma(\lambda) = \sum_{k=0}^{\infty} p_k \varphi^k(\lambda) = P(\varphi(\lambda)).$$

This principle can be extended to arbitrary power series with positive coefficients. We turn to specific applications.

(*d*) *Bessel function densities.* In example II,7(*c*) we saw that for $r = 1, 2, \ldots$ the density

$$(3.8) \qquad v_r(x) = e^{-x}\frac{r}{x}I_r(x)$$

corresponds to a distribution of the form (3.6) where F is exponential with $\varphi(\lambda) = 1/(\lambda+1)$, and $\{p_k\}$ is the distribution of the first-passage epoch through the point $r > 0$ in an ordinary symmetric random walk. The generating function of this distribution is

$$(3.9) \qquad P(s) = \left(\frac{1 - \sqrt{1 - s^2}}{s}\right)^r$$

[see **1**; XI,(3.6)]. Substituting $s = (1 + \lambda)^{-1}$ we conclude that *the ordinary Laplace transform of the probability density* (3.8) *is given by*

$$(3.10) \qquad [\lambda + 1 - \sqrt{(\lambda+1)^2 - 1}\,]^r.$$

That v_r *is a probability density and* (3.10) *its transform* has been proved only for $r = 1, 2, \ldots$. However, the statement is true[5] for *all* $r > 0$. It is of probabilistic interest because it implies the convolution formula $v_r * v_s = v_{r+s}$ and thus the *infinite divisibility of* v_r. (See section 7.)

(*e*) *Another Bessel density*. In (3.6) choose for F the exponential distribution with $\varphi(\lambda) = 1/(\lambda+1)$ and for $\{p_k\}$ the Poisson distribution with $P(s) = e^{-t+ts}$. It is easy to calculate G explicitly, but fortunately this task was already accomplished in example II,7(*a*). We saw there that the density

$$w_\rho(x) = e^{-t-x}\sqrt{(x/t)^\rho}\, I_\rho(2\sqrt{tx}) \tag{3.11}$$

defined in II,(7.2) is the convolution of our distribution G with a gamma density $f_{1,\rho+1}$. It follows that the ordinary Laplace transform of w_ρ is the product of our γ with the transform of $f_{1,\rho+1}$, namely $(\lambda+1)^{\rho+1}$. Accordingly, *the probability density* (3.11) *has the Laplace transform*

$$\frac{1}{(\lambda+1)^{\rho+1}}\, e^{-t+t/(\lambda+1)}. \tag{3.12}$$

For $t = 1$ we see using the translation rule (1.9) that $\sqrt{x}_\rho\, I_\rho(2\sqrt{x})$ *has the ordinary transform* $\lambda^{-\rho-1}e^{1/\lambda}$.

(*f*) *Mixtures of exponential densities*. Let the density f be of the form

$$f(x) = \sum_{k=1}^{n} p_k a_k e^{-a_k x}, \quad p_k > 0, \qquad \sum_{k=1}^{n} p_k = 1 \tag{3.13}$$

where for definiteness we assume $0 < a_1 < \cdots < a_n$. The corresponding Laplace transform is given by

$$\varphi(\lambda) = \sum_{k=1}^{n} p_k \frac{a_k}{\lambda + a_k} = \frac{Q(\lambda)}{P(\lambda)} \tag{3.14}$$

where P is a polynomial of degree n with roots $-a_k$, and Q is a polynomial of degree $n - 1$. Conversely, for any polynomial Q of degree $n - 1$ the ratio $Q(\lambda)/P(\lambda)$ admits of a partial fraction expansion of the form (3.14) with

$$a_r\, p_r = \frac{Q(-a_r)}{P'(-a_r)} \tag{3.15}$$

[see **1**; XI,(4.5)]. For (3.14) to correspond to a mixture (3.13) it is necessary and sufficient that $p_r > 0$ and that $Q(0)/P(0) = 1$. From the graph of P it is clear that $P'(-a_r)$ and $P'(-a_{r+1})$ are of opposite signs, and hence the same must be true of $Q(-a_r)$ and $Q(-a_{r+1})$. In other words, it is necessary that Q has a root $-b_r$ between $-a_r$ and $-a_{r+1}$. But as Q

[5] This result is due to H. Weber. The extremely difficult analytic proof is now replaced by an elementary proof in J. Soc. Industr. Appl. Math., vol. 14 (1966) pp. 864–875.

cannot have more than $n - 1$ roots $-b_r$ we conclude that these must satisfy

$$0 < a_1 < b_1 < a_2 < b_2 < \cdots < b_{n-1} < a_n. \tag{3.16}$$

This guarantees that all p_r are of the same sign, and we reach the following conclusion: Let P and Q be polynomials of degree n and $n - 1$, respectively, and $Q(0)/P(0) = 1$. *In order that* $Q(\lambda)/P(\lambda)$ *be the Laplace transform of a mixture* (3.13) *of exponential densities it is necessary and sufficient that the roots* $-a_r$ *of* P *and* $-b_r$ *of* Q *be distinct and* (*with proper numbering*) *satisfy* (3.16). ►

4. COMPLETELY MONOTONE FUNCTIONS. INVERSION FORMULAS

As we saw in VII,2 a function f in $\overline{0, 1}$ is a generating function of a positive sequence $\{f_n\}$ iff f is absolutely monotone, that is, if f possesses positive derivatives $f^{(n)}$ of all orders. An analogous theorem holds for Laplace transforms, except that now the derivatives alternate in sign.

Definition 1. *A function* φ *on* $\overline{0, \infty}$ *is completely monotone if it possesses derivatives* $\varphi^{(n)}$ *of all orders and*

$$(-1)^n \varphi^{(n)}(\lambda) \geq 0, \qquad \lambda > 0. \tag{4.1}$$

As $\lambda \to 0$ the values $\varphi^{(n)}(\lambda)$ approach finite or infinite limits which we denote by $\varphi^{(n)}(0)$. Typical examples are $1/\lambda$ and $1/(1+\lambda)$.

The following beautiful theorem due to S. Bernstein (1928) was the starting point of much research, and the proof has been simplified by stages. We are able to give an extremely simple proof because the spade work was laid by the characterization of generating functions derived in theorem 2 of VII,2 as a consequence of the law of large numbers.

Theorem 1. *A function* φ *on* $\overline{0, \infty}$ *is the Laplace transform of a probability distribution* F, *iff it is completely monotone, and* $\varphi(0) = 1$.

We shall prove a version of this theorem which appears more general in form, but can actually be derived from the restricted version by an appeal to the translation principle explained in connection with (1.9).

Theorem 1a. *The function* φ *on* $\overline{0, \infty}$ *is completely monotone iff it is of the form*

$$\varphi(\lambda) = \int_0^\infty e^{-\lambda x}\, F\{dx\}, \qquad \lambda > 0, \tag{4.2}$$

where F *is not a necessarily finite measure on* $\overline{0, \infty}$.

(By our initial convention the interval of integration is *closed:* a possible atom of F at the origin has the effect that $\varphi(\infty) > 0$.)

Proof. The necessity of the condition follows by formal differentiation as in (2.4). Assuming φ to be completely monotone consider $\varphi(a-as)$ for fixed $a > 0$ and $0 < s < 1$ as a function of s. Its derivatives are evidently positive and by theorem 2 of VII,2 the Taylor expansion

$$\varphi(a-as) = \sum_{n=0}^{\infty} \frac{(-a)^n \varphi^{(n)}(a)}{n!} s^n \tag{4.3}$$

is valid for $0 \leq s < 1$. Thus

$$\varphi_a(\lambda) = \varphi(a-ae^{-\lambda/a}) = \sum_{n=0}^{\infty} \frac{(-a)^n \varphi^{(n)}(a)}{n!} e^{-n\lambda/a} \tag{4.4}$$

is the Laplace transform of an arithmetic measure attributing mass $(-a)^n\varphi^{(n)}(a)/n!$ to the point n/a (where $n = 0, 1, \ldots$). Now $\varphi_a(\lambda) \to \varphi(\lambda)$ as $a \to \infty$. By the extended continuity theorem there exists therefore a measure F such that $F_a \to F$ and φ is its Laplace transform. ▶

We have not only proved theorem 1*a*, but the relation $F_a \to F$ may be restated in the form of the important

Theorem 2. (*Inversion formula.*) *If* (4.2) *holds for* $\lambda > 0$, *then at all points of continuity*[6]

$$F(x) = \lim_{a\to\infty} \sum_{n \leq ax} \frac{(-a)^n}{n!} \varphi^{(n)}(a). \tag{4.5}$$

This formula is of great theoretical interest and permits various conclusions. The following boundedness criterion may serve as an example of particular interest for semi-group theory. (See problem 13.)

Corollary. *For* φ *to be of the form*

$$\varphi(\lambda) = \int_0^\infty e^{-\lambda x} f(x)\, dx \qquad \textit{where} \quad 0 \leq f \leq C \tag{4.6}$$

it is necessary and sufficient that

$$0 \leq \frac{(-a)^n \varphi^{(n)}(a)}{n!} \leq \frac{C}{a} \tag{4.7}$$

for all $a > 0$.

[6] The inversion formula (4.5) was derived in VII,(6.4) as a direct consequence of the law of large numbers. In VII,(6.6) we have an analogous *inversion formula for integrals of the form* (4.6) *with continuous* f (not necessarily positive).

Proof. Differentiating (4.6) under the integral we get (4.7) [see (2.4)]. Conversely, (4.7) implies that φ is completely monotone and hence the transform of a measure F. Substituting from (4.7) into (4.5) we conclude that

$$F(x_2) - F(x_1) \leq C(x_2 - x_1)$$

for any pair $x_1 < x_2$. This means that F has bounded difference ratios and hence F is the integral of a function $f \leq C$ (see V,3). ►

Theorem 1 leads to simple *tests* that a given function is the Laplace transform of a probability distribution. The standard technique is illustrated by the proof of

Criterion 1. *If φ and ψ are completely monotone so is their product $\varphi\psi$.*

Proof. We show by induction that the derivatives of $\varphi\psi$ alternate in sign. Assume that for *every* pair φ, ψ of completely monotone functions the first n derivatives of $\varphi\psi$ alternate in sign. As $-\varphi'$ and $-\psi'$ are completely monotone the induction hypothesis applies to the products $-\varphi'\psi$ and $-\varphi\psi'$, and we conclude from $-(\varphi\psi)' = -\varphi'\psi - \varphi\psi'$ that in fact the first $n + 1$ derivatives of $\varphi\psi$ alternate in sign. Since the hypothesis is trivially true for $n = 1$ the criterion is proved. ►

The same proof yields the useful

Criterion 2. *If φ is completely monotone and ψ a positive function with a completely monotone derivative then $\varphi(\psi)$ is completely monotone. (In particular, $e^{-\psi}$ is completely monotone.)*

Typical applications are given in section 6 and in the following example, which occurs frequently in the literature with unnecessary complications.

Example. (*a*) *An equation occurring in branching processes.* Let φ be the Laplace transform of a probability distribution F with expectation $0 < \mu \leq \infty$, and let $c > 0$. We prove that *the equation*

$$\beta(\lambda) = \varphi(\lambda + c - c\beta(\lambda)) \tag{4.8}$$

has a unique root $\beta(\lambda) \leq 1$ and β is the Laplace transform of a distribution B which is proper iff $\mu c \leq 1$, defective otherwise.

(See XIV,4 for applications and references.)

Proof. Consider the equation

$$\varphi(\lambda + c - cs) - s = 0 \tag{4.9}$$

for fixed $\lambda > 0$ and $0 \leq s \leq 1$. The left side is a convex function which assumes a negative value at $s = 1$ and a positive value at $s = 0$. It follows that there exists a unique root.

To prove that the root $\beta(\lambda)$ is a Laplace transform put $\beta_0 = 0$ and recursively $\beta_{n+1} = \varphi(\lambda+c-c\beta_n)$. Then $\beta_0 \leq \beta_1 \leq 1$ and since φ is decreasing this implies $\beta_1 \leq \beta_2 \leq 1$, and by induction $\beta_n \leq \beta_{n+1} \leq 1$. The limit of the bounded monotone sequence $\{\beta_n\}$ satisfies (4.8) and hence $\beta = \lim \beta_n$. Now $\beta_1(\lambda) = \varphi(\lambda+c)$ is completely monotone and criterion 2 shows recursively that $\beta_2, \beta_3, \ldots$ are completely monotone. By the continuity theorem the same is true of the limit β, and hence β is the Laplace transform of a measure B. Since $\beta(\lambda) \leq 1$ for all λ the total mass of B is $\beta(0) \leq 1$. It remains to decide under what conditions $\beta(0) = 1$.

By construction $s = \beta(0)$ is the *smallest* root of the equation

$$\varphi(c-cs) - s = 0. \tag{4.10}$$

Considered as a function of s the left side is convex; it is positive for $s = 0$ and vanishes for $s = 1$. A second root $s < 1$ exists therefore iff at $s = 1$ the derivative is positive, that is iff $-c\varphi'(0) > 1$. Otherwise $\beta(0) = 1$ and β is the Laplace transform of a proper probability distribution B. Hence B is proper iff $-c\varphi'(0) = c\mu \leq 1$. ►

5. TAUBERIAN THEOREMS

Let U be a measure concentrated on $\overline{0, \infty}$ and such that its Laplace transform

$$\omega(\lambda) = \int_0^\infty e^{-\lambda x}\, U\{dx\} \tag{5.1}$$

exists for $\lambda > 0$. It will be convenient to describe the measure U in terms of its improper distribution function defined for $x \geq 0$ by $U\{\overline{0, x}\}$. We shall see that under fairly general conditions the behavior of ω near the origin uniquely determines the asymptotic behavior of $U(x)$ as $x \to \infty$ and vice versa. Historically any relation describing the asymptotic behavior of U in terms of ω is called a Tauberian theorem, whereas theorems describing the behavior of ω in terms of U are usually called Abelian. We shall make no distinction between these two classes because our relations will be symmetric.

To avoid unsightly formulas involving reciprocals we introduce two positive variables t and τ related by

$$t\tau = 1. \tag{5.2}$$

Then $\tau \to 0$ when $t \to \infty$.

To understand the background of the Tauberian theorems note that for fixed t the change of variables $x = ty$ in (5.1) shows that $\omega(\tau\lambda)$ is the Laplace transform corresponding to the improper distribution function

$U(ty)$. Since ω decreases it is possible to find a sequence $\tau_1, \tau_2, \ldots \to 0$ such that as τ runs through it

$$\frac{\omega(\tau\lambda)}{\omega(\tau)} \to \gamma(\lambda) \tag{5.3}$$

with $\gamma(\lambda)$ finite at least for $\lambda > 1$. By the extended continuity theorem the limit γ is the Laplace transform of a measure G and as t runs through the reciprocals $t_k = 1/\tau_k$

$$\frac{U(tx)}{\omega(\tau)} \to G(x) \tag{5.4}$$

at all points of continuity of G. For $x = 1$ it is seen that the asymptotic behavior of $U(t)$ as $t \to \infty$ is intimately connected with the behavior of $\omega(t^{-1})$.

In principle we could formulate this fact as an all-embracing Tauberian theorem, but it would be too clumsy for practical use. To achieve reasonable simplicity we consider only the case where (5.3) is valid for *any* approach $\tau \to 0$, that is, when ω varies regularly at 0. The elementary lemma[7] 1 of VIII,8 states that the limit γ is necessarily of the form $\gamma(\lambda) = \lambda^{-\rho}$ with $\rho \geq 0$. The corresponding measure is given by $G(x) = x^\rho/\Gamma(\rho+1)$, and (5.4) implies that U varies regularly and the exponents of ω and U are the same in absolute value. We formulate this important result together with its converse in

Theorem 1. *Let U be a measure with a Laplace transform ω defined for $\lambda > 0$. Then each of the relations*

$$\frac{\omega(\tau\lambda)}{\omega(\tau)} \to \frac{1}{\lambda^\rho}, \qquad \tau \to 0 \tag{5.5}$$

and

$$\frac{U(tx)}{U(t)} \to x^\rho, \qquad t \to \infty \tag{5.6}$$

implies the other as well as

$$\omega(\tau) \sim U(t)\,\Gamma(\rho+1). \tag{5.7}$$

Proof. (*a*) Assume (5.5). The left side is the Laplace transform corresponding to $U(tx)/\omega(\tau)$, and by the extended continuity theorem this implies

$$\frac{U(tx)}{\omega(\tau)} \to \frac{x^\rho}{\Gamma(\rho+1)}. \tag{5.8}$$

For $x = 1$ we get (5.7), and substituting this back into (5.8) we get (5.6).

[7] This lemma is used *only* to justify the otherwise artificial form of the relations (5.5) and (5.6). The theory of regular variation is *not* used in this section [except for the side remark that (5.18) implies (5.16)].

(*b*) Assume (5.6). Taking Laplace transforms we get

$$\frac{\omega(\tau\lambda)}{U(t)} \to \frac{\Gamma(\rho+1)}{\lambda^\rho} \tag{5.9}$$

provided the extended continuity theorem is applicable, that is, *provided* the left-side remains bounded for some λ. As under (*a*) it is seen that (5.9) implies (5.7) and (5.5), and to prove the theorem it suffices to verify that $\omega(\tau)/U(t)$ remains bounded.

On partitioning the domain of integration by the points $t, 2t, 4t, \ldots$ it is clear that

$$\omega(\tau) \leq \sum_0^\infty e^{-2^{n-1}}\, U(2^n t). \tag{5.10}$$

In view of (5.7) there exists a t_0 such that $U(2t) < 2^{\rho+1}\, U(t)$ for $t > t_0$. Repeated application of this inequality yields

$$\frac{\omega(\tau)}{U(t)} \leq \sum_0^\infty 2^{n(\rho+1)} e^{-2^{n-1}} \tag{5.11}$$

and so the left side indeed remains bounded as $t \to \infty$. ►

Examples. (*a*) $U(x) \sim \log^2 x$ as $x \to \infty$ iff $\omega(\lambda) \sim \log^2 \lambda$ as $\lambda \to 0$. Similarly $U(x) \sim \sqrt{x}$ iff $\omega(\lambda) \sim \frac{1}{2}\sqrt{\pi/\lambda}$.

(*b*) Let F be a probability distribution with Laplace transform φ. The measure $U\{dx\} = x\, F\{dx\}$ has the transform $-\varphi'$. Hence if $-\varphi'(\lambda) \sim \mu\lambda^{-\rho}$ as $\lambda \to \infty$ then

$$U(x) = \int_0^x y\, F\{dy\} \sim \frac{\mu}{\Gamma(\rho+1)}\, x^\rho, \qquad x \to \infty$$

and vice versa. This generalizes the differentiation rule (2.4) which is contained in (5.7) for $\rho = 0$. ►

It is sometimes useful to know to what extent the theorem remains valid in the limit $\rho \to \infty$. We state the result in the form of a

Corollary. *If for some* $a > 1$ *as* $t \to \infty$

$$\text{either} \quad \frac{\omega(\tau a)}{\omega(\tau)} \to 0 \quad \text{or} \quad \frac{U(ta)}{U(t)} \to \infty \tag{5.12}$$

then

$$\frac{U(t)}{\omega(\tau)} \to 0. \tag{5.13}$$

Proof. The first relation in (5.12) implies that $\omega(\tau\lambda)/\omega(\tau) \to 0$ for $\lambda > a$ and by the extended continuity theorem $U(tx)/\omega(\tau) \to 0$ for all $x > 0$. The second relation in (5.12) entails (5.13) because

$$\omega(\tau) \geq \int_0^{at} e^{-x/t}\, U\{dx\} \geq e^{-a}\, U(ta). \qquad \blacktriangleright$$

In applications it is more convenient to express theorem 1 in terms of slow variation. We recall that a positive function L defined on $\overline{0, \infty}$ *varies slowly at* ∞ if for every fixed x

$$\frac{L(tx)}{L(t)} \to 1, \qquad t \to \infty. \tag{5.14}$$

L varies slowly at 0 if this relation holds as $t \to 0$, that is, if $L(1/x)$ varies slowly at ∞. Evidently U satisfies (5.6) iff $U(x)/x^\rho$ varies slowly at ∞ and similarly (5.5) holds iff $\lambda^\rho \omega(\lambda)$ varies slowly at 0. Consequently theorem 1 may be rephrased as follows.

Theorem 2. *If L is slowly varying at infinity and $0 \leq \rho < \infty$, then each of the relations*

$$\omega(\tau) \sim \tau^{-\rho}\, L\left(\frac{1}{\tau}\right), \qquad \tau \to 0, \tag{5.15}$$

and

$$U(t) \sim \frac{1}{\Gamma(\rho+1)}\, t^\rho\, L(t), \qquad t \to \infty \tag{5.16}$$

implies the other.

Theorem 2 has a glorious history. The implication (5.16) $\to$ (5.15) (from the measure to the transform) is called an Abelian theorem; the converse (5.15) $\to$ (5.16) (from transform to measure), a Tauberian theorem. In the usual setup, the two theorems are entirely separated, the Tauberian part causing the trouble. In a famous paper G. H. Hardy and J. E. Littlewood treated the case $\omega(\lambda) \sim \lambda^{-\rho}$ by difficult calculations. In 1930, J. Karamata created a sensation by a simplified proof for this special case. (This proof is still found in texts on complex variables and Laplace transforms.) Soon afterwards he introduced the class of regularly varying functions and proved theorem 2; the proof was too complicated for textbooks, however. The notion of slow variation was introduced by R. Schmidt about 1925 in the same connection. Our proof simplifies and unifies the theory and leads to the little-known, but useful, corollary.

A great advantage of our proof is that it applies without change when the roles of infinity and zero are interchanged, that is, if $\tau \to \infty$ while $t \to 0$. In this way we get the dual theorem connecting the behavior of ω at infinity with that of U at the origin. [It will not be used in this book except to derive (6.2).]

Theorem 3. *The last two theorems and the corollary remain valid when the roles of the origin and infinity are interchanged, that is, for $\tau \to \infty$ and $t \to 0$.*

Theorem 2 represents the main result of this section, but for completeness we derive two useful complements. First of all, when U has a density $U' = u$ it is desirable to obtain estimates for u. This problem cannot be treated in full generality, because a well-behaved distribution U can have an extremely ill-behaved density u. In most applications, however, the density u will be *ultimately monotone*, that is, monotone in some interval $\overline{x_0, \infty}$. For such densities we have

Theorem 4.[8] *Let* $0 < \rho < \infty$. *If* U *has an ultimately monotone derivative* u *then as* $\lambda \to 0$ *and* $x \to \infty$, *respectively,*

$$\omega(\lambda) \sim \frac{1}{\lambda^\rho} L\left(\frac{1}{\lambda}\right) \quad \textit{iff} \quad u(x) \sim \frac{1}{\Gamma(\rho)} x^{\rho-1} L(x). \tag{5.17}$$

(For a formally stronger version see problem 16.)

Proof. The assertion is an immediate consequence of theorem 2 and the following

Lemma. *Suppose that* U *has an ultimately monotone density* u. *If* (5.16) *holds with* $\rho > 0$ *then*

$$u(x) \sim \rho U(x)/x, \qquad x \to \infty. \tag{5.18}$$

[Conversely, (5.18) implies (5.16) even if u is not monotone. This is contained in VIII,(9.6) with $Z = u$ and $p = 0$.]

Proof. For $0 < a < b$

$$\frac{U(tb) - U(ta)}{U(t)} = \int_a^b \frac{u(ty)t}{U(t)}\, dy. \tag{5.19}$$

As $t \to \infty$ the left side tends to $b^\rho - a^\rho$. For sufficiently large t the integrand is monotone, and then (5.16) implies that it remains bounded as $t \to \infty$. By the selection theorem of VIII,6 there exists therefore a sequence $t_1, t_2, \ldots \to \infty$ such that as t runs through it

$$\frac{u(ty)t}{U(t)} \to \psi(y) \tag{5.20}$$

at all points of continuity. It follows that the integral of ψ over $\overline{a, b}$ equals $b^\rho - a^\rho$, and so $\psi(y) = \rho y^{\rho-1}$. This limit being independent of the sequence $\{t_k\}$ the relation (5.20) is true for an arbitrary approach $t \to \infty$, and for $y = 1$ it reduces to (5.18). ▶

[8] This includes the famous Tauberian theorem of E. Landau. Our proof serves as a new example of how the selection theorem obviates analytical intricacies.

Example. (*c*) For a probability distribution F with characteristic function φ we have [see (2.7)]

$$\int_0^\infty e^{-\lambda x}[1 - F(x)]\, dx = [1 - \varphi(\lambda)]/\lambda. \tag{5.21}$$

Since $1 - F$ is monotone each of the relations

$$1 - \varphi(\lambda) \sim \lambda^{1-\rho}L(1/\lambda) \quad \textit{and} \quad 1 - F(x) \sim \frac{1}{\Gamma(\rho)}\, x^{\rho-1}L(x) \tag{5.22}$$

($\rho > 0$) implies the other. The next section will illustrate the usefulness of this observation. ►

The use of this theorem is illustrated in the next section. In conclusion we show how theorem 2 leads to a *Tauberian theorem for power series.* [It is used in XII,(8.10) and in XVII,5.]

Theorem 5. *Let* $q_n \geq 0$ *and suppose that*

$$Q(s) = \sum_{n=0}^{\infty} q_n s^n \tag{5.23}$$

converges for $0 \leq s < 1$. *If* L *varies slowly at infinity and* $0 \leq \rho < \infty$ *then each of the two relations*

$$Q(s) \sim \frac{1}{(1-s)^\rho}\, L\left(\frac{1}{1-s}\right), \qquad s \to 1- \tag{5.24}$$

and

$$q_0 + q_1 + \cdots + q_{n-1} \sim \frac{1}{\Gamma(\rho+1)}\, n^\rho\, L(n), \qquad n \to \infty \tag{5.25}$$

implies the other.

Furthermore, if the sequence $\{q_n\}$ *is monotonic and* $0 < \rho < \infty$, *then* (5.24) *is equivalent to*

$$q_n \sim \frac{1}{\Gamma(\rho)}\, n^{\rho-1}\, L(n), \qquad n \to \infty. \tag{5.26}$$

Proof. Let U be the measure with density u defined by

$$u(x) = q_n \qquad \text{for} \quad n \leq x < n + 1. \tag{5.27}$$

The left side in (5.25) equals $U(n)$. The Laplace transform ω of U is given by

$$\omega(\lambda) = \frac{1 - e^{-\lambda}}{\lambda} \sum_{n=0}^{\infty} q_n e^{-n\lambda} = \frac{1 - e^{-\lambda}}{\lambda}\, Q(e^{-\lambda}). \tag{5.28}$$

It is thus seen that the relations (5.24) and (5.25) are equivalent to (5.15) and (5.16), respectively, and hence they imply each other by virtue of theorem 2. Similarly, (5.26) is an immediate consequence of theorem 4. ►

Example. (*d*) Let $q_n = n^{\rho-1}\log^a n$ where $\rho > 0$ and a is arbitrary. The sequence $\{q_n\}$ is ultimately monotone and so (5.24) holds with $L(t) = \Gamma(\rho)\log^a t$. ▶

*6. STABLE DISTRIBUTIONS

To show the usefulness of the Tauberian theorems we now derive the most general stable distributions concentrated on $\overline{0, \infty}$ and give a complete characterization of their domains of attraction. The proofs are straightforward and of remarkable simplicity when compared with the methods required for distributions not concentrated on $\overline{0, \infty}$.

Theorem 1. *For fixed* $0 < \alpha < 1$ *the function* $\gamma_\alpha(\lambda) = e^{-\lambda^\alpha}$ *is the Laplace transform of a distribution* G_α *with the following properties:*

G_α *is stable; more precisely, if* $\mathbf{X}_1, \ldots, \mathbf{X}_n$ *are independent variables with the distribution* G_α, *then* $(\mathbf{X}_1+\cdots+\mathbf{X}_n)/n^{1/\alpha}$ *has again the distribution* G_α.

$$x^\alpha[1 - G_\alpha(x)] \to \frac{1}{\Gamma(1-\alpha)}, \qquad x \to \infty, \tag{6.1}$$

$$e^{x^{-\alpha}}G_\alpha(x) \to 0, \qquad x \to 0. \tag{6.2}$$

Proof. The function γ_α is completely monotone by the second criterion of section 4, because $e^{-\lambda}$ is completely monotone and λ^α has a completely monotone derivative. Since $\gamma_\alpha(0) = 1$, the measure G_α with Laplace transform γ_α has total mass 1. The asserted stability property is obvious since $\gamma_\alpha^n(\lambda) = \gamma_\alpha(n^{1/\alpha}\lambda)$.

(6.1) is a special case of (5.22), and (6.2) is an immediate consequence of theorem 3 and the corollary to theorem 1 of the preceding section. ▶

Theorem 2. *Suppose that* F *is a probability distribution concentrated on* $\overline{0, \infty}$ *such that*

$$F^{n\star}(a_n x) \to G(x) \tag{6.3}$$

(*at points of continuity*) *where* G *is a proper distribution not concentrated at a single point. Then*

(*a*) *There exists a function* L *that varies slowly at infinity*[9] *and a constant* α *with* $0 < \alpha < 1$ *such that*

$$1 - F(x) \sim \frac{x^{-\alpha}L(x)}{\Gamma(1-\alpha)} \qquad x \to \infty. \tag{6.4}$$

* Except for (6.2) the results of this section are derived independently in chapters IX and XVII. Stable distributions were introduced in VI,1.

[9] That is, L satisfies (5.14). The norming factor $\Gamma(1-\alpha)$ in (6.4) is a matter of convenience and affects only notations.

(*b*) *Conversely, if* F *is of the form* (6.4) *it is possible to choose* a_n *such that*

$$\frac{nL(a_n)}{a_n^\alpha} \to 1, \tag{6.5}$$

and in this case (6.3) *holds with* $G = G_\alpha$.

This implies that the possible limits G in (6.3) differ only by scale factors from some G_α. It follows, in particular, that *there are no other stable distributions concentrated on* $\overline{0, \infty}$. (See lemma 1 of VIII,2.)

Proof. If φ and γ are the Laplace transforms of F and G, then (6.3) is equivalent to

$$-n \log \varphi(\lambda/a_n) \to -\log \gamma(\lambda). \tag{6.6}$$

By the simple theorem of VIII,8 this implies that $-\log \varphi$ varies regularly at the origin, that is

$$-\log \varphi(\lambda) \sim \lambda^\alpha L(1/\lambda), \qquad \lambda \to 0, \tag{6.7}$$

with L varying slowly at infinity and $\alpha \geq 0$. From (6.6) then $-\log \gamma(\lambda) = C\lambda^\alpha$. Since G is not concentrated at a single point we have $0 < \alpha < 1$.

Now (6.7) implies

$$\frac{1 - \varphi(\lambda)}{\lambda} \sim \lambda^{\alpha-1} L\left(\frac{1}{\lambda}\right), \qquad \lambda \to 0. \tag{6.8}$$

In view of (5.22) the two relations (6.4) and (6.8) imply each other. Accordingly, (6.4) is necessary for (6.1) to hold.

For the converse part we start from (6.4) which was just shown to imply (6.8). For fixed n define a_n as the lower bound of all x such that $n[1 - F(x)]\ 1/\Gamma(1 - \alpha)$. Then (6.5) holds. Using this and the slow variation of L we conclude from (6.8) that

$$1 - \varphi(\lambda/a_n) \sim \lambda^\alpha a_n^{-\alpha} L(\sigma_n/\lambda) \sim \lambda^\alpha/n. \tag{6.9}$$

It follows that the left side in (6.6) tends to λ^α, and this concludes the proof. ▶

(See problem 26 for the influence of the maximal term.)

*7. INFINITELY DIVISIBLE DISTRIBUTIONS

According to the definition in VI,3 a probability distribution U with Laplace transform ω is infinitely divisible iff for $n = 1, 2, \ldots$ the positive nth root $\omega_n = \omega^{1/n}$ is the Laplace transform of a probability distribution.

* Not used in the sequel.

Theorem 1. *The function* ω *is the Laplace transform of an infinitely divisible probability distribution iff* $\omega = e^{-\psi}$ *where* ψ *has a completely monotone derivative and* $\psi(0) = 0$.

Proof. Using the criterion 2 of section 4, it is seen that when $\psi(0) = 0$ and ψ' is completely monotone then $\omega_n = e^{-\psi/n}$ is the Laplace transform of a probability distribution. The condition is therefore sufficient.

To prove the necessity of the condition assume that $\omega_n = e^{-\psi/n}$ is, for each n, the Laplace transform of a probability distribution and put

$$\psi_n(\lambda) = n[1 - \omega_n(\lambda)]. \tag{7.1}$$

Then $\psi_n \to \psi$ and the derivative $\psi_n' = -n\omega_n'$ is completely monotone. By the mean value theorem $\psi_n(\lambda) = \lambda\psi_n'(\theta\lambda) \geq \lambda\psi_n'(\lambda)$, and since $\psi_n \to \psi$ this implies that the sequence $\{\psi_n'(\lambda)\}$ is bounded for each fixed $\lambda > 0$. It is therefore possible to find a convergent subsequence, and the limit is automatically completely monotone by the extended continuity theorem. Thus ψ is an integral of a completely monotone function, and this completes the proof. ▶

An alternative form of this theorem is as follows.

Theorem 2. *The function* ω *is the Laplace transform of an infinitely divisible distribution iff it is of the form* $\omega = e^{-\psi}$ *where*

$$\psi(\lambda) = \int_0^\infty \frac{1 - e^{-\lambda x}}{x} P\{dx\} \tag{7.2}$$

and P *is a measure such that*

$$\int_1^\infty x^{-1} P\{dx\} < \infty. \tag{7.3}$$

Proof. In view of the representation theorem for completely monotone functions the conditions of theorem 1 may be restated to the effect that we must have $\psi(0) = 0$ and

$$\psi'(\lambda) = \int_0^\infty e^{-\lambda x} P\{dx\} \tag{7.4}$$

where P is a measure. Truncating the integral at a changes the equality sign into $\geq$, and this implies that

$$\psi(\lambda) \geq \int_0^a \frac{1 - e^{-\lambda x}}{x} P\{dx\} \tag{7.5}$$

for each $a > 0$ (the integrand being bounded). It follows that (7.2) makes sense and the condition (7.3) is satisfied. Formal differentiation now shows that (7.2) represents the integral of (7.4) vanishing at zero. ▶

[See problems 17–23 and example 9(*a*).]

Examples. (*a*) The compound Poisson distribution

$$U = e^{-c} \sum_{0}^{\infty} \frac{c^n}{n!} F^{n\star} \tag{7.6}$$

has the Laplace transform $e^{-c+c\varphi}$ and (7.2) is true with $P\{dx\} = cx\, F\{dx\}$.

(*b*) *The gamma density* $x^{a-1}\, e^{-x}/\Gamma(a)$ has transform $\omega(\lambda) = 1/(\lambda+1)^a$. Here

$$\psi(\lambda) = a \int_0^{\infty} \frac{1 - e^{-\lambda x}}{x}\, e^{-x}\, dx \tag{7.7}$$

because $\psi'(\lambda) = a(\lambda+1)^{-1} = \omega'(\lambda)/\omega(\lambda)$.

(*c*) *Stable distributions.* For the transform $\omega(\lambda) = e^{-\lambda^\alpha}$ of section 6 we have $\psi(\lambda) = \lambda^\alpha$ and

$$\lambda^\alpha = \frac{\alpha}{\Gamma(1-\alpha)} \int_0^{\infty} \frac{1 - e^{-\lambda x}}{x^{\alpha+1}}\, dx \tag{7.8}$$

as is again seen by differentiation.

(*d*) *Bessel functions.* Consider the density v_r of example 3(*d*) with Laplace transform (3.10). It is obvious from the form of the latter that v_r is the n-fold convolution of $v_{r/n}$ with itself, and hence infinitely divisible. Formal differentiation shows that in this case $\psi'(\lambda) = r/\sqrt{(\lambda+1)^2 - 1}$ and it is easily shown (see problem 6) that this ψ' is of the form (7.4) with

$$P\{dx\} = re^{-x}\, I_0(x)\, dx.$$

(*e*) *Subordination.* It is easily seen from the criteria in section 4 that if ψ_1 and ψ_2 are positive functions with completely monotone derivatives, the composite function $\psi(\lambda) = \psi_1(\psi_2(\lambda))$ has the same property. The corresponding infinitely divisible distribution is of special interest. To find it, denote by $Q_t^{(i)}$ the probability distribution with Laplace transform $e^{-t\psi_i(\lambda)}$ (where $i = 1, 2$), and put

$$U_t(x) = \int_0^{\infty} Q_s^{(2)}(x) Q_t^{(1)}\{ds\}. \tag{7.9}$$

(The distribution U_t is thus obtained by randomization of the parameter s in $Q_s^{(2)}$.) The Laplace transform of U_t is

$$\omega_t(\lambda) = \int_0^{\infty} e^{-s\psi_2(\lambda)} Q_t^{(1)}\{ds\} = e^{-t\psi(\lambda)}. \tag{7.10}$$

Readers of X,7 will recognize in (7.9) the *subordination of processes:* U_t is subordinated to $Q_t^{(2)}$ by the directing process $Q_t^{(1)}$. It is seen with what ease we get the Laplace transforms of the new process although only for the special case that the distributions $Q_t^{(2)}$ are concentrated on the positive half-axis.

A special case deserves attention: if $\psi_1(\lambda) = \lambda^\alpha$ and $\psi_2(\lambda) = \lambda^\beta$ then $\psi(\lambda) = \lambda^{\alpha\beta}$. Thus *a stable α-process directed by a stable β-process leads to a stable $\alpha\beta$-process.* Readers should verify that this statement in substance repeats the assertion of problem 10. For a more general proposition see example VI,2(h).

(f) *Every mixture of exponential distributions is infinitely divisible.*[10] The most general such distribution has a density of the form

$$f(x) = \int_0^\infty se^{-sx}\, U\{ds\} \tag{7.11}$$

where U is a probability distribution. In the special case where U is concentrated on finitely many points $0 < a_1 < \cdots < a_n$ it was shown in example 3(f) that the Laplace transform is of the form

$$\varphi(\lambda) = C \cdot \frac{\lambda + b_1}{\lambda + a_1} \cdots \frac{\lambda + b_{n-1}}{\lambda + a_{n-1}} \cdot \frac{1}{\lambda + a_n} \tag{7.12}$$

with $a_k < b_k < a_{k+1}$. Now

$$-\frac{d}{d\lambda} \log \frac{\lambda + b_k}{\lambda + a_k} = \frac{1}{\lambda + a_k} - \frac{1}{\lambda + b_k} = \frac{b_k - a_k}{(\lambda + a_k)(\lambda + b_k)} \tag{7.13}$$

is the product of two completely monotone functions, and therefore itself completely monotone. It follows that each factor in (7.12) is infinitely divisible and therefore the same is true of φ. For general mixtures the assertion follows by a simple passage to the limit (see problems 20–23). ▶

*8. HIGHER DIMENSIONS

The generalization to higher dimensions is obvious: not even the definition (1.1) requires a change if x is interpreted as column matrix $(x_1, \ldots, x_n)$ and λ as row matrix $(\lambda_1, \ldots, \lambda_n)$. Then

$$\lambda x = \lambda_1 x_1 + \cdots + \lambda_n x_n$$

is the inner product of λ and x. Within probability theory the use of multidimensional transforms is comparatively restricted.

Examples. (a) *Resolvent equation.* Let f be a continuous function in *one* dimension with ordinary Laplace transform $\varphi(\lambda)$. Consider the function

[10] This surprising observation is due to F. W. Steutel, Ann. Math. Statist., vol. 40 (1969), pp. 1130–1131 and vol. 38 (1967), pp. 1303–1305.

* Not used in the sequel.

$f(s+t)$ of the two variables s, t. Its two-dimensional transform is given by

$$\omega(\lambda, \nu) = \int_0^\infty \int_0^\infty e^{-\lambda s - \nu t} f(s+t)\, ds\, dt. \tag{8.1}$$

After the change of variables $s + t = x$ and $-s + t = y$ the integral reduces to

$$\frac{1}{2}\int_0^\infty e^{-\frac{1}{2}(\lambda+\nu)x} f(x)\, dx \int_{-x}^{x} e^{\frac{1}{2}(\lambda-\nu)y}\, dy = \frac{1}{\lambda - \nu}\int_0^\infty (e^{-\nu x} - e^{-\lambda x}) f(x)\, dx.$$

Thus

$$\omega(\lambda, \nu) = -\frac{\varphi(\lambda) - \varphi(\nu)}{\lambda - \nu}. \tag{8.2}$$

We shall encounter this relation in more dignified surroundings as the basic *resolvent equation* for semi-groups [see (10.5) and the concluding remarks to section 10].

(*b*) *Mittag-Leffler functions.* This example illustrates the use of higher dimensions as a technical tool for evaluating simple transforms. We shall prove the following *proposition*:

If F *is stable with Laplace transform* $e^{-\lambda^\alpha}$, *the distribution*

$$G_t(x) = 1 - F(t/x^{1/\alpha}), \qquad x > 0, \tag{8.3}$$

(t *fixed*) *has as Laplace transform the Mittag-Leffler function*

$$\sum_{k=0}^{\infty} \frac{(-\lambda)^k}{\Gamma(1+k\alpha)}\, t^{k\alpha}. \tag{8.4}$$

This result is of considerable interest because in various limit theorems the distribution G appears in company with F [see, for example, XI,(5.6)]. A direct calculation seems difficult, but it is easy to proceed as follows. First keep x fixed and take t as variable. The ordinary Laplace transform $\gamma_t(\nu)$ (with ν as variable) of $G_t(x)$ is obviously $(1-e^{-\nu^\alpha x})/\nu$. Except for the norming factor ν this is a distribution function in x, and *its* Laplace transform is evidently

$$\frac{\nu^{\alpha-1}}{\lambda + \nu^\alpha}. \tag{8.5}$$

This, then, is the bivariate transform of (8.3). In theory it could have been calculated by taking first the transform with respect to x, then t, and so (8.5) is the transform with respect to t of the transform which we seek. But expanding (8.5) into a geometric series one sees that (8.5) is in fact the transform of (8.4) and thus the proposition is proved.

The Mittag-Leffler function (8.4) is a generalization of the exponential to which it reduces when $\alpha = 1$. ►

9. LAPLACE TRANSFORMS FOR SEMI-GROUPS

The notion of Laplace integrals can be generalized to abstract-valued functions and integrals,[11] but we shall consider only Laplace transforms of semi-groups of transformations associated with Markov processes.[12] We return to the basic conventions and notations of X,8.

Let Σ be a space (for example, the line, an interval, or the integers), and $\mathscr{L}$ a Banach space of bounded functions on it with the norm $\|u\| = \sup |u(x)|$. It will be assumed that if $u \in \mathscr{L}$ then also $|u| \in \mathscr{L}$. Let $\{\mathfrak{Q}(t), t > 0\}$ be a continuous semi-group of contractions on $\mathscr{L}$. In other words we assume that for $u \in \mathscr{L}$ there exists a function $\mathfrak{Q}(t)u \in \mathscr{L}$ and that $\mathfrak{Q}(t)$ has the following properties: $0 \leq u \leq 1$ implies $0 \leq \mathfrak{Q}(t)u \leq 1$; furthermore $\mathfrak{Q}(t+s) = \mathfrak{Q}(t)\,\mathfrak{Q}(s)$, and $\mathfrak{Q}(h) \to \mathfrak{Q}(0) = \mathbf{1}$, the identity operator.[13]

We begin by defining integration. Given an arbitrary probability distribution F on $\overline{0, \infty}$ we want to define a contraction operator E from $\mathscr{L}$ to $\mathscr{L}$, to be denoted by

$$E = \int_0^\infty \mathfrak{Q}(s)\, F\{ds\}, \tag{9.1}$$

such that

$$\mathfrak{Q}(t)E = E\mathfrak{Q}(t) = \int_0^\infty \mathfrak{Q}(t+s)\, F\{ds\}. \tag{9.2}$$

(The dependence of E on the distribution F should be kept in mind.)

For a semi-group associated with a Markov process with transition probabilities $Q_t(x, \Gamma)$ this operator E will be induced by the stochastic or substochastic kernel

$$\int_0^\infty Q_s(x, \Gamma)\, F\{ds\}. \tag{9.3}$$

A natural (almost trivial) definition of the operator E presents itself if F is atomic and a simple limiting procedure leads to the desired definition as follows.

---

[11] A fruitful theory covering transforms of the form (9.6) was developed by S. Bochner, *Completely monotone functions in partially ordered spaces*, Duke Math. J., vol. 9 (1942) 519–526. For a generalization permitting an arbitrary family of operators see the book by E. Hille and R. S. Phillips (1957).

[12] The construction of the minimal solution in XIV,7 may serve as a typical example for the present methods.

[13] Recall from X,8 that strong convergence $T_n \to T$ of endomorphisms means $\|T_n u - Tu\| \to 0$ for all $u \in \mathscr{L}$. Our "continuity" is an abbreviation for "strong continuity for $t \geq 0$."

Let $p_j \geq 0$ and $p_1 + \cdots + p_r = 1$. The linear combination

$$E = p_1\mathfrak{Q}(t_1) + \cdots + p_r(\mathfrak{Q}t_r) \tag{9.4}$$

is again a contraction and may be interpreted as the expectation of $\mathfrak{Q}(t)$ with respect to the probability distribution attaching weight p_j to t_j. This defines (9.1) for the special case of finite discrete distributions, and (9.2) is true. The general expectation (9.1) is defined by a passage to the limit just as a Riemann integral: partition $\overline{0, \infty}$ into intervals $I_1, \ldots, I_n$, choose $t_j \in I_j$, and form the Riemann sum $\sum \mathfrak{Q}(t_k)\, F\{I_k\}$ which is a contraction. In view of the uniform continuity property X,(8.7) the familiar convergence proof works without change. This defines (9.1) as a special case of a Bochner integral.

If the semi-group consists of transition operators, that is, if $\mathfrak{Q}(t)1 = 1$ for all t, then $E1 = 1$. The notation (9.1) will be used for E, and for the function Ew we shall use the usual symbol

$$Ew = \int_0^\infty \mathfrak{Q}(s)w \cdot F\{ds\} \tag{9.5}$$

(although it would be logically more consistent to write w outside the integral). The value $Ew(x)$ at a given point x is the ordinary expectation with respect to F of the numerical function $\mathfrak{Q}(s)\, w(x)$.

In the special case $F\{ds\} = e^{-\lambda s}\, ds$ the operator E is called the *Laplace integral of the semi-group*, or *resolvent*. It will be denoted by

$$\mathfrak{R}(\lambda) = \int_0^\infty e^{-\lambda s}\, \mathfrak{Q}(s)\, ds, \qquad \lambda > 0. \tag{9.6}$$

In view of (9.2) the resolvent operators $\mathfrak{R}(\lambda)$ commute with the operators $\mathfrak{Q}(t)$ of the semi-group. In order that $\lambda\mathfrak{R}(\lambda)1 = 1$ it is necessary and sufficient that $\mathfrak{Q}(t)1 = 1$ for all t, and thus the contraction $\lambda\mathfrak{R}(\lambda)$ is a transition operator iff all $\mathfrak{Q}(s)$ are transition operators.

Lemma. *The knowledge of* $\mathfrak{R}(\lambda)w$ *for all* $\lambda > 0$ *and* $w \in \mathscr{L}$ *uniquely determines the semi-group.*

Proof. The value

$$\mathfrak{R}(\lambda)\, w(x) = \int_0^\infty e^{-\lambda t}\mathfrak{Q}(t)\, w(x) \cdot dt$$

at a given point x is the ordinary Laplace transform of the numerical function of t defined by $\mathfrak{Q}(t)\, w(x)$. This function being continuous, it is uniquely determined by its Laplace transform (see the corollary in section 1). Thus $\mathfrak{Q}(t)w$ is uniquely determined for all t and all $w \in \mathscr{L}$. ▶

The Laplace transform (9.6) leads to a simple characterization of the *infinitesimal generator* $\mathfrak{A}$ of the semi-group. By the definition of this operator in X,10 we have

$$\frac{\mathfrak{Q}(h) - \mathbf{1}}{h} u \to \mathfrak{A}u, \qquad \to 0+, \tag{9.7}$$

if $\mathfrak{A}u$ exists (that is, if the norm of the difference of the two sides tends to zero).

Theorem 1. *For fixed* $\lambda > 0$

$$u = \mathfrak{R}(\lambda)w \tag{9.8}$$

iff u *is in the domain of* $\mathfrak{A}$ *and*

$$\lambda u - \mathfrak{A}u = w. \tag{9.9}$$

Proof. (i) Define u by (9.8). Referring to the property (9.2) of expectations we have

$$\frac{\mathfrak{Q}(h) - \mathbf{1}}{h} u = \frac{1}{h}\int_0^\infty e^{-\lambda s}\mathfrak{Q}(s+h)w \cdot ds - \frac{1}{h}\int_0^\infty e^{-\lambda s}\mathfrak{Q}(s)w \cdot ds. \tag{9.10}$$

The change of variable $s + h = t$ in the first integral reduces this to

$$\begin{aligned}\frac{\mathfrak{Q}(h) - \mathbf{1}}{h} u &= \frac{e^{\lambda h} - 1}{h}\int_0^\infty e^{-\lambda t}\mathfrak{Q}(t)w \cdot dt - \frac{1}{h}\int_0^h e^{\lambda(h-t)}\mathfrak{Q}(t)w \cdot dt \\ &= \frac{e^{\lambda h} - 1}{h}(u - \lambda^{-1}w) - \frac{1}{h}\int_0^h e^{\lambda(h-t)}(\mathfrak{Q}(t)w - w)\, dt.\end{aligned} \tag{9.11}$$

Since $\|\mathfrak{Q}(t)w - w\| \to 0$ as $t \to 0$ the second term on the right tends in norm to 0, and the whole right side therefore tends to $\lambda(u - \lambda^{-1}w)$. Thus (9.9) is true.

(ii) Conversely, assume that $\mathfrak{A}u$ exists, that is, (9.7) holds. Since $\lambda\mathfrak{R}(\lambda)$ is a contraction commuting with the semi-group, (9.7) implies

$$\frac{\mathfrak{Q}(h) - \mathbf{1}}{h}\mathfrak{R}(\lambda)u \to \mathfrak{R}(\lambda)\mathfrak{A}u. \tag{9.12}$$

But we have just seen that the left side tends to $\lambda\mathfrak{R}(\lambda)u - u$, and the resulting identity exhibits u as the Laplace transform of the function w in (9.9). ►

Corollary 1. *For given* $w \in \mathscr{L}$ *there exists exactly one solution* u *of* (9.9).

Corollary 2. *Two distinct semi-groups cannot have the same generator* $\mathfrak{A}$.

Proof. The knowledge of the generator $\mathfrak{Q}$ permits us to find the Laplace transform $\mathfrak{R}(\lambda)w$ for all $w \in \mathscr{L}$ and by the above lemma this uniquely determines all operators of the semi-group. ►

It is tempting to derive Tauberian theorems analogous to those of section 5, but we shall be satisfied with the rather primitive

Theorem 2. *As* $\lambda \to \infty$

$$\lambda\mathfrak{R}(\lambda) \to \mathbf{1}. \tag{9.13}$$

Proof. For arbitrary $w \in \mathscr{L}$ we have

$$\|\lambda\mathfrak{R}(\lambda)w - w\| \leq \int_0^\infty \|\mathfrak{Q}(t)w - w\| \cdot \lambda e^{-\lambda t}\, dt. \tag{9.14}$$

As $\lambda \to \infty$ the probability distribution with density $\lambda e^{-\lambda t}$ tends to the distribution concentrated at the origin. The integrand is bounded and tends to 0 as $t \to 0$, and so the integral tends to 0 and (9.13) is true. ►

Corollary 3. *The generator* $\mathfrak{A}$ *has a domain which is dense in* $\mathscr{L}$.

Proof. It follows from (9.13) that every $w \in \mathscr{L}$ is the strong limit of a sequence of elements $\lambda\mathfrak{R}(\lambda)w$, and by theorem 1 these elements are in the domain of $\mathfrak{A}$. ►

Examples. (*a*) *Infinitely divisible semi-groups.* Let U be the infinitely divisible distribution with Laplace transform $\omega = e^{-\varphi}$ described in (7.2). The distributions U_t with Laplace transforms

$$\int_0^\infty e^{-\lambda x}\, U_t\{dx\} = e^{-t\varphi(\lambda)} = \exp\left(-t\int_0^\infty \frac{1 - e^{-\lambda x}}{x}\, P\{dx\}\right) \tag{9.15}$$

are again infinitely divisible, and the associated convolution operators $\mathfrak{U}(t)$ form a semi-group. To find its generator[14] choose a bounded continuously differentiable function v. Then clearly

$$\frac{\mathfrak{U}(t) - \mathbf{1}}{t}\, v(x) = \int_0^\infty \frac{v(x-y) - v(x)}{y} \cdot \frac{1}{t}\, y\, U_t\{dy\}. \tag{9.16}$$

Differentiation of (9.15) shows that the measure $t^{-1}y\, U_t\{dy\}$ has the transform $\varphi'(\lambda)e^{-t\varphi(\lambda)}$ which tends to $\varphi'(\lambda)$ as $t \to 0$. But φ' is the transform of the measure P, and so our measures tend to P. Since the fraction under the last integral is (for fixed x) a bounded continuous function of y,

[14] This derivation is given for purposes of illustration. The generator is already known from chapter IX and can be obtained by a passage to the limit from compound Poisson distributions.

we get

$$\mathfrak{A}v(x) = \int_0^\infty \frac{v(x-y) - v(x)}{y} P\{dy\} \tag{9.17}$$

and have thus an interpretation of the measure P in the canonical representation of infinitely divisible distributions.

(*b*) *Subordinated semi-groups.* Let $\{\mathfrak{Q}(t)\}$ stand for an arbitrary Markovian semi-group, and let U_t be the infinitely divisible distribution of the preceding example. As explained in X,7 a new Markovian semigroup $\{\mathfrak{Q}^*(t)\}$ may be obtained by randomization of the parameter t. In the present notation

$$\mathfrak{Q}^*(t) = \int_0^\infty \mathfrak{Q}(s)\, U_t\{ds\}. \tag{9.18}$$

Putting for abbreviation

$$V(s, x) = \frac{\mathfrak{Q}(s) - 1}{s} v(x) \tag{9.19}$$

we have

$$\frac{\mathfrak{Q}^*(t) - 1}{t} v(x) = \int_0^\infty V(s, x) \cdot \frac{1}{t} s\, U_t\{ds\}. \tag{9.20}$$

For a function v in the domain of $\mathfrak{A}$ and for x fixed the function V is continuous everywhere including the origin since $V(s, x) \to \mathfrak{A}v(x)$ as $s \to 0$. We saw in the last example that $t^{-1}s\, U_t\{ds\} \to P\{ds\}$ if $t \to 0$. Thus the right side in (9.20) tends to a limit and hence $\mathfrak{A}^*v$ exists and is given by

$$\mathfrak{A}^*v(x) = \int_0^\infty V(s, x)\, P\{ds\}. \tag{9.21}$$

The conclusion is that *the domains of* $\mathfrak{A}$ *and* $\mathfrak{A}^*$ *coincide*, and

$$\mathfrak{A}^* = \int_0^\infty \frac{\mathfrak{Q}(s) - \mathbf{1}}{s} P\{ds\} \tag{9.22}$$

in the sense that (9.21) holds for v in the domain of $\mathfrak{A}$. ►

10. THE HILLE-YOSIDA THEOREM

The famous and exceedingly useful Hille-Yosida theorem characterizes generators of arbitrary semi-groups of transformations, but we shall specialize it to our contraction semi-groups. The theorem asserts that the properties of generators found in the last section represent not only necessary but also sufficient conditions.

Theorem 1. (*Hille-Yosida.*) *An operator* $\mathfrak{A}$ *with domain* $\mathscr{L}' \subset \mathscr{L}$ *is the generator of a continuous semi-group of contractions* $\mathfrak{Q}(t)$ *on* $\mathscr{L}$ (*with* $\mathfrak{Q}(0) = \mathbf{1}$) *iff it has the following properties.*

(i) *The equation*

$$\lambda u - \mathfrak{A}u = w, \qquad \lambda > 0, \tag{10.1}$$

has for each $w \in \mathscr{L}$ *exactly one solution* u;

(ii) *if* $0 \leq w \leq 1$ *then* $0 \leq \lambda u \leq 1$;

(iii) *the domain* $\mathscr{L}'$ *of* $\mathfrak{A}$ *is dense in* $\mathscr{L}$.

We know already that every generator possesses these properties, and so the conditions are necessary. Furthermore, if the solution u is denoted by $u = \mathfrak{R}(\lambda)w$, we know that $\mathfrak{R}(\lambda)$ coincides with the Laplace transform (9.6). Accordingly the conditions of the theorem may be restated as follows.

(i′) The operator $\mathfrak{R}(\lambda)$ satisfies the identity

$$\lambda\mathfrak{R}(\lambda) - \mathfrak{A}\mathfrak{R}(\lambda) = \mathbf{1}. \tag{10.2}$$

The domain of $\mathfrak{R}(\lambda)$ is $\mathscr{L}$; the range coincides with the domain $\mathscr{L}'$ of $\mathfrak{A}$

(ii′) The operator $\lambda\mathfrak{R}(\lambda)$ is a contraction.

(iii′) The range of $\mathfrak{R}(\lambda)$ is dense in $\mathscr{L}$.

From theorem 2 in section 9 we know that $\mathfrak{R}(\lambda)$ must satisfy the further condition

$$\lambda\mathfrak{R}(\lambda) \to \mathbf{1}, \qquad \lambda \to \infty. \tag{10.3}$$

This implies that every u is the limit of its own transforms and hence that the range $\mathscr{L}'$ of $\mathfrak{R}(\lambda)$ is dense. It follows that (10.3) can serve as replacement for (iii′), and thus *the three conditions of the theorem are fully equivalent to the set* (i′), (ii′), (10.3).

We now suppose that we are given a family of operators $\mathfrak{R}(\lambda)$ with these properties and proceed to construct the desired semi-group as the limit of a family of pseudo-Poisson semi-groups. The construction depends on

Lemma 1. *If* w *is in domain* $\mathscr{L}'$ *of* $\mathfrak{A}$ *then*

$$\mathfrak{A}\mathfrak{R}(\lambda)w = \mathfrak{R}(\lambda)\mathfrak{A}w. \tag{10.4}$$

The operators $\mathfrak{R}(\lambda)$ *and* $\mathfrak{R}(\nu)$ *commute and satisfy the resolvent equation*

$$\mathfrak{R}(\lambda) - \mathfrak{R}(\nu) = (\nu-\lambda)\mathfrak{R}(\lambda)\mathfrak{R}(\nu). \tag{10.5}$$

Proof. Put $v = \mathfrak{A}u$. Since both u and w are in the domain $\mathscr{L}'$ of $\mathfrak{A}$ it follows from (10.1) that the same is true of v and

$$\lambda v - \mathfrak{A}v = \mathfrak{A}w.$$

Thus $v = \mathfrak{R}(\lambda)\mathfrak{A}w$ which is the same as (10.4).

Next, define z as the unique solution of $\nu z - \mathfrak{A}z = w$. Subtracting this from (10.1) we get after a trite rearrangement

$$\lambda(u-z) - \mathfrak{A}(u-z) = (\nu-\lambda)z,$$

which is the same as (10.5). The symmetry of this identity implies that the operators commute. ▶

For the construction of our semi-group we recall from theorem 1 of X,9 that to an arbitrary contraction T and $a > 0$ there corresponds a semi-group of contractions defined by

$$e^{at(T-1)} = e^{-at}\sum_{n=0}^{\infty}\frac{(at)^n}{n!}T^n. \tag{10.6}$$

The generator of this semi-group is $a(T-\mathbf{1})$, which is an endomorphism.

We apply this result to $T = \lambda_{\mathfrak{R}(\lambda)}$. Put for abbreviation

$$\mathfrak{A}_\lambda = \lambda[\lambda\mathfrak{R}(\lambda) - \mathbf{1}] = \lambda\mathfrak{A}\mathfrak{R}(\lambda), \qquad \mathfrak{Q}_\lambda(t) = e^{t\mathfrak{A}_\lambda}. \tag{10.7}$$

These operators defined for $\lambda > 0$ commute with each other, and for fixed λ the operator $\mathfrak{A}_\lambda$ generates the quasi-Poissonian semi-group of contractions $\mathfrak{Q}_\lambda(t)$.

It follows from (10.4) that $\mathfrak{A}_\lambda u \to \mathfrak{A}u$ for all u in the domain $\mathscr{L}'$ of the given operator $\mathfrak{A}$. We can forget about the special definition of $\mathfrak{A}_\lambda$ and consider the remaining assertion of the Hille-Yosida theorem as a special case of a more general limit theorem which is useful in itself. In it λ may be restricted to the sequence of integers.

Approximation lemma 2. *Let $\{\mathfrak{Q}_\lambda(t)\}$ be a family of pseudo-Poissonian semi-groups commuting with each other and generated by the endomorphisms $\mathfrak{A}_\lambda$.*

If $\mathfrak{A}_\lambda u \to \mathfrak{A}u$ for all u of a dense set $\mathscr{L}'$, then

$$\mathfrak{Q}_\lambda(t) \to \mathfrak{Q}(t), \qquad \lambda \to \infty, \tag{10.8}$$

where $\{\mathfrak{Q}(t)\}$ is a semi-group of contractions whose generator agrees with $\mathfrak{A}$ for all $u \in \mathscr{L}'$.

Furthermore, for $u \in \mathscr{L}'$

$$\|\mathfrak{Q}(t)u - \mathfrak{Q}_\lambda(t)u\| \le t\,\|\mathfrak{A}u - \mathfrak{A}_\lambda u\|. \tag{10.9}$$

Proof. For two commuting contractions we have the identity

$$S^n - T^n = (S^{n-1} + \cdots + T^{n-1})(S-T)$$

and hence

$$\|S^n u - T^n u\| \le n\,\|Su - Tu\|. \tag{10.10}$$

Applied to operators $\mathfrak{Q}_\lambda(t/n)$ this inequality yields after a trite rearrangement

$$\|\mathfrak{Q}_\lambda(t)u - \mathfrak{Q}_\nu(t)u\| \le t\left\|\frac{\mathfrak{Q}_\lambda(t/n) - \mathbf{1}}{t/n}u - \frac{\mathfrak{Q}_\nu(t/n) - \mathbf{1}}{t/n}u\right\|. \tag{10.11}$$

Letting $n \to \infty$ we get

$$\|\mathfrak{Q}_\lambda(t)u - \mathfrak{Q}_\nu(t)u\| \le t\,\|\mathfrak{A}_\lambda u - \mathfrak{A}_\nu u\|. \tag{10.12}$$

This shows that for $u \in \mathscr{L}'$ the sequence $\{\mathfrak{Q}_\lambda(t)u\}$ is uniformly convergent as $\lambda \to \infty$. Since $\mathscr{L}'$ is dense in $\mathscr{L}$ this uniform convergence extends to all u, and if we denote the limit by $\mathfrak{Q}(t)u$ we have a contraction $\mathfrak{Q}(t)$ for which (10.8) is true. The semi-group property is obvious. Also, letting $\nu \to \infty$ in (10.12) we get (10.9). Rewriting the left side as in (10.11) we have

$$\left\| \frac{\mathfrak{Q}(t) - \mathbf{1}}{t} u - \frac{\mathfrak{Q}_\lambda(t) - \mathbf{1}}{t} u \right\| \leq \|\mathfrak{A}u - \mathfrak{A}_\lambda u\|. \tag{10.13}$$

Choose λ large enough to render the right side $<\epsilon$. For sufficiently small t the second difference ratio on the left differs in norm from $\mathfrak{A}_\lambda u$ by less than ϵ, and hence from $\mathfrak{A}u$ by less than 3ϵ. Thus for $u \in \mathscr{L}'$

$$\frac{\mathfrak{Q}(t) - \mathbf{1}}{t} u \to \mathfrak{A}u \tag{10.14}$$

and this concludes the proof. ►

Examples. *Diffusion.* Let $\mathscr{L}$ be the family of continuous functions on the line vanishing at $\pm\infty$. To use familiar notations we replace λ by h^{-2} and let $h \to 0$. Define the difference operator ∇_h by

$$\nabla_h u(x) = \frac{1}{h^2}\left[\frac{u(x+h) + u(x-h)}{2} - u(x)\right]. \tag{10.15}$$

This is of the form $h^{-2}(T - \mathbf{1})$ where T is a transition operator, and hence ∇_h generates a semi-group $e^{t\nabla_h}$ of transition operators (a Markovian semi-group). The operators ∇_h commute with each other, and for functions with three bounded derivatives $\nabla_h u \to \frac{1}{2}u''$ uniformly. The lemma implies the existence of a limiting semi-group $\{\mathfrak{Q}(t)\}$ generated by an operator $\mathfrak{A}$ such that $\mathfrak{Q}u = \frac{1}{2}u''$ at least when u is sufficiently smooth.

In this particular case we know that $\{\mathfrak{Q}(t)\}$ is the semi-group of convolutions with normal distributions of variance t and we have not obtained new information. The example reveals nevertheless how easy it can be (sometimes) to establish the existence of semi-group with given generators. The argument applies, for example, to more general differential operators and also to boundary conditions. (See problems 24, 25.) ►

Note *on the resolvent and complete monotonicity.* The Hille-Yosida theorem emphasizes properties of the generator $\mathfrak{A}$, but it is possible to reformulate the theorem so as to obtain a characterization of the family $\{\mathfrak{R}(\lambda)\}$.

Theorem 2. (*Alternative form of the Hille-Yosida theorem.*) *In order that a family* $\{\mathfrak{R}(\lambda);\, \lambda > 0\}$ *of endomorphisms be the resolvent of a semi-group* $\{\mathfrak{Q}(t)\}$ *of contractions it is necessary and sufficient* (*a*) *that the resolvent equation*

$$\mathfrak{R}(\lambda) - \mathfrak{R}(\nu) = (\nu-\lambda)\mathfrak{R}(\lambda)\mathfrak{R}(\nu). \tag{10.16}$$

be satisfied, (*b*) *that* $\lambda\mathfrak{R}(\lambda)$ *be a contraction, and* (c) *that* $\lambda\mathfrak{R}(\lambda) \to \mathbf{1}$ *as* $\lambda \to \infty$.

Proof. (10.16) is identical with (10.5), while conditions (*b*) and (*c*) appear above as (ii′) and (10.3). All three conditions are therefore necessary.

Assuming the conditions to hold we define an operator $\mathfrak{A}$ as follows. Choose some $\nu > 0$ and define $\mathscr{L}'$ as the range of $\mathfrak{R}(\nu)$, that is: $u \in \mathscr{L}'$ iff $u = \mathfrak{R}(\nu)w$ for some $w \in \mathscr{L}$. For such u we put $\mathfrak{A}u = \lambda u - w$. This defines an operator $\mathfrak{A}$ with domain $\mathscr{L}'$ and satisfying the identity

$$\nu\mathfrak{R}(\nu) - \mathfrak{A}\mathfrak{R}(\nu) = \mathbf{1}. \tag{10.17}$$

We show that this identity extends to all λ, that is

$$\lambda\mathfrak{R}(\lambda) - \mathfrak{A}\mathfrak{R}(\lambda) = \mathbf{1}. \tag{10.18}$$

The left side may be rewritten in the form

$$(\lambda-\nu)\mathfrak{R}(\lambda) + (\nu-\mathfrak{A})\mathfrak{R}(\lambda). \tag{10.19}$$

Using (10.16) and the fact that $(\nu-\mathfrak{A})\mathfrak{R}(\nu) = \mathbf{1}$ we get

$$(\nu-\mathfrak{A})\mathfrak{R}(\lambda) = \mathbf{1} + (\nu-\lambda)\mathfrak{R}(\lambda). \tag{10.20}$$

Using (10.19) the identity (10.18) follows. It shows that all conditions of the Hille-Yosida theorem are satisfied, and this accomplishes the proof. ▶

The preceding theorem shows that the whole semi-group theory hinges on the resolvent equation (10.16), and it is therefore interesting to explore its meaning in terms of ordinary Laplace transforms of functions. It is clear from (10.16) that $\mathfrak{R}(\lambda)$ depends continuously on λ in the sense that $\mathfrak{R}(\nu) \to \mathfrak{R}(\lambda)$ as $\nu \to \lambda$. However, we can go a step farther and define a derivative $\mathfrak{R}'(\lambda)$ by

$$\mathfrak{R}'(\lambda) = \lim_{\nu\to\lambda} \frac{\mathfrak{R}(\nu) - \mathfrak{R}(\lambda)}{\nu - \lambda} = -\mathfrak{R}^2(\lambda). \tag{10.21}$$

The same procedure now shows that the right side has a derivative given by $-2\mathfrak{R}(\lambda)\mathfrak{R}'(\lambda)$. Proceeding by induction it is seen that $\mathfrak{R}(\lambda)$ has derivatives $\mathfrak{R}^{(n)}(\lambda)$ of all orders and

$$(-1)^n\mathfrak{R}^{(n)}(\lambda) = n!\mathfrak{R}^{n+1}(\lambda). \tag{10.22}$$

Let now u be an arbitrary function in $\mathscr{L}$ such that $0 \leq u \leq 1$. Choose an arbitrary point x and put $\omega(\lambda) = \mathfrak{R}(\lambda)u(x)$. The right side in (10.22) is a positive operator of norm $\leq n!/\lambda^{n+1}$ and therefore ω is completely monotone and $|\omega^{(n)}(\lambda)| \leq n!/\lambda^{n+1}$. From the corollary in section 4 it follows now that ω is the ordinary Laplace transform with values lying between 0 and 1. If we denote this function by $\mathfrak{Q}(t)\, u(x)$, this defines $\mathfrak{Q}(t)$ as a contraction operator. Comparing the resolvent equation (10.16) with (8.2) it is now clear that it implies the semi-group property

$$\mathfrak{Q}(t+s)u(x) = \mathfrak{Q}(t)\mathfrak{Q}(s)u(x). \tag{10.23}$$

We see thus that the essential features of the semi-group theory could have been derived from (10.16) using only the classical Laplace transforms of ordinary functions. In particular, the resolvent equation turns out to be merely an abstract paraphrasing of the elementary example 8(*a*).

To emphasize further that the present abstract theory merely paraphrases the theorems concerning ordinary Laplace transforms we prove an *inversion formula.*

Theorem 3. *For fixed* $t > 0$ *as* $\lambda \to \infty$

$$\frac{(-1)^{n-1}}{(n-1)!}\mathfrak{R}^{(n-1)}(n/t)(n/t)^n \to \mathfrak{Q}(t). \tag{10.24}$$

Proof. From the definition (9.6) of $\mathfrak{R}_{(\lambda)}$ as a Laplace transform of $\mathfrak{Q}(t)$ it follows that

$$(-1)^n \mathfrak{R}^{(n)}(\lambda) = \int_0^\infty e^{-\lambda s} s^n \mathfrak{Q}(s)\, ds. \tag{10.25}$$

The left side of (10.24) is the integral of $\mathfrak{Q}(s)$ with respect to the density $[e^{-ns/t}(ns/t)^{n-1}/t(n-1)!]\, n$ which has expected value t and variance t^2/n. As $n \to \infty$ this measure tends to the distribution concentrated at t, and because of the continuity of $\mathfrak{Q}(s)$, this implies (10.24) just as in the case of functions [formula (10.24) is the same as VII,(1.6)]. ▶

11. PROBLEMS FOR SOLUTION

1. Let F_q be the geometric distribution attributing weight qp^n to the point nq $(n = 0, 1, \ldots)$. As $q \to 0$ show that F_q tends to the exponential distribution $1 - e^{-x}$ and that its Laplace transform tends to $1/(\lambda+1)$.

2. Show that the ordinary Laplace transforms of $\cos x$ and $\sin x$ are $\lambda/(\lambda^2+1)$ and $1/(\lambda^2+1)$. Conclude that $(1+a^{-2})e^{-x}(1-\cos ax)$ is a probability density with Laplace transform $(1+a^2)(\lambda+1)^{-1}[(\lambda+1)^2 + a^2]^{-1}$. *Hint:* Use $e^{ix} = \cos x + i \sin x$ or, alternatively, two successive integrations by parts.

3. Let ω be the transform of a measure U. Then ω is integrable over $\overline{0, 1}$ and $\overline{1, \infty}$ iff $1/x$ is integrable with respect to U over $\overline{1, \infty}$ and $\overline{0, 1}$, respectively.

4. *Parseval relation.* If $\mathbf{X}$ and $\mathbf{Y}$ are independent random variables with distributions F and G, and transforms φ and γ, the transform of $\mathbf{XY}$ is

$$\int_0^\infty \varphi(\lambda y)\, G\{dy\} = \int_0^\infty \gamma(\lambda y)\, F\{dy\}.$$

5. Let F be a distribution with transform φ. If $a > 0$ then $\varphi(\lambda+a)/\varphi(a)$ is the transform of the distribution $e^{-ax} F\{dx\}/\varphi(a)$. For fixed $t > 0$ conclude from example 3(*b*) that[15] $\exp[-t\sqrt{2\lambda + a^2} + at]$ is the transform of an infinitely divisible distribution with density $\dfrac{t}{\sqrt{2\pi x^3}} \exp\left[-\dfrac{1}{2}\left(\dfrac{t}{\sqrt{x}} - a\sqrt{x}\right)^2\right]$.

6. From the definition II,(7.1) show that *the ordinary Laplace transform of* $I_0(x)$ *is* $\omega_0(\lambda) = 1/\sqrt{\lambda^2 - 1}$ for $\lambda > 1$. $\left[\text{Recall the identity } \binom{2n}{n} = \binom{-\frac{1}{2}}{n}(-4)^n.\right]$

7. *Continuation.* Show that $I_0' = I_1$, and hence that I_1 has the ordinary Laplace transform $\omega_1(\lambda) = \omega_0(\lambda)\, R(\lambda)$ where $R(\lambda) = \lambda - \sqrt{\lambda^2 - 1}$.

8. *Continuation.* Show that $2I_n' = I_{n-1} + I_{n+1}$ for $n = 1, 2, \ldots$ and hence by induction that I_n has the ordinary transform $\omega_n(\lambda) = \omega_0(\lambda)\, R^n(\lambda)$.

9. From example 3(*e*) conclude by integration that $e^{1/\lambda} - 1$ *is the ordinary transform* of $I_1(2\sqrt{x})/\sqrt{x}$.

10. Let $\mathbf{X}$ and $\mathbf{Y}$ be independent random variables with Laplace transforms φ and $e^{-\lambda^\alpha}$, respectively. Then $\mathbf{YX}^{1/\alpha}$ has the Laplace transform $\varphi(\lambda^\alpha)$.

---

[15] This formula occurs in applications and has been derived repeatedly by lengthy calculations.

11. The density f of a probability distribution is completely monotone iff it is a mixture of exponential densities [that is, if it is of the form (7.11)]. *Hint:* Use problem 3.

12. Verify the inversion formula (4.5) by direct calculation in the special cases $\varphi(\lambda) = 1/(\lambda+1)$ and $\varphi(\lambda) = e^{-\lambda}$.

13. Show that the corollary in section 4 remains valid if f and $\varphi^{(n)}$ are replaced by their absolute values.

14. Assuming $e^{-x}I_n(x)$ monotone at infinity, conclude from problem 8 that

$$e^{-x}I_n(x) \sim \frac{1}{\sqrt{2\pi x}} \qquad x \to \infty.$$

15. Suppose that $1 - \varphi(\lambda) \sim \lambda^{1-\rho} L(\lambda)$ as $\lambda \to 0$ where $\rho > 0$. Using example 5(*c*) show that $1 - F^{n\star}(x) \sim nx^{\rho-1} L(1/x)/\Gamma(\rho)$ as $x \to \infty$. [Compare this with example VIII,8(*c*).]

16. In theorem 4 of section 5 it suffices that $u(x) \sim v(x)$ where v is ultimately monotone.

17. Every infinitely divisible distribution is the limit of compound Poisson distributions.

18. If in the canonical representation (7.2) for infinitely divisible distributions $P(x) \sim x^c L(x)$ as $x \to \infty$ with $0 < c < 1$, prove that $1 - F(x) \sim (c/1 - c)x^{c-1} L(x)$. [Continued in example XVII,4(*d*).]

19. Let P be the generating function of an infinitely divisible integral-valued random variable and φ the Laplace transform of a probability distribution. Prove that $P(\varphi)$ is infinitely divisible.

20. The infinitely divisible Laplace transforms φ_n converge to the Laplace transform φ of a probability distribution iff the corresponding measure P_n in the canonical representation (7.2) converges to P. Hence: the limit of a sequence of infinitely divisible distributions is itself infinitely divisible.

21. Let F_n be a mixture (7.11) of exponential distributions corresponding to a mixing distribution U_n. The sequence $\{F_n\}$ converges to a probability distribution F iff the U_n converge to a probability distribution U. In this case F is a mixture corresponding to U.

22. A probability distribution with a completely monotone density is infinitely divisible. *Hint:* Use problems 11 and 21 as well as example 7(*f*).

23. Every mixture of geometric distributions is infinitely divisible. *Hint:* Follow the pattern of example 7(*f*).

24. *Diffusion with an absorbing barrier.* In the example of section 10 restrict x to $x > 0$ and when $x - h \leq 0$ put $u(x-h) = 0$ in the definition of ∇_h. Show that the convergence proof goes through if $\mathscr{L}$ is the space of continuous functions with $u(\infty) = 0$, $u(0) = 0$, but not if the last condition is dropped. The resulting semi-group is given in example X,5(*b*).

25. *Reflecting barriers.* In the example of section 10 restrict x to $x > 0$ and when $x - h < 0$ put $u(x-h) = u(x+h)$ in the definition of ∇_h. Then $\nabla_h u$ converges for every u with three bounded derivatives such that $u'(0) = 0$. The domain $\mathscr{L}'$ of $\mathfrak{A}$ is restricted by this boundary condition. The semi-group is described in example X,5(*e*).

26. *The influence of the maximal term in the convergence to stable distributions.* Let $\mathbf{X}_1, \mathbf{X}_2, \ldots$ be independent variables with the common distribution F satisfying (6.4), that is, belonging to the domain of attraction of the stable distribution G_α. Put $\mathbf{S}_n = \mathbf{X}_1 + \cdots + \mathbf{X}_n$ and $\mathbf{M}_n = \max[\mathbf{X}_1, \ldots, \mathbf{X}_n]$. Prove that the ratio $\mathbf{S}_n/\mathbf{M}_n$ has a Laplace transform $\omega_n(\lambda)$ converging to[16]

$$(*) \qquad \omega(\lambda) = \frac{e^{-\lambda}}{1 + \alpha \int_0^1 (1 - e^{-\lambda t}) t^{-\alpha-1}\, dt}.$$

Hence $\mathbf{E}(\mathbf{S}_n/\mathbf{M}_n) \to 1/(1-\alpha)$.

Hint: Evaluating the integral over the region $\mathbf{X}_j \leq \mathbf{X}_1$ one gets

$$\omega_n(\lambda) = ne^{-\lambda} \int_0^\infty F\{dx\} \left(\int_0^\infty e^{-\lambda y/x}\, F\{dy\} \right)^{n-1}.$$

Substitute $y = tx$ and then $x = a_n s$ where a_n satisfies (6.5). The inner integral is easily seen to be

$$1 - \frac{1 - F(a_n s)}{n[1 - F(a_n)]} - \frac{1}{n}\int_0^1 (1 - e^{-\lambda t}) \frac{F\{a_n\, dt\}}{1 - F(a_n)} + o\left(\frac{1}{n}\right) = 1 - \frac{s^{-\alpha}\psi(\lambda)}{n} - o\left(\frac{1}{n}\right),$$

where $\psi(\lambda)$ stands for the denominator in (*). Thus

$$\omega_n(\lambda) \to e^{-\lambda} \int_0^\infty e^{-s^{-\alpha}\psi(\lambda)} \cdot \frac{\alpha\, ds}{s^{\alpha+1}} = \omega(\lambda).$$

---

[16] This result and its analogue for stable distributions with exponent $\alpha > 0$ was derived by D. A. Darling in terms of characteristic functions. See Trans. Amer. Math. Soc., vol. 73 (1952) pp. 95–107.

CHAPTER XIV

Applications of Laplace Transforms

This chapter can serve as collateral reading to chapter XIII. It covers several independent topics ranging from practical problems (sections 1, 2, 4, 5) to the general existence theorem in section 7. The limit theorem of section 3 illustrates the power of the methods developed in connection with regular variation. The last section serves to describe techniques for the analysis of asymptotic properties and first-passage times in Markov processes.

1. THE RENEWAL EQUATION: THEORY

For the probabilistic background the reader is referred to VI,6–7. Although the whole of chapter XI was devoted to renewal theory, we give here an independent and much less sophisticated approach. A comparison of the methods and results is interesting. Given the rudiments of the theory of Laplace transforms, the present approach is simpler and more straightforward, but the precise result of the basic renewal theorem is at present not obtainable by Laplace transforms. On the other hand, Laplace transforms lead more easily to the limit theorems of section 3 and to explicit solutions of the type discussed in section 2.

The object of the present study is the integral equation

$$V(t) = G(t) + \int_0^t V(t-x)\, F\{dx\} \tag{1.1}$$

in which F and G are given monotone right continuous functions vanishing for $t < 0$. We consider them as improper distribution functions of measures and suppose F is not concentrated at the origin and that their Laplace transforms

$$\varphi(\lambda) = \int_0^\infty e^{-\lambda t}\, F\{dt\}, \qquad \gamma(\lambda) = \int_0^\infty e^{-\lambda t}\, G\{dt\} \tag{1.2}$$

exist for $\lambda > 0$. As in the preceding chapter all intervals of integration are taken *closed*. It will be shown that there exists exactly one solution V; it is an improper distribution function whose Laplace transform ψ exists for all $\lambda > 0$. If G has a density g, then V has a density v satisfying the integral equation

$$v(t) = g(t) + \int_0^t v(t-x)\, F\{dx\} \tag{1.3}$$

obtained by differentiation from (1.1).

Recalling the convolution rule we get for the Laplace transform ψ of the distribution V (or the ordinary transform of its density) $\psi = \gamma + \psi\varphi$, whence formally

$$\psi(\lambda) = \frac{\gamma(\lambda)}{1 - \varphi(\lambda)}. \tag{1.4}$$

To show that this formal solution is the Laplace transform of a measure (or density) we distinguish three cases (of which only the first two are probabilistically significant).

Case (a). F is a probability distribution, not concentrated at the origin. Then $\varphi(0) = 1$ and $\varphi(\lambda) < 1$ for $\lambda > 0$. Accordingly

$$\omega = \frac{1}{1 - \varphi} = \sum_0^\infty \varphi^n \tag{1.5}$$

converges for $\lambda > 0$. Obviously ω is completely monotone and therefore the Laplace transform of a measure U (theorem 1 of XIII,4). Now $\psi = \omega\gamma$ is the Laplace transform of the convolution $V = U \bigstar G$, that is

$$V(t) = \int_0^t G(t-x)\, U\{dx\}. \tag{1.6}$$

Finally, if G has a density g then V possesses a density $v = U \bigstar g$. We have thus proved the *existence* and the *uniqueness* of the desired solution of our integral equations.

The asymptotic behavior of V at infinity is described by the Tauberian theorem 2 of XIII,4. Consider the typical case where $G(\infty) < \infty$ and F has a finite expectation μ. Near the origin $\psi(\lambda) \sim \mu^{-1}G(\infty)\lambda^{-1}$ which implies that

$$V(t) \sim \mu^{-1}G(\infty) \cdot t, \qquad t \to \infty. \tag{1.7}$$

The renewal theorems in XI,1 yield the more precise result that

$$V(t+h) - V(t) \to \mu^{-1}G(\infty)h,$$

but this cannot be derived from Tauberian theorems. [These lead to better results when F has no expectation; (section 3).]

Case (*b*). *F is a defective distribution*, $F(\infty) < 1$. Assume for simplicity that also $G(\infty) < \infty$. The preceding argument applies with the notable simplification that $\varphi(0) = F(\infty) < 1$ and so $\omega(0) < \infty$: *the measure V is now bounded.*

Case (*c*). *The last case is* $F(\infty) > 1$. For small values of λ the denominator in (1.4) is negative, and for such values $\omega(\lambda)$ cannot be a Laplace transform. Fortunately this fact causes no trouble. To avoid trivialities assume that F has no atom at the origin so that $\varphi(\lambda) \to 0$ as $\lambda \to \infty$. In this case there exists a unique root $\kappa > 0$ of the equation $\varphi(\kappa) = 1$ and the argument under (*a*) applies without change for $\lambda > \kappa$. In other words, there exists a unique solution V, but its Laplace transform ω converges only for $\lambda > \kappa$. For such values ω is still given by (1.4).[1]

2. RENEWAL-TYPE EQUATIONS: EXAMPLES

(*a*) *Waiting times for gaps in a Poisson process.* Let V be the distribution of the waiting time to the completion of the first gap of length ξ in a Poisson process with parameter c (that is, in a renewal process with exponential interarrival times). This problem was treated analytically in example XI,7(*b*). Empirical interpretations (delay of a pedestrian or car trying to cross a stream of traffic, locked times in type II Geiger counters, etc.) are given in VI,7. We proceed to set up the renewal equation afresh.

The waiting time commencing at epoch 0 necessarily exceeds ξ. It terminates before $t > \xi$ if no arrival occurs before epoch ξ (probability $e^{-c\xi}$) or else if the first arrival occurs at an epoch $x < \xi$ and the residual waiting time is $\leq t - x$. Because of the inherent lack of memory the probability $V(t)$ of a waiting time $\leq t$ is therefore

$$V(t) = e^{-c\xi} + \int_0^\xi V(t-x) \cdot e^{-cx} c\, dx \tag{2.1}$$

for $t \geq \xi$ and $V(t) = 0$ for $t < \xi$. Despite its strange appearance (2.1) is a renewal equation of the standard type (1.1) in which F has the density $f(x) = ce^{-ex}$ concentrated on $0 < x < \xi$, while G is concentrated at the point ξ. Thus

$$\varphi(\lambda) = \frac{c}{c + \lambda}(1 - e^{-(c+\lambda)\xi}), \qquad \gamma(\lambda) = e^{-(c+\lambda)\xi}, \tag{2.2}$$

[1] The solution V is of the form $V\{dx\} = e^{\kappa x} V^{\#}\{dx\}$ where $V^{\#}$ is the solution [with Laplace transform $\psi^{\#}(\lambda) = \psi(\lambda + \kappa)$] of a standard renewal equation (1.1) with F replaced by the proper probability distribution $F^{\#}\{dx\} = e^{-\kappa x} F\{dx\}$ and G by $G^{\#}\{dx\} = e^{-\kappa x} G\{dx\}$.

and hence the transform ψ of V is given by

$$\psi(\lambda) = \frac{(c+\lambda)e^{-(c+\lambda)\xi}}{\lambda + ce^{-(c+\lambda)\xi}}. \tag{2.3}$$

The expressions XI,(7.8) for the expectation and variance are obtained from this by simple differentiations[2] and the same is true of the higher moments.

It is instructive to derive from (2.3) *an explicit formula* for the solution. For reasons that will become apparent we switch to the *tail* $1 - V(t)$ of the distribution. Its ordinary Laplace transform is $[1-\psi(\lambda)]/\lambda$ [see XIII,(2.7)] which admits of an expansion into a geometric series

$$\frac{1-\psi(\lambda)}{\lambda} = \xi \sum_{n=1}^{\infty} c^{n-1}\xi^{n-1}\left\{\frac{1-e^{-(c+\lambda)\xi}}{(c+\lambda)\xi}\right\}^n. \tag{2.4}$$

The expression within braces differs from the Laplace transform $(1-e^{-\lambda})/\lambda$ of the uniform distribution merely by a scale factor ξ and by the change from λ to $\lambda + c$. As was observed repeatedly, this change corresponds to a multiplication of the densities by e^{-ct}. Thus

$$1 - V(t) = e^{-ct} \sum_{n=1}^{\infty} c^{n-1}\xi^{n-1}f^{n*}(t/\xi) \tag{2.5}$$

where f^{n*} is the density of the n-fold convolution of the uniform distribution with itself. Using I,(9.6) we get finally

$$1 - V(t) = e^{-ct} \sum_{n=1}^{\infty} \frac{(ct)^{n-1}}{(n-1)!} \sum_{k=0}^{\infty} (-1)^k \binom{n}{k}\left(1 - k\frac{\xi}{t}\right)_+^{n-1}. \tag{2.6}$$

The relation to *covering theorems* is interesting. As was shown in I,(9.9) the inner sum represents (for t, ξ fixed) the probability that $n-1$ *points chosen at random in* $\overline{0,t}$ *partition this interval into* n *parts each of which is* $\leq \xi$. Now the waiting time exceeds t iff every subinterval of $\overline{0,t}$ contains at least one arrival and so (2.6) states that if in a Poisson process exactly $n-1$ arrivals occur in $\overline{0,t}$ *their conditional distribution is uniform.* If one starts from this fact one can take (2.6) as a consequence of the covering theorem; alternatively, (2.6) represents a new proof of the covering theorem *by randomization.*

(*b*) *Ruin problem in compound Poisson processes.* As a second illustrative example we treat the *integro-differential equation*

$$R'(t) = (\alpha/c)R(t) - (\alpha/c)\int_0^t R(t-x)\,F\{dx\} \tag{2.7}$$

in which F is a probability distribution with finite expectation μ. This equation was derived in VI,5, where its relevance for collective risk theory, storage problems, etc., is discussed. Its solution and asymptotic properties are derived by different methods in example XI,7(*a*).

[2] It saves labor first to clear the denominator to avoid tedious differentiations of fractions.

The problem is to find a *probability distribution* R satisfying (2.7). This equation is related to the renewal equation and can be treated in the same way. Taking ordinary Laplace transforms and noticing that

$$\rho(\lambda) = \int_0^\infty e^{-\lambda x} R(x)\, dx = \lambda^{-1}\int_0^\infty e^{-\lambda x} R'(x)\, dx + \lambda^{-1}R(0) \tag{2.8}$$

we get

$$\rho(\lambda) = \frac{R(0)}{1 - \dfrac{\alpha}{c}\dfrac{1-\varphi(\lambda)}{\lambda}} \cdot \frac{1}{\lambda} \tag{2.9}$$

where φ is the Laplace transform of F. Recalling that $[1-\varphi(\lambda)]/\lambda$ is the ordinary Laplace transform of $1 - F(x)$ we note that the first fraction on the right is of the form (1.4) and hence the Laplace-Stieltjes transform of a measure R. The factor $1/\lambda$ indicates an integration, and hence $\rho(\lambda)$ is the *ordinary* Laplace transform of the improper distribution function $R(x)$ [as indicated in (2.8)]. Since $R(x) \to 1$ as $x \to \infty$ it follows from theorem 4 in XIII,5 that $\rho(\lambda) \to 1$ as $\lambda \to 0$. From (2.9) we get therefore for the unknown constant $R(0)$

$$R(0) = 1 - (\alpha/c)\mu. \tag{2.10}$$

Accordingly, our problem admits of *a unique solution if* $\alpha\mu < c$ *and admits of no solution if* $\alpha\mu \geq c$. This result was to be anticipated from the probabilistic setup.

Formula (2.9) appears also in queuing theory under the name Khintchine-Pollaczek formula [see example XII,5(*a*)]. Many papers derive explicit expressions in special cases. In the case of the *pure Poisson process*, F is concentrated at the point 1, and $\varphi(\lambda) = e^{-\lambda}$. The expression for ρ is now almost the same as in (2.3) and the same method leads easily to the *explicit solution*

$$R(x) = \left(1 - \frac{\alpha}{c}\right)\sum_{k=0}^{\infty}\left(\frac{-a}{c}\right)^k \frac{(x-k)_+^k}{k!}\exp\left(\frac{\alpha}{c}(x-k)_+\right). \tag{2.11}$$

Although of no practical use, this formula is interesting because of the presence of *positive* exponents which must cancel out in curious ways. It has been known in connection with collective risk[3] theory since 1934 but was repeatedly rediscovered.

3. LIMIT THEOREMS INVOLVING ARC SINE DISTRIBUTIONS

It has become customary to refer to distributions concentrated on $\overline{0, 1}$ with density

$$q_\alpha(x) = \frac{\sin \pi\alpha}{\pi} x^{-\alpha}(1-x)^{\alpha-1}, \qquad 0 < \alpha < 1 \tag{3.1}$$

[3] An explicit solution for ruin before epoch t is given by R. Pyke, *The supremum and infimum of the Poisson process*, Ann. Math. Statist., vol. 30 (1959) pp. 568–576.

as "*generalized arc sine distributions*" although they are special beta distributions. The special case $\alpha = \frac{1}{2}$ corresponds to the distribution function $2\pi^{-1} \arc \sin \sqrt{x}$ which plays an important role in the fluctuation theory for random walks. An increasing number of investigations are concerned with limit distributions related to q_α, and their intricate calculations make the occurrence of q_α seem rather mysterious. The deeper reason lies in the intimate connection of q_α to distribution functions with regularly varying tails, that is, distributions of the form

$$(3.2) \qquad 1 - F(x) = x^{-\alpha} L(x), \qquad 0 < \alpha < 1$$

where, $L(tx)/L(t) \to 1$ as $t \to \infty$. For such functions the renewal theorem may be supplemented to the effect that the renewal function $U = \sum F^{n\star}$ satisfies

$$(3.3) \qquad U(t) \sim \frac{1}{\Gamma(1-\alpha)\,\Gamma(1+\alpha)} \frac{t^\alpha}{L(t)}, \qquad t \to \infty.$$

In other words, if F varies regularly, so does U. It is known (but not obvious) that the constant in (3.3) equals $(\sin \pi\alpha)/\pi\alpha$ and so (3.3) may be rewritten in the form

$$(3.4) \qquad [1-F(x)]\, U(x) \to \frac{\sin \pi\alpha}{\pi\alpha}, \qquad x \to \infty.$$

Lemma. *If* F *is of the form* (3.2) *then* (3.4) *holds.*

Proof. By the Tauberian theorem 4 of XIII,5

$$1 - \varphi(\lambda) \sim \Gamma(1-\alpha)\lambda^\alpha L(1/\lambda) \qquad \lambda \to 0.$$

The Laplace transform of U is $\sum \varphi^n = 1/(1-\varphi)$ and (3.3) is true by virtue of theorem 2 of XIII,5. ►

Consider now a sequence of positive independent variables $\mathbf{X}_k$ with the common distribution F and their partial sums $\mathbf{S}_n = \mathbf{X}_1 + \cdots + \mathbf{X}_n$. For fixed $t > 0$ denote by $\mathbf{N}_t$ the chance-dependent index for which

$$(3.5) \qquad \mathbf{S}_{\mathbf{N}_t} \leq t < \mathbf{S}_{\mathbf{N}_t+1}.$$

We are interested in the two subintervals

$$\mathbf{Y}_t = t - \mathbf{S}_{\mathbf{N}_t} \quad \text{and} \quad \mathbf{Z}_t = \mathbf{S}_{\mathbf{N}_{t+1}} - t.$$

They were introduced in VI,7 as "*spent waiting time*" and "*residual waiting time*" at epoch t. The interest attached to these variables was explained in various connections, and in XI,4 it was proved that as $t \to \infty$ the variables $\mathbf{Y}_t$ and $\mathbf{Z}_t$ have a common proper limit distribution iff F has a finite expectation. Otherwise, however, $\mathbf{P}\{\mathbf{Y}_t \leq x\} \to 0$ for each fixed $x > 0$, and similarly for $\mathbf{Z}_t$. The following interesting theorem emerges as a by-product of our results, but the original proof presented formidable analytical difficulties.[4]

Theorem. *If* (3.2) *is true, then the normed variable* $\mathbf{Y}_t/t$ *has the limit density* q_α *of* (3.1), *and* $\mathbf{Z}_t/t$ *has the limit density given by*[5]

$$p_\alpha(x) = \frac{\sin \pi\alpha}{\pi} \cdot \frac{1}{x^\alpha(1+x)}, \qquad x > 0. \tag{3.6}$$

Proof. The inequality $tx_1 < \mathbf{Y}_t < tx_2$ occurs iff $\mathbf{S}_n = ty$ and

$$\mathbf{X}_{n+1} > t(1-y)$$

for some combination n, y such that $1 - x_2 < y < 1 - x_1$. Summing over all n and possible y we get

$$\mathbf{P}\{tx_1 < \mathbf{Y}_t < tx_2\} = \int_{1-x_2}^{1-x_1} [1 - F(t(1-y))]\, U\{t\, dy\} \tag{3.7}$$

and hence using (3.4)

$$\mathbf{P}\{tx_1 < \mathbf{Y}_t \le tx_2\} \sim \frac{\sin \pi\alpha}{\pi\alpha} \int_{1-x_2}^{1-x_1} \frac{1 - F(t(1-y))}{1 - F(t)} \cdot \frac{U\{t\, dy\}}{U(t)}. \tag{3.8}$$

Now $U(ty)/U(t) \to y^\alpha$ and so the measure $U\{t\, dy\}/U(t)$ tends to the measure with density $\alpha y^{\alpha-1}$ while the first factor approaches $(1-y)^{-\alpha}$. Because of the monotonicity the approach is uniform, and so

$$\mathbf{P}\{tx_1 < \mathbf{Y}_t < tx_2\} \to \frac{\sin \pi\alpha}{\pi} \int_{1-x_2}^{1-x_1} y^{\alpha-1}(1-y)^{-\alpha}\, dy, \tag{3.9}$$

which proves the first assertion. For $\mathbf{P}\{\mathbf{Z}_t > ts\}$ we get the same integral between the limits 0 and $1/(1+s)$ and by differentiation one gets (3.6). ▶

It is a remarkable fact that the density q_α becomes infinite near the endpoints 0 and 1. The most probable values for $\mathbf{Y}_t/t$ are therefore near 0 and 1.

It is easy to amend our argument to obtain converses to the lemma and the theorem. The condition (3.2) is then seen to be *necessary* for the existence of a limit distribution for $\mathbf{Y}_t/t$. On the other hand, (3.2) characterizes the domain of attraction of stable distributions, and this explains the frequent occurrence of q_α in connection with such distributions.

---

[4] E. B. Dynkin, *Some limit theorems for sums of independent random variables with infinite mathematical expectations.* See Selected Trans. in Math. Statist. and Probability, vol. 1 (1961) IMS-AMS, pp. 171–189.

[5] Since $\mathbf{S}_{\mathbf{N}_t+1} = \mathbf{Z}_t + t$ the distribution of $\mathbf{Z}_t/\mathbf{S}_{\mathbf{N}_t+1}$ is obtained from (3.6) by the change of variable $x = y/(1-y)$. It is thus seen that *also* $\mathbf{Z}_t/\mathbf{S}_{\mathbf{N}_t+1}$ *has the limit density* q_α.

4. BUSY PERIODS AND RELATED BRANCHING PROCESSES

It was shown in example XIII,4(*a*) that, if φ is the Laplace transform of a probability distribution F with expectation μ, the equation

$$\beta(\lambda) = \varphi(\lambda + c - c\beta(\lambda)), \qquad \lambda > 0, \tag{4.1}$$

possesses a unique solution β: furthermore *β is the Laplace transform of a distribution B which is proper if $c\mu \leq 1$ and defective otherwise.* This simple and elegant theory is being applied with increasing frequency and it is therefore worthwhile to explain the probabilistic background of (4.1) and its applications.

The derivation of (4.1) and similar equations is simple if one gets used to expressing probabilistic relations directly in terms of Laplace transforms. A typical situation is as follows. Consider a random sum $\mathbf{S_N} = \mathbf{X}_1 + \cdots + \mathbf{X_N}$ where the $\mathbf{X}_j$ are independent with Laplace transform $\gamma(\lambda)$, and $\mathbf{N}$ is an independent variable with generating function $P(s)$. The Laplace transform of $\mathbf{S_N}$ is obviously $P(\gamma(\lambda))$ [see example XIII,3(*c*)]. For a Poisson variable $\mathbf{N}$ this Laplace transform is of the form $e^{-\alpha[1-\gamma(\lambda)]}$. As we have seen repeatedly, in applications the parameter α is often taken as a random variable subject to a distribution U. Adapting the terminology of distribution functions we can then say that $e^{-\alpha[1-\gamma(\lambda)]}$ is the conditional Laplace transform of $\mathbf{S_N}$ given the value α of the parameter. The absolute Laplace transform is obtained by integration with respect to U. Due to the peculiar form of the integrand the result is obviously $\omega(1-\gamma(\lambda))$ where ω stands for the Laplace transform of U.

Examples. (*a*) *Busy periods.*[6] Customers (or calls) arrive at a server (or trunkline) in accordance with a Poisson process at a rate c. The successive service times are supposed to be independent variables with the common distribution F. Suppose that at epoch 0 a customer arrives and the server is free. His service time commences immediately: the customers arriving during his service time join a queue, and the service times continue without interruption as long as a queue exists. By *busy period* is meant the interval from 0 to the first epoch when the server again

[6] That (4.1) governs the busy periods was pointed out by D. G. Kendall, *Some problems in the theory of queues*, J. Roy. Statist. Soc. (B), vol. 13 (1951) pp. 151–185. The elegant reduction to branching processes was contributed by I. J. Good. Equation (4.1) is equivalent to

$$B(t) = \sum \int_0^t e^{-cx} \frac{(cx)^n}{n!} B^{n\star}(t-x)\, F\{dx\}$$

which is frequently referred to as *Takacs' integral equation.* The intrinsic simplicity of the theory is not always understood.

becomes free. Its duration is a random variable and we denote by B and β its distribution and Laplace transform, respectively.

In the terminology of branching processes the customer initiating the busy period is the "ancestor," the customers arriving during his service time are his direct descendents, and so on. Given that the progenitor departs at epoch x the number $\mathbf{N}$ of his direct descendants is a Poisson variable with expectation cx. Denote by $\mathbf{X}_j$ the total service time of the jth *direct descendant and all of his progeny.* Although these service times are not necessarily consecutive their total duration has clearly the same distribution as the busy period. The total service time required by all (direct and indirect) descendants is therefore $\mathbf{S}_\mathbf{N} = \mathbf{X}_1 + \cdots + \mathbf{X}_\mathbf{N}$ where the $\mathbf{X}_j$ have the Laplace transform β and all the variables are independent. For the busy period we have to add the service time x of the ancestor himself. Accordingly, given the length of the ancestor's service time the busy period $x + \mathbf{S}_\mathbf{N}$ has the (conditional) Laplace transform $e^{-x[\lambda+c-c\beta(\lambda)]}$. The parameter x has the distribution F and integration with respect to x yields (4.1).

If B is defective the defect $1 - B(\infty)$ represents the probability of a never-ending busy period (congestion). The condition $c\mu \leq 1$ expresses that the expected total service time of customers arriving per time unit must not exceed unity. It is easy from (4.1) to calculate the expectation and variance of B.

In the special case of exponential service times $F(t) = 1 - e^{-\alpha t}$ and $\varphi(\lambda) = \alpha/(\lambda+\alpha)$. In this case (4.1) reduces to a quadratic equation one of whose roots is unbounded at infinity. The solution β therefore agrees with the other root, namely

$$\beta(\lambda) = \sqrt{\frac{\alpha}{c}}\left[\frac{\lambda + \alpha + c}{2\sqrt{\alpha c}} - \sqrt{\left(\frac{\lambda + \alpha + c}{2\sqrt{\alpha c}}\right)^2 - 1}\,\right]. \tag{4.2}$$

This Laplace transform occurs in example XIII,3(c). Taking into account the changed scale parameter and the translation principle we find that the corresponding density is given by

$$\sqrt{\alpha/c}\; e^{-(\alpha+c)x} x^{-1}\, I_1(2\sqrt{\alpha c}\; x). \tag{4.3}$$

The same result will be derived by another method in example 6(b); it was used in example VI,9(e).

(b) *Delays in traffic.*[7] Suppose that cars passing a given point of the road conform to a Poisson process at a rate c. Let the traffic be stopped (by a red light or otherwise) for a duration δ. When traffic is resumed $\mathbf{K}$ cars will wait in line, where $\mathbf{K}$ is a Poisson variable with parameter $c\delta$.

[7] This example is inspired by J. D. C. Little's treatment of the number of cars delayed. [Operations Res., vol. 9 (1961) pp. 39–52.]

Because the rth car in the line cannot move before the $r-1$ cars ahead of it, each car in the line causes a delay for all following cars. It is natural to assume that the several delays are independent random variables with a common distribution F. For the duration of a waiting line newly arriving cars are compelled to join the line, thus contributing to the total delay. The situation is the same as in the preceding example except that we have $\mathbf{K}$ "ancestors." The total delay caused by each individual car and its direct and indirect descendants has the Laplace transform β satisfying (4.1), and the total "busy period"—the interval from the resumption of traffic to the first epoch where no car stands waiting—has the Laplace transform

$$e^{-c\delta}\sum\frac{(c\delta)^k}{k!}\,\beta^k(\lambda)=e^{-c\delta[1-\beta(\lambda)]}.$$

It is easy to calculate the expected delay and one can use this result for the discussion of the effect of successive traffic lights, etc. (See problems 6, 7.) ▶

5. DIFFUSION PROCESSES

In the one-dimensional Brownian motion the transition probabilities are normal and the first passage times have a stable distribution with index $\frac{1}{2}$ [see example VI,2(*e*)]. Being in possession of these explicit formulas we must not expect new information from the use of Laplace transforms. The reason for starting afresh from the diffusion equation is that the method is instructive and applicable to the most general diffusion equation (except that no *explicit* solutions can be expected when the coefficients are arbitrary). To simplify writing we take it for granted that the transition probabilities Q_t have densities q_t (although the method to be outlined would lead to this result without special assumptions).

We begin with the special case of Brownian motion. For a given bounded continuous function f put

$$u(t,x)=\int_{-\infty}^{+\infty}q_t(x,y)f(y)\,dy. \tag{5.1}$$

Our starting point is the fact derived in example X,4(*a*) that (at least for f sufficiently smooth) u will satisfy the diffusion equation

$$\frac{\partial u(t,x)}{\partial t}=\frac{1}{2}\,\frac{\partial^2u(t,x)}{\partial x^2} \tag{5.2}$$

with the initial condition $u(t,x)\to f(x)$ as $t\to 0$. In terms of the ordinary Laplace transform

$$\omega_\lambda(x)=\int_0^\infty e^{-\lambda t}u(t,x)\,dt \tag{5.3}$$

we conclude from (5.2) that[8]

$$\lambda\omega_\lambda - \tfrac{1}{2}\omega''_\lambda = f \tag{5.4}$$

and from (5.1) that

$$\omega_\lambda(x) = \int_{-\infty}^{+\infty} K_\lambda(x, s) f(s)\, ds \tag{5.5}$$

where $K_\lambda(x, y)$ is the ordinary Laplace transform of $q_t(x, y)$. In the theory of differential equations K_λ is called the Green function of (5.4). We shall show that

$$K_\lambda(x, y) = \frac{1}{\sqrt{2\lambda}}\, e^{-\sqrt{2\lambda}\,|x-y|}. \tag{5.6}$$

The truth of this formula can be verified by checking that (5.5) represents the desired solution of the differential equation (5.4), but this does not explain how the formula was found.

We propose to derive (5.6) by a probabilistic argument applicable to more general equations and leading to explicit expressions for the basic first passage times. (Problem 9.) We take it as known that the path variables $\mathbf{X}(t)$ depend continuously on t. Let $\mathbf{X}(0) = x$ and denote by $F(t, x, y)$ the probability that the point y will be reached before epoch t. We call F the distribution of the *first-passage epoch* from x to y and denote its Laplace transform by $\varphi_\lambda(x, y)$.

For $x < y < z$ the event $\mathbf{X}(t) = z$ takes place iff a first passage through y occurs at some epoch $\tau < t$ and is followed by a transition from y to z within time $t - \tau$. Thus $q_t(x, z)$ represents the convolution of $F(t, x, y)$ and $q_t(y, z)$, whence

$$K_\lambda(x, z) = \varphi_\lambda(x, y)\, K_\lambda(y, z), \qquad x < y < z. \tag{5.7}$$

Fix a point y and choose for f a function concentrated on $\overline{y, \infty}$. Multiply (5.7) by $f(z)$ and integrate with respect to z. In view of (5.5) the result is

$$\omega_\lambda(x) = \varphi_\lambda(x, y)\, \omega_\lambda(y), \qquad x > y, \tag{5.8}$$

while (5.4) requires that for y fixed $\varphi_\lambda(x, y)$ satisfy the differential equation

$$\lambda\varphi_\lambda - \frac{1}{2}\frac{\partial^2 \varphi_\lambda}{\partial x^2} = 0, \qquad x < y. \tag{5.9}$$

A solution which is bounded at $-\infty$ is necessarily of the form $C_\lambda e^{-\sqrt{2\lambda}\,x}$.

[8] Readers of the sections on semi-groups will notice that we are concerned with a Markovian semi-group generated by the differential operator $\mathfrak{A} = \frac{1}{2}d^2/dx^2$. The differential equation (5.4) is a special case of the basic equation XIII,(10.1) occurring in the Hille-Yosida theorem.

Since (5.8) shows that $\varphi_\lambda(x, y) \to 1$ as $x \to y$, we have $\varphi_\lambda(x, y) = e^{\sqrt{2\lambda}(x-y)}$ provided $x < y$. A similar argument applies when $x > y$ and it is clear that for reasons of symmetry the *Laplace transform of the first-passage time from* x *to* y *is given by*

$$\varphi_\lambda(x, y) = e^{-\sqrt{2\lambda}\,|x-y|}. \tag{5.10}$$

Letting $z = y$ in (5.7) we see therefore that

$$K_\lambda(x, y) = e^{-\sqrt{2\lambda}\,|x-y|} K_\lambda(y, y),$$

and since K must depend symmetrically on x and y it follows that $K(y, y)$ reduces to a constant C_λ depending only on λ. We have thus determined K_λ up to a multiplicative constant C_λ; that $\sqrt{2\lambda}C_\lambda = 1$ follows easily from the fact that to $f = 1$ there corresponds the solution $\omega_\lambda(x) = 1/\lambda$. This proves the truth of (5.6).

The following examples show how to calculate the probability that a point $y_1 > x$ will be reached before another point $y_2 < x$. At the same time they illustrate the treatment of *boundary conditions.*

Examples. (*a*) *One absorbing barrier*. The Brownian motion on $\overline{0, \infty}$ with an absorbing barrier at the origin is obtained by stopping an ordinary Brownian motion with $\mathbf{X}(0) = x > 0$ when it reaches the origin. We denote its transition densities by $q_t^{\text{abs}}(x, y)$ and adapt similarly the other notations.

In the unrestricted Brownian motion the probability density of a passage from $x > 0$ to $y > 0$ with an intermediate passage through 0 is the convolution of the first passage from x to 0 and $q_t(0, y)$. The corresponding Laplace transform is $\varphi_\lambda(x, 0)\, K_\lambda(0, y)$ and hence we must have

$$K_\lambda^{\text{abs}}(x, y) = K_\lambda(x, y) - \varphi_\lambda(x, 0)\, K_\lambda(0, y), \tag{5.11}$$

where $x > 0, y > 0$. This is equivalent to

$$K_\lambda^{\text{abs}}(x, y) = [e^{-\sqrt{2\lambda}\,|x-y|} - e^{-\sqrt{2\lambda}(x+y)}]/\sqrt{2\lambda} \tag{5.12}$$

or

$$q_t^{\text{abs}}(x, y) = q_t(x, y) - q_t(x, -y) \tag{5.13}$$

in agreement with the solution X,(5.5) obtained by the *reflection principle.*

The argument leading to (5.7) applies without change to the absorbing barrier process and we conclude from (5.12) that for $0 < x < y$

$$\varphi_\lambda^{\text{abs}}(x, y) = \frac{e^{\sqrt{2\lambda}\,x} - e^{-\sqrt{2\lambda}\,x}}{e^{\sqrt{2\lambda}\,y} - e^{-\sqrt{2\lambda}\,y}}. \tag{5.14}$$

This[9] is the Laplace transform of *the probability that in an unrestricted Brownian motion with* $\mathbf{X}(0) = x$ *the point* $y > x$ *is reached before epoch* t *and before a passage through the origin.* Letting $\lambda \to 0$ we conclude that *the probability that* y *will be reached before the origin equals* x/y. just as in the symmetric Bernoulli random walk (see the ruin problem in **1**; XIV,2). (Continued in problem 8.)

(*b*) *Two absorbing barriers.* Consider now a Brownian motion starting at a point x in $\overline{0, 1}$ and terminating when either 0 or 1 is reached. It is easiest to derive this process from the preceding absorbing barrier process by introducing an additional absorbing barrier at 1 so that the reasoning leading to (5.11) applies without change. The transition densities $q_t^{\#}(x, y)$ of the new process have therefore the Laplace transform $K_\lambda^{\#}$ given by

$$K_\lambda^{\#}(x, y) = K_\lambda^{\text{abs}}(x, y) - \varphi_\lambda^{\text{abs}}(x, 1)\, K_\lambda^{\text{abs}}(1, y) \tag{5.15}$$

with x and y restricted to $\overline{0, 1}$. [Note that the boundary conditions $K_\lambda^{\#}(0, y) = K_\lambda^{\#}(1, y)$ are satisfied.] Simple arithmetic shows that

$$K_\lambda^{\#}(x, y) = \frac{e^{-\sqrt{2\lambda}\,|x-y|} + e^{-\sqrt{2\lambda}(2-|x-y|)} - e^{-\sqrt{2\lambda}(x+y)} - e^{-\sqrt{2\lambda}(2-x-y)}}{\sqrt{2\lambda}(1-e^{-2\sqrt{2\lambda}})}. \tag{5.16}$$

Expanding $1/[1-e^{-2\sqrt{2\lambda}}]$ into a geometric series, one is led to the alternative representation

$$K_\lambda^{\#}(x, y) = \frac{1}{\sqrt{2\lambda}} \sum_{n=-\infty}^{+\infty} [e^{-\sqrt{2\lambda}\,|x-y+2n|} - e^{-\sqrt{2\lambda}\,|x+y+2n|}], \tag{5.17}$$

which is equivalent to the solution X,(5.7) obtained by the *reflection principle.* ►

The same argument applies to the more *general diffusion equation*

$$\frac{\partial u(t, x)}{\partial t} = \tfrac{1}{2}a(x)\frac{\partial^2 u(t, x)}{\partial x^2} + b(x)\frac{\partial u(t, x)}{\partial x}, \qquad a > 0, \tag{5.18}$$

in a finite or infinite interval. Instead of (5.4) we get

$$\lambda\omega_\lambda - \tfrac{1}{2}a\omega_\lambda'' - b\omega_\lambda' = f \tag{5.19}$$

and the solution is again of the form (5.5) with a Green function K_λ of the form (5.7) where $\varphi_\lambda(x, y)$ is the transform of the *first-passage density* from x to $y > x$. For fixed y, this function must satisfy the differential equation corresponding to (5.9), namely

$$\lambda\varphi_\lambda - \tfrac{1}{2}a\varphi_\lambda'' - b\varphi_\lambda' = 0. \tag{5.20}$$

[9] For y fixed, φ_2^{abs} represents the solution of the differential equation (5.9) which reduces to 0 when $x = 0$ and to 1 when $x = y$. In this form the result applies to arbitrary triples of points $a < x < b$ and $a > x > b$ and to more general differential equations.

It must be bounded at the left endpoint and $\varphi_\lambda(y, y) = 1$. These conditions determine φ_λ uniquely except if (5.20) possesses a bounded solution, in which case (as in the above examples) appropriate boundary conditions must be imposed. (See problems 9, 10.)

6. BIRTH-AND-DEATH PROCESSES AND RANDOM WALKS

In this section we explore the connection between the birth-and-death processes of **1**; XVII,5 and the randomized random walk of II,7. The main purpose is to illustrate the techniques involving Laplace transforms and the proper use of boundary conditions.

Consider a simple random walk starting at the origin in which the individual steps equal 1 or -1 with respective probabilities p and q. The times between successive steps are supposed to be independent random variables with an exponential distribution with expectation $1/c$. The probability $P_n(t)$ of the position n at epoch t was found in II,(7.7), but we start afresh from a new angle. To derive an equation for $P_n(t)$ we argue as follows. The position $n \neq 0$ at epoch t is possible only if a jump has occurred before t. Given that the first jump occurred at $t - x$ and led to 1, the (conditional) probability of the position n at epoch t is $P_{n-1}(x)$. Thus for $n = \pm 1, \pm 2, \ldots$.

$$(6.1a) \qquad P_n(t) = \int_0^t ce^{-c(t-x)}[pP_{n-1}(x) + qP_{n+1}(x)]\, dx.$$

For $n = 0$ the term e^{-ct} must be added to account for the possibility of no jump up to epoch t. Thus

$$(6.1b) \qquad P_0(t) = e^{-ct} + \int_0^t ce^{-c(t-x)}[pP_{-1}(x) + qP_1(x)]\, dx.$$

Accordingly, the P_n must satisfy the infinite system of *convolution equations* (6.1). A simple differentiation leads to the infinite system of *differential equations*[10]

$$(6.2) \qquad P_n'(t) = -cP_n(t) + cpP_{n-1}(t) + cqP_{n+1}(t)$$

together with the initial conditions $P_0(0) = 1$, $P_n(0) = 0$ for $n \neq 0$.

The two systems (6.1) and (6.2) are equivalent, but the latter has the formal advantage that the special role of $n = 0$ is noticeable only in the initial conditions. For the use of Laplace transforms it does not matter where we start.

---

[10] They are a special case of the equations **1**; XVII,(5.2) for general birth-and-death processes and may be derived in like manner.

We pass to Laplace transforms putting

$$\pi_n(\lambda) = \int_0^\infty e^{-\lambda t}\, P_n(t)\, dt. \tag{6.3}$$

Since convolutions correspond to multiplication of Laplace transforms and e^{-cx} has the transform $1/(c+\lambda)$ the system (6.1) is equivalent to

$$\pi_n(\lambda) = \frac{c}{c+\lambda}\,[p\pi_{n-1}(\lambda) + q\pi_{n+1}(\lambda)], \qquad n \neq 0 \tag{6.4a}$$

$$\pi_0(\lambda) = \frac{1}{c+\lambda} + \frac{c}{c+\lambda}\,[p\pi_{-1}(\lambda) + q\pi_1(\lambda)], \tag{6.4b}$$

[The same result could have been obtained from (6.2) since

$$\int_0^\infty e^{-\lambda t} P_n'(t)\, dt = -P_n(0) + \lambda\pi_n(\lambda)$$

which follows on integration by parts.]

The system of linear equations (6.4) is of the type encountered in connection with random walks in **1**; XIV, and we solve it by the same method. The quadratic equation

$$cqs^2 - (c+\lambda)s + cp = 0 \tag{6.5}$$

has the roots

$$s_\lambda = \frac{c+\lambda-\sqrt{(c+\lambda)^2 - 4c^2pq}}{2cq} \quad \text{and} \quad \sigma_\lambda = (p/q)s_\lambda^{-1}. \tag{6.6}$$

It is easily verified that with arbitrary constants A_λ, B_λ the linear combinations $\pi_n(\lambda) = A_\lambda s_\lambda^n + B_\lambda \sigma_\lambda^n$ satisfy (6.4*a*) for $n = 1, 2, \ldots$, and the coefficients can be chosen so as to yield the correct values for $\pi_0(\lambda)$ and $\pi_1(\lambda)$. Given π_0 and π_1 it is possible from (6.4*a*) to calculate recursively $\pi_2, \pi_3, \ldots$, and so for $n \geq 0$ *every* solution is of the form $\pi_n(\lambda) = A_\lambda s_\lambda^n + B_\lambda \sigma_\lambda^n$. Now $s_\lambda \to 0$ but $\sigma_\lambda \to \infty$ as $\lambda \to \infty$. As our $\pi_n(\lambda)$ remain bounded at infinity we must have $B_\lambda = 0$, and hence

$$\pi_n(\lambda) = \pi_0(\lambda)s_\lambda^n, \qquad n = 0, 1, 2, \ldots. \tag{6.7a}$$

For $n \leq 0$ we get analogously

$$\pi_n(\lambda) = \pi_0(\lambda)\sigma_\lambda^n = (p/q)^n \pi_0(\lambda) s_\lambda^{-n}, \qquad n = 0, -1, -2, \ldots. \tag{6.7b}$$

Substituting into (6.4*b*) we get finally

$$\pi_0(\lambda) = \frac{1}{\sqrt{(c+\lambda)^2 - 4c^2pq}} \tag{6.8}$$

and so all $\pi_n(\lambda)$ are uniquely determined.

Much information can be extracted from these Laplace transforms without knowledge of explicit formulas for the solution itself. For example, since multiplication of Laplace transforms corresponds to convolutions, the form (6.7) suggests that for $n \geq 0$ the probabilities P_n are of the form $P_n = F^{n\star} \star P_0$ where F is a (possibly defective) probability distribution with transform s_λ. That this is so can be seen probabilistically as follows. If at epoch t the random walk is at the point n, the first passage through n must have occurred at some epoch $\tau \leq t$. In this case the (conditional) probability of being at epoch t again at n equals $P_0(t-\tau)$. Thus P_n is the convolution of P_0 and the distribution F_n of the first passage time through n. Again, this first passage time is the sum of n identically distributed independent random variables, namely the waiting times between successive passages through $1, 2, \ldots$. This explains the form (6.7) and shows at the same time that s_λ^n *is (for* $n > 0$*) the transform of the distribution* F_n *of the first passage time through* n. This distribution is defective unless $p = q = \frac{1}{2}$, for only in this case is $s_0 = 1$.

In the present case we are fortunately able to invert the transforms s_λ. It was shown in example XIII,3(*d*) that $(\lambda - \sqrt{\lambda^2 - 1})^r$ is (for $\lambda > 1$) the ordinary Laplace transform of $(r/x)I_r(x)$. Changing λ into $\lambda/2c\sqrt{pq}$ merely changes a scale factor, and replacing λ by $\lambda + c$ reflects multiplication of the density by e^{-cx}. It follows that s_λ^n *(with* $n > 0$*) is the ordinary Laplace transform of a distribution* F_n *with density*

$$f_n(t) = \sqrt{(p/q)^n}\, nt^{-1}\, I_n(2c\sqrt{pq}\, t)e^{-ct}. \tag{6.9}$$

This is the density of the first passage time through $n > 0$. This fact was established by direct methods in II,(7.13) [and so the present argument may be viewed as a new derivation of the Laplace transform of $x^{-1} I_n(x)$].

An explicit expression for the probabilities $P_n(t)$ can be obtained similarly. In problem 8 of XIII,11 we found the Laplace transform of I_n, and the adjustment of parameters just described leads directly to the explicit formulas

$$P_n(t) = \sqrt{(p/q)^n}\, e^{-ct}\, I_n(2c\sqrt{pq}\, t), \qquad n = 0, \pm 1, \pm 2, \ldots. \tag{6.10}$$

Again, this result was derived by direct methods in II,(7.7).

As we have seen in **1**; XVII,7 various trunking and servicing problems lead to the same system of differential equations (6.2) except that n is restricted to $n \geq 0$ and that a different equation corresponds to the boundary state $n = 0$. Two examples will show how the present method operates in such cases.

Examples. (*a*) *Single-server queues.* We consider a single server in which newly arriving customers join the queue if the server is busy. The state of the system is given by the number $n \geq 0$ of customers in the queue including the customer being served. The interarrival times and the service times are mutually independent and have the exponential densities $\lambda e^{-\lambda t}$ and $\mu e^{-\mu t}$, respectively. This is a special case of the multi-server example (*b*) in **1**; XVII,7, but we derive the differential equations afresh in order to elucidate the intimate connection with our present random walk model.

Suppose that at present there are $n \geq 1$ customers in the queue. The next change of state will be $+1$ if it is due to a new arrival, and -1 if it is due to the termination of the present service time. The waiting time $\mathbf{T}$ for this change is the smaller of the waiting times for these two contingencies and so $\mathbf{P}\{\mathbf{T} > t\} = e^{-ct}$ where we put $c = \lambda + \mu$. When a change occurs it is $+1$ with probability $p = \lambda/c$, and -1 with probability $q = \mu/c$. In other words, as long as the queue lasts our process conforms to our random walk model, and hence the differential equations (6.2) hold for $n \geq 1$. However, when no service is going on, a change can be caused only by new arrivals, and so for $n = 0$ the differential equation takes on the form

$$P_0'(t) = -cpP_0(t) + cqP_1(t). \tag{6.11}$$

We solve these differential equations assuming that originally the server was free, that is $P_0(0) = 1$. For $n \geq 1$ the Laplace transforms $\pi_n(\lambda)$ again satisfy the equations (6.4*a*), but for $n = 0$ we get from (6.11)

$$(cp+\lambda)\pi_0(\lambda) = 1 + cq\pi_1(\lambda). \tag{6.12}$$

As in the general random walk we get $\pi_n(\lambda) = \pi_0(\lambda)s_\lambda^n$ for $n \geq 1$, but in view of (6.12)

$$\pi_0(\lambda) = \frac{1}{cp + \lambda - cqs_\lambda} = \frac{1 - s_\lambda}{\lambda}. \tag{6.13}$$

Thus

$$\pi_n(\lambda) + \pi_{n+1}(\lambda) + \cdots = \frac{1 - s_\lambda}{\lambda}(s_\lambda^n + s_\lambda^{n+1} + \cdots) = s_\lambda^n/\lambda.$$

We found that s^n is the Laplace transform of the distribution F_n with density (6.9); the factor $1/\lambda$ corresponds to integration, and so for $n > 0$

$$P_n(t) + P_{n+1}(t) + \cdots = F_n(t) \tag{6.14}$$

where F_n *is the distribution with density* (6.9). For $n = 0$ the left side is, of course, unity.

(*b*) *Fluctuations during a busy period.* We consider the same server, but only during a busy period. In other words, it is assumed that at epoch 0 a customer arrives at the empty server, and we let the process *terminate* when the server becomes empty. Analytically this implies that n is now restricted to $n \geq 1$, and the initial condition is $P_1(0) = 1$. Nothing changes in the differential equations (6.2) for $n \geq 2$, but in the absence of a zero state the term $cpP_0(t)$ drops out in equation number one. Thus the Laplace transforms $\pi_n(\lambda)$ satisfy (6.4a) for $n \geq 2$ and

$$(\lambda+c)\pi_1(\lambda) = 1 + cq\pi_2(\lambda). \tag{6.15}$$

As before we get $\pi_n(\lambda) = \pi_1(\lambda)s_\lambda^{n-1}$ for $n \geq 2$, but $\pi_1(\lambda)$ is to be determined from (6.15). A routine calculation shows that $\pi_1(\lambda) = s_\lambda/(cp)$, and hence $\pi_n(\lambda) = s_\lambda^n/(cp)$. Using the preceding example we have thus the final result that $P_n(t) = f_n(t)/(cp)$ with f_n given by (6.9).

To ensure that the busy period has a finite duration we assume that $p < q$. Denote the *duration of the busy period* by **T**. Then $\mathbf{P}\{\mathbf{T} > t\} = P(t) = \sum P_n(t)$. Now $P'(t) = -cqP_1(t)$ as can be seen summing the differential equations, or probabilistically as follows. Neglecting events of negligible probabilities the busy time terminates between t and $t+h$ iff at epoch t there is only one customer in the queue and his service terminates within the next time interval of duration h. The two conditions have probabilities $P_1(t)$ and $cqh + o(h)$, and so the density of **T** satisfies the condition $-P'(t) = cqP_1(t)$. Accordingly, *the duration of the busy period has the density*

$$-P'(t) = \sqrt{q/p}\, t^{-1} I_1(2c\sqrt{pq}\, t)e^{-ct}. \tag{6.16}$$

This result was derived by a different method in example 4(a) and was used in the queuing process VI,9(e). See problem 13. ▶

7. THE KOLMOGOROV DIFFERENTIAL EQUATIONS[11]

We return to the Markovian processes restricted to the integers $1, 2, \ldots$. The Kolmogorov differential equations were derived in **1**; XVII,9 and again in X,3. This section contains an independent treatment by means of Laplace transforms. To render the exposition self-contained we give a new derivation of the basic equations, this time in the form of *convolution* equations.

The basic assumption is that if $\mathbf{X}(\tau) = i$ at some epoch τ, the value $\mathbf{X}(t)$ will remain constant for an interval $\tau \leq t < \tau+\mathbf{T}$ whose duration has the exponential density $c_i e^{-c_i x}$; the probability of a jump to j_{is} then p_{ij}. Given that $\mathbf{X}(0) = i$ the probability $P_{ik}(t)$ that $\mathbf{X}(t) = k \neq i$ can now be

[11] The theory developed in this section applies without essential change to the general jump processes of X,3. It is a good exercise to reformulate the proofs in terms of the probabilities themselves without using Laplace transforms. Some elegance is lost, but the theory then easily generalizes to the non-stationary case where the coefficients c_j and p_{jk} depend on t. In this form the theory is developed (for general jump processes) in W. Feller, Trans. Amer. Math. Soc., vol. 48 (1940), pp. 488–515 [erratum vol. 58, p. 474].

For a probabilistic treatment based on the study of sample paths see Chung (1967). For generalizations to *semi-Markov processes* see problem 14.

calculated by summing over all possible epochs and results of the *first* jump:

$$P_{ik}(t) = \sum_{j=1}^{\infty} \int_0^t c_i e^{-c_i x} p_{ij} P_{jk}(t-x)\, dx \qquad (k \neq i). \tag{7.1a}$$

For $k = i$ we must add a term accounting for the possibility of no jump:

$$P_{ii}(t) = e^{-c_i t} + \sum_{j=1}^{\infty} \int_0^t c_i e^{-c_i x} p_{ij} P_{ji}(t-x)\, dx. \tag{7.1b}$$

These equations can be unified by introducing the Kronecker symbol δ_{ik} which equals 1 or 0 according as $k = i$ or $k \neq i$.

The backward equations (7.1) *are our point of departure;*[12] given arbitrary $c_i > 0$ and a stochastic matrix $\mathbf{p} = (p_{ik})$ we seek stochastic matrices $P(t) = (P_{ik}(t))$ satisfying (7.1).

Alternatively, if we suppose that any finite time interval contains only finitely many jumps we can modify the argument by considering the epoch x of the *last* jump preceding t. The probability of a jump from j to k has density $\sum P_{ij}(x) c_j p_{jk}$, while the probability of no jump between x and t equals $e^{-c_j(t-x)}$. Instead of (7.1) we get *the forward equations*

$$P_{ik}(t) = \delta_{ik} e^{-c_i t} + \int_0^t \sum_{j=1}^{\infty} P_{ij}(x) c_j p_{jk} e^{-c_k(t-x)}\, dx. \tag{7.2}$$

As will be seen, however, there exist processes with *infinitely many jumps* satisfying the backward equations, and hence the forward equations are not implied by the basic assumptions underlying the process. This phenomenon was discussed in X,3 and also in **1**; XVII,9.

In terms of the Laplace transforms

$$\Pi_{ik}(\lambda) = \int_0^\infty e^{-\lambda t} P_{ik}(t)\, dt \tag{7.3}$$

the backward equations (7.1) take on the form

$$\Pi_{ik}(\lambda) = \frac{\delta_{ik}}{\lambda + c_i} + \frac{c_i}{\lambda + c_i} \sum_{j=1}^{\infty} p_{ij} \Pi_{jk}(\lambda). \tag{7.4}$$

We now switch to a more convenient matrix notation. (The rules of matrix

---

[12] The change of variables $y = t - x$ makes differentiation easy, and it is seen that the convolution equations (7.1) are equivalent to the system of differential equations

$$P'_{ik}(t) = -c_i P_{ik}(t) + c_i \sum_j p_{ij} P_{jk}(t)$$

together with the initial conditions $P_{ii}(0) = 1$ and $P_{ik}(0) = 0$ for $k \neq i$. This system agrees with **1**; XVII,(9.14), except that there the coefficients c_i and p_{ij} depend on time, and hence P_{ik} is a function of two epochs τ and t rather than of the duration $t - \tau$.

calculus apply equally to infinite matrices with non-negative elements.) We introduce the matrices $\Pi(\lambda) = (\Pi_{ik}(\lambda))$ and similarly $P(t) = (P_{ik}(t))$, $\mathbf{p} = (p_{ik})$, and the diagonal matrix $\mathbf{c}$ with elements c_i. By $\mathbf{1}$ we denote the column vector all of whose elements equal 1. The row sums of a matrix A are then given by $A\mathbf{1}$. Finally, I is the identity matrix.

It is then clear from (7.4) that the *backward equations* (7.1) are transformed into

$$(\lambda+\mathbf{c})\,\Pi(\lambda) = I + \mathbf{cp}\,\Pi(\lambda), \tag{7.5}$$

and the *forward equations* into

$$\Pi(\lambda)(\lambda+\mathbf{c}) = I + \Pi(\lambda)\mathbf{cp}. \tag{7.6}$$

To construct the *minimal solution* we put recursively

$$(\lambda+\mathbf{c})\,\Pi^{(0)}(\lambda) = I, \qquad (\lambda+\mathbf{c})\,\Pi^{(n+1)}(\lambda) = I + \mathbf{cp}\,\Pi^{(n)}(\lambda). \tag{7.7}$$

For the row sums of $\lambda\Pi^{(n)}(\lambda)$ we introduce the notation

$$\lambda\Pi^{(n)}(\lambda)\mathbf{1} = \mathbf{1} - \xi^{(n)}(\lambda). \tag{7.8}$$

Substituting into (7.7) and remembering that $\mathbf{p1} = \mathbf{1}$ it is seen that

$$(\lambda+\mathbf{c})\xi^{(n+1)}(\lambda) = \mathbf{cp}\xi^{(n)}(\lambda). \tag{7.9}$$

Since $\xi^{(0)} \geq 0$ it follows that $\xi^{(n)}(\lambda) \geq 0$ for all n, and so the matrices $\lambda\Pi^{(n)}(\lambda)$ are substochastic. Their elements are non-decreasing functions of n and therefore there exists a finite limit

$$\Pi^{(\infty)}(\lambda) = \lim_{n\to\infty} \Pi^{(n)}(\lambda) \tag{7.10}$$

and $\lambda\Pi^{(\infty)}(\lambda)$ is substochastic or stochastic.

Obviously $\Pi^{(\infty)}(\lambda)$ satisfies the backward equation (7.5) and for any other non-negative solution $\Pi(\lambda)$ one has trivially $\Pi(\lambda) \geq \Pi^{(0)}(\lambda)$, and by induction $\Pi(\lambda) \geq \Pi^{(n)}(\lambda)$ for all n. Thus

$$\Pi(\lambda) \geq \Pi^{(\infty)}(\lambda). \tag{7.11}$$

Less obvious is that $\Pi^{(\infty)}(\lambda)$ satisfies also the forward equation (7.6). To show it we prove by induction that

$$\Pi^{(n)}(\lambda)(\lambda+\mathbf{c}) = I + \Pi^{(n-1)}(\lambda)\mathbf{cp}. \tag{7.12}$$

This is true for $n = 1$. Assuming the truth of (7.12), substitution into (7.7) leads to

$$(\lambda+\mathbf{c})\Pi^{(n+1)}(\lambda)(\lambda+\mathbf{c}) = \lambda I + \mathbf{c} + [I + \mathbf{cp}\Pi^{(n-1)}(\lambda)]\mathbf{cp}. \tag{7.13}$$

The expression within brackets equals $(\lambda+\mathbf{c})\Pi^{(n)}(\lambda)$. Premultiplication of

(7.13) by $(\lambda+\mathbf{c})^{-1}$ yields (7.12) with n replaced by $n+1$. This relation is therefore true for all n, and hence $\Pi^{(\infty)}(\lambda)$ satisfies the forward equation.

Repeating the argument that led to (7.11) we see equally that any non-negative solution of the forward equations (7.6) satisfies $\Pi(\lambda) \geq \Pi^{(\infty)}(\lambda)$. For this reason $\Pi^{(\infty)}(\lambda)$ is called the *minimal* solution.

We have thus proved

Theorem 1. *There exists a matrix* $\Pi^{(\infty)}(\lambda) \geq 0$ *with row sums* $\leq \lambda^{-1}$ *satisfying both* (7.5) *and* (7.6) *and such that for every non-negative solution of either* (7.5) *or* (7.6) *the inequality* (7.11) *holds.*

Theorem 2. *The minimal solution is the Laplace transform of a family of substochastic or stochastic matrices* $P(t)$ *satisfying the Chapman-Kolmogorov equation*

$$P(s+t) = P(s)\,P(t) \tag{7.14}$$

and both the backward and forward equations (7.1)–(7.2). *Either all matrices* $P(t)$ *and* $\lambda\Pi^{(\infty)}(\lambda)$ $(t > 0, \lambda > 0)$ *are strictly stochastic or none is.*

Proof. We drop the superscript ∞ and write $\Pi(\lambda)$ for $\Pi^{(\infty)}(\lambda)$. From the definition (7.7) it is clear that $\Pi_{ik}^{(n)}(\lambda)$ is the transform of a positive function $P_{ik}^{(n)}$ which is the convolution of finitely many exponential distributions. Because of (7.8) the row sums of $P^{(n)}(t)$ form a monotone sequence bounded by $\mathbf{1}$ and so it follows that $\Pi(\lambda)$ is the transform of a matrix $P(t)$ which is substochastic or stochastic. From (7.5)–(7.6) it is clear that $P(t)$ satisfies the original forward and backward equations. These imply that $P(t)$ depends continuously on t. It follows that if the ith row sum is < 1 for some t the ith row sum of $\Pi(\lambda)$ is $< \lambda^{-1}$ for all λ and conversely.

To restate (7.14) in terms of Laplace transforms multiply it by $e^{-\lambda t-\nu s}$ and integrate over s and t. The right side leads to the matrix product $\Pi(\lambda)\,\Pi(\nu)$, and the left side is easily evaluated by the substitution $x = t + s$, $y = -t + s$. The result is

$$-\frac{\Pi(\nu) - \Pi(\lambda)}{\nu - \lambda} = \Pi(\lambda)\,\Pi(\nu); \tag{7.15}$$

conversely (7.15) implies (7.14). [This argument was used in example XIII,8(a).]

To prove (7.15) consider the matrix equation

$$(\lambda+\mathbf{c})Q = A + \mathbf{cp}Q. \tag{7.16}$$

If A and Q are non-negative then obviously $Q \geq (\lambda+\mathbf{c})^{-1}A = \Pi^{(0)}(\lambda)A$ and by induction $Q \geq \Pi^{(n)}(\lambda)A$ for all n. Thus $Q \geq \Pi(\lambda)A$. Now $\Pi(\nu)$ satisfies (7.16) with $A = I + (\lambda-\nu)\Pi(\nu)$ and hence for $\lambda > \nu$

$$\Pi(\nu) \geq \Pi(\lambda) + (\lambda-\nu)\Pi(\lambda)\,\Pi(\nu). \tag{7.17}$$

On the other hand, the right-hand member satisfies the forward equation (7.6) with λ replaced by ν. It follows that it is $\geq \Pi(\nu)$ and thus the equality sign holds in (7.17). This concludes the proof.[13] ▶

To see whether the matrix $\lambda\Pi^{(\infty)}(\lambda)$ is strictly stochastic[14] we return to the relations (7.8) and (7.9). Since the elements $\xi_i^{(n)}(\lambda)$ are non-increasing functions of n there exists a limit $\xi(\lambda) = \lim \xi^{(n)}(\lambda)$ such that

$$\lambda\Pi^{(\infty)}(\lambda)\mathbf{1} = \mathbf{1} - \xi(\lambda) \tag{7.18}$$

and

$$(\lambda+\mathbf{c})\xi(\lambda) = \mathbf{cp}\xi(\lambda), \qquad 0 \leq \xi(\lambda) \leq \mathbf{1}. \tag{7.19}$$

On the other hand, we have

$$(\lambda+\mathbf{c})\xi^{(0)}(\lambda) = \mathbf{c1} = \mathbf{cp1} \tag{7.20}$$

and therefore $\xi^{(0)}(\lambda) \geq \xi(\lambda)$ for *any* vector $\xi(\lambda)$ satisfying (7.19). From (7.9) it follows by induction that $\xi^{(n)}(\lambda) \geq \xi(\lambda)$ for all n, and so the vector $\xi(\lambda)$ in (7.18) represents the *maximal* vector satisfying (7.19). We have thus

Theorem 3. *The row defects of the minimal solution are represented by the well-defined maximal vector* $\xi(\lambda)$ *satisfying* (7.19).

Thus $\lambda\Pi^{(\infty)}(\lambda)$ is strictly stochastic iff (7.19) implies $\xi(\lambda) = 0$.

Corollary 1. *If* $c_i \leq M < \infty$ *for all* i *the minimal solution is strictly stochastic (so that neither the forward nor the backward equations possess other admissible solutions).*

Proof. Since $c/(\lambda+c)$ is an increasing function of c it follows from (7.19) by induction that

$$\xi(\lambda) \leq \left(\frac{M}{\lambda+M}\right)^n \cdot \mathbf{1} \tag{7.21}$$

for all n, and hence $\xi(\lambda) = 0$. ▶

If $A(\lambda)$ is a matrix of elements of the form $\xi_i(\lambda)\,\eta_k(\lambda)$ with arbitrary $\eta_k(\lambda)$ then $\Pi(\lambda) + A(\lambda)$ is again a solution of the backward equation (7.5). It is always possible to choose $A(\lambda)$ so as to obtain admissible matrices $P(t)$ satisfying the Chapman-Kolmogorov equation. The procedure is illustrated in the next section. The corresponding

[13] (7.15) is the *resolvent equation* for the family of contractions $\lambda\Pi(\lambda)$ on the Banach space of bounded column vectors. We saw in XIII,10 that it holds iff the range of these transformations is independent of λ, and minimal character guarantees this. (In terms of boundary theory the range is characterized by the vanishing of the vectors at the "active exit boundary.")

[14] *Warning:* A formal multiplication of the forward equations by the column vector $\mathbf{1}$ would seem to lead to the identity $\lambda\Pi(\lambda)\mathbf{1} = \mathbf{1}$, but the series involved may diverge. The procedure is legitimate if the c_i are bounded (corollary 1).

processes are characterized by transitions involving infinitely many jumps in a finite time interval. Curiously enough, the forward equations may be satisfied even though their interpretation in terms of a last jump is false.

These are the main results. We conclude with a criterion that is useful in applications and interesting because its proof introduces notions of potential theory; the kernel Γ of (7.25) is a typical *potential*.

We assume $c_i > 0$ and rewrite (7.19) in the form

$$\xi(\lambda) + \lambda \mathbf{c}^{-1}\xi(\lambda) = \mathbf{p}\xi(\lambda). \tag{7.22}$$

Multiplying by $\mathbf{p}^k$ and adding over $k = 0, \ldots, n-1$ we get

$$\xi(\lambda) + \lambda \sum_{k=0}^{n-1} \mathbf{p}^k \mathbf{c}^{-1}\xi(\lambda) = \mathbf{p}^n \xi(\lambda). \tag{7.23}$$

This implies that $\mathbf{p}^n\xi(\lambda)$ depends monotonically on n and so $\mathbf{p}^n\xi(\lambda) \to x$ where x is the minimal column vector satisfying[15]

$$\mathbf{p}x = x, \qquad \xi(\lambda) \leq x \leq 1. \tag{7.24}$$

Now define a matrix (with possibly infinite elements) by

$$\Gamma = \sum_{k=1}^{\infty} \mathbf{p}^k \mathbf{c}^{-1}. \tag{7.25}$$

Letting $n \to \infty$ in (7.23) we get

$$\xi(\lambda) + \lambda\Gamma\xi(\lambda) = x, \tag{7.26}$$

which implies in particular that $\xi_k(\lambda) = 0$ for each k such that $\Gamma_{kk} = \infty$. This is the case if k is a persistent state for the Markov chain with matrix $\mathbf{p}$ and hence we have

Corollary 2. *The minimal solution is strictly stochastic (and hence unique) whenever the discrete Markov chain with matrix* $\mathbf{p}$ *has only persistent states.*

8. EXAMPLE: THE PURE BIRTH PROCESS

Instead of pursuing the general theory we consider in detail processes in which only transitions $i \to i+1$ are possible, for they furnish good illustrations for the types of processes arising from non-uniqueness. To avoid trivalities we suppose $c_i > 0$ for all i. By definition $p_{i,i+1} = 1$ whence $p_{ik} = 0$ for all other combinations. The backward and forward equations now reduce to

$$(\lambda + c_i)\Pi_{ik}(\lambda) - c_i\Pi_{i+1,k}(\lambda) = \delta_{ik} \tag{8.1}$$

[15] It is not difficult to see that x is independent of λ and $\lambda\Pi^{(\infty)}(\lambda) = x - \xi(\lambda)$.

and

$$(\lambda+c_k)\Pi_{ik}(\lambda) - c_{k-1}\Pi_{i,k-1}(\lambda) = \delta_{jk}, \tag{8.2}$$

where δ_{ik} equals 1 for $i = k$ and 0 otherwise. We put for abbreviation

$$\rho_i = \frac{c_i}{c_i + \lambda}, \qquad r_i = \frac{1}{c_i + \lambda}. \tag{8.3}$$

ρ_i is the Laplace transform of the (exponential) sojourn time distribution at i, and r_i is the ordinary Laplace transform of the probability that this sojourn time extends beyond t. The dependence of r_j and ρ_j on λ should be borne in mind.

(*a*) *The minimal solution.* It is easily verified that

$$\Pi_{ik}(\lambda) = \begin{matrix} \rho_i\rho_{i+1}\cdots\rho_{k-1}r_k & \text{for} \quad k \geq i \\ 0 & \text{for} \quad k < i \end{matrix} \tag{8.4}$$

is the minimal solution for both (8.1) and (8.2). It reflects the fact that transitions from i to $k < i$ are impossible, and that the epoch of the *arrival* at $k > i$ is the sum of the k independent sojourn times at $i, i+1, \ldots, k-1$.

Let $P_{ik}(t)$ stand for the transition probabilities of the process defined by (8.4). We prove the following important result derived by other methods in **1**; XVII,4.

Lemma. *If*

$$\sum 1/c_n = \infty \tag{8.5}$$

then

$$\sum_{k=i}^{\infty} P_{ik}(t) = 1 \tag{8.6}$$

for all i *and* $t > 0$. *Otherwise* (8.6) *is false for all* i.

Proof. Note that $\lambda r_k = 1 - \rho_k$, whence

$$\lambda[\Pi_{ii}(\lambda) + \cdots + \Pi_{i,i+n}(\lambda)] = 1 - \rho_i \cdots \rho_{i+n}. \tag{8.7}$$

Thus (8.6) holds iff for all $\lambda > 0$

$$\rho_i\rho_{i+1}\cdots\rho_n \to 0 \qquad n \to \infty. \tag{8.8}$$

Now if $c_n \to \infty$ then $\rho_n \sim e^{-\lambda/c_n}$ and hence in this case (8.5) is necessary and sufficient for (8.8). On the other hand, if c_n does not tend to infinity there exists a number $q < 1$ such that $\rho_n < q$ for infinitely many n, and hence both (8.8) and (8.5) hold. ►

In the case where the series in (8.5) diverges there are no surprises: the c_n determine uniquely a birth process satisfying the basic postulates from which we started. From now on we assume therefore that $\sum c_n^{-1} < \infty$.

The defect $1 - \sum_k P_{ik}(t)$ is the probability that by epoch t the system has passed through *all* states, or has "arrived at the boundary ∞." The epoch of the arrival is the sum of the sojourn times at $i, i+1, \ldots$. The series converges with probability one because the sum of the mean sojourn times $1/c_n$ converges.

In a process starting at i *the lifetime of the process up to the epoch of the arrival at* ∞ *has the Laplace transform*

$$\xi_i = \lim_{n\to\infty} \rho_i \rho_{i+1} \cdots \rho_{i+n} \tag{8.9}$$

and the ξ_i satisfy the equations (7.19), namely,

$$(\lambda + c_i)\xi_i = c_i \xi_{i+1}. \tag{8.10}$$

For the row sums we get from (8.7)

$$\lambda \sum_{k=i}^{\infty} \Pi_{ik}(\lambda) = 1 - \xi_i. \tag{8.11}$$

(*b*) *Return processes.* Starting from the process (8.4) new processes may be defined as follows. Choose numbers q_i such that $q_i \geq 0$, $\sum q_i = 1$. We stipulate that on arrival at ∞ *with probability*[16] q_i *the state of the system passes instantaneously to* i. The original process now starts afresh until a second arrival at ∞ takes place. The time elapsed between the two arrivals at ∞ is a random variable with Laplace transform

$$\tau(\lambda) = \sum q_i \xi_i. \tag{8.12}$$

The Markovian character of the process requires that on the second arrival at ∞ the process recommences in the same manner. We now describe the transition probabilities $P_{ik}^{\text{ret}}(t)$ of the new process in terms of its Laplace transforms $\Pi_{ik}^{\text{ret}}(\lambda)$. The probability of a transition from i at epoch 0 to k at epoch t *without* an intervening passage through ∞ has the transform (8.4). The probability to reach k after exactly one passage through ∞ has therefore the Laplace transform $\xi_i \sum_j q_j \Pi_{jk}(\lambda)$, and the epoch of the *second* arrival at ∞ has transform $\xi_i \tau(\lambda)$. Considering further returns we see in this way that we must have

$$\Pi_{ik}^{\text{ret}}(\lambda) = \Pi_{ik}(\lambda) + \xi_i \frac{1}{1 - \tau(\lambda)} \sum_j q_j \Pi_{jk}(\lambda) \tag{8.13}$$

[16] Variants of the return processes are obtained by letting $\Sigma q_i < 1$; on arrival at ∞ the process terminates with probability $1 - \Sigma q_i$.

where $[1-\tau(\lambda)]^{-1} = \sum \tau^n(\lambda)$ counts the number of passages through ∞. A trite calculation using (8.11) shows that the row sums in (8.13) equal $1/\lambda$, and so the $\Pi_{ik}^{\text{ret}}(\lambda)$ are the transforms of a strictly stochastic matrix of transition probabilities $P^{\text{ret}}(t)$.

It is easily verified that *the new process satisfies the backward equations* (8.1) *but not the forward equations* (8.2). This is as should be: the postulates leading to the forward equations are violated since no last jump need exist.

(*c*) *The bilateral birth process.* To obtain a process satisfying both the forward and the backward equations we modify the birth process by letting the states of the system run through $0, \pm 1, \pm 2, \ldots$. Otherwise the conventions remain the same: the constants $c_i > 0$ are defined for all integers, and transitions from i are possible only to $i + 1$. We assume again that $\sum 1/c_n < \infty$, the summation now extending from $-\infty$ to ∞.

Nothing changes for the *minimal* solution which is still given by (8.4). The limit

$$\eta_k = \lim_{i\to-\infty} \Pi_{ik}(\lambda) = r_k\rho_{k-1}\rho_{k-2}\rho_{k-3}\cdots \tag{8.14}$$

exists and may be interpreted as the transform of "the probability $P_{-\infty,k}(t)$ of a transition from $-\infty$ at epoch 0 to k at epoch t." With this starting point the process will run through all states from $-\infty$ to ∞ and "arrive at ∞" at an epoch with Laplace transform $\xi_{-\infty} = \lim_{n\to-\infty} \xi_n$. We now define a new process as follows. It starts as the process corresponding to the minimal solution (8.4) but on reaching ∞ it recommences at $-\infty$, and in this way the process continues forever. By the construction used in (*b*) we get for the transition probabilities

$$\Pi_{ik}^{\#}(\lambda) = \Pi_{ik}(\lambda) + \frac{\xi_i\eta_k}{1-\xi_{-\infty}}. \tag{8.15}$$

It is easily verified that *the* $\Pi_{ik}^{\#}$ *satisfy both the backward and the forward equations* (8.1) *and* (8.2). The process satisfies the hypotheses leading to the backward equations, but *not* those for the forward equations.

9. CALCULATION OF ERGODIC LIMITS AND OF FIRST-PASSAGE TIMES

As can be expected, the behavior as $t \to \infty$ of the transition probabilities $P_{ij}(t)$ of Markov processes on integers is similar to that of higher transition probabilities in discrete chains with the pleasing simplification, however, that the nuisance of periodic chains disappears. Theorem 1 establishes this fact as a simple consequence of the ergodic theorem of **1**; XV. Our main concern will then be to calculate the limits for the general processes of section

7 and to show how first passage times can be found. The methods used are of wide applicability.

Theorem 1. *Suppose that for the family of stochastic matrices* $P(t)$

$$P(s+t) = P(s)\,P(t) \tag{9.1}$$

and $P(t) \to I$ *as* $t \to 0$. *If no* P_{ik} *vanishes*[17] *identically, then as* $t \to \infty$

$$P_{ik}(t) \to u_k \tag{9.2}$$

where either $u_k = 0$ *for all* k *or else*

$$u_k > 0, \qquad \sum_k u_k = 1, \tag{9.3}$$

and

$$\sum_j u_j\, P_{jk}(t) = u_k. \tag{9.4}$$

The second alternative occurs whenever there exists a probability vector $(u_1, u_2, \ldots)$ *satisfying* (9.4) *for some* $t > 0$. *In this case* (9.4) *holds for all* $t > 0$, *and the probability vector* u *is unique.*

(As explained in **1**; XVII,6 the important feature is that the limits do not depend on i, which indicates that the influence of the initial conditions is asymptotically negligible.)

Proof. For a fixed $\delta > 0$ consider the discrete Markov chain with matrix $P(\delta)$ and higher transition probabilities given by $P^n(\delta) = P(n\delta)$. If all elements $P_{ik}(n\delta)$ are ultimately positive the chain is irreducible and aperiodic, and by the ergodic theorem **1**; XV,7 the assertions are true for t restricted to the sequence $\delta, 2\delta, 3\delta, \ldots$. Since two rationals have infinitely many multiples in common the limit as $n \to \infty$ of $P_{ik}(n\delta)$ is the same for all rational δ. To finish the proof it suffices to show that $P_{ik}(t)$ is a uniformly continuous function of t and is positive for large t. Now by (9.1)

$$P_{ii}(s)\,P_{ik}(t) \le P_{ik}(s+t) \le P_{ik}(t) + [1 - P_{ii}(s)] \tag{9.5}$$

[the first inequality is trivial, the second follows from the fact that the terms $P_{ij}(s)$ with $j \neq i$ add up to $1 - P_{ii}(s)$]. For s sufficiently small we have $1 - \epsilon \le P_{ii}(s) \le 1$ and so (9.5) shows the uniform continuity of P_{ik}. It follows from (9.5) also that if $P_{ik}(t) > 0$ then $P_{ik}(t+s) > 0$ in some s-interval of fixed length and hence P_{ik} is either identically zero or ultimately positive. ▶

[17] This condition is introduced only to avoid trivialities that may be circumvented by restrictions to appropriate sets of states. It is not difficult to see that our conditions imply strict positivity of $P_{ik}(t)$ for all t.

We now apply this result to the minimal solution of section 7 assuming that it is strictly stochastic, and hence *unique*. In matrix notation (9.4) reads $uP(t) = u$; for the corresponding ordinary Laplace transform this implies

$$u\lambda\Pi(\lambda) = u. \tag{9.6}$$

If a vector u satisfies (9.6) for some particular value $\lambda > 0$ the resolvent equation (7.15) entails the truth of (9.6) for *all* $\lambda > 0$, and hence the truth of (9.4) for all $t > 0$. Introducing (9.6) into the forward equation (7.6) we get

$$u\mathbf{c}\mathbf{p} = u\mathbf{c}; \tag{9.7}$$

the components $u_k c_k$ are finite though possibly unbounded. On the other hand, if u is a probability vector satisfying (9.7) it follows by induction from (7.12) that $u\lambda\Pi^{(n)}(\lambda) \leq u$ for all n, and hence $u\lambda\Pi(\lambda) \leq u$. But the matrix $\lambda\Pi(\lambda)$ being strictly stochastic the sums of the components on either side must be equal and hence (9.6) is true. We have thus

Theorem 2. *If the minimal solution is strictly stochastic (and hence unique) the relations* (9.2) *hold with* $u_k > 0$ *iff there exists a probability vector* u *such that* (9.7) *holds.*

This implies in particular that the solution u of (9.7) is unique.

Probabilistic interpretation. To fix ideas consider the simplest case where the discrete chain with transition probabilities p_{ij} is ergodic. In other words, we assume that there exists a strictly positive probability vector $\alpha = (\alpha_1, \alpha_2, \ldots)$ such that $\alpha\mathbf{p} = \alpha$ and $p_{ik}^{(n)} \to \alpha_k$ as $n \to \infty$. It is then clear that *if* $\sigma = \sum \alpha_k c_k^{-1} < \infty$, *the probability vector with components* $u_k = \alpha_k c_k^{-1}/\sigma$ *satisfies* (9.7) whereas no solution exists if $\sigma = \infty$.

Now it is intuitively obvious that the *transitions* in our process are the same as in the discrete Markov chain with matrix $\mathbf{p}$, but their timing is different. For an orientation consider a particular state and label it with the index 0. The successive *sojourn times* at 0 alternate with *off times* during which the system is at states $j > 0$. The number of visits to the state j is regulated by $\mathbf{p}$, their duration depends on c_j. In the discrete Markov chain the long-run frequencies of j and 0 are in the ratio α_j/α_0 and hence α_j/α_0 should be *the expected number* of visits to j during an off interval. The expected duration of each visit being $1/c_j$ we conclude that in the long run the probabilities of the states j and 0 should stand in the proportion $\alpha_j c_j^{-1} : \alpha_0 c_0^{-1}$ or $u_j : u_0$.

This argument can be made rigorous even in the case where $P_{ij}(t) \to 0$. According to a theorem of C. Derman mentioned in **1**; XV,11, if $\mathbf{p}$ induces an irreducible and persistent chain there exists a vector α such that $\alpha\mathbf{p} = \alpha$ and α is unique up to a multiplicative constant; here $\alpha_k \geq 0$, but the series $\sum \alpha_k$ may diverge. Even in this case the *ratio*

$\alpha_j:\alpha_0$ have the relative frequency interpretation given above and the argument holds generally. *If* $\sum \alpha_k c_k^{-1} < \infty$ *then* (9.2)–(9.4) *are true with* u_k *proportional to* $\alpha_k c_k^{-1}$, *and otherwise* $P(t) \to 0$ *as* $t \to \infty$. The interesting feature is that the limits u_k may be positive even if the discrete chain has only null states.

The existence of the limits $P_{ik}(\infty)$ can be obtained also by a renewal argument intimately connected with the recurrence times. To show how the distribution of recurrence and first passage times may be calculated we number the states $0, 1, 2, \ldots$ and use 0 as pivotal state. Consider a new process which coincides with the original process up to the random epoch of the first visit to 0 but with the state fixed at 0 forever after. In other words, the new process is obtained from the old one by making 0 *an absorbing state.* Denote the transition probabilities of the modified process by ${}^0P_{ik}(t)$. Then ${}^0P_{00}(t) = 1$. In terms of the original process ${}^0P_{i0}(t)$ is the probability of a *first passage from* $i \neq 0$ *to* 0 *before epoch* t, and ${}^0P_{ik}(t)$ gives the probability of a transition from $i \neq 0$ to $k \neq 0$ without intermediate passage through 0. It is probabilistically clear that the matrix ${}^0P(t)$ should satisfy the same backward and forward equations as $P(t)$ except that c_0 is replaced by 0. We now proceed the inverse way: *we modify the backward and forward equations by changing* c_0 *to* 0 *and show that the unique solution of this absorbing-state process has the predicted properties.*

If ξ is the vector represented by the zeroth column of $\Pi(\lambda)$, the backward equations show that the vector

$$(9.8) \qquad (\lambda + \mathbf{c} - \mathbf{cp})\xi = \eta$$

has components $1, 0, 0, \ldots$. Now the backward equations for ${}^0\Pi(\lambda)$ are obtained on replacing c_0 by 0, and so it ξ stands for the zeroth column of ${}^0\Pi(\lambda)$ the vector (9.8) has components $\eta_1 = \eta_2 = \cdots = 0$, but $\eta_0 = \rho \neq 0$. It follows that the vector with components $\xi_k = \Pi_{k0}(\lambda) - \rho\,{}^0\Pi_{k0}(\lambda)$ satisfies (9.8) with $\eta = 0$, and as $\lambda\Pi(\lambda)$ is strictly stochastic this implies $\xi_k = 0$ for all k (theorem 3 of section 7). Since ${}^0\Pi_{00}(\lambda) = 1/\lambda$ we have therefore for $k \geq 0$

$$(9.9) \qquad \Pi_{k0}(\lambda) = \lambda\,{}^0\Pi_{k0}(\lambda)\Pi_{00}(\lambda).$$

Referring to the first equation in (9.8) we see also that

$$(9.10) \qquad \Pi_{00}(\lambda) = \frac{1}{\lambda + c_0} + \frac{c_0}{\lambda + c_0} \sum_j p_{0j}\lambda\,{}^0\Pi_{j0}(\lambda)\Pi_{00}(\lambda).$$

(9.9) and (9.10) are *renewal equations* with obvious probabilistic interpretation. In fact, let the process start at $k > 0$. Then ${}^0\Pi_{k0}$ is the ordinary Laplace transform of the probability ${}^0P_{k0}(t)$ that the first entry to 0 occurs before t, and hence $\lambda\,{}^0\Pi_{k0}(\lambda)$ is the Laplace transform of the distribution F_k of the *epoch of first entry to* 0. Thus (9.9) states that $P_{k0}(t)$ is the convolution of F_k and P_{00}; the event $\mathbf{X}(t) = 0$ takes place iff the first entry occurs at some epoch $x < t$ and $t - x$ time units later the system is again at 0.

Similarly, $\sum p_{0j}\lambda^0\Pi_{j0}$ represents the distribution F_0 of an off time, that is, the interval between two consecutive sojourn times at 0. The factor of $\Pi_{00}(\lambda)$ on the right in (9.10) therefore represents the *waiting time for a first return to* 0 if the system is initially at 0. (This is also the distribution of a complete period = sojourn time plus off time.) The renewal equation (9.10) expresses $P_{00}(t)$ as the sum of the probability that the sojourn time at 0 extends beyond t and the probability of $\mathbf{X}(t) = 0$ after a first return at epoch $x < t$. If 0 is persistent (9.10) implies by the renewal theorem that

$$P_{00}(\infty) = \frac{1}{1 + c_0\mu} \tag{9.11}$$

where μ is the expected duration of an off time and $c_0^{-1} + \mu$ is the expected duration of a complete cycle.

10. PROBLEMS FOR SOLUTION

1. In the renewal equation (1.3) let $F'(t) = g(t) = e^{-t}t^{p-1}/\Gamma(p)$. Then

$$\psi(\lambda) = \frac{1}{(\lambda+1)^p - 1}. \tag{10.1}$$

By the method of partial fractions show that for integral[18] p

$$v(t) = \frac{1}{p}\sum_{0=k}^{p-1} a_k e^{-(1-a_k)t} \tag{10.2}$$

where $a_k = e^{-i2\pi k/p}$ and $i^2 = -1$.

2. A server has Poisson incoming traffic with parameter α and a holding time distribution G with Laplace transform γ. Let $H(t)$ be the probability that the duration of a holding time does not exceed t and that no new call arrives during it. Show that H is a defective distribution with Laplace-Stieltjes transform $\gamma(\lambda+\alpha)$.

3. *Lost calls.* Suppose that the server of the preceding example is free at epoch 0. Denote by $U(t)$ the probability that up to epoch t all arriving calls find the server free. Derive a renewal equation for U and conclude that the ordinary Laplace transform ω of U satisfies the linear equation

$$\omega(\lambda) = \frac{1}{\lambda + \alpha} + \alpha\frac{1 - \gamma(\lambda+\alpha)}{(\lambda+\alpha)^2} + \frac{\alpha}{\lambda + \alpha}\gamma(\lambda+\alpha)\,\omega(\lambda).$$

The expected waiting time for the first call arriving during a busy period is

$$\alpha^{-1} + \alpha^{-1}[1 - \gamma(\alpha)]^{-1}.$$

[18] The roots of the denominator are the same for $p = n$ and $p = n/2$, but the solutions are entirely different. This shows that the popular "expansion according to the roots of the denominator" requires caution when ψ is an irrational function.

Hint: Use the preceding problem.

4. (*Continuation.*) Solve the preceding problem by the method described in problem 17 of VI,13 considering U as the distribution of the total lifetime of a delayed terminating renewal process.

5. If F has expectation μ and variance σ^2 and if $c\mu < 1$, the solution of the busy-period equation (4.1) has variance $(\sigma^2 + c\mu^3)/(1 - c\mu)$.

6. In example 4(*b*) the generating function of the total numbers of cars delayed is $e^{c\delta[\psi(s)-1]}$ where

$$\psi(s) = s\varphi(c - c\psi(s)). \tag{10.3}$$

7. Show that if φ is the Laplace transform of a proper distribution the solution ψ of (10.3) is the generating function of a possibly defective distribution. The latter is proper iff F has an expectation $\mu \leq 1/c$.

8. In the *absorbing barrier* process of example 5(*a*) denote by $F(t, x)$ the probability that (starting from x) absorption will take place within time t. (Thus F stands for the distribution of the total lifetime of the process.) Show that the ordinary Laplace transform of $1 - F$ is given by the integral of $K^{\text{abs}}(x, y)$ over $0 < y < \infty$. Conclude that the Laplace-Stieltjes transform of F is given by $e^{-\sqrt{2\lambda}\, x}$, in agreement with the fact that it must satisfy the differential equation (5.9).

9. Starting from (5.7) show that the Green function of the general diffusion equation (5.19) in any interval is necessarily of the form

$$K_\lambda(x, y) = \begin{cases} \dfrac{\xi_\lambda(x)\, \eta_\lambda(y)}{W(y)} & \text{for } x \leq y \\[2ex] \dfrac{\eta_\lambda(x)\, \xi_\lambda(y)}{W(y)} & \text{for } x \geq y, \end{cases} \tag{10.4}$$

where ξ_λ and η_λ are solutions of the homogeneous equation

$$\lambda\varphi - \tfrac{1}{2}a\varphi'' - b\varphi' = 0 \tag{*}$$

bounded, respectively, at the left and the right boundary. If (*) has no bounded solution then ξ_λ and η_λ are determined up to arbitrary multiplicative constants which can be absorbed in W. (Otherwise appropriate boundary conditions must be imposed.)

Show that ω_λ defined by (10.4) and (5.5) satisfies the differential equation (5.19) iff W is the Wronskian

$$W(y) = [\xi_\lambda'(y)\, \eta_\lambda(y) - \xi_\lambda(y)\, \eta_\lambda'(y)]a/2. \tag{10.5}$$

The solutions ξ_λ and η_λ are necessarily monotonic, and hence $W(y) \neq 0$.

10. *Continuation.* For $x < y$ the first-passage epoch from x to y has the Laplace transform $\xi_\lambda(x)/\xi_\lambda(y)$. For $x > y$ it is given by $\eta_\lambda(x)/\eta_\lambda(y)$.

11. Show that the method described in section 5 for diffusion processes applies equally to a general *birth-and-death process.*[19]

12. Adjust example 6(*a*) to the case of $a > 1$ channels. (*Explicit* calculations of the a constants are messy and not recommended.)

[19] For details and boundary conditions see W. Feller, *The birth and death process as diffusion process*, Journal Mathématiques Pures Appliquées, vol. 38 (1959), pp. 301–345.

13. In example 6(*b*) show directly from the differential equations that the busy time has expectation $\dfrac{1}{c(q-p)}$ and variance $\dfrac{1}{c^2(q-p)^3}$.

14. *Semi-Markov processes.* A semi-Markov process on $1, 2, \ldots$ differs from a Markov process in that the sojourn times may depend on the terminal state: given that the state i was *entered* at τ the probability that the sojourn time ends before $\tau + t$ by a jump to k is $F_{ik}(t)$. Then $\sum_k F_{ik}(t)$ gives the distribution of the sojourn time and $p_{ik} = F_{ik}(\infty)$ is the probability of a jump to k. Denote by $P_{ik}(t)$ the probability of k at epoch $t + \tau$ given that i was *entered* at epoch τ. Derive an analogue to the Kolmogorov *backward* equations. With self-explanatory notations the transformed version is given by

$$\Pi(\lambda) = \gamma(\lambda) + \Phi(\lambda)\,\Pi(\lambda)$$

where $\gamma(\lambda)$ is the diagonal matrix with elements $[1 - \sum_k \varphi_{ik}(\lambda)]/\lambda$. For $F_{ik}(t) = p_{ik}(1 - e^{-c_i t})$ this reduces to the backward equations (7.5). The construction of the minimal solution of section 7 goes through.[20]

[20] For details see W. Feller, *On semi-Markov processes*, Proc. National Acad. of Sciences, vol. 51 (1964) pp. 653–659. Semi-Markov processes were introduced by P. Lévy and W. L. Smith, and were investigated in particular by R. Pyke.

CHAPTER XV

Characteristic Functions

This chapter develops the elements of the theory of characteristic functions and is entirely independent of chapters VI, VII, IX–XIV. A refined Fourier analysis is deferred to chapter XIX.

1. DEFINITION. BASIC PROPERTIES

The generating function of a non-negative integral-valued random variable $\mathbf{X}$ is the function defined for $0 \leq s \leq 1$ by $\mathbf{E}(s^{\mathbf{X}})$, the expectation of $s^{\mathbf{X}}$. As was shown in chapter XIII, the change of variable $s = e^{-\lambda}$ makes this useful tool available for the study of arbitrary non-negative random variables. The usefulness of these transforms derives largely from the multiplicative property $s^{x+y} = s^x s^y$ and $e^{-\lambda(x+y)} = e^{-\lambda x}e^{-\lambda y}$. Now this property is shared by the exponential function with a purely imaginary argument, that is, by the function defined for real x by

$$(1.1) \qquad e^{i\zeta x} = \cos \zeta x + i \sin \zeta x$$

where ζ is a real constant and $i^2 = -1$. This function being bounded, its expectation exists under any circumstances. The use of $\mathbf{E}(e^{i\zeta\mathbf{X}})$ as a substitute for generating functions provides a powerful and universally applicable tool, but it is bought at the price of introducing complex-valued functions and random variables. Note, however, that the independent variable remains restricted to the real line, (or, later on, to $\mathcal{R}^r$).

By a complex-valued function $w = u + iv$ is meant the pair of *real* functions u and v defined for *real* x. The expectation $\mathbf{E}(w)$ is merely an abbreviation for $\mathbf{E}(u) + i\mathbf{E}(v)$. We write, as usual, $\bar{w} = u - iv$ for the conjugate function, and $|w|$ for the absolute value (that is, $|w|^2 = w\bar{w} = u^2 + v^2$). The elementary properties of expectation remain valid, and only the *mean value theorem* requires comment: *if* $|w| \leq a$ *then* $|\mathbf{E}(w)| \leq a$. In fact, by Schwarz' inequality

$$(1.2) \qquad |\mathbf{E}(w)|^2 = (\mathbf{E}(u))^2 + (\mathbf{E}(v))^2 \leq \mathbf{E}(u^2) + \mathbf{E}(v^2) = \mathbf{E}(|w|^2) \leq a^2.$$

Two complex-valued random variables $\mathbf{W}_j = \mathbf{U}_j + i\mathbf{V}_j$ are called *independent* iff the pairs $(\mathbf{U}_1, \mathbf{V}_1)$ and $(\mathbf{U}_2, \mathbf{V}_2)$ are independent. That the multiplicative property $\mathbf{E}(\mathbf{W}_1\mathbf{W}_2) = \mathbf{E}(\mathbf{W}_1)\,\mathbf{E}(\mathbf{W}_2)$ holds as usual is seen by decomposition into real and imaginary parts. (This formula illustrates the advantage of the complex notation.) With these preparations we define an analogue to generating functions as follows.

Definition. *Let* $\mathbf{X}$ *be a random variable with probability distribution* F. *The characteristic function of* F *(or of* $\mathbf{X}$*) is the function* φ *defined for real* ζ *by*

$$\varphi(\zeta) = \int_{-\infty}^{+\infty} e^{i\zeta x}\, F\{dx\} = u(\zeta) + iv(\zeta) \tag{1.3}$$

where

$$u(\zeta) = \int_{-\infty}^{+\infty} \cos \zeta x \cdot F\{dx\}, \qquad v(\zeta) = \int_{-\infty}^{+\infty} \sin \zeta x \cdot F\{dx\}. \tag{1.4}$$

For distributions F with a density f, of course,

$$\varphi(\zeta) = \int_{-\infty}^{+\infty} e^{i\zeta x} f(x)\, dx. \tag{1.5}$$

Terminological note. In the accepted terminology of Fourier analysis φ is the *Fourier-Stieltjes transform of* F. Such transforms are defined for all bounded measures and the term "characteristic function" emphasizes that the measure has unit mass. (No other measures have characteristic functions.) On the other hand, integrals of the form (1.5) occur in many connections and we shall say that (1.5) defines the *ordinary Fourier transform of* f. The characteristic function of F is the ordinary Fourier transform of the density f (when the latter exists), but the term Fourier transform applies also to other functions. ▸

For ease of reference we list some basic properties of characteristic functions.

Lemma 1. *Let* $\varphi = u + iv$ *be the characteristic function of a random variable* $\mathbf{X}$ *with distribution* F. *Then*

(*a*) φ *is continuous.*

(*b*) $\varphi(0) = 1$ *and* $|\varphi(\zeta)| \leq 1$ *for all* ζ.

(*c*) $a\mathbf{X} + b$ *has the characteristic function*

$$\mathbf{E}(e^{i\zeta(a\mathbf{X}+b)}) = e^{ib\zeta}\varphi(a\zeta). \tag{1.6}$$

In particular, $\bar{\varphi} = u - iv$ *is the characteristic function of* $-\mathbf{X}$.

(*d*) u *is even and* v *is odd. The characteristic function is real iff* F *is symmetric.*

(*e*) *For all* ζ

$$0 \leq 1 - u(2\zeta) \leq 4(1-u(\zeta)). \tag{1.7}$$

(For variants see problems 1–3.)

Proof. (*a*) Note that $|e^{i\zeta x}| = 1$ and hence

$$|e^{i\zeta(x+h)} - e^{i\zeta x}| = |e^{i\zeta h} - 1|. \tag{1.8}$$

The right side is independent of x and is arbitarily small for h sufficiently close to 0. Thus φ is, in fact, *uniformly* continuous. Property (*b*) is obvious from the mean value theorem, and (*c*) requires no comment. For the proof of (*d*) we anticipate the fact that distinct distributions have distinct characteristic functions. Now φ is real iff $\varphi = \bar{\varphi}$, that is, if $\mathbf{X}$ and $-\mathbf{X}$ have the same characteristic function. But then $\mathbf{X}$ and $-\mathbf{X}$ have the same distribution, and so F is symmetric. Finally, to prove (*e*) consider the elementary trigonometric relation

$$1 - \cos 2\zeta x = 2(1-\cos^2 \zeta x) \leq 4(1-\cos \zeta x) \tag{1.9}$$

valid because $0 \leq 1 + \cos \zeta x \leq 2$. Taking expectations we get (1.7). ▶

Consider now two random variables $\mathbf{X}_1, \mathbf{X}_2$ with distributions F_1, F_2 and characteristic functions φ_1, φ_2. If $\mathbf{X}_1$ and $\mathbf{X}_2$ are independent, the multiplicative property of the exponential entails

$$\mathbf{E}(e^{i\zeta(\mathbf{X}_1+\mathbf{X}_2)}) = \mathbf{E}(e^{i\zeta\mathbf{X}_1})\, \mathbf{E}(e^{i\zeta\mathbf{X}_2}). \tag{1.10}$$

This simple result is used frequently and we record it therefore as

Lemma 2. *The convolution* $F_1 \star F_2$ *has the characteristic function* $\varphi_1\varphi_2$. In other words: *to the sum* $\mathbf{X}_1 + \mathbf{X}_2$ *of two independent random variables there corresponds the product* $\varphi_1\varphi_2$ *of their characteristic functions.*[1]

If $\mathbf{X}_2$ has the same distribution as $\mathbf{X}_1$, then the sum $\mathbf{X}_1 - \mathbf{X}_2$ represents the symmetrized variable (see V,5). We have therefore the

Corollary. $|\varphi|^2$ *is the characteristic function of the symmetrized distribution* ${}^0F$.

The following lemma gives a characterization of *arithmetic* distributions.

Lemma 3. *If* $\lambda \neq 0$ *the following three statements are equivalent:*

(*a*) $\varphi(\lambda) = 1$.

(*b*) φ *has period* λ, *that is* $\varphi(\zeta+n\lambda) = \varphi(\zeta)$ *for all* ζ *and* n.

---

[1] The converse is false, for it was shown in II,4(*e*) and again in problem 1 of III,9 that in some exceptional cases the sum of two *dependent* variables may have the distribution $F_1 \star F_2$, and consequently the characteristic function $\varphi_1\varphi_2$.

(*c*) *All points of increase of* F *are among* $0, \pm h, \pm 2h, \ldots$ *where* $h = 2\pi/\lambda$.

Proof. If (*c*) is true and F attributes weight p_n to nh then

$$\varphi_n(\zeta) = \sum p_n e^{inh\zeta}.$$

This function has period $2\pi/h$, and so (*c*) implies (*b*), which in turn is stronger than (*a*).

Conversely, if (*a*) holds, the expectation of the non-negative function $1 - \cos \lambda x$ vanishes, and this is possible only if $1 - \cos \lambda x = 0$ at every point x that is a point of increase for F. Thus F is concentrated on the multiples of $2\pi/\lambda$, and hence (*c*) is true.

Technically this lemma covers the extreme case of a distribution F concentrated at the origin. Then $\varphi(\zeta) = 1$ for all ζ, and so every number is a period of φ. In general, if λ is a period of φ the same is true of all multiples $\pm \lambda, \pm 2\lambda, \ldots$, but for a non-constant periodic function φ there exists a smallest positive period, and this is called the *true period.* Similarly, for an arithmetic F there exists a *largest* positive h for which property (*c*) holds, and this is called the *span* of F. It follows from lemma 3 that the span h and the period λ are related by $\lambda h = 2\pi$. Thus unless either $\varphi(\zeta) \neq 1$ for all $\zeta \neq 0$, or $\varphi(\zeta) = 1$ identically, there exists a smallest $\lambda > 0$ such that $\varphi(\lambda) = 1$ but $\varphi(\zeta) \neq 1$ for $0 < \zeta < \lambda$.

All this can be restated in a form of more general appearance. Instead of $\varphi(\lambda) = 1$ assume only that $|\varphi(\lambda)| = 1$. There exists then a real b such that $\varphi(\lambda) = e^{ib\lambda}$, and we can apply the preceding result to the variable $\mathbf{X} - b$ with characteristic function $\varphi(\zeta)e^{-ib\lambda}$ which equals 1 at $\zeta = \lambda$. Every period of this characteristic function is automatically a period of $|\varphi|$, and we have thus proved

Lemma 4. *There exist only the following three possibilities:*

(*a*) $|\varphi(\zeta)| < 1$ *for all* $\zeta \neq 0$.

(*b*) $|\varphi(\lambda)| = 1$ *and* $|\varphi(\zeta)| < 1$ *for* $0 < \zeta < \lambda$. *In this case* $|\varphi|$ *has period* λ *and there exists a real number* b *such that* $F(x+b)$ *is arithmetic with span* $h = 2\pi/\lambda$.

(*c*) $|\varphi(\zeta)| = 1$ *for all* ζ. *In this case* $\varphi(\zeta) = e^{ib\zeta}$ *and* F *is concentrated at the point* b.

Example. Let F be concentrated on 0 and 1 attributing probability $\frac{1}{2}$ to each. Then F is arithmetic with span 1, and its characteristic function $\varphi(\zeta) = (1+e^{i\zeta})/2$ has period 2π. The distribution $F(x+\frac{1}{2})$ is concentrated on $\pm\frac{1}{2}$. It has span $\frac{1}{2}$ and its characteristic function $\cos \zeta/2$ has period 4π. ►

2. SPECIAL DISTRIBUTIONS. MIXTURES

For ease of reference we give a table of the characteristic functions of ten common densities and describe the method of deriving them.

Notes to table 1. (1) *Normal density.* If one is not afraid of complex integration the result is obvious by the substitution $y = x - i\zeta$. To prove the formula in the real domain use differentiation and integration by parts to obtain $\varphi'(\zeta) = -\zeta\varphi(\zeta)$. Since $\varphi(0) = 1$ it follows that $\log \varphi(\zeta) = -\frac{1}{2}\zeta^2$, as asserted.

(2)–(3) *Uniform densities.* The calculation in (2) and (3) is obvious. The two distributions differ only by location parameters, and the relation between the characteristic functions illustrates the rule (1.6).

(4) *Triangular density.* Direct calculation is easy using integration by parts. Alternatively, observe that our triangular density is the convolution of the uniform density in $-\frac{1}{2}a < x < \frac{1}{2}a$ with itself and in view of (3) its characteristic function is therefore $\left(\frac{2}{a\zeta} \cdot \sin \frac{a\zeta}{2}\right)^2$.

(5) This is obtained by application of the inversion formula (3.5) to the triangular density (4). See also problem 4. This formula is of great importance because many Fourier-analytic proofs depend on the use of a characteristic function vanishing outside a finite interval.

(6) *Gamma densities.* Use the substitution $y = x(1-i\zeta)$. If one prefers to stay in the real domain expand e^{ix} into a power series. For the characteristic function one gets in this way

$$\frac{1}{\Gamma(t)} \sum_{n=0}^{\infty} \frac{(i\zeta)^n}{n!} \int_0^{\infty} e^{-x} x^{n+t-1}\, dx = \sum_{n=0}^{\infty} \frac{\Gamma(n+t)}{n!\,\Gamma(t)} (i\zeta)^n = \sum_{n=0}^{\infty} \binom{-t}{n} (-i\zeta)^n$$

which is the binomial series for $(1-i\zeta)^{-t}$. For the special case $t = 1$ (exponential distribution) the calculation can be performed in the real by repeated integration by parts. The same is true (by recursion) for all integral values of t.

(7) *The bilateral exponential* is obtained by symmetrization from the exponential distribution, and so the characteristic function follows from (6) with $t = 1$. A direct verification is easy using repeated integrations by parts.

(8) *Cauchy distribution.* Again the formula follows from the preceding one by the use of the inversion formula (3.5). The direct verification of this formula is a standard exercise in the calculus of residues.

(9) *Bessel density.* This is the Fourier version of the Laplace transform derived in XIII,3(d), and may be proved in the same way.

(10) *Hyperbolic cosine.* The corresponding distribution function is $F(x) = 1 - 2\pi^{-1} \arctan e^{-x}$. Formula 10 is of no importance, but it has a

TABLE 1

No.	Name	Density	Interval	Characteristic Function
1	Normal	$\frac{1}{\sqrt{2\pi}} e^{-\frac{1}{2}x^2}$	$-\infty < x < \infty$	$e^{-\frac{1}{2}\zeta^2}$
2	Uniform	$\frac{1}{a}$	$0 < x < a$	$\frac{e^{ia\zeta} - 1}{ia\zeta}$
3	Uniform	$\frac{1}{2a}$	$\|x\| < a$	$\frac{\sin a\zeta}{a\zeta}$
4	Triangular	$\frac{1}{a}\left(1 - \frac{\|x\|}{a}\right)$	$\|x\| < a$	$2\frac{1 - \cos a\zeta}{a^2\zeta^2}$
5	—	$\frac{1}{\pi}\frac{1 - \cos ax}{ax^2}$	$-\infty < x < \infty$	$\begin{cases} 1 - \frac{\|\zeta\|}{a} & \text{for } \|\zeta\| \leq a \\ 0 & \text{for } \|\zeta\| > a \end{cases}$
6	Gamma	$\frac{1}{\Gamma(t)} x^{t-1}e^{-x}$	$x > 0,\quad t > 0$	$\frac{1}{(1 - i\zeta)^t}$
7	Bilateral exponential	$\frac{1}{2}e^{-\|x\|}$	$-\infty < x < \infty$	$\frac{1}{1 + \zeta^2}$
8	Cauchy	$\frac{1}{\pi}\frac{t}{t^2 + x^2}$	$-\infty < x < \infty$, $t > 0$	$e^{-t\|\zeta\|}$
9	Bessel	$e^{-x}\frac{t}{x}I_t(x)$	$x > 0,\quad t > 0$	$[1 - i\zeta - \sqrt{(1 - i\zeta)^2 - 1}]^t$
10	Hyperbolic cosine[a]	$\frac{1}{\pi \cosh x}$	$-\infty < x < \infty$	$\frac{1}{\cosh(\pi\zeta/2)}$

[a] $\cosh x = \frac{1}{2}(e^x + e^{-x})$.

curiosity value in that it exhibits a "*self-reciprocal pair*": the density and its characteristic function differ only by scale parameters. (The normal density is the prime example for this phenomenon.) To calculate the characteristic function expand the density into the geometric series

$$\frac{1}{2\pi}\sum(-1)^k e^{-(2k+1)|x|}.$$

Applying number 7 to the individual term one gets the canonical partial fraction expansion for the characteristic function. ▶

(For further examples see problems 5–8.)

Returning to the general theory, we give a method of constructing new characteristic functions out of given ones. The principle is extremely simple, but example (*b*) will show that it can be exploited to avoid lengthy calculations.

Lemma. *Let* $F_0, F_1, \ldots$ *be probability distributions with characteristic functions* $\varphi_0, \varphi_1, \ldots$. *If* $p_k \geq 0$ *and* $\sum p_k = 1$ *the mixture*

$$U = \sum p_k F_k \tag{2.1}$$

is a probability distribution with characteristic function

$$\omega = \sum p_k \varphi_k. \tag{2.2}$$

Examples. (*a*) *Random sums.* Let $\mathbf{X}_1, \mathbf{X}_2, \ldots$ be independent random variables with a common distribution F and characteristic function φ. Let $\mathbf{N}$ be an integral-valued random variable with generating function $P(s) = \sum p_k s^k$ and independent of the $\mathbf{X}_j$. The random sum $\mathbf{X}_1 + \cdots + \mathbf{X}_\mathbf{N}$ has then the distribution (2.1) with $F_k = F^{k\star}$, and the corresponding characteristic function is

$$\omega(\zeta) = P(\varphi(\zeta)). \tag{2.3}$$

The most noteworthy special case is that of the *compound Poisson distribution.* Here $p_k = e^{-t}t^k/k!$ and

$$\omega(\zeta) = e^{-t+t\varphi(\zeta)}. \tag{2.4}$$

The ordinary Poisson distribution represents the special case where F is concentrated at the point 1, that is, when $\varphi(\zeta) = e^{i\zeta}$.

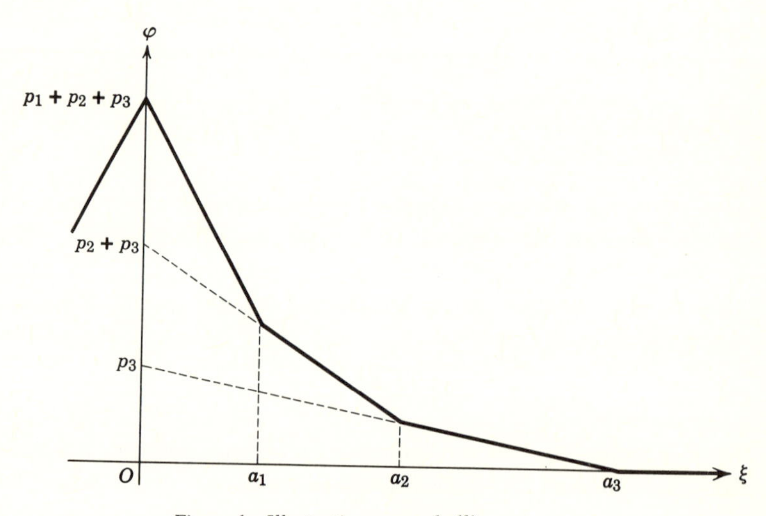

Figure 1. *Illustrating example* (*b*).

(*b*) *Convex polygons.* From number 5 in table 1 we know that

$$\varphi(\zeta) = \begin{cases} 1 - |\zeta| & \text{for } |\zeta| \leq 1 \\ 0 & \text{for } |\zeta| \geq 1 \end{cases} \tag{2.5}$$

is a characteristic function. If $a_1, \ldots, a_n$ are arbitrary positive numbers, the mixture

$$\omega(\zeta) = p_1\varphi\left(\frac{\zeta}{a_1}\right) + \cdots + p_n\varphi\left(\frac{\zeta}{a_n}\right) \tag{2.6}$$

is an even characteristic function whose graph in $\overline{0, \infty}$ is a convex polygon (fig. 1). In fact, without loss of generality assume $a_1 < a_2 < \cdots < a_n$. In the interval $0 < \zeta < a_1$ the graph of ω is a segment of a line with slope $-\left(\frac{p_1}{a_1} + \cdots + \frac{p_n}{a_n}\right)$. Between a_1 and a_2 the term p_1/a_1 drops out, and so on, until between a_{n-1} and a_n the graph coincides with a segment of slope $-p_n/a_n$. In $\overline{0, \infty}$ the graph is therefore a polygon consisting of n finite segments with decreasing slopes and the segment $\overline{a_n, \infty}$ of the ζ-axis. It is easily seen that every polygon with these properties may be represented in the form (2.6) (the n sides intercepting the ω-axis at the points p_n, $p_n+p_{n-1}, \ldots, p_n + \cdots + p_1 = 1$). We conclude that *every even function* $\omega \geq 0$ *with* $\omega(0) = 1$ *whose graph in* $\overline{0, \infty}$ *is a convex polygon is a characteristic function.*

A simple passage to the limit will lead to the famous Polya criterion [example 3(*b*)] and reveals its natural source. Even the present special criterion leads to surprising and noteworthy results.

2a. SOME UNEXPECTED PHENOMENA

We digress somewhat to introduce certain special types of characteristic functions with surprising and interesting properties. We begin by a preliminary remark concerning arithmetic distributions.

Suppose that the distribution G is concentrated on the multiples $n\pi/L$ of some fixed number $\pi/L > 0$, the point $n\pi/L$ carrying probability p_n; here $n = 0, \pm 1, \ldots$. The characteristic function γ is given by

$$\gamma(\zeta) = \sum_{n=-\infty}^{+\infty} p_n e^{in\pi\zeta/L} \tag{2.7}$$

and has period $2L$. It is usually not easy to find simple explicit expressions for γ, whereas it is easy to express the probabilities p_r in terms of the characteristic function γ. Indeed, multiply (2.7) by $e^{-ir\pi\zeta/L}$. The probability p_n appears as the coefficient of the periodic function $e^{i(n-r)\pi\zeta/L}$ whose integral over $\overline{-L, L}$ vanishes except when $n = r$. It follows that

$$p_r = \frac{1}{2L}\int_{-L}^{L} \gamma(\zeta)e^{-ir\pi\zeta/L}\, d\zeta \qquad r = 0, \pm 1, \ldots. \tag{2.8}$$

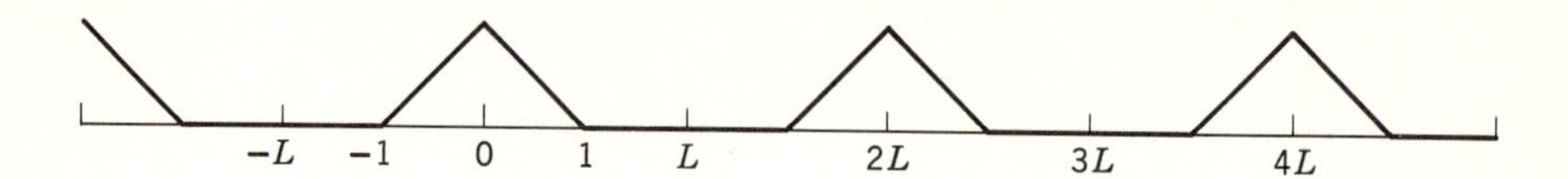

Figure 2. Illustrating example (c).

We now anticipate the following criterion of theorem 1 in XIX,4. Let γ be a continuous function with period $L > 0$ and normed by $\gamma(0) = 1$. Then γ is *a characteristic function iff all the numbers* p_r *in* (2.8) *are* ≥ 0. In this case $\{p_r\}$ is automatically a probability distribution and (2.7) holds.

Example. (*c*) Choose $L \geq 1$ arbitrary and let γ be the function with period $2L$ which for $|\zeta| \leq L$ agrees with the characteristic function φ of (2.5). Then γ *is the characteristic function* of an arithmetic distribution concentrated on the multiples of π/L. In fact, for reasons of symmetry (2.8) reduces to

$$p_r = \frac{1}{L}\int_0^L \gamma(\zeta)\cos r\pi\zeta/L\,d\zeta = \frac{1}{L}\int_0^L (1-\zeta)\cos r\pi\zeta/L\,d\zeta, \tag{2.9}$$

and a simple integration by parts shows that

$$p_0 = 1/(2L), \qquad p_r = L\pi^{-2}(1-\cos r\pi/L) \geq 0, \qquad r \neq 0. \tag{2.10}$$

We have thus obtained a *whole family of periodic characteristic functions whose graphs consist of a periodic repetition of a right triangle with* bases $2nL - 1 < x < 2nL + 1$ and intermittent sections of the ζ-axis (see fig. 2). (We shall return to this example in the more general setting of Poisson's summation formula in XIX,5.) ►

Curiosities. (i) *Two distinct characteristic functions can coincide within a finite interval* $\overline{-a, a}$. This obvious corollary to examples (*b*) or (*c*) shows a marked contrast between characteristic functions and generating functions (or Laplace transforms).

(ii) *The relation* $F \star F_1 = F \star F_2$ *between three probability distributions does not imply*[2] *that* $F_1 = F_2$. Indeed, with φ defined by (2.5) we have $\varphi\varphi_1 = \varphi\varphi_2$ for any two characteristic functions that coincide within the interval $\overline{-1, 1}$. In particular, we have $\varphi^2 = \varphi\gamma$ for each of the periodic characteristic functions of example (*c*).

(iii) Even more surprising is that *there exist two real characteristic functions* φ_j *such that* $|\varphi_2| = \varphi_1 \geq 0$ *everywhere*. In fact, consider the characteristic function γ of example (*c*) with $L = 1$. Its graph is shown by the heavy polygonal line in fig. 3. We saw that the corresponding distribution attributes to the origin weight $\frac{1}{2}$. Eliminating this atom and doubling all the other weights yields a distribution with characteristic function $2\gamma - 1$. Its graph is given by a polygonal line oscillating between ± 1 whose slopes are ± 2. It follows that $2\gamma(\frac{1}{2}\zeta) - 1$ is a characteristic function whose graph is obtained from that of γ by

[2] Statisticians and astronomers sometimes ask whether a given distribution has a normal component. This problem makes sense because the characteristic function of a normal distribution $\mathfrak{N}_a$ has no zeros and therefore $\mathfrak{N}_a \star F_1 = \mathfrak{N}_a \star F_2$ does imply $\varphi_1 = \varphi_2$ and hence, by the uniqueness theorem, $F_1 = F_2$.

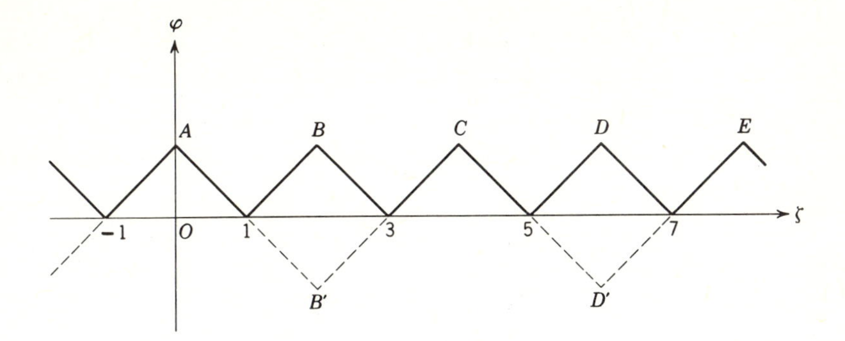

Figure 3. *Illustrating curiosity (iii).*

mirroring every second triangle along the ζ-axis (fig. 3). Thus $\gamma(\zeta)$ and $2\gamma(\frac{1}{2}\zeta) - 1$ are two real characteristic functions that differ only by their signs. (For a similar construction relating to fig. 2 see problem 9.)

3. UNIQUENESS. INVERSION FORMULAS

Let F and G be two distributions with characteristic functions φ and γ. Then

$$e^{-i\zeta t}\varphi(\zeta) = \int_{-\infty}^{+\infty} e^{i\zeta(x-t)}\, F\{dx\}. \tag{3.1}$$

Integrating with respect to $G\{d\zeta\}$ one gets

$$\int_{-\infty}^{+\infty} e^{-i\zeta t}\varphi(\zeta)\, G\{d\zeta\} = \int_{-\infty}^{+\infty} \gamma(x-t)\, F\{dx\}. \tag{3.2}$$

This identity is known as the *Parseval relation* (which, however, can be written in many equivalent forms; we shall return to it in chapter XIX).

We shall use only the special case where $G = \mathfrak{N}_a$ is the normal distribution with density $a\mathfrak{n}(ax)$. Its characteristic function is given by $\gamma(\zeta) = \sqrt{2\pi}\,\mathfrak{n}(\zeta/a)$, and so (3.2) takes on the form

$$\int_{-\infty}^{+\infty} e^{-i\zeta t}\varphi(\zeta) a\mathfrak{n}(a\zeta)\, d\zeta = \sqrt{2\pi}\int_{-\infty}^{+\infty} \mathfrak{n}\left(\frac{x-t}{a}\right) F\{dx\} \tag{3.3}$$

which is the same as

$$\frac{1}{\sqrt{2\pi}}\int_{-\infty}^{+\infty} e^{-i\zeta t}\varphi(\zeta) e^{-\frac{1}{2}a^2\zeta^2}\, d\zeta = \int_{-\infty}^{+\infty} \frac{1}{a}\,\mathfrak{n}\left(\frac{t-x}{a}\right) F\{dx\}. \tag{3.4}$$

Surprisingly many conclusions can be drawn from this identity. To begin with, the right side is the density of the convolution $\mathfrak{N}_a \star F$ of F with a normal distribution of zero expectation and variance a^2. Thus the knowledge of φ enables us in principle to calculate the distributions $\mathfrak{N}_a \star F$ for all a. But $\mathfrak{N}_a$ has variance a^2, and hence $\mathfrak{N}_a \star F \to F$ as $a \to 0$. It follows that the knowledge of φ uniquely determines the distribution F. We have thus the important

Theorem 1. *Distinct probability distributions have distinct characteristic functions.*

Suppose then that we are given a sequence of probability distributions F_n with characteristic functions φ_n, such that $\varphi_n(\zeta) \to \varphi(\zeta)$ for all ζ. By the selection theorem of VIII,6 there exists a sequence $\{n_k\}$ and a possibly defective distribution F such that $F_{n_k} \to F$. We apply (3.4) to the pair (φ_{n_k}, F_{n_k}) and let $k \to \infty$. In the limit we get again the identity (3.4) [the left side by bounded convergence, the right side because the integrand $\mathfrak{n}((t - x)a)$ vanishes at infinity]. But we have seen that for given φ the identity (3.4) determines F uniquely, and hence the limit F is the same for all convergent subsequences $\{F_{n_k}\}$. We have thus the

Lemma. *Let F_n be a probability distribution with characteristic function φ_n. If $\varphi_n(\zeta) \to \varphi(\zeta)$ for all ζ then there exists a possibly defective distribution F such that $F_n \to F$.*

Example. (*a*) Let U be a probability distribution with a real, non-negative characteristic function ω. Let $F_n = U^{n\star}$ so that $\varphi_n(\zeta) = \omega^n(\zeta)$. Then $\varphi_n(\zeta) \to 0$ except at the points where $\omega(\zeta) = 1$, and by lemma 4 of section 1, this set consists of all points of the form $\pm n\lambda$, where $\lambda \geq 0$ is a fixed number. It follows that the left side in (3.4) is identically zero, and so $U^{n\star} \to 0$. By symmetrization we conclude that $G^{n\star} \to 0$ *for any probability distribution* G *not concentrated at zero.* ►

The next theorem states essentially that the limit F is defective iff the limit φ is discontinuous at the origin.

Theorem 2. (*Continuity theorem.*) *In order that a sequence $\{F_n\}$ of probability distributions converges properly to a probability distribution F it is necessary and sufficient that the sequence $\{\varphi_n\}$ of their characteristic functions converges pointwise to a limit φ, and that φ is continuous in some neighborhood of the origin.*

In this case φ is the characteristic function of F. (Hence φ is continuous everywhere and the convergence $\varphi_n \to \varphi$ is uniform in every finite interval.)

Proof. (*a*) Assume that $F_n \to F$ where F is a proper probability distribution. By the corollary in VIII,1 the characteristic functions φ_n converge to the characteristic function φ of F, and the convergence is uniform in finite intervals.

(*b*) Assume $\varphi_n(\zeta) \to \varphi(\zeta)$ for all ζ. By the preceding lemma the limit $F = \lim F_n$ exists and the identity (3.3) applies. The left side is the expectation of the bounded function $e^{-\zeta t}\varphi(\zeta)$ with respect to a normal distribution with zero expectation and variance a^{-2}. As $a \to \infty$ this distribution concentrates near the origin and so the left side tends to $\varphi(0)$ whenever φ is continuous in some neighborhood of the origin. But since $\varphi_n(0) = 1$ we have $\varphi(0) = 1$. On the other hand, $\sqrt{2\pi}\,\mathfrak{n}(x) \leq 1$ for all x, and so the right side is $\leq F\{\overline{-\infty, \infty}\}$. Thus $F\{\overline{-\infty, \infty}\} \geq 1$, and hence F is proper. ►

Corollary. *A continuous function which is the pointwise limit of a sequence of characteristic functions is itself a characteristic function.*

Example. (*b*) *Polya's criterion. Let* ω *be a real even function with* $\omega(0) = 1$ *and a graph that is convex in* $\overline{0, \infty}$. *Then* ω *is a characteristic function.* Indeed, we saw in example 2(*b*) that the assertion is true when the graph is a convex polygon. Now the inscribed polygons to a concave curve are convex, and hence the general assertion is an immediate consequence of the corollary. The criterion (together with a tricky proof) had a surprise value in the early days. G. Polya used it in 1920 to prove that $e^{-|\zeta|^\alpha}$ for $0 < \alpha \leq 1$ is a characteristic function of a stable distribution. (Cauchy is said to have been aware of this fact, but gave no proof.) Actually $e^{-|\zeta|^\alpha}$ is a characteristic function even for $1 < \alpha \leq 2$, but the criterion breaks down. ►

We defer to chapter XIX a full use of the method developed for the proof of theorem 1. We use it here, however, to derive an important theorem which was used in numbers 5 and 8 of the table in section 2. For abbreviation we write $\varphi \in L$ iff $|\varphi|$ is integrable over $\overline{-\infty, \infty}$.

Theorem 3. (*Fourier inversion.*) *Let* φ *be the characteristic function of the distribution* F *and suppose that* $\varphi \in L$. *Then* F *has a bounded continuous density* f *given by*

$$f(x) = \frac{1}{2\pi}\int_{-\infty}^{+\infty} e^{-i\zeta x}\varphi(\zeta)\, d\zeta. \tag{3.5}$$

Proof. Denote the right side in (3.4) by $f_a(t)$. Then f_a is the density of the convolution $F_a = \mathfrak{N}_a \bigstar F$ of F with the normal distribution $\mathfrak{N}_a$ of zero expectation and variance a^2. As already mentioned, this implies that $F_a \to F$ as $a \to 0$. From the representation on the left it is clear that $f_a(t) \to f(t)$ boundedly, where f is the bounded continuous function defined in (3.5). Thus for every bounded interval I

$$F_a\{I\} = \int_I f_a(t)\, dt \to \int_I f(x)\, dx. \tag{3.6}$$

But if I is an interval of continuity for F the leftmost member tends to $F\{I\}$, and so f is indeed the density of F. ▶

Corollary. *If* $\varphi \geq 0$ *then* $\varphi \in L$ *iff the corresponding distribution* F *has a bounded density.*

Proof. By the last theorem the integrability of φ entails that F has a bounded continuous density. Conversely, if F has a density $f < M$ we get from (3.4) for $t = 0$

$$\frac{1}{2\pi}\int_{-\infty}^{+\infty} \varphi(\zeta)e^{-\frac{1}{2}a^2\zeta^2}\, d\zeta = \frac{1}{\sqrt{2\pi}\cdot a}\int_{-\infty}^{+\infty} e^{-x^2/(2a^2)}f(x)\, dx < M. \tag{3.7}$$

The integrand on the left is ≥ 0 and if φ were not integrable the integral would tend to ∞ as $a \to 0$. ▶

Examples. (*c*) *Plancherel identity.* Let the distribution F have a density f and characteristic function φ. Then $|\varphi|^2 \in L$ iff $f^2 \in L$ and in this case

$$\int_{-\infty}^{+\infty} f^2(y)\, dy = \frac{1}{2\pi}\int_{-\infty}^{+\infty} |\varphi(\zeta)|^2\, d\zeta. \tag{3.8}$$

Indeed, $|\varphi|^2$ is the characteristic function of the symmetrized distribution ${}^0F$. If $|\varphi|^2 \in L$ it follows that the density

$${}^0f(x) = \int_{-\infty}^{+\infty} f(y+x)f(y)\, dy \tag{3.9}$$

of ${}^0F$ is bounded and continuous. The left side in (3.8) equals ${}^0f(0)$, and the inversion formula (3.5) applied to ${}^0f$ shows the same is true of the right side. Conversely, if $f^2 \in L$ an application of Schwarz' inequality to (3.9) shows that ${}^0f$ is bounded, and hence $|\varphi|^2 \in L$ by the last corollary. We shall return to the relation (3.8) in XIX,7.

(*d*) *Continuity theorem for densities.* Let φ_n and φ be *integrable* characteristic functions such that

$$\int_{-\infty}^{+\infty} |\varphi_n(\zeta) - \varphi(\zeta)|\, d\zeta \to 0. \tag{3.10}$$

By the last corollary the corresponding distributions F_n and F have bounded continuous densities f_n and f, respectively. From the inversion formula (3.5) we see that

$$|f_n(x) - f(x)| \leq (2\pi)^{-1}\int_{-\infty}^{+\infty} |\varphi_n(\zeta) - \varphi(\zeta)|\, d\zeta.$$

Therefore $f_n \to f$ uniformly. (See also problem 12.)

(*e*) *Inversion formula for distribution functions.* Let F be a distribution with characteristic function φ, and let $h > 0$ be arbitrary, but fixed. We prove that

$$\frac{F(x+h)-F(x)}{h} = \frac{1}{2\pi}\int_{-\infty}^{+\infty} \varphi(\zeta)\,\frac{1-e^{-i\zeta h}}{i\zeta h}\,e^{-i\zeta x}\,d\zeta \tag{3.11}$$

whenever the integrand is integrable (for example, if it is $O(1/\zeta^2)$, that is, if $|\varphi(\zeta)| = O(1/\zeta)$ as $\zeta \to \infty$). Indeed, the left side is the *density* of the convolution of F with the uniform distribution concentrated on $\overline{-h, 0}$; by the product rule the factor of $e^{-i\zeta x}$ under the integral is the characteristic function of this convolution. Thus (3.11) represents but a special case of the general inversion formula (3.5). ►

Note *on so-called inversion formulas.* Formula (3.11) is applicable only when $|\varphi(\zeta)/\zeta|$ is integrable near infinity, but trite variations of this formula are generally applicable. For example, let F_a again denote the convolution of F with the symmetric normal distribution with variance a^2. Then by (3.11)

$$\frac{F_a(x+h)-F_a(x)}{h} = \frac{1}{2\pi}\int_{-\infty}^{+\infty} \varphi(\zeta)e^{-\frac{1}{2}a^2\zeta^2}\,\frac{1-e^{-i\zeta h}}{i\zeta h}\,e^{-i\zeta k}\,d\zeta. \tag{3.12}$$

The statement that if x and $x+h$ are points of continuity of F the right side tends to $[F(x+h)-F(x)]/h$ as $a \to 0$ is a typical "inversion theorem." An infinite variety of equivalent formulas may be written down. The traditional form consists in replacing in (3.12) the normal distribution by the uniform distribution in $\overline{-t, t}$ and letting $t \to \infty$. By force of tradition such inversion formulas remain a popular topic even though they have lost much of their importance; their derivation from the Dirichlet integral detracts from the logical structure of the theory.

From distributions with integrable characteristic functions we turn to *lattice distributions.* Let F attribute weight p_k to the point $b + kh$, where $p_k \geq 0$ and $\sum p_k = 1$. The characteristic function φ is then given by

$$\varphi(\zeta) = \sum_{-\infty}^{+\infty} p_k e^{i(b+kh)\zeta}. \tag{3.13}$$

We suppose $h > 0$.

Theorem 4. *If* φ *is a characteristic function of the form* (3.13) *then*

$$p_r = \frac{h}{2\pi}\int_{-\pi/h}^{\pi/h} \varphi(\zeta)e^{-i(b+rh)\zeta}\,d\zeta. \tag{3.14}$$

Proof. The integrand is a series in which the factor of p_k equals $e^{i(k-r)h\zeta}$. Its integral equals 0 or $2\pi/h$ according as $k \neq r$ or $k = r$, and so (3.14) is true. ►

4. REGULARITY PROPERTIES

The main result of this section may be summarized roughly to the effect that the smaller the tails of a distribution F, the smoother is its characteristic function φ; conversely, the smoother F, the better will φ behave at infinity. (Lemmas 2 and 4.) Most estimates connected with characteristic functions depend on an appraisal of the error committed in approximating e^{it} by finitely many terms of its Taylor expansion. The next lemma states that this error is dominated by the first omitted term.

Lemma 1.[3] *For* $n = 1, 2, \ldots$ *and* $t > 0$

$$\left| e^{it} - 1 - \frac{it}{1!} - \cdots - \frac{(it)^{n-1}}{(n-1)!} \right| \leq \frac{t^n}{n!}. \tag{4.1}$$

Proof. Denote the expression within the absolute value signs by $\rho_n(t)$. Then

$$\rho_1(t) = i \int_0^t e^{ix}\, dx, \tag{4.2}$$

whence $|\rho_1(t)| \leq t$. Furthermore for $n > 1$

$$\rho_n(t) = i \int_0^t \rho_{n-1}(x)\, dx, \tag{4.3}$$

and (4.1) now follows by induction. ▶

In the sequel F is an arbitrary distribution function, and φ its characteristic function. For the moments and absolute moments of F (when they exist) we write

$$m_n = \int_{-\infty}^{+\infty} x^n\, F\{dx\}, \qquad M_n = \int_{-\infty}^{+\infty} |x|^n\, F\{dx\}. \tag{4.4}$$

Lemma 2. *If* $M_n < \infty$, *the* nth *derivative of* φ *exists and is a continuous function given by*

$$\varphi^{(n)}(\zeta) = i^n \int_{-\infty}^{+\infty} e^{i\zeta x} x^n\, F\{dx\}. \tag{4.5}$$

Proof. The difference ratios of φ are given by

$$\frac{\varphi(\zeta+h) - \varphi(\zeta)}{h} = \int_{-\infty}^{+\infty} e^{i\zeta x}\, \frac{e^{ihx} - 1}{h}\, F\{dx\}. \tag{4.6}$$

According to the last lemma the integrand is dominated by $|x|$ and so for $n = 1$ the assertion (4.5) follows by dominated convergence. The general case follows by induction. ▶

Corollary. *If* $m_2 < \infty$ *then*

$$\varphi'(0) = im_1, \qquad \varphi''(0) = -m_2. \tag{4.7}$$

[3] The same proof shows that when the Taylor development for either $\sin t$ or $\cos t$ is stopped after finitely many terms, *the error is of the same sign, and smaller in absolute value, than the first omitted term.* For example, $1 - \cos t \leq t^2/2$.

The *converse*[4] of the last relation is also true: *If* $\varphi''(0)$ *exists, then* $m_2 < \infty$.

Proof. Denoting the real part of φ by u we have

$$\frac{1-u(h)}{h^2} = \int_{-\infty}^{+\infty} \frac{1-\cos hx}{h^2x^2} \cdot x^2 F\{dx\}. \tag{4.8}$$

Proof. The existence of $u''(0)$ implies that u' exists near the origin and is continuous there. In particular, $u'(0) = 0$ because u is even. By the mean value theorem there exists a θ such that $0 < \theta < 1$ and

$$\left|\frac{u(h)-1}{h^2}\right| = \left|\frac{u'(\theta h)}{h}\right| \le \left|\frac{u'(\theta h)}{\theta h}\right|. \tag{4.9}$$

As $h \to 0$ the right side tends to $u''(0)$. But the fraction under the integral in (4.8) tends to $\frac{1}{2}$, and so the integral approaches ∞ if $m_2 = \infty$. For a generalization see problem 15. ▶

Examples. (*a*) A non-constant function ψ such that $\psi''(0) = 0$ cannot be a characteristic function, since the corresponding distribution would have a vanishing second moment. For example, $e^{-|\zeta|^\alpha}$ is not a characteristic function when $\alpha > 2$.

(*b*) *The weak law of large numbers.* Let $\mathbf{X}_1, \mathbf{X}_2, \ldots$ be independent random variables with $\mathbf{E}(\mathbf{X}_j) = 0$ and the common characteristic function φ. Put $\mathbf{S}_n = \mathbf{X}_1 + \cdots + \mathbf{X}_n$. The average $\mathbf{S}_n/n$ has the characteristic function $\varphi^n(\zeta/n)$. Now near the origin $\varphi(h) = 1 + o(h)$, and hence $\varphi(\zeta/n) = 1 + o(1/n)$ as $n \to \infty$. Taking logarithms we see, therefore, that $\varphi^n(\zeta/n) \to 1$. By the continuity theorem 2 of section 3 this implies that *the distribution* of $\mathbf{S}_n/n$ tends to the distribution concentrated at the origin. This is the weak law of large numbers. The simple and straightforward nature of the proof is typical for characteristic functions; a variant will lead to the central limit theorem.[5] ▶

Lemma 3. (*Riemann-Lebesgue.*) *If* g *is integrable and*

$$\gamma(\zeta) = \int_{-\infty}^{+\infty} e^{i\zeta x} g(x)\, dx, \tag{4.10}$$

then $\gamma(\zeta) \to 0$ *as* $\zeta \to \pm\infty$.

---

[4] The argument does not apply to the first derivative. The long outstanding problem of finding conditions for the existence of $\varphi'(0)$ is solved in section XVII,2*a*.

[5] It was shown in VII,7 that the weak law of large numbers can hold even when the variables $\mathbf{X}_j$ have no expectations. The proof of the text shows the existence of a derivative $\varphi'(0)$ is a sufficient condition. It is actually also necessary (see section XVII,2*a*).

Proof. The assertion is easily verified for finite step functions g. For an arbitrary integrable function g and $\epsilon > 0$ there exists by the mean approximation theorem of IV,2 a finite step function g_1 such that

$$\int_{-\infty}^{+\infty} |g(x) - g_1(x)|\, dx < \epsilon. \tag{4.11}$$

The transform (4.10) γ_1 of g_1 vanishes at infinity, and in consequence of the last two relations we have $|\gamma(\zeta) - \gamma_1(\zeta)| < \epsilon$ for all ζ. Accordingly $|\gamma_1(\zeta)| < 2\epsilon$ for all sufficiently large $|\zeta|$, and as ϵ is arbitrary this means that $\gamma_1(\zeta) \to 0$ as $\zeta \to \pm\infty$. ▶

As a simple corollary we get

Lemma 4. *If F has a density f, then $\varphi(\zeta) \to 0$ as $\zeta \to \pm\infty$. If f has integrable derivatives $f', \ldots, f^{(n)}$, then $|\varphi(\zeta)| = o(|\zeta|^{-n})$ as $|\zeta| \to \infty$.*

Proof. The first assertion is contained in lemma 3. If f' is integrable, an integration by parts shows that

$$\varphi(\zeta) = \frac{1}{i\zeta} \int_{-\infty}^{+\infty} e^{i\zeta x} f'(x)\, dx, \tag{4.12}$$

and hence $|\varphi(\zeta)| = o(|\zeta|^{-1})$, and so on. ▶

Appendix: The Taylor Development of Characteristic Functions

The inequality (4.1) may be rewritten in the form

$$\left| e^{i\zeta x}\left(e^{itx} - 1 - \frac{itx}{1!} - \cdots - \frac{(itx)^{n-1}}{(n-1)!}\right)\right| \le \frac{|tx|^n}{n!}. \tag{4.13}$$

From this we get using (4.5)

$$\left| \varphi(\zeta+t) - \varphi(\zeta) - \frac{t}{1!}\varphi'(\zeta) - \cdots - \frac{t^{n-1}}{(n-1)!}\varphi^{(n-1)}(\zeta) \right| < M_n \frac{|t|^n}{n!}. \tag{4.14}$$

If $M_n < \infty$ this inequality is valid for arbitrary ζ and t and provides an upper bound for the difference between φ and the first terms of its Taylor development. In the special case when F is concentrated at the point 1 the inequality (4.14) reduces to (4.1).

Suppose now that all moments exist and that

$$\limsup_{n\to\infty} \frac{1}{n} M_n^{1/n} = \lambda < \infty. \tag{4.15}$$

Stirling's formula for $n!$ then shows trivially that for $|t| < 1/(3\lambda)$ the right side in (4.14) tends to zero as $n \to \infty$, and so the Taylor series for φ converges in some interval about ζ. It follows that φ is analytic in a neighborhood of any point of the real axis, and hence completely determined by its power series about the origin. But $\varphi^{(n)}(0) = (i)^n m_n$, and thus φ is completely determined by the moments m_n of F. Accordingly, *if* (4.15) *holds then F is uniquely determined by its moments*, and φ is analytic in a neighborhood of the real axis. This uniqueness criterion is weaker than Carleman's sufficient condition $\sum M_n^{-1/n} = \infty$ mentioned in VII,(3.14), but the two criteria are not very far apart. (For an example of a distribution not determined by its moments see VII,3.)

5. THE CENTRAL LIMIT THEOREM FOR EQUAL COMPONENTS

Work connected with the central limit theorem has greatly influenced the development and sharpening of the tools now generally used in probability theory, and a comparison of different proofs is therefore illuminating. Until recently the method of characteristic functions (first used by P. Lévy) was incomparably simpler than the direct approach devised by Lindeberg (not to mention other approaches). The streamlined modern version of the latter (presented in VIII,4) is not more complicated and has, besides, other merits. On the other hand, the method of characteristic functions leads to refinements which are at present not attainable by direct methods. Among these are the local limit theorem in this section as well as the error estimates and asymptotic expansions developed in the next chapter. We separate the case of variables with a common distribution, partly because of its importance, and partly to explain the essence of the method in the simplest situation.

Throughout this section $\mathbf{X}_1, \mathbf{X}_2, \ldots$ are mutually independent variables with the common distribution F and characteristic function φ. We suppose

$$\mathbf{E}(\mathbf{X}_j) = 0, \qquad \mathbf{E}(\mathbf{X}_j^2) = 1 \tag{5.1}$$

and put $\mathbf{S}_n = \mathbf{X}_1 + \cdots + \mathbf{X}_n$.

Theorem[6] **1.** *The distribution of* $\mathbf{S}_n/\sqrt{n}$ *tends to the normal distribution* $\mathfrak{N}$.

By virtue of the continuity theorem 2 in section 3 the assertion is equivalent to the statement that as $n \to \infty$

$$\varphi^n(\zeta/\sqrt{n}) \to e^{-\frac{1}{2}\zeta^2} \qquad \text{for all } \zeta. \tag{5.2}$$

Proof. By lemma 2 of the preceding section has a continuous second derivative, and hence by Taylor's formula

$$\varphi(x) = \varphi(0) + x\varphi'(0) + \tfrac{1}{2}x^2\varphi''(0) + o(x^2), \qquad x \to 0. \tag{5.3}$$

Choose ζ arbitrary and let $x = \zeta/n$ to conclude that

$$\varphi\left(\frac{\zeta}{\sqrt{n}}\right) = 1 - \frac{1}{2n}\zeta^2 + o\left(\frac{1}{n}\right) \qquad n \to \infty. \tag{5.4}$$

Taking nth powers we get (5.2). ▶

[6] The existence of a variance is not necessary for the asymptotic normality of $\mathbf{S}_n$. For the necessary and sufficient conditions see corollary 1 in XVII,5.

It is natural to expect that when F possesses a density f, the *density* of $\mathbf{S}_n/\sqrt{n}$ should tend to the normal density $\mathfrak{n}$. This is not always true, but the exceptions are fortunately rather "pathological." The following theorem covers the situations occurring in common practice.

Theorem 2. *If* $|\varphi|$ *is integrable, then* $\mathbf{S}_n/\sqrt{n}$ *has a density* f_n *which tends uniformly to the normal density* $\mathfrak{n}$.

Proof. The fourier inversion formula (3.5) holds both for f_n and $\mathfrak{n}$, and therefore

$$|f_n(x) - \mathfrak{n}(x)| \le \frac{1}{2\pi}\int_{-\infty}^{\infty}\left|\varphi^n\left(\frac{\zeta}{\sqrt{n}}\right) - e^{-\frac{1}{2}\zeta^2}\right| d\zeta. \tag{5.5}$$

The right side is independent of x and we have to show that it tends to 0 as $n \to \infty$. In view of (5.3) it is possible to choose $\delta > 0$ such that

$$|\varphi(\zeta)| \le e^{-\frac{1}{4}\zeta^2} \qquad \textit{for} \quad |\zeta| < \delta. \tag{5.6}$$

We now split the integral into three parts and prove that each is $< \epsilon$ for n sufficiently large. (1) As we have seen in the last proof, within a fixed interval $-a \le \zeta \le a$ the integrand tends uniformly to zero and so the contribution of $\overline{-a, a}$ tends to zero. (2) For $a < |\zeta| < \delta\sqrt{n}$ the integrand is $< 2e^{-\frac{1}{4}\zeta^2}$ and so the contribution of this interval is $< \epsilon$ if a is chosen sufficiently large. (3) We know from lemma 4 of section 1 that $|\varphi(\zeta)| < 1$ for $\zeta \ne 0$, and from lemma 3 of the last section that $\varphi(\zeta) \to 0$ as $|\zeta| \to \infty$. It follows that the maximum of $|\varphi(\zeta)|$ for $|\zeta| \ge \delta$ equals a number $\eta < 1$. The contribution of the intervals $|\zeta| > \delta\sqrt{n}$ to the integral (5.5) is then less than

$$\eta^{n-1}\int_{-\infty}^{+\infty}\left|\varphi\left(\frac{\zeta}{\sqrt{n}}\right)\right| d\zeta + \int_{|\zeta|>\delta\sqrt{n}} e^{-\frac{1}{2}\zeta^2}\, d\zeta. \tag{5.7}$$

The first integral equals the integral of $\sqrt{n}\,|\varphi|$, and so the quantity (5.7) tends to zero. ►

Actually the proof yields[7] the somewhat stronger result that if $|\varphi|^r \in L$ *for some integer* r *then* $f_n \to \mathfrak{n}$ *uniformly.* On the other hand, the corollary to theorem 3.3 shows that if no $|\varphi|^r$ is integrable, then every f_n is unbounded. Because of their curiosity value we insert examples showing that such pathologies can in fact occur.

Examples. (*a*) For $x > 0$ and $p \ge 1$ put

$$u_p(x) = \frac{1}{x \log^{2p} x}. \tag{5.8}$$

[7] The only change is that in (5.7) the factor η^{n-1} is replaced by η^{n-r}, and φ by φ^r.

Let g be a density concentrated on $\overline{0, 1}$ such that $g(x) > u_p(x)$ in some interval $\overline{0, h}$. There exists an interval $\overline{0, \delta}$ in which u_p decreases monotonically, and within this interval

$$g^{2*}(x) \geq \int_0^x u_p(x-y)\, u_p(y)\, dy > x\, u_p^2(x) = u_{2p}(x). \tag{5.9}$$

By induction it follows that for $n = 2^k$ there exists an interval $\overline{0, h_n}$ in which $g^{n*} \geq u_{np}$, and hence $g^{n*}(x) \to \infty$ as $x \to 0+$. Thus no convolution g^{n*} is bounded.

(*b*) A variant of the preceding example exhibits the same pathology in a more radical form. Let v be the density obtained by symmetrization of g and put

$$f(x) = \tfrac{1}{2}[v(x+1) + v(x-1)]. \tag{5.10}$$

Then f is an even probability density concentrated on $\overline{-2, 2}$, and we may suppose that it has unit variance. The analysis of the last example shows that v is continuous except at the origin, where it is unbounded. The same statement is true of all convolutions v^{n*} Now $f^{2n*}(x)$ is a linear combination of values $v^{2n*}(x+k)$ with $k = 0, \pm 1, \pm 2, \ldots,$ $\pm n$, and is therefore unbounded at all these points. The density of the normalized sum $\mathbf{S}_{2n}/\sqrt{2n}$ of variables $\mathbf{X}_j$ with the density f is given by $f_{2n}(x) = \sqrt{2n}\, f^{2n*}(x\sqrt{2n})$. It is continuous except at the $2n + 1$ points of the form $k/\sqrt{2n}$ $(k = 0, \pm 1, \ldots, \pm n)$, where it is unbounded. Since to every rational point t there correspond infinitely many pairs k, n such that $k/\sqrt{2n} = t$ it follows that *the distribution of $\mathbf{S}_n/\sqrt{n}$ tends to $\mathfrak{N}$, but the densities f_n do not converge at any rational point, and the sequence $\{f_n\}$ is unbounded in every interval.*

(*c*) *See problem* 20. ▶

To round off the picture we turn to *lattice distributions*, that is, we suppose that the variables $\mathbf{X}_j$ are restricted to values of the form $b, b\pm h, b\pm 2h, \ldots.$ We assume that h is the *span* of the distribution F, that is, h is the largest positive number with the stated property. Lemma 4 in section 1 states that $|\varphi|$ has period $2\pi/h$, and hence $|\varphi|$ is not integrable. Theorem 2, however, has a perfect analogue for the weights of the atoms of the distribution of $\mathbf{S}_n/\sqrt{n}$. All these atoms are among the points of the form $x = (nb+kh)/\sqrt{n}$, where $k = 0, \pm 1, \pm 2, \ldots.$ For such x we put

$$p_n(x) = \mathbf{P}\left\{\frac{\mathbf{S}_n}{\sqrt{n}} = x\right\} \tag{5.11}$$

and we leave $p_n(x)$ undefined for all other x. In (5.12) therefore x is restricted to the smallest lattice containing all atoms of $\mathbf{S}_n/\sqrt{n}$.

Theorem 3. *If F is a lattice distribution with span h, then as $n \to \infty$*

$$\frac{\sqrt{n}}{h}\, p_n(x) - \mathfrak{n}(x) \to 0 \tag{5.12}$$

uniformly in x.

Proof. By (3.14)

$$\frac{\sqrt{n}}{h}\,p_n(x) = \frac{1}{2\pi}\int_{-\sqrt{n}\pi/h}^{\sqrt{n}\pi/h} \varphi^n\left(\frac{\zeta}{\sqrt{n}}\right)e^{-ix\zeta}\,d\zeta. \tag{5.13}$$

Using again the Fourier inversion formula (3.5) for the normal density $\mathfrak{n}$ we see that the left side in (5.12) is dominated by

$$\int_{-\sqrt{n}\pi/h}^{\sqrt{n}\pi/h}\left|\varphi^n\left(\frac{\zeta}{\sqrt{n}}\right) - e^{-\frac{1}{2}\zeta^2}\right|d\zeta + \int_{|\zeta|>\sqrt{n}\pi/h} e^{-\frac{1}{2}\zeta^2}\,d\zeta. \tag{5.14}$$

It was shown in the proof of theorem 2 that the first integral tends to zero. The second integral trivially tends to zero and this completes the proof. ►

6. THE LINDEBERG CONDITIONS

We consider now a sequence of independent variables $\mathbf{X}_k$ such that

$$\mathbf{E}(\mathbf{X}_k) = 0, \qquad \mathbf{E}(\mathbf{X}_k^2) = \sigma_k^2. \tag{6.1}$$

We denote the distribution of $\mathbf{X}_k$ by F_k, its characteristic function by φ_k, and as usual we put $\mathbf{S}_n = \mathbf{X}_1 + \cdots + \mathbf{X}_n$ and $s_n^2 = \text{Var}(\mathbf{S}_n)$. Thus

$$s_n^2 = \sigma_1^2 + \cdots + \sigma_n^2. \tag{6.2}$$

We say that the *Lindeberg condition* is satisfied if

$$\frac{1}{s_n^2}\sum_{k=1}^{n}\int_{|x|>ts_n} x^2\,F_k\{dx\} \to 0, \qquad n \to \infty, \tag{6.3}$$

for each fixed $t > 0$. Roughly speaking, this condition requires that the variance σ_k^2 be due mainly to masses in an interval whose length is small in comparison with s_n. It is clear that σ_k^2/s_n^2 is less than t^2 plus the left side in (6.3) and, t being arbitrary, (6.3) implies that for arbitrary $\epsilon > 0$ and n sufficiently large

$$\frac{\sigma_k}{s_n} \leq \epsilon, \qquad k = 1, \ldots, n. \tag{6.4}$$

This, of course, implies that $s_n \to \infty$.

The ratio σ_n/s_n may be taken as a measure for the contribution of the component $\mathbf{X}_n$ to the weighted sum $\mathbf{S}_n/s_n$ and so (6.10) may be described as stating that asymptotically $\mathbf{S}_n/s_n$ is the sum of "*many individually negligible components*." The Lindeberg condition was introduced in VIII, (4.15) and the following theorem coincides with theorem 3 of VIII,4. Each proof has its advantages. The present one permits us to prove that the Lindeberg conditions are, in a certain sense, necessary; it leads also to the asymptotic expansions in chapter XVI, and to convergence theorems for densities (problem 28).

Theorem 1. *If the Lindeberg condition* (6.3) *holds, the distribution of the normalized sums* $\mathbf{S}_n/s_n$ *tends to the standard normal distribution* $\mathfrak{N}$.

Proof. Choose $\zeta > 0$ arbitrary, but fixed. We have to show that

$$\varphi_1(\zeta/s_n)\cdots\varphi_n(\zeta/s_n) \to e^{-\frac{1}{2}\zeta^2}. \tag{6.5}$$

Since $\varphi_k'(0) = 0$ and $|\varphi_k''(x)| \le \sigma_k^2$ for all x it follows from the two-term Taylor expansion and (6.4) that for n sufficiently large

$$|\varphi_k(\zeta/s_n) - 1| \le \tfrac{1}{2}\zeta^2\sigma_k^2/s_n^2 \le \epsilon\zeta^2. \tag{6.6}$$

We show that if this is true (6.5) is equivalent to

$$\sum_{k=1}^{n}[\varphi_k(\zeta/s_n) - 1] + \tfrac{1}{2}\zeta^2 \to 0. \tag{6.7}$$

In fact, we saw in (2.4) that $e^{\varphi_k - 1}$ is the characteristic function of a compound Poisson distribution and therefore $|e^{\varphi_k-1}| \le 1$. Now for any complex numbers such that $|a_k| \le 1$ and $|b_k| \le 1$

$$|a_1\cdots a_n - b_1\cdots b_n| \le \sum_{k=1}^{n}|a_k - b_k| \tag{6.8}$$

as can be seen by induction from the identity

$$x_1x_2 - y_1y_2 = (x_1-y_1)x_2 + (x_2-y_2)y_1.$$

For any $\delta > 0$ we have $|e^z - 1 - z| < \delta\,|z|$ if $|z|$ is sufficiently small, and hence we get from (6.6) for large n

$$|e^{\Sigma[\varphi_k(\zeta/s_n)-1]} - \varphi_1(\zeta/s_n)\cdots\varphi_n(\zeta/s_n)| \le \sum_{k=1}^{n}|e^{\varphi_k(\zeta/s_m)-1} - \varphi_k(\zeta/s_n)| \tag{6.9}$$

$$\le \delta\sum_{k=1}^{n}|\varphi_k(\zeta/s_n) - 1| \le \delta(\zeta^2/s_n^2)\sum_{k=1}^{n}\sigma^{k2} = \delta\zeta^2.$$

Since δ is arbitrary this means that the left side tends to zero and hence (6.5) holds iff (6.7) is true.

Now (6.7) may be rewritten in the form

$$\sum_{k=1}^{n}\int_{-\infty}^{+\infty}\left[e^{ix\zeta/s_n} - 1 - \frac{ix\zeta}{s_n} + \frac{x^2\zeta^2}{2s_n^2}\right]F_k\{dx\} \to 0. \tag{6.10}$$

From the basic inequality (4.1) it follows that for $|x| \le ts_n$ the integrand is dominated by $|x\zeta/s_n|^3 < t\zeta^3x^2/s_n^2$ and for $|x| > ts_n$ by $x^2\zeta^2/s_n^2$. The left side in (6.10) is therefore in absolute value

$$< t\zeta^3 + \zeta^2 s_n^{-2} \sum_{k=1}^{n} \int_{|x|>ts_n} x^2 F_k\{dx\}. \tag{6.11}$$

In consequence of the Lindeberg condition (6.3) the second term tends to zero, and as t can be chosen arbitrarily small it follows that (6.10) is true. ▶

Illustrative examples are given in VIII,4, in problems 17–20 of VIII,10, and in problems 26–27 below.

The next theorem contains a partial converse to theorem 1.

Theorem 2. *Suppose that* $s_n \to \infty$ *and* $\sigma_n/s_n \to 0$. *Then the Lindeberg condition* (6.3) *is necessary for the convergence of the distribution of* $\mathbf{S}_n/s_n$ *to* $\mathfrak{N}$.

Warning. We shall presently see that even when the Lindeberg condition fails the distribution of $\mathbf{S}_n/s_n$ can tend to a normal distribution with variance < 1.

Proof. We begin by showing that (6.4) holds. By assumption there exists a ν such that $\sigma_n/s_n < \epsilon$ for $n > \nu$. For $\nu < k \leq n$ we have then $\sigma_k/s_n \leq \sigma_k/s_k < \epsilon$, and the ν ratios σ_k/s_n with $k \leq \nu$ tend to 0 because $s_n \to \infty$.

Assume then that the distribution of $\mathbf{S}_n/s_n$ tends to $\mathfrak{N}$, that is, assume (6.5). We saw in the preceding proof that when (6.4) holds, this relation implies (6.10). Since $\cos z - 1 + \frac{1}{2}z^2 \geq 0$ the real part of the integrand is non-negative and so the real part of the left side is

$$\geq \sum_{k=1}^{n} \int_{|x|>ts_n} \left(\frac{x^2\zeta^2}{2s_n^2} - 2\right) F_k\{dx\} \geq (\tfrac{1}{2}\zeta^2 - 2t^{-2}) \frac{1}{s_n^2} \sum_{k=1}^{n} \int_{|x|>ts_n} x^2 F_k\{dx\}. \tag{6.12}$$

Thus for arbitrary ζ and t the right side tends to zero, and hence (6.3) is true. ▶

The condition $\sigma_n/s_n \to 0$ is not strictly necessary as can be seen in the special case where all the distributions F_k are normal: the σ_k may then be chosen arbitrarily and yet the distribution of $\mathbf{S}_n/s_n$ coincides with $\mathfrak{N}$. (See also problem 27.) However, the condition σ_n/s_n is a natural way to ensure that the influence of the individual terms $\mathbf{X}_k$ is in the limit negligible, and without this condition there is a radical change in the character of the problem. Even if $\sigma_n/s_n \to 0$ and $s_n \to \infty$ the Lindeberg condition is not necessary in order that there exist some norming constants a_n such that the distribution of $\mathbf{S}_n/a_n$ tends to $\mathfrak{N}$. The following example will clarify the situation.

Example. Let $\{\mathbf{X}_n\}$ be a sequence of variables satisfying the conditions of theorem 1 including the norming (6.1). Let the $\mathbf{X}'_n$ be independent of each other and of the $\mathbf{X}_k$ and such that

$$\sum_{n=1}^{\infty} \mathbf{P}\{\mathbf{X}'_n \neq 0\} < \infty. \tag{6.13}$$

Put $\bar{\mathbf{X}}_n = \mathbf{X}_n + \mathbf{X}'_n$ and denote the partial sums of $\{\mathbf{X}'_n\}$ and $\{\bar{\mathbf{X}}_n\}$ by $\mathbf{S}'_n$ and $\bar{\mathbf{S}}_n$. By the first Borel-Cantelli lemma with probability one only finitely many $\mathbf{X}'_n$ will differ from 0, and hence with probability one $\mathbf{S}'_n = O(s_n)$. It follows easily that the distributions of $\bar{\mathbf{S}}_n/s_n$ and $\mathbf{S}_n/s_n$ have the same asymptotic behavior. Thus *the distribution of* $\bar{\mathbf{S}}_n/s_n$ *tends to* $\mathfrak{N}$ *even though* s_n^2 *is not the variance of* $\bar{\mathbf{S}}_n$*; in fact, the* $\bar{\mathbf{X}}_n$ *need not have a finite expectation.* If $\mathbf{E}(\bar{\mathbf{S}}_n) = 0$ and $\mathbf{E}(\bar{\mathbf{S}}_n^2) = \bar{s}_n^2 < \infty$, the distribution of $\bar{\mathbf{S}}_n/\bar{s}_n$ converges only if $s_n/\bar{s}_n$ tends to a limit p. In this case the limit distribution is normal with a variance $p \leq 1$. ▶

This example shows that the partial sums $\mathbf{S}_n$ can be asymptotically normally distributed even when the components $\mathbf{X}_n$ have no expectations, and also that the variances are not always the appropriate norming constants. We shall not pursue this topic here for two reasons. First, the whole theory will be covered in chapter XVII. More importantly, generalizations of the above theorems provide excellent exercises, and problems 29–32 are designed to lead by easy stages to necessary and sufficient conditions for the central limit theorem.

7. CHARACTERISTIC FUNCTIONS IN HIGHER DIMENSIONS

The theory of characteristic functions in higher dimensions is so closely parallel to the theory in $\mathcal{R}^1$ that a systematic exposition appears unnecessary. To describe the basic ideas and notations it suffices to consider the case of two dimensions. Then $\mathbf{X}$ stands for a pair of two real random variables $\mathbf{X}_1$ and $\mathbf{X}_2$ with a given joint probability distribution F. We treat $\mathbf{X}$ *as a row vector* with components $\mathbf{X}_1$ and $\mathbf{X}_2$; similarly, in $F(x)$ the variable x should be interpreted as row vector with components x_1, x_2. On the other hand the variable ζ of the corresponding characteristic function stands for a column *vector* $\zeta = (\zeta_1, \zeta_2)$. This convention has the advantage that $x\zeta$ now denotes the inner product $x\zeta = \zeta_1 x_1 + \zeta_2 x_2$. *The characteristic function* φ of $\mathbf{X}$ *(or of* F*) is defined by*

$$\varphi(\zeta) = \mathbf{E}(e^{i\mathbf{X}\zeta}). \tag{7.1}$$

This definition is formally the same as in one dimension, but the exponent has a new interpretation and the integration is with respect to a bivariate distribution.

The main properties of bivariate characteristic functions are self-evident. For example, the choice $\zeta_2 = 0$ reduces the inner product $x\zeta$ to $x_1\zeta_1$, and hence $\varphi(\zeta_1, 0)$ *represents the characteristic function of the (marginal) distribution of* $\mathbf{X}_1$. For any *fixed choice* of the parameters ζ_1, ζ_2 the linear combination $\zeta_1\mathbf{X}_1 + \zeta_2\mathbf{X}_2$ *is a (one-dimensional) random variable and its characteristic function is given by*

$$\mathbf{E}(e^{i\lambda(\zeta_1\mathbf{X}_1+\zeta_2\mathbf{X}_2)}) = \varphi(\lambda\zeta_1, \lambda\zeta_2); \tag{7.2}$$

here ζ_1 and ζ_2 are fixed and λ serves as independent variable. In particular, the characteristic function of the sum $\mathbf{X}_1 + \mathbf{X}_2$ is given by $\varphi(\lambda, \lambda)$. In this manner the bivariate characteristic function yields the univariate characteristic function of all linear combinations $\zeta_1\mathbf{X}_1 + \zeta_2\mathbf{X}_2$. Conversely if we know the distributions of all such combinations, we can calculate all expressions $\varphi(\lambda\zeta_1, \lambda\zeta_2)$, and hence the bivariate characteristic function.[8] The next example shows the usefulness and flexibility of this approach. It uses the notations introduced in III,5.

Example. (*a*) *Multivariate normal characteristic functions.* Let $\mathbf{X} = (\mathbf{X}_1, \mathbf{X}_2)$ (thought of as a row vector!) have a non-degenerate normal distribution. For the sake of simplicity we suppose that $\mathbf{E}(\mathbf{X}) = 0$ and denote the *covariance matrix* $\mathbf{E}(\mathbf{X}^T\mathbf{X})$ by C. Its elements are $c_{kk} = \text{Var}(\mathbf{X}_k)$ and $c_{12} = c_{21} = \text{Cov}(\mathbf{X}_1, \mathbf{X}_2)$. For fixed ζ_1 and ζ_2 the linear combination $\zeta\mathbf{X} = \zeta_1\mathbf{X}_1 + \zeta_2\mathbf{X}_2$ has zero expectation and variance

$$\sigma^2 = \zeta^T C\zeta = c_{11}\zeta_1^2 + 2c_{12}\zeta_1\zeta_2 + c_{22}\zeta_2^2. \tag{7.3}$$

With λ as independent variable, the characteristic function of the variable $\mathbf{X}\zeta = \zeta_1\mathbf{X}_1 + \zeta_2\mathbf{X}_2$ is therefore given by $e^{-\frac{1}{2}\sigma^2\lambda^2}$. Accordingly, the bivariate characteristic function of $\mathbf{X} = (\mathbf{X}_1, \mathbf{X}_2)$ is given by

$$\varphi(\zeta) = e^{-\frac{1}{2}\zeta^T C\zeta}. \tag{7.4}$$

This formula holds also in r dimensions except that then $\zeta^T C\zeta$ is a quadratic form in r variables $\zeta_1, \ldots, \zeta_r$. Thus (7.4) *represents the characteristic function of the r-dimensional normal distribution with zero expectation and covariance matrix* C.

It is occasionally desirable to change both pairs $(\mathbf{X}_1, \mathbf{X}_2)$ and (ζ_1, ζ_2) to polar coordinates, that is, to introduce new variables by

$$\mathbf{X}_1 = \mathbf{R}\cos\Theta, \qquad \mathbf{X}_2 = \mathbf{R}\sin\Theta, \qquad \zeta_1 = \rho\cos\alpha, \qquad \zeta_2 = \rho\sin\alpha. \tag{7.5}$$

[8] This proves incidentally that a *probability distribution in* $\mathfrak{R}^2$ *is uniquely determined by the probabilities of all half-planes.* This fact (noted by H. Cramér and H. Wold) does not seem to be accessible by elementary methods. For an application to moments see problem 21.

(For such transformations, see III,1.) Then

$$\varphi(\zeta) = \mathbf{E}(e^{i\rho\mathbf{R}\cos(\Theta-\alpha)}) \tag{7.6}$$

but it must be borne in mind that this is *not* the characteristic function of the pair $(\mathbf{R}, \Theta)$; the latter is given by $\mathbf{E}(e^{i(\zeta_1\mathbf{R}+\zeta_2\Theta)})$.

Examples. (*b*) *Rotational symmetry.* When the pair $(\mathbf{X}_1, \mathbf{X}_2)$ represents a "vector issued in a random direction" (see I,10) the joint distribution of $(\mathbf{R}, \Theta)$ factors into the distribution G of $\mathbf{R}$ and the uniform distribution over $-\pi < \theta < \pi$. The expectation in (7.6) is then independent of α and takes on the form

$$\varphi(\zeta_1, \zeta_2) = (2\pi)^{-1}\int_0^\infty G\{dr\}\int_{-\pi}^{\pi} e^{i\rho r\cos\theta}\,d\theta. \tag{7.7}$$

The change of variable $\cos\theta = x$ reduces the inner integral to that discussed in problem 6, and thus

$$\varphi(\zeta_1,\zeta_2) = \int_0^\infty J_0(\rho r)G\{dr\} \qquad (\rho = \sqrt{\zeta_1^2 + \zeta_2^2}), \tag{7.8}$$

where

$$J_0(x) = I_0(ix) = \sum_{k=0}^{\infty} \frac{(-1)^k}{k!k!}\left(\frac{x}{2}\right)^{2k}. \tag{7.9}$$

(The Bessel function I_0 was introduced in II,7.)

A unit vector in a random direction has the distribution G concentrated at the point 1. Thus $J_0^n(\sqrt{\zeta_1^2 + \zeta_2^2})$ *is the characteristic function of the resultant of* n *independent unit vectors issued in random directions.* This result was derived by Raleigh in connection with random flights.

(*c*) We consider the special case where $(\mathbf{X}_1, \mathbf{X}_2)$ has a bivariate density f given by

$$f(x_1, x_2) = (2\pi)^{-1}a^2e^{-ar}, \qquad r = \sqrt{x_1^2 + x_2^2} \tag{7.10}$$

where a is a positive constant. Then (7.8) takes on the form[9]

$$\varphi(\zeta_1, \zeta_2) = a^2\int_0^\infty e^{-ar}J_0(\rho r)r\,dr = (1+\rho^2/a^2)^{-\frac{3}{2}}. \tag{7.11}$$

(*d*) *Rotational symmetry in* $\mathcal{R}^3$. Example (*b*) carries over to three dimensions except that we have now two polar angles: the geographic longitude

---

[9] Substituting for J_0 its expansion (7.9) one gets

$$\varphi(\zeta_1, \zeta_2) = \sum \frac{(2k+1)!}{k!k!}(-1)^k\left(\frac{\rho}{2a}\right)^{2k} = \sum \binom{-\frac{3}{2}}{k}\left(\frac{\rho^2}{a^2}\right)^k$$

which is the binomial series for $(1+\rho^2/a^2)^{-\frac{3}{2}}$.

ω and the polar distance θ. The inner integral in (7.7) takes on the form

(7.12)

$$\frac{1}{4\pi}\int_{-\pi}^{\pi}d\omega\int_0^{\pi}e^{ir\rho\cos\theta}\sin\theta\,d\theta = \tfrac{1}{2}\int_0^{\pi/2}(e^{ir\rho\cos\theta}+e^{-ir\rho\cos\theta})\sin\theta\,d\theta$$

where $\rho^2 = \zeta_1^2+\zeta_2^2+\zeta_3^2$. The substitution $\cos\theta = x$ reduces this expression to $(r\rho)^{-1}\sin r\rho$, and so (7.8) has the analogue

$$\varphi(\zeta_1, \zeta_2, \zeta_3) = \int_0^{\infty}\frac{\sin r\rho}{r\rho}\,G\{dr\}. \tag{7.13}$$

In particular, for a *unit* random vector the integral reduces to $\rho^{-1}\sin\rho$. Letting $\zeta_2 = \zeta_3$ we see that the characteristic function of the $\mathbf{X}_1$-component of a unit random vector is given by $\zeta_1^{-1}\sin\zeta_1$. We have thus a new proof for the fact established in I,10 that this component is distributed uniformly over $\overline{-1, 1}$. ▶

It may be left to the reader to verify that the main theorems concerning characteristic functions in one dimension carry over without essential change. *The Fourier inversion theorem in $\mathcal{R}^2$ states that if φ is (absolutely) integrable over the entire plane, then* $\mathbf{X}$ *has a bounded continuous density given by*

$$f(x_1, x_2) = \frac{1}{(2\pi)^2}\iint_{-\infty}^{+\infty}e^{-i(x_1\zeta_1+x_2\zeta_2)}\varphi(\zeta_1, \zeta_2)\,d\zeta_1\,d\zeta_2. \tag{7.14}$$

Example. (*e*) *Bivariate Cauchy distribution.* When the inversion formula (7.14) is applied to the density f in example (*c*) it is seen upon division by $f(0, 0) = (2\pi)^{-1}a^2$ that

$$\gamma(\zeta_1, \zeta_2) = e^{-a\sqrt{\zeta_1^2+\zeta_2^2}} \tag{7.15}$$

represents the characteristic function of a bivariate density g defined by

$$g(x_1, x_2) = \frac{a}{2\pi(a^2+x_1^2+x_2^2)^{\frac{3}{2}}}. \tag{7.16}$$

It follows that this density shares the main properties of the ordinary Cauchy density. In particular, it is *strictly stable*: if $\mathbf{X}^{(1)}, \ldots, \mathbf{X}^{(n)}$ are mutually independent vector variables with the density (7.16), their average $(\mathbf{X}^{(1)}+\cdots+\mathbf{X}^{(n)})/n$ has the same density. ▶

*8. TWO CHARACTERIZATIONS OF THE NORMAL DISTRIBUTION

We begin by a famous theorem conjectured by P. Lévy and proved in 1936 by H. Cramér. Unfortunately its proof depends on analytic function theory and is therefore not quite in line with our treatment of characteristic functions.

Theorem 1. *Let* $\mathbf{X}_1$ *and* $\mathbf{X}_2$ *be independent random variables whose sum is normally distributed. Then both* $\mathbf{X}_1$ *and* $\mathbf{X}_2$ *have normal distributions.*

In other words, the normal distribution cannot be decomposed except in the trivial manner. The proof will be based on the following lemma of some independent interest.

Lemma. *Let* F *be a probability distribution such that*

$$f(\eta) = \int_{-\infty}^{+\infty} e^{\eta^2 x^2} F\{dx\} < \infty \tag{8.1}$$

for some $\eta > 0$. *The characteristic function* φ *is then an entire function (defined for all complex* ζ). *If* $\varphi(\zeta) \neq 0$ *for all complex* ζ, *then* F *is normal.*

Proof *of the lemma.* For all complex ζ and real x, η one has $|x\zeta| \leq \eta^2 x^2 + \eta^{-2} |\zeta|^2$ and so the integral defining φ converges for all complex ζ and

$$|\varphi(\zeta)| \leq e^{\eta^{-2}|\zeta|^2} \cdot f(\eta). \tag{8.2}$$

This means that φ is an entire function of order ≤ 2, and if such a function has no zeros, then $\log \varphi(\zeta)$ is quadratic polynomial.[10] Hence, $\varphi(\zeta) = e^{-\frac{1}{2}a\zeta^2 + ib\zeta}$ where a and b are (possibly complex) numbers. But φ is a characteristic function and hence $-i\varphi'(0)$ equals the expectation, and $-\varphi''(0)$ the second moment of the distribution. It follows that b is real and $a \geq 0$, and so F is indeed normal. ▶

Proof *of theorem* 1. Without loss of generality we may assume the variables $\mathbf{X}_1$ and $\mathbf{X}_2$ centered so that the origin is a median for each. Then

$$\mathbf{P}\{|\mathbf{X}_1 + \mathbf{X}_2| > t\} \geq \tfrac{1}{2}\mathbf{P}\{|\mathbf{X}_1| > t\}. \tag{8.3}$$

Now the usual integration by parts [see V,6] shows that

$$f(\eta) \leq \eta^2 \int_0^{\infty} x \cdot e^{\eta^2 x^2} [1 - F(x) + F(-x)]\, dx, \tag{8.4}$$

and therefore the functions f_k corresponding to $\mathbf{X}_k$ satisfy the inequalities

---

* This section treats special topics and is used only in problem 27.

[10] See, for example, E. Hille, *Analytic function theory*, Boston, 1962, vol. II, p. 199 (Hadamard's factorization theorem).

$f_k(\eta) \leq 2f(\eta) < \infty$. Since $\varphi_1(\zeta)\,\varphi_2(\zeta) = e^{-\frac{1}{2}a\zeta^2+ib\zeta}$ neither φ_1 nor φ_2 can have a zero, and so $\mathbf{X}_1$ and $\mathbf{X}_2$ are normal. ►

We turn to a proof of the following characterization of the normal distribution enunciated and discussed in III,4.

Theorem 2. *Let* $\mathbf{X}_1$ *and* $\mathbf{X}_2$ *be independent variables and*

$$\mathbf{Y}_1 = a_{11}\mathbf{X}_1 + a_{12}\mathbf{X}_2, \qquad \mathbf{Y}_2 = a_{21}\mathbf{X}_1 + a_{22}\mathbf{X}_2. \tag{8.5}$$

If also $\mathbf{Y}_1$ *and* $\mathbf{Y}_2$ *are independent of each other then either all four variables are normal, or else the transformation* (8.5) *is trivial in the sense that either* $\mathbf{Y}_1 = a\mathbf{X}_1$ *and* $\mathbf{Y}_2 = b\mathbf{X}_2$ *or* $\mathbf{Y}_1 = a\mathbf{X}_2$ *and* $\mathbf{Y}_2 = b\mathbf{X}_1$.

Proof. For the special case of variables $\mathbf{X}_j$ with continuous densities the theorem was proved in III,4. The proof depended on the general solution of the functional equation III,(4.4), and we shall now show that an equation of the same type is satisfied by the characteristic functions φ_j of the variables $\mathbf{X}_j$. We show first that it suffices to consider real characteristic functions. This argument illustrates the usefulness of theorem 1.

(*a*) *Reduction to symmetric distributions.* Introduce a pair of variables $\mathbf{X}_1^-$ and $\mathbf{X}_2^-$ that are independent of each other and of the $\mathbf{X}_j$, and distributed as $-\mathbf{X}_1$ and $-\mathbf{X}_2$, respectively. The linear transformation (8.5) changes the symmetrized variables ${}^0\mathbf{X}_j = \mathbf{X}_j + \mathbf{X}_j^-$ into a pair $({}^0\mathbf{Y}_1, {}^0\mathbf{Y}_2)$ of *symmetric* independent variables. If the theorem is true for such variables then ${}^0\mathbf{X}_j$ is normal, and by theorem 1 this implies that also $\mathbf{X}_j$ is normal.

(*b*) *The functional equation.* Because of the assumed independence of $\mathbf{Y}_1$ and $\mathbf{Y}_2$ the bivariate characteristic function of $(\mathbf{Y}_1, \mathbf{Y}_2)$ must factor:

$$\mathbf{E}(e^{i(\zeta_1\mathbf{Y}_1+\zeta_2\mathbf{Y}_2)}) = \mathbf{E}(e^{i\zeta_1\mathbf{Y}_1})\mathbf{E}(e^{i\zeta_2\mathbf{Y}_2}). \tag{8.6}$$

Substituting from (8.5) we see that this relation implies the following identity for the characteristic functions of $\mathbf{X}_1$ and $\mathbf{X}_2$

$$\varphi_1(a_{11}\zeta_1+a_{21}\zeta_2)\,\varphi_2(a_{12}\zeta_1+a_{22}\zeta_2) = \tag{8.7}$$
$$= \varphi_1(a_{11}\zeta_1)\,\varphi_2(a_{12}\zeta_1)\,\varphi_1(a_{21}\zeta_2)\,\varphi_2(a_{22}\zeta_2).$$

This identity coincides with III,(4.4) except that the roles of a_{12} and a_{21} are interchanged. By assumption the φ_j are real and continuous, and as in III,4 it is seen that all a_{jk} may be assumed to be different from zero. By the lemma of III,4 therefore $\varphi_j(\zeta) = e^{-a_j\zeta^2}$, and so the $\mathbf{X}_j$ are normal. ►

9. PROBLEMS FOR SOLUTION

1. From the inequality (1.7) conclude (without calculations) that for every characteristic function φ

$$|\varphi(\zeta)|^2 \leq 1 - \frac{1 - |\varphi(2\zeta)|}{4} \leq e^{-\frac{1}{4}(1-|\varphi(2\zeta)|)}. \tag{9.1}$$

2. If $\varphi = u + iv$ is a characteristic function show that

$$u^2(\zeta) \leq \tfrac{1}{2}(1 + u(2\zeta)). \tag{9.2}$$

This in turn implies

$$|\varphi(\zeta)|^2 \leq \tfrac{1}{2}(1 + |\varphi(2\zeta)|). \tag{9.3}$$

Hint: For (9.2) use Schwarz' inequality, for (9.3) consider characteristic functions of the form $e^{i\alpha\zeta}\varphi(\zeta)$.

3. With the same notations

$$|\varphi(\zeta_2) - \varphi(\zeta_1)|^2 \leq 2[1 - u(\zeta_2 - \zeta_1)]. \tag{9.4}$$

The inequality (1.7) is contained herein when $\zeta_2 = -\zeta_1$.

4. From elementary formulas prove (without explicit integrations) that the characteristic function φ of the density $(1/\pi)[(1 - \cos x)/x^2]$ differs only by a constant factor from $2|\zeta| - |\zeta+1| - |\zeta-1|$. Conclude that $\varphi(\zeta) = 1 - |\zeta|$ for $|\zeta| \leq 1$.

5. From the characteristic function of the density $\frac{1}{2}ae^{-a|x|}$ derive a new characteristic function by simple differentiation with respect to a. Use the result to show that the convolution of the given distribution with itself has density $\frac{1}{4}ae^{-a|x|}(1 + a|x|)$.

6. Let f be the density concentrated on $\overline{-1, 1}$ and defined by

$$f(x) = \frac{1}{\pi\sqrt{1 - x^2}}. \tag{9.5}$$

Show that its characteristic function is given by

$$\varphi(\zeta) = \sum_{k=0}^{\infty} \frac{(-1)^k}{k!\,k!} (\tfrac{1}{2}\zeta)^{2k} = J_0(\zeta). \tag{9.6}$$

Note that $J_0(\zeta) = I_0(i\zeta)$ where I_0 is the Bessel function defined in II,(7.1). *Hint:* Expand $e^{i\zeta x}$ into a power series. The coefficient of ζ^n is given by an integral, and (9.6) can be verified by induction on n using an integration by parts.

7. The *arc sine* distribution with density $1/[\pi\sqrt{x(1-x)}]$ concentrated on 0, 1 has the characteristic function $e^{i\zeta/2}J_0(\zeta/2)$. *Hint:* Reduce to the preceding problem.

8. Using the entry 10 of the table in section 2 show that $2\pi^2 x \cdot (\sinh x)^{-1}$ is a density with characteristic function $2/[1 + \cosh(\pi\zeta)]$. *Hint:* Use problem 6 of II,9.

9. Let γ_L stand for the characteristic function with period $2L$ described in example 2(*c*). Show that $2\gamma_{2L} - \gamma_L$ is again a characteristic function of an arithmetic distribution. Its graph is obtained from fig. 2 by reflecting every second triangle about the ζ-axis.

10.[11] Let **X** and **Y** be independent random variables with distributions F and G, and characteristic functions φ and γ, respectively. Show that *the product* **XY** *has the characteristic function*

$$\int_{-\infty}^{\infty} \gamma(\zeta x)\, F\{dx\} = \int_{-\infty}^{\infty} \varphi(\zeta x)\, G\{dx\}. \tag{9.7}$$

[11] Combining (9.7) with the theorem in the footnote to example V,9(*b*) one gets the following criterion due to A. Khintchine. A function ω *is the characteristic function of a unimodal distribution iff* $\omega(\zeta) = \int_0^1 \varphi(\zeta x)\, dx$ *where* φ *is a characteristic function.*

11. If $\{\varphi_n\}$ is a sequence of characteristic functions such that $\varphi_n(\zeta) \to 1$ for $-\delta < \zeta < \delta$, then $\varphi_n(\zeta) \to 1$ for all ζ.

12. Let g be an even density with a strictly positive characteristic function γ. Then

$$g_a(x) = \frac{g(x)[1 - \cos ax]}{1 - \gamma(a)} \tag{9.8}$$

is a probability density with characteristic function

$$\gamma_a(\zeta) = \frac{2\gamma(\zeta) - \gamma(\zeta+a) - \gamma(\zeta-a)}{2[1 - \gamma(a)]}. \tag{9.9}$$

As $a \to \infty$ *we have* $\gamma_a \to \gamma$ *but not* $g_a \to g$. This shows that in the continuity theorem for densities the condition (3.10) is essential.

13. If γ is a real characteristic function and $\gamma \geq 0$, there exist even densities g_n with strictly positive characteristic functions γ_n such that $\gamma_n \to \gamma$. *Hint:* Consider mixtures $(1 - \epsilon)G + \epsilon F$ and convolutions.

14. If γ is a characteristic function such that $\gamma \geq 0$ and $\gamma(a) \neq 1$, then (9.9) defines a characteristic function. *Hint:* Use the preceding two problems.

15. *Generalization of the converse to* (4.7). Considering the distributions $(1/m_{2k})x^{2k} F\{dx\}$ (when they exist) prove by induction: the distribution F possesses a finite moment m_{2r} iff the $2r$th derivative of the characteristic function φ exists at the origin.

16. Let f be a probability density with a *positive and integrable* characteristic function. Then f has a unique maximum at the origin. If a second derivative f'' exists, then

$$f(0) > f(x) > f(0) - \frac{x^2}{2} f''(0); \tag{9.10}$$

analogous expansions hold for the first $2r$ terms of the Taylor development. [Note that f is even and hence $f^{(2k+1)}(0) = 0$.]

17. Let φ be a real characteristic function with continuous second derivative φ''. Then [unless $\varphi(\zeta) = 1$ for all ζ]

$$\psi(\zeta) = \frac{1 - \varphi(\zeta)}{\zeta^2} \frac{2}{|\varphi''0)|} \tag{9.11}$$

is a characteristic function belonging to an even density f_2 defined for $x > 0$ by

$$\frac{2}{|\varphi''(0)|} \int_x^\infty [1 - F(t)]\, dt. \tag{9.12}$$

Generalize to higher moments.

18. Let f be an even density with characteristic function φ. For $x > 0$ put

$$g(x) = \int_x^\infty \frac{f(s)\, ds}{s}, \qquad g(-x) = g(x). \tag{9.13}$$

Then g is again an even density and its characteristic function is

$$\gamma(\zeta) = \frac{1}{\zeta} \int_0^\zeta \varphi(s)\, ds. \tag{9.14}$$

19. Let γ be a characteristic function such that $\limsup |\gamma(\zeta)| = 1$ as $\zeta \to \infty$. The corresponding distribution F is purely singular (with respect to Lebesgue measure).

20. Suppose that $c_k > 0$, $\sum c_k = 1$ but $\sum c_k 2^k = \infty$. Let u be an even continuous density concentrated on $\overline{-1, 1}$ and let ω be its characteristic function. Then

$$f(x) = \sum c_k 2^k\, u(2^k x) \tag{9.15}$$

defines a density that is continuous except at the origin and has the characteristic function

$$\varphi(\zeta) = \sum c_k \omega(2^{-k}\zeta). \tag{9.16}$$

Show that $|\varphi|^n$ is not integrable for any n. *Hint:* For $x \neq 0$ the series in (9.15) is finite. Use the trivial inequality $(\sum c_k p_k)^n \geq \sum c_k^n p_k^n$ valid for $p_k \geq 0$.

21. *Moment problem in* $\mathfrak{R}^2$. Let $\mathbf{X}_1$ and $\mathbf{X}_2$ be two random variables with a joint distribution F. Put $A_k = \mathbf{E}(|\mathbf{X}_1|^k) + \mathbf{E}(|\mathbf{X}_2|^k)$. Show that F is uniquely determined by its moments if $\limsup k^{-1} A_k^{1/k} < \infty$. *Hint:* As pointed out in the footnote 8 to section 7 it suffices to prove that the distributions of all linear combinations $a_1\mathbf{X}_1 + a_2\mathbf{X}_2$ are uniquely determined. Use the criterion (4.15).

22. *Degenerate bivariate distributions.* If φ is a *univariate* characteristic function and a_1, a_2 arbitrary constants, show that $\varphi(a_1\zeta_1 + a_2\zeta_2)$ as a function of ζ_1, ζ_2 represents the *bivariate* characteristic function of a pair $(\mathbf{X}_1, \mathbf{X}_2)$ such that identically $a_2\mathbf{X}_1 = a_1\mathbf{X}_2$. Formulate the converse. Consider the special case $a_2 = 0$.

23. Let $\mathbf{X}, \mathbf{Y}, \mathbf{U}$ be mutually independent random variables with characteristic functions φ, γ, ω. Show that the product $\varphi(\zeta_1)\, \gamma(\zeta_2)\, \omega(\zeta_1 + \zeta_2)$ represents the bivariate characteristic function of the pair $(\mathbf{U} + \mathbf{X}, \mathbf{U} + \mathbf{Y})$. *Hint:* Use a trivariate characteristic function.

Examples and complements to the central limit theorem

24. Prove the central limit theorem 4 of VIII,4 for random sums by the method of characteristic functions.

25. Let $\mathbf{X}_k$ have the density $e^{-x} x^{a_k - 1}/\Gamma(a_k)$ where $a_k \to \infty$. The variance of $\mathbf{S}_n$ is $s_n^2 = (a_1 + \cdots + a_n)$. Show that the Lindeberg condition is satisfied if

$$s_n^{-2} \sum_{k=1}^{n} a_k^2 \to 0.$$

26. Let $\mathbf{P}\{\mathbf{X}_k = \pm 1\} = (k-1)/2k$ and $\mathbf{P}\{\mathbf{X}_k = \pm\sqrt{k}\} = 1/2k$. Show that there do *not* exist norming constants a_n such that the distribution of $\mathbf{S}_n/a_n$ tends to $\mathfrak{N}$. *Hint:* Pass to exponentials using

$$1 - \frac{\zeta^2}{a_2^n} \leq \varphi_k\left(\frac{\zeta}{a_n}\right) \leq 1 - \frac{k-1}{2k}\,\frac{\zeta^2}{a_n^2}.$$

27. If the distribution of $\mathbf{S}_n/s_n$ tends to $\mathfrak{N}$, but $\sigma_n/s_n \to p > 0$ then the distribution of $\mathbf{X}_n/s_n$ tends to a normal distribution with variance p^2. *Hint:* By the Cramér-Lévy theorem in section 8 if $\mathfrak{N} = U \star V$, then both U and V are normal. Use convergent subsequences for the distributions of $\mathbf{X}_n/s_n$ and $\mathbf{S}_{n-1}/s_n$.

28. (*Central limit theorem for densities.*) Show that theorem 2 of section 5 generalizes to sequences with variable densities f_k provided sufficient uniformity conditions are imposed on the characteristic functions. It suffices, for example, that the third absolute moments remain bounded and that the f_k have derivatives such that $|f_k'| < M$ for all k.

29. (*Central limit theorem for triangular arrays.*) For each n let $\mathbf{X}_{1,n}, \ldots, \mathbf{X}_{n,n}$ be n independent variables with distributions $F_{k,n}$. Let $\mathbf{T}_n = \mathbf{X}_{1,n} + \cdots + \mathbf{X}_{n,n}$. Suppose that $\mathbf{E}(\mathbf{X}_{k,n}) = 0$ and $\mathbf{E}(\mathbf{T}_n^2) = 1$, and that

$$\sum_{k=1}^{n} \int_{|x|>t} x^2 \, F_{k,n}\{dx\} \to 0 \tag{9.17}$$

for each $t > 0$. Show that the distribution of $\mathbf{T}_n$ tends to $\mathfrak{N}$. *Hint:* Adapt the proof of theorem 1 in section 6.

Note. The Lindeberg theorem represents the special case $\mathbf{X}_{k,n} = \mathbf{X}_k/s_n$ and $\mathbf{T}_n = \mathbf{S}_n/s_n$. Then (9.17) reduces to the Lindeberg condition (6.3). For triangular arrays see VI,3.

30. (*Truncation.*) Let $\{\mathbf{X}_k\}$ be a sequence of independent variables with *symmetric* distributions. For each n and $k \leq n$ let $\mathbf{X}_{k,n}$ be the variable obtained by truncating $\mathbf{X}_k$ at $\pm a_n$. Suppose that $\sum_{k=1}^{n} \mathbf{P}\{|\mathbf{X}_k| > a_n\} \to 0$ and that (9.17) holds. Show that the distribution of $\mathbf{S}_n/a_n$ tends to $\mathfrak{N}$.

31. (*Generalized central limit theorem.*) Suppose the distributions F_k are symmetric and that for every $t > 0$

$$\sum_{k=1}^{n} \int_{|x|>ta_n} F_k\{dx\} \to 0, \qquad a_n^{-2} \sum_{k=1}^{n} \int_{|x|<ta_n} x^2 \, F_k\{dx\} \to 1. \tag{9.18}$$

Prove that the distribution of $\mathbf{S}_n/a_n$ tends to $\mathfrak{N}$ (*a*) using the last two problems, (*b*) directly, by adapting the proof of theorem 1 in section 6.[12]

32. (*Continuation.*) The condition of symmetry may be replaced by the weaker condition

$$\sum_{k=1}^{n} \left| \int_{|x|<a_n} x \, F_k\{dx\} \right| \to 0. \tag{9.19}$$

33. In order that there exist norming constants a_n for which the conditions (9.18) are satisfied it is necessary and sufficient that there exists a sequence of numbers $t_n \to \infty$ such that

$$\sum_{k=1}^{n} \int_{|x|<t_n} F_k\{dx\} \to 0, \qquad \frac{1}{t_n^2} \sum_{k=1}^{a} \int_{|x|<t_n} x^2 \, F_k\{dx\} \to \infty.$$

In this case one can take

$$a_n^2 = \sum_{k=1}^{n} \int_{|x|<t_n} x^2 \, F_k\{dx\}.$$

(This criterion usually can be applied without difficulty.)

[12] Theorem 2 generalizes similarly but requires a different proof.

CHAPTER XVI*

Expansions Related to the Central Limit Theorem

The topics of this chapter are highly technical and may be divided into two classes. One problem is to obtain estimates for the error in the central limit theorem and to improve on this result by providing asymptotic expansions. A problem of an entirely different nature is to supplement the central limit theorem for large values of the independent variable, where the classical formulation becomes empty.

In order to facilitate access to important theorems, and to explain the basic ideas, we separate the case of identically distributed variables. Section 7 on large deviations is independent of the first five sections. The theory developed in these sections depends essentially on two techniques: direct estimation of absolutely convergent Fourier integrals, and smoothing methods. At the cost of some repetitions and some loss of elegance we separate the two main ideas by first treating expansions for densities.

The chapter culminates in the Berry-Esseen theorem of section 5. The smoothing method described in section 3 was first used by A. C. Berry in the proof of this theorem. An endless variety of smoothing procedures are in general use. In fact, the long and glorious history of the subject matter of this chapter has the unfortunate effect that accidents of historical development continue to influence the treatment of individual topics. The resulting diversity of tools and abundance of ad hoc methods has rendered the field proverbial for its messiness. The following systematic exploitation of Berry's method and of modern inequalities fortunately permits an amazing unification and simplification of the whole theory.[1]

* This chapter treats special topics and should be omitted at first reading.

[1] The best-known introduction to the asymptotic expansions is H. Cramér (1962). It contains the expansion theorems of sections 2 and 4 for equally distributed variables and a slightly sharper version of the theorems of section 7. The first rigorous treatment of the expansion theorems is due to Cramér, but his methods are no longer useful. Gnedenko and Kolmogorov (1954) treat the material of sections 1–5.

1. NOTATIONS

Except in the last section (which deals with unequal components) we shall denote by F a one-dimensional probability distribution with characteristic function φ. When the kth moment exists, it will be denoted by:

$$\mu_k = \int_{-\infty}^{+\infty} x^k\, F\{dx\}. \tag{1.1}$$

We suppose $\mu_1 = 0$ and put, as usual, $\mu_2 = \sigma^2$. For the normalized n-fold convolution we write F_n. Thus

$$F_n(x) = F^{n\star}(x\sigma\sqrt{n}\,). \tag{1.2}$$

When a density of F_n exists we shall denote it by f_n.

Except in section 6 (concerned with large deviations) we shall have to deal with functions of the form

$$u(x) = \frac{1}{2\pi}\int_{-\infty}^{+\infty} e^{-i\zeta x}v(\zeta)\, d\zeta, \tag{1.3}$$

and the obvious estimate

$$|u(x)| \le \frac{1}{2\pi}\int_{-\infty}^{+\infty} |v(\zeta)|\, d\zeta. \tag{1.4}$$

Both u and v will be integrable. If u is a probability density, then v is its characteristic function. To simplify expressions we introduce the

Convention. *The function* v *in* (1.3) *will be called the Fourier transform of* u *and the right side of* (1.4) *will be called the Fourier norm of* u.

As always, the normal density is denoted by

$$\mathfrak{n}(x) = \frac{1}{\sqrt{2\pi}}\, e^{-\frac{1}{2}x^2}. \tag{1.5}$$

Its Fourier transform is the characteristic function $e^{-\frac{1}{2}\zeta^2}$. By repeated differentiation we get therefore the identity

$$\frac{d^k}{dx^k}\,\mathfrak{n}(x) = \frac{1}{2\pi}\int_{-\infty}^{+\infty} e^{-i\zeta x}(-i\zeta)^k e^{-\frac{1}{2}\zeta^2}\, d\zeta \tag{1.6}$$

valid for $k = 1, 2, \ldots$. Obviously the left side is of the form

$$\frac{d^k}{dx^k}\,\mathfrak{n}(x) = (-1)^k\, H_k(x)\,\mathfrak{n}(x) \tag{1.7}$$

where H_k is a polynomial of degree k. The H_k are called *Hermite polynomials*.[2] In particular,

$$H_1(x) = x, \qquad H_2(x) = x^2 - 1, \qquad H_3(x) = x^3 - 3x. \tag{1.8}$$

The characteristic property of H_k is, then, that $H_k(x)\,\mathfrak{n}(x)$ *has the Fourier transform* $(i\zeta)^k e^{-\frac{1}{2}\zeta^2}$.

2. EXPANSIONS FOR DENSITIES

The central limit theorem 2 of XV,5 for densities can be strengthened considerably when some higher moments μ_k exist. The important assumption is that[3]

$$\int_{-\infty}^{+\infty} |\varphi(\zeta)|^\nu \, d\zeta < \infty \tag{2.1}$$

for some $\nu \geq 1$. The proof given in XV,5 may be summarized roughly as follows. The difference $u_n = f_n - \mathfrak{n}$ has the Fourier transform

$$v_n(\zeta) = \varphi^n\left(\frac{\zeta}{\sigma\sqrt{n}}\right) - e^{-\frac{1}{2}\zeta^2}. \tag{2.2}$$

The integral of $|v_n|$ tends to zero for two reasons. Given an arbitrarily small but fixed $\delta > 0$ the contribution of the intervals $|\zeta| > \delta\sigma\sqrt{n}$ tends to zero because of (2.1). Within $|\zeta| < \delta\sigma\sqrt{n}$ the integrand v_n is small by virtue of the behavior of φ near the origin. The latter conclusion depends only on the fact that $\mu_1 = 0$ and $\mu_2 = \sigma^2$. When higher moments exist we can use more terms in the Taylor development for φ and thus obtain more precise information concerning the speed of convergence $f_n \to \mathfrak{n}$. Unfortunately the problem becomes notationally involved when more than three terms are involved, and we therefore separate out the simplest and most important special case.

Theorem 1. *Suppose that* μ_3 *exists and that* $|\varphi|^\nu$ *is integrable for some* $\nu \geq 1$. *Then* f_n *exists for* $n \geq \nu$ *and as* $n \to \infty$

$$f_n(x) - \mathfrak{n}(x) - \frac{\mu_3}{6\sigma^3\sqrt{n}}(x^3 - 3x)\,\mathfrak{n}(x) = o\left(\frac{1}{\sqrt{n}}\right) \tag{2.3}$$

uniformly in x.

Proof. By the Fourier inversion theorem of XV,3 the left side in (2.3) exists for $n \geq \nu$ and has the Fourier norm

$$N_n = \frac{1}{2\pi}\int_{-\infty}^{+\infty} \left| \varphi^n\left(\frac{\zeta}{\sigma\sqrt{n}}\right) - e^{-\frac{1}{2}\zeta^2} - \frac{\mu_3}{6\sigma^3\sqrt{n}}(i\zeta)^3 e^{-\frac{1}{2}\zeta^2} \right| d\zeta. \tag{2.4}$$

[2] Sometimes Chebyshev-Hermite polynomials. The terminology is not unique. Various norming factors are in use and frequently e^{-x^2} replaces our $e^{-\frac{1}{2}x^2}$.

[3] Concerning this condition see examples XV,5(*a–b*) and problem 20 in XV,9.

Choose $\delta > 0$ arbitrary, but fixed. Since φ^n is the characteristic function of a density we have $|\varphi(\zeta)| < 1$ for $|\zeta| \neq 0$ and $\varphi(\zeta) \to 0$ as $|\zeta| \to \infty$ (lemmas 4 of XV,1 and 3 of XV,4). There exists therefore a number $q_\delta < 1$ such that $|\varphi(\zeta)| < q_\delta$ for $|\zeta| \geq \delta$. The contribution of the intervals $|\zeta| > \delta\sigma\sqrt{n}$ to the integral in (2.4) is then

$$(2.5) \qquad < q_\delta^{n-\nu} \int_{-\infty}^{+\infty} \left| \varphi\left(\frac{\zeta}{\sigma\sqrt{n}}\right) \right|^\nu d\zeta + \int_{|\zeta|>\delta\sigma\sqrt{n}} e^{-\frac{1}{2}\zeta^2} \left(1 + \left|\frac{\mu_3\zeta^3}{\sigma^3}\right|\right) d\zeta$$

and this tends to zero more rapidly than any power of $1/n$.

With the abbreviation[4]

$$(2.6) \qquad \psi(\zeta) = \log \varphi(\zeta) + \tfrac{1}{2}\sigma^2\zeta^2$$

we have therefore

(2.7)

$$N_n = \frac{1}{2\pi} \int_{|\zeta|<\delta\sigma\sqrt{n}} e^{-\frac{1}{2}\zeta^2} \left| \exp\left(n\psi\left(\frac{\zeta}{\sigma\sqrt{n}}\right)\right) - 1 - \frac{\mu_3}{6\sigma^3\sqrt{n}}(i\zeta)^3 \right| d\zeta + o\left(\frac{1}{n}\right)$$

The integrand will be estimated using the following general scheme

$$(2.8) \qquad |e^\alpha - 1 - \beta| = |(e^\alpha - e^\beta) + (e^\beta - 1 - \beta)| \leq (|\alpha - \beta| + \tfrac{1}{2}\beta^2)e^\gamma,$$

where $\gamma \geq \max(|\alpha|, |\beta|)$. (That this inequality is valid for arbitrary real or complex α and β becomes evident on replacing e^α and e^β by their power series.)

The function ψ is thrice differentiable and $\psi(0) = \psi'(0) = \psi''(0) = 0$ while $\psi'''(0) = i^3\mu_3$. Since ψ''' is continuous it is possible to find a neighborhood $|\zeta| < \delta$ of the origin in which ψ''' varies by less than ϵ. From the three-term Taylor expansion we conclude that

$$(2.9) \qquad |\psi(\zeta) - \tfrac{1}{6}\mu_3(i\zeta)^3| < \epsilon\sigma^3\,|\zeta|^3 \qquad \text{for } |\zeta| < \delta.$$

Here we choose δ so small that also

$$(2.10) \qquad |\psi(\zeta)| < \tfrac{1}{4}\sigma^2\zeta^2, \qquad |\tfrac{1}{6}\mu_3(i\zeta)^3| \leq \tfrac{1}{4}\sigma^2\zeta^2 \qquad \text{for } |\zeta| < \delta.$$

With this choice of δ it is seen using (2.8) that the integrand in (2.7) is less than

$$(2.11) \qquad e^{-\frac{1}{4}\zeta^2}\left(\frac{\epsilon}{\sqrt{n}}|\zeta|^3 + \frac{\mu_3^2}{72n}\zeta^6\right),$$

and as ϵ is arbitrary we have $N_n = o(1/\sqrt{n})$ and so (2.3) is true. ►

---

[4] All logarithms of complex numbers used in the sequel are defined by the Taylor series $\log(1+z) = \sum (-z)^n/n$ valid for $|z| < 1$. No other values of z will occur.

The same argument leads to higher-order expansions, but their terms cannot be expressed by simple explicit formulas. We therefore postpone the explicit construction of the polynomials involved.

Theorem 2. *Suppose that the moments* $\mu_3, \ldots, \mu_r$ *exist and that* $|\varphi|^\nu$ *is integrable for some* $\nu \geq 1$. *Then* f_n *exists for* $n \geq \nu$ *and as* $n \to \infty$

$$f_n(x) - \mathfrak{n}(x) - \mathfrak{n}(x) \sum_{k=3}^{r} n^{-\frac{1}{2}k+1} P_k(x) = o(n^{-\frac{1}{2}r+1}) \tag{2.12}$$

uniformly in x. *Here* P_k *is a real polynomial depending only on* $\mu_1, \ldots, \mu_k$ *but not on* n *and* r *(or otherwise on* F*).*

The first two terms are given by

$$P_3 = \frac{\mu_3}{6\sigma^3} H_3, \qquad P_4 = \frac{\mu_3^2}{72\sigma^6} H_3 + \frac{\mu_4 - 3\sigma^4}{24\sigma^4} H_4, \tag{2.13}$$

where H_k stands for the Hermite polynomial defined in (1.7). The expansion (2.12) is called (or used to be called) the *Edgeworth expansion* for f_n.

Proof. We adhere to the notation (2.6). If p is a polynomial with *real* coefficients $p_1, p_2, \ldots$ then

$$f_u - \mathfrak{n} - \mathfrak{n}\Sigma p_k H_k \tag{2.14}$$

has the Fourier norm

$$N_n = \frac{1}{2\pi} \int_{-\infty}^{+\infty} e^{-\frac{1}{2}\zeta^2} \left| \exp\left(n\psi\left(\frac{\zeta}{\sigma\sqrt{n}}\right)\right) - 1 - p(i\zeta) \right| d\zeta. \tag{2.15}$$

The theorem will be proved by exhibiting appropriate polynomials p. (Their dependence on n is not stressed in order not to encumber the notations.)

We begin by estimating the integrand. The procedure is as in the last proof except that we use the Taylor approximation for ψ up to and including the term of degree r. This approximation will be denoted by $\zeta^2 \psi_r(\zeta)$. Thus ψ_r is a polynomial of degree $r - 2$ with $\psi_r(0) = 0$; it is uniquely determined by the property that

$$\psi(\zeta) - \zeta^2\psi_r(\zeta) = o(|\zeta|^r) \qquad \zeta \to 0.$$

We now put

$$p(\zeta) = \sum_{k=1}^{r-2} \frac{1}{k!} \left[\zeta^2 \psi_r\left(\frac{\zeta}{\sigma\sqrt{n}}\right)\right]^k. \tag{2.16}$$

Then $p(i\zeta)$ is a polynomial with *real* coefficients depending on n. For fixed ζ, on the other hand, p is a polynomial in $1/\sqrt{n}$ whose coefficients can be calculated explicitly as polynomials in $\mu_1, \mu_2, \ldots, \mu_r$. As in the last proof it is obvious that for fixed $\delta > 0$ the contribution of $|\zeta| > \delta\sigma\sqrt{n}$ to the integral in (2.15) tends to zero more rapidly than any power of $1/n$, and thus we are concerned only with the integrand for $|\zeta| < \delta\sigma\sqrt{n}$. To estimate it we use [instead of (2.8)] the inequality

$$\left| e^\alpha - 1 - \sum_1^{r-2} \beta^k/k! \right| \leq |e^\alpha - e^\beta| + \left| e^\beta - 1 - \sum_1^{r-2} \beta^k/k! \right| \leq e^\gamma \left(|\alpha - \beta| + \frac{1}{(r-1)!} |\beta|^{r-1} \right) \tag{2.17}$$

valid when $|\alpha| < \gamma$ and $|\beta| < \gamma$.

By analogy to (2.9) we now determine δ such that for $|\zeta| < \delta$

$$|\psi(\zeta) - \zeta^2\psi_r(\zeta)| \le \epsilon\sigma^r |\zeta|^r. \tag{2.18}$$

The coefficient of ζ in ψ_r being $i^3\mu_3/6$, we can suppose that for $|\zeta| < \delta$ also

$$|\psi_r(\zeta)| < a\,|\zeta| < \tfrac{1}{4}\sigma^2 \tag{2.19}$$

provided $a > 1 + |\mu_3|$. Finally we require that for $|\zeta| < \delta$

$$|\psi(\zeta)| < \tfrac{1}{4}\sigma^2\zeta^2. \tag{2.20}$$

For $|\zeta| < \delta\sigma\sqrt{n}$ the integrand in (2.15) is then less than

$$e^{-\frac{1}{4}\zeta^2}\left(\frac{\epsilon\,|\zeta|^r}{n^{\frac{1}{2}r-1}} + \frac{a^{r-1}}{(r-1)!}\,\frac{|\zeta|^{3(r-1)}}{(\sigma\sqrt{n})^{r-1}}\right). \tag{2.21}$$

As ϵ is arbitrary we have $N_n = o(n^{-\frac{1}{2}r+1})$.

We have now found real coefficients p_k depending on n such that the left side in (2.14) is $o(n^{-\frac{1}{2}r+1})$ uniformly in x. For fixed ζ the left side is a polynomial in $1/\sqrt{n}$. Rearranging it according to ascending powers of $1/\sqrt{n}$ we get an expression of the form postulated in the theorem except that the summation extends beyond r. But the terms involving powers $1/n^k$ with $k > \frac{1}{2}r - 1$ can be dropped, and we get then the desired expansion (2.12). ►

The explicit definition of the polynomials P_k is thus as follows. A polynomial ψ_r of degree $r - 2$ is uniquely determined by the Taylor formula

$$\log \varphi(\zeta) = \zeta^2[-\tfrac{1}{2}\sigma^2 + \psi_r(\zeta)] + o(|\zeta|^r) \tag{2.22}$$

valid near the origin. Rearrange (2.16) *according to powers of* $1/\sqrt{n}$. *Denote the coefficient of $n^{-\frac{1}{2}k+1}$ by $q_k(i\zeta)$. Then P_k is the polynomial such that $\mathfrak{n}(x)\,P_k(x)$ has the inverse Fourier transform $e^{-\frac{1}{2}\zeta^2}q_k(i\zeta)$.*

3. SMOOTHING

Every expansion for the densities f_n leads by integration to an analogous expansion for the distributions F_n, but this simple procedure is not available when the integrability condition (2.1) fails. To cope with this situation we shall proceed indirectly (following A. C. Berry). To estimate the discrepancy $F_n - \mathfrak{N}$ or a similar function Δ we shall use the Fourier methods of the last section to estimate an approximation ${}^T\Delta$ to Δ, and then appraise the error ${}^T\Delta - \Delta$ by direct methods. In this section we develop the basic tools for this procedure.

Let V_T be the probability distribution with density

$$v_T(x) = \frac{1}{\pi}\,\frac{1 - \cos Tx}{Tx^2}, \tag{3.1}$$

and characteristic function ω_T. For $|\zeta| \le T$ we have

$$\omega_T(\zeta) = 1 - \frac{|\zeta|}{T}, \tag{3.2}$$

but this explicit form is of no importance. What matters is that $\omega_T(\zeta)$ *vanishes for* $|\zeta| \geq T$, for this circumstance will eliminate all questions of convergence.

We shall be interested in bounds for $F_n - \mathfrak{N}$ and, more generally, for functions of the form $\Delta_n = F_n - G_n$. Such functions will be approximated by their convolutions with V_T and we put generically ${}^T\Delta = V_T \bigstar \Delta$. In other words, given any function Δ we define

$$(3.3) \qquad {}^T\Delta(t) = \int_{-\infty}^{+\infty} \Delta(t-x)\, v_T(x)\, dx.$$

If Δ is bounded and continuous, then ${}^T\Delta \to \Delta$ as $T \to \infty$. Our main problem is to estimate the maximum of $|\Delta|$ in terms of the maximum of $|{}^T\Delta|$.

Lemma 1. *Let* F *be a probability distribution and* G *a function such that* $G(-\infty) = 0$, $G(\infty) = 1$, *and* $|G'(x)| \leq m < \infty$.

Put

$$(3.4) \qquad \Delta(x) = F(x) - G(x)$$

and

$$(3.5) \qquad \eta = \sup_x |\Delta(x)|, \qquad \eta_T = \sup_x |{}^T\Delta(x)|.$$

Then

$$(3.6) \qquad \eta_T \geq \frac{\eta}{2} - \frac{12m}{\pi T}.$$

Proof. The function Δ vanishes at infinity and the one-sided limits $\Delta(x+)$ and $\Delta(x-)$ exist everywhere, and so it is clear that at some point x_0 either $|\Delta(x_0+)| = \eta$ or $|\Delta(x_0-)| = \eta$. We may assume $\Delta(x_0) = \eta$. As F does not decrease and G grows at a rate $\leq m$ this implies

$$(3.7) \qquad \Delta(x_0+s) \geq \eta - ms \qquad \textit{for} \quad s > 0.$$

Putting

$$(3.8) \qquad h = \frac{\eta}{2m}, \qquad t = x_0 + h, \qquad x = h - s,$$

we have then

$$(3.9) \qquad \Delta(t-x) \geq \frac{\eta}{2} + mx \qquad \textit{for} \quad |x| \leq h.$$

We now estimate the convolution integral in (3.3) using (3.9) and the bound $\Delta(t-x) \geq -\eta$ for $|x| > h$. The contribution of the linear term vanishes for reasons of symmetry; since the density v_T attributes to $|x| > h$ a mass $\leq 4/(\pi Th)$ we get

$$(3.10) \qquad \eta_T \geq {}^T\Delta(x_0) \geq \frac{\eta}{2}\left[1 - \frac{4}{\pi Th}\right] - \eta \cdot \frac{4}{\pi Th} = \frac{\eta}{2} - \frac{6\eta}{\pi Th} = \frac{\eta}{2} - \frac{12m}{\pi T}. \qquad \blacktriangleright$$

In our applications G will have a derivative g coinciding either with the normal density $\mathfrak{n}$ or with one of the finite expansions described in the last section. In every case g will have a Fourier transform γ with two continuous derivatives such that $\gamma(0) = 1$ and $\gamma'(0) = 0$. Obviously then the convolution ${}^Tg = V_T \bigstar g$ has the Fourier transform $\gamma\omega_T$. Similarly, by the Fourier inversion theorem of XV,3 the product $\varphi\omega_T$ is the Fourier transform of the density ${}^Tf$ of $V_T \bigstar F$. In other words,

$$(3.11) \qquad {}^Tf(x) - {}^Tg(x) = \frac{1}{2\pi}\int_{-T}^{T} e^{-i\zeta x}[\varphi(\zeta) - \gamma(\zeta)]\,\omega_T(\zeta)\,d\zeta.$$

Integrating with respect to x we obtain

$$(3.12) \qquad {}^T\Delta(x) = \frac{1}{2\pi}\int_{-T}^{T} e^{-i\zeta x}\,\frac{\varphi(\zeta) - \gamma(\zeta)}{-i\zeta}\,\omega_T(\zeta)\,d\zeta.$$

No integration constant appears because both sides tend to 0 as $|x| \to \infty$, the left because $F(x) - G(x) \to 0$, the right by the Riemann-Lebesgue lemma 4 of XV,4. Note that $\varphi(0) = \gamma(0) = 1$ and $\varphi'(0) = \gamma'(0) = 0$; hence the integrand is a continuous function vanishing at the origin, and so no problem of convergence arises.

From (3.12) we get an upper bound for η_T which, combined with (3.6), yields an upper bound for η, namely

$$(3.13) \qquad |F(x) - G(x)| \le \frac{1}{\pi}\int_{-T}^{T}\left|\frac{\varphi(\zeta) - \gamma(\zeta)}{\zeta}\right| d\zeta + \frac{24m}{\pi T}.$$

As this inequality will be the basis for all estimates in the next two sections we recapitulate the conditions of its validity.

Lemma 2. *Let F be a probability distribution with vanishing expectation and characteristic function φ. Suppose that $F - G$ vanishes at $\pm\infty$ and that G has a derivative g such that $|g| \le m$. Finally, suppose that g has a continuously differentiable Fourier transform γ such that $\gamma(0) = 1$ and $\gamma'(0) = 0$. Then* (3.13) *holds for all x and $T > 0$.*

We shall give two *independent* applications of this inequality: In the next section we derive integrated versions of the expansion theorems of section 2. In section 5 we derive the famous Berry-Esseen bound for the discrepancy $F_n - \mathfrak{N}$.

4. EXPANSIONS FOR DISTRIBUTIONS

From the expansion (2.3) for densities we get by simple integration

$$(4.1) \qquad F_n(x) - \mathfrak{N}(x) - \frac{\mu_3}{6\sigma^3\sqrt{n}}(1-x^2)\,\mathfrak{n}(x) = o\left(\frac{1}{\sqrt{n}}\right).$$

For this expansion to hold it is not necessary that F has a density. In fact, we shall now prove that (4.1) holds for all distributions with the sole exception of lattice distributions (that is, when F is concentrated on the set of points of the form $b \pm nh$). For a lattice distribution the inversion formula XV,(5.12) shows that the largest jump of F_n is of the order of magnitude $1/\sqrt{n}$, and hence (4.1) cannot be true of any lattice distribution. However, even for lattice distributions the following theorem applies with a minor amendment. For convenience we separate the two cases.

Theorem 1. *If F is not a lattice distribution and if the third moment μ_3 exists, then* (4.1) *holds uniformly for all x.*

Proof. Put

$$G(x) = \mathfrak{N}(x) - \frac{\mu_3}{6\sigma^3\sqrt{n}}(x^2-1)\mathfrak{n}(x). \tag{4.2}$$

Then G satisfies the conditions of the last lemma with

$$\gamma(\zeta) = e^{-\frac{1}{2}\zeta^2}\left[1 + \frac{\mu_3}{6\sigma^3\sqrt{n}}(i\zeta)^3\right]. \tag{4.3}$$

We use the inequality (3.13) with $T = a\sqrt{n}$ where the constant a is chosen so large that $24\,|G'(x)| < \epsilon a$ for all x. Then

$$|F_n(x) - G(x)| \le \int_{-a\sqrt{n}}^{a\sqrt{n}} \left| \frac{\varphi^n\left(\dfrac{\zeta}{\sigma\sqrt{n}}\right) - \gamma(\zeta)}{\zeta} \right| d\zeta + \frac{\epsilon}{\sqrt{n}}. \tag{4.4}$$

As the domain of integration is finite we can use the argument of section 2 even when $|\varphi|$ is not integrable over the whole line. We partition the interval of integration into two parts. First, since F is not a lattice distribution the maximum of $|\varphi(\zeta)|$ for $\delta \le |\zeta| \le a\sigma$ is strictly less than 1 owing to lemma 4 of XV,1. As in section 2 it follows that the contribution of $|\zeta| > \delta\sigma\sqrt{n}$ tends to zero faster than any power of $1/n$. Second, by the estimate (2.11) for $|\zeta| \le \delta\sigma\sqrt{n}$ the integrand in (4.4) is

$$< e^{-\frac{1}{4}\zeta^2}\left(\frac{\epsilon}{\sqrt{n}}|\zeta| + \frac{\mu_3^2}{72n}|\zeta|^5\right)$$

and so for large n the right side in (4.4) is $< 1000\epsilon/\sqrt{n}$. Since ϵ is arbitrary this concludes the proof. ►

This argument breaks down for lattice distributions because their characteristic functions are periodic (and so the contribution of $|\zeta| > \delta\sigma\sqrt{n}$ does not tend to zero). The theorem can nevertheless be saved by a natural reformulation which takes into account the lattice character. The distribution function F is a stepfunction, but we shall approximate it by a continuous distribution function $F^\#$ with polygonal graph.

Definition. *Let F be concentrated on the lattice of points $b \pm nh$, but on no sublattice (that is, h is the span of F).*

The polygonal approximant $F^\#$ to F is the distribution function with a polygonal graph with vertices at the midpoints $b \pm (n+\frac{1}{2})h$ lying on the graph of F.

Thus

$$F^\#(x) = F(x) \qquad \text{if } x = b \pm (n+\tfrac{1}{2})h \tag{4.5}$$

$$F^\#(x) = \tfrac{1}{2}[F(x) + F(x-)] \qquad \text{if } x = b \pm nh. \tag{4.6}$$

Now F_n is a lattice distribution with span

$$h_n = \frac{h}{\sigma\sqrt{n}}, \tag{4.7}$$

and hence for large n the polygonal approximant $F_n^\#$ is very close to F_n.

Theorem 2.[5] *For lattice distributions the expansion* (4.1) *holds with F_n replaced by its polygonal approximant $F_n^\#$.*

In particular, (4.1) *is true at all midpoints of the lattice for F_n (with span h_n), while at the points of the lattice* (4.1) *holds with $F_n(x)$ replaced by*

$$\tfrac{1}{2}[F_n(x) + F_n(x-)].$$

Proof. The approximant $F^\#$ is easily seen to be identical with the convolution of F with the uniform distribution over $-\frac{1}{2}h < x < \frac{1}{2}h$. Accordingly, $F_n^\#$ is the convolution of F_n with the uniform distribution over $-\frac{1}{2}h_n < x < \frac{1}{2}h_n$, and we denote by $G^\#$ the convolution of this distribution with G, that is

$$G^\#(x) = h_n^{-1}\int_{-h_n/2}^{h_n/2} G(x-y)\,dy. \tag{4.8}$$

If M denotes the maximum of $|G''|$ it follows from the two-term Taylor expansion of G about the point x that

$$|G^\#(x) - G(x)| < \tfrac{1}{3}Mh_n^2 = O(1/n), \tag{4.9}$$

and to prove the theorem it suffices therefore to show that

$$|F_n^\#(x) - G^\#(x)| = o(1/\sqrt{n}). \tag{4.10}$$

[5] Instead of replacing F_n by $F_n^\#$ one can expand $F_n^\# - F_n$ into a Fourier series and add it to the right side in (4.1). In this way one arrives formally at a form of the theorem proved by Esseen by intricate formal calculations. See, for example, the book by Gnedenko and Kolmogorov.

Since taking convolutions corresponds to multiplying the transforms, we conclude from (4.4) that

$$(4.11)\quad |F_n^{\#}(x) - G^{\#}(x)| \leq \int_{-a\sqrt{n}}^{a\sqrt{n}} \left| \frac{\varphi^n(\zeta/\sigma n) - \gamma(\zeta)}{\zeta} \right| |\omega_n(\zeta)|\, d\zeta + \frac{\epsilon}{\sqrt{n}}$$

where $\omega_n(\zeta) = (\sin \frac{1}{2}h_n\zeta)/(\frac{1}{2}h_n\zeta)$ is the characteristic function of the uniform distribution. The estimates used for (4.4) apply except that a new argument is required to show that

$$(4.12)\quad \int_{\delta\sigma\sqrt{n}}^{a\sqrt{n}} |\varphi^n(\zeta/\sigma n)\omega_n(\zeta)|\, \zeta^{-1}\, d\zeta = \frac{2}{h} \int_{\delta}^{a/\sigma} \left| \varphi^n(y) \sin \frac{hy}{2} \right| y^{-2}\, dy = o\left(\frac{1}{n}\right).$$

By Lemma 4 of XV,1 the characteristic function φ has period $2\pi/h$, and the same is obviously true of $|\sin \frac{1}{2}hy|$. It suffices therefore to prove that

$$(4.13)\quad \int_0^{\pi/h} |\varphi^n(y)|\, y\, dy = o\left(\frac{1}{\sqrt{n}}\right).$$

But this is trivially true because within a neighborhood of the origin $|\varphi(y)| < e^{-\frac{1}{4}\sigma y^2}$ while outside this neighborhood $|\varphi(y)|$ is bounded away from 0 and hence the integrand in (4.13) decreases faster than any power of n. The integral is therefore actually $O(1/n)$. ▶

We turn to higher-order expansions. The proof of (4.1) differs from the proof of (2.3) only by the smoothing, which accounts for the finite limits in the integral (4.4). The same smoothing can be applied to the higher expansions (2.12), but it is obvious that to achieve an error term of the order of magnitude $n^{-\frac{1}{2}r+1}$ we shall have to take $T \sim an^{\frac{1}{2}r-1}$. Here one difficulty arises. The proof of (4.1) depended on the fact that the maximum of $|\varphi(\zeta/(\sigma\sqrt{n}))|$ in $\delta\sigma\sqrt{n} < |\zeta| < T$ is less than one. For non-lattice distributions this is always true when $T = a\sqrt{n}$, but not necessarily when T increases as some higher power of n. For higher-order expansions we are therefore compelled to introduce the assumption that

$$(4.14)\quad \limsup_{|\zeta|\to\infty} |\varphi(\zeta)| < 1$$

which for non-lattice distributions implies that the maximum q_δ of $|\varphi(\zeta)|$ for $|\zeta| > \delta$ is less than 1. With this additional assumption the method of proof given in detail for (4.1) applies *without change* to the expansions (2.12) and leads to

Theorem 3. *If* (4.14) *holds and the moments* $\mu_3, \ldots, \mu_r$ *exist, then as* $n \to \infty$,

$$(4.15)\quad F_n(x) - \mathfrak{N}(x) - \mathfrak{n}(x) \sum_{k=3}^{r} n^{-\frac{1}{2}k+1} R_k(x) = o(n^{-\frac{1}{2}r+1})$$

uniformly in x. *Here* R_k *is a polynomial depending only on* $\mu_1, \ldots, \mu_r$ *but not on* n *and* r *(or otherwise on* F*).*

The expansion (4.15) is simply the integrated version of (2.12) and the polynomials R_k are related to those in (2.12) by

$$(4.16)\quad \mathfrak{n}(x)\, P_k(x) = \frac{d}{dx}\, \mathfrak{n}(x)\, R_k(x).$$

There is therefore no need to repeat their construction. The condition (4.14) is satisfied by every non-singular F.

(4.15) is called the *Edgeworth expansion* of F. If F has moments of all orders, one is tempted to let $r \to \infty$, but the resulting infinite series need not converge for any n. (Cramér showed that it converges for all n iff $e^{\frac{1}{4}x^2}$ is integrable with respect to F.) The formal Edgeworth series should not be confused with the Hermite polynomial expansion

$$F_n(x) - \mathfrak{N}(x) = \sum_{k=1}^{r} c_k H_k(x) e^{-\frac{1}{2}x^2} \tag{4.17}$$

which is convergent whenever F has a finite expectation, but is without deeper probabilistic meaning. For example even if it is possible to expand each F_n into a series of the form (4.17) the coefficients are not indicative of the speed of convergence $F_n \to \mathfrak{N}$.

5. THE BERRY-ESSEEN THEOREMS[6]

The following important theorem was discovered (with radically different proofs) by A. C. Berry (1941) and C. G. Essen (1942).

Theorem 1. *Let the* $\mathbf{X}_k$ *be independent variables with a common distribution* F *such that*

$$\mathbf{E}(\mathbf{X}_k) = 0, \qquad \mathbf{E}(\mathbf{X}_k^2) = \sigma^2 > 0, \qquad \mathbf{E}(|\mathbf{X}_k|^3) = \rho < \infty, \tag{5.1}$$

and let F_n *stand for the distribution of the normalized sum*

$$(\mathbf{X}_1 + \cdots + \mathbf{X}_n)/\sigma\sqrt{n}.$$

Then for all x *and* n

$$|F_n(x) - \mathfrak{N}(x)| \le \frac{3\rho}{\sigma^3\sqrt{n}}. \tag{5.2}$$

The striking feature of the inequality is that it depends only on the first three moments. The expansion (4.1) provides a better asymptotic estimate, but the speed of convergence depends on more delicate properties of the underlying distribution. The factor 3 on the right could be replaced by a better upper bound C but no attempt is made in our setup to achieve optimal results.[7]

---

[6] This section uses the smoothing inequality (3.13) (with G standing for the normal distribution) but is otherwise independent of the preceding sections.

[7] Berry gives a bound $C \le 1.88$, but his calculations were found in error. Esseen gives $C \le 7.59$. Unpublished calculations are reported to yield $C \le 2.9$ (Esseen 1956) and $C \le 2.05$ (D. L. Wallace 1958). Our streamlined method yields a remarkably good bound even though it avoids the usual messy numerical calculations. No substantial improvement can be expected without improving the error term $24m/\pi$ in (3.13).

Proof. The proof will be based on the smoothing inequality (3.13) with $F = F_n$ and $G = \mathfrak{N}$. For T we choose

$$T = \frac{4}{3} \cdot \frac{\sigma^3}{\rho} \sqrt{n} \leq \tfrac{4}{3}\sqrt{n}, \tag{5.3}$$

the last inequality being a consequence of the moment inequality $\sigma^3 < \rho$. [See V,8(*c*).] Since the normal density $\mathfrak{n}$ has a maximum $m < \frac{2}{5}$ we get

$$\pi\, |F_n(x) - \mathfrak{N}(x)| \leq \int_{-T}^{T} |\varphi^n(\zeta/\sigma\sqrt{n}) - e^{-\frac{1}{2}\zeta^2}| \frac{d\zeta}{|\zeta|} + \frac{9.6}{T}. \tag{5.4}$$

To appraise the integrand we note that the familiar expansion for $\alpha^n - \beta^n$ leads to the inequality

$$|\alpha^n - \beta^n| \leq n\, |\alpha - \beta| \cdot \gamma^{n-1} \qquad \textit{if} \quad |\alpha| \leq \gamma, \quad |\beta| \leq \gamma. \tag{5.5}$$

We use this with $\alpha = \varphi(\zeta/\sigma\sqrt{n})$ and $\beta = e^{-\frac{1}{2}\zeta^2/n}$. From the inequality XV,(4.1) for e^{it} we have

$$|\varphi(t) - 1 + \tfrac{1}{2}\sigma^2 t^2| = \left| \int_{-\infty}^{\infty} (e^{itx} - 1 - itx + \tfrac{1}{2}t^2x^2)\, F\{dx\} \right| \leq \tfrac{1}{6}\rho\, |t|^3 \tag{5.6}$$

and hence

$$|\varphi(t)| < 1 - \tfrac{1}{2}\sigma^2 t^2 + \tfrac{1}{6}\rho\, |t|^3 \quad \textit{if} \quad \tfrac{1}{2}\sigma^2 t^2 \leq 1. \tag{5.7}$$

We conclude that for $|\zeta| \leq T$

$$|\varphi(\zeta/\sigma\sqrt{n})| \leq 1 - \frac{1}{2n}\zeta^2 + \frac{\rho}{6\sigma^3 n^{\frac{3}{2}}} |\zeta|^3 \leq 1 - \frac{5}{18n}\zeta^2 \leq e^{-\frac{5}{18}\zeta^2/n}. \tag{5.8}$$

Since $\sigma^3 < \rho$ the assertion of the theorem is trivially true for $\sqrt{n} \leq 3$ and hence we may assume $n \geq 10$. Then

$$|\varphi(\zeta/\sigma\sqrt{n})|^{n-1} \leq e^{-\frac{1}{4}\zeta^2}, \tag{5.9}$$

and the right side may serve for the bound γ^{n-1} in (5.5). Noting that $e^{-x} - 1 + x \leq \frac{1}{2}x^2$ for $x > 0$ we get from (5.6)

$$\begin{aligned} n \left| \varphi\left(\frac{\epsilon}{\sigma\sqrt{n}}\right) - e^{-\frac{1}{2}\zeta^2/n} \right| &\leq n \left| \varphi\left(\frac{\zeta}{\sigma\sqrt{n}}\right) - 1 + \frac{\zeta^2}{2n} \right| \\ &\quad + n \left| 1 - \frac{\zeta^2}{2n} - e^{-\frac{1}{2}\zeta^2/n} \right| \leq \frac{\zeta}{6\sigma^3\sqrt{n}} |\zeta|^3 + \frac{1}{8n}\zeta^4. \end{aligned} \tag{5.10}$$

Since $\sqrt{n} > 3$ it follows from (5.5) and (5.9) that the integrand in (5.4) is

$$\leq \frac{1}{T}\left(\tfrac{2}{9}\zeta^2 + \tfrac{1}{18}|\zeta|^3\right)e^{-\frac{1}{4}\zeta^2}. \tag{5.11}$$

This function is integrable over $-\infty < \zeta < \infty$, and simple integrations by parts now show that

$$\pi T\,|F_n(x) - \mathfrak{N}(x)| \le \tfrac{8}{9}\sqrt{\pi} + \tfrac{8}{9} + 10. \tag{5.12}$$

Since $\sqrt{\pi} < \frac{9}{5}$ the right side is $< \frac{113}{9} < 4\pi$, and so (5.2) is true. ▶

The theorem and its proof can be generalized to sequences $\{\mathbf{X}_k\}$ with varying distributions as follows.

Theorem 2.[8] *Let the* $\mathbf{X}_k$ *be independent variables such that*

$$\mathbf{P}(\mathbf{X}_k) = 0, \qquad \mathbf{P}(\mathbf{X}_k^2) = \sigma_k^2, \qquad \mathbf{P}(|\mathbf{X}_k^3|) = \rho_k \tag{5.13}$$

Put

$$s_n^2 = \sigma_1^2 + \cdots + \sigma_n^2, \qquad r_n = \rho_1 + \cdots + \rho_n \tag{5.14}$$

and denote by F_n *the distribution of the normalized sum* $(\mathbf{X}_1+\cdots+\mathbf{X}_n)/s_n$. *Then for all* x *and* n

$$|F_n(x) - \mathfrak{N}(x)| \le 6\,\frac{r_n}{s_n^3}. \tag{5.15}$$

Proof. If ω_k stands for the characteristic function of $\mathbf{X}_k$ the starting inequality (5.4) is now replaced by

$$\pi\,|F_n(x) - \mathfrak{N}(x)| \le \int_{-T}^{T} \left|\omega_1\!\left(\frac{\zeta}{s_n}\right)\cdots\omega_n\!\left(\frac{\zeta}{s_n}\right) - e^{-\frac{1}{2}\zeta^2}\right| \frac{d\zeta}{|\zeta|} + \frac{9.6}{T}. \tag{5.16}$$

This time we choose

$$T = \frac{8}{9}\cdot\frac{s_n^3}{r_n}. \tag{5.17}$$

Instead of (5.5) we now use

$$|\alpha_1\cdots\alpha_n - \beta_1\cdots\beta_n| \le \sum_{k=1}^{n} \gamma_1\cdots\gamma_{k-1}\alpha_k - \beta_k\gamma_{k+1}\cdots\gamma_n \tag{5.18}$$

valid if $|\alpha_k| \le \gamma_k$ and $|\beta_k| \le \gamma_k$. This inequality will be applied to

$$\alpha_k = \omega_k(\zeta/s_n), \qquad \beta_k = e^{-\frac{1}{2}(\sigma_k^2/s_n^2)\zeta^2}, \qquad |\zeta| < T. \tag{5.19}$$

By analogy with (5.8) we have

$$|\omega_k(\zeta/s_n)| \le 1 - \frac{1}{2}\frac{\sigma_k^2}{s_n^2}\zeta^2 + \frac{\rho_k}{6s_n^3}|\zeta|^3 \le \exp\left(-\frac{\sigma_k^2}{2s_n^2} + \frac{\rho_k T}{6s_n^3}\right)\zeta^2 \tag{5.20}$$

provided $\sigma_k T < s_n\sqrt{2}$. To obtain a bound γ_k applicable for all k we change

[8] Due (with an entirely different proof) to Esseen.

the coefficient $\frac{1}{6}$ to $\frac{3}{8}$ and put

$$(5.21) \qquad \gamma_k = \exp\left(-\frac{\sigma_k^2}{2s_n^2} + \frac{3}{8}\frac{\rho_k T}{s_n^3}\right)\zeta^2.$$

Obviously $|\beta_k| \leq \gamma_k$, and from (5.20) also $|\alpha_k| \leq \gamma_k$ for k such that $\sigma_k T \leq \frac{3}{4}s_n$. But from the moment inequality $\rho_k \geq \sigma_k^3$ it follows that $\gamma_k > 1$ if $\sigma_k T > \frac{4}{3}s_n$, and hence $|\alpha_k| \leq \gamma_k$ for all k.

The theorem is trivially true when the right side in (5.15) is ≥ 1, that is, if $r_n/s_n^3 \geq \frac{1}{6}$. Accordingly we assume from now on that $r_n/s_n^3 < \frac{1}{6}$ or $T > \frac{16}{3}$. The minimum value of γ_k is assumed for some k such that $\sigma_k/s_n < 4/3T < \frac{1}{4}$, and hence $\gamma_k \geq e^{-\zeta^2/32}$ for all k. Thus finally

$$(5.22) \quad |\gamma_1 \cdots \gamma_{k-1}\beta_{k+1} \cdots \gamma_n| \leq \exp \zeta^2\left(-\frac{1}{2} + \frac{3r_n T}{8s_n^2} + \frac{1}{32}\right) < e^{-\zeta^2/8}.$$

By analogy with (5.10) we get

$$(5.23) \qquad \sum_{k=1}^{n} |\alpha_k - \beta_k| \leq \frac{r_n}{6s_n^3}|\zeta|^3 + \frac{\zeta_4}{8s_n^4}\sum_{k=1}^{n}\sigma_k^4.$$

To appraise the last sum we recall that $\sigma_k^4 \leq \rho_k^{\frac{4}{3}} \leq r_n^{\frac{1}{3}} \cdot \rho_k$ whence

$$(5.24) \qquad \frac{1}{s_n^4}\sum_{k=1}^{n}\sigma_k^4 \leq \left(\frac{r_n}{s_n^3}\right)^{\frac{4}{3}} \leq \frac{1}{6^{\frac{1}{3}}}\frac{r_n}{s_n^3} \leq \frac{5}{9}\frac{r_n}{s_n^3}.$$

These inequalities show that the integrand in (5.16) is

$$(5.25) \qquad < \frac{8}{9T}\left(\tfrac{1}{6}\zeta^2 + \tfrac{5}{72}|\zeta|^3\right)e^{-\zeta^2/8},$$

and hence finally

$$(5.26) \qquad \pi T\,|F_n(x) - \mathfrak{N}(x)| \leq \tfrac{32}{27}\sqrt{2\pi} + \tfrac{5}{81}\cdot 64 + 9.6.$$

The right side is $<16\pi/3$, and thus (5.26) implies (5.15). ►

Recently much attention has been paid to generalizations of the Berry-Esseen theorem to variables without third moment; the upper bound is then replaced by a fractional moment or some related quantity. The first step in this direction was taken by M. L. Katz (1963). The usual calculations are messy, and no attempt has been made to develop unified methods applicable to the several variants. Our proof was developed for this purpose and can be restated so as to cover a much wider range. Indeed, the third moment occurs in the proof only because of the use of the inequality

$$|e^{itx} - 1 - itx + \tfrac{1}{2}t^2x^2| \leq \tfrac{1}{6}|tx|^3.$$

Actually it would have sufficed to use this estimate in some finite interval $|x| < a$ and to be otherwise satisfied with the bound t^2x^2. In this way one obtains the following theorem obtained by different methods and with an unspecified constant by L. V. Osipov and V. V. Petrov.

Theorem 3. *Assume the conditions of theorem 2 except that no third moments need exist. Then for arbitrary* $\tau_k > 0$

$$(5.27)\quad |F_n(x) - \mathfrak{N}(x)| \leq 6\left(s_n^{-3}\int_{|x|\leq\tau_k} |x|^3\, F\{dx\} + s_n^{-2}\int_{|x|>\tau_k} x^2\, F\{dx\}\right).$$

Simple truncation methods permit one to extend this result to variables without moments.[9]

6. EXPANSIONS IN THE CASE OF VARYING COMPONENTS

The theory of sections 2 and 4 is easily generalized to sequences $\{\mathbf{X}_k\}$ of independent variables with varying distributions U_k. In fact, our notations and arguments were intended to prepare for this task and were therefore not always the simplest.

Let $\mathbf{E}(\mathbf{X}_k) = 0$ and $\mathbf{E}(\mathbf{X}_k^2) = \sigma_k^2$. As usual we put $s_n^2 = \sigma_1^2 + \cdots + \sigma_n^2$. To preserve continuity we let F_n again stand for the distribution of the normalized sum $(\mathbf{X}_1 + \cdots + \mathbf{X}_n)/s_n$.

To fix ideas, let us consider the one-term expansion (4.1). The left side has now the obvious analogue

$$(6.1)\qquad D_n(x) = F_n(x) - \mathfrak{N}(x) - \frac{\mu_3^{(n)}}{6s_n^2}\,\mathfrak{n}(x)$$

where

$$(6.2)\qquad \mu_3^{(n)} = \sum_{k=1}^{n} \mathbf{E}(\mathbf{X}_k^3).$$

In the case of equal components it was shown that $D_n(x) = o(1/\sqrt{n})$. Now D_n is the sum of various error terms which in the present situation need not be of comparable magnitude. In fact, if the $\mathbf{X}_k$ have fourth moments it can be shown that under mild further conditions

$$(6.3)\qquad |D_n(x)| = O(n^2 s_n^{-6}) + O(n s_n^{-4}).$$

Here either of the two terms can preponderate depending on the behavior of the sequence ns_n^{-2} which may fluctuate between 0 and ∞. In theory it would be possible to find universal bounds for the error,[10] but these would

[9] For details see W. Feller, *On the Berry-Esseen theorem*, Zs. Wahrscheinlichkeitstheorie verw. Gebiete, vol. 10 (1968) pp. 261–268. It is surprising that the unified general method actually simplifies the argument even in the classical case and, moreover, leads without effort to better numerical estimates.

[10] For example, in Cramér's basic theory the bound is of the form

$$D_n(x) = O\left(n^{\frac{1}{2}} s_n^{-6}\left(\sum_{k=1}^{n} E(\mathbf{X}_k^4)\right)^{\frac{3}{2}}\right)$$

which may be worse than (6.3).

be messy and too pessimistic in individual cases arising in practice. It is therefore more prudent to consider only sequences $\{\mathbf{X}_k\}$ with some typical pattern of behavior, but to keep the proofs so flexible as to be applicable in various situations.

For a typical pattern we shall consider sequences such that the ratios s_n^2/n remain bounded between two positive constants.[11] We show that under mild additional restrictions the expansion (4.1) remains valid and its proof requires no change. In other situations the error term may take on different forms. For example, if $s_n^2 = o(n)$ it can be said only that $|D_n(x)| = o(n/s_n^2)$. However, the proof is adaptable to this situation.

The proof of (4.1) depended on taking the Fourier transform of $D_n(x)$. If ω_k denotes the characteristic function of $\mathbf{X}_k$ this transform may be written in the form

$$(6.4) \qquad e^{nv_n(\zeta/s_n)} - e^{-\frac{1}{2}\zeta^2} - \frac{nv_n'''(0)}{6s_n^3}\zeta^3 e^{-\frac{1}{2}\zeta^2}$$

where

$$(6.5) \qquad v_n(\zeta) = n^{-1}\sum_{k=1}^{n}\log\omega_k(\zeta)$$

Now this is exactly the same form as used in the proof of (4.1), except that there $v_n(\zeta) = \log\varphi(\zeta)$ was independent of n. Let us now see how this dependence on n influences the proof. Only two properties of v were used.

(*a*) We used the continuity of the third derivative φ''' to find an interval $|\zeta| < \delta$ within which v_n''' varies by less than ϵ. To assure that this δ can be chosen independent of n we have now to assume some uniformity condition concerning the derivatives ω_k''' near the origin. To avoid uninteresting technical discussions we shall suppose that the moments $\mathbf{E}(\mathbf{X}_k^4)$ exist and remain bounded. Then the ω_k''' have uniformly bounded derivatives and the same is true of v_n'''.

(*b*) The proof of (4.1) depended on the fact that $|\varphi^n(\zeta)| = o(1/\sqrt{n})$ uniformly for all $\zeta > \delta$. The analogue now would be

$$(6.6) \qquad |\omega_1(\zeta)\cdots\omega_n(\zeta)| = o(1/\sqrt{n}) \quad \textit{uniformly in} \quad \zeta > \delta > 0.$$

This condition eliminates the possibility that all $\mathbf{X}_k$ have lattice distributions with the same span, in which case the product in (6.6) would be periodic function of ζ. Otherwise this condition is so mild as to be trivially satisfied in most cases. For example, if the $\mathbf{X}_k$ have densities each factor $|\omega_k|$ remains bounded away from 1, and the left side in (6.6) decreases faster

[11] Then (6.3) gives $|D_n(x)| = O(1/n)$ which is sharper than the bound $o(1/\sqrt{n})$ obtained in (4.1). The improvement is due to the assumption that fourth moments exist.

than any power of $1/n$ unless $|\omega_n(\zeta)| \to 1$; that is, unless the $\mathbf{X}_n$ tend to be concentrated at one point. Thus in general the stronger condition

$$(6.7) \qquad |\omega_1(\zeta) \cdots \omega_n(\zeta)| = o(n^{-a}) \qquad \textit{uniformly in} \quad \zeta > \delta$$

will be satisfied and easily verified for all $a > 0$.

Under our two additional assumptions the proof of (4.1) goes through without change and we have thus

Theorem 1. *Suppose that with some positive constants*

$$(6.8) \qquad cn < s_n^2 < Cn, \qquad \mathbf{E}(\mathbf{X}_n^2) < M$$

for all n and that (6.6) *holds. Then* $|D_n(x)| = o(1/\sqrt{n})$ *uniformly for all* x.

As mentioned before, the proof applies equally to other situations. For example, suppose that

$$(6.9) \qquad s_n^2/n \to 0 \qquad but \qquad s_n^3/n \to \infty.$$

The proof of (4.1) carries through with $T = as_n^3/n$, and since $T = o(s_n)$ the condition (6.6) becomes unnecessary. In this way one arrives at the following variant.

Theorem 1a. *If* (6.9) *holds and the* $\mathbf{E}(\mathbf{X}_k^4)$ *are uniformly bounded then* $|D_n(x)| = o(n/s_n^3)$ *uniformly in* x.

The other theorems of sections 2 and 4 generalize in like manner. For example, the proof of theorem 3 in section 4 leads without essential change to the following general expansion theorem.[12]

Theorem 2. *Suppose that*

$$(6.10) \qquad 0 < c < \mathbf{E}(|\mathbf{X}|^\nu) < C < \infty, \qquad \nu = 1, \ldots, r+1,$$

and that (6.7) *holds with* $a = r + 1$. *Then the asymptotic expansion* (4.15) *holds uniformly in* x.

The polynomials R_j depend on the moments occurring in (6.10) but for fixed x the sequence $\{R_j(x)\}$ is bounded.

7. LARGE DEVIATIONS[13]

We begin again by considering our general problem in the special case of variables with a common distribution F such that $\mathbf{E}(\mathbf{X}_k) = 0$ and

[12] With slightly milder uniformity conditions this theorem is contained in Cramér's pioneer work. Cramér's methods, however, are now obsolete.

[13] This section is entirely independent of the preceding sections in this chapter.

$\mathbf{E}(\mathbf{X}_k^2) = \sigma^2$. As before, F_n stands for the distribution of the normalized sums $(\mathbf{X}_1+\cdots+\mathbf{X}_n)/\sigma\sqrt{n}$. Then F_n tends to the normal distribution $\mathfrak{N}$. This information is valuable for moderate values of x, but for large x both $F_n(x)$ and $\mathfrak{N}(x)$ are close to unity and the statement of the central limit theorem becomes empty. Similarly most of our expansions and approximations become redundant: One needs an estimate of the *relative* error in approximating $1 - F_n$ by $1 - \mathfrak{N}$. Many times we would like to use the relation

$$\frac{1 - F_n(x)}{1 - \mathfrak{N}(x)} \to 1 \tag{7.1}$$

in situations where both x and n tend to infinity. This relation cannot be true generally since for the symmetric binomial distribution the numerator vanishes for all $x > \sqrt{n}$. We shall show, however that (7.1) is true if x varies with n in such a way that $xn^{-\frac{1}{6}} \to 0$ provided that the integral

$$f(\zeta) = \int_{-\infty}^{+\infty} e^{\zeta x}\, F\{dx\} \tag{7.2}$$

exists for all ζ in some interval $|\zeta| < \zeta_0$. [This amounts to saying that the characteristic function $\varphi(\zeta) = f(i\zeta)$ is analytic in a neighborhood of the origin, but it is preferable to deal with the real function f.]

Theorem 1. *If the integral* (7.2) *converges in some interval about the origin, and if* x *varies with* n *in such a way that* $x \to \infty$ *and* $x = o(n^{\frac{1}{6}})$, *then* (7.1) *is true.*

Changing x into $-x$ we obtain the dual theorem for the left tail. The theorem is presumably general enough to cover "all situations of practical interest," but the method of proof will lead to much stronger results.

For the proof we switch from f to its logarithm. In a neighborhood of the origin

$$\psi(\zeta) = \log f(\zeta) = \sum \frac{\psi_k}{k!}\zeta^k \tag{7.3}$$

defines an analytic function. The coefficient ψ_k depends only on the moments $\mu_1, \ldots, \mu_k$ of the distribution F and is called the *semi-invariant of order* k of F. In general $\psi_1 = \mu_1$, $\psi_2 = \sigma^2, \ldots.$ In the present case $\mu_1 = 0$ and therefore $\psi_1 = 0$, $\psi_2 = \sigma^2$, $\psi_3 = \mu_3, \ldots.$

The proof is based on the technique of associated distributions.[14] With the distribution F we associate the new probability distribution V such that

$$V\{dx\} = e^{-\psi(s)}e^{sx}\, F\{dx\}, \tag{7.4}$$

[14] It was employed in renewal theory XI,6 and for random walks XII,4.

where the parameter s is chosen within the interval of convergence of ψ. The function

$$(7.5) \qquad v(\zeta) = \frac{f(\zeta+s)}{f(s)}$$

plays for V the same role as does f for the original distribution F. In particular, it follows by differentiation of (7.5) that V has expectation $\psi'(s)$ and variance $\psi''(s)$.

The idea of the proof can now be explained roughly as follows: It is readily seen from either (7.4) or (7.5) that the distributions $F^{n\star}$ and $V^{n\star}$ again stand in the relationship (7.4) except that the norming constant $e^{-\psi(s)}$ is replaced by $e^{-n\psi(s)}$. Inverting this relation we get

$$(7.6) \qquad 1 - F_n(x) = 1 - F^{n\star}(x\sigma\sqrt{n}) = e^{n\psi(s)} \int_{x\sigma\sqrt{n}}^{\infty} e^{-sy} V^{n\star}\{dy\}.$$

In view of the central limit theorem it seems natural here to replace $V^{n\star}$ by the corresponding normal distribution with expectation $n\psi'(s)$ and variance $n\psi''(s)$. The relative error committed in this approximation will be small if the lower limit of the integral is close to the expectation of $V^{n\star}$, that is, if x is close to $\psi'(s)\sqrt{n}/\sigma$. In this way one can derive good approximations to $1 - F_n(x)$ for certain large values of x, and (7.1) is among them.

Proof. In a neighborhood of the origin ψ is an analytic function with a power series of the form

$$(7.7) \qquad \psi(s) = 1 + \tfrac{1}{2}\sigma^2 s^2 + \tfrac{1}{6}\mu_3 s^3 + \cdots.$$

ψ is a convex function with $\psi'(0) = 0$, and hence increases for $s > 0$. The relation

$$(7.8) \qquad \sqrt{n}\psi'(s) = \sigma x \qquad s > 0, \quad x > 0,$$

therefore establishes a one-to-one correspondence between the variables s and x as long as s and $x/\sqrt{n}$ are restricted to a suitable neighborhood of the origin. Each variable may be considered as an analytic function of the other, and clearly

$$(7.9) \qquad s \sim \frac{x}{\sigma\sqrt{n}} \quad \textit{if} \quad \frac{x}{\sqrt{n}} \to 0.$$

We now proceed in two steps:

(*a*) We begin by calculating the quantity A_s obtained on replacing $V^{n\star}$ in (7.6) by the normal distribution having the same expectation $n\psi'(s)$ and the same variance $n\psi''(s)$. The standard substitution $y = n\psi'(s) + t\sqrt{n\psi''(s)}$ yields

$$(7.10) \qquad A_s = e^{n[\psi(s)-s\psi'(s)]} \frac{1}{\sqrt{2\pi}} \int_0^{\infty} e^{-ts\sqrt{n\psi''(s)}-\frac{1}{2}t^2}\, dt.$$

Completing the square in the exponent we get

$$(7.11)\quad A_s = \exp\left(n[\psi(s) - s\psi'(s) + \tfrac{1}{2}s^2\psi''(s)]\right) \cdot [1 - \mathfrak{N}(s\sqrt{n\psi''(s)})].$$

The exponent and its first two derivatives vanish at the origin and so its power series starts with cubic terms. Thus

$$(7.12)\qquad A_s = [1 - \mathfrak{N}(s\sqrt{n\psi''(s)})] \cdot [1 + O(ns^3)], \qquad s \to 0.$$

If $ns^3 \to 0$ or, what amounts to the same, if $x = o(n^{\frac{1}{6}})$ we may rewrite (7.12) in the form

$$(7.13)\qquad A_s = [1 - \mathfrak{N}(\bar{x})][1 + O(x^3/\sqrt{n}\,)],$$

where we put for abbreviation

$$(7.14)\qquad \bar{x} = s\sqrt{n\psi''(s)}.$$

It remains to show that in (7.13) we may replace $\bar{x}$ by x. The power series for $(\bar{x}-x)/n$ is independent of n and a trite calculation shows that it starts with cubic terms. Accordingly,

$$(7.15)\qquad |\bar{x} - x| = O(\sqrt{n}\,s^3) = O(x^3/n).$$

From **1**; VII,(1.8) we know that as $t \to \infty$

$$(7.16)\qquad \frac{\mathfrak{n}(t)}{1 - \mathfrak{N}(t)} \sim t.$$

Integrating between x and $\bar{x}$ we get for $x \to \infty$

$$(7.17)\qquad \left|\log\frac{1 - \mathfrak{N}(\bar{x})}{1 - \mathfrak{N}(x)}\right| = O(x \cdot |\bar{x}-x|) = O(x^4/n)$$

and hence

$$(7.18)\qquad \frac{1 - \mathfrak{N}(\bar{x})}{1 - \mathfrak{N}(x)} = 1 + O(x^4/n).$$

Substituting into (7.13) we get finally if $x \to \infty$ so that $x = o(n^{\frac{1}{6}})$

$$(7.19)\qquad A_s = [1 - \mathfrak{N}(x)][1 + O(x^3/\sqrt{n}\,)].$$

(*b*) If $\mathfrak{N}_s$ denotes the normal distribution with expectation $n\psi'(s)$ and variance $n\psi''(s)$ then A_s stands for the right side in (7.6) when $V^{n\star}$ is replaced by $\mathfrak{N}_s$. We now proceed to appraise the error committed by this replacement. By the Berry-Esseen theorem (section 5)

$$(7.20)\qquad |V^{n\star}(y) - \mathfrak{N}_s(y)| < 3M_s\sigma^{-3}/\sqrt{n}$$

for all y, where M_s denotes the third absolute moment of the distribution

V. After a simple integration by parts it is therefore seen that

(7.21)

$$|1-F_n(x)-A_s| < \frac{3M_s}{\sigma^3\sqrt{n}}\, e^{n\psi(s)}\left[e^{-s\psi'(s)} + s\int_{n\psi'(s)}^{\infty} e^{-sy}\,dy\right] = \frac{6M_s}{\sigma^3\sqrt{n}}\, e^{n[\psi(s)-s\psi'(s)]}.$$

But by (7.11)

$$A_s = e^{n[\psi(s)-s\psi'(s)]}\cdot e^{\frac{1}{2}\bar{x}^2}[1-\mathfrak{N}(\bar{x})] \sim \frac{1}{x}\, e^{n[\psi(s)-s\psi'(s)]} \tag{7.22}$$

and hence the right side in (7.21) is $A_s \cdot O(x/\sqrt{n})$. Thus

$$1 - F_n(x) = A_s[1 + O(x/\sqrt{n})]. \tag{7.23}$$

In combination with (7.19) this not only proves the theorem, but also the stronger

Corollary. If $x \to \infty$ so that $x = o(n^{\frac{1}{6}})$ then

$$\frac{1 - F_n(x)}{1 - \mathfrak{N}(x)} = 1 + O\left(\frac{x^3}{\sqrt{n}}\right). \tag{7.24}$$

We have indirectly derived a much further-going result applicable whenever x varies with n in such a way that $x \to \infty$ but $x = 0(\sqrt{n})$. Indeed, by (7.23) we have then $1 - F_n(x) \sim A_s$ with A_s given by (7.11). Here the argument of $\mathfrak{N}$ is $\bar{x}$, but (7.18) shows that $\bar{x}$ may be replaced by x. We get therefore the general approximation formula

$$1 - F_n(x) = \exp\,(n[\psi(s)-s\psi'(s)+\tfrac{1}{2}\psi'^2(s)])[1-\mathfrak{N}(x)]\cdot[1+O(x/\sqrt{n})]. \tag{7.25}$$

The exponent is a power series in s commencing with the term of third order. As in (7.8) we now define an analytic function s of the variable z by $\psi'(s) = \sigma z$. With this function we define a *power series* λ *such that*

$$z^2\,\lambda(z) = \lambda_1 z^3 + \lambda_2 z^4 + \cdots = \psi(x) - s\psi'(s) + \tfrac{1}{2}\psi'^2(s). \tag{7.26}$$

In terms of this series we have

Theorem 2.[15] *If in theorem* 1 *the condition* $x = o(n^{\frac{1}{6}})$ *is replaced by* $x = o(\sqrt{n})$, *then*

$$\frac{1 - F_n(x)}{1 - \mathfrak{N}(x)} = \exp\left(x^2\lambda\left(\frac{x}{\sqrt{n}}\right)\right)\left[1 + O\left(\frac{x}{\sqrt{n}}\right)\right]. \tag{7.27}$$

---

[15] The use of the transformation (7.4) in connection with the central limit theorem seems due to F. Esscher (1932). The present theorem is due to H. Cramér (1938), and was generalized to variable components by Feller (1943). For newer results in this case see V. V. Petrov, Uspekhi Matem. Nauk, vol. 9 (1954) (in Russian), and W. Richter, *Local limit theorems for large deviations*, Theory of Probability and Its Applications (transl.), vol. 2 (1957) pp. 206–220. The latter author treats densities rather than distributions. For a different approach leading to approximations of the form $1 - F_n(x) = \exp\,[v(x)+o(v(x))]$ see W. Feller, Zs. Wahrscheinlichkeitstheorie verw. Gebiete vol. 14 (1969), pp. 1–20.

In particular, if $x = o(n^{\frac{1}{6}})$, only the first term in the power series matters, and we get

$$\frac{1 - F_n(x)}{1 - \mathfrak{N}(x)} \sim \exp\left(\frac{\lambda_1 x^3}{\sqrt{n}}\right), \qquad \lambda_1 = \frac{\mu_3}{6\sigma^3}. \tag{7.28}$$

For an increase such that $x = o(n^{\frac{3}{10}})$ we get

$$\frac{1 - F_n(x)}{1 - \mathfrak{N}(x)} \sim \exp\left(\lambda_1 \frac{x^3}{\sqrt{n}} + \lambda_2 \frac{x^4}{n}\right), \qquad \lambda_2 = \frac{\sigma^2\psi_4 - 3\psi_3^2}{24\sigma^6}, \tag{7.29}$$

and so on. Note that the right sides may tend to 0 or ∞, and hence these formulas do *not* imply an asymptotic equivalence of $1 - F_n(x)$ and $1 - \mathfrak{N}(x)$. Such an equivalence exists only if $x = o(n^{\frac{1}{6}})$ [or, in the case of a vanishing third moment, if $x = o(n^{\frac{1}{4}})$]. Under any circumstances we have the following interesting

Corollary. *If* $x = o(\sqrt{n})$ *then for any* $\epsilon > 0$ *ultimately*

$$\exp(-(1+\epsilon)x^2/2) < 1 - F_n(x) < \exp(-(1-\epsilon)x^2/2). \tag{7.30}$$

The preceding theory may be generalized to cover partial sums of random variables $\mathbf{X}_k$ with varying distributions and characteristic functions ω_k. The procedure may be illustrated by the following generalization of theorem 1 in which the uniformity conditions are unnecessarily severe. In it F_n stands again for the distribution of the normalized sum $(\mathbf{X}_1 + \cdots + \mathbf{X}_n)/s_n$.

Theorem 3. *Suppose that there exists an interval* $\overline{-a, a}$ *in which all the characteristic functions* ω_k *are analytic, and that*

$$\mathbf{E}(|\mathbf{X}_n|^3) \leq M\sigma_n^2, \tag{7.31}$$

where M *is independent of* n. *If* s_n *and* x *tend to* ∞ *so that* $x = o(s_n^{\frac{1}{3}})$, *then*

$$\frac{1 - F_n(x)}{1 - \mathfrak{N}(x)} \to 1 \tag{7.32}$$

with an error $O(x^3/s_n)$.

The proof is the same except that ψ is now replaced by the real-valued analytic function ψ_n defined for $-a < s < a$ by

$$\psi_n(s) = \frac{1}{n}\sum_{k=1}^{n} \log \omega_k(-is). \tag{7.33}$$

In the formal calculations now xs_n replaces $x\sigma\sqrt{n}$. The basic equation (7.8) takes on the form $\psi_n'(s) = xs_n/n$.

CHAPTER XVII

Infinitely Divisible Distributions

This chapter presents the core of the now classical limit theorems of probability theory—the reservoir created by the contributions of innumerably many individual streams and developments. The most economical treatment of the subject would start from the theory of triangular arrays developed in section 7, but once more we begin by a discussion of simple special cases in order to facilitate access to various important topics.

The notions of infinite divisibility, stability, etc., and their intuitive meaning were discussed in chapter VI. The main results of the present chapter were derived in a different form, and by different methods, in chapter IX, but the present chapter provides more detailed information. It is self-contained, and may be studied as a sequel to chapter XV (characteristic functions) independently of the preceding chapters.

1. INFINITELY DIVISIBLE DISTRIBUTIONS

We continue the practice of using descriptive terms interchangeably for distributions and their characteristic functions. With this understanding the definition of infinite divisibility given in VI,3 may be rephrased as follows.

Definition. *A characteristic function* ω *is infinitely divisible iff for each* n *there exists a characteristic function* ω_n *such that*

$$\omega_n^n = \omega. \tag{1.1}$$

We shall presently see that infinite divisibility can be characterized by other striking properties which explain why the notion plays an important role in probability theory.

Note *concerning roots and logarithms of characteristic functions.* It is tempting to refer to ω_n in (1.1) as the nth root of ω, but to make this meaningful we have to show that this root is essentially unique. To discuss the indeterminacy of roots and logarithms in the complex domain it is convenient to start from the polar representation $a = re^{i\theta}$ of the complex number $a \neq 0$. The positive number r is uniquely determined, but the argument θ is determined only up to multiples of 2π. In principle this indeterminancy is inherited by $\log a = \log r + i\theta$ and by $a^{1/n} = r^{1/n}e^{i\theta/n}$ (here $r^{1/n}$ stands for the positive root, and $\log r$ for the familiar real logarithm). Nevertheless, in any interval $|\zeta| < \zeta_0$ in which $\omega(\zeta) \neq 0$ the characteristic function ω admits of a *unique* polar representation $\omega(\zeta) = r(\zeta)e^{i\theta(\zeta)}$ such that θ is continuous and $\theta(0) = 0$. In such an interval we can write without ambiguity $\log \omega(\zeta) = \log r(\zeta) + i\theta(\zeta)$ and $\omega^{1/n}(\zeta) = r^{1/n}(\zeta)e^{i\theta(\zeta)/n}$; these determinations are the only ones that render $\log \omega$ and $\omega^{1/n}$ continuous functions that are real at the point $\zeta = 0$. In this sense $\log \omega$ *and* $\omega^{1/n}$ *are uniquely determined in any interval* $|\zeta| < \zeta_0$ *free of zeros of* ω. We shall use the symbols $\log \omega$ and $\omega^{1/n}$ only in this sense, but it must be borne in mind that this definition breaks down[1] as soon as $\omega(\zeta_0) = 0$. ►

Let F be an arbitrary probability distribution and φ its characteristic function. We recall from XV,(2.4) that F generates the family of *compound Poisson distributions*

$$e^{-c}\sum_{k=0}^{\infty} \frac{c^k}{k!} F^{k\star} \tag{1.2}$$

with characteristic functions $e^{c(\varphi-1)}$. Here $c > 0$ is arbitrary. Obviously $\omega = e^{c(\varphi-1)}$ is infinitely divisible (the root $\omega^{1/n}$ being of the same form with c replaced by c/n). The normal and the Cauchy distribution show that an infinitely divisible distribution need not be of the compound Poisson type, but we shall now show that every infinitely divisible distribution is the limit of a sequence of compound Poisson distributions. Basic for the whole theory is

Theorem 1. *Let* $\{\varphi_n\}$ *be a sequence of characteristic functions. In order that there exist a continuous limit*

$$\omega(\zeta) = \lim \varphi_n^n(\zeta) \tag{1.3}$$

it is necessary and sufficient that

$$n[\varphi_n(\zeta) - 1] \to \psi(\zeta) \tag{1.4}$$

with ψ *continuous. In this case*

$$\omega = e^{-\psi}. \tag{1.5}$$

---

[1] At the end of XV,2 (as well as in problem 9 of XV,9) we found pairs of real characteristic functions such that $\varphi_2^1 = \varphi_2^2$. This shows that in the presence of zeros even real characteristic functions may possess two real roots that are again characteristic functions.

Proof. We recall from the continuity theorem in XV,3 that if a sequence of characteristic functions converges to a continuous function, the latter represents a characteristic function and the convergence is automatically uniform in every finite interval.

(*a*) We begin with the easy part of the theorem. Assume (1.4) where ψ is continuous. This implies that $\varphi_n(\zeta) \to 1$ for every ζ, and the convergence is automatically uniform in finite intervals. This means that in any interval $|\zeta| < \zeta_1$ we have $|1 - \varphi_n(\zeta)| < 1$ for all n sufficiently large. For such n we conclude then from the Taylor expansion for $\log(1-z)$ that

$$(1.6)\qquad n \log \varphi_n(\zeta) = n \log [1 - [1-\varphi_n(\zeta)] =$$

$$= -n[1 - \varphi_n(\zeta)] - \frac{n}{2}[1 - \varphi_n(\zeta)]^2 - \cdots.$$

Because of (1.4) the first term on the right tends to $-\psi(\zeta)$, and since $\varphi_n(s) \to 1$ this implies that all other terms tend to zero. Thus $n \log \varphi_n \to -\psi$ or $\varphi_n^n \to e^{-\psi}$, as asserted.

(*b*) The converse is equally simple if it is known that the limit ω in (1.3) has no zeros. Indeed, consider an arbitrary finite interval $|\zeta| \leq \zeta_1$. In it the convergence in (1.3) is uniform and the absence of zeros of ω implies that also $\varphi_n(\zeta) \neq 0$ for $|\zeta| \leq \zeta_1$ and all n sufficiently large. We can therefore pass to logarithms and conclude that $n \log \varphi_n \to \log \omega$, and hence $\log \varphi_n \to 0$. This implies that $\varphi_n(\zeta) \to 1$ for each fixed ζ, and the convergence is automatically uniform in every finite interval. As under (*a*) therefore we conclude that the expansion (1.6) is valid, and since

$$1 - \varphi_n(\zeta) \to 0$$

this implies that

$$(1.7)\qquad n \log \varphi_n(\zeta) = -n[1-\varphi_n(\zeta)](1+O(1))$$

where $O(1)$ stands for a quantity that tends to 0 as $n \to \infty$. By assumption the left side tends to $\log \omega(\zeta)$, and hence obviously $n[1-\varphi_n] \to -\log \omega$ as asserted.

To validate this argument we have to show that $\omega(\zeta)$ cannot vanish for any ζ. For that purpose we can replace ω and φ_n by the characteristic functions $|\omega|^2$ and $|\varphi_n|^2$, respectively, and hence it suffices to consider the special case of (1.3) where all φ_n are real and $\varphi_n \geq 0$. Let then $|\zeta| \leq \zeta_1$ be an interval in which $\omega(\zeta) > 0$. Within this interval $-n \log \varphi_n(\zeta)$ is positive and remains bounded. On the other hand, for $|\zeta| \leq \zeta_1$, the expansion (1.6) is valid, and since all terms are of the same sign it follows that $n[1 - \varphi_n(\zeta)]$ remains bounded for all $|\zeta| \leq \zeta_1$. But by the basic inequality XV,(1.7) for characteristic functions

$$n[1-\varphi_n(2\zeta)] \leq 4n[1-\varphi_n(\zeta)],$$

and so $n[1-\varphi_n(\zeta)]$ remains bounded for all $|\zeta| \leq 2\zeta_1$. It follows that this interval can contain no zero of ω. But then the initial argument applies to this interval and leads to the conclusion that $\omega(\zeta) > 0$ for all $|\zeta| \leq 4\zeta_1$. Continued doubling shows that $\omega(\zeta) > 0$ for all ζ, and this concludes the proof. ▶

Theorem 1 has many consequences. On multiplying (1.4) by $t > 0$ it is seen that this relation is equivalent to

$$e^{tn[\varphi_n(\zeta)-1]} \to e^{t\psi(\zeta)} = \omega^t(\zeta). \tag{1.8}$$

The left side represents a characteristic function of the compound Poisson type, and therefore $e^{t\psi(\zeta)}$ is a characteristic function for every $t > 0$. We conclude in particular that $\omega = e^{\psi}$ is necessarily infinitely divisible. In other words, every characteristic function ω appearing as the limit of a sequence $\{\varphi_n^n\}$ of characteristic functions is infinitely divisible. This may be regarded as a widening of the definition of infinite divisibility in that it replaces the identity (1.1) by the more general limit relation (1.3). It will be seen in section 7 that this result may be further extended to more general triangular arrays, but we record our preliminary result in the form of

Theorem 2. *A characteristic function ω is infinitely divisible iff there exists a sequence $\{\varphi_n\}$ of characteristic functions such that $\varphi_n^n \to \omega$.*

In this case ω^t is a characteristic function for every $t > 0$, and $\omega(\zeta) \neq 0$ for all ζ.

Corollary. *A continuous limit of a sequence $\{\omega_n\}$ of infinitely divisible characteristic functions is itself infinitely divisible.*

Proof. By assumption $\varphi_n = \omega_n^{1/n}$ is again a characteristic function, and so the relation $\omega_n \to \omega$ may be rewritten in the form $\varphi_n^n \to \omega$. ▶

Every compound Poisson distribution is infinitely divisible, and theorem 1 tells us that every infinitely divisible distribution can be represented as a limit of a sequence of compound Poisson distributions [see (1.8) with $t = 1$]. In this way we get a new characterization of infinite divisibility.

Theorem 3. *The class of infinitely divisible distributions coincides with the class of limit distributions of compound Poisson distributions.*

Application to processes with independent increments. As explained in VI,3 such processes can be described by a family $\{\mathbf{X}(t)\}$ of random variables with the property that for any partition $t_0 < t_1 < \cdots < t_n$ the increments $\mathbf{X}(t_k) - \mathbf{X}(t_{k-1})$ represent n mutually independent variables. The increments are *stationary* if the distribution of $\mathbf{X}(s+t) - \mathbf{X}(s)$ depends only on the length t of the interval, but not on its position on the time axis. In this case $\mathbf{X}(s+t) - \mathbf{X}(s)$ is the sum of n independent variables distributed as $\mathbf{X}(s + t/n) - \mathbf{X}(s)$, and hence *the distribution of* $\mathbf{X}(s + t) - \mathbf{X}(s)$ *is infinitely divisible.*

Conversely, every family of infinitely divisible distributions with characteristic functions of the form $e^{t\psi}$ can regulate a process with independent stationary increments. The results concerning triangular arrays in section 7 will generalize this result to processes with *non-stationary* independent increments. The increment $\mathbf{X}(t+s) - \mathbf{X}(s)$ is then the sum of increments $\mathbf{X}(t_{k+1}) - \mathbf{X}(t_k)$ and these are mutually independent random variables. The theorem of section 7 then applies provided the process is continuous in the sense that $\mathbf{X}(t+h) - \mathbf{X}(t)$ tends in probability to zero as $h \to 0$. For such processes *the distribution of the increments* $\mathbf{X}(t+s) - \mathbf{X}(t)$ *are infinitely divisible.* (Discontinuous processes of this type exist, but the discontinuities are of a trite nature and, in a certain sense, removable. See the discussion in IX,5*a* and IX,9.)

Compound Poisson processes admit of a particularly simple probabilistic interpretation (see VI,3 and IX,5) and the fact that every infinitely divisible distribution appears as limit of compound Poisson distributions helps to understand the nature of the more general processes with independent increments.

2. CANONICAL FORMS. THE MAIN LIMIT THEOREM

We saw that to find the most general form of infinitely divisible characteristic functions $\omega = e^{\psi}$ it suffices to determine the general form of possible limits of sequences of characteristic functions $\exp c_n(\zeta_n - 1)$ of the compound Poisson type. For various applications it is desirable to state the problem more generally by permitting arbitrary centerings, and hence we seek the possible limits of characteristic functions of the form $\omega_n = e^{\psi_n}$, where we put for abbreviation

$$\psi_n(\zeta) = c_n[\varphi_n(\zeta) - 1 - i\beta_n\zeta]. \tag{2.1}$$

The ω_n are infinitely divisible, and the same is therefore true of their continuous limits.

Our problem is to find conditions under which there exists a *continuous* limit

$$\psi(\zeta) = \lim \psi_n(\zeta). \tag{2.2}$$

It is understood that φ_n is the characteristic function of a probability distribution F_n, the c_n are positive constants, and the centering constants β_n are real.

For distributions with expectations the natural centering is to zero expectation, and whenever possible we shall choose β_n accordingly. However, we need a universally applicable centering with similar properties. As it turns out, the simplest such centering is obtained by the requirement that for $\zeta = 1$ the value of ψ_n be real. If u_n and v_n stand for the real and imaginary part of φ_n our condition requires that

$$\beta_n = v_n(1) = \int_{-\infty}^{+\infty} \sin x \, F_n\{dx\}. \tag{2.3}$$

This shows that our centering is always possible. With it

$$\psi_n(\zeta) = c_n \int_{-\infty}^{+\infty} [e^{i\zeta x} - 1 - i\zeta \sin x]\, F_n\{dx\}. \tag{2.4}$$

Near the origin the integrand behaves like $-\frac{1}{2}\zeta^2 x^2$, just as is the case with the more familiar centering to zero expectation. The usefulness of the centering (2.3) is due largely to the following

Lemma. *Let* $\{c_n\}$ *and* $\{\varphi_n\}$ *be given. If there exist centering constants* β_n *such that* ψ_n *tends to a continuous limit* ψ, *then* (2.3) *will achieve the same goal.*

Proof. Define ψ_n by (2.1) with arbitrary β_n, and suppose that $\psi_n \to \psi$. If b denotes the imaginary part of $\psi(1)$ we conclude for $\zeta = 1$ that

$$c_n(v_n(1) - \beta_n) \to b. \tag{2.5}$$

Multiplying by $i\zeta$ and subtracting from $\psi_n \to \psi$ we see that

$$c_n[\varphi_n(\zeta) - 1 - iv_n(1)\zeta] \to \psi(\zeta) - ib\zeta \tag{2.6}$$

and this proves the assertion. ►

We begin by treating our convergence problem in a special case in which the solution is particularly simple. Suppose that the functions ψ_n and ψ are twice continuously differentiable (which means that the corresponding distributions have variances; see XV,4). Suppose that not only $\psi_n \to \psi$, but also $\psi_n'' \to \psi''$. In view of (2.1) this means that

$$c_n \int_{-\infty}^{+\infty} e^{i\zeta x} x^2\, F_n\{dx\} \to -\psi''(\zeta). \tag{2.7}$$

By assumption $c_n x^2\, F_n\{dx\}$ defines a finite measure, and we denote its total mass by μ_n. For $\zeta = 0$ we see from (2.7) that $\mu_n \to -\psi''(0)$. On dividing (2.7) by μ_n we get on the left the characteristic function of a proper probability distribution, and as $n \to \infty$ it tends to $\psi''(\zeta)/\psi''(0)$. It follows that $\psi''(\zeta)/\psi''(0)$ is the characteristic function of a probability distribution, and hence

$$-\psi''(\zeta) = \int_{-\infty}^{+\infty} e^{i\zeta x}\, M\{dx\} \tag{2.8}$$

where M is a finite measure. From this we obtain ψ by repeated integration. Bearing in mind that $\psi(0) = 0$ and that with our centering condition $\psi(1)$ must be real, we get

$$\psi(\zeta) = \int_{-\infty}^{+\infty} \frac{e^{i\zeta x} - 1 - i\zeta \sin x}{x^2}\, M\{dx\}. \tag{2.9}$$

This integral makes sense, the integrand being a bounded continuous function assuming at the origin the value $-\frac{1}{2}\zeta^2$.

Under our differentiability conditions the limit ψ is necessarily of the form (2.9). We show next that with an arbitrarily chosen finite measure M the integral (2.9) defines an infinitely divisible characteristic function e^{ψ}. However, we can go a step further. For the integral to make sense it is not necessary that the measure M be finite. It suffices that M attributes finite masses to finite intervals and that $M\overline{\{-x, x\}}$ increases sufficiently slowly for the integrals

$$M^+(x) = \int_x^{\infty} y^{-2}\, M\{dy\}, \qquad M^-(-x) = \int_{-\infty}^{-x} y^{-2}\, M\{dy\} \tag{2.10}$$

to converge for all $x > 0$. (For definiteness we take the intervals of integration closed.) Measures defined by the densities $|x|^p\, dx$ with $0 < p < 1$ are typical examples. We show that if M has these properties (2.9) defines an infinitely divisible characteristic function, and that *all* such characteristic functions are obtained in this manner. For this reason it is convenient to introduce a special term for our measures.

Definition 1. *A measure M will be called canonical if it attributes finite masses to finite intervals and the integrals* (2.10) *converge for some (and therefore all) $x > 0$.*

Lemma 2. *If M is a canonical measure and ψ defined by* (2.9) *then e^{ψ} is an infinitely divisible characteristic function.*

Proof. We consider two important special cases.

(*a*) Suppose that M is concentrated at the origin and attributes mass $m > 0$ to it. Then $\psi(\zeta) = -m\zeta^2/2$, and so e^{ψ} is a normal characteristic function with variance m^{-1}.

(*b*) Suppose that M is concentrated on $|x| > \eta$ where $\eta > 0$. In this case (2.9) may be rewritten in a simpler form. Indeed, $x^{-2}\, M\{dx\}$ now defines a finite measure with total mass $\mu = M^+(\eta) + M^-(-\eta)$. Accordingly, $x^{-2}\, M\{dx\}/\mu = F\{dx\}$ defines a probability measure with characteristic function φ, and obviously $\psi(\zeta) = \mu[\varphi(\zeta) - 1 - ib\zeta]$, where b is a real constant. Thus in this case e^{ψ} is the characteristic function of the compound Poisson type, and hence infinitely divisible.

(*c*) In the general case, let $m \geq 0$ be the mass attributed by M to the origin, and put

$$\psi_\eta(\zeta) = \int_{|x|>\eta} \frac{e^{i\zeta x} - 1 - i\zeta \sin x}{x^2}\, M\{dx\}. \tag{2.11}$$

Then

$$\psi(\zeta) = -\frac{m}{2}\zeta^2 + \lim_{\eta \to 0} \psi_\eta(\zeta) \tag{2.12}$$

We saw that $e^{\psi_\eta(\zeta)}$ is the characteristic function of an infinitely divisible distribution U_η. If $m > 0$ the addition of $-m\zeta^2/2$ to $\psi_\eta(\zeta)$ corresponds to a convolution of U_η with a normal distribution. Thus (2.12) represents e^ψ as the limit of a sequence of infinitely divisible characteristic functions, and hence e^ψ is infinitely divisible as asserted. ►

We show next that the representation (2.9) is unique in the sense that distinct canonical measures give rise to distinct integrals.

Lemma 3. *The representation* (2.9) *of* ψ *is unique.*

Proof. In the special case of a finite measure M it is clear that the second derivative ψ'' exists and that $-\psi''(\zeta)$ coincides with the expectation of $e^{i\zeta x}$ with respect to M. The uniqueness theorem for characteristic functions guarantees that M is uniquely determined by ψ'', and hence by ψ.

This argument can be adapted to unbounded canonical measures, but it is necessary to replace the second derivative by an operation with a similar effect and applicable to arbitrary continuous functions. Such operations can be chosen in various ways (see problems 1–3). We choose the operation that transforms ψ into the function $\psi^\star$ defined by

$$\psi^\star(\zeta) = \psi(\zeta) - \frac{1}{2h}\int_{-h}^{h} \psi(\zeta + s)\, ds, \tag{2.13}$$

where $h > 0$ is arbitrary, but fixed. For the function ψ defined by (2.9) we get

$$\psi^\star(\zeta) = \int_{-\infty}^{+\infty} e^{i\zeta x} \cdot K(x)\, M\{dx\} \tag{2.14}$$

where we put for abbreviation

$$K(x) = x^{-2}\left[1 - \frac{\sin xh}{xh}\right]. \tag{2.15}$$

This is a strictly positive continuous function assuming at the origin the value $h^2/6$ and as $x \to \pm\infty$ we have $K(x) \sim x^{-2}$. The measure $M^\star$ defined by $M^\star\{dx\} = K(x)\, M\{dx\}$ is therefore finite, and (2.14) states that $\psi^\star$ is its Fourier transform. By the uniqueness theorem for characteristic functions the knowledge of $\psi^\star$ uniquely determines the measure $M^\star$. But then $M\{dx\} = K^{-1}(x)\, M^\star\{dx\}$ is uniquely determined, and so the knowledge of ψ enables us to calculate the corresponding canonical measure (cf. problem 3). ►

Our next goal should be to prove that lemma 2 describes the totality of all infinitely divisible characteristic functions, but to do this we must first solve the convergence problem described at the beginning of this section. We put it now in the following slightly more general form: Let $\{M_n\}$ be a sequence of canonical measures and

$$\psi_n(\zeta) = \int_{-\infty}^{+\infty} \frac{e^{i\zeta x} - 1 - i\zeta \sin x}{x^2} M_n\{dx\} + ib_n\zeta \tag{2.16}$$

where b_n is real. We ask for the necessary and sufficient conditions that $\psi_n \to \psi$ where ψ is a continuous function. Note that functions ψ_n defined by (2.1) or (2.4) are a special case of (2.16) with

$$M_n\{dx\} = c_n x^2 F_n\{dx\} \tag{2.17}$$

Assume $\psi_n \to \psi$, the limit ψ being continuous. The transforms defined by (2.13) then satisfy $\psi_n^\star \to \psi^\star$, that is

$$\int_{-\infty}^{+\infty} e^{i\zeta x} K(x)\, M_n\{dx\} \to \psi^\star(\zeta), \tag{2.18}$$

where K is the strictly positive continuous function defined by (2.15). On the left we recognize the Fourier transform of a finite measure with total mass

$$\mu_n = \int_{-\infty}^{+\infty} K(x)\, M_n\{dx\}. \tag{2.19}$$

Clearly $\mu_n \to \psi^\star(0)$. It is easily seen that $\mu_n \to 0$ would imply $\psi(\zeta) = 0$ for all ζ, and hence we may suppose $\psi^\star(0) = \mu > 0$. Then the measures $M_n^\star$ defined by

$$M_n^\star\{dx\} = \frac{1}{\mu_n} K(x)\, M_n\{dx\} \tag{2.20}$$

are probability measures, and (2.18) states that their characteristic functions tend to the continuous function $\psi^\star(\zeta)/\psi_\star(0)$. It follows that

$$M_n^\star \to M^\star \tag{2.21}$$

where $M^\star$ is the probability distribution with characteristic function $\psi^\star(\zeta)/\psi_\star(0)$. But ψ_n may be written in the form

$$\psi_n(\zeta) = \mu_n \int_{-\infty}^{+\infty} \frac{e^{i\zeta x} - 1 - i\zeta \sin x}{x^2} K^{-1}(x)\, M_n^\star\{dx\} + ib_n\zeta. \tag{2.22}$$

The integrand is a bounded continuous function of x, and so (2.21) implies that the integral converges. It follows that $b_n \to b$ and our limit ψ is of the form

$$\psi(\zeta) = \mu \int_{-\infty}^{+\infty} \frac{e^{i\zeta x} - 1 - i\zeta \sin x}{x^2} K^{-1}(x)\, M^\star\{dx\} + ib\zeta. \tag{2.23}$$

Since $M^\star$ is a probability measure it is clear that the measure M defined by

$$M\{dx\} = \mu K^{-1}(x)\, M^\star\{dx\} \tag{2.24}$$

is canonical, and

$$\psi(\zeta) = \int_{-\infty}^{+\infty} \frac{e^{i\zeta x} - 1 - i\zeta \sin x}{x^2} M\{dx\} + ib\zeta. \tag{2.25}$$

This shows that, except for the irrelevant centering term $ib\zeta$, all our limits are of the form described by lemma 2. As already remarked, functions ψ_n defined by (2.1) are a special case of (2.16), and hence we have solved the convergence problem formulated at the beginning of this section. We state the result as

Theorem 1. *The class of infinitely divisible characteristic functions is identical with the class of functions of the form* e^{ψ} *with* ψ *defined by* (2.25) *in terms of a canonical measure* M *and a real number* b.

In other words, except for the arbitrary centering there is a one-to-one correspondence between canonical measures and infinitely divisible distributions.

In the preceding section we have emphasized that the conditions $M_n^\star \to M^\star$ and $b_n \to b$ are necessary for the relation $\psi_n \to \psi$. Actually we have also shown the sufficiency of these conditions, since ψ_n could be written in the form (2.22) which makes it obvious that ψ_n tends to the limit defined by (2.23). We have thus found a useful limit theorem, but it is desirable to express the condition $M_n^\star \to M^\star$ in terms of the canonical measures M_n and M. The relationship between M_n and $M_n^\star$ is defined (2.22). In finite intervals K remains bounded away from 0 and ∞, and so for every finite I the relations $M_n^\star\{I\} \to M^\star\{I\}$ and $M_n\{I\} \to M\{I\}$ imply each other. As $x \to \infty$ the behavior of K is nearly the same as that of x^{-2}, and hence $M_n^\star\{\overline{x, \infty}\} \sim M_n^+(x)$ where M_n^+ stands for the integral occurring in the definition (2.10) of canonical measures. Thus proper convergence $M_n^\star \to M^\star$ is fully equivalent to the conditions

$$M_n\{I\} \to M\{I\} \tag{2.26}$$

for all finite intervals of continuity for M, and

$$M_n^+(x) \to M^+(x), \qquad M_n^-(x) \to M^-(x) \tag{2.27}$$

at all points $x > 0$ of continuity. In the special case of canonical measures of the form $M_n\{dx\} = c_n x^2 F_n\{dx\}$ (with F_n a probability distribution) these relations take on the form

$$c_n \int_I x^2 F_n\{dx\} \to M\{I\} \tag{2.28}$$

and

$$c_n[1 - F_n(x)] \to M^+(x), \qquad c_n F_n(-x) \to M^-(-x). \tag{2.29}$$

If $I = \overline{a, b}$ is a finite interval of continuity not containing the origin, then

(2.28) obviously implies $c_n F_n\{I\} \to M^+(a) - M^+(b)$ and so (2.29) may be taken as an extension of (2.28) to semi-infinite intervals. An equivalent condition is that no masses flow out to infinity in the sense that to each $\epsilon > 0$ there corresponds a τ such that

$$c_n[1 - F_n(\tau) + F_n(-\tau)] < \epsilon \tag{2.30}$$

at least for all n sufficiently large. In the presence of (2.28) the conditions (2.29) and (2.30) imply each other. [Note that the left side of (2.30) is a decreasing function of τ.] Sequences of canonical measures $M_n\{dx\} = c_n x^2 F_n\{dx\}$ will occur so frequently that it is desirable to introduce a convenient term for reference.

Definition 2. *A sequence* $\{M_n\}$ *of canonical measures is said to converge properly to the canonical measure* M *if the conditions* (2.26) *and* (2.27) *are satisfied. We write* $M_n \to M$ *iff this is the case.*

With this terminology we can restate our finding concerning the convergence $\psi_n \to \psi$ as follows.

Theorem 2. *Let* M_n *be a canonical measure and* ψ_n *be defined by* (2.16). *In order that* ψ_n *tends to a continuous limit* ψ *it is necessary and sufficient that there exist a canonical measure* M *such that* $M_n \to M$, *and that* $b_n \to b$. *In this case* ψ *is given by* (2.25).

In the following we shall use this theorem only in the special case

$$\psi_n(\zeta) = c_n[\varphi_n(\zeta) - 1 - ib_n\zeta] \tag{2.31}$$

where φ_n is the characteristic function of a probability distribution F_n. Our conditions then take on the form

$$c_n x^2 F_n\{dx\} \to M\{dx\}, \qquad c_n(\beta_n - b_n) \to b \tag{2.32}$$

where we put again

$$\beta_n = \int_{-\infty}^{+\infty} \sin x \cdot F_n\{dx\}. \tag{2.33}$$

By virtue of theorem 1 of section 1 our conditions apply not only to sequences of compound Poisson distributions, but also to more general sequences of the form $\{\varphi_n^n\}$.

Note on other canonical representations. The measure M is not the one encountered in the literature. In his pioneer work P. Lévy used the measure Λ defined outside the origin by $\Lambda\{dx\} = x^{-2}M\{dx\}$ and which represents the limit of $nF_n\{dx\}$. It is finite on intervals $|x| > \delta > 0$ but unbounded near the origin. It does not take into account the atom of M at the origin, if any. In terms of this measure (2.9) takes on the form

$$\psi(\zeta) = -\tfrac{1}{2}\sigma^2\zeta^2 + ib\zeta + \lim_{\delta \to 0} \int_{|x|>\delta} [e^{i\zeta x} - 1 + i\zeta \sin x]\, \Lambda\{dx\}. \tag{2.34}$$

This is (except for a different choice of the centering function) P. Lévy's original canonical representation. Its main drawback is that it requires many words for a full description of all required properties of the measure Λ.

Khintchine introduced the bounded measure K defined by $K\{dx\} = (1+x^2)^{-1} M\{dx\}$ This bounded measure may be chosen arbitrarily, and Khintchine's canonical representation is given by

$$x(\zeta) = ib\zeta + \int_{-\infty}^{+\infty} \left[e^{i\zeta x} - 1 - \frac{i\zeta x}{1+x^2} \right] \frac{1+x^2}{x^2} K\{dx\}. \tag{2.35}$$

It is easiest to describe since it avoids unbounded measures. This advantage is counterbalanced by the fact that the artificial nature of the measure K complicates many arguments unnecessarily. Stable distributions and the example 3(f) illustrate this difficulty.

2a. DERIVATIVES OF CHARACTERISTIC FUNCTIONS

Let F be a probability distribution with characteristic function φ. It was shown in XV,4 that if F has an expectation μ then φ has a derivative φ' with $\varphi'(0) = i\mu$. The converse is false. The differentiability of φ is closely connected with the law of large numbers for a sequence $\{\mathbf{X}_n\}$ of independent random variables with the common distribution F, and hence many studies were concerned with conditions on F that will ensure the existence of φ'. This problem was solved by E. J. G. Pitman in 1956 following a partial answer by A. Zygmund (1947) who had still to impose smoothness conditions on φ. In view of the formidable difficulties of a direct attack on the problem it is interesting to see that its solution follows as a simple corollary of the last theorem.

Theorem. *Each of the following three conditions implies the other two.*

(i) $\varphi'(0) = i\mu$.

(ii) As $t \to \infty$.

$$t[1 - F(t) + F(t)] \to 0, \qquad \int_{-t}^{t} x\, F\{dx\} \to \mu. \tag{2.36}$$

(iii) *The average* $(\mathbf{X}_1 + \cdots + \mathbf{X}_n)/n$ *tends in probability to* μ.

Proof. The real part of φ being even, the derivative $\varphi'(0)$ is necessarily purely imaginary. To see the connection between our limit theorem and the relation $\varphi'(0) = i\mu$ it is best to write the latter in the form

$$t[\varphi(\zeta/t) - 1] \to i\mu\zeta, \qquad t \to \infty. \tag{2.37}$$

If t runs through a sequence $\{c_n\}$ this becomes a special case of (2.31) with $\varphi_n(\zeta) = \varphi(\zeta/c_n)$ and $F_n(x) = F(c_n x)$. Thus theorem 2 asserts that (2.37) holds iff

$$tx^2\, F\{t\, dx\} \to 0, \qquad t\int_{-\infty}^{+\infty} \sin x\, F\{t\, dx\} \to \mu. \tag{2.38}$$

(a) Assume (2.36). An integration by parts shows that for arbitrary $a > 0$

$$t\int_{-a}^{a} x^2\, F\{t\, dx\} \le 4\int_0^a tx[1 - F(tx) + F(-tx)]\, dx. \tag{2.39}$$

As $t \to \infty$ the integrand tends to 0, and hence the same is true of the integral. Since

$|t \sin x/t - x| < Cx^2$ it follows easily that (2.38) is true and this entails (2.37). Conversely (2.38) clearly implies (2.37). Thus conditions (i) and (ii) are equivalent.

(*b*) According to theorem 1 of section 1 we have $\varphi^n(\zeta/n) \to e^{i\mu\zeta}$ iff

$$n[\varphi(\zeta/n) - 1] \to i\mu\zeta. \tag{2.40}$$

In other words, the law of large numbers applies iff (2.37) holds when t runs through the sequence of positive integers. Since the convergence of characteristic functions is automatically uniform in finite intervals it is clear that (2.40) implies (2.37) and so the conditions (i) and (iii) are equivalent.

That (2.36) represents the necessary and sufficient conditions for the law of large numbers (iii) was shown by different methods in theorem 1 of VII,7. ▶

3. EXAMPLES AND SPECIAL PROPERTIES

We list a few special distributions and turn then to properties such as existence of moments and positivity. They are listed as "examples" partly for clarity of exposition, partly to emphasize that the individual items are not connected. None of the material of this section is used in the sequel. Further examples are found in problems 6, 7, and 19.

Examples. (*a*) *Normal distribution.* If M is concentrated at the origin and attributes weight σ^2 to it, then (2.25) leads to $\psi(\zeta) = -\frac{1}{2}\sigma^2\zeta^2$ and e^ψ is normal with zero expectation and variance σ^2.

(*b*) *Poisson distribution.* The standard Poisson distribution with expectation α has the characteristic function $\omega = e^\psi$ with $\psi(\zeta) = \alpha(e^{i\zeta}-1)$. We change the location parameters so as to obtain a distribution concentrated at points of the form $-b + nh$. This changes the exponent into $\rho(\zeta) = \alpha(e^{i\zeta h}-1) - ib\zeta$, which is a special case of (2.25) with M concentrated at the single point h. The property that the measure M is concentrated at a single point is therefore characteristic of the normal and the Poisson distributions with arbitrary scale parameters. Convolutions of finitely many such distributions correspond to canonical measures with finitely many atoms. The most general measure M may be obtained as the limit of a sequence of such measures and so *all infinitely divisible distributions are limits of convolutions of finitely many Poisson and normal distributions.*

(*c*) *Randomized random walks.* In II,(7.7) we encountered the family of arithmetic distributions attributing to $r = 0, \pm 1, \pm 2, \ldots$ probability

$$a_r(t) = \sqrt{\left(\frac{p}{q}\right)^r} e^{-t} I_r(2\sqrt{pq}\, t); \tag{3.1}$$

here the parameters p, q, t are positive, $p + q = 1$, and I_r is the Bessel function defined in II,(7.1). The fact that $\{a_r\}$ satisfies the Chapman-Kolmogorov equation shows that it is infinitely divisible. Its characteristic function $\omega = e^\psi$ is easily calculated because it differs from the Schlömilch expansion II,(7.8) merely by the change of variable $u = \sqrt{p/q}\, e^{-i\zeta}$. The result

$$\psi(\zeta) = -t + t(pe^{i\zeta}+qe^{-i\zeta}) \tag{3.2}$$

shows that $\{a_r(t)\}$ *is the distribution of the difference of two independent Poisson variables* with expectations pt and qt. The canonical measure is concentrated at the points ± 1.

(*d*) *Gamma distributions.* The distribution with density

$$g_t(x) = e^{-x}x^{t-1}/\Gamma(t)$$

for $x > 0$ has the characteristic function $\gamma_t(\zeta) = (1-i\zeta)^{-t}$ which is clearly infinitely divisible. To put it into the canonical form note that

$$(\log \gamma_t(\zeta))' = it(1-i\zeta)^{-1} = it\int_0^\infty e^{i\zeta x-x}\, dx. \tag{3.3}$$

Integration shows that

$$\log \gamma_t(\zeta) = t\int_0^\infty \frac{e^{i\zeta x}-1}{x}\, e^{-x}\, dx. \tag{3.4}$$

Thus the canonical measure M is defined by the density txe^{-x} for $x > 0$. Here no centering term is necessary since the integral converges without it.

(*e*) *Hyperbolic cosine density.* We saw in XV,2 that the density $f(x) = 1/\pi \cosh x$ has the characteristic function $\omega(\zeta) = 1/\cosh(\pi\zeta/2)$. To show that it is infinitely divisible we note that $(\log \omega)'' = -(\pi^2/4)\omega^2$. Now ω^2 is the characteristic function of the density $f^{2\star}$ which was calculated in problem 6 of II,9. Thus

$$\frac{d^2}{d\zeta^2}\log \omega(\zeta) = -\int_{-\infty}^{+\infty} e^{i\zeta x}\frac{x}{e^x - e^{-x}}\, dx. \tag{3.5}$$

Since $(\log \omega)'$ vanishes at the origin we get

$$\log \omega(\zeta) = \int_{-\infty}^{+\infty} \frac{e^{i\zeta x}-1-i\zeta x}{x^2}\,\frac{x}{e^x-e^{-x}}\, dx. \tag{3.6}$$

The canonical measure therefore has density $x/(e^x - e^{-x})$. For reasons of symmetry the contribution of the term $i\zeta x$ vanishes, and since the integral converges without it, this term may be omitted from the numerator.

(*f*) *P. Lévy's example.* The function

$$\psi(\zeta) = 2\sum_{k=-\infty}^{+\infty} 2^{-k}[\cos 2^k\zeta - 1] \tag{3.7}$$

is of the form (2.9) with M symmetric and attributing weight 2^k to the points $\pm 2^k$ with $k = 0, \pm 1 \pm 2, \ldots$. (The series converges because $1 - \cos 2^k\zeta \sim 2^{2k-1}\zeta^2$ as $k \to -\infty$.) The characteristic function $\omega = e^\psi$

has the curious property that $\omega^2(\zeta) = \omega(2\zeta)$, and hence $\omega^{2^k}(\zeta) = \omega(2^k\zeta)$. For stability (in the sense introduced in VI,1) one should have $\omega^n(\zeta) = \omega(a_n\zeta)$ for all n, but this requirement is satisfied only for $n = 2, 4, 8, \cdots$. In the terminology of section 9 this ω belongs to its own domain of *partial attraction*, but it does not have a domain of attraction. (See problem 10.)

(*g*) *One-sided stable densities.* We proceed to calculate the characteristic functions corresponding to the canonical measures concentrated on $\overline{0, \infty}$ such that

$$M\{\overline{0, x}\} = Cx^{2-\alpha} \qquad 0 < \alpha < 2, \quad C > 0. \tag{3.8}$$

This example is of great importance because it will turn out that from it we may derive the general form of stable characteristic functions.

(i) If $0 < \alpha < 1$ we consider the characteristic function $\omega_\alpha = e^{\psi_\alpha}$ with

$$\psi_\alpha(\zeta) = C(2-\alpha)\int_0^\infty \frac{e^{i\zeta x} - 1}{x^{\alpha+1}}\, dx. \tag{3.9}$$

This differs from the canonical form (2.9) by the omission of the centering term, which is dispensable since the integral converges without it. To evaluate the integral we suppose $\zeta > 0$, and consider it as the limit as $\lambda \to 0+$ of

$$\int_0^\infty \frac{e^{-(\lambda - i\zeta)x} - 1}{x^{\alpha+1}}\, dx = \frac{1}{\alpha}(\lambda - i\zeta)\int_0^\infty e^{-(\lambda-i\zeta)x}x^{-\alpha}\, dx = \tag{3.10}$$

$$= -\frac{1}{\alpha}\Gamma(1-\alpha)(\lambda - i\zeta)^\alpha$$

(for the characteristic function of gamma densities see XV,2). Now

$$(\lambda - i\zeta)^\alpha = (\lambda^2 + \zeta^2)^{\alpha/2}e^{i\theta\alpha}$$

where θ is the argument of $\lambda - i\zeta$, that is, $\tan\theta = -\zeta/\lambda$. Obviously $\theta \to -\pi/2$ as $\lambda \to 0+$, and hence $(\lambda - i\zeta)^\alpha \to \zeta^\alpha e^{-i\alpha\pi/2}$. We write the final result in the form

$$\psi_\alpha(\zeta) = \zeta^\alpha \cdot C \cdot \frac{\Gamma(3-\alpha)}{(\alpha-1)\alpha}\, e^{-i\pi\alpha/2}, \qquad \zeta > 0. \tag{3.11}$$

For $\zeta < 0$ one gets $\psi_\alpha(\zeta)$ as the conjugate of $\psi_\alpha(-\zeta)$.

(ii) When $1 < \alpha < 2$ we put

$$\psi_\alpha(\zeta) = C\int_0^\infty \frac{e^{i\zeta x} - 1 - i\zeta x}{x^{\alpha+1}}\, dx. \tag{3.12}$$

This differs from the canonical form (2.9) by the more convenient centering to zero expectation. An integration by parts reduces the exponent in the denominator and enables us to use the preceding result. A routine calculation shows that ψ_α is again given by (3.11). (The real part is again negative, because now $\cos \pi/2 < 0$.)

(iii) When $\alpha = 1$ we use the standard form

$$\psi_1(\zeta) = C\int_0^\infty \frac{e^{i\zeta x} - 1 - i\zeta \sin x}{x^2}\,dx. \tag{3.13}$$

We know from XV,2 that $(1-\cos x)/(\pi x^2)$ is a probability density, and hence the real part of $\psi_1(\zeta)$ equals $-\frac{1}{2}\pi\zeta$. For the imaginary part we get

$$\int_0^\infty \frac{\sin \zeta x - \zeta \sin x}{x^2}\,dx = \lim_{\epsilon\to 0}\left[\int_\epsilon^\infty \frac{\sin \zeta x}{x^2}\,dx - \zeta\int_\epsilon^\infty \frac{\sin x}{x^2}\,dx\right] \tag{3.14}$$

When $\zeta > 0$ the substitution $\zeta x = y$ reduces the first integral to the form of the second, and the whole reduces to

$$-\zeta \lim_{\epsilon\to 0}\int_\epsilon^{\epsilon\zeta} \frac{\sin x}{x^2}\,dx = -\zeta \lim_{\epsilon\to 0}\int_1^\zeta \frac{\sin \epsilon y}{\epsilon y}\cdot\frac{dy}{y} = -\zeta \log \zeta. \tag{3.15}$$

Thus finally

$$\psi_1(\zeta) = C(-\tfrac{1}{2}\pi\zeta - i\zeta\log\zeta), \qquad \zeta > 0. \tag{3.16}$$

Of course, $\psi_1(-\zeta)$ is the conjugate to $\psi_1(\zeta)$.

When $\alpha \neq 1$ the characteristic function $\omega = e^{\psi_\alpha}$ enjoys the property that $\omega^n(\zeta) = \omega(n^{1/\alpha}\zeta)$. This means that ω is *strictly stable* according to the definition of VI,1: The sum of independent variables $\mathbf{X}_1, \ldots, \mathbf{X}_n$ with characteristic function ω has the same distribution as $n^{1/\alpha}\mathbf{X}_1$. When $\alpha = 1$ we have $\omega^n(\zeta) = \omega(n\zeta)e^{-i\zeta \log n}$, and hence the distribution of the sum differs from that of $n^{1/\alpha}\mathbf{X}_1$ by its centering. Thus ψ_1 is *stable* in the wide sense.

[For various properties of, and examples for, stable distributions see VI,1–2. Additional properties will be derived in section 5. In section 4*c* we shall see that when $\alpha < 1$ the distribution is concentrated on the positive half-axis. This is not true for $\alpha \neq 1$.]

(*h*) *General stable densities.* To each ψ_α of the preceding example there corresponds an analogous characteristic function induced by the canonical measure with the same density, but concentrated on the negative half-axis. To obtain these characteristic functions we have merely to change i into $-i$ in our formulas. One can derive more general stable characteristic functions by taking linear combinations of the two extreme cases, that is, using a canonical measure M such that for $x > 0$

$$M\overline{\{0, x\}} = Cpx^{2-\alpha}, \qquad M\overline{\{-x, 0\}} = Cqx^{2-\alpha}. \tag{3.17}$$

Here $p \geq 0$, $q \geq 0$ and $p + q = 1$. From what was said it is clear that the corresponding characteristic function $\omega = e^{\psi}$ is given by

$$\psi(\zeta) = |\zeta|^{\alpha}\, C\, \frac{\Gamma(3-\alpha)}{\alpha(\alpha-1)}\left[\cos\frac{\pi\alpha}{2} \pm i(p-q)\sin\frac{\pi\alpha}{2}\right] \tag{3.18}$$

if $0 < \alpha < 1$ or $1 < \alpha \le 2$, while for $\alpha = 1$

$$\psi(\zeta) = -|\zeta| \cdot C[\tfrac{1}{2}\pi \pm i(p-q)\log|\zeta|]; \tag{3.19}$$

here the upper sign applies when $\zeta > 0$, the lower for $\zeta < 0$. Note that for $\alpha = 2$ we get $\psi(\zeta) = -\frac{1}{2}(p+q)\zeta^2$, that is, the *normal distribution.* It corresponds to a measure M concentrated at the origin.

It will be shown in section 5 that (neglecting arbitrary centerings) these formulas yield the *most general stable characteristic functions.* In particular, all symmetric stable distributions have characteristic functions of the form $e^{-a|\zeta|^{\alpha}}$ with $a > 0$. ►

4. SPECIAL PROPERTIES

In this section $\omega = e^{\psi}$ stands for an infinitely divisible characteristic function with ψ given in the standard form

$$\psi(\zeta) = \int_{-\infty}^{+\infty} \frac{e^{i\zeta x} - 1 - i\zeta \sin x}{x^2}\, M\{dx\} + ib\zeta \tag{4.1}$$

where M is a canonical measure and b a real constant. By the definition of canonical measures the integral

$$M^{+}(x) = \int_{x}^{\infty} y^{-2}\, M\{dy\} \tag{4.2}$$

converges for all $x > 0$, and a similar statement holds for $x < 0$.

The probability distribution with characteristic function ω will be denoted by U.

(*a*) *Existence of moments.* It was shown in XV,4 that the second moment of U is finite iff ω is twice differentiable, that is, iff ψ'' exists. The same argument shows that this is the case iff the measure M is finite. In other words, *for a second moment of U to exist it is necessary and sufficient that the measure M be finite.*

A similar reasoning (see problem 15 of XV,9) shows more generally that for any integer $k \ge 1$ *the* $2k$th *moment of U exists iff M has a moment of order* $2k-2$.

(*b*) *Decompositions.* Every representation of $M = M_1 + M_2$ of M as the sum of two measures induces in an obvious manner a factorization $\omega = e^{\psi_1}e^{\psi_2}$ of ω into two infinitely divisible characteristic functions. If

M is concentrated at a single point the same is true of M_1 and M_2; in other words, if ω is normal or Poisson,[2] the same will be true of the two factors e^{ψ_1} and e^{ψ_2}. But any other infinitely divisible ω can be split into two essentially different components. In particular, any non-normal stable characteristic function can be factorized into non-stable infinitely divisible characteristic functions.

A particularly useful decomposition $\omega = e^{\psi_1}e^{\psi_2}$ is obtained by representing M as a sum of two measures concentrated on the intervals $|x| \leq \eta$ and $|x| > \eta$, respectively. For the latter we express M in terms of the measure N defined by $N\{dx\} = x^{-2} M\{dx\}$. Thus we write

$$\psi(\zeta) = \psi_1(\zeta) + \psi_2(\zeta) + i\beta\zeta \tag{4.3}$$

where

$$\psi_1(\zeta) = \int_{|x|\leq\eta} \frac{e^{i\zeta x} - 1 - i\zeta x}{x^2} M\{dx\} \tag{4.4}$$

$$\psi_2(\zeta) = \int_{|x|>\eta} (e^{i\zeta x}-1)N\{dx\} \tag{4.5}$$

and the difference $b - \beta$ accounts for the changed centering terms in (4.4) and (4.5).

Note that e^{ψ_2} is the characteristic function of a compound Poisson distribution generated by a probability distribution F such that $F\{dx\} = c\, N\{dx\}$, or

$$1 - F(x) = c\, M^+(x), \qquad x > 0. \tag{4.6}$$

The function e^{ψ_1} is infinitely differentiable. We see thus that *every infinitely divisible distribution U is the convolution of a distribution U_1 possessing moments of all orders and a compound Poisson distribution U_2* generated by a probability distribution F with tails proportional to M^+ and M^-. It follows in particular that U possesses a kth moment iff the kth moment of F exists.

(*c*) *Positive variables.* We proceed to prove that *U is concentrated on $\overline{0, \infty}$ iff*[3]

$$\psi(\zeta) = \int_0^\infty \frac{e^{i\zeta x} - 1}{x} P\{dx\} + ib\zeta \tag{4.7}$$

---

[2] By theorem 1 of XV,8 the normal characteristic function does not admit of *any* factorization into non-normal characteristic functions. An analogous statement holds for the Poisson distribution (Raikov's theorem).

[3] This remark was made by P. Lévy and is also an immediate consequence of the Laplace transform version XIII,(7.2). It is interesting that without probabilistic arguments a formal verification of the assertion is cumbersome (see G. Baxter and J. M. Shapiro, Sankhya, vol. 22.)

where $b \geq 0$ *and* P *is a measure such that* $(1+x)^{-1}$ *is integrable with respect to* P. (In the original notation of (4.1) we have $P\{dx\} = x^{-1} M\{dx\}$.)

Assume U concentrated on $[0, \infty)$ and consider the decomposition described by (4.3)–(4.5). The origin is a point of increase for the compound Poisson distribution U_2. The distribution U_1 has zero expectation, and therefore some point of increase $s \leq 0$. It follows that $s + \beta$ is a point of increase for U, and hence $\beta \geq 0$. The same argument shows that U_1 can have no normal component, and therefore the contribution of U_1 must tend to 0 as $\eta \to 0$. Finally, if t is a point of increase for the probability distribution F generating U_2, then nt is a point of increase for U_2 itself. It follows that F, and hence N, are concentrated on the positive half-axis, and so the integral in (4.5) actually extends only over $x > \eta$. The integrand vanishes at the origin, and therefore in the passage to the limit $\eta \to 0$ the measure N need not remain bounded. However, for $x > \eta$ we can switch from the measure N to $P\{dx\} = x\, N\{dx\}$ (which is the same as $x^{-1} M\{dx\}$). Near the origin the new integrand $(e^{i\zeta x}-1\, x^{-1}$ is bounded away from 0, and hence P must assign finite values to neighborhoods of the origin. In this way we obtain the representation (4.7).

Conversely, if ψ is defined by (4.7) then our argument shows e^{ψ} to be the limit of characteristic functions of compound Poisson distributions concentrated on $\overline{0, \infty}$. The same is therefore true of the limit distribution U.

(*d*) *Asymptotic behavior.* The result concerning the existence of moments appears to indicate that the asymptotic behavior of the distribution function U as $x \to \pm\infty$ depends only on the behavior of the canonical measure M near $\pm\infty$ or, what amounts to the same, on the asymptotic behavior of the functions M^+ and M^-. Rather than attempting to prove this conjecture in the greatest possible generality we consider a typical situation.

Suppose that M^+ *varies regularly at* ∞, *that is,*

$$M^+(x) = x^{-\zeta} L(x) \tag{4.8}$$

where $\zeta \geq 0$ *and* L *is slowly varying. Then*

$$1 - U(x) \sim M^+(x), \qquad x \to \infty. \tag{4.9}$$

Proof. Let $\mathbf{S}$ be a random variable having U for distribution function. Consider the canonical measure M as a sum $M_1 + M_2 + M_3$ of three measures concentrated on the intervals $\overline{1, \infty}$, $\overline{-1, 1}$, and $\overline{-\infty, -1}$. As shown under (*b*), this induces a representation $\mathbf{S} = \mathbf{X}_1 + \mathbf{X}_2 + \mathbf{X}_3 + \beta$ as a sum of three independent random variables such that: $\mathbf{X}_1$ has a compound Poisson distribution U_1 generated by a probability distribution F concentrated on $\overline{1, \infty}$ and defined by (4.6); the canonical measure corresponding to $\mathbf{X}_2$ is concentrated on $\overline{-1, 1}$; finally $\mathbf{X}_3$ is defined as $\mathbf{X}_1$ except that $\overline{-\infty, 1}$ takes over the role of $\overline{1, \infty}$. It is not difficult to show that

$$\mathbf{P}\{\mathbf{X}_1 > x\} \sim M^+(x), \qquad x \to \infty \tag{4.10}$$

(see theorem 2 of VIII,9). To prove the assertion (4.9) it suffices therefore to show that

$$\mathbf{P}\{\mathbf{S} > x\} \sim \mathbf{P}\{\mathbf{X}_1 > x\}, \qquad x \to \infty. \tag{4.11}$$

In this connection the centering constant β plays no role and we assume $\beta = 0$. Then for every $\epsilon > 0$

$$\mathbf{P}\{\mathbf{S} > x\} \geq \mathbf{P}\{\mathbf{X}_1 > (1+\epsilon)x\} \cdot \mathbf{P}\{\mathbf{X}_2 + \mathbf{X}_3 > -\epsilon x\}. \tag{4.12}$$

On the other hand, since $\mathbf{X}_3 \leq 0$

$$\mathbf{P}\{\mathbf{S} > x\} \leq \mathbf{P}\{\mathbf{X}_1 > (1-\epsilon)x\} + \mathbf{P}\{\mathbf{X}_2 > \epsilon x\}. \tag{4.13}$$

As $x \to \infty$ the last probability in (4.12) tends to 1, while the last probability in (4.13) decreases faster than any power x^{-a} because $\mathbf{X}_2$ has moments of all orders. Thus (4.12) and (4.13) imply the truth of (4.11).

(*e*) *Subordination.* If e^{ψ} is infinitely divisible, so is $e^{s\psi}$ for every $s > 0$. By randomization of the parameter s we obtain a new characteristic function of the form

$$\varphi(\zeta) = \int_0^\infty e^{s\psi(\zeta)} G\{ds\} \tag{4.14}$$

where G is an arbitrary probability distribution concentrated on $\overline{0, \infty}$. The characteristic function φ need not be infinitely divisible, but it is easily verified that if $G = G_1 \star G_2$ is the convolution of two probability distributions then (with obvious notations) $\varphi = \varphi_1\varphi_2$. It follows that *if* G *is infinitely divisible then* (4.14) *defines an infinitely divisible characteristic function.*

This result has a simple probabilistic interpretation. Let $\{\mathbf{X}(t)\}$ stand for the variables of a process with independent increments such that $\mathbf{X}(t)$ has the characteristic function $e^{t\psi}$. If $\mathbf{T}$ is a positive variable with distribution G then φ may be interpreted as the characteristic function of the composite random variable $\mathbf{X}(\mathbf{T})$. Suppose then that G is infinitely divisible with characteristic function e^{γ}. We may envisage a second process $\{\mathbf{T}(t)\}$ with independent increments such that $\mathbf{T}(t)$ has the characteristic function $e^{t\gamma}$. For each $t > 0$ we get a new variable $\mathbf{X}(\mathbf{T}(t))$, and these are again the variables of a process with independent increments.[4] Thus $\mathbf{T}(t)$ serves as

[4] If G has the Laplace transform $e^{-\rho(\lambda)}$ then $\mathbf{T}(t)$ corresponds to the Laplace transform $e^{-t\rho(\lambda)}$ and it is easily verified that the characteristic function of $\mathbf{X}(\mathbf{T}(t))$ is given by $e^{-t\rho(-\psi)}$.

operational time. In the terminology of X,7 the new process $\{\mathbf{X}(\mathbf{T}(b))\}$ is obtained by *subordination*, with $\{\mathbf{T}(t)\}$ as *directing process*. We have now found a purely analytic proof that the subordination process always leads to infinitely divisible distributions.

5. STABLE DISTRIBUTIONS AND THEIR DOMAINS OF ATTRACTION

Let $\{\mathbf{X}_n\}$ be a sequence of mutually independent random variables with a common distribution F, and put $\mathbf{S}_n = \mathbf{X}_1 + \cdots + \mathbf{X}_n$. Let U be a distribution not concentrated at one point. According to the terminology introduced in VI,1 we say that F belongs to the domain of attraction of U iff there exist constants $a_n > 0$ and b_n such that the distribution of $a_n^{-1}\mathbf{S}_n - nb_n$ tends to U. The exclusion of limit distributions concentrated at a single point serves to eliminate the trivial situation where $b_n \to b$ while a_n increases so rapidly that $a_n^{-1}\mathbf{S}_n$ tends in probability to zero.

We wish to rephrase the definition in terms of the characteristic functions φ and ω of the distributions F and U. According to lemma 4 of XV,1 the distribution U is concentrated at one point iff $|\omega(\zeta)| = 1$ for all ζ. Accordingly, *φ belongs to the domain of attraction of the characteristic function ω if $|\omega|$ is not identically one, and there exist constants $a_n > 0$ and b_n such that*

$$(\varphi(\zeta/a_n)e^{-ib_n\zeta})^n \to \omega(\zeta). \tag{5.1}$$

It was shown in VI,1 that the limit ω is necessarily stable, but we shall now develop the whole theory anew as a simple consequence of the basic limit theorem of section 2. In conformity with the notations used there we put

$$\varphi_n(\zeta) = \varphi(\zeta/a_n)e^{-ib_n\zeta}, \qquad F_n(x) = F(a_n(x+b_n)). \tag{5.2}$$

According to theorem 1 of section 1 the relation (5.1) holds iff

$$n[\varphi_n(\zeta)-1] \to \psi(\zeta) \tag{5.3}$$

for all ζ, where $\omega = e^\psi$.

Consider first the special case of a *symmetric* F. Then $b_n = 0$. We know from theorem 1 of section 2 that (5.3) implies the existence of a canonical measure M such that $nx^2\, F_n\{dx\} \to M\{dx\}$. To express this we introduce the truncated moment function

$$\mu(x) = \int_{-x}^{x} y^2\, F\{dy\}, \qquad x > 0. \tag{5.4}$$

Then at all points of continuity

$$\frac{n}{a_n^2}\,\mu(a_n x) \to M\{\overline{-x,\, x}\} \tag{5.5}$$

and

$$n[1-F(a_nx)] \to M^+(x) \tag{5.6}$$

where

$$M^+(x) = \int_x^\infty y^{-2}\, M\{dy\}. \tag{5.7}$$

The relation $\varphi(\zeta/a_n) \to 1$ implies $a_n \to \infty$, and therefore $\mathbf{S}_n/a_n$ and $\mathbf{S}_n/a_{n+1}$ have the same limit distribution U. It follows that the ratio a_{n+1}/a_n tends to 1, and hence lemma 3 of VIII,8 applies to (5.5). We conclude that μ *varies regularly* and the canonical measure M is of the form

$$M\{\overline{-x, x}\} = Cx^{2-\alpha}, \qquad x > 0 \tag{5.8}$$

with $\alpha \leq 2$. (The exponent is denoted by $2 - \alpha$ in conformity with a usage introduced by P. Lévy.) If $\alpha = 2$ the measure M is concentrated at the origin. The convergence of the integral in (5.7) requires that $\alpha > 0$; for $0 < \alpha < 2$ we find

$$M^+(x) = C\frac{2-\alpha}{\alpha}\, x^{-\alpha}, \qquad x > 0. \tag{5.9}$$

A similar argument applies to unsymmetric distributions F, but instead of (5.6) we get the less appealing relations

$$n[1 - F(a_n(x+b_n))] \to M^+(x), \qquad nF(a_n(-x+b_n)) \to M^-(-x), \tag{5.10}$$

and an analogous modification applies to (5.5). However, the fact that $\varphi_n(\zeta) \to 1$ and $a_n \to \infty$ implies $b_n \to 0$, and so (5.10) is actually fully equivalent to (5.6) and the analogous relation for the left tail.

We see thus that (5.6) holds whenever F belongs to a domain of attraction In view of lemma 3 of VIII,8 this means that either M^+ vanishes identically, or else the tail $1 - F$ *varies regularly* and $M^+(x) = Ax^{-\alpha}$. Then (5.7) shows that on the positive half-axis the measure M has the *density* $A\alpha x^{1-\alpha}$. The same argument applies to the left tail and also to the tail sum; the exponent α must therefore be common to the two tails.

If both tails vanish identically M is concentrated at the origin. In no other case can M have an atom at the origin. This is so because the canonical measure ${}^0M$ corresponding to the symmetrized distribution ${}^0U$ is the sum of M and its mirror image with respect to the origin, and we saw that ${}^0M$ is either atomless or concentrated at the origin. Accordingly, when $\alpha < 2$ the canonical measure M is uniquely determined by its densities on the two half-axes and these are proportional to $|x|^{1-\alpha}$. For intervals $\overline{-y, x}$ containing the origin we have therefore

$$M\{\overline{-y, x}\} = C(qy^{2-\alpha} + px^{2-\alpha}), \tag{5.11}$$

where $0 < \alpha \leq 2$, $C > 0$, and $p + q = 1$. For $\alpha = 2$ the measure is concentrated at the origin. In accordance with (5.7) this is equivalent to

$$M^+(x) = Cp\frac{2-\alpha}{\alpha}x^{-\alpha}, \qquad M^-(-x) = Cq\frac{2-\alpha}{\alpha}x^{-\alpha}. \tag{5.12}$$

The characteristic function corresponding to these measures are given by (3.18) and (3.19). They show clearly that our distributions are *stable* in the sense that $U^{n\star}$ differs from U only by location parameters. This means that each stable distribution belongs to its own domain of attraction, and we have therefore solved the problem of finding *all* distributions possessing a domain of attraction. We record this in

Theorem 1. *A distribution possesses a domain of attraction iff it is stable.*

(i) *The class of stable distributions coincides with the class of infinitely divisible distributions with canonical measures given by* (5.11).

(ii) *The corresponding characteristic functions are of the form* $\omega(\zeta) = e^{\psi(\zeta)+ib\zeta}$ *with* ψ *defined by* (3.18)–(3.19), *and* $0 < \alpha \leq 2$.

(iii) *As* $x \to \infty$ *the tails of the corresponding distribution* U *satisfy*

$$x^\alpha[1-U(x)] \to Cp\frac{2-\alpha}{\alpha}, \qquad x^\alpha U(-x) \to Cq\frac{2-\alpha}{\alpha}. \tag{5.13}$$

The last statement is a direct corollary of (5.6) if one remembers that U belongs to its own domain of attraction with norming constants given by $a_n = n^{1/\alpha}$. [Alternatively, (5.13) represents a special case of the result obtained in 4(*d*).]

Note that each of the three descriptions in the theorem determines U uniquely up to an arbitrary centering.

Before returning to the investigation of the conditions under which a distribution F belongs to the domain of attraction of a stable distribution we recall a basic result concerning regular variation. According to the definition in VIII,8 a function L varies *slowly* at infinity if for each fixed $x > 0$

$$\frac{L(tx)}{L(t)} \to 1, \qquad t \to \infty. \tag{5.14}$$

In this case we have for arbitrary $\delta > 0$ and all x sufficiently large

$$x^{-\delta} < L(x) < x^\delta. \tag{5.15}$$

A function μ varies regularly if it is of the form $\mu(x) = x^\rho L(x)$. We consider in particular the truncated moment function μ defined by (5.4). Applying theorem 2 of VIII,9 with $\zeta = 2$ and $\eta = 0$ to the distribution function on $\overline{0, \infty}$ defined by $F(x) - F(-x)$, we obtain the following important result:

If μ *varies regularly with exponent* $2 - \alpha$ (*where* $0 < \alpha \leq 2$) *then*

$$\frac{x^2[1 - F(x) + F(-x)]}{\mu(x)} \to \frac{2 - \alpha}{\alpha} \tag{5.16}$$

Conversely, if (5.16) *is true with* $\alpha < 2$, *then* μ *and the tailsum*

$$1 - F(x) + F(-x)$$

vary regularly with exponents $2 - \alpha$ *and* $-\alpha$, *respectively. If* (5.16) *holds with* $\alpha = 2$ *then* μ *varies slowly.*

In deriving (5.7) we saw that for a *symmetric* F to belong to a domain of attraction it is necessary that the truncated moment function μ varies regularly:

$$\mu(x) \sim x^{2-\alpha} L(x), \qquad x \to \infty, \tag{5.17}$$

where L varies slowly. We shall now see that this is true also for unsymmetric distributions. When $\alpha = 2$ this condition turns out to be sufficient, but when $\alpha < 2$ the canonical measure (5.11) attributes to the positive and negative half-axes weights in the proportion $p{:}q$ and it turns out the two tails of F must be similarly balanced.

We are now in a position to prove the basic

Theorem 2. (*a*) *In order that a distribution* F *belong to some domain of attraction it is necessary that the truncated moment function* μ *varies regularly with an exponent* $2 - \alpha$ $(0 < \alpha \leq 2)$. [*That is,* (5.17) *holds.*]

(*b*) *If* $\alpha = 2$, *this condition is also sufficient provided* F *is not concentrated at one point.*

(c) *If* (5.17) *holds with* $0 < \alpha < 2$ *then* F *belongs to some domain of attraction iff the tails are balanced so that as* $x \to \infty$

$$\frac{1 - F(x)}{1 - F(x) + F(-x)} \to p, \qquad \frac{F(-x)}{1 - F(x) + F(-x)} \to q. \tag{5.18}$$

Note that nothing is assumed concerning the centering of F. The theorem therefore implies that (5.17) either holds with an arbitrary centering or with none. The truth of this is easily verified directly except that when F is concentrated at a single point t the left side in (5.17) vanishes identically for the centering at t, and varies regularly for all other centerings.

The theorem was formulated so as to cover also convergence to the normal distributions. When $\alpha < 2$ it appears more natural to express the main condition in terms of the tailsum of F rather than ∞. The following corollaries restate the theorem in equivalent forms.

Corollary 1. *A distribution* F *not concentrated at one point belongs to the domain of attraction of the normal distribution iff* μ *varies slowly.*
This is the case iff (5.16) *holds with* $\alpha = 2$.

Needless to say, μ varies slowly whenever F has a finite variance.

Corollary 2. *A distribuition* F *belongs to the domain of attraction of a stable distribution with exponent* $\alpha < 2$ *iff its tails satisfy the balancing condition* (5.18) *and the tailsum varies regularly with exponent* α.
The latter condition is fully equivalent to (5.16).

Proof. (*a*) *Necessity.* Suppose that the canonical measure of the limit distribution U is given by (5.11). In the process of deriving this relation we saw that a distribution belonging to the domain of attraction of U satisfies (5.6) and its analogue for the left tail, and so

$$n[1 - F(a_n x) + F(-a_n x)] \to M^+(x) + M^-(-x.) \tag{5.19}$$

Assume first $\alpha < 2$, so that the right side is not identically zero. As already mentioned, lemma 3 of VIII,8 then guarantees that the tailsum $1 - F(x) + F(-x)$ varies regularly with exponent $-\alpha$. But then (5.16) holds and so μ varies regularly with exponent $2 - \alpha$. The balancing condition (5.18) is now an immediate consequence of (5.6).

There remains the case $\alpha = 2$. The left side in (5.19) then tends to zero, and thus the probability that $|\mathbf{X}_k| > a_n$ for some $k \leq n$ tends to zero. In order that $\mathbf{S}_n/a_n$ does not tend in probability to zero it is therefore necessary that the sum of the truncated second moments of $\mathbf{X}_k a_n^{-1}$ be bounded away from 0. But

$$\frac{\mu(a_n)}{1 - F(a_n) + F(-a_n)} \to \infty \tag{5.20}$$

and hence (5.16) holds with $\alpha = 2$. This implies the slow variation of μ, and so our conditions are necessary.

(*b*) *Sufficiency.* We shall not only prove that our conditions are sufficient, but shall at the same time specify norming constants a_n and b_n that will guarantee convergence to a prescribed stable distribution. This is done in theorem 3. ▶

The formulation of theorem 3 assumes knowledge of the fact that distributions in any domain of attraction with $\alpha > 1$ possess expectations. In the proof we shall require additional information concerning the truncated first moments. It is natural to formulate these results in a more general setting, although we shall require only the special case $\beta = 1$.

Lemma. *A distribution* F *belonging to a domain of attraction with index* α *possesses absolute moments* m_β *of all orders* $\beta < \alpha$. *If* $\alpha < 2$ *no moments of order* $\beta > \alpha$ *exist.*

More precisely, if $\beta < \alpha$ *then as* $t \to \infty$

$$\frac{t^{2-\beta}}{\mu(t)} \int_{|x|>t} |x|^\beta\, F\{dx\} \to \frac{2-\alpha}{\alpha-\beta}, \tag{5.21}$$

while for $\alpha < 2$ *and* $\beta > \alpha$

$$\int_{|x|<t} |x|^\beta\, F\{dx\} \sim \frac{\alpha}{\beta-\alpha}\, t^\beta[1 - F(t) + F(-t)]. \tag{5.22}$$

(Note that in each case the integral is a regularly varying function with exponent $\beta - \alpha$.)

Proof. The relations (5.21) and (5.22) represent the general form (5.15) and are direct consequences of theorem 2 in VIII,9 applied to the distribution defined on $\overline{0, \infty}$ by $F(x) + F(-x)$. For (5.21) set $\zeta = 2$ and $\eta = \beta$ and $\eta = 0$. ▶

It is implicit in the proof of theorem 2 that the norming constants a_n must satisfy the condition

$$\frac{n\mu(a_n)}{a_n^2} \to C. \tag{5.23}$$

If μ varies regularly [satisfies (5.17)] such a_n exist: one may define a_n as the lower bound of all x for which $nx^{-2}\mu(x) \le C$. Because of the regular variation we have then for $x > 0$

$$\frac{n\mu(a_n x)}{a_n^2} \to Cx^{2-\alpha}. \tag{5.24}$$

This means that the mass attributed by the measure $nx^2\, F\{a_n\, dx\}$ to any symmetric interval $\overline{-x, x}$ tends to $M\{\overline{-x, x}\}$. In view of (5.16) the relation (5.24) automatically entails the analogous relation (5.19) for the tailsum of F. When $\alpha = 2$ the right side is identically zero; when $\alpha < 2$ the balancing condition (5.18) guarantees that also the individual tails satisfy the required conditions

$$n[1 - F(a_n x)] \to Cp\,\frac{2-\alpha}{\alpha}\, x^{-\alpha}, \qquad nF(-a_n x) \to C_q\, \frac{2-\alpha}{\alpha}\, x^{-\alpha}, \tag{5.25}$$

the right sides being identical with $M^+(x)$ and $M^-(x)$. [Incidentally, when $\alpha < 2$ the relations (5.25) in turn imply (5.24).]

We have thus shown that the measures $nx^2\, F\{a_n\, dx\}$ tend properly to the canonical measure M. By theorem 2 of section 2 this implies that

(5.26)

$$\int_{-\infty}^{+\infty} \frac{e^{i\zeta x} - 1 - i\zeta \sin x}{x^2}\, nx^2\, F\{a_n\, dx\} \to \int_{-\infty}^{+\infty} \frac{e^{i\zeta x} - 1 - i\zeta \sin x}{x^2}\, M\{dx\}.$$

From this it is now easy to derive

Theorem 3. *Let U be the stable distribution determined (including centering) by the characteristic function* (3.18) *if $\alpha \neq 1$ or* (3.19) *if $\alpha = 1$.*

Let the distribution F satisfy the conditions of theorem 2, *and let $\{a_n\}$ satisfy* (5.23).

(i) *If $0 < \alpha < 1$ then $\varphi^n(\zeta/a_n) \to \omega(\zeta) = e^{\psi(\zeta)}$.*

(ii) *If $1 < \alpha \leq 2$ the same is true provided F is centered to zero expectation.*

(iii) *If $\alpha = 1$ then*

$$(\varphi(\zeta/a_n)e^{-ib_n\zeta})^n \to \omega(\zeta) = e^{\psi(\zeta)}, \tag{5.27}$$

where

$$b_n = \int_{-\infty}^{+\infty} \sin \frac{x}{a_n}\, F\{dx\}. \tag{5.28}$$

We have thus the pleasing result that when $\alpha < 1$ no centering procedure is required, while for $\alpha > 1$ the natural centering to zero expectation suffices.

Proof. (i) Let $\alpha < 1$. The integral defining $\psi(\zeta)$ in (3.18) differs from the right side in (5.26) in that the term $i\zeta \sin x$ is missing. We show that these terms may be omitted also in (5.26) so that

$$\int_{-\infty}^{+\infty} \frac{e^{i\zeta x} - 1}{x^2} \cdot nx^2\, F\{a_n\, dx\} \to \int_{-\infty}^{+\infty} \frac{e^{i\zeta x} - 1}{x^2} \cdot M\{dx\}. \tag{5.29}$$

Outside a neighborhood of the origin the integrand is continuous, and since $nx^2\, F\{a_n\, dx\} \to M\{dx\}$ the relation (5.29) holds if an interval $|x| < \delta$ is cut out of the domain of integration. It suffices therefore to show that the contribution of $|x| < \delta$ to the integral on the left can be made arbitrarily small by choosing δ sufficiently small. Now this contribution is dominated by

$$n\int_{|x|<\delta} |x|\, F\{a_n\, dx\} = \frac{n}{a_n}\int_{|y|<a_n\delta} |y|\, F\{dy\}, \tag{5.30}$$

and (5.22) with $\beta = 1$ shows that the right side is $\sim(2 - \alpha/(1 - \alpha))C\delta^{1-\alpha}$ which tends to zero with δ.

Thus (5.29) holds. It can be rewritten in the form $n[\varphi(\zeta/a_n) - 1] \to \psi(\zeta)$, and by theorem 1 of section 1 this is equivalent to the assertion $\varphi^n(\zeta/a_n) \to \log \psi(\zeta)$.

(ii) Let $\alpha > 1$. The argument used under (i) carries over except that the modified version of (5.26) now takes the form

$$\int_{-\infty}^{+\infty} \frac{e^{i\zeta x} - 1 - i\zeta x}{x^2}\, nx^2\, F\{a_n\, dx\} \to \int_{-\infty}^{+\infty} \frac{e^{i\zeta x} - 1 - i\zeta x}{x^2}\, M\{dx\}. \tag{5.31}$$

To justify it we have to show that the contribution of $|x| > t$ to the integral on the left can be made arbitrarily small by choosing t sufficiently large. This follows directly from (5.21).

(iii) Let $\alpha = 1$. No modification is required in (5.26), but to show that this relation is equivalent to the assertion (5.27) it is necessary to prove that for fixed ζ

$$\varphi^n(\zeta/a_n) \sim e^{n[\varphi(\zeta/a_n)-1]}, \tag{5.32}$$

or, what amounts to the same, that

$$n\,|\varphi(\zeta/a_n) - 1|^2 \to 0. \tag{5.33}$$

For $\beta < 1$ the absolute moment m_β of F is finite. From the obvious inequality $|e^{it} - 1| < 2\,|t|^\beta$ we conclude that $|\varphi(\zeta/a_n) - 1| < 2m_\beta\,|\zeta|^\beta a_n^{-\beta}$ and so the left side in (5.33) is $O(na_n^{-2\beta})$. But the defining relation (5.23) shows that $n = O(a_n^{1+\epsilon})$ for every $\epsilon > 0$, and so (5.33) is true. ►

Concluding remark. The domain of attraction of the normal distribution must not be confused with the notion of the "domain of normal attraction of a stable distribution U with exponent a^α introduced by B. V. Gnedenko. A distribution F is said to belong to this domain if it belongs to the domain of attraction of U with norming coefficients $a_n = n^{1/\alpha}$. The delimitation of this domain originally posed a serious problem, but within the present setup the solution is furnished by the condition (5.23) on the norming constants. A distribution F belongs to the "normal" domain of attraction of U iff $x^\alpha[1 - F(x)] \to Cp$ and $x^\alpha F(-x) \to Cq$ as $x \to \infty$. Here $C > 0$ is a constant. (Note, incidentally, that in this terminology the normal distribution possesses a domain of non-normal attraction.)

*6. STABLE DENSITIES

It seems impossible to express stable densities in a closed form, but series expansions were given independently by Feller (1952) and H. Bergström (1953). They contain implicitly results discovered later by more complicated methods, and they provide a good example for the use of the Fourier inversion formula (although complex integration is used). We shall not consider the exponent $\alpha = 1$.

For $\zeta > 0$ we can put the stable characteristic functions in the form $e^{-a\zeta^\alpha}$, where a is a complex constant. Its absolute value affects only a scale parameter so that we are free to let a have a unit modulus and write $a = e^{i\pi\gamma/2}$ with γ real. Thus we put

$$\psi(\zeta) = -|\zeta|^\alpha \cdot e^{\pm i\pi\gamma/2} \tag{6.1}$$

where in $\pm$ the upper sign prevails for $\zeta > 0$, the lower for $\zeta < 0$. [See the canonical form (3.18).] The ratio of the real and imaginary parts are

* This section treats a special topic and should be omitted at first reading.

subject to inequalities evident in (3.18); with the present notations *for* e^{ψ} *to be stable it is necessary and sufficient that*

$$|\gamma| \leq \begin{cases} \alpha & \textit{if} \quad 0 < \alpha < 1 \\ 2-\alpha & \textit{if} \quad 1 < \alpha < 2. \end{cases} \tag{6.2}$$

Since e^{ψ} is absolutely integrable the corresponding distribution has a density. It will be denoted by $p(x;\alpha,\gamma)$, and we proceed to calculate it from the Fourier inversion formula XV,(3.5). Knowing that p is real and that $\psi(-\zeta)$ is the conjugate of $\psi(\zeta)$ we get

$$p(x;\alpha,\gamma) = \pi^{-1}\,\mathrm{Re}\int_0^{\infty} e^{-ix\zeta-\zeta^{\alpha}e^{i\pi\gamma/2}}\,d\zeta. \tag{6.3}$$

It suffices to calculate this function for $x > 0$ since

$$p(-x;\alpha,\gamma) = p(x;\alpha,-\gamma). \tag{6.4}$$

(*a*) *The case* $\alpha < 1$. Consider the integrand as a function of the complex variable ζ. When $x > 0$ and $\mathrm{Im}\,\zeta \to -\infty$ the integrand tends to 0 owing to the dominance of the linear term in the exponent. This enables one to move the path of integration to the negative imaginary axis, which amounts to using the substitution $\zeta = (t/x)e^{-i\frac{1}{2}\pi}$ and proceeding as if all coefficients were real. The new integrand is of the form $e^{-t-ct^{\alpha}}$. The exponential expansion for $e^{-ct^{\alpha}}$ and the familiar gamma integral lead without further artifice to

$$p(-x;\alpha,\gamma) = \mathrm{Re}\,\frac{-i}{\pi x}\sum_{k=0}^{\infty}\frac{\Gamma(k\alpha+1)}{k!}\left(-x^{-\alpha}\exp\left[i\frac{\pi}{2}(\gamma-\alpha)\right]\right)^k. \tag{6.5}$$

(*b*) *The case* $1 < \alpha < 2$. The use of the formal substitution

$$\zeta = t^{\alpha-1}\exp\left(-\tfrac{1}{2}i\pi\gamma/\alpha\right)$$

can be justified as in the case $\alpha < 1$. The new integrand is of the form $e^{-t-ct^{\alpha^{-1}}}t^{\alpha^{-1}-1}$. Expanding $e^{-ct^{\alpha^{-1}}}$ into an exponential series we get

$$p(-x;\alpha,\gamma) =$$

$$= \frac{1}{\alpha\pi}\,\mathrm{Re}\exp\left(-i\frac{\pi\gamma}{2\alpha}\right)\sum_{n=0}^{\infty}\frac{\Gamma((n+1)/\alpha)}{n!}\left(-ix\exp\left[-i\frac{\pi\gamma}{2\alpha}\right]\right)^n. \tag{6.6}$$

Changing the summation index n to $k-1$ and using the familiar recursion formula $\Gamma(s+1) = s\,\Gamma(s)$ leads to

$$p(-x;\alpha,\gamma) = \frac{1}{\pi x}\,\mathrm{Re}\,i\sum_{k=1}^{\infty}\frac{\Gamma(1+k/\alpha)}{k!}\left(-x\exp\left[-i\frac{\pi}{2\alpha}(\gamma-\alpha)\right]\right)^k. \tag{6.7}$$

We have thus proved

Lemma 1. *For* $x > 0$ *and* $0 < \alpha < 1$

$$(6.8) \qquad p(x;\alpha,\gamma) = \frac{1}{\pi x}\sum_{k=1}^{\infty}\frac{\Gamma(k\alpha+1)}{k!}(-x^{-\alpha})^k \sin\frac{k\pi}{2}(\gamma-\alpha).$$

For $x > 0$ *and* $1 < \alpha < 2$

$$(6.9) \qquad p(x;\alpha,\gamma) = \frac{1}{\pi x}\sum_{k=1}^{\infty}\frac{\Gamma(1+k/\alpha)}{k!}(-x)^k \sin\frac{k\pi}{2\alpha}(\gamma-\alpha).$$

The values for $x < 0$ *are given by* (6.4).

Note that (6.8) provides asymptotic estimates for $x \to \infty$. A curious by-product of these formulas is as follows:

Lemma 2. *If* $\frac{1}{2} < \alpha < 1$ *and* $x > 0$ *then*

$$(6.10) \qquad \frac{1}{x^{\alpha+1}}\,p\left(\frac{1}{x^{\alpha}};\frac{1}{\alpha},\gamma\right) = p(x;\alpha,\gamma^*)$$

where $\gamma^* = \alpha(\gamma+1) - 1$.

A trite check shows that γ^* falls within the range prescribed by (6.2). The identity (6.10) was first noticed (with a complicated proof) by V. M. Zolotarev.

7. TRIANGULAR ARRAYS

The notion of a triangular array was explained in VI,3 as follows. For each n we are given finitely many, say r_n, independent random variables $\mathbf{X}_{k,n}$ $(k = 1, 2, \ldots, r_n)$ with distributions $F_{k,n}$ and characteristic functions $\varphi_{k,n}$. We form the row sum $\mathbf{S}_n = \mathbf{X}_{l,n} + \cdots + \mathbf{X}_{r_n,n}$, and denote its distribution and it characteristic function by U_n and ω_n, respectively. For reasons explained in VI,3 we are interested primarily in arrays where the influence of individual components is asymptotically negligible. To ensure this we imposed the condition VI,(3.2) that the variables $\mathbf{X}_{k,n}$ tend in probability to zero uniformly in $k = 1, \ldots, r_n$. In terms of characteristic functions this means that given $\epsilon > 0$ and $\zeta_0 > 0$ one has for all n sufficiently large

$$(7.1) \qquad |1 - \varphi_{k,n}(\zeta)| < \epsilon \qquad \textit{for} \quad |\zeta| < \zeta_0, \quad k = 1, \ldots, r_n.$$

Such an array is called a *null array*.

In effect sections 1 and 2 are concerned with triangular arrays in which the distributions $F_{k,n}$ do not depend on k, and such arrays are automatically null arrays. The condition (7.1) enables us to use the theory developed in the first two sections. In particular, it will be now shown that the main result carries over to arbitrary null-arrays: *if the distributions of the row-sums* $\mathbf{S}_n$

tend to a limit, the latter is infinitely divisible.[5] We shall find precise criteria for the convergence to a specified infinitely divisible distribution.

For the reader's convenience we recall that a measure M is *canonical* if it attributes finite masses $M\{I\}$ to *finite* intervals and is such that the integrals

$$(7.2) \qquad M^+(x) = \int_x^\infty y^{-2}\, M\{dy\}, \qquad M^-(-x) = \int_{-\infty}^{-x} y^{-2}\, M\{dy\}$$

exist for each $x > 0$. (Definition 1 of section 1.)

To simplify notations we introduce a measure M_n defined by

$$(7.3) \qquad M_n\{dx\} = \sum_{k=1}^{r_n} x^2\, F_{k,n}\{dx\}.$$

This is the analogue to the measures $nx^2\, F_n\{dx\}$ in the preceding sections. By analogy to (7.2) we put for $x > 0$

$$(7.4) \qquad M_n^+(x) = \sum_{k=1}^{r_n} [1 - F_{k,n}(x)], \qquad M_n^-(x) = \sum_{k=1}^{r_n} F_{k,n}(-x).$$

Extensive use will be made of truncated variables, but the standard truncation procedure will be modified slightly in order to avoid trite complications resulting from the use of discontinuous functions. The modified procedure will replace the random variable $\mathbf{X}$ by the truncated variable $\tau(x)$ where τ is the *continuous monotone function such that*

$$(7.5) \qquad \tau(x) = x \quad \textit{for} \quad |x| \le a, \qquad \tau(x) = \pm a \quad \text{for} \quad |x| \le a$$

[obviously $\tau(-x) = -\tau(x)$]. For the expectations of the truncated variables we write

$$(7.6) \qquad \begin{gathered} \beta_{k,n} = \mathbf{E}(\tau(\mathbf{X}_{k,n})) \\ b_n = \sum_{k=1}^{r_n} \beta_{k,n}, \qquad B_n = \sum_{k=1}^{r_n} \beta_{k,n}^2. \end{gathered}$$

Theoretically it would be possible to center the $\mathbf{X}_{k,n}$ in such a way that all $\beta_{k,n}$ vanish. This would simplify arguments, but the resulting criterion would not be directly applicable in many concrete situations. However, it is usually possible to center the $\mathbf{X}_{k,n}$ so as to render the $\beta_{k,n}$ small enough that $B_n \to 0$. In this case the conditions of the following theorem reduce to the condition $M_n \to M$ familiar from the preceding sections. In the general case we still have $M_n\{I\} \to M\{I\}$ for intervals I at a positive distance from the origin, but neighborhoods of the origin are affected by B_n. (The choice of the truncation point a has no effect.)

[5] Concerning the implications of this result for processes with independent but non-stationary increments, see the concluding remarks to section 1.

Theorem. *Let* $\{\mathbf{X}_{k,n}\}$ *be a null array. If it is possible to find constants* b_n *such that the distributions of* $\mathbf{S}_n - b_n$ *tend to a limit distribution* U, *the* b_n *of* (7.6) *will do.*[6] *The limit distribution* U *is infinitely divisible.*[7]

In order that convergence takes place to a limit U *with canonical measure* M *it is necessary and sufficient that at all points of continuity* $x > 0$

$$M_n^+(x) \to M^+(x), \qquad M_n^-(-x) \to M^-(x) \tag{7.7}$$

and that for some $s > 0$

$$M_n\overline{\{-s, s\}} - B_n \to M\overline{\{-s, s\}}. \tag{7.8}$$

In this case the distribution of $\mathbf{S}_n - b_n$ *tends to the distribution with characteristic function* $\omega = e^{\psi}$ *defined by*

$$\psi(\zeta) = \int_{-\infty}^{-\infty} \frac{e^{i\zeta x} - 1 - i\zeta\tau(x)}{x^2}\, M\{dx\}. \tag{7.9}$$

[The condition (7.8) will automatically hold at *all* points of continuity.]

Proof. We proceed by steps.

(*a*) Suppose first that all variables $\mathbf{X}_{k,n}$ are symmetric so that the distributions of $\mathbf{S}_n$ must converge without preliminary centering. The characteristic functions $\varphi_{k,n}$ are real, and in view of (7.1) the Taylor expansion

$$-\log \varphi_{k,n}(\zeta) = [1 - \varphi_{k,n}(\zeta)] + \tfrac{1}{2}[1 - \varphi_{k,n}(\zeta)]^2 + \cdots \tag{7.10}$$

holds for arbitrary ζ provided only that n is sufficiently large. The question is whether

$$\sum_{k=1}^{r_n} \log \varphi_{k,n}(\zeta) \to \psi(\zeta). \tag{7.11}$$

All the terms of the expansion in (7.10) are positive, and hence (7.11) requires that the sum of the linear terms remains bounded. In view of (7.1) this implies that the contribution of the higher-order terms is asymptotically negligible and we conclude that (7.11) holds iff

$$\sum_{k=1}^{r_n} [\varphi_{k,n}(\zeta) - 1] \to \psi(\zeta). \tag{7.12}$$

The left side may be written in the form $r_n[\varphi_n - 1]$ where φ_n is the characteristic function of the arithmetic mean of the distributions $F_{k,n}$. We are thus

[6] The theorem remains valid also with standard truncation, that is, if τ is replaced by the truncation function vanishing outside $|x| > a$. To avoid notational complications it is then necessary to assume that there are no atoms of M at $\pm a$. [Part (*b*) of the proof gets more involved since it may not be possible to find θ such that $\mathbf{E}(\tau(\mathbf{X}+\theta) = 0$.]

[7] A distribution concentrated at a single point is infinitely divisible, the corresponding canonical measure being identically zero.

concerned with a special case of theorem 2 of section 2 and conclude that a relation of the form (7.12) holds iff there exists a canonical measure M such that $M_n \to M$. When $B_n = 0$ the conditions (7.7)–(7.8) are equivalent to $M_n \to M$ because for an interval I at a positive distance from the origin the relation $M_n\{I\} \to M\{I\}$ is implied by (7.7). This proves the theorem for symmetric distributions.

(*b*) Suppose next that $\beta_{k,n} = 0$ for all k and n. We shall prove that (7.11) cannot take place unless

$$\sum_{k=1}^{r_n} |\varphi_{k,n}(\zeta) - 1| < C(\zeta), \tag{7.13}$$

that is, unless the sum on the left remains bounded. In this case (7.11) and (7.12) are again equivalent, and the concluding argument used under (*a*) again reduces the assertion to theorem 2 of section 2. It is true that this theorem refers to a centering b'_n using $\sin x$ instead of $\tau(x)$, but this is compensated by the corresponding change for the limit distribution since

$$b_n - b'_n = \int_{-\infty}^{+\infty} \frac{\tau(x) - \sin x}{x^2}\, M_n\{dx\} \to \int_{-\infty}^{+\infty} \frac{\tau(x) - \sin x}{x^2}\, M\{dx\}. \tag{7.14}$$

To derive (7.13) from (7.11) we start from the identity

$$\varphi_{k,n}(\zeta) - 1 = \int_{-\infty}^{+\infty} [e^{i\zeta x} - 1 - i\zeta\tau(x)]F_{k,n}\{dx\} \tag{7.15}$$

valid because $\beta_{k,n} = 0$. For $|x| < a$ the integrand equals $e^{i\zeta x} - 1 - i\zeta x$ and is dominated by $\frac{1}{2}\zeta^2 x^2$. Since $|\tau(x)| \leq a$ it follows that

$$\sum_{k=1}^{r_n} |\varphi_{k,n}(\zeta) - 1| \leq \tfrac{1}{2}\zeta^2 M_n\{\overline{-a, a}\} + (2+a\,|\zeta|)(M_n^+(a)+M_n^-(-a)). \tag{7.16}$$

To show that $M_n^+(a)$ must remain bounded we consider the array $\{{}^0\mathbf{X}_{k,n}\}$ obtained by symmetrization of $\{\mathbf{X}_{k,n}\}$. The condition (7.1) for null arrays implies that for n sufficiently large the probability of the event

$${}^0\mathbf{X}_{k,n} > \mathbf{X}_{k,n} - \epsilon$$

exceeds $\frac{1}{2}$ for all $k \leq r_n$. Thus $\frac{1}{2}M_n^+(a) < {}^0M_n^+(a)$, and we know that the latter quantity remains bounded if convergence takes place. We conclude that in case of convergence $M_n^+(a) + M_n^-(-a)$ remains bounded, and hence

$$\sum_{k=1}^{r_n} |\varphi_{k,n}(\zeta) - 1|^2 = M_n\{\overline{-a, a}\} \cdot \epsilon_n(\zeta), \tag{7.17}$$

where ϵ_n stands for a quantity tending to zero. On the other hand, the real part of the integrand in (7.15) does not change sign. For $|x| < a$ and ζ

sufficiently small it is in absolute value $>\frac{1}{4}\zeta^2x^2$, and hence

$$-\text{Re}\sum_{k=1}^{r_n}(\varphi_{k,n}(\zeta)-1) \geq \tfrac{1}{4}\zeta^2 M_n\{\overline{-a,a}\}. \tag{7.18}$$

The last two inequalities show that the left side in (7.11) cannot remain bounded unless $M_n\{\overline{-a,a}\}$ remains bounded, and in this case (7.13) is implied by (7.16).

(*c*) We turn finally to an arbitrary null-array $\{\mathbf{X}_{k,n}\}$. Since $\mathbf{E}(\tau(\mathbf{X}_{k,n}-\theta))$ is a continuous monotone function of θ going from a to $-a$ there exists a unique value $\theta_{k,n}$ such that the variable $\mathbf{Y}_{k,n}=\mathbf{X}_{k,n}-\theta_{k,n}$ satisfies the condition $\mathbf{E}(\tau(\mathbf{Y}_{k,n}))=0$. Clearly $\{\mathbf{Y}_{k,n}\}$ is a null-array and hence the theorem applies to it.

We have thus found the general form of the possible limit distributions, but the conditions for convergence are expressed in terms of the measure $N_n\{dx\}=\sum x^2F_{k,n}\{\theta_{k,n}+dx\}$ of the artificially centered distributions of the $\mathbf{Y}_{k,n}$. In other words, we have proved the theorem with M_n replaced by N_n in (7.7) and (7.8), and B_n replaced by 0.

To eliminate the centering constants $\theta_{k,n}$ we recall that they tend uniformly to 0 and so ultimately

$$M_n^+(x+\epsilon) \leq N_n^+(x) \leq M_n^+(x-\epsilon).$$

It follows that the condition (7.7) applies interchangeably to both arrays.

Before turning to condition (7.8) we show that the arrays $\{\mathbf{Y}_{k,n}\}$ and $\{\mathbf{X}_{k,n}-\beta_{k,n}\}$ have the same limit distribution, that is,

$$b_n-\sum_{k=1}^{r_n}\theta_{k,n}\to 0. \tag{7.19}$$

Let $\mathbf{Z}_{k,n}=\tau(\mathbf{Y}_{k,n})-\tau(\mathbf{X}_{k,n})+\theta_{k,n}$. From the definition of τ it is clear that $\mathbf{Z}_{k,n}$ vanishes unless $\mathbf{X}_{k,n}\,|>a-1\theta_{k,n}|$ and even there $|\mathbf{Z}_{k,n}|\leq|\theta_{k,n}|\to 0$. Condition (7.7) therefore guarantees that

$$\sum_{k=1}^{r_n}|\mathbf{E}(\mathbf{Z}_{k,n})|=\sum_{k=1}^{r_n}|\beta_{k,n}-\theta_{k,n}|\to 0, \tag{7.20}$$

and this is stronger than (7.19).

Finally, we turn to condition (7.8). We use the sign $\approx$ to indicate that the difference of the two sides tends to 0 as $n\to\infty$. When (7.7) and (7.20) hold it is easily seen that

$$\begin{aligned} M_n\{\overline{-a,a}\}-B_n &\approx \sum_{k=1}^{r_n}\int_{|x|<a}(x-\theta_{k,n})^2\,F_{k,n}\{dx\} \\ &\approx \sum_{k=1}^{r_n}\int_{|y|<a}y^2\,F_{k,n}\{dy+\theta_{k,n}\}\approx N_n\{\overline{-a,a}\}, \end{aligned} \tag{7.21}$$

and thus (7.8) is equivalent to the corresponding condition for the array $\{\mathbf{Y}_{k,n}\}$. ▶

Example. *The role of centering.* For $k = 1, \ldots, n$ let $\mathbf{X}_{k,n}$ be normally distributed with expectation $n^{-\frac{1}{4}}$ and variance n^{-1}. With the centering to zero expectations the limit distribution exists and is normal. But with the centering constants $\beta_{k,n} = n^{-\frac{1}{4}}$ we have $B_n \sim 2\sqrt{n} \to \infty$. It follows that $\overline{M_n\{-a, a\}} \to \infty$. This example shows that the non-linear form of the theorem is unavoidable if arbitrary centerings are permitted. It shows also that in this case it does *not* suffice to consider the linear term in the expansion (7.10) for $\log \varphi_{k,n}$. ▶

For further results see problems 17 and 18.

†8. THE CLASS L

As an illustration of the power of the last theorem we give a simple proof of a theorem discovered by P. Lévy. We are once more concerned with partial sums $\mathbf{S}_n = \mathbf{X}_1 + \cdots + \mathbf{X}_n$ of a sequence of mutually independent random variables but, in contrast to section 5, the distribution F_n of $\mathbf{X}_n$ is permitted to depend on n. We put $\mathbf{S}_n^* = (\mathbf{S}_n - b_n)/a_n$ and wish to characterize the possible limit distributions of $\{\mathbf{S}_n^*\}$, under the assumption that

$$(8.1) \qquad a_n \to \infty, \qquad \frac{a_{n+1}}{a_n} \to 1.$$

The first condition eliminates convergent series $\sum \mathbf{X}_k$ which are treated in section 10. Situations avoided by the second condition are best illustrated by the

Example. Let $\mathbf{X}_n$ have an exponential distribution with expectation $n!$. Put $a_n = n!$ and $b_n = 0$. Obviously the distribution of $\mathbf{S}_n^*$ tends to the exponential distribution with expectation 1, but the convergence is due entirely to the preponderance of the term $\mathbf{X}_n$. ▶

Following Khintchine it is usual to say that *a distribution belongs to the class L if it is the limit distribution of a sequence $\{\mathbf{S}_n^*\}$ satisfying the conditions* (8.1).

In this formulation it is not clear that all distributions of the class L are infinitely divisible, but we shall prove this as a consequence of the

Lemma. *A characteristic function ω belongs to the class L iff for each $0 < s < 1$ the ratio $\omega(\zeta)/\omega(s\zeta)$ is a characteristic function.*

---

† This section treats a special topic.

Proof. (*a*) *Necessity.* Denote the characteristic function of $\mathbf{S}_n^*$ by ω_n and let $n > m$. The variable $\mathbf{S}_n^*$ is the sum of $(a_m/a_n)\mathbf{S}_m^*$ and a variable depending only on $\mathbf{X}_{m+1}, \ldots, \mathbf{X}_n$. Therefore

$$\omega_n(\zeta) = \omega_m(\zeta a_m/a_n) \cdot \varphi_{m,n}(\zeta) \tag{8.2}$$

where $\varphi_{m,n}$ is a characteristic function. Now let $n \to \infty$ and $m \to \infty$ in such a way that $a_m/a_n \to s < 1$. [This is possible on account of (8.1).] The left side tends to $\omega(\zeta)$ and the first factor on the right tends to $\omega(s\zeta)$ because the convergence of characteristic functions is uniform in finite intervals. (Theorem 2 of XV,3.) We conclude first that ω has no zeros. In fact, since $\varphi_{m,n}$ remains bounded $\omega(\zeta_0) = 0$ would imply $\omega(s\zeta_0) = 0$, and hence $\omega(s^k\zeta_0)$ for all $k > 0$, whereas actually $\omega(s^k\zeta_0) \to 1$. Accordingly, the ratio $\omega(\zeta)/\omega(s\zeta)$ appears as the continuous limit of the characteristic functions $\varphi_{m,n}$, and is therefore a characteristic function.

(*b*) *Sufficiency.* The above argument shows that ω has no zeros, and hence we have the identity

$$\omega(n\zeta) = \omega(\zeta) \cdot \frac{\omega(2\zeta)}{\omega(\zeta)} \cdots \frac{\omega(n\zeta)}{\omega((n-1)\zeta)}. \tag{8.3}$$

Under the conditions of the lemma the factor $\omega(k\zeta)/\omega((k-1)\zeta)$ is the characteristic function of a random variable $\mathbf{X}_k$ and hence $\omega(\zeta)$ is the characteristic function of $(\mathbf{X}_1 + \cdots + \mathbf{X}_n)/n$. ▶

We have not only proved the theorem but have found that ω is the characteristic function of the nth row sum in a triangular array. The condition (7.1) for null arrays is trivially satisfied, and hence ω is infinitely divisible. To find the canonical measure M determining ω we note that the ratio $\omega(\zeta)/\omega(s\zeta)$ is infinitely divisible as can be seen from the factorization (8.3). The canonical measure N determining $\omega(\zeta)/\omega(s\zeta)$ is related to M by the identity

$$N\{dx\} = M\{dx\} - s^2 M\{s^{-1}\,dx\}. \tag{8.4}$$

In terms of the functions M^+ and M^- this relation reads

$$N^+(x) = M^+(x) - M^+(x/s), \qquad N^-(-x) = M^-(-x) - M^-(-x/s). \tag{8.5}$$

We have shown that if the canonical measure M determines a characteristic function ω of class L, then the functions N^+ and N^- defined in (8.5) must be monotone for each $0 < s < 1$. Conversely, if this is true then (8.4) defines a canonical measure determining $\omega(\zeta)/\omega(s\zeta)$. We have thus proved the

Theorem. *A characteristic function* ω *belongs to the class* L *iff it is infinitely divisible and its determining canonical measure* M *is such that the two functions in* (8.5) *are monotone for every fixed* $0 < s < 1$.

Note. It is easily verified that the functions are monotone iff $M^+(e^x)$ and $M^-(-e^x)$ are *convex* functions.

*9. PARTIAL ATTRACTION. "UNIVERSAL LAWS"

As we have seen, a distribution F need not belong to any domain of attraction, and the question arises whether there exist general patterns in the asymptotic behavior of the sequence $\{F^{n\star}\}$ of its successive convolutions. The sad answer is that practically every imaginable behavior occurs and no general regularity properties are discernible. We describe a few of the possibilities principally for their curiosity value.

The characteristic function φ *is said to belong to the domain of partial attraction of* γ *iff there exist norming constants* a_r, b_r *and a sequence of integers* $n_r \to \infty$ *such that*

$$(9.1) \qquad [\varphi(\zeta/a_r)e^{-ib_r\zeta}]^{n_r} \to \gamma(\zeta).$$

Here it is understood that $|\gamma|$ is not identically 1, that is, the corresponding distribution is not concentrated at one point. Thus (9.1) generalizes the notion of domains of attraction by considering limits of subsequences.

The limit γ *is necessarily infinitely divisible* by virtue of theorem 2 of section 1. The following examples will show that both extremes are possible: *there exist distributions that belong to no domain of partial attraction and others that belong to the domain of partial attraction of every infinitely divisible distribution.*

Examples. (*a*) Example 3(*f*) exhibits a characteristic function φ which is not stable but belongs to its own domain of partial attraction.

(*b*) *A symmetric distribution with slowly varying tails belongs to no domain of partial attraction.* Suppose that $L(x) = 1 - F(x) + F(-x)$ varies slowly at infinity. By theorem 2 of VIII,9 in this case

$$(9.2) \qquad U(x) = \int_{-x}^{x} y^2\, F\{dy\} = o(x^2L(x)), \qquad x \to \infty.$$

By the theorem of section 7, for F to belong to some domain of partial attraction it is necessary that as n runs through an appropriate sequence $n[1 - F(a_nx) + F(-a_nx)]$ and $na_n^{-2}\, U(a_nx)$ converge at all points of continuity. The first condition requires that $n\, L(a_n) \sim 1$, the second that $n\, L(a_n) \to \infty$.

(*c*) *An infinitely divisible* γ *need not belong to its own domain of partial attraction.* Indeed, it follows from theorem 1 of section 1 that if φ belongs to the domain of attraction of γ so does the characteristic function $e^{\varphi-1}$,

---

* This section treats special topics.

which is infinitely divisible. The last example shows that $e^{\zeta-1}$ need not belong to any domain of partial attraction.

(*d*) As a preparation to the oddities in the subsequent examples we prove the following proposition. Consider an arbitrary sequence of infinitely divisible characteristic functions $\omega_r = e^{\psi_r}$ with bounded exponents. Put

$$\lambda(\zeta) = \sum_{k=1}^{\infty} \psi_k(a_k\zeta)/n_k. \tag{9.3}$$

It is possible to choose the constants $a_k > 0$ and integers n_k such that as $r \to \infty$

$$n_r\lambda(\zeta/a_r) - \psi_r(\zeta) \to 0 \tag{9.4}$$

for all ζ.

Proof. Choose for $\{n_k\}$ a monotone sequence of integers increasing so rapidly that $n_k/n_{k-1} > 2^k \max |\psi_k|$. The left side in (9.4) is then dominated by

$$n_r\sum_{k=1}^{r-1} |\psi_k(\zeta a_k/a_r)| + \sum_{k=r+1}^{\infty} 2^{-k}. \tag{9.5}$$

We choose the coefficients a_r recursively as follows. Put $a_1 = 1$. Given $a_1, \ldots, a_{r-1}$ choose a_r so large that the quantity (9.5) is $<1/r$ for all $|\zeta| < r$. This is possible because the first sum depends continuously on ζ and vanishes for $\zeta = 0$.

(*e*) *Every infinitely divisible characteristic function* $\omega = e^{\psi}$ *possesses a domain of partial attraction.* Indeed, we know that ω is the limit of a sequence of characteristic functions $\omega_k = e^{\psi_k}$ of the compound Poisson type. Define λ by (9.3) and put $\varphi = e^{\lambda}$. Then φ is a characteristic function and (9.4) states that

$$\lim \varphi^{n_r}(\zeta/a_r) = \lim e^{\psi_r(\zeta)} = \omega(\zeta). \tag{9.6}$$

(*f*) *Variants.* Let e^{α} and e^{β} be two infinitely divisible characteristic functions and choose the terms in (9.3) such that $\psi_{2k} \to \alpha$ and $\psi_{2k+1} \to \beta$. It follows from (9.4) easily that if a sequence $\nu_r\lambda(\zeta/a_{k_r})$ converges, the limit is necessarily a linear combination of α and β. In other words, e^{λ} *belongs to the domain of partial attraction of all characteristic functions of the form* $e^{p\alpha+q\beta}$, *and to no others.* This example generalizes easily. In the terminology of convex sets it shows that a distribution F may belong to the domains of partial attraction of all distributions in the *convex hull of n prescribed infinitely divisible distributions.*

(*g*) *Given a sequence of infinitely divisible characteristic functions* e^{α_1}, $e^{\alpha_2}, \ldots$ *there exists a* $\varphi = e^{\lambda}$ *belonging to the domain of partial attraction of each of them.* Partition the integers into infinitely many subsequences. (For example, let the *n*th subsequence contain all those integers that are divisible by 2^{n-1} but not by 2^n.) We can then choose the ψ_r in example (*d*) such that $\psi_r \to \alpha_n$ when r runs through the nth subsequence. With this choice (9.4) shows that $\varphi = e^\lambda$ has the desired property.

(*h*) *Doblin's "universal laws." It is possible that* φ *belongs to the domain of partial attraction of every infinitely divisible* ω. Indeed, it is obvious that if φ belongs to the domain of partial attraction of $\omega_1, \omega_2, \ldots$ and $\omega_n \to \omega$, then φ belongs also to the domain of partial attraction of ω. Now there exist only countably many infinitely divisible characteristic functions whose canonical measures are concentrated at finitely many rational points and have only rational weights. We can therefore order these functions in a simple sequence $e^{\alpha_1}, e^{\alpha_2}, \ldots$. Then *every* infinitely divisible ω is the limit of a subsequence of $\{e^{\alpha_k}\}$. The characteristic function φ of the last example belongs to the domain of partial attraction of each α_k, and therefore also of ω.

[**Note.** The last result was obtained by W. Doblin in a masterly study in 1940, following previous work by A. Khintchine in 1937. The technical difficulties presented by the problem at that time were formidable. The phenomenon of example (*b*) was discovered in special cases by B. V. Gnedenko, A. Khintchine, and P. Lévy. It is interesting to observe the complications encountered in a special example when the underlying phenomenon of regular variation is not properly understood.]

*10. INFINITE CONVOLUTIONS

Let $\mathbf{X}_1, \mathbf{X}_2, \ldots$ be independent random variables with characteristic functions $\varphi_1, \varphi_2, \ldots$. As in (7.5) we denote by τ the monotone continuous truncation function defined by $\tau(x) = x$ for $|x| \leq a$ and $\tau(x) = \pm a$ for $|x| \geq a$. The basic theorem on infinite convolutions states that *the distributions of the partial sums* $\mathbf{X}_1 + \cdots + \mathbf{X}_n$ *converge to a probability distribution* U *iff*

$$(10.1) \qquad \sum_{k=1}^{\infty} \mathrm{Var}(\tau(\mathbf{X}_k)) < \infty, \qquad \sum_{k=1}^{\infty} \mathbf{P}\{|\mathbf{X}_k| > a\} < \infty$$

and

$$(10.2) \qquad \sum_{k=1}^{n} \mathbf{E}(\tau(\mathbf{X}_k)) \to b$$

where b is a number.

The special case of finite variance was treated in VIII,5 together with examples and applications. In full generality the theorem appears in IX,9 where the result is also extended by proving the convergence of the series $\sum \mathbf{X}_n$ (the "three-series theorem"). The theorem was shown to be a simple corollary to the basic theorems concerning triangular arrays, and it is not

* This section treats a special topic.

necessary to repeat the argument.[8] We shall therefore be satisfied with examples illustrating the use of characteristic functions.

Examples. (*a*) *Factorization of the uniform distribution.* Let $\mathbf{X}_k = \pm 2^{-k}$ with probability $\frac{1}{2}$. It was shown in example I,11(*c*) informally that $\sum \mathbf{X}_k$ may be interpreted as "a number chosen at random between 1 and -1." This amounts to the assertion that the characteristic function $(\sin \zeta)/\zeta$ of the uniform distribution is the infinite product of the characteristic functions $\cos(\zeta/2^k)$. For an analytic proof we start from the identity

$$\frac{\sin \zeta}{\zeta} = \cos\frac{\zeta}{2} \cdot \cos\frac{\zeta}{4} \cdots \cos\frac{\zeta}{2^n} \cdot \frac{\sin(\zeta/2^n)}{\zeta/2^n} \tag{10.3}$$

which is proved by induction using the formula $\sin 2\alpha = 2 \sin \alpha \cos \alpha$. As $n \to \infty$ the last factor tends to 1 uniformly in every finite interval.

Note that the product of the even-numbered terms again corresponds to a sum of independent random variables. We know from example I,11(*d*) that this sum has a *singular distribution* of the Cantor type.[9]

(See problems 5, 7, and 19.)

(*b*) Let $\mathbf{Y}_k$ have density $\frac{1}{2}e^{-|x|}$ with characteristic function $1/(1 + \zeta^2)$. Then $\sum \mathbf{Y}_k/k$ converges. For the characteristic function we get the canonical product representation for $\pi\zeta/\sinh \pi\zeta$ where sinh denotes the hyperbolic sine. Using problem 8 in XV,9 we find that *the density of* $\sum \mathbf{Y}_k/k$ *is given by* $1/(2 + e^x + e^{-x}) = 1/4(\cosh(x/2))^2$.

11. HIGHER DIMENSIONS

The theory developed in this chapter carries over without essential changes to higher dimensions, and we shall not give all the details. In the canonical form for infinitely divisible distributions it is best to separate the normal component and consider only canonical measures without atom at the origin. The formulas then require no change provided ζx is interpreted as an inner product in the manner described in XV,7. For definiteness we spell out the formula in two dimensions.

A measure without an atom at the origin is canonical if it attributes finite masses to finite intervals and if $1/(1+x_1^2+x_2^2)$ is integrable with respect

[8] It is a good exercise to verify *directly* that the conditions (10.1)–(10.2) assure that the products $\varphi_1 \cdots \varphi_n$ converge uniformly in every finite interval. The *necessity* of the conditions is less obvious, but follows easily on observing that the triangular array whose nth row is $\mathbf{X}_n, \mathbf{X}_{n+1}, \ldots, \mathbf{X}_{n+r_n}$ must satisfy the conditions of the theorem of section 7 with $M = 0$.

[9] G. Choquet gave a charming geometric proof applicable to more general infinite convolutions. It is given in A. Tortrat, J. Math. Pures. Appl., vol. 39 (1960) pp. 231–273.

to it, and if it has no atom at the origin. Choose an appropriate centering function in *one* dimension, say $\tau(x) = \sin x$ or the one defined in (7.5). Put

$$(11.1)\qquad \psi(\zeta_1, \zeta_2) = \int \frac{e^{i(\zeta_1 x_1 + \zeta_2 x_2)} - 1 - i\zeta_1\tau(x_1) - i\zeta_2\tau(x_2)}{x_1^2 + x_2^2} M\{dx\},$$

the integral extending over the whole plane. Then $\omega = e^{\psi}$ is an infinitely divisible bivariate characteristic function. The most general infinitely divisible characteristic function is obtained by multiplication by a normal characteristic function.

A reformulation in polar coordinates may render the situation more intuitive. Put

$$(11.2)\qquad \zeta_1 = \rho\cos\varphi, \qquad \zeta_2 = \rho\sin\varphi, \qquad x = r\cos\theta, \qquad y = r\sin\theta.$$

Define the canonical measure in polar coordinates as follows. For each θ with $-\pi < \theta \leq \pi$ choose a one-dimensional canonical measure Λ_θ concentrated on $\overline{0, \infty}$; furthermore, choose a finite measure W on $-\pi < \theta \leq \pi$ (the circle). Then M may be defined by randomization of the parameter θ, and (with a trite change in centering) (11.1) may be recast in the form

(11.3)

$$\psi(\zeta_1, \zeta_2) = \int_{-\pi}^{\pi} W\{d\theta\} \int_{0+}^{\infty} \frac{e^{i\rho r\cos(\varphi-\theta)} - 1 - i\rho\tau(r)\cos(\varphi-\theta)}{r^2} \Lambda_\theta\{dr\}.$$

(This form permits one to absorb the normal component by adding an atom at the origin to Λ_θ.)

Example. *Stable distributions.* By analogy with one dimension we put $\Lambda_\theta\{dr\} = r^{-\alpha+1}\,dr$. One could add an arbitrary factor C_θ, but this would merely change the measure W. As we have seen in example 3(*g*), with this measure (11.3) takes on the form

$$(11.4)\qquad \psi(\zeta_1, \zeta_2) = -C\rho^\alpha \int_{-\pi}^{\pi} |\cos(\varphi-\theta)|^\alpha \left(1 \mp \tan\frac{\pi}{2}\alpha\right) W\{d\theta\},$$

where the upper or lower sign prevails according as $\varphi - \theta > 0$ or $\varphi - \theta < 0$. This shows that e^{ψ} is a strictly stable characteristic function, and as in section 5 one sees that there are no others. However, just as in one dimension, the exponent $\alpha = 1$ leads to characteristic functions that are stable only in the wide sense and have a logarithmic term in the exponent.

When $\alpha = 1$ and W is the uniform distribution we get the characteristic function $e^{-a\rho}$ of *the symmetric Cauchy distribution in* $\mathfrak{R}^2$ [see example XV,7(*e*) and problems 21–23]. ▶

12. PROBLEMS FOR SOLUTION

1. It was shown in section 2 that if ψ is the logarithm of an infinitely divisible characteristic function, then

$$\psi(\zeta) - \frac{1}{2h}\int_{-h}^{h} \psi(\zeta - s)\,dx = \chi(\zeta) \tag{12.1}$$

is a real multiple of a characteristic function. Prove the converse: Suppose that ψ is a continuous function such that $\chi(\zeta)/\chi(0)$ is a characteristic function for every choice of $h > 0$. Then ψ differs only by a linear function from the logarithm of an infinitely divisible characteristic function. Furthermore, ψ is such a logarithm if it satisfies the further conditions $\psi(0) = 0$ and $\psi(-\zeta) = \overline{\psi(\zeta)}$.

[*Hint:* Prove that the solutions of the homogeneous equation (with $\chi = 0$) are linear.]

2. Show that problem 1 and the argument of section 2 remain valid if in (12.1) or (2.13) the uniform distribution is replaced by a distribution concentrated at the points:

$$\psi(\zeta) - \tfrac{1}{2}[\psi(\zeta+h) + \psi(\zeta-h)] = \chi(\zeta). \tag{12.2}$$

However, there arises a slight complication from the fact that the density corresponding to χ is not strictly positive.

3. *Generalization.* Let R be an arbitrary even probability distribution with finite variance. If e^{ψ} is an infinitely divisible characteristic function and

$$\chi = \psi - R \bigstar \psi, \tag{12.3}$$

then $\chi(\zeta)/\chi(0)$ is a characteristic function. The argument of section 2 goes through using (12.3) instead of (2.13).

In particular, if R has the density $\frac{1}{2}e^{-|x|}$ one is led directly to Khintchine's normal form for ψ. (See the concluding note to section 2.) However, some care is required by the fact that ψ is unbounded.

4. If ω is an infinitely divisible characteristic function then there exist constants a and b such that $|\log \omega(\zeta)| < a + b\zeta^2$ for all ζ.

5. *Shot noise in vacuum tubes.* In example VI,3(h) we considered a triangular array in which $\mathbf{X}_{k,n}$ had the characteristic function

$$\varphi_{k,n}(\zeta) = 1 + \alpha h[e^{i\zeta I(kh)} - 1],$$

where $h = n^{-\frac{1}{2}}$. Show that the characteristic functions of $\mathbf{S}_n = \mathbf{X}_{1,n} + \cdots + \mathbf{X}_{n,n}$ tend to e^{ψ} where

$$\psi(\zeta) = \alpha \int_0^{\infty} [e^{i\zeta I(x)} - 1]\,dx;$$

e^{ψ} is the characteristic function of the random variable $\mathbf{X}(t)$, and by differentiation one gets *Campbell's theorem* VI,(3.4).

6. Let $U = \sum \mathbf{X}_n/n$ where the variables $\mathbf{X}_k$ are independent and have the common density $\frac{1}{2}e^{-|x|}$. Show that[10] U is infinitely divisible with the canonical

[10] The characteristic function ω is defined by an infinite product which happens to be the well-known canonical product of $2\pi\,|\zeta|/e^{\pi|\zeta|} - e^{-\pi|\zeta|}$.

measure $M\{dx\} = |x| \dfrac{e^{-|x|}}{1 - e^{-|x|}} dx$. [No calculations beyond summing a geometric series are required.]

7. Let $P(s) = \Sigma p_k s^k$ where $p_k \geq 0$ and $\Sigma p_k = 1$. Assume $P(0) > 0$ and that $\log \dfrac{P(s)}{P(0)}$ is a power series with positive coefficients. If φ is the characteristic function of an arbitrary distribution F show that $P(\varphi)$ is an infinitely divisible characteristic function. Find its canonical measure M in terms of $F^{n\star}$.

Special case of interest: If $0 \leq a < b < 1$ then $\dfrac{1-b}{1-a} \cdot \dfrac{1-a\varphi}{1-b\varphi}$ is an infinitely divisible characteristic function. (See also problem 19.)

8. *Continuation.* Interpret $P(\varphi)$ in terms of randomization and subordinated processes using the fact that P is the generating function of an infinitely divisible integral-valued random variable.

9. Let $\mathbf{X}$ be stable with characteristic function $e^{-|\zeta|^\alpha}$ $(0 < \alpha \leq 2)$ and let $\mathbf{Y}$ be independent of $\mathbf{X}$. If $\mathbf{Y}$ is positive with a distribution G (concentrated on $\overline{0, \infty}$ show that the characteristic function of $\mathbf{X}\mathbf{Y}^{1/\alpha}$ is given by

$$\int_0^\infty e^{-|\zeta|^\alpha y}\, G\{dy\}.$$

Conclude: *If* $\mathbf{X}$ *and* $\mathbf{Y}$ *are independent strictly stable variables with exponents* α *and* β *and if* $\mathbf{Y} > 0$, *then* $\mathbf{X}\mathbf{Y}^{1/\alpha}$ *is strictly stable with exponent* $\alpha\beta$.

10. Let ω be a characteristic function such that $\omega^2(\zeta) = \omega(a\zeta)$ and $\omega^3(\zeta) = \omega(b\zeta)$. Then ω is stable.

[Example 3(f) shows that the first relation does not suffice. The exponents 2, 3 may be replaced by any two relatively prime integers.]

11. Show that the simple lemma 3 of VIII,8 applies (not only to monotone functions but also) to logarithms of characteristic functions. Conclude that if $\omega_n(\zeta) = \omega(a_n\zeta)$ for all n then $\log \omega(\zeta) = A\zeta^\alpha$ for $\zeta > 0$, where A is a complex constant.

12. *Continuation.* Using the result of problem 28 in VIII,10 show *directly* that if ω is a stable characteristic function then for $\zeta > 0$ either $\log \omega(\zeta) = A\zeta^\alpha + ib\zeta$ or else $\log \omega(\zeta) = A\zeta + ib\zeta \log \zeta$ with b real.

13. Let F be carried by $\overline{0, \infty}$ and $1 - F(x) = x^{-\alpha}L(x)$ with $0 < \alpha < 1$ and L slowly varying at infinity. Prove that $1 - \varphi(\zeta) \sim A\zeta^\alpha L(1/\zeta)$ as $\zeta \to 0+$

14. *Continuation.* From the results of section 5 prove the *converse*, and also that $A = \Gamma(1-\alpha)e^{-i\pi\alpha/2}$.

15. *Continuation.* By induction on k prove: in order that $1 - F(x) \sim ax^{-\alpha}L(x)$ as $x \to \infty$ with L slowly varying and $k < \alpha < k+1$ it is necessary and sufficient that as $\zeta \to 0+$

$$(*) \qquad \varphi(\zeta) - 1 - \frac{\mu_1(i\zeta)}{1!} - \cdots - \frac{\mu_k(i\zeta)^k}{k!} \sim A\zeta^\alpha L\left(\frac{1}{\zeta}\right).$$

Then automatically $A = -a\Gamma(k-\alpha)e^{-i\frac{1}{2}\pi\alpha}$.

16. Formulate the weak law for triangular arrays as a special case of the general theorem of section 7.

17. Let $\{\mathbf{X}_{k,n}\}$ be a null array whose row sums have a limit distribution determined by the canonical measure M. Show that for $x > 0$

$$\mathbf{P}\{\max [\mathbf{X}_{1,n}, \ldots, \mathbf{X}_{r_n,n}] \leq x\} \to e^{-M^+(x)}.$$

Formulate a converse.

18. Let $\{\mathbf{X}_{k,n}\}$ be a null array of symmetric variables whose row sums have a limit distribution determined by the canonical measure M with an atom of weight σ^2 at the origin. Show that the distribution of $\mathbf{S}_n^\# = \Sigma \mathbf{X}_{k,n}^2 - \sigma^2$ converges to a distribution determined by a measure $M_\#$ without atom at the origin and such that $M_\#^+(x) = 2M^+(\sqrt{x})$ for $x > 0$.

19. Let $0 < r_j < 1$ and $\Sigma r_j < \infty$. For arbitrary real a_j the infinite product

$$\frac{1 - r_1}{1 - r_1 e^{ia_1\zeta}} \cdot \frac{1 - r_2}{1 - r_2 e^{ia_2\zeta}} \cdots$$

converges and represents an infinitely divisible characteristic function. (*Hint:* Each factor is infinitely divisible by problem 7.)

20. Use the method of example 9(*d*) to construct a distribution F such that $\limsup F^{n\star}(x) = 1$ and $\liminf F^{n\star}(x) = 0$ at all points.

21. In (11.4) let W stand for the uniform distribution. Then

$$\psi(\zeta_1, \zeta_2) = -c[\zeta_1^2 + \zeta_2^2]^{\frac{1}{2}\alpha},$$

and e^ψ is a symmetric stable distribution.

22. In (11.4) let W attribute weight $\frac{1}{4}$ to each of the four points $0, \pi, \frac{1}{2}\pi, -\frac{1}{2}\pi$. Then (11.4) represents the bivariate characteristic function of two *independent* one-dimensional stable variables.

23. In (11.4) let W be concentrated on the two points σ and $\sigma + \pi$. Then (11.4) represents a *degenerate* characteristic function of a pair such that

$$\mathbf{X}_1 \sin \sigma - \mathbf{X}_2 \cos \sigma = 0.$$

More generally, any discrete W leads to a convolution of degenerate distributions. Explain (11.4) by a limiting process.

CHAPTER XVIII

Applications of Fourier Methods to Random Walks

To a large extent this chapter treats topics already covered in chapter XII, for which reason applications are kept to a minimum. A serious attempt has been made to make it self-contained and accessible with a minimum of previous knowledge except the Fourier analysis of chapter XV. The theory is entirely independent of the last two chapters. Section 6 is independent of the preceding ones.

1. THE BASIC IDENTITY

Throughout this chapter $\mathbf{X}_1, \mathbf{X}_2, \ldots$ are mutually independent random variables with a common distribution F and characteristic function φ. As usual we put $\mathbf{S}_0 = 0$ and $\mathbf{S}_n = \mathbf{X}_1 + \cdots + \mathbf{X}_n$; the sequence $\{\mathbf{S}_n\}$ constitutes the random walk generated by F.

Let A be an arbitrary set on the line and A' its complement. (In most applications A' will be a finite or infinite interval.) If I is a subset (interval) of A' and if

$$\mathbf{S}_1 \in A, \ldots, \mathbf{S}_{n-1} \in A, \mathbf{S}_n \in I \qquad (I \subset A') \tag{1.1}$$

we say that the set A' *is entered (for the first time) at epoch* n *and at a point of* I. Since A' need not be entered at all the *epoch* $\mathbf{N}$ *of the entry* is a possibly defective random variable, and the same is true of *the point* $\mathbf{S}_\mathbf{N}$ *of first entry*. For the joint distribution of the pair $(\mathbf{N}, \mathbf{S}_\mathbf{N})$ we write

$$\mathbf{P}\{\mathbf{N} = n, \mathbf{S}_\mathbf{N} \in I\} = H_n\{I\}, \qquad n = 1, 2, \ldots. \tag{1.2}$$

Thus $H_n\{I\}$ is the probability of the event (1.1), but the distribution (1.2) is defined for all sets I on the line by the convention that $H_n\{I\} = 0$ if $I \subset A$. The probabilities (1.2) will be called *hitting probabilities*. Their study is intimately connected with the study of the random walk prior to the first entry into A', that is, *the random walk restricted to* A. For $I \subset A$ and $n = 1, 2, \ldots$ put

$$(1.3) \qquad G_n\{I\} = \mathbf{P}\{\mathbf{S}_1 \in A, \ldots, \mathbf{S}_{n-1} \in A, \mathbf{S}_n \in I\};$$

in words, this is the probability that at epoch n the set $I \subset A$ is visited and up to epoch n no entry into A' took place. We extend this definition to all sets on the line by letting $G_n\{I\} = 0$ if $I \subset A'$.

$$(1.4) \qquad G_n\{A\} = 1 - \mathbf{P}\{\mathbf{N} \leq n\}.$$

The variable $\mathbf{N}$ is not defective iff this quantity tends to 0 as $n \to \infty$.

Considering the position $\mathbf{S}_n$ of the random walk at epochs $n = 1, 2, \ldots$ it is obvious that for $I \subset A'$

$$(1.5a) \qquad H_{n+1}\{I\} = \int_A G_n\{dy\}\, F\{I-y\}$$

whereas for $I \subset A$

$$(1.5b) \qquad G_{n+1}\{I\} = \int_A G_n\{dy\}\, F\{I-y\}.$$

We now agree to let G_0 *stand for the probability distribution concentrated at the origin*. Then the relations (1.5) hold for $n = 0, 1, 2, \ldots$ and determine recursively all the probabilities H_n and G_n. The two relations can be combined in one. Given an arbitrary set I on the line we split it into the components IA' and IA and apply (1.5) to these components. Recalling that H_n and G_n are concentrated, respectively, on A' and A we get

$$(1.6) \qquad H_{n+1}\{I\} + G_{n+1}\{I\} = \int_A G_n\{dy\}\, F\{I-y\}$$

for $n = 0, 1, \ldots$ and arbitrary I.

The special case $A = \overline{0, \infty}$ was treated in XII,3, the relation XII,(3.5) being the same as the present (1.5). We could retrace our steps and derive an integral equation of the Wiener-Hopf type analogous to XII,(3.9) and again possessing only one probabilistically possible solution (though the uniqueness is not absolute). It is preferable, however, to rely this time on the powerful method of Fourier analysis.

We are concerned with the distribution of the pair $(\mathbf{N}, \mathbf{S_N})$. Since $\mathbf{N}$ is integral-valued we use generating functions for $\mathbf{N}$ and characteristic functions for $\mathbf{S_N}$. Accordingly we put

$$(1.7) \qquad \chi(s, \zeta) = \sum_{n=1}^{\infty} s^n \int_{A'} e^{i\zeta x} H_n\{dx\}, \qquad \gamma(s, \zeta) = \sum_{n=0}^{\infty} s^n \int_A e^{i\zeta x} G_n\{dx\}.$$

(The zero terms of the two series equal 0 and 1, respectively.) These series converge at least for $|s| < 1$, but usually in a wider interval.

The effective domains of integration are inserted for clarity, but the limits of integration may be given as well as $-\infty$ and $+\infty$. In particular, the integral in (1.6) is an ordinary convolution. On taking Fourier-Stieltjes transforms the relation (1.6) therefore takes on the form

$$\chi_{n+1}(\zeta) + \gamma_{n+1}(\zeta) = \gamma_n(\zeta)\, \varphi(\zeta). \tag{1.8}$$

Multiplying by s^{n+1} and adding over $n = 0, 1, \ldots$ we get

$$\chi(s, \zeta) + \gamma(s, \zeta) - 1 = s\, \gamma(\zeta)\, \varphi(\zeta)$$

for all s for which the series in (1.7) converge. We have thus established *the basic identity*

$$1 - \chi = \gamma[1 - s\varphi]. \tag{1.9}$$

(For an alternative proof see problem 6.)

In principle χ and γ can be calculated recursively from (1.5), and the identity (1.9) appears at first glance redundant. In reality direct calculations are rarely feasible, but much valuable information can be extracted directly from (1.9).

Example. Let F stand for the bilateral exponential distribution with density $\frac{1}{2}e^{-|x|}$ and characteristic function $\varphi(\zeta) = 1/(1+\zeta^2)$, and let $A = \overline{-a, a}$. For $x > a$ we get from (1.5*a*)

$$H_{n+1}\{\overline{x, \infty}\} = H_{n+1}\{\overline{-\infty, -x}\} = \tfrac{1}{2}\int_{-a}^{a} G_n\{dy\}e^{-(x-y)} = c_n e^{-x} \tag{1.10}$$

with c_n independent of x. It follows that the point $\mathbf{S_N}$ of first entry into $|x| > a$ is independent of the epoch of this entry and has a density proportional to $e^{-|x|}$ (for $|x| > a$). This result accords intuitively with the lack of memory of the exponential distribution described in chapter I. The independence means that the joint characteristic function χ must factor, and from the form of the density for $\mathbf{S_N}$ we conclude that

$$\chi(s, \zeta) = \tfrac{1}{2}P(s)\left[\frac{e^{ia\zeta}}{1 - i\zeta} + \frac{e^{-ia\zeta}}{1 + i\zeta}\right] \tag{1.11}$$

where P is the generating function of the epoch $\mathbf{N}$ of the first entry into $|x| > a$. [The proportionality factor is deduced from the fact that $\chi(1, 0) = 1$.]

A direct calculation of $P(s)$ would be cumbersome, but an explicit expression can be easily deduced from (1.9). In fact, with our form of the characteristic function the right side in (1.9) vanishes for $\zeta = \pm i\sqrt{1 - s}$, and so for this value $\chi(s, \zeta)$ must reduce to 1. Thus

$$P(s) = 2\left[\frac{e^{-a\sqrt{1-s}}}{1+\sqrt{1-s}} + \frac{e^{a\sqrt{1-s}}}{1-\sqrt{1-s}}\right]^{-1}. \tag{1.12}$$

From this it follows that the epoch $\mathbf{N}$ of the first entry into $|x| > a$ has expectation $1 + a + \frac{1}{2}a^2$.

(For further examples see problems 1–5.) ►

*2. FINITE INTERVALS. WALD'S APPROXIMATION

Theorem. *Let* $A = \overline{-a, b}$ *be a finite interval containing the origin and let* $(\mathbf{N}, \mathbf{S_N})$ *be the hitting point for the complement* A'.

The variables $\mathbf{N}$ *and* $\mathbf{S_N}$ *are proper. The generating function*

$$\sum_{n=0}^{\infty} s^n\, \mathbf{P}\{\mathbf{N} > n\} = \sum_{n=0}^{\infty} s^n\, G_n\{A\} \tag{2.1}$$

converges for some[1] $s > 1$ *and hence* $\mathbf{N}$ *has moments of all orders. The hitting point* $\mathbf{S_N}$ *has an expectation iff the random-walk distribution* F *has an expectation* μ, *in which case*

$$\mathbf{E}(\mathbf{S_N}) = \mu \cdot \mathbf{E}(\mathbf{N}). \tag{2.2}$$

The identity (2.2) was first discussed by A. Wald. In the special case $A = \overline{0, \infty}$ it reduces to XII,(2.8).

Proof. As was already pointed out, $G_n\{A\}$ and $\mathbf{P}\{\mathbf{N} > n\}$ are different notations for the probability that the random walk lasts for more than n steps, and so the two sides in (2.1) are identical.

Choose an integer r such that $\mathbf{P}\{|\mathbf{S}_r| < a + b\} = \eta < 1$. The event $\{\mathbf{N} > n + r\}$ cannot occur unless

$$\mathbf{N} > n \quad \text{and} \quad |\mathbf{X}_{n+1} + \cdots + \mathbf{X}_{n+r}| < a + b$$

(These two events are independent because $\{\mathbf{N} > n\}$ depends only on the variables $\mathbf{X}_1, \ldots, \mathbf{X}_n$. Since $\mathbf{X}_{n+1} + \cdots + \mathbf{X}_{n+r}$ has the same distribution as $\mathbf{S}_r$ we conclude that

$$\mathbf{P}\{\mathbf{N} > n + r\} \leq \mathbf{P}\{\mathbf{N} > n\}\eta.$$

Hence by induction

$$\mathbf{P}\{\mathbf{N} > kr\} \leq \eta^k, \tag{2.3}$$

* This section is included because of its importance in statistics; it should be omitted at first reading.

[1] This is known to statisticians as C. Stein's lemma. For an alternative proof see problem 8.

which shows that the sequence $\mathbf{P}\{\mathbf{N} > n\}$ decreases at least as fast as a geometric sequence with ratio $\eta^{1/r}$. It follows that $\mathbf{N}$ is a proper variable and that the series in (2.1) converge at least for $|s| < \eta^{-1/r}$. This proves the first assertion.

It follows also that (1.9) is meaningful for $|s| < \eta^{-1/r}$. For $s = 1$ we conclude

$$1 - \chi(1, \zeta) = \gamma(1, \zeta)[1-\varphi(\zeta)]. \tag{2.4}$$

But $\chi(1, \zeta)$ is the characteristic function of $\mathbf{S_N}$, and the fact that $\chi(1, 0) = 1$ shows that $\mathbf{S_N}$ is proper.

The event $|\mathbf{S_N}| > t + a + b$ cannot occur unless for some n one has $\mathbf{N} > n - 1$ and $|\mathbf{X}_n| > t$. As already remarked, these two events are independent, and since the $\mathbf{X}_n$ are identically distributed we conclude that

$$\mathbf{P}\{|\mathbf{S_N}| > t + a + b\}$$
$$\leq \sum_{n=1}^{\infty} \mathbf{P}\{\mathbf{N} > n - 1\} \cdot \mathbf{P}\{|\mathbf{X}_1| > t\} = \mathbf{E}(\mathbf{N}) \cdot \mathbf{P}\{|\mathbf{X}_1| > t\}.$$

The expectation $\mu = \mathbf{E}(\mathbf{X}_1)$ exists iff the right side is integrable over $\overline{0, \infty}$. In this case the same is true of the left side, and then $\mathbf{E}(\mathbf{S_N})$ exists. On the other hand,

$$\mathbf{P}\{|\mathbf{S_N}| > t\} \geq \mathbf{P}\{|\mathbf{X}_1| > t + a + b\}$$

because the occurrence of the event on the right implies $\mathbf{S_N} = \mathbf{X}_1$. Thus the existence of $\mathbf{E}(\mathbf{S_N})$ implies the existence of $\mu = \mathbf{E}(\mathbf{X}_1)$. When these expectations exist we can differentiate (2.4) to obtain

$$i\,\mathbf{E}(\mathbf{S_N}) = \frac{\partial\chi(1, 0)}{\partial\zeta} = \varphi'(0)\,\gamma(1, 0) = i\mu\,\mathbf{E}(\mathbf{N}). \tag{2.5}$$ ►

We proceed now to derive a variant of the basic identity (1.9) known as Wald's identity. To avoid the use of imaginary arguments we put

$$f(\lambda) = \int_{-\infty}^{+\infty} e^{-\lambda x}\, F\{dx\}. \tag{2.6}$$

Suppose that this integral converges in some interval $-\lambda_0 < \lambda < \lambda_1$ about the origin. The characteristic function is then given by $\varphi(i\lambda) = f(\lambda)$, this function being analytic in a complex neighborhood of each λ in the given interval. Wald's identity is obtained formally from (1.9) letting $\zeta = i\lambda$ and $s = 1/\varphi(i\lambda)$. For these particular values the right side vanishes and hence $\chi(s, \zeta) = 1$. In view of the definition of χ this relation may be restated in probabilistic terms as follows.

Wald's lemma.[2] *If the integral* (2.6) *converges for* $-\lambda_0 < \lambda < \lambda_1$, *then in this interval*

$$\mathbf{E}(f^{-\mathbf{N}}(\lambda)e^{-\lambda \mathbf{S_N}}) = 1. \tag{2.7}$$

Proof. We repeat the argument leading to (1.9). As the measures G_n are concentrated on a finite interval their Fourier transforms χ_n converge for all ζ in the complex plane. By assumption $\varphi(i\lambda) = f(\lambda)$ exists, and hence it is seen that the Fourier version (1.8) of (1.6) is valid for $\zeta = i\lambda$. On multiplication by $f^{-n-1}(\lambda)$ this relation takes on the form

$$f^{-n-1}(\lambda)\,\chi_{n+1}(i\lambda) = f(\lambda)^{-n}\,\gamma_n(i\lambda) - f^{-n-1}(\lambda)\,\gamma_{n+1}(i\lambda). \tag{2.8}$$

If $f^{-n}(\lambda)\,\gamma_n(i\lambda) \to 0$ the right sides add to unity due to the obvious cancellation of terms. In this case summation of (2.8) leads to the assertion (2.7) and hence it suffices to show that

$$f^{-n}(\lambda)\,G_n\{A\} \to 0. \tag{2.9}$$

Now if $f(\eta) < \infty$

$$G_n\{A\} \le \mathbf{P}\{-a < \mathbf{S}_n < b\} \le e^{(a+b)|\eta|} \cdot \int_{-a}^{b} e^{-\eta x}\, F^{n\star}\,\{dx\}$$
$$\le e^{(a+b)|\eta|} \cdot f^n(\eta).$$

Thus (2.9) is true if $f(\lambda) > f(\eta)$. As we are free to choose η this proves (2.7) for all λ excepting values where f assumes its minimum. But being convex f has at most one minimum, and at it (2.7) follows by continuity. ▶

Example. *Estimates concerning* **N**. Wald was led to his lemma from problems in sequential analysis where it was required to find approximations to the distribution of the epoch **N** of the first exit from A, as well as estimates for the probabilities that this exit takes place to the right or left of this interval. Wald's method is a generalization of the procedure described in **1**; XIV,8 for arithmetic distributions with finitely many jumps. (There it is also shown how strict inequalities can be obtained.) Put

$$p_k = \mathbf{P}\{\mathbf{N}{=}k, \mathbf{S_N}{\ge}b\}, \qquad q_k = \mathbf{P}\{\mathbf{N}{=}k, \mathbf{S_N}{\le}{-a}\} \tag{2.10}$$

and write for the corresponding generating functions $P(s)$ and $Q(s)$. (Then $P + Q$ is the generating function for **N**.) Suppose now that a and b are large in comparison with the expectation and variance of F. The

[2] Wald used (2.7) in connection with sequential analysis. This was before 1945 and before the general random walks were systematically explored. It is therefore natural that his conditions were severe and his methods difficult, but unfortunately they still influence the statistical literature. The argument of the text utilizes an idea of H. D. Miller (1961).

hitting point $\mathbf{S_N}$ is then likely to be relatively close to either b or $-a$. If these were the only possible values of $\mathbf{S_N}$ the identity (2.7) would take on the form

$$P(1/f(\lambda))e^{-\lambda b} + Q(1/f(\lambda))e^{\lambda a} = 1, \tag{2.11}$$

and one expects naturally that under the stated assumptions (2.11) will be satisfied at least approximately. The function f is convex and it is usually possible to find an interval $s_0 < s < s_1$ such that in it the equation

$$sf(\lambda) = 1 \tag{2.12}$$

admits of two roots $\lambda_1(s)$ and $\lambda_2(s)$ depending continuously on s. Substituting into (2.11) we get *two linear equations for the generating functions* P and Q, and thus we get (at least approximately) the distribution of $\mathbf{N}$ and the probabilities for an exodus to the right and left. ▶

3. THE WIENER-HOPF FACTORIZATION

In this section we derive by purely analytical methods various consequences of the basic identity (1.9). It turns out that they contain, in a more flexible and sharper form, many of the results derived in chapter XII by combinatorial methods. This may produce the false impression of a superiority of the Fourier methods, but in reality it is the interplay of the two methods that characterizes the recent progress of the theory. Each method leads to results which seem inaccessible to the other. (For examples in one direction see section 5; the arc sine law for the number of positive partial sums as well as generalizations of the whole theory to exchangeable variables illustrate advantages of the combinatorial approach.)

From now on $\mathbf{N}$ and $\mathbf{S_N}$ will denote the epoch and the point of *first entry into the open half-line* $\overline{0, \infty}$. Their joint distribution

$$\mathbf{P}\{\mathbf{N} = n, \mathbf{S_N} \in I\} = H_n\{I\}$$

is given by

$$H_n\{I\} = \mathbf{P}\{\mathbf{S}_1 \leq 0, \ldots, \mathbf{S}_{n-1} \leq 0, \mathbf{S}_n \in I\}, \qquad I \subset \overline{0, \infty} \tag{3.1}$$

with the understanding that $H_0 = 0$ and that H_n is concentrated on $\overline{0, \infty}$. Instead of the bivariate characteristic function we introduce as before the more convenient combination of generating and characteristic function

$$\chi(s, \zeta) = \mathbf{E}(s^{\mathbf{N}}e^{i\zeta\mathbf{S_N}}), \tag{3.2}$$

namely,

$$\chi(s, \zeta) = \sum_{n=1}^{\infty} s^n \int_0^{\infty} e^{i\zeta x}\, H_n\{dx\}. \tag{3.3}$$

(The integration is over the open half-axis, but nothing changes if the lower limit is replaced by $-\infty$.) For brevity we shall refer to χ as "*the transform*" of the sequence of measures H_n.

For the epoch and point of first entry into the *open negative* half-axis we write $\mathbf{N}^-$ and $\mathbf{S}_{\mathbf{N}^-}$; then $\{H_n^-\}$ and χ^- denote the corresponding distribution and transform.

When the underlying distribution F is discontinuous we must distinguish between first entries into open and closed half-axes. It is therefore necessary to consider the event of a *return to the origin through negative values.* Its probability distribution $\{f_n\}$ is given by

$$f_n = \mathbf{P}\{\mathbf{S}_1 < 0, \ldots, \mathbf{S}_{n-1} < 0, \mathbf{S}_n = 0\}, \qquad n \geq 1, \tag{3.4}$$

and we put $f(s) = \sum_{n=1}^{\infty} f_n s^n$. It will be seen presently that the right side in (3.4) remains unchanged if all the inequalities are reversed. Clearly $\sum f_n \leq \mathbf{P}\{\mathbf{X}_1 < 0\} < 1$.

With these notations we can now formulate the basic

Wiener-Hopf factorization theorem. *For* $|s| \leq 1$ *one has the identity*

$$1 - s\varphi(\zeta) = [1-f(s)] \cdot [1-\chi(s, \zeta)] \cdot [1-\chi^-(s, \zeta)]. \tag{3.5}$$

The proof will lead to explicit expressions for f and χ which we state in the form of separate lemmas.[3]

Lemma 1. *For* $0 \leq s < 1$

$$\log \frac{1}{1 - \chi(s, \zeta)} = \sum_{n=1}^{\infty} \frac{s^n}{n} \int_{0+}^{\infty} e^{i\zeta x}\, F^{n\star}\{dx\}. \tag{3.6}$$

An analogous formula for χ^- follows by symmetry.

Lemma 2. *For* $0 \leq s \leq 1$

$$\log \frac{1}{1 - f(s)} = \sum_{n=1}^{\infty} \frac{s^n}{n} \mathbf{P}\{\mathbf{S}_n = 0\}. \tag{3.7}$$

Since no inequalities enter the right side it follows that (3.4) remains valid with all inequalities reversed. This result was obtained in example XII,2(*a*) as a consequence of the duality principle.

The remarkable feature of the factorization (3.5) is that it represents an arbitrary characteristic function φ in terms of two (possibly defective)

[3] For $\zeta = 0$ lemma 1 reduces to theorem 1 of XII,7. The generalized version XII,(9.3) is equivalent to lemma 1, but is clumsy by comparison. Lemma 2 restates XII,(9.6). It is due to G. Baxter. A greatly simplified (but still rather difficult) proof was given by F. Spitzer, Trans. Amer. Math. Soc., vol. 94 (1960), pp. 150–169.

distributions concentrated on the two half-axes. Lemma 1 shows that this representation is *unique*.

The proof is straightforward for continuous distributions, but for the general case we require the analogue to lemma 1 for the entrance probabilities into the closed interval $\overline{\vert 0, \infty}$. These will be denoted by $R_n\{I\}$, that is,

$$R_n\{I\} = \mathbf{P}\{\mathbf{S}_1 < 0, \ldots, \mathbf{S}_{n-1} < 0, \mathbf{S}_n \in I\} \tag{3.8}$$

for any interval I in $\overline{\vert 0, \infty}$. Of course, $R_0 = 0$ and $R_n\{\overline{-\infty, 0}\} = 0$.

Lemma 3. *For* $0 \leq s < 1$ *the transform* ρ *of* $\{R_n\}$ *is given by*

$$\log \frac{1}{1 - \rho(s, \zeta)} = \sum_{n=1}^{\infty} \frac{s^n}{n} \int_{0-}^{\infty} e^{i\zeta x} F^{n\star}\{dx\}. \tag{3.9}$$

Proof. We start from the basic identity (1.9) applied to $A = \overline{\vert 0, \infty}$. With our present notation the entrance probabilities are R_n rather than H_n, and so (1.9) reads

$$1 - \rho(s, \zeta) = \gamma(s, \zeta)[1 - s\rho(\zeta)]. \tag{3.10}$$

Here γ is the transform of the sequence of probabilities G_n defined on $\overline{-\infty, 0}$ by

$$G_n\{I\} = \mathbf{P}\{\mathbf{S}_1 < 0, \ldots, \mathbf{S}_{n-1} < 0, \mathbf{S}_n < 0, \mathbf{S}_n \in I\}, \tag{3.11}$$

that is

$$\gamma(s, \zeta) - 1 = \sum_{n=1}^{\infty} s^n \int_{-\infty}^{0-} e^{i\zeta x} G_n\{dx\}. \tag{3.12}$$

For fixed $|s| < 1$ the functions $1 - s\varphi(\zeta)$ and $1 - \chi(s, \zeta)$ can have no zeros, and hence (see XVII,1) their logarithms are uniquely defined as continuous functions of ζ vanishing at the origin. We can therefore rewrite (3.10) in the form

$$\log \frac{1}{1 - s\varphi(\zeta)} = \log \frac{1}{1 - \rho(s, \zeta)} + \log \gamma(s, \zeta) \tag{3.13}$$

or

$$\sum_{n=1}^{\infty} \frac{s^n}{n} \int_{-\infty}^{+\infty} e^{i\zeta x} F^{n\star}\{dx\} = \sum_{n=1}^{\infty} \frac{s^n}{n} \rho^n(s, \zeta) + \sum_{n=1}^{\infty} \frac{(-1)^n}{n} [\gamma(s, \zeta) - 1]^n. \tag{3.14}$$

Consider this relation for a fixed value $0 < s < 1$. Then $\rho^n(s, \zeta)$ is the characteristic function of a defective probability distribution concentrated on $\overline{\vert 0, \infty}$, and hence the first series on the right is the Fourier-Stieltjes transform of a finite measure concentrated on $\overline{\vert 0, \infty}$. Similarly, (3.12) shows that $\rho(s, \zeta) - 1$ is the Fourier-Stieltjes transform of a finite measure concentrated on $\overline{-\infty, 0}$. The same is therefore true of $[\rho(s, \zeta)-1]^n$, and so the last series is the transform of the difference of two measures on $\overline{-\infty, 0}$. It follows that, when restricted to sets in $\overline{0, \infty}$, the first two series in (3.14) represent the same finite measure, namely $\sum (s^n/n)F^{n\star}$. The assertion (3.9) restates this fact in terms of the corresponding transforms. ▶

Proof of lemma 2. This lemma is contained in lemma 3 inasmuch as the two sides in (3.7) are the weights of the atoms at the origin of the measures whose transforms appear in (3.9). This is obviously true of the right sides. As for the left sides, by the definition (3.8) the atom of R_n at the origin has weight f_n. Thus $f(s)$ is the weight attributed to the origin by the measure $\sum s^n R_n$ with transform $\rho(s, \zeta)$. The measure with transform $\sum \rho^n(s, \zeta)/n$ therefore attributes to the origin the weight $\sum f^n(s)/n = \log\,(1-f(s))^{-1}$.

Proof of lemma 1. We may proceed in two ways.

(i) Lemma 1 is the analogue of lemma 3 for the open half-axis and exactly the same proof applies. If both lemmas are considered known we may subtract (3.6) from (3.9) to conclude that

$$\rho(s, \zeta) = f(s) + [1-f(s)]\chi(s, \zeta). \tag{3.15}$$

[This identity states that the first entry into $\overline{0, \infty}$ can be a return to the origin through negative values and that, when such a return does not take place, the (conditional) distribution of the point of first entry into $\overline{0, \infty}$ reduces to the distribution $\{H_n\}$ of the first entry into $\overline{0, \infty}$.]

(ii) Alternatively we may prove (3.15) directly from the definitions (3.1) and (3.8) of H_n and R_n. [For that it suffices in (3.8) to consider the *last* index $k \leq n$ for which $S_k = 0$ and take $(k, 0)$ as new origin.] Substituting (3.15) into (3.9) we get lemma 1 as a corollary to lemmas 2 and 3. ▶

Proof of the factorization theorem. Adding the identities of lemmas 1–2 and the analogue of lemma 1 relating to $\overline{-\infty, 0}$, we get (3.5) in its logarithmic form. That (3.5) holds also for $s = 1$ follows by continuity.

Corollary.

$$\gamma(s, \zeta) = \frac{1}{1 - \chi^{-}(s, \zeta)} \tag{3.16}$$

Proof. In view of (3.13) and lemma 3

$$\gamma(s, \zeta) = \exp\left(\sum_{n=1}^{\infty} \frac{s^n}{n} \int_{-\infty}^{0-} e^{i\zeta x}\, F^{n\star}\{dx\}\right), \tag{3.17}$$

and by lemma 1 the right sides in (3.16) and (3.17) are identical. ▶

Examples. (*a*) *Binomial random walk.* Let

$$\mathbf{P}\{\mathbf{X}_1 = 1\} = p \quad \text{and} \quad \mathbf{P}\{\mathbf{X}_1 = -1\} = q.$$

The first entries into the two half-axes necessarily take place at ± 1, and hence

$$\chi(s, \zeta) = P(s)e^{i\zeta}, \qquad \chi^-(s, \zeta) = Q(s)e^{-i\zeta} \tag{3.18}$$

where P and Q are the generating functions of the epochs of first entry. The two sides of the factorization formula (3.5) are therefore linear combinations of three exponentials $e^{ik\zeta}$ with $k = 0, \pm 1$. Equating the coefficients one gets the three equations

$$\begin{gathered}[1-f(s)][1+P(s)Q(s)] = 1, \qquad [1-f(s)]P(s) = sp, \\ [1-f(s)]Q(s) = sq. \end{gathered} \tag{3.19}$$

This leads to a quadratic equation for $1 - f(s)$, and the condition $f(0) = 0$ implies that f is given by

$$f(s) = \tfrac{1}{2}(1 - \sqrt{1-4pqs^2}). \tag{3.20}$$

The generating functions P and Q now follow from (3.19). If $p > q$ we have $f(1) = q$ and hence $Q(1) < 1$. In this way the factorization theorem leads directly to the first passage and recurrence time distributions found by other methods in **1**; XI and **1**; XIV.

(*b*) *Finite arithmetic distributions.* In principle the same method applies if F is concentrated on the integers between $-a$ and b. The transforms χ and χ^- together with f are now determined by $a + b + 1$ equations, but explicit solutions are hard to come by [see example XII,4(*c*)].

(*c*) *Let F be the convolution of exponential distributions* concentrated on the two half-axes, that is, let

$$\varphi(\zeta) = \frac{a}{a + i\zeta} \cdot \frac{b}{b - i\zeta}, \qquad a > 0, \quad b > 0. \tag{3.21}$$

Because of the continuity of F we have $f(s) = 0$ identically. The left side in the factorization formula (3.5) has a pole at $\zeta = -ib$, but $\chi^-(s, \zeta)$ is regular around any point ζ with negative imaginary part. (This is so because χ^- is the transform of a measure concentrated on $\overline{-\infty, 0}$.) It follows that χ must be of the form $\chi(s, \zeta) = (b-i\zeta)^{-1}\, U(s, \zeta)$, with U regular for all ζ. One may therefore surmise that U will be independent of ζ, that is, that χ and χ^- will be of the form

$$\chi(s, \zeta) = \frac{P(s)}{b - i\zeta}, \qquad \chi^-(s, \zeta) = \frac{Q(s)}{a + i\zeta}. \tag{3.22}$$

For this to be so we must have

$$(3.23)\qquad 1 - s\frac{ab}{(a+i\zeta)(b-i\zeta)} = \left(1 - \frac{P(s)}{b - i\zeta}\right)\left(1 - \frac{Q(s)}{a + i\zeta}\right).$$

Clearing the denominators and equating the coefficients we find that $P(s) = Q(s)$ and that $P(s)$ satisfies a quadratic equation. The condition $P(0) = 0$ eliminates one of the two roots, and we find finally

$$(3.24)\qquad P(s) = Q(s) = \tfrac{1}{2}[a + b - \sqrt{(a+b)^2 - 4abs}].$$

Assume $a > b$. Then $P(1) = b$, and hence $P(s)/b$ and $Q(s)/a$ are generating functions of a proper and a defective probability distribution. The function χ defined in (3.22) is therefore the transform of a pair $(\mathbf{N}, \mathbf{S_N})$ such that $\mathbf{S_N}$ is independent of $\mathbf{N}$ and has the characteristic function $b/(b-i\zeta)$. A similar statement holds for χ^-, and because of the uniqueness of the factorization $P(s)/b$ and $Q(s)/a$ are indeed the generating functions of the epochs $\mathbf{N}$ and $\mathbf{N}^-$ of first entries. [That $\mathbf{S_N}$ and $\mathbf{S_{N^-}}$ are exponentially distributed was found also in example XII,4(*a*). Recall from example VI,9(*e*) that distributions of the form (3.21) play an important role in queueing theory.] ▶

For further examples see problems 9–11.

4. IMPLICATIONS AND APPLICATIONS

We proceed to analyze the preceding section from a probabilistic point of view and to relate it to certain results derived in chapter XII.

(i) *The duality principle*. We begin by showing that the corollary (3.16) is equivalent to

Lemma 1. *For any interval* I *in* $\overline{0, \infty}$

$$(4.1)\qquad \mathbf{P}\{\mathbf{S}_1 < \mathbf{S}_n, \ldots, \mathbf{S}_{n-1} < \mathbf{S}_n, \mathbf{S}_n \in I\} = \mathbf{P}\{\mathbf{S}_1 > 0, \ldots, \mathbf{S}_{n-1} > 0, \mathbf{S}_n \in I\}.$$

This fact was derived in XII,(2.1) by considering the variables $\mathbf{X}_1, \ldots, \mathbf{X}_n$ in reverse order. Viewed in this way the lemma appears almost self-evident, but we saw that many important relations are simple consequences of it. In the Fourier analytic treatment it plays no role, but it is remarkable that it comes as a byproduct of a purely analytic theory.[4] [For a reminder of the

[4] Our Fourier analytic arguments are rather elementary, but historically the original Wiener-Hopf theory served as point of departure. Most of the literature therefore uses deep complex variable techniques which are really out of place in probability theory because even the original Wiener-Hopf techniques simplify greatly by a restriction to positive kernels. See the discussion in XII,3*a*.

fantastic consequences of lemma 1 concerning fluctuations the reader may consult example XII,2(*b*).]

Proof. The corollary (3.16) refers to the negative half-axis and for a direct comparison all inequalities in (4.1) should therefore be reversed. The probability on the right in (4.1) then coincides with the probability $G_n\{I\}$ introduced in (3.11), and $\gamma(s, \zeta)$ is simply the corresponding transform. To prove the lemma we have therefore to show that $[1-\chi(s, \zeta)]^{-1}$ is the transform of the sequence of probabilities appearing on the left side in (4.1).

Now $\chi(s, \zeta)$ was defined as the transform of the distribution of the point $(\mathbf{N}, \mathbf{S_N})$ of first entry into $\overline{0, \infty}$, and hence χ^r is the transform of the rth ladder point $(\mathbf{N}_r, \mathbf{S}_{\mathbf{N}_r})$. It follows that

$$\frac{1}{1-\chi} - 1 = \chi + \chi^2 + \cdots \tag{4.2}$$

is the transform of the sequence of probabilities that n be a ladder epoch and $\mathbf{S}_n \in I$. But these are the probabilities appearing on the left in (4.1), and this concludes the proof. ►

(ii) *The epoch* $\mathbf{N}$ *of the first entry* into $\overline{0, \infty}$ has the generating function τ given by $\tau(s) = \chi(s, 0)$. Thus by (3.6)

$$\log \frac{1}{1-\tau(s)} = \sum_{n=1}^{\infty} \frac{s^n}{n} \mathbf{P}\{\mathbf{S}_n > 0\}. \tag{4.3}$$

This formula was derived by combinatorial methods in XII,7 where various consequences were discussed. For example, letting $s \to 1$ in (4.3) it is seen that the variable $\mathbf{N}$ is proper iff the series $\sum n^{-1}\mathbf{P}\{\mathbf{S}_n > 0\}$ diverges; in case of convergence the random walk drifts to $-\infty$. On adding $\log(1-s) = -\sum s^n/n$ to (4.3) and letting $s \to 1$ one finds that

$$\log \mathbf{E}(\mathbf{N}) = \log \tau'(1) = \sum_{n=1}^{\infty} \frac{1}{n} \mathbf{P}\{\mathbf{S}_n \le 0\} \tag{4.4}$$

provided only that $\mathbf{N}$ is proper. But we have just observed that the last series converges iff the random walk drifts to ∞, and hence we have

Lemma 2. *A necessary and sufficient condition that* $\mathbf{N}$ *be proper and* $\mathbf{E}(\mathbf{N}) < \infty$ *is that the random walk drifts to* ∞.

This result was derived by different methods in XII,2. For further properties of the distribution of $\mathbf{N}$ the reader is referred to XII,7.

(iii) *On the expectation of the point* $\mathbf{S_N}$ *of first entry.* With the methods of chapter XII not much could be said about the distribution of $\mathbf{S_N}$, but now we get the characteristic function of $\mathbf{S_N}$ by setting $s = 1$ in (3.6). However, it is preferable to derive some pertinent information directly from the factorization formula.

Lemma 3. *If both* $\mathbf{S_N}$ *and* $\mathbf{S_{N^-}}$ *are proper and have finite expectations then* F *has zero expectation and a variance* σ^2 *given by*

$$\tfrac{1}{2}\sigma^2 = -[1-f(1)]\cdot \mathbf{E}(\mathbf{S_N})\cdot \mathbf{E}(\mathbf{S_{N^-}}). \tag{4.5}$$

Theorem 1 of the next section shows that the converse is also true. The surprising implication is that the existence of a second moment of F is necessary to ensure a finite expectation for $\mathbf{S_N}$.

Proof. For $s = 1$ we get from (3.5)

$$\frac{\varphi(\zeta)-1}{\zeta^2} = [1-f(1)]\,\frac{\chi(1,\zeta)-1}{\zeta}\cdot\frac{\chi^-(1,\zeta)-1}{\zeta}. \tag{4.6}$$

As $\zeta\to 0$ the fractions on the right tend to the derivatives of the characteristic functions χ and χ^-, that is, to $i\,\mathbf{E}(\mathbf{S_N})$ and $i\,\mathbf{E}(\mathbf{S_{N^-}})$. The left side has therefore a finite limit σ^2, which means that $\varphi'(0) = 0$ and $\varphi''(0) = \frac{1}{2}\sigma^2$. It follows that σ^2 is the variance of F (see the corollary in XV,4). ▶

We turn to the case of a drift toward ∞. It follows from lemmas 1–2 of section 3, together with (4.5) that in this case as $s\to 1$ and $\zeta\to 0$

$$[1-f(s)]^{-1}\cdot[1-\tilde{\chi}(s,\zeta)]^{-1}\to \exp\left(\sum_{n=1}^{\infty}\frac{1}{n}\mathbf{P}\{\mathbf{S}_n\le 0\}\right) = \mathbf{E}(\mathbf{N}) < \infty. \tag{4.7}$$

Now by the factorization theorem

$$\frac{\chi(1,\zeta)-1}{\zeta} = \frac{\varphi(\zeta)-1}{\zeta}\cdot\frac{1}{[1-f(1)]\cdot[1-\chi^-(1,\zeta)]}. \tag{4.8}$$

Letting $\zeta\to 0$ we get the important result that

$$\mathbf{E}(\mathbf{S_N}) = \mathbf{E}(\mathbf{X}_1)\cdot\mathbf{E}(\mathbf{N}) \tag{4.9}$$

provided $\mathbf{E}(\mathbf{S_N})$ and $\mathbf{E}(\mathbf{X}_1)$ exist (the latter is positive because of the assumed drift to ∞).

We can go a step further. Our argument shows that the left side in (4.8) tends to a finite limit iff $\varphi'(0) = i\mu$ exists. Now it was shown in XVII,2*a* that this is the case iff our random walk obeys the generalized weak law of large numbers, namely iff

$$\frac{1}{n}\mathbf{S}_n \xrightarrow{\text{P}} \mu \tag{4.10}$$

($\xrightarrow{\text{P}}$ signifying convergence in probability). It was shown also that for positive variables this implies that μ coincides with their expectation. Thus in (4.8) the left side approaches a limit iff $\mathbf{E}(\mathbf{N}) < \infty$, and the right iff $\varphi'(0)$ exists. We have thus

Lemma 4. *When the random walk drifts to* ∞, *then* $\mathbf{E}(\mathbf{N}) < \infty$ *iff there exists a number* $\mu > 0$ *such that* (4.10) *holds.*

In XII,8 we could only show that the existence of $\mathbf{E}(\mathbf{X}_1)$ suffices, and even for this weaker result we required the strong law of large numbers together with its converse.

5. TWO DEEPER THEOREMS

To illustrate the use of more refined methods we derive two theorems of independent interest. The first refines lemma 3 of the preceding section; the second has applications in queueing theory. The proofs depend on deep Tauberian theorems, and the second uses Laplace transforms.

Theorem 1. *If* F *has zero expectation and variance* σ^2 *the series*

$$\sum_{n=1}^{\infty} \frac{1}{n} [\mathbf{P}\{\mathbf{S}_n > 0\} - \tfrac{1}{2}] = c \tag{5.1}$$

converges at least conditionally, and

$$\mathbf{E}(\mathbf{S}_\mathbf{N}) = \frac{\sigma}{\sqrt{2}} e^{-c}. \tag{5.2}$$

This theorem is due to F. Spitzer. The convergence of the series played a role in theorems 1a of XII,7 and 8.

Proof. Differentiating (3.6) with respect to ζ and setting $\zeta = 0$ one gets

$$-i \cdot \frac{d\,\chi(s, 0)}{d\zeta} = \sum_{n=1}^{\infty} \frac{s^n}{n} \int_0^{\infty} x\, F^{n\star}\{dx\} \cdot \exp\left[-\sum_{n=1}^{\infty} \frac{s^n}{n} \mathbf{P}\{\mathbf{S}_n > 0\}\right]. \tag{5.3}$$

Both series converge absolutely for $|s| < 1$ since the coefficients of s^n remain bounded. Indeed, by the central limit theorem the moments of order ≤ 2 of $\mathbf{S}_n/\sigma\sqrt{n}$ tend to the corresponding moments of the normal distribution, which means that as $n \to \infty$

$$\int_0^{\infty} x\, F^{n\star}\{dx\} \sim \sigma \sqrt{\frac{n}{2\pi}}. \tag{5.4}$$

Accordingly, by the easy part of theorem 5 of XIII,5 as $s \to 1$

$$\sum_{n=1}^{\infty} \frac{s^n}{n} F^{n\star}\{dx\} \sim \frac{\sigma}{\sqrt{2\pi}} \sum_{n=1}^{\infty} \frac{s^n}{\sqrt{n}} \sim \frac{\sigma}{\sqrt{2}} (1-s)^{-\frac{1}{2}}. \tag{5.5}$$

The left side in (5.3) tends to $\mathbf{E}(\mathbf{S}_\mathbf{N})$ which may be finite or infinite, but cannot be zero. Combining (5.3) and (5.5) we get therefore

$$\mathbf{E}(\mathbf{S}_\mathbf{N}) = \frac{\sigma}{\sqrt{2\pi}} \lim_{s \to 1} \exp\left[\sum_{n=1}^{\infty} \frac{s^n}{n} (\tfrac{1}{2} - \mathbf{P}\{\mathbf{S}_n > 0\})\right]. \tag{5.6}$$

The exponent tends to a finite number or to $+\infty$. The same argument applies to $\mathbf{N}^-$, that is, to the exponent with $\mathbf{S}_n > 0$ replaced by $\mathbf{S}_n < 0$. But the sum of the two exponents equals $\sum (s^n/n)\,\mathbf{P}\{\mathbf{S}_n = 0\}$ and remains bounded as $s \to 1$. It follows that the exponent in (5.6) remains bounded, and hence tends to a finite limit $-c$. Since its coefficients are $o(n^{-1})$ this implies[5] that for $s = 1$ the series converges to $-c$. This concludes the proof. ►

Next we consider random walks with a drift to $-\infty$ and put

$$\mathbf{M}_n = \max\{0, \mathbf{S}_1, \ldots, \mathbf{S}_n\}. \tag{5.7}$$

It may be recalled from VI,9 that in applications to queueing theory $\mathbf{M}_n$ represents the waiting time of the nth customer. However, the proof of the following limit theorem is perhaps more interesting than the theorem itself.

Theorem 2. *If the random walk drifts to $-\infty$ the distributions U_n of $\mathbf{M}_n$ tend to a limit distribution U with characteristic function ω given by*

$$\omega(\zeta) = \exp\left[\sum_{n=1}^{\infty} \frac{1}{n}\int_0^{\infty} (e^{i\zeta x}-1)\, F^{n\star}\{dx\}\right]. \tag{5.8}$$

Note that $\sum n^{-1}\mathbf{P}\{\mathbf{S}_n > 0\} < \infty$ in consequence of (4.3), and so the series in (5.8) converges absolutely for all ζ with positive imaginary part.

Proof. Let ω_n denote the characteristic function of U_n. We begin by showing that for $|s| < 1$

$$\sum_{n=0}^{\infty} s^n \omega_n(\zeta) = \frac{1}{1-s}\exp\left[\sum_{n=1}^{\infty} \frac{s^n}{n}\int_0^{\infty} (e^{i\zeta x} - 1)\, F^{n\star}\{dx\}\right]. \tag{5.9}$$

The event $\{\mathbf{M}_\nu \in I\}$ occurs iff the following two conditions are satisfied. First, for some $0 \le n \le \nu$ the point $(n, \mathbf{S}_n)$ is a ladder point with $\mathbf{S}_n \in I$; second, $\mathbf{S}_k - \mathbf{S}_n \le 0$ for all $n < k \le \nu$. The first condition involves only $\mathbf{X}_1, \ldots, \mathbf{X}_n$, the second only $\mathbf{X}_{n+1}, \ldots, \mathbf{X}_\nu$. The two events are therefore independent and so

$$\mathbf{P}\{\mathbf{M}_\nu \in I\} = a_0 b_\nu + \cdots + a_\nu b_0 \tag{5.10}$$

where

$$a_n = \mathbf{P}\{\mathbf{S}_1 < \mathbf{S}_n, \ldots, \mathbf{S}_{n-1} < \mathbf{S}_n, \mathbf{S}_n \in I\}, \qquad b_n = \mathbf{P}\{\mathbf{N} > n\}. \tag{5.11}$$

The probabilities a_n occur on the left in (4.1), and we saw that their transform is given by $[1-\chi(s, \zeta)]^{-1}$. The generating function of $\{b_n\}$ is given by $[1-\tau(s)]/(1-s)$ with τ defined in (4.3). In view of the convolution property

[5] By the elementary (original) theorem of Tauber. See, for example, E. C. Titchmarsh, *Theory of Functions*, 2nd ed., Oxford 1939, p. 10.

(5.10) the product of these functions represents the transform of the probabilities $\mathbf{P}\{\mathbf{M}_n \in I\}$, and (5.9) merely records this fact.

We have already noticed that the exponents in (5.8) and (5.9) are regular for all ζ with positive imaginary part. For $\lambda > 0$ we may therefore put $\zeta = i\lambda$ which leads us to the Laplace transforms

$$\omega_n(i\lambda) = \int_0^\infty e^{-\lambda x}\, U_n\{dx\} = \lambda \int_0^\infty e^{-\lambda x}\, U_n(x)\, dx. \tag{5.12}$$

From the monotone character of the sequence of maxima $\mathbf{M}_n$ it follows that for fixed x the sequence $\{U_n(x)\}$ decreases, and hence for fixed λ the Laplace transforms $\omega_n(i\lambda)$ form a decreasing sequence. In view of (5.9) we have as $s \to 1$

$$\sum_{n=0}^\infty s^n \omega_n(i\lambda) \sim \frac{1}{1-s}\, \omega(i\lambda), \tag{5.13}$$

and by the last part of the Tauberian theorem 5 of XIII,5 this implies that $\omega_n(i\lambda) \to \omega(i\lambda)$. This implies the asserted convergence $U_n \to U$. ►

6. CRITERIA FOR PERSISTENCY

The material of this section is independent of the preceding theory. It is devoted to the method developed by K. L. Chung and W. H. J. Fuchs (1950) to decide whether a random walk is persistent or transient. Despite the criteria and methods developed in chapters VI and XII the Fourier-analytic method preserves its methodological and historical interest and is at present the only method applicable in higher dimensions. In the following F stands for a one-dimensional distribution with characteristic function $\varphi(\zeta) = u(\zeta) + iv(\zeta)$.

For $0 < s < 1$ we introduce the finite measure

$$U_s = \sum_{n=0}^\infty s^n F^{n\star}. \tag{6.1}$$

According to the theory developed in VI,10 the distribution F is transient iff for some open interval I about the origin $U_s\{I\}$ remains bounded as $s \to 1$; in this case $U_s\{I\}$ remains bounded for *every* open interval I. Non-transient distributions are called persistent.

Criterion. *The distribution F is transient iff for some $a > 0$*

$$\int_0^a \frac{1 - su}{(1-su)^2 + s^2 v^2}\, d\zeta \tag{6.2}$$

remains bounded as $s \to 1$ from below.

(It will be seen that in the contrary case the integral tends to ∞.)

Proof. (i) Assume that the integral (6.2) remains bounded for some fixed $a > 0$. The Parseval relation XV,(3.2) applied to $F^{n\star}$ and a triangular density (number 4 in XV,2) reads

$$2\int_{-\infty}^{+\infty} \frac{1 - \cos ax}{a^2x^2}\, F^{n\star}\{dx\} = \frac{1}{a}\int_{-a}^{a}\left(1 - \frac{|\zeta|}{a}\right) \varphi^n(\zeta)\, d\zeta. \tag{6.3}$$

Multiplying by s^n and adding over n we get

$$\begin{aligned} 2\int_{-\infty}^{+\infty} \frac{1 - \cos ax}{a^2x^2}\, U_s\{dx\} &= \frac{1}{a}\int_{-a}^{a}\left(1 - \frac{|\zeta|}{a}\right)\frac{d\zeta}{1 - s\varphi(\zeta)} \\ &= \frac{2}{a}\int_{0}^{a}\left(1 - \frac{\zeta}{a}\right)\cdot\frac{1 - su}{(1-su)^2 + s^2v^2}\, d\zeta \end{aligned} \tag{6.4}$$

(because the real part of φ is even and the imaginary part is odd). Let I stand for the interval $|x| < 2/a$. For $x \in I$ the integrand on the left is $> \frac{1}{3}$ and so $U_s\{I\}$ remains bounded. The condition of the theorem is therefore *sufficient*.

(ii) To prove the necessity of the condition we use Parseval's relation with the distribution number 5 in XV,2 which has the characteristic function $1 - |\zeta|/a$ for $|\zeta| < a$. This replaces (6.4) by

$$\int_{-a}^{a}\left(1 - \frac{|x|}{a}\right) U_s\{dx\} = \frac{2}{\pi}\int_{0}^{\infty} \frac{1 - \cos a\zeta}{a\zeta^2}\cdot\frac{1 - su}{(1-su)^2 + s^2v^2}\, d\zeta. \tag{6.5}$$

For a transient F the left side remains bounded, and so the integral (6.2) remains bounded. ►

As an application we prove a lemma which was proved by different methods in theorem 4 of VI,10. For further examples see problems 13–16.

Lemma 1.[6] *A probability distribution with vanishing expectation is persistent.*

Proof. The characteristic function has a derivative vanishing at the origin, and hence we can choose a so small that

$$0 \leq 1 - u(\zeta) \leq \epsilon\zeta, \qquad \textit{for}\quad 0 \leq \zeta < a.$$

Then $1 - su(\zeta) \leq 1 - s + \epsilon\zeta$ and using the inequality $2\,|xy| \leq x^2 + y^2$ it is seen that the integral in (6.2) is

$$\geq \frac{1}{3}\int_0^a \frac{(1-s)\, d\zeta}{(1-s)^2 + \epsilon^2\zeta^2} = \frac{1}{3\epsilon}\arctan\frac{a\epsilon}{1 - s} \to \frac{\pi}{6\epsilon}.$$

[6] The fact that $\mathbf{E}(\mathbf{X}_j) = 0$ implies persistency was first established by Chung and Fuchs. It is interesting to reflect that in 1950 this presented a serious problem and many attempts to solve it had ended in failures. Attention on this problem was focused by the surprise discovery of the unfavorable "fair" random walk in which $\mathbf{P}\{\mathbf{S}_n > n/\log n\} \to 1$. See **1**; X,3 and problem 15 in **1**; X,8. For a related phenomenon see the footnote to problem 13.

The right side can be made arbitrarily large, and so the integral (6.2) tends to ∞. ▶

The passage to the limit involved in the criterion is rather delicate, and it is therefore useful to have the simpler sufficient conditions stated in the following

Corollary. *The probability distribution* F *is persistent if*

$$\int_0^a \frac{1-u}{(1-u)^2+v^2}\,d\zeta = \infty \tag{6.6}$$

for every $a > 0$, and transient if for some $a > 0$

$$\int_0^a \frac{d\zeta}{1-u} < \infty\,. \tag{6.7}$$

Proof. The integrand in (6.2) is decreased when $1 - su$ in the numerator is replaced by $1 - u$, and sv in the denominator by v. Then (6.6) follows by monotone convergence. Similarly, the integrand in (6.2) increases when the term s^2v^2 is dropped, and then (6.7) follows by monotone convergence. ▶

These criteria apply without change in higher dimensions except that then u and v become functions of several variables ζ_j, and the integrals are extended over spheres centered at the origin. In this way we prove the following criteria.

Lemma 2. *A truly two-dimensional probability distribution with zero expectations and finite variances is persistent.*

Proof. The characteristic function is twice continuously differentiable, and from the two-term Taylor expansion it is seen that in a neighborhood of the origin the integrand of (6.6) is $>\delta/(\zeta_1^2+\zeta_2^2)$. The integral corresponding to (6.6) therefore diverges. ▶

Lemma 3. *Every truly three-dimensional distribution is transient.*

Proof. On considering the Taylor expansion of $\cos(x_1\zeta_1+x_2\zeta_2+x_3\zeta_3)$ in some x-neighborhood of the origin one sees that for any characteristic function there exists a neighborhood of the origin in which

$$1 - u(\zeta_1, \zeta_2, \zeta_3) \geq \delta(\zeta_1^2+\zeta_2^2+\zeta_3^2).$$

The three-dimensional analogue to (6.7) is therefore dominated by an integral of $(\zeta_1^2+\zeta_2^2+\zeta_3^2)^{-1}$ extended over a neighborhood of the origin, and in three dimensions this integral converges. ▶

7. PROBLEMS FOR SOLUTION

1. Do the example of section 1 for the case of an unsymmetric interval $\overline{-a, b}$. (Derive two linear equations for two generating functions corresponding to the two boundaries. Explicit solutions are messy.)

Problems 2–5 *refer to a symmetric binomial random walk*, that is, $\varphi(\zeta) = \cos\zeta$. The notations are those of section 1.

2. Let A consist of the two points $0, 1$. Show by elementary considerations that $\chi(s,\zeta) = \dfrac{1}{1-\frac{1}{4}s^2}\left(\dfrac{s}{2}e^{-i\zeta}+\dfrac{s^2}{4}e^{2i\zeta}\right)$ and $\gamma(s,\zeta) = \dfrac{1}{1-\frac{1}{4}s^2}\left(1+\dfrac{s}{2}e^{i\zeta}\right)$. Verify (1.9).

3. If in the preceding problem the roles of A and A' are interchanged one gets

$$\chi(s,\zeta) = \frac{s}{2}e^{i\zeta} + \tfrac{1}{2}(1 - \sqrt{1-s^2}), \qquad \gamma(s,\zeta) = \left[1 - \frac{1-\sqrt{1-s^2}}{s}e^{-i\zeta}\right]^{-1}.$$

Interpret probabilistically.

4. If A' consists of the origin alone χ depends only on s and γ must be the sum of two power series in $e^{i\zeta}$ and $e^{-i\zeta}$, respectively. Using this information derive χ and γ directly from (1.9).

5. If A' consists of the origin alone one has $\chi = s\varphi$ and $\gamma = 1$.

6. *Alternative proof of the identity* (1.9). With the notations of section 1 show that

$$(*) \qquad F^{n\star}\{I\} = \sum_{k=1}^{n}\int_{A'} H_k\{dy\}\, F^{(n-k)\star}\{I-y\} + G_n\{I\}$$

(*a*) by a direct probabilistic argument, and (*b*) by induction. Show that (*) is equivalent to (1.9).

7. In the case of a (not necessarily symmetric) binomial random walk Wald's approximation in section 2 leads to a rigorous solution. Show that (2.12) reduces to a quadratic equation for $\tau = e^{-\lambda}$ and that one is led to the solution known from **1**; XIV,(4.11). Specifically, $Q(s)$ agrees with U_z except that the latter refers to a basic interval $\overline{0, a}$ rather than $\overline{-a, b}$, and to a starting point z.

8. As in section 2 let $G_n\{I\}$ be the probability that $\mathbf{S}_n \in I \subset A$ and that no exit from $A = \overline{-a, b}$ has taken place previously. Show that if two distributions F and $F^\#$ agree within the interval $|x| < a + b$ they lead to the same probabilities G_n. Use this and an appropriate truncation for an alternative proof that the series (2.1) converges for some $s > 1$.

9. Random walks in which the distribution F is concentrated on finitely many integers were treated in example XII,4(*c*). Show that the formulas derived there contain implicitly the Wiener-Hopf factorization for $1 - \varphi$.

10. (*Khintchine-Pollaczek formula.*) Let F be the convolution of an exponential with expectation $1/a$ concentrated on $\overline{0, \infty}$ and a distribution B concentrated on $\overline{-\infty, 0}$. Denote the characteristic function of B by β, its expectation by $-b < 0$. We suppose that the expectation $a^{-1} - b$ of F is positive. Then

$$1 - \varphi(\zeta) = 1 - \frac{a}{a - i\zeta}\beta(\zeta) = \left(1 - \frac{a}{a - i\zeta}\right)\left(1 - a\frac{1 - \beta(\zeta)}{i\zeta}\right).$$

Note: This formula plays an important role in queueing theory. For alternative treatments see examples XII,5(*a–b*) and XIV,2(*b*).

11. (*Continuation.*) If $ab > 1$ show that there exists a unique positive number κ between 0 and a such that

$$a\beta(-i\kappa) = a - \kappa.$$

Prove that $\chi^-(1,\zeta) = \dfrac{a - \kappa - a\beta(\zeta)}{i\zeta - \kappa}$. *Hint:* Apply problem 10 to the associated random walk with characteristic function ${}^a\varphi(\zeta) = \varphi(\zeta - i\kappa)$. Recall that $\chi^-(1,\zeta) = {}^a\chi^-(1, \zeta - i\kappa)$. [See example XII,4(*b*).]

12. Let $\mathbf{U}_n = \max[0, \mathbf{S}_1, \ldots, \mathbf{S}_n]$ and $\mathbf{V}_n = \mathbf{S}_n - \mathbf{U}_n$. By a very slight change of the argument used for (5.9) show that the bivariate characteristic function of the pair $(\mathbf{U}_n, \mathbf{V}_n)$ is the coefficient of s^n in[7]

$$\frac{1}{1-s} \exp \sum_{n=1}^{\infty} \frac{s^n}{n} \left[\int_0^{\infty} (e^{i\zeta_1 x} - 1) F^{n\star}\{dx\} + \int_{-\infty}^{0} (e^{i\zeta_2 x} - 1) F^{n\star}\{dx\} \right].$$

13.[8] Suppose that in a neighborhood of the origin $|1 - \varphi(\zeta)| < A \cdot |\zeta|$. Then F is persistent unless it has an expectation $\mu \neq 0$.

Hint: The integral in (6.6) exceeds $\int_0^{\alpha} d\zeta \int_{-1/\zeta}^{1/\zeta} x^2 \, F\{dx\}$. Substitute $\zeta = 1/t$ and interchange the order of integration to see that this integral diverges unless μ exists.

14. Using the criterion (6.7) show that if $t^{-1-\rho} \int_{-t}^{t} x^2 \, F\{dx\} \to \infty$ for some $\rho > 0$ as $t \to \infty$ the distribution F is transient.[9]

15. The distribution with characteristic function $\varphi(\zeta) = e^{-1} \sum 1/n! \cos(n!\zeta)$ is transient.

Hint: Use (6.7) and the change of variable $\zeta = (1/n!)t$.

16. The unsymmetric stable distributions with characteristic exponent $\alpha = 1$ are transient, but the Cauchy distribution is persistent.

---

[7] First derived analytically by F. Spitzer, Trans. Amer. Math. Soc., vol. 82 (1956) pp. 323–339.

[8] This problem commands theoretical interest. It applies whenever φ has a derivative at the origin. We saw in XVII,2*a* that this is possible even without F having an expectation, and that in this case the weak law of large numbers applies nevertheless. Thus we get examples of random walks in which for each sufficiently large n there is an overwhelming probability that $\mathbf{S}_n > (1 - \epsilon)n\mu$ with $\mu > 0$, and yet there is no drift to ∞: the random walk is persistent.

[9] This shows that under slight regularity conditions F is transient whenever an absolute moment of order $\rho < 1$ diverges. The intricacies of the problem without any regularity conditions are shown in L. A. Shepp, Bull. Amer. Math. Soc., vol. 70 (1964) pp. 540–542.

CHAPTER XIX

Harmonic Analysis

This chapter supplements the theory of characteristic functions presented in chapter XV and gives applications to stochastic processes and integrals. The discussion of Poisson's summation formula in section 5 is practically independent of the remainder. The whole theory is independent of chapters XVI–XVIII.

1. THE PARSEVAL RELATION

Let U be a probability distribution with characteristic function

$$\omega(\zeta) = \int_{-\infty}^{+\infty} e^{i\zeta x}\, U\{dx\}. \tag{1.1}$$

Integrating this relation with respect to some other probability distribution F we get

$$\int_{-\infty}^{+\infty} \omega(\zeta)\, F\{d\zeta\} = \int_{-\infty}^{+\infty} \varphi(x)\, U\{dx\}, \tag{1.2}$$

where φ is the characteristic function of F. This is one form of the *Parseval relation* from which the basic results of XV,3 were derived. Surprisingly enough, a wealth of new information can be obtained by rewriting Parseval's formula in equivalent forms and considering special cases. A simple example of independent interest may illustrate this method, which will be used repeatedly.

Example. The formula

$$\int_{-\infty}^{+\infty} e^{-ia\zeta}\omega(\zeta)\, F\{d\zeta\} = \int_{-\infty}^{+\infty} \varphi(x)\, U\{a + dx\} \tag{1.3}$$

differs from (1.2) only notationally. We apply the special case where F is the uniform distribution in $\overline{-t, t}$ and $\varphi(x) = \sin tx/tx$. This function does not exceed 1 in absolute value and as $t \to \infty$ it tends to 0 at all points

$x \neq 0$. By bounded convergence we get therefore

$$U(a) - U(a-) = \lim_{t\to\infty} \frac{1}{2t} \int_{-t}^{t} e^{-ia\zeta}\omega(\zeta)\, d\zeta. \tag{1.4}$$

This formula makes it possible to decide whether a is a point of continuity and to find the weight of the atom at a, if any. The most interesting result is obtained by applying (1.4) to the symmetrized distribution ${}^0U$ with characteristic function $|\omega|^2$. If $p_1, p_2, \ldots$ are the weights of the atoms of U then ${}^0U$ has an atom of weight $\sum p_k^2$ at the origin (problem 11 in V,12) and so

$$\frac{1}{2t} \int_{-t}^{t} |\omega(\zeta)|^2\, d\zeta \to \sum p_k^2. \tag{1.5}$$

This formula shows, in particular, that the characteristic functions of continuous distributions are, on the average, small. ▶

A versatile and useful variant of the Parseval formula (1.2) is as follows. *If A and B are arbitrary probability distributions with characteristic functions α and β, respectively, then*

$$\iint_{-\infty}^{+\infty} \omega(s-t)\, A\{ds\}\, B\{dt\} = \int_{-\infty}^{+\infty} \alpha(x)\, \overline{\beta(x)}\, U\{dx\} \tag{1.6}$$

where $\bar\beta$ is the conjugate of β. For a direct verification it suffices to integrate

$$\omega(s-t) = \int_{-\infty}^{+\infty} e^{i(s-t)x}\, U\{dx\} \tag{1.7}$$

with respect to A and B. This argument produces the erroneous impression that (1.6) is more general than (1.2), whereas the relation (1.6) *is in reality the special case of the Parseval relation* (1.2) *corresponding to* $F = A \star {}^-B$ where ${}^-B$ is the distribution with characteristic function $\bar\beta$ [that is, ${}^-B(x) = 1 - B(-x)$ at all points of continuity]. Indeed, F has the characteristic function $\varphi = \alpha\bar\beta$, and so the right sides in (1.2) and (1.6) are identical. That the left sides differ only notationally is best seen using two independent random variables $\mathbf{X}$ and $\mathbf{Y}$ with distributions A and B, respectively. The left side in (1.6) represents the direct definition of the expectation $\mathbf{E}(\omega(\mathbf{X}-\mathbf{Y}))$, whereas the left side in (1.2) expresses this expectation in terms of the distribution F of $\mathbf{X} - \mathbf{Y}$.

(We return to Parseval's formula in section 7.)

2. POSITIVE DEFINITE FUNCTIONS

An important theorem due to S. Bochner (1932) makes it possible to describe the class of characteristic functions by intrinsic properties. The following simple criterion will point the way.

Lemma 1. *Let* ω *be a bounded continuous (complex-valued) function that is integrable*[1] *over* $\overline{-\infty, \infty}$. *Define* u *by*

$$u(x) = \frac{1}{2\pi}\int_{-\infty}^{+\infty} e^{-i\zeta x}\,\omega(\zeta)\,d\zeta. \tag{2.1}$$

In order that ω *be a characteristic function it is necessary and sufficient that* $\omega(0) = 1$ *and that* $u(x) \geq 0$ *for all* x. *In this case* u *is the probability density corresponding to* ω.

Proof. The Fourier inversion formula XV,(3.5) shows that the conditions are necessary. Now choose an arbitrary even density f with integrable characteristic function $\varphi \geq 0$. Multiply (2.1) by $\varphi(tx)e^{iax}$ and integrate with respect to x. Since the inversion formula XV,(3.5) applies to the pair f, φ the result is

$$\int_{-\infty}^{+\infty} u(x)\,\varphi(tx)e^{iax}\,dx = \int_{-\infty}^{+\infty}\omega(\zeta)f\left(\frac{\zeta - a}{t}\right)\frac{d\zeta}{t}. \tag{2.2}$$

The right side is the expectation of ω with respect to a probability distribution, and hence it is bounded by the maximum of $|\omega|$. For the particular value $a = 0$ the integrand on the left is non-negative and tends to $u(x)$ as $t \to 0$. The boundedness of the integral therefore implies that u is *integrable*. Letting $t \to 0$ in (2.2) we get therefore

$$\int_{-\infty}^{+\infty} u(x)e^{iax}\,dx = \omega(a) \tag{2.3}$$

(the left side by bounded convergence, the right side because the probability distribution involved tends to the distribution concentrated at the point a). For $a = 0$ we see that u is a probability density, and ω is indeed its characteristic function. ►

The integrability condition of the lemma looks more restrictive than it is. In fact, by the continuity theorem a continuous function ω is characteristic iff $\omega(\zeta)e^{-\epsilon\zeta^2}$ is a characteristic function for every fixed $\epsilon > 0$. It follows that a bounded continuous function with $\omega(0) = 1$ is characteristic iff for all x and $\epsilon > 0$

$$\int_{-\infty}^{+\infty} e^{-i\zeta x}\omega(\zeta)e^{-\epsilon\zeta^2}d\zeta \geq 0. \tag{2.4}$$

This criterion is perfectly general, but it is not easy to apply in individual situations; moreover, the arbitrary choice of the convergence factor $e^{-\epsilon\zeta^2}$ is a drawback. For this reason we restate the criterion in a form in which the condition is sharpened.

---

[1] As elsewhere this means *absolute* integrability.

Lemma 2. *A bounded continuous function* ω *is characteristic iff* $\omega(0) = 1$ *and if for every probability distribution* A *and all* x

$$\int_{-\infty}^{+\infty} e^{-i\zeta x}\omega(\zeta)\,{}^0A\{d\zeta\} \geq 0 \tag{2.5}$$

where ${}^0A = A \bigstar {}^-A$ *is the distribution obtained by symmetrization.*

Proof. (*a*) *Necessity.* If α is the characteristic function of A then ${}^0A$ has the characteristic function $|\alpha|^2$ and the necessity of (2.5) is implicit in the Parseval relation (1.3).

(*b*) *Sufficiency.* It was shown in (2.4) that the condition is sufficient if A is restricted to normal distributions with arbitrary variances. ▶

We have seen that (2.5) may be rewritten in the form (1.6) with $B = A$. In particular, if A is concentrated at finitely many points $t_1, t_2, \ldots, t_n$ with corresponding weights $p_1, p_2, \ldots, p_n$, then (2.5) takes on the form

$$\sum_{j,k} \omega(t_j - t_k)e^{-ix(t_j - t_k)}p_jp_k \geq 0. \tag{2.6}$$

If this inequality is valid for all choices of t_j and p_j then (2.5) is satisfied for all discrete distributions A with finitely many atoms. As every distribution is the limit of a sequence of such discrete distributions the condition (2.6) is necessary and sufficient. With the change of notation $z_j = p_je^{-ixt_j}$ it takes on the form

$$\sum_{j,k} \omega(t_j - t_k)z_j\bar{z}_k \geq 0. \tag{2.7}$$

For the final formulation of our criterion we introduce a frequently used term.

Definition. *A complex-valued function* ω *of the real variable* t *is called positive definite iff* (2.7) *holds for every choice of finitely many real numbers* $t_1, \ldots, t_n$ *and complex numbers* $z_1, \ldots, z_n$.

Theorem. (*Bochner.*) *A continuous function* ω *is the characteristic function of a probability distribution iff it is positive definite and* $\omega(0) = 1$.

Proof. We have shown that the condition is necessary, and also that it is sufficient when ω is bounded. The proof is completed by the next lemma which shows that all positive definite functions are bounded. ▶

Lemma 3. *For any positive definite* ω

$$\omega(0) \geq 0, \qquad |\omega(t)| \leq \omega(0) \tag{2.8}$$

$$\omega(-t) = \overline{\omega(t)}. \tag{2.9}$$

Proof. We use (2.7) with $n = 2$ letting $t_2 = 0$ and $z_2 = 1$. Dropping the unnecessary subscripts we get

$$\omega(0)[1+|z|^2] + \omega(t)z + \omega(-t)\bar{z} \geq 0. \tag{2.10}$$

For $z = 0$ it is seen that $\omega(0) \geq 0$. For positive z we get (2.9) and it follows that $\omega \equiv 0$ if $\omega(0) = 0$. Finally, if $\omega(0) \neq 0$ and $z = -\overline{\omega(t)}/\omega(0)$ then (2.10) reduces to $|\omega(t)|^2 \leq \omega^2(0)$. ►

3. STATIONARY PROCESSES

The last theorem has important consequences for stochastic processes with stationary covariances. By this is meant a family of random variables $\{\mathbf{X}_t\}$ defined for $-\infty < t < \infty$ and having covariances such that

$$\text{Cov}\,(\mathbf{X}_{s+t}, \mathbf{X}_s) = \rho(t) \tag{3.1}$$

is independent of s. So far we have considered only real random variables, but now the notations will become simpler and more symmetric if we admit *complex-valued* random variables. A complex random variable is, of course, merely a pair of real variables written in the form $\mathbf{X} = \mathbf{U} + i\mathbf{V}$ and nothing need be assumed concerning the joint distribution of $\mathbf{U}$ and $\mathbf{V}$. The variable $\bar{\mathbf{X}} = \mathbf{U} - i\mathbf{V}$ is called the conjugate of $\mathbf{X}$ and the product $\mathbf{X}\bar{\mathbf{X}}$ takes over the role of $\mathbf{X}^2$ in the real theory. This necessitates a slight unsymmetry in the definition of variances and covariances:

Definition. *For complex random variables with*

$$\mathbf{E}(\mathbf{X}) = \mathbf{E}(\mathbf{Y}) = 0$$

we define

$$\text{Cov}\,(\mathbf{X}, \mathbf{Y}) = \mathbf{E}(\mathbf{X}\bar{\mathbf{Y}}). \tag{3.2}$$

Then $\text{Var}\,(\mathbf{X}) = \mathbf{E}(|\mathbf{X}|^2) \geq 0$, but $\text{Cov}\,(\mathbf{Y}, \mathbf{X})$ is the conjugate of $\text{Cov}\,(\mathbf{X}, \mathbf{Y})$.

Theorem. *Let $\{\mathbf{X}_t\}$ be a family of random variables such that*

$$\rho(t) = \mathbf{E}(\mathbf{X}_{t+s}\bar{\mathbf{X}}_s) \tag{3.3}$$

is a continuous function[2] *independent of s. Then ρ is positive definite, that is,*

$$\rho(t) = \int_{-\infty}^{+\infty} e^{i\lambda t} R\{d\lambda\} \tag{3.4}$$

where R is a measure on the real line with total mass $\rho(0)$.

[2] Continuity is important: for mutually independent variables $\mathbf{X}_t$ one has $\rho(t) = 0$ except when $t = 0$, and this covariance function is not of the form (3.4). See problem 4.

If the variables $\mathbf{X}_t$ *are real the measure* R *is symmetric and*

$$\rho(t) = \int_{-\infty}^{+\infty} \cos \lambda t \, R\{d\lambda\}. \tag{3.5}$$

Proof. Choose arbitrary real points $t_1, \ldots, t_n$ and complex constants $z_1, \ldots, z_n$. Then

$$\begin{aligned} \sum \rho(t_j - t_k) z_j \bar{z}_k = \sum \mathbf{E}(\mathbf{X}_{t_j} \bar{\mathbf{X}}_{t_k}) z_j \bar{z}_k = \\ = \mathbf{E}(\sum \mathbf{X}_{t_j} z_j \bar{\mathbf{X}}_{t_k} \bar{z}_k) = \mathbf{E}(|\sum \mathbf{X}_{t_j} z_j|^2) \geq 0 \end{aligned} \tag{3.6}$$

and so (3.4) is true by the criterion of the last section. When ρ is real the relation (3.4) holds also for the mirrored measure obtained by changing x to $-x$, and because of the uniqueness R is symmetric. ►

The measure R is called the *spectral measure*[3] of the process; the set formed by its points of increase is called the *spectrum* of $\{\mathbf{X}_t\}$. In most applications the variables are centered so that $\mathbf{E}(\mathbf{X}_t) = 0$, in which case $\rho(t) = \text{Cov}\,(\mathbf{X}_{t+s}, \mathbf{X}_s)$. For this reason ρ is usually referred to as *the covariance function* of the process. Actually the centering of $\mathbf{X}_t$ has no influence on the properties of the process with which we shall be concerned.

Examples. (*a*) Let $\mathbf{Z}_1, \ldots, \mathbf{Z}_n$ be mutually uncorrelated random variables with zero expectation and variances $\sigma_1^2, \ldots, \sigma_n^2$. Put

$$\mathbf{X}_t = \mathbf{Z}_1 e^{i\lambda_1 t} + \cdots + \mathbf{Z}_n e^{i\lambda_n t} \tag{3.7}$$

with $\lambda_1, \ldots, \lambda_n$ real. Then

$$\rho(t) = \sigma_1^2 e^{i\lambda_1 t} + \cdots + \sigma_n^2 e^{i\lambda_n t} \tag{3.8}$$

and so R is concentrated at the n points $\lambda_1, \ldots, \lambda_n$. We shall see that the most general stationary process may be treated as a limiting case of this example.

If the process (3.7) is real it can be put into the form

$$\mathbf{X}_t = \mathbf{U}_1 \cos \lambda_1 t + \cdots + \mathbf{U}_r \cos \lambda_r t + \mathbf{V}_1 \sin \lambda_1 t + \cdots + \mathbf{V}_r \sin \lambda_r t \tag{3.9}$$

where the $\mathbf{U}_j$ and $\mathbf{V}_j$ are real uncorrelated random variables and

$$\mathbf{E}(\mathbf{U}_j^2) = \mathbf{E}(\mathbf{V}_j^2) = \sigma_j^2.$$

A typical example occurs in III,(7.23). The corresponding covariances are $\rho(t) = \sigma_1^2 \cos \lambda_1 t + \cdots + \sigma_r^2 \cos \lambda_r t$.

(*b*) *Markovian processes.* If the variables $\mathbf{X}_t$ are normal and the process is Markovian, then $\rho(t) = e^{-a|t|}$ [see III,(8.14)]. The spectral measure is proportional to a Cauchy density.

[3] In communication engineering, also called the "power spectrum."

(*c*) Let $\mathbf{X}_t = \mathbf{Z}e^{it\mathbf{Y}}$ where $\mathbf{Y}$ and $\mathbf{Z}$ are independent *real* random variables, $\mathbf{E}(\mathbf{Z}) = 0$. Then

$$\rho(t) = \mathbf{E}(\mathbf{Z}\bar{\mathbf{Z}})\mathbf{E}(e^{it\mathbf{Y}})$$

which shows that the spectral measure R is given by the probability distribution of $\mathbf{Y}$ multiplied by the factor $\mathbf{E}(\mathbf{Z}\bar{\mathbf{Z}})$. ▶

Theoretically it matters little whether a process is described in terms of its covariance function ρ or, equivalently, in terms of the corresponding spectral measure R, but in practice the description in terms of the spectral measure R is usually simpler and preferable. In applications to communication engineering the spectral analysis has technical advantages in instrumentation and measurement, but we shall not dwell on this point. Of greater importance from our point of view is that linear operations (often called "filters") on the variables $\mathbf{X}_t$ are more readily described in terms of R than of ρ.

Example. (*d*) *Linear operations.* As the simplest example consider the family of random variables $\mathbf{Y}_t$ defined by

$$\mathbf{Y}_t = \sum c_k \mathbf{X}_{t-\tau_k} \tag{3.10}$$

where the c_k and τ_k are constants (τ_k real) and the sum is finite. The covariance function of $\mathbf{Y}_t$ is given by the double sum

$$\rho_{\mathbf{Y}}(t) = \sum c_j \bar{c}_k\, \rho(t-\tau_j+\tau_k). \tag{3.11}$$

Substituting into (3.4) one finds

$$\rho_{\mathbf{Y}}(t) = \int_{-\infty}^{+\infty} |\sum c_j e^{-i\tau_j\lambda}|^2 \cdot e^{it\lambda}\, R\{d\lambda\}.$$

This shows that the spectral measure $R_{\mathbf{Y}}$ is determined by

$$R_{\mathbf{Y}}\{d\lambda\} = |\sum c_j e^{-i\tau_j\lambda}|^2\, R\{d\lambda\}. \tag{3.12}$$

In contrast to (3.11) this relationship admits of an intuitive interpretation: the "frequency" λ is affected by a "frequency response factor" $f(\lambda)$ which depends on the given transformation (3.10).

This example is of much wider applicability than appears at first sight because integrals and derivatives are limits of sums of the form (3.10) and therefore a similar remark applies to them. For example, if $\mathbf{X}_t$ serves as input to a standard electric circuit, the output $\mathbf{Y}_t$ can be represented by integrals involving $\mathbf{X}_t$; the spectral measure $R_{\mathbf{Y}}$ is again expressible by R and a frequency response. The latter depends on the characteristics of the network, and our result can be used in two directions; namely, to describe the output process and also to construct networks which will yield an output with certain prescribed properties. ▶

We turn to the converse of our theorem and show that *given an arbitrary measure* R *on the line there exists a stationary process* $\{\mathbf{X}_t\}$ *with spectral measure* R. Since the mapping $\mathbf{X}_t \to a\mathbf{X}_t$ changes R into a^2R there is no loss of generality in assuming that R is a probability measure. We take the λ-axis equipped with the probability measure R as sample space and denote by $\mathbf{X}_t$ the random variable defined by $\mathbf{X}_t(\lambda) = e^{it\lambda}$. Then

$$\rho(t) = \mathbf{E}(\mathbf{X}_{t+s}\overline{\mathbf{X}}_s) = \int_{-\infty}^{+\infty} e^{it\lambda}\, R\{d\lambda\}, \tag{3.13}$$

and so the spectral measure of our process is given by R. We have thus constructed *an explicit model of a stationary process with the prescribed spectral measure.* That such a model is possible with the *real line as sample space* is surprising and gratifying. We shall return to it in section 8.

It is easy to modify the model so as to obtain variables with zero expectation. Let $\mathbf{Y}$ be a random variable that is independent of all the $\mathbf{X}_t$ and assumes the values ± 1 each with probability $\frac{1}{2}$. Put $\mathbf{X}'_t = \mathbf{Y}\mathbf{X}_t$. Then $\mathbf{E}(\mathbf{X}'_t) = 0$ and $\mathbf{E}(\mathbf{X}'_{t+s}\mathbf{X}'_s) = \mathbf{E}(\mathbf{Y}^2)\,\mathbf{E}(\mathbf{X}_{t+s}\overline{\mathbf{X}}_s)$. Thus $\{\mathbf{X}'_t\}$ *is a stationary process with zero expectations, and* (3.13) *represents its true covariance function.*

4. FOURIER SERIES

An arithmetic distribution attributing probability φ_n to the point n has the characteristic function

$$\varphi(\zeta) = \sum_{-\infty}^{+\infty} \varphi_n e^{in\zeta} \tag{4.1}$$

with period 2π. The probabilities φ_n can be expressed by the inversion formula

$$\varphi_k = \frac{1}{2\pi}\int_{-\pi}^{\pi} e^{-ik\zeta}\,\varphi(\zeta)\,d\zeta \tag{4.2}$$

which is easily verified from (4.1) [see XV,(3.14)].

We now start from an arbitrary function φ with period 2π and define φ_k by (4.2). Our problem is to decide whether φ is a characteristic function, that is, whether $\{\varphi_n\}$ is a probability distribution. The method depends on investigating the behavior of the family of functions f_r defined for $0 < r < 1$ by

$$f_r(\zeta) = \sum_{-\infty}^{+\infty} \varphi_n r^{|n|} e^{in\zeta}. \tag{4.3}$$

Despite its simplicity the same argument will yield important results concerning Fourier series and characteristic functions of distributions concentrated on finite intervals.

In what follows it is best to interpret the basic interval $\overline{-\pi, \pi}$ as a *circle* (that is, to identify the points π and $-\pi$). For an integrable φ the number φ_k *will be called the kth Fourier coefficient of* φ. The series occurring in (4.1) is the corresponding "*formal Fourier series*." It need not converge, but the sequence $\{\varphi_n\}$ being bounded, the series (4.3) converges to a continuous (even differentiable) function f_r. When $r \to 1$ it is possible for f_r to tend to a limit ψ even when the series in (4.1) diverges. In this case one says that the series is "*Abel summable*" to ψ.

Examples. (*a*) Let $\varphi_n = 1$ for $n = 0, 1, 2, \ldots$, but $\varphi_n = 0$ for $n < 0$. Each term of the series in (4.1) has absolute value 1, and so the series cannot converge for any value ζ. On the other hand, the right side in (4.3) reduces to a geometric series which converges to

$$f_r(\zeta) = \frac{1}{1 - re^{i\zeta}}. \tag{4.4}$$

As $r \to 1$ a limit exists at all points except $\zeta = 0$.

(*b*) An important special case of (4.3) is represented by the functions

$$p_r(t) = \frac{1}{2\pi} \sum_{-\infty}^{+\infty} r^{|n|} e^{int} \tag{4.5}$$

obtained when $\varphi_n = 1/(2\pi)$ for all n. The contribution of the terms $n \geq 0$ was evaluated in (4.4). For reasons of symmetry we get

$$2\pi p_r(t) = \frac{1}{1 - re^{it}} + \frac{1}{1 - re^{-it}} - 1 \tag{4.6}$$

or

$$p_r(t) = \frac{1}{2\pi} \cdot \frac{1 - r^2}{1 + r^2 - 2r\cos t}. \tag{4.7}$$

This function is of constant use in the theory of harmonic functions where $p_r(t-\zeta)$ is called the "*Poisson kernel.*" For reference we state its main property in the next lemma.

Lemma. *For fixed* $0 < r < 1$ *the function* p_r *is the density of a probability distribution* P_r *on the circle. As* $r \to 1$ *the latter tends to the probability distribution concentrated at the origin.*

Proof. Obviously $p_r \geq 0$. That the integral of p_r over $\overline{-\pi, \pi}$ equals one is evident from (4.5) because for $n \neq 0$ the integral of e^{int} vanishes. For $\delta \leq t \leq \pi$ the denominator in (4.7) is bounded away from zero. As $r \to 1$ it follows that in every open interval excluding the origin $p_r(t) \to 0$ boundedly as $r \to 1$, and so P_r has a limit distribution concentrated at the origin. ▶

Theorem 1. *A continuous function* φ *with period* 2π *is a characteristic function iff its Fourier coefficients* (4.2) *satisfy* $\varphi_k \geq 0$ *and* $\varphi(0) = 1$. *In this case* φ *is represented by the uniformly convergent Fourier series* (4.1).

[In other words, a formal Fourier series with non-negative coefficients φ_k converges to a continuous function iff $\sum \varphi_k < \infty$. In this case (4.1) holds.]

Proof. In view of (4.2) and (4.5) the function f_r of (4.3) may be put into the form

$$f_r(\zeta) = \int_{-\pi}^{\pi} \varphi(t) \cdot p_r(\zeta - t)\, dt. \tag{4.8}$$

On the right we recognize the convolution of φ and the probability distribution P_r, and we conclude

$$f_r(\zeta) \to \varphi(\zeta), \qquad r \to 1. \tag{4.9}$$

Furthermore, if m is an upper bound for $|\varphi|$ then by (4.8)

$$f_r(0) = \sum_{-\infty}^{+\infty} \varphi_n r^{|n|} \leq m. \tag{4.10}$$

The terms of the series being non-negative it follows for $r \to 1$ that $\sum \varphi_n \leq m$. Therefore $\sum \varphi_n e^{in\zeta}$ converges uniformly and it is evident from (4.3) that $f_r(\zeta)$ tends to this value. Thus (4.1) is true, and this concludes the proof. ▶

Note that (4.9) is a direct consequence of the convergence properties of convolutions and hence independent of the positivity of the coefficients φ_n. As a by-product we thus have

Theorem 2.[4] *If* φ *is continuous with period* 2π, *then* (4.9) *holds uniformly in* ζ.

(For generalizations to discontinuous functions see corollary 2 and problems 6–8.)

Corollary 1. (*Féjer.*) *A continuous periodic function* φ *is the uniform limit of a sequence of trigonometric polynomials.*

---

[4] The theorem may be restated as follows: *The Fourier series of a continuous periodic function* φ *is Abel summable to* φ. The theorem (and the method of proof) apply equally to other methods of summability.
The phenomenon was first discovered by L. Féjer using Cesaro summability (see problem 9) at a time when divergent series still seemed mysterious. The discovery therefore came as a sensation, and for historical reasons texts still use Cesaro summability although the Abel method is more convenient and unifies proofs.

In other words, given $\epsilon > 0$ there exist numbers $a_{-N}, \ldots, a_N$ such that

$$\left|\varphi(\zeta) - \sum_{n=-N}^{N} a_n e^{in\zeta}\right| < \epsilon \tag{4.11}$$

for all ζ.

Proof. For arbitrary N and $0 < r < 1$

$$\left|\varphi(\zeta) - \sum_{n=-N}^{N} \varphi_n r^{|n|} e^{in\zeta}\right| \le |\varphi(\zeta) - f_r(\zeta)| + \sum_{|n|>N} |\varphi_n| \cdot r^{|n|}. \tag{4.12}$$

The claim is that we can choose r so close to 1 that the first term on the right will be $<\epsilon/2$ for all ζ. Having chosen r we can choose N so large that the last series in (4.12) adds to $< \epsilon/2$. Then (4.11) holds with $a_n = \varphi_n r^{|n|}$. ►

The following result is mentioned for completeness only. It is actually contained in lemma 1 of section 6.

Corollary 2. *Two integrable periodic functions with identical Fourier coefficients differ at most on a set of measure zero* (*that is, their indefinite integrals are the same*).

Proof. For an integrable periodic φ with Fourier coefficients φ_n put

$$\Phi(x) = \int_{-\pi}^{x} [\varphi(t) - \varphi_0]\, dt. \tag{4.13}$$

This Φ is a continuous periodic function, and an integration by parts shows that for $n \neq 0$ its nth Fourier coefficient equals $-i\varphi_n/n$.

The relations (4.9) and (4.3) together show that a continuous function φ is uniquely determined by its Fourier coefficients φ_n. The coefficients φ_n with $n \neq 0$ therefore determine φ up to an additive constant. For an arbitrary integrable φ it follows that its Fourier coefficients determine the integral Φ, and hence φ is determined up to values on a set of measure zero. ►

*5. THE POISSON SUMMATION FORMULA

In this section φ stands for a characteristic function such that $|\varphi|$ is integrable over the whole line. By the Fourier inversion formula XV,(3.5) this implies the existence of a *continuous density* f. By the Riemann-Lebesgue lemma 3 of XV,4 both f and φ vanish at infinity. If φ tends to zero sufficiently fast it is possible to use it to construct periodic functions by a method that may be described roughly as wrapping the ζ-axis around a

* This section treats important special topics. It is not used in the sequel, and it is independent of the preceding sections except that it uses theorem 1 of section 4.

circle of length 2λ. The new function may be presented in the form

$$\psi(\zeta) = \sum_{k=-\infty}^{+\infty} \varphi(\zeta+2k\lambda). \tag{5.1}$$

(The sum of a doubly infinite series $\sum_{k=-\infty}^{+\infty} a_k$ is here defined as $\lim \sum_{k=-N}^{N} a_k$ when this limit exists.) In case of convergence the function ψ is obviously periodic with period 2λ. In its simplest probabilistic form the Poisson summation formula asserts that whenever ψ is continuous $\psi(\zeta)/\psi(0)$ *is a characteristic function of an arithmetic probability distribution with atoms of weight proportional* to $f(n\pi/\lambda)$ *at the points* $n\pi/\lambda$. (Here $n = 0, \pm 1, \pm 2, \ldots.$) At first sight this result may appear as a mere curiosity, but it contains the famous sampling theorem of communication theory and many special cases of considerable interest.

Poisson summation formula.[5] *Suppose that the characteristic function* φ *is absolutely integrable, and hence the corresponding probability density* f *continuous. Then*

$$\sum_{-\infty}^{+\infty} \varphi(\zeta+2k\lambda) = \frac{\pi}{\lambda} \sum_{-\infty}^{+\infty} f(n\,\pi/\lambda) e^{in(\pi/\lambda)\zeta} \tag{5.2}$$

provided the series on the left converges to a continuous function ψ.

For $\zeta = 0$ this implies that

$$\sum_{-\infty}^{+\infty} \varphi(2k\lambda) = (\pi/\lambda) \sum f(n\pi/\lambda) \tag{5.3}$$

is a positive number A, and so $\psi(\zeta)/A$ is a characteristic function.

Proof. It suffices to show that the right side in (5.2) is the formal Fourier series of the periodic function ψ on the left, that is

$$\frac{1}{2\lambda} \int_{-\lambda}^{\lambda} \psi(\zeta) e^{-in(\pi/\lambda)\zeta}\, d\zeta = \frac{\pi}{\lambda} f(n\pi^{\lambda}). \tag{5.4}$$

In fact, these Fourier coefficients are non-negative, and ψ was assumed continuous; by theorem 1 of the preceding section the Fourier series therefore converges to ψ, and so (5.2) is true.

The contribution of the kth term of the series (5.1) for ψ to the left side in (5.4) equals

$$\frac{1}{2\lambda} \int_{-\lambda}^{\lambda} \varphi(\zeta+2k\lambda) e^{-in(\pi/\lambda)\zeta}\, d\zeta = \frac{1}{2\lambda} \int_{(2k-1)\lambda}^{(2k+1)\lambda} \varphi(s) e^{-in(\pi/\lambda)s}\, ds \tag{5.5}$$

[5] The identity (5.2) is usually established under a variety of subsidiary conditions. Our simple formulation, as well as the greatly simplified proof, are made possible by a systematic exploitation of the positivity of f. For variants of the theorem and its proof see problems 12 and 13.

and is in absolute value less than

$$(5.6)\qquad \int_{(2k-1)\lambda}^{(2k+1)\lambda} |\varphi(s)|\, ds.$$

The intervals $(2k-1)\lambda < s \leq (2k+1)\lambda$ cover the real axis without overlap, and so the quantities (5.6) add up to the integral of $|\varphi|$ which is finite. Summing (5.5) over $-N < k < N$ and letting $N \to \infty$ we get therefore by dominated convergence

$$(5.7)\qquad \frac{1}{2\lambda}\int_{-\lambda}^{\lambda} \psi(\zeta) e^{-in(\pi/\lambda)\zeta}\, d\zeta = \frac{1}{2\lambda}\int_{-\infty}^{+\infty} \varphi(s)\, e^{-in(\pi/\lambda)s}\, ds.$$

The right side equals $(\pi/\lambda) f(n\pi/\lambda)$ by the Fourier inversion theorem and this concludes the proof. ▶

The most interesting special case arises when φ vanishes identically for $|\zeta| \geq a$ where $a < \lambda$. The infinite series in (5.1) reduces to a single term, and ψ is simply the periodic continuation of φ with period 2λ. Then (5.2) holds. In (5.3) the left side reduces to 1 which shows that ψ is the characteristic function of a probability distribution. We have thus the

Corollary. *If a characteristic function vanishes for* $|\zeta| \geq a$ *then all its periodic continuations with period* $2\lambda > 2a$ *are again characteristic functions.*

This corollary is actually somewhat sharper[6] than the "sampling theorem" as usually stated in texts on communication engineering and variously ascribed to H. Nyquist or C. Shannon.

"Sampling theorem." *A probability density* f *whose characteristic function* φ *vanishes outside* $\overline{-a, a}$ *is uniquely determined*[7] *by the values* $(\pi/\lambda) f(n\pi/\lambda)$

[6] Usually unnecessary conditions are introduced because the proofs rely on standard Fourier theory which neglects the positivity of f germane to probability theory.

[7] An explicit expression for $f(x)$ may be derived as follows. For $|\zeta| \leq \lambda$ we have $\varphi(\zeta) = \psi(\zeta)$ and hence by the Fourier inversion formula

$$f(x) = \frac{1}{2\pi}\int_{-\infty}^{+\infty} \psi(\zeta) e^{-ix\zeta}\, d\zeta.$$

Now ψ is given by the right side in (5.2), and a trite integration leads to the final formula

$$f(x) = \frac{1}{2\pi}\frac{\pi}{\lambda}\sum_{-\infty}^{+\infty}\int_{-\lambda}^{\lambda} f\left(n\frac{\pi}{\lambda}\right) e^{in(\pi/\lambda)\zeta - i\zeta x}\, d\zeta$$

$$= \frac{\sin \lambda x}{\lambda}\sum_{-\infty}^{+\infty} f\left(n\frac{\pi}{\lambda}\right)\frac{(-1)^n}{x - n\pi/\lambda}.$$

This expansion is sometimes referred to as "cardinal series." It has many applications. [See theorem 16 in J. M. Whittaker, *Interpolatory function theory*, Cambridge Tracts No. 33, 1935. For an analogue in higher dimensions see D. P. Petersen and D. Middleton, Information and Control, vol. 5 (1962) pp. 279–323.]

for any fixed $\lambda > a$. *(Here* $n = 0, \pm 1, \ldots$.) *(These values induce a probability distribution whose characteristic function is the periodic continuation of* φ *with period* 2λ.)

Examples. (*a*) Consider the density $f(x) = (1-\cos x)/(\pi x^2)$ with the characteristic function $\varphi(\zeta) = 1 - |\zeta|$ vanishing for $|\zeta| > 1$. For $\lambda = 1$ and $\zeta = 1$ we get from (5.3) remembering that $f(0) = 1/(2\pi)$

$$\frac{1}{2} + \frac{4}{\pi^2} \sum_{\nu=0}^{\infty} \frac{1}{(2\nu+1)^2} = 1. \tag{5.8}$$

The periodic continuation of φ with period $2\lambda = 2$ is graphed in figure 2 of XV,2, a continuation with period $\lambda > 2$ in figure 3.

(*b*) For a simple example for (5.2) see problem 11. ▶

As usual in similar situations, formula (5.2) may be rewritten in a form that *looks* more general. Indeed, applying (5.2) to the density $f(x+s)$ we get *the alternative form of the Poisson summation formula*

$$\sum_{-\infty}^{+\infty} \varphi(\zeta+2k\lambda) e^{-is(\zeta+2k\lambda)} = \frac{\pi}{\lambda} \sum_{-\infty}^{+\infty} f\left(n\frac{\pi}{\lambda} + s\right) e^{in(\pi/\lambda)\zeta}. \tag{5.9}$$

Examples. (*c*) Applying (5.9) to the normal density and using only the special value $\zeta = \lambda$ one gets

$$\sum_{-\infty}^{+\infty} e^{-\frac{1}{2}(2k+1)^2\lambda^2} \cos(2k+1)\lambda s = \frac{\pi}{\lambda} \sum_{-\infty}^{+\infty} (-1)^k \mathfrak{n}\left(\frac{(2k+1)\pi}{\lambda} + s\right). \tag{5.10}$$

This is a famous formula from the theory of theta functions which was proved in X,5 by more elementary methods. In fact, differentiation with respect to x shows that the identity of X (5.8) and X (5.9) is equivalent to (5.10) with $\lambda = (\pi/a)\sqrt{t}$ and $s = x/\sqrt{t}$.

(*d*) For the density $f(x) = \pi^{-1}(1+x^2)^{-1}$ with characteristic function $\varphi(\zeta) = e^{-|\zeta|}$ we get from (5.2) for $\zeta = 0$

$$\frac{e^{\lambda} + e^{-\lambda}}{e^{\lambda} - e^{-\lambda}} = \sum_{n=-\infty}^{+\infty} \frac{\lambda}{\lambda^2 + n^2\pi^2}. \tag{5.11}$$

This is the *partial fraction decomposition for the hyperbolic cotangent*.

(*e*) *Densities on the circle* of length 2π may be obtained by wrapping the real axis around the circle as described in II,8. To a given density f on the line there corresponds on the circle the density given by the series $\sum f(2\pi n + s)$. From (5.9) with $\zeta = 0$ we get a new representation of this density in terms of the original characteristic function. In the special case

$f = \mathfrak{n}$ we get the analogue to *the normal density on the unit circle* in the form

$$\text{(5.12)} \qquad \frac{1}{\sqrt{2\pi t}} \sum_{-\infty}^{+\infty} \exp\left(-\frac{1}{2t}(s+2n\pi)^2\right) = \frac{1}{2\pi} \sum_{-\infty}^{+\infty} e^{-\frac{1}{2}n^2 t} \cos ns.$$

The second representation shows clearly that the convolution of two normal densities with parameters t_1 and t_2 is a normal density with parameter $t_1 + t_2$. ▶

6. POSITIVE DEFINITE SEQUENCES

This section is concerned with probability distributions on a finite interval; for definiteness its length will be taken to be 2π. As in section 4 we identify the two endpoints of the interval and interpret the latter as a circle of unit radius. Thus we consider F as a *probability distribution on the unit circle* and define its *Fourier coefficients* by

$$\text{(6.1)} \qquad \varphi_k = \frac{1}{2\pi} \int_{-\pi}^{\pi} e^{-ikt} F\{dt\}, \qquad k = 0, \pm 1, \ldots .$$

Note that $\bar{\varphi}_k = \varphi_{-k}$. It will now be shown that the coefficients φ_k uniquely determine the distribution. Allowing for a trivial change of scale the assertion is equivalent to the following: *A distribution concentrated on* $\overline{-\lambda, \lambda}$ *is uniquely determined by the knowledge of the values* $\varphi(n\pi/\lambda)$ assumed by its characteristic function at the multiples of π/λ. The assertion represents the dual to the sampling theorem of the preceding section according to which a characteristic function vanishing outside $\overline{-\lambda, \lambda}$ is uniquely determined by the values $f(n\pi/\lambda)$ of the density.

Theorem 1. *A distribution F on the circle is uniquely determined by its Fourier coefficients φ_k.*

Proof. As in (4.3) we put for $0 \leq r < 1$

$$\text{(6.2)} \qquad f_r(\zeta) = \sum_{-\infty}^{+\infty} \varphi_n \cdot r^{|n|} \cdot e^{in\zeta}.$$

The trite calculation that led to (4.8) shows that now

$$\text{(6.3)} \qquad f_r(\zeta) = \int_{-\pi}^{\pi} p_r(\zeta - t)\, F\{dt\},$$

where p_r stands for the Poisson kernel defined in (4.7). We know that p_r is a probability density, and we denote the corresponding distribution P_r. Then f_r is the density of the convolution $P_r \star F$ which tends to F as $r \to 1$, and so F is actually calculable in terms of f_r. ▶

Lemma 1. *Let* $\{\varphi_n\}$ *be an arbitrary bounded sequence of complex numbers. In order that there exists a measure* F *on the circle with Fourier coefficients* φ_n *it is necessary and sufficient that for each* $r < 1$ *the function* f_r *defined in* (6.2) *be non-negative.*

Proof. The necessity is obvious from (6.3) and the strict positivity of p_r. Multiply (6.2) by $e^{-ik\zeta}$ and integrate to obtain

$$\varphi_k \cdot r^{|k|} = \frac{1}{2\pi}\int_{-\pi}^{\pi} f_r(\zeta)e^{-ik\zeta}d\zeta. \tag{6.4}$$

For the particular value $k = 0$ it is seen that $\varphi_0 > 0$ and without loss of generality we may assume that $\varphi_0 = 1/(2\pi)$. With this norming f_r is the density of a probability distribution F_r on the circle, and (6.5) states that $\varphi_k r^{|k|}$ is the kth Fourier coefficient of F_r. By the selection theorem it is possible to let $r \to 1$ in such a manner that F_r converges to a probability distribution F. From (6.5) it is obvious that φ_k satisfies (6.1), and this completes the proof. ▶

Note that this lemma is stronger than corollary 2 in section 4. We proceed as in section 2 and derive a counterpart to Bochner's theorem; it is due to G. Herglotz.

Definition. *A sequence* $\{\varphi_k\}$ *is called positive definite if for every choice of finitely many complex numbers* $z_1, \ldots, z_n$

$$\sum_{j,k} \varphi_{j-k}\, z_j \bar{z}_k \geq 0. \tag{6.5}$$

Lemma 2. *If* $\{\varphi_n\}$ *is positive definite then* $\varphi_0 \geq 0$ *and* $|\varphi_n| \leq \varphi_0$.

Proof. The proof of lemma 3 of section 2 applies. (See also problem 14.)

Theorem 2. *A sequence* $\{\varphi_n\}$ *represents the Fourier coefficients of a measure* F *on the circle iff it is positive definite.*

Proof. (*a*) A trite calculation shows that if the φ_k are given by (6.1) the left side in (6.5) equals the integral of $(1/2\pi)\,|\sum e^{-ijt}z_j|^2$ with respect to F. The condition is therefore necessary.

(*b*) To show its sufficiency choose $z_k = r^k e^{ikt}$ for $k \geq 0$ and $z_k = 0$ for $k < 0$. With this sequence the sum in (6.5) takes on the form

$$\sum_{j=0}^{\infty}\sum_{k=0}^{\infty} \varphi_{j-k} r^{j+k} e^{i(j-k)t} = \sum_{n=-\infty}^{+\infty} \varphi_n e^{int} \sum_{k=0}^{\infty} r^{|n|+2k} = (1-r^2)^{-1} \sum_{n=-\infty}^{+\infty} \varphi_n r^{|n|} e^{int}, \tag{6.6}$$

and by the definition (6.2) the last sum equals $f_r(t)$. It is true that the inequality (6.5) was postulated only for finite sequences $\{z_n\}$, but a simple passage to the limit shows that it applies also to our infinite sequence, and we have thus proved that $f_r(t) \geq 0$. By lemma 1 this implies that the φ_n are indeed the Fourier coefficients of a measure on the unit circle. ▶

From this criterion we derive an analogue to the theorem of section 3:

Theorem 3. *Let* $\{\mathbf{X}_n\}$ *be a sequence of random variables defined on some probability space such that*

$$\rho_n = \mathbf{E}(\mathbf{X}_{n+\nu}\bar{\mathbf{X}}_\nu) \tag{6.7}$$

is independent of ν. *Then there exists a unique measure* R *on the circle* $\overline{-\pi, \pi}$ *such that* ρ_n *is its nth Fourier coefficient.*

Proof. Clearly

$$\sum \rho_{j-k} z_j \bar{z}_k = \sum \mathbf{E}(\mathbf{X}_j z_j \bar{\mathbf{X}}_k \bar{z}_k) = \mathbf{E}(|\sum \mathbf{X}_j z_j|^2) \tag{6.8}$$

which shows that the sequence $\{\rho_n\}$ is positive definite. ▶

The converse is also true: to any measure on the circle there exists a sequence $\{\mathbf{X}_n\}$ such that (6.7) yields its Fourier coefficients. This can be seen by the construction used at the end of section 3, but we shall return to this point in section 8.

Examples. (*a*) Let the $\mathbf{X}_n$ be real identically distributed independent variables with $\mathbf{E}(\mathbf{X}_n) = \mu$ and $\operatorname{Var}(\mathbf{X}_n) = \sigma^2$. Then $\rho_0 = \sigma^2 + \mu^2$ and $\rho_k = \mu^2$ for all $k \neq 0$. The spectral measure is the sum of an atom of weight μ^2 at the origin plus a uniform distribution with density $\sigma^2/(2\pi)$.

(*b*) The construction in section 4 shows that the density defined for fixed $r < 1$ and θ by $p_r(t-\theta)$ has Fourier coefficients $\rho_n = r^{|n|}e^{in\theta}$.

(*c*) *Markov processes.* It was shown in III,8 that stationary Markov sequences of real normal variables have covariances of the form $\rho_n = r^{|n|}$ with $0 \leq r \leq 1$. A similar argument shows that the covariances of arbitrary complex stationary Markov sequences are of the form $r^{|n|}e^{in\theta}$. When $r < 1$ the spectral measure has density $p_r(t-\theta)$; when $r = 1$ it is concentrated at the point θ. ▶

7. L^2 THEORY

For purposes of probability theory it was necessary to introduce characteristic functions as transforms of measures, but other approaches to harmonic analysis are equally natural. In particular, it is possible to define Fourier transforms of functions (rather than measures) and the Fourier inversion formula makes it plausible that greater symmetry can be achieved in this way. It turns out that the greatest simplicity and elegance is attained when only square integrable functions are admitted. This theory will now be developed for its intrinsic interest and because it is extensively used in the harmonic analysis of stochastic processes.

For a complex-valued function u of the real variable x we define the *norm* $\|u\| \geq 0$ by

$$\|u\|^0 = \int_{-\infty}^{+\infty} |u(x)|^2 \, dx. \tag{7.1}$$

Two functions differing only on a set of measure zero will be considered identical. (In other words, we are actually dealing with equivalence classes of functions, but indulge in a usual harmless abuse of language.) With this convention $\|u\| = 0$ iff $u = 0$. The class of all functions with finite norm will be denoted by L^2. The *distance* of two functions u, v in L^2 is defined by $\|u - v\|$. With this definition L^2 is a metric space and a sequence of functions u_n in L^2 converges in this metric to u iff $\|u_n - u\| \to 0$. This convergence[8] will be indicated by $u = \text{l.i.m.}\, u_n$ or $u_n \xrightarrow{\text{l.m.}} u$. It is also called "convergence in the mean square". $\{u_n\}$ is a *Cauchy sequence* iff

$$\|u_n - u_m\| \to 0 \quad \text{as} \quad n, m \to \infty.$$

We mention without proof that the metric space L^2 is complete in the sense that every Cauchy sequence $\{u_n\}$ possesses a unique limit $u \in L^2$.

Examples. (*a*) A function u in L_2 is integrable over every *finite* interval because $|u(x)| < |u(x)|^2 + 1$ at all points. The statement is *not* true for infinite intervals since $(1+|x|)^{-1}$ is in L^2 but not integrable.

(*b*) Every *bounded* integrable function is in L^2 because $|u| \leq M$ implies $|u|^2 \leq M\,|u|$. The statement is false for unbounded functions since $x^{-\frac{1}{2}}$ is integrable over $\overline{0, 1}$, but not square integrable. ▶

The *inner product* (u, v) of two functions is defined by

$$(u, v) = \int_{-\infty}^{+\infty} u(x)\, \overline{v(x)} \, dx. \tag{7.2}$$

It exists for every pair of functions in L^2 since by Schwarz' inequality

$$\int_{-\infty}^{\infty} |uv| \, dx \leq \|u\| \cdot \|v\|. \tag{7.3}$$

In particular, $(u, u) = \|u\|^2$. With this definition of the inner product L^2

---

[8] Pointwise convergence of u_n to a limit v does not imply that $u_n \xrightarrow{\text{l.m.}} v$ [see example IV,2(*e*)]. However, if it is known that if also $u = \text{l.i.m.}\, u_n$ exists, then $u = v$. In fact, by Fatou's lemma

$$\int_{-\infty}^{+\infty} |u(x) - v(x)|^2 \, dx \leq \lim \int_{-\infty}^{+\infty} |u(x) - u_n(x)|^2 \, dx = 0.$$

becomes a *Hilbert space*. The analogy of the inner product (7.2) with the covariance of two random variables with zero expectations is manifest and will be exploited later on.

After these preparations we turn to our main object, namely to define transforms of the form

$$\hat{u}(\zeta) = \frac{1}{\sqrt{2\pi}} \int_{-\infty}^{+\infty} u(x) e^{i\zeta x}\, dx. \tag{7.4}$$

When u is a probability density, $\hat{u}$ differs by the factor $\sqrt{2\pi}$ from the characteristic function and to avoid confusion $\hat{u}$ will be called *the Plancherel transform of* u. The definition (7.4) applies only to integrable functions u but we shall extend it to all L^2. The following examples may facilitate an understanding of the procedure and of the nature of the generalized transform.

Examples. (*c*) Any function u in L^2 is integrable over finite intervals, and hence we may define the truncated transforms

$$\hat{u}^{(n)}(\zeta) = \frac{1}{\sqrt{2\pi}} \int_{-n}^{n} u(x) e^{i\zeta x}\, dx.$$

Note that $\hat{u}^{(n)}$ is the true Plancherel transform of the function $u^{(n)}$ defined by $u^{(n)}(x) = u(x)$ for $|x| < n$ and $u^{(n)}(x) = 0$ for all other x. As $n \to \infty$ the values $\hat{u}^{(n)}(\zeta)$ need not converge for any particular ζ, but we shall show that $\{\hat{u}^{(n)}\}$ is a Cauchy sequence and hence there exists an element $\hat{u}$ of L^2 such that $\hat{u} = \text{l.i.m.}\, \hat{u}^{(n)}$. This $\hat{u}$ will be defined to be the Plancherel transform of u even though the integral in (7.4) need not converge. The particular mode of truncation plays no role, and the same $\hat{u}$ might have been obtained by taking any other sequence of integrable functions $u^{(n)}$ converging in the mean to u.

(*d*) If u stands for the uniform density in $\overline{-1, 1}$ then its Plancherel transform $\hat{u} = \sin x/(x\sqrt{2\pi})$ is not integrable. However, $\hat{u}$ is in L^2 and we shall see that its Plancherel transform coincides with the original density u. In this way we get a generalization of the Fourier inversion formula XV,(3.5) applicable to densities whose characteristic functions are not integrable. ►

We proceed to the definition of the general Plancherel transform. For any integrable function u the transform $\hat{u}$ is defined by (7.4). Such $\hat{u}$ is continuous (by the principle of dominated convergence) and $|\hat{u}|$ is bounded by $(2\pi)^{-\frac{1}{2}}$ times the integral of $|u|$. In general $\hat{u}$ is not integrable. For brevity we now agree to call a function u "*good*" if it is bounded and continuous, and $\hat{u}$ as well as u is integrable. Then also $\hat{u}$ is good, and both u and $\hat{u}$ belong to L^2 [see example (*b*)].

First we show that the *inversion formula*

$$(7.5)\qquad u(t) = \frac{1}{\sqrt{2\pi}} \int_{-\infty}^{+\infty} \hat{u}(\zeta) e^{-i\zeta t}\, d\zeta$$

holds for good functions. We repeat the argument used for the inversion formula in XV,3. Multiply (7.4) by $(1/\sqrt{2\pi})e^{-i\zeta t - \frac{1}{2}\epsilon^2\zeta^2}$ and integrate to obtain

$$(7.6)\qquad \frac{1}{\sqrt{2\pi}} \int_{-\infty}^{+\infty} \hat{u}(\zeta) e^{-i\zeta t - \frac{1}{2}\epsilon^2\zeta^2}\, d\zeta = \int_{-\infty}^{+\infty} u(x)\, \mathfrak{n}\left(\frac{t-x}{\epsilon}\right) \frac{dx}{\epsilon}$$

where $\mathfrak{n}$ denotes the standard normal density. As $\epsilon \to 0$ the integral on the left tends to the integral in (7.5) while the convolution on the right tends to $u(t)$. Thus (7.5) holds for good functions.

Now let v be another good function. Multiply (7.5) by the conjugate $\bar{v}(t)$ and integrate over $-\infty < t < \infty$. The left side equals the inner product (u, v) and after interchanging the order of integration the right side reduces to $(\hat{u}, \hat{v})$. Thus good functions satisfy the identity

$$(7.7)\qquad (\hat{u}, \hat{v}) = (u, v)$$

which will be referred to as the *Parseval relation* for L^2. For $v = u$ it reduces to

$$(7.8)\qquad \|\hat{u}\| = \|u\|.$$

It follows that the distance of two transforms $\hat{u}$ and $\hat{v}$ is the same as the distance between u and v. We express this by saying that among good functions the Plancherel transform is an *isometry*.

Next we show that the relations (7.7) and (7.8) remain valid for arbitrary integrable functions u and v belonging to L^2. The transforms are not necessarily integrable, but (7.8) implies that they belong to L^2 [compare examples (*a*) and (*d*)].

First we observe that an integrable function w with two integrable derivatives is necessarily good; indeed, from lemma 4 of XV,4 one concludes that $|\hat{w}(\zeta)| = o(\zeta^{-2})$ as $\zeta \to \pm\infty$, and so $\hat{w}$ is certainly integrable.

Suppose now that u is bounded and integrable (and hence in L^2). By the mean approximation theorem of IV,2 it is possible to find a sequence of good functions u_n such that

$$(7.9)\qquad \int_{-\infty}^{+\infty} |u(x) - u_n(x)|\, dx \to 0.$$

If $|u| < M$ these u_n may be chosen such that also $|u_n| < M$. Then u_n tends in the mean to u because $\|u - u_n\|^2$ cannot exceed $2M$ times the integral in (7.9). The isometry (7.8) for good functions therefore guarantees that $\{\hat{u}_n\}$ is a Cauchy sequence. On the other hand, $\hat{u}_n(\zeta) \to \hat{u}(\zeta)$ for every fixed ζ because $|\hat{u}(\zeta) - \hat{u}_n(\zeta)|$ cannot exceed the integral in (7.9). But, as pointed out in the last footnote, the pointwise convergence of the elements of a Cauchy sequence entails the convergence of the sequence itself, and thus we have

$$\hat{u} = \text{l.i.m. } \hat{u}_n. \tag{7.10}$$

Applying (7.7) to the pair u_n and v and letting $n \to \infty$ we see now that (7.7) remains valid whenever u is bounded and integrable while v is good. Another such passage to the limit shows that (7.7) remains valid for any pair of bounded integrable functions.

It remains to show that (7.7) is valid also for unbounded functions u and v provided they are integrable and belong to L^2. For the proof we repeat the preceding argument with the sole change that the approximating functions u_n are now defined by truncation: $u_n(x) = u(x)$ if $|u(x)| < n$ and $u_n(x) = 0$ for all other x. Then (7.10) holds and $u_n \xrightarrow{\text{l.i.m.}} u$, and the proof applies without further change.

We are now ready for the final step, namely to extend the definition of the Plancherel transform to the whole of L^2. As shown in example (*c*), every function u in L^2 is the limit of a Cauchy sequence of integrable functions u_n in L^2. We have just shown that the transforms $\hat{u}_n$ defined by (7.4) form a Cauchy sequence, and we now *define* $\hat{u}$ as the limit of this Cauchy sequence. Since two Cauchy sequences may be combined into one the limit $\hat{u}$ is independent of the choice of the approximating sequence $\{u_n\}$. Also, if u happens to be integrable we may take $u_n = u$ for all n, and thus it is seen that the new definition is consistent with (7.4) whenever u is integrable. To summarize:

A Plancherel transform $\hat{u}$ *is defined for every* u *in* L^2; *for integrable* u *it is given by* (7.4), *and in general by the rule that*

$$\textit{if} \quad u = \text{l.i.m. } u_n \quad \textit{then} \quad \hat{u} = \text{l.i.m. } u_n. \tag{7.11}$$

The Parseval relation (7.7) *and the isometry* (7.8) *apply generally. The mapping* $u \to \hat{u}$ *is one-to-one, the transform of* $\hat{u}$ *being given by* $u(-x)$.

The last statement is a version of the Fourier inversion formula (7.5) applicable when u or $\hat{u}$ are not integrable so that the integrals in (7.4) and (7.5) are not defined in the usual sense. This complete symmetry in the relationship between the original functions and their transforms represents the main advantage of Fourier theory in Hilbert spaces.

The theory as outlined is widely used in prediction theory. As an example of a probabilistic application we mention a criterion usually ascribed to A.

Khintchine although it appears in the classical work of N. Wiener. For reasons of historical tradition even newer texts fail to notice that it is really merely a special case of the Parseval formula and requires no elaborate proof.

Wiener-Khintchine criterion. *In order that a function* φ *be the characteristic function of a probability density* f *it is necessary and sufficient that there exist a function* u *such that* $\|u\|^2 = 1$ *and*

$$\varphi(\lambda) = \int_{-\infty}^{+\infty} u(x)\, \overline{u(x+\lambda)}\, dx \tag{7.12}$$

In this case $f = \|\hat{u}\|^2$.

Proof. For fixed λ put $v(x) = u(x+\lambda)$. Then $\hat{v}(\zeta) = \hat{u}(\zeta)e^{-i\lambda\zeta}$. The Parseval relation (7.7) reduces to

$$\int_{-\infty}^{+\infty} |\hat{u}(x)|^2\, e^{i\lambda x}\, dx = \int_{-\infty}^{+\infty} u(x)\, \overline{u(x+\lambda)}\, dx. \tag{7.13}$$

and since $\|\hat{u}\|^2 = 1$ the left side represents the characteristic function of a probability density. Conversely, given a probability density f it is possible to choose u such that $f = |\hat{u}|^2$, and then (7.12) holds. The choice of u is not unique. (One problem of prediction theory concerns the possibility of choosing u vanishing on a half-line.) ►

The L^2 theory for Fourier integrals carries over to Fourier series. The functions are now defined on the circle, but to the Fourier (or Plancherel) transforms there correspond now sequences of Fourier coefficients. Except for this formal difference the two theories run parallel, and a brief summary will suffice.

To our L^2 there corresponds now the Hilbert space $L^2(\overline{-\pi, \pi})$ of square integrable functions on the circle. The norm and inner product are now defined by

$$\|u\|^2 = \frac{1}{2\pi}\int_{-\pi}^{\pi} |u(x)|^2\, dx, \qquad (u, v) = \frac{1}{2\pi}\int_{-\pi}^{\pi} u(x)\, \overline{v(x)}\, dx \tag{7.14}$$

it being understood that the integrals extend over the whole circle (the points $-\pi$ and π being identified). The role of "good functions" is played by finite trigonometric polynomials of the form

$$u(x) = \sum u_n e^{inx}, \tag{7.15}$$

whose Fourier coefficients u_n are given by

$$u_n = \frac{1}{2\pi}\int_{-\pi}^{\pi} u(x) e^{-inx}\, dx. \tag{7.16}$$

To a good function there corresponds the finite sequence $\{u_n\}$ of its coefficients, and, conversely, to every finite sequence of complex numbers there corresponds a good function. The relations (7.16) and (7.15) define the Fourier transform $\hat{u} = \{u_n\}$ and its inverse. Formal multiplication and integration shows that for two good functions

$$\frac{1}{2\pi}\int_{-\pi}^{\pi} u(x)\,\overline{v(x)}\,dx = \sum u_k \bar{v}_k. \tag{7.17}$$

We now consider the Hilbert space $\mathfrak{H}$ of infinite sequences $\hat{u} = \{u_n\}$, $\hat{v} = \{v_n\}$, etc., with norm and inner product defined by

$$\|\hat{u}\| = \sum |u_n|^2, \qquad (\hat{u}, \hat{v}) = \sum u_n\overline{v_n}. \tag{7.18}$$

This space enjoys properties analogous to L^2; in particular, finite sequences are dense in the whole space. It follows that there exists a one-to-one correspondence between the sequences $\hat{u} = \{u_n\}$ in $\mathfrak{H}$ and the functions u in $L^2(\overline{-\pi, \pi})$. *To each sequence* $\{u_n\}$ *such that* $\sum |u_n|^2 < \infty$ *there corresponds a square integrable function* u *with Fourier coefficients* u_n *and conversely*. The mapping $u \leftrightarrow \{u_n\}$ is again an *isometry*, and the *Parseval relation* $(u, v) = (\hat{u}, \hat{v})$ holds. The Fourier series need not converge but the partial sums $\sum_{-n}^{n} u_k e^{ikx}$ form a sequence of continuous functions that converges to u in the L^2 metric. The same statement is true of other continuous approximations. Thus $\sum u_k r^{|k|} e^{ikx}$ tends to u as $r \to 1$.

As above, we consider the special case of the Parseval relation represented by

$$\frac{1}{2\pi}\int_{-\pi}^{\pi} u(x)\,\bar{v}(x) e^{-inx}\,dx = \sum_k u_{k+n}\overline{v_k}. \tag{7.19}$$

Choosing $v = u$ one sees again that *a sequence* $\{\varphi_n\}$ *represents the Fourier coefficients of a probability density on* $\overline{-\pi, \pi}$ *iff it is of the form*

$$\varphi_n = \sum u_{k+n}\overline{u_k} \quad \textit{where} \quad \sum |u_k|^2 = 1. \tag{7.20}$$

A covariance of this form occurs in III,(7.4). (See also problem 17.)

8. STOCHASTIC PROCESSES AND INTEGRALS

For notational simplicity we refer in this section to *sequences* $\{\mathbf{X}_n\}$ of random variables, but it will be evident that the exposition applies to families depending on a continuous time parameter with the sole change that the spectral measure is not confined to a finite interval and that series are replaced by integrals.

Let, then, $\{\mathbf{X}_n\}$ stand for a doubly infinite sequence of random variables defined on some probability space $\mathfrak{S}$ and having finite second moments. The sequence is assumed stationary in the restricted sense that

$$\mathbf{E}(\mathbf{X}_{n+\nu}\bar{\mathbf{X}}_\nu) = \rho_n$$

is independent of ν. According to theorem 3 of section 6 there exists a unique measure R on the circle $\overline{-\pi, \pi}$ such that

$$\rho_n = \frac{1}{2\pi}\int_{-\pi}^{\pi} e^{-inx}\, R\{dx\}, \qquad n = 0, \pm 1, \ldots. \tag{8.1}$$

We shall now elaborate on the idea mentioned in section 3 that the circle equipped with the spectral measure R may be used to construct *a concrete representation for the stochastic process* $\{\mathbf{X}_n\}$ (at least for all properties depending only on second moments). The description uses Hilbert space terminology, and we shall consider two Hilbert spaces.

(*a*) *The space* L_R^2. We construct a space of functions on the circle by literal repetition of the definition of L^2 in section 7, except that the line is replaced by the circle $\overline{-\pi, \pi}$ and Lebesgue measure by the measure R. The norm and the inner product of (complex-valued) functions on the circle are defined by

$$\|u\|^2 = \frac{1}{2\pi}\int_{-\pi}^{\pi} |u(x)|^2\, R\{dx\}, \qquad (u, v) = \frac{1}{2\pi}\int_{-\pi}^{\pi} u(x)\, \overline{v(x)}\, R\{dx\}, \tag{8.2}$$

respectively. The basic convention now is that *two functions are considered identical if they differ only on a set of R-measure zero.* The impact of this convention is serious. If R is concentrated on the two points 0 and 1 then a "function" (in our sense) is completely determined by its values at these two points. For example, $\sin n\pi x$ is the zero function. Even in such radical cases no harm is done in using the customary formulas for continuous functions and in referring to their graphs. Thus reference to a "step function" is always meaningful and it simplifies the language.

The Hilbert space L_R^2 *consists of all functions on the circle with finite norm.* If $\|u_n - u\| \to 0$ the sequence $\{u_n\}$ is said to converge to u in our metric (or in mean square with respect to the weight distribution R). The Hilbert space L_R^2 is a complete metric space in which the continuous functions are dense. (For definitions see section 7.)

(*b*) *The Hilbert space* $\mathfrak{H}$ *spanned by* $\{\mathbf{X}_n\}$. Denote by $\mathfrak{H}_0$ the family of random variables with finite second moments defined in the arbitrary, but fixed, sample space $\mathfrak{S}$. By Schwarz's inequality $\mathbf{E}(\mathbf{U}\overline{\mathbf{V}})$ exists for any pair of such variables, and it is natural to generalize (8.2) from the circle to the sample space $\mathfrak{S}$ using the underlying probability measure instead of R. We accordingly agree again to identify two random variables if they differ only on a set of probability zero and define inner products by $\mathbf{E}(\mathbf{U}\overline{\mathbf{V}})$; the norm of $\mathbf{U}$ is the positive root of $\mathbf{E}(\mathbf{U}\overline{\mathbf{U}})$. With these conventions $\mathfrak{H}_0$ again becomes a Hilbert space; it is a complete metric space in which a sequence of random variables $\mathbf{U}_n$ is said to converge to $\mathbf{U}$ if $\mathbf{E}(|\mathbf{U}_n - \mathbf{U}|^2) \to 0$.

In dealing with a sequence $\{\mathbf{X}_n\}$ one is usually interested only in random variables that are functions of the $\mathbf{X}_k$ and in many connections one considers only linear functions. This restricts the consideration to finite linear combinations $\sum a_k \mathbf{X}_k$ and limits of sequences of such finite linear combinations. Random variables of this kind form a subspace $\mathfrak{H}$ of $\mathfrak{H}_0$, called the *Hilbert space spanned by the* $\mathbf{X}_k$. In it inner products, norms, and convergence are defined as just described and $\mathfrak{H}$ is a complete metric space.

In the present context the expectations $\mathbf{E}(\mathbf{X}_n)$ play no role whatever, but as "covariance" sounds better than "inner product" we introduce the usual convention that $\mathbf{E}(\mathbf{X}_n) = 0$. The sole purpose of this is to establish ρ_n as a covariance, and no centering is necessary if one agrees to call $\mathbf{E}(\mathbf{X}\bar{\mathbf{Y}})$ the covariance of $\mathbf{X}$ and $\mathbf{Y}$. We come now to the crucial point, namely that for our purposes the intuitively simple space L^2_R *may serve as concrete model for* $\mathfrak{H}$. Indeed, by definition the covariance $\rho_{j-k} = \text{Cov}\,(\mathbf{X}_j, \mathbf{X}_k)$ of any pair equals the inner product of the functions e^{ijx} and e^{ikx} in L^2_R. It follows that the covariance of two finite linear combinations $\mathbf{U} = \sum a_j \mathbf{X}_j$ and $\mathbf{V} = \sum b_k \mathbf{X}_k$ equals the inner product of the corresponding linear combinations $u = \sum a_j e^{ijx}$ and $v = \sum b_k e^{ikx}$. By the very definition of convergence in the two spaces this mapping now extends to all random variables. We have thus the important result that the mapping $\mathbf{X}_k \leftrightarrow e^{ikx}$ *induces a one-to-one correspondence between the random variables in* $\mathfrak{H}$ *and the functions in* L^2_R, *and this correspondence preserves inner products and norms* (and hence limits). In technical language the two spaces are isometric.[9] We are in a position to study $\mathfrak{H}$ and $\{\mathbf{X}_n\}$ referring explicitly only to the concrete space L^2_R. This procedure has theoretical advantages in addition to being an aid to intuition. Since functions on the circle are a familiar object it is relatively easy to discover sequences $\{u^{(n)}\}$ of functions in L^2_R with desirable structural properties. To $u^{(n)}$ there corresponds a random variable $\mathbf{Z}_n$ on the original sample space $\mathfrak{S}$; if the Fourier coefficients of $u^{(n)}$ are known it is possible to represent $\mathbf{Z}_n$ explicitly as a limit of finite linear

---

[9] Readers acquainted with Hilbert space theory should note the connection with the standard *spectral theorem for unitary operators*. The linear operator which maps $\mathfrak{H}$ into itself in such a way that $\mathbf{X}_n \to \mathbf{X}_{n+1}$ is called a *shift operator*, and L^2_R serves as a model in which the action of this shift operator becomes multiplication by e^{ix}. Conversely, given an arbitrary unitary operator T on a Hilbert space $\mathfrak{H}_0$ and an arbitrary element $\mathbf{X}_0 \in \mathfrak{H}_0$, the sequence of elements $\mathbf{X}_n = T^n \mathbf{X}_0$ may be treated as a stationary sequence and T as the shift operator on the subspace $\mathfrak{H}$ spanned by this sequence. If $\mathbf{X}_0$ can be chosen such that $\mathfrak{H} = \mathfrak{H}_0$ we have obtained the standard spectral theorem for T except that we have a concrete representation of the "resolution of the identity" based on the choice of $\mathbf{X}_0$. If $\mathfrak{H} \subset \mathfrak{H}_0$, then $\mathfrak{H}_0$ is the direct sum of two invariant subspaces, and the presentation applies to each of them. By a simple change of notations one derives the general spectral representation, including the theory of multiplicity for the spectrum.

combinations of the variables $\mathbf{X}_k$. If the joint distributions of the $\mathbf{X}_k$ are normal the same is true for those of the $\mathbf{Z}_n$.

In practice this procedure is usually reversed. Given a complicated process $\{\mathbf{X}_k\}$ our aim is to express it in terms of the variables $\mathbf{Z}_n$ of a simpler process. A practical way to achieve this is to proceed in L_R^2 rather than in the original space. A few examples will explain this better than a theoretical discourse.

Examples. (*a*) *Representation of* $\{\mathbf{X}_n\}$ *by independent variables.* As elsewhere in this section $\{\mathbf{X}_n\}$ stands for a given process with covariances ρ_n and spectral measure R defined by (8.1). In our mapping $\mathbf{X}_n$ corresponds to the function e^{inx} on the circle. We show now that for certain functions γ the random variables corresponding to $e^{inx}/\gamma(x)$ are uncorrelated. We consider only the situation when the spectral measure R has an ordinary density r. For simplicity[10] r will be assumed strictly positive and continuous. Choose a function γ such that

$$|\gamma(x)|^2 = r(x). \tag{8.3}$$

The Fourier series of γ converges in the L^2 norm as explained in section 7. Denoting the Fourier coefficients of γ by γ_k we have $\sum |\gamma_k|^2 < \infty$ and by the Parseval relation (7.20)

$$\rho_n = \sum_{k=-\infty}^{+\infty} \gamma_{k+n}\overline{\gamma}_k. \tag{8.4}$$

Consider now the doubly infinite sequence of functions $u^{(n)}$ defined by

$$u^{(n)}(x) = \frac{e^{inx}}{\gamma(x)}. \tag{8.5}$$

Substituting into (8.2) it is seen that

$$\|u^{(n)}\| = 1, \qquad (u^{(n)}, u^{(m)}) = 0 \tag{8.6}$$

for $m \neq n$. For the random variables $\mathbf{Z}_n$ corresponding to the functions $u^{(n)}$ this implies that they are uncorrelated and of unit variance. In particular, *if the* $\mathbf{X}_k$ *are normal the* $\mathbf{Z}_k$ *are mutually independent.*

It is interesting that the space spanned by the variables $\mathbf{X}_k$ contains a stationary sequence $\{\mathbf{Z}_n\}$ of uncorrelated variables. An explicit expression of $\mathbf{Z}_n$ in terms of the $\mathbf{X}_k$ can be obtained from the Fourier expansion of the function $u^{(n)}$, but it is more profitable to proceed in the opposite direction: the structure of $\{\mathbf{Z}_n\}$ being simpler than that of $\{\mathbf{X}_k\}$ it is preferable to express the $\mathbf{X}_k$ in terms of the $\mathbf{Z}_n$. Now

$$\sum_{n=-N}^{N} \gamma_n\, u^{(n+k)}(x) = \frac{e^{ikx}}{\gamma(x)} \sum_{n=-N}^{N} \gamma_n e^{inx}. \tag{8.7}$$

---

[10] The restriction is not used except to avoid trite explanations of what is meant by $r(x)/\gamma(x)$ when $r(x) = \gamma(x) = 0$ and of how series converge.

The sum on the right is a section of the Fourier series for γ and tends in the Hilbert space metric to γ. It follows that the quantity (8.7) tends to e^{ikx}. In our mapping $u^{(n+k)}$ corresponds to $\mathbf{Z}_{n+k}$ and so the series $\sum \gamma_n \mathbf{Z}_{n+k}$ converges, and we can write

$$\mathbf{X}_k = \sum_{n=-\infty}^{+\infty} \gamma_n \mathbf{Z}_{n+k}. \tag{8.8}$$

We have thus obtained *an explicit representation of* $\mathbf{X}_k$ *as a "moving average" in a stationary sequence of uncorrelated variables* $\mathbf{Z}_k$.

The representation (8.8) is evidently not unique. The natural question arises whether it is possible to express $\mathbf{X}_k$ solely by the variables $\mathbf{Z}_k$, $\mathbf{Z}_{k-1}, \mathbf{Z}_{k-2}, \ldots$ (representing the "past"), that is, whether the function γ in (8.3) can be chosen such that $\gamma_n = 0$ for $n \geq 1$. This problem is fundamental in prediction theory, but lies outside the scope of the present volume. For typical examples see III,(7.5), and problem 18.

(*b*) *The associated process with uncorrelated increments.* For each t with $-\pi < t \leq \pi$ define y_t by

$$y_t(x) = \begin{cases} 1 & \textit{for} \quad x \leq t \\ 0 & \textit{for} \quad x > t \end{cases} \tag{8.9}$$

and denote by $\mathbf{Y}_t$ the corresponding random variable in $\mathfrak{H}$. The increments $\mathbf{Y}_t - \mathbf{Y}_s$ for non-overlapping intervals have obviously covariance 0; furthermore $\text{Var}\,(\mathbf{Y}_t) = R\{\overline{-\pi, t}\}$. Thus $\{\mathbf{Y}_t\}$ is a process *with uncorrelated increments and variances given by* R. If the $\mathbf{X}_t$ are normal, the increments of the $\mathbf{Y}_t$ process are actually independent.

With every stationary sequence $\{\mathbf{X}_k\}$ there is in this way associated a process with uncorrelated increments. An explicit expression of $\mathbf{Y}_t$ in terms of the $\mathbf{X}_k$ is obtainable in the standard way by expanding the function y_t in (8.9) into a Fourier series. Once more it is preferable to proceed in the opposite direction. This will be done in the next example.

(*c*) *Stochastic integrals.* The representation of a random variable $\mathbf{U}$ in terms of the $\mathbf{X}_k$ depends (as we have seen) on the Fourier expansion of the function corresponding to $\mathbf{U}$. By contrast, the following representation in terms of the variables $\mathbf{Y}_t$ is almost too simple for comfort. It refers to the graph of the function, and for simplicity we assume the latter continuous.

Consider first a step function w, that is, a function of the form

$$w = a_1 y_{t_1} + a_2(y_{t_2} - y_{t_1}) + \cdots + a_n(y_\pi - y_{t_{n-1}}) \tag{8.10}$$

where the a_j are constants and $-\pi < t_1 < t_2 < \cdots < t_{n-1} < \pi$. The associated random variable $\mathbf{W}$ is obtained on replacing in this expression each y_{t_j} by $\mathbf{Y}_{t_j}$. Now an arbitrary continuous function w can be approximated uniformly by step functions $w^{(n)}$ of the form (8.10). Uniform convergence of $w^{(n)}$ to w implies the convergence in the norm of L^2_R and hence also the convergence of the corresponding random variables $\mathbf{W}^{(n)}$ to $\mathbf{W}$. This gives us a prescription for finding the image $\mathbf{W}$ of an arbitrary continuous function w by a simple limiting procedure: approximate w by step functions of the form (8.10) and replace y_{t_j} by $\mathbf{Y}_{t_j}$. Remember that (8.10) is a function, and not a number, just as the limit w of $w^{(n)}$ is a function rather than a number. But (8.10) *looks* like a Riemann sum and our procedure is formally reminiscent of the definition of the Riemann integral. It has therefore become standard practice to use the notation

$$\mathbf{W} = \int_{-\pi}^{\pi} w(t)\, d\mathbf{Y}_t \tag{8.11}$$

to indicate the described limiting process. The random variable (8.11) is called *the stochastic integral* of the continuous function w. The name is arbitrary and the notation mere shorthand for the limiting procedure which we have rigorously defined. By definition the function e^{int} corresponds to the random variable $\mathbf{X}_n$ and hence we can write

$$\mathbf{X}_n = \int_{-\pi}^{\pi} e^{int}\, d\mathbf{Y}_t. \tag{8.12}$$

This is the basic *spectral representation of the arbitrary stationary sequence* $\{\mathbf{X}_n\}$ *in terms of the associated process with uncorrelated increments.*

The notation for stochastic integrals is, perhaps, more suggestive than logical but we are not concerned with this usage. Our aim was to show that this useful concept and the important representation (8.12) are easily established by means of Fourier analysis. This illustrates the power of the canonical mapping used in this section and first introduced by Cramér. ▶

The theory depends only on the second moments of $\{\mathbf{X}_n\}$ and is in practice applicable only when these moments are truly significant. Such is the case when the process is normal, because normal distributions are completely determined by their covariances. In other applications one may trust that the process is "not too far off a normal process" just as the oldest regression analysis trusted in a universal applicability of methods developed for normal variables. Unfortunately the mere existence of a beautiful theory in no way justifies this trust. In example (3.*c*) the sample functions of the process are strictly periodic. The future of an individual path is completely determined by the data for a full period, but the prediction theory based on the L^2 theory takes no account of this fact and identifies all processes with the same spectral measure. One who observes the sequence $1, -1, 1, -1, \ldots$ going on since time immemorial can safely predict the next observation, but L^2 methods will lead him to predict the miraculous occurrence of 0. This example shows that the L^2 methods are not universally applicable, but they are the ideal tool for treating normal processes.

9. PROBLEMS FOR SOLUTION

1. *Curious characteristic functions.* Let $\tau_k(x) = 1 - |x|/h$ for $|x| \leq h$ and $\tau_h(x) = 0$ for $|x| \geq h$. Put

$$\alpha(x) = \sum_{n=-\infty}^{+\infty} a_n \, \tau_k(x-n). \tag{9.1}$$

When the a_n are real and $h = 1$ *the graph of* α *is the polygonal line with vertices* (n, a_n). When $h < \frac{1}{2}$ the graph of α consists of segments of the x-axis and the sloping sides of isosceles triangles with vertices (n, a_n).

Using the criterion of lemma 1 in section 2 show that $\alpha(\zeta)/\alpha(0)$ is a characteristic function if the a_n are real and such that $a_{-n} = a_n$ and $|a_1| + |a_2| + \cdots \leq \frac{1}{2}a_0$.

2. (*Generalization.*) The last statement remains true when τ_k is replaced by an arbitrary even integrable characteristic function. (Using the characteristic functions exhibited in fig. 1 of XV,2 you may construct characteristic functions with exceedingly weird polygonal graphs.)

3. (*Continuation.*) The preceding result is a special case of the following: Let τ be an even integrable characteristic function; if the λ_n are real, and the a_n complex constants with $\sum |a_n| < \infty$, then

$$\alpha(\zeta) = \sum a_k \tau(\zeta - \lambda_k) \tag{9.2}$$

is a characteristic function iff $\alpha(0) = 1$ and $\sum a_k e^{-i\lambda_k\zeta} \geq 0$ for all ζ. (It suffices actually that the last series be Abel summable to a positive function.)

4. The covariance function ρ defined in (3.3) is continuous everywhere iff it is continuous at the origin. This is the case iff $\mathbf{E}(\mathbf{X}_t - \mathbf{X}_0)^2 \to 0$ as $t \to 0$.

5. (*Difference ratios and derivatives*). Let $\{\mathbf{X}_t\}$ be a stationary process with $\rho(t) = \mathbf{E}(\mathbf{X}_{t+s}\overline{\mathbf{X}}_s)$ and spectral measure R. For $h > 0$ a new process is defined by $\mathbf{X}_t^{(h)} = (\mathbf{X}_{t+h} - \mathbf{X}_t)/h$.

(*a*) Show that the spectral measure $R^{(h)}$ of the new process is given by $R^{(h)}\{dx\} = 2h^{-2}[1 - \cos hx]R\{dx\}$. The covariances $\rho^{(h)}(t)$ tend to a limit as $h \to 0$ iff a continuous second derivative $\rho''(t)$ exists, that is, if the measure $x^2 R\{dx\}$ is finite.

(*b*) In the latter case $\mathbf{E}(|\mathbf{X}_t^{(\epsilon)} - \mathbf{X}_t^{(\delta)}|^2) \to 0$ as $\epsilon \to 0$ and $\delta \to 0$.

Note: In the Hilbert space terminology of section 7 this means that for fixed t as $\epsilon_n \to 0$ the sequence $\{\mathbf{X}_t^{(\epsilon_n)}\}$ is Cauchy, and hence a derivative $\mathbf{X}_t' = \text{l.i.m.}\mathbf{X}_t^{(h)}$ exists.

6. (*See theorem* 2 *of section* 4.) If φ is continuous except for a jump at the origin, then $f_r(\zeta) \to \varphi(\zeta)$ uniformly outside a neighborhood of the origin and $f_r(0) \to \frac{1}{2}[\varphi(0+) - \varphi(0-)]$.

7. *Continuation.* If φ is the difference between two monotone functions then $f_r(\zeta) \to \frac{1}{2}[\varphi(\zeta+) - \varphi(\zeta-)]$ at all points.

8. A bounded periodic function φ with non-negative Fourier coefficients φ_n is necessarily continuous and $\sum \varphi_n < \infty$. The example $\varphi_n = 1/n$ shows that this is false if φ is only supposed to be integrable. *Hint:* Use the main idea of the proof of theorem 1 in section 4.

9. *Cesaro summability.* Replace (4.3) by

$$f_r(\zeta) = \sum \varphi_n a_n e^{in\zeta}$$

where $a_n = 1 - |n|(2N+1)^{-1}$ for $|n| \leq 2N$ and $a_n = 0$ for $|n| > 2N$. Show that the theory of section 4 goes through with $p_r(t)$ replaced by

$$q_N(t) = \frac{1}{2N+1}\,\frac{\sin^2 (N+\frac{1}{2})t}{\sin^2 \frac{1}{2}t},$$

which is again a probability density.

10. *Continuation.* Show, more generally, that the theory goes through if the a_n are the Fourier coefficients of a symmetrized probability density on the circle.

11. Use the Poisson summation formula (5.2) to show that

$$\sum_{-\infty}^{+\infty}\left\{\mathfrak{n}\left(\frac{y-x+2k\lambda}{\sqrt{t}}\right)+\mathfrak{n}\left(\frac{y+x+2k\lambda}{\sqrt{t}}\right)\right\} = \frac{1}{\lambda}\sum_{-\infty}^{+\infty}\exp\left(-\tfrac{1}{2}tn^2\frac{\pi^2}{\lambda^2}\right)\cos\frac{n\pi}{\lambda}x\cdot\cos\frac{n\pi}{\lambda}y,$$

where $\mathfrak{n}$ stands for the standard normal density. [This is the solution of the reflecting barrier problem in example X,5(*e*).]

12. In the *Poisson summation formula* (5.2) the condition that $\sum \varphi(\zeta + 2k\lambda)$ converges to a continuous function ψ may be replaced by the condition that $\sum f(n\pi/\lambda) < \infty$. *Hint:* ψ is under any circumstances an integrable function. Use corollary 2 in section 4.

13. *Alternative derivation of the Poisson summation formula.* Let φ be the characteristic function of an arbitrary probability distribution F. If p_r stands for the Poisson kernel of (4.5) and (4.7) show (without further calculations) that for $0 \leq r < 1$

$$\frac{1}{2\pi}\sum_{n=-\infty}^{+\infty}\varphi(\zeta+n)r^{|n|}e^{in\lambda} = \int_{-\infty}^{+\infty} e^{i\zeta x}\,p_r(x+\lambda)\,F\{dx\}. \tag{9.3}$$

Hence the left side is a characteristic function. Letting $r \to 1$ conclude that if F has a density f then

$$\frac{1}{2\pi}\sum_{n=-\infty}^{+\infty}\varphi(\zeta+n)e^{in\lambda} = \sum_{k=-\infty}^{+\infty} e^{i\zeta(-\lambda+2k\pi)}f(-\lambda+2k\pi) \tag{9.4}$$

whenever $\sum f(-\lambda+2k\pi) < \infty$. Show that (9.4) is equivalent to the general version (5.9) of the summation formula.

Note: The result may be restated to the effect that the left side in (9.4) is *Abel summable* to the right side whenever the latter is continuous.

14. The sequence $\{\varphi_n\}$ is positive definite iff $\{\varphi_n r^{|n|}\}$ is positive definite for every $0 < r < 1$. Necessary and sufficient is that $\sum \varphi_n r^{|n|}e^{in\lambda} \geq 0$ for all λ and $0 < r < 1$.

15. From problems 1 and 14 derive (without calculations) the following theorem (observed by L. Shepp): *Let $\{\varphi_n\}$ be positive definite and denote by α the piecewise linear function with vertices at (n, φ_n). Then α is positive definite.*

16. *Let φ be a characteristic function and denote by α the piecewise linear function with vertices $(n, \varphi(n))$. Then α is a characteristic function.* (This merely paraphrases problem 15 and is a special case of problem 1 for $h = 1$.) Use the

other cases of problem 1 to describe other curious characteristic functions obtainable from φ.

17. If $r < 1$ the covariances $\rho_n = r^{|n|}e^{in\theta}$ of Markov sequences satisfy (7.20) with $u_k = \sqrt{1 - r^2}\, r^k e^{ik\theta}$ for $k \geq 0$ and $u_k = 0$ for $k < 0$. Find alternative representations.

18. *Continuation.* Let $\{\mathbf{X}_n\}$ be Markovian with covariances $\rho_n = r^{|n|}e^{in\theta}$. If $r < 1$ one has $\mathbf{X}_n = \sqrt{1 - r^2} \sum_{k=0}^{\infty} r^k e^{ik\theta}\mathbf{Z}_{n-k}$ where the $\mathbf{Z}_k$ are uncorrelated. If $r = 1$ one has $\mathbf{X}_n = e^{in\theta}\mathbf{Z}_0$.

Answers to Problems

CHAPTER I

1. (i) $\dfrac{\alpha}{3}\dfrac{2}{\sqrt[3]{x^2}}\,e^{-\alpha x^{\frac{1}{3}}}$ (ii) $\dfrac{\alpha}{2}\,e^{-\alpha(x-3)/2}$ for $x > 3$

   (iii) $\dfrac{\alpha}{2}\,e^{-\alpha|x|}$, all x (iv) $\alpha e^{-\alpha x}$, $x > 0$

   (v) $\alpha\left(1 + \dfrac{1}{4}\dfrac{1}{\sqrt[3]{x^2}}\right)e^{-\alpha x-\alpha x^{\frac{1}{3}}}$

   (vi) $\alpha e^{-\alpha x} + \dfrac{\alpha}{3}\dfrac{1}{\sqrt[3]{x^2}}\,e^{-\alpha x^{\frac{1}{3}}} - \alpha\left(1 + \dfrac{1}{3}\dfrac{1}{\sqrt[3]{x^2}}\right)e^{-\alpha x-\alpha x^{\frac{1}{3}}}$

2. (i) $\dfrac{1}{6}\dfrac{1}{\sqrt[3]{x^2}}$ for $|x| < 1$ (ii) $\frac{1}{4}$ for $1 < t < 5$

   (iii) $\dfrac{1}{2}\left(1 - \dfrac{|x|}{2}\right)$ for $|x| < 2$ (iv) $1 - \dfrac{x}{2}$ for $0 < x < 2$

   (v) $\frac{1}{4} - \frac{1}{3}x^{\frac{1}{3}} + \frac{1}{12}x^{-\frac{2}{3}}$ for $|x| < 1$ (vi) $\frac{1}{4} + \frac{1}{3}x^{\frac{1}{3}} + \frac{1}{12}x^{-\frac{1}{3}}$ for $|x| < 1$.

3. (i) $h^{-1}(1-e^{-\alpha x})$ for $0 < x < h$, and $h^{-1}(e^{\alpha h}-1)e^{-\alpha x}$ for $x > h$.
   (ii) $h^{-1}(1-e^{-\alpha(x+h)})$ for $-h < x < 0$, and $h^{-1}(1-e^{-\alpha h})e^{-\alpha x}$ for $x > 0$.

4. (i) $h/3$ if $h \le 1$ and $1/(3\sqrt{h})$ if $h \ge 1$.

   (ii) $\sqrt{\alpha\pi}\,e^{\frac{1}{4}\alpha}(1-\mathfrak{N}(\sqrt{\alpha/2}))$.

5. (i) $1 - x^{-1}$ for $x > 1$; (ii) $x^2(x+1)^{-2}$.

7. $\mathbf{P}\{\mathbf{Z} \le x\} = 1 - e^{-\alpha x}$ for $x < t$ and $=1$ for $x > t$.

10. (*b*) The platoon together with cars directly ahead or trailing form a sample in which the smallest element is at the last place, the next smallest at the second.

16. $p = \displaystyle\sum_{0}^{m-1}\binom{n+k-1}{k}2^{-n-k}$. For $m = 1, n = 2$ one gets $p = \frac{1}{4}$.

18. $nt^{n-1} - (n-1)t^n$.

19. (i) $2\int_0^1 dx \int_x^1 (1-z)\, dz = \frac{1}{3}$

(ii) The density is $2t - t^2$ for $0 < t \leq 1$ and $(2-t)^2$ for $1 \leq t < 2$

(iii) The density is $2t^2$ for $0 < t < 1$ and again $(2-t)^2$ for $1 \leq t < 2$.

20. $2\int_0^1 x(1-x)\, dx = \frac{1}{3}$. Two out of six permutations produce an intersection.

21. $\mathbf{X}_{11}$: $4 \log \frac{1}{4x}$ for $x < \frac{1}{4}$; $\mathbf{X}_{12}$ and $\mathbf{X}_{21}$: $4 \log 2$ for $x < \frac{1}{4}$, $4 \log \frac{1}{2x}$ for $\frac{1}{4} < x < \frac{1}{2}$; $\mathbf{X}_{22}$: $4 \log 4x$ for $\frac{1}{4} < x < \frac{1}{2}$, $4 \log \frac{1}{x}$ for $\frac{1}{2} < x < 1$. The expectations are $\frac{1}{16}, \frac{3}{16}, \frac{9}{16}$.

27. Distributions $\frac{2}{\pi} \arcsin \frac{1}{2}x$ and $\frac{1}{4}x^2$; densities $\frac{2}{\pi} \frac{1}{\sqrt{4 - x^2}}$ and $\frac{1}{2}x$ for $0 < x < 2$.

28. $2\pi^{-1} \arcsin \frac{1}{2}x$.

30. (*a*) $\log \frac{1}{x}$, (*b*) $\frac{2}{\pi} \log \frac{1 + \sqrt{1 - x^2}}{x}$ where $0 < x < 1$.

31. $\frac{4}{\pi} \int_{0<\cos\theta<x} \sin^2\theta\, d\theta = \frac{4}{\pi} \int_0^x \sqrt{1-y^2}\, dy = 2\pi^{-1}[\arcsin x + x\sqrt{1-x^2}\,]$ where $0 < x < 1$.

32. $F(t) = \frac{\pi}{2} \int_0^{\pi/2} V\left(\frac{t}{\cos\theta}\right)(1 - \cos 2\theta)\, d\theta$.

35. The substitution $s = F(y)$ reduces the integral to the corresponding integral for the uniform distribution. Note that $F(m + x) \approx \frac{1}{2} + f(m)x$ for small x.

CHAPTER II

4. $g \star g(x) = \frac{1}{4}e^{-|x|}(1 + |x|)$

$g^{3\star}(x) = \frac{1}{16}e^{-|x|}(3 + 3\,|x| + x^2)$

$g^{4\star}(x) = \frac{1}{32}e^{-|x|}(5 + 5\,|x| + 2x^2 + \frac{1}{3}\,|x|^3)$.

10. (*a*) $\lambda\mu(e^{-\lambda t} - e^{-\mu t})(\mu - \lambda)$ as a convolution of two exponentials. (*b*) Using (*a*) one gets $\lambda e^{-\lambda t}$.

12. For a person arriving at random the density is $1 - \frac{1}{2}t^2$ for $0 < t < 1$ and $\frac{1}{2}(2-t)^2$ for all $1 < t < 2$. The expectation equals $\frac{7}{12}$.

13. $\binom{j+k+\gamma-1}{j}\binom{m+n+\mu-j-k-1}{m-j}\Big/\binom{m+n+\mu+\gamma-1}{m}$.

CHAPTER III

7. (*a*) e^{-x} and $1 - e^{-x} - e^{-y} + e^{-x-y-axy}$ for $x > 0, y > 0$.

(*b*) $\mathbf{E}(\mathbf{Y} \mid \mathbf{X}) = \frac{1 + a + ax}{(1 + ax)^2}$,

$$\mathrm{Var}\,(\mathbf{Y} \mid \mathbf{X}) = \frac{1}{(1+ax)^2} + \frac{2a}{(1+ax)^3} - \frac{a^2}{(1+ax)^4}.$$

8. If f has expectation μ and variance σ^2 then $\mathbf{E}(\mathbf{X}) = \mathbf{E}(\mathbf{Y}) = \frac{1}{2}\mu$, $\text{Var}(\mathbf{X}) = \text{Var}(\mathbf{Y}) = \frac{1}{3}\sigma^2 + \frac{1}{12}\mu^2$, $\text{Cov}(\mathbf{X}, \mathbf{Y}) = \frac{1}{6}\sigma^2 - \frac{1}{12}\mu^2$.
9. Density $2x_2$ in unit square. In n variables $(n-1)!\mathbf{X}_2\mathbf{X}_3^2 \cdots \mathbf{X}_{n-1}^{n-2}$.
10. $\frac{1}{3}e^{-(x+y)}$ for $y > x > 0$ and $\frac{1}{3}e^{-y+2x}$ for $y > x$, $x < 0$. Interchange x and y when $y < x$.
11. (a) $4\int_{2x}^{\frac{1}{2}} f(s) f\left(\frac{x}{s}\right) \frac{ds}{s}$, $\quad 0 < x < \frac{1}{4}$.

    (b) $8\int f(s) f\left(\frac{x}{s}\right) f\left(\frac{y}{1-s}\right) \frac{ds}{s(1-s)}$

    where $0 < x < \frac{1}{4} < y < 1$ and the domain of integration satisfies the conditions that $2x < s < \frac{1}{2}$ and also $1 - 2y < s < 1 - y$.
12. Bivariate normal with variances m, n and covariance $\sqrt{m/n}$. Conditional density has expectation $\frac{m}{n}t$ and variance $m \cdot \frac{n-m}{n}$ as is clear intuitively.
13. $\mathbf{X}_1^2 + \cdots + \mathbf{X}_n^2$ has the gamma density $f_{1/2,n/2}$ [see II,(2.2)]. From (3.1) therefore

$$u_t = \frac{\Gamma\left(\frac{n}{2}\right)}{\Gamma\left(\frac{m}{2}\right)\Gamma\left(\frac{n-m}{2}\right)} \left(\frac{x}{t}\right)^{\frac{1}{2}m-1} \left(1 - \frac{x}{t}\right)^{\frac{1}{2}(n-m)-1} \frac{1}{t}.$$

    For $m = 2, n = 4$ we get example 3(*a*).
15. (a) $4xy$ when $x + y < 1$, $x > 0, y > 0$

    $4xy - 4(x+y-1)^2$ when $x + y > 1$, $0 < x$, $y < 1$

    $4x(2-x-y)$ when $y > 1$, $x + y < 2$, $x > 0$

    $4y(2-x-y)$ when $x > 1$, $x + y < 2$, $y > 0$.

    (b) $2(1-x-y)^2$ for $0 < x$, $y < 1$, $x + y < 1$

    $2(1-x)^2$ for $x > 0$, $y < 0$, $x + y > 0$.

    $2(1+y)^2$ for $x > 0$, $y < 0$, $x + y < 0$.

    For $x < 0$ by symmetry.
16. $2\frac{1}{\pi^2}\left(\arccos\frac{r}{2} - \frac{r}{2}\sqrt{1 - \frac{r^2}{4}}\right)$.
17. $\int_0^\infty f(\rho)\rho \, d\rho \int_0^{2\pi} g(\sqrt{r^2 + \rho^2 - 2r\rho\cos\theta}) \, d\theta$.
20. (a) $\mathbf{X}_n = \mathbf{U}\cos\frac{1}{2}\pi n + \mathbf{V}\sin\frac{1}{2}\pi n$

    (b) $\mathbf{U} + \mathbf{V}(-1)^n$

    (c) $\mathbf{U}\cos\frac{1}{2}\pi n + \mathbf{V}\sin\frac{1}{2}\pi n + \mathbf{W}$.
21. (a) $\text{Var}(\mathbf{Y}_{n+1}) - \text{Var}(\mathbf{Y}_n) = \text{Var}(\mathbf{C}_n) - 2\,\text{Cov}(\mathbf{Y}_n, \mathbf{C}_n) + 1$ whence

    (b) $\alpha^2 - 2\alpha\sigma\rho + 1 = 0$

    (c) $\sigma = \frac{1}{2}\left(\alpha + \frac{1}{\alpha}\right)$, $\mathbf{Y}_n = \sum_{k=0}^{n-1} q^k\mathbf{X}_{n-1-k} + q^n\mathbf{Y}_0 + (bp - a)(1 - q^n)/p$ where $q = 1 - p$.
22. $\sigma^2 \geq \frac{1}{4}\left(\alpha + \frac{1}{\alpha}\right)^2 + N$.

CHAPTER VI

11. Not necessarily. It is necessary that $n[1-F(\epsilon n)] \to 0$.
12. For $x > 0$ the densities are given by 1 and $\frac{1}{2}(1-e^{-2x})$.
13. $qU(x) = 1 - qe^{-pct}$. The number of renewal epochs is *always* geometrically distributed.
19. $Z = z + F \bigstar Z$ where $z(t) = 1 - e^{-ct}$ for $t \leq \xi$, $z(t) = z(\xi)$ for $t \geq \xi$ and $F(t) = e^{-c\xi} - c^{-ct}$ for $t > \xi$.
20. $V = A + B \bigstar V$ where $A\{dx\} = [1-G(x)]\,F\{dx\}$ and $B\{dx\} = G(x)\,F\{dx\}$.
23. The arc sine density $g(y) = \dfrac{1}{\pi}\dfrac{1}{\sqrt{y(1-y)}}$.

CHAPTER VII

6. (*a*) $\binom{n}{k} p^k(1-p)^{n-k}$ with F concentrated at p.

   (*b*) $\dfrac{1}{n+1}$, density $f(x) = 1$.

   (*c*) $\dfrac{2(k+1)}{(n+1)(n+2)}$, density $2x$.

Some Books on Cognate Subjects

A. INTRODUCTORY TEXTS

Krickeberg, K. [1965], *Probability Theory*. (Translated from the German 1963) Addison-Wesley, Reading, Mass. 230 pp.

Loève, M. [1963], *Probability Theory*. 3rd ed. Van Nostrand, Princeton, N.J. 685 pp.

Neveu, J. [1965], *Mathematical Foundations of the Calculus of Probability*. (Translated from the French 1964.) Holden Day, San Francisco, Calif. 233 pp.

B. SPECIFIC SUBJECTS

Bochner, S. [1955], *Harmonic Analysis and the Theory of Probability*. Univ. of California Press. 176 pp.

Grenander, U. [1963], *Probabilities on Algebraic Structures*. John Wiley, New York. 218 pp.

Lukacs, E. [1960], *Characteristic Functions*. Griffin, London. 216 pp.

Lukacs, E. and R. G. Laha [1964], *Applications of Characteristic Functions*. Griffin, London. 202 pp.

C. STOCHASTIC PROCESSES WITH EMPHASIS ON THEORY

Chung, K. L. [1967], *Markov Chains with Stationary Transition Probabilities*. 2nd ed. Springer, Berlin. 301 pp.

Dynkin, E. B. [1965], *Markov Processes*. Two vols. (Translation from the Russian 1963) Springer, Berlin. 174 pp. 365 + 271 pp.

Ito, K. and H. P. McKean Jr. [1965], *Diffusion Processes and Their Sample Paths*. Springer, Berlin. 321 pp.

Kemperman, J. H. B. [1961], *The Passage Problem for a Stationary Markov Chain*. University of Chicago Press. 127 pp.

Lévy, Paul [1965], *Processus Stochastiques et Mouvement Brownien*. 2nd ed. Gauthier-Villars, Paris. 438 pp.

Spitzer, Frank [1964], *Principles of Random Walk*. Van Nostrand, Princeton. 406 pp.

Skorokhod, A. V. [1965], *Studies in the Theory of Random Processes*. (Translation from the Russian 1961.) Addison-Wesley, Reading, Mass. 199 pp.

Yaglom, A. M. [1962], *Stationary Random Functions.* (Translation from the Russian.) Prentice-Hall, Englewood Cliffs, N.J. 235 pp.

D. STOCHASTIC PROCESSES WITH EMPHASIS ON APPLICATIONS OR EXAMPLES

Barucha-Reid, A. T. [1960], *Elements of the Theory of Stochastic Processes and Their Applications.* McGraw-Hill, New York. 468 pp.

Beneš, V. E. [1963], *General Stochastic Processes in the Theory of Queues.* Addison-Wesley, Reading, Mass. 88 pp.

Grenander, U. and M. Rosenblatt [1957], *Statistical Analysis of Stationary Time Series.* John Wiley, New York. 300 pp.

Khintchine, A. Y. [1960], *Mathematical Methods in the Theory of Queueing.* (Translation from the Russian.) Griffin, London. 120 pp.

Prabhu, N. U. [1965], *Stochastic Processes.* Macmillan, New York. 233 pp.

——— [1965], *Queues and Inventories.* John Wiley, New York. 275 pp.

Riordan, J. [1962], *Stochastic Service Systems.* John Wiley, New York. 139 pp.

Wax, N. (editor) [1954], *Selected Papers on Noise and Stochastic Processes.* Dover, New York. 337 pp.

E. BOOKS OF HISTORICAL INTEREST

Cramér, H. [1962], *Random Variables and Probability Distributions.* 2nd ed. (The first appeared in 1937.) Cambridge Tracts. 119 pp.

Doob, J. L. [1953], *Stochastic Processes.* John Wiley, New York. 654 pp.

Gnedenko, B. V. and A. N. Kolmogorov [1954], *Limit Distributions for Sums of Independent Random Variables.* (Translated from the Russian 1949) Addison-Wesley, Reading, Mass. 264 pp.

Kolmogorov, A. N. [1950], *Foundations of the Theory of Probability.* Chelsea Press, New York. 70 pp. (The German original appeared in 1933.)

Lévy, P. [1925], *Calcul des Probabilités.* Gauthier-Villars, Paris. 350 pp.

Lévy, P. [1937 and 1954], *Théorie de l'Addition des Variables Aléatoires.* Gauthier-Villars, Paris. 384 pp.

F. SEMI-GROUPS AND GENERAL ANALYSIS

Hille, E. and R. S. Phillips [1957], *Functional Analysis and Semi-groups.* (Revised edition.) Amer. Math. Soc. 808 pp.

Karlin, S. and W. Studden [1966], *Tchebycheff Systems: With Applications in Analysis and Statistics.* Interscience, New York. 586 pp.

Yosida, K. [1965], *Functional Analysis.* Springer, Berlin. 458 pp.

Index